Retinal Degenerative Diseases and Experimental Therapy

Retinal Degenerative Diseases and Experimental Therapy

Edited by

Joe G. Hollyfield

Cole Eye Institute
The Cleveland Clinic Foundation
Cleveland, Ohio

Robert E. Anderson

Dean A. McGee Eye Institute
University of Oklahoma Health Sciences Center
Oklahoma City, Oklahoma

and

Matthew M. LaVail

Beckman Vision Center
University of California, San Francisco
San Francisco, California

Springer Science+Business Media, LLC

Library of Congress Cataloging-in-Publication Data

Retinal degenerative diseases and experimental therapy / edited by Joe G. Hollyfield, Robert E. Anderson, and Matthew M. LaVail.
p. cm.
Includes bibliographic references and index.
ISBN 978-1-4757-7224-1 ISBN 978-0-585-33172-0 (eBook)
DOI 10.1007/978-0-585-33172-0
1. Retina—Degeneration—Congresses. 2. Photereceptors—Congresses. 3. Retina—Degeneration—Treatment—Congresses. I. Hollyfield, Joe G. II. Anderson, Robert E. (Robert Eugene) III. LaVail, Matthew, M.
[DNLM]: 1. Retinal Degeneration—genetics—Congresses. 2. Photoreceptors—Congresses. 3. Retinal Degeneration—physiopathology—Congresses 4. Retinal Degeneration—therapy—Congresses. WW 270 R438282 1999]
RE661.D3 R482 1999
617.7'35 21—dc21

99-040819

Proceedings of the Eighth International Symposium on Retinal Degeneration, held July 20–25, 1998 in Schluchsee, Germany

ISBN 978-1-4757-7224-1

Originally published by Kluwer Academic/Plenum Publishers, New York in 1999
Softcover reprint of the hardcover 1st edition 1999

10 9 8 7 6 5 4 3 2 1

A C.I.P. record for this book is available from the Library of Congress.

Edward H. Gollob

This book is dedicated to Edward H. Gollob, President of the Foundation Fighting Blindness, Hunt Valley, Maryland, USA, for his devotion, commitment, and untiring efforts to bring us closer to the day when there will be no blindness from inherited retinal disorders.

PREFACE

To create a forum for scientists and clinicians interested in degenerative retinal diseases, we began in 1984 to organize a biennial symposium on Retinal Degeneration as a satellite meeting of the International Congress of Eye Research. The timing and varying location of these meetings provides an important assembly for investigators from throughout the world to convene for presentation of their new findings on the causes and potential therapies for degenerative retinal disorders.

The VIII International Symposium on Retinal Degeneration was held from July 20–25, 1998, at the Hotel Vier Jahreszeiten in Schluchsee, a small town in the Black Forest of southwestern Germany. Most of the participants in this meeting contributed to this volume, and we are appreciative of the efforts of each author in making this publication possible.

The research presented at the meeting, and described in this proceedings volume, reflects a strong emphasis on the molecular genetic approach to understanding these disorders. Several of the papers provide important new insights into the mechanism of photoreceptor degeneration and cell death. A number of the studies are targeted at retarding or reversing the degeneration process. Included for the first time are presentations from all the principal laboratories involved in the field of visual prostheses—implant (chip) technology—in which investigations are targeted at restoring vision in eyes that have lost photoreceptor cells. A variety of diagnostic, clinical, histopathological, and physiological assessments of retinal degeneration in patients are also included. The scope, depth, and variety of approaches toward investigating these disorders, their prevention, and progress toward sight restoration, makes this volume the most up-to-date compendium covering this rapidly progressing field.

The symposium received financial support from several agencies. We are very appreciative of the funds provided by the Foundation Fighting Blindness, Hunt Valley, Maryland, for this and all previous symposia, without which we could not have convened these important meetings. We also thank the Retinal Preservation Foundation, South Africa; Deutsche Forschungsgemeinschaft, Germany; Pro Retina, Germany; and CIBA Vision, Switzerland, for their important financial help.

To Professor Eberhart and Mrs. Claudia Zrenner, our very generous local host and hostess, we extend our heartfelt thanks for a job well done. All the local arrangements in Germany were made under their very capable guidance. The Zrenners were assisted by a number of their colleagues in Tübingen, including: Christine Augustin, Eckart Apfelstedt-Sylla, Dorothea Besch, Klaus-Dieter Miliczek, Marcus Reichl, Stephanie Schaerer, Matthias Seeliger, Elke Guenther, Jens Hartmann, Konrad Kohler, Moritz Meins, Janina Kramer, Ian Giddings, Thomas Wheeler-Schilling, and Bernd

Wissinger. We thank all of these members of the Local Organizing Committee for their efforts in making this an outstanding symposium.

We thank Ms. Roni Isom for her assistance with all the correspondence generated during the two years that the symposium was being organized, and for her help in collating the manuscripts included in this publication. Thanks also go to Kluwer Academic / Plenum Publishers for publishing this volume.

Joe G. Hollyfield
Robert E. Anderson
Matthew M. LaVail

CONTENTS

IV Animal Models with Photoreceptor Degeneration

V Molecular and Cell Biological Studies

Retinal Degenerative Diseases and Experimental Therapy

Section I

SPECIFIC GENE DEFECTS CAUSE PHOTORECEPTOR DEGENERATION

1

EVIDENCE FOR MYOSIN VIIa-DRIVEN TRANSPORT OF RHODOPSIN IN THE PLASMA MEMBRANE OF THE PHOTORECEPTOR-CONNECTING CILIUM

Uwe Wolfrum and Angelika Schmitt

Institut für Zoologie
Johannes Gutenberg-Universität Mainz
D-55099 Mainz, Germany

SUMMARY

Defects in the gene encoding for the unconventional myosin VIIa leads to human Usher syndrome 1B, the most common form of hereditary combined blindness and deafness. To determine cellular function of myosin VIIa, we have investigated the subcellular localization of myosin VIIa in spacial relation relationship to potentially interacting proteins in mammalian photoreceptor cells. Western blot analysis of the axonemal fraction of photoreceptor cells by Western blot show that myosin VIIa and actin, as well as opsin, were present in the ciliary portion of the photoreceptors. Improved immunoelectron microscopy revealed that in mammalian photoreceptor cells, myosin VIIa was localized at the membrane of the connecting cilium linking photoreceptor inner segments with their outer segments. In addition, actin, the functional partner of myosins, actin, was detected at the ciliary membrane providing considerable evidence that myosin VIIa also interacts with actin filaments in the connecting cilium. Applying the highly-sensitive silver-enhanced indirect nanogold™-labeling for immunoelectron microscopy, intense anti-opsin immunoreactivity was recognized in the membrane of the connecting cilium indicating that rhodopsin trafficing to the outer segment occurs through the ciliary membrane. Co-localization of rhodopsin with the myosin VIIa-actin systems at the membrane also provides evidence for mysion VIIa driven ciliary transport of rhodopsin. Absence of myosin VIIa or presence of

Retinal Degenerative Diseases and Experimental Therapy, edited by Hollyfield *et al.*
Kluwer Academic / Plenum Publishers, New York, 1999.

non-functioning myosin VIIa may lead to defects in the ciliary transport and cause *retinitis pigmentosa* in Usher syndrome 1B patients.

INTRODUCTION

Usher syndrome is the most common form of inherited blindness combined with deafness.[1–3] Usher syndrome 1B is caused by defects in the gene encoding myosin VIIa.[4] Myosin VIIa is a large unconventional myosin possessing all the main characteristics of a myosin motor domain in its N-terminal domain.[5,6] Further regions of functional interest are present in the long tail domain of myosin VIIa: A coiled coil structure indicates that it is likely that myosin VIIa molecules dimerize, and two talin-like sequences are most probably involved in membrane-binding.[5,6] Although, direct biochemical experimental data on myosin VIIa function do not exist, available information indicate that in cells, dimers of myosin VIIa act as a membrane-associated, actin-based motor protein. In an effort to better understand the process of sensory degeneration in Usher 1B at the cellular function level, we have investigated the subcellular localization of myosin VIIa and its potential functional partner proteins in photoreceptor cells of mammalian retinae (see also: Liu et al.[7]).

Vertebrate photoreceptor cells are highly-polarized neuronal cells which are composed of a photosensitive outer segment and an inner segment. Photoreceptor segments are linked via the so-called connecting cilium. The outer segment of rod cells contain over a thousand membrane disks and all proteins involved in phototransduction including the transmembranal visual pigment rhodopsin are found here. The phototransductive membranes of the outer segment are continually renewed throughout life by forming new disk membranes at the base of the outer segments and shedding the old disks at the distal tip of the outer segment which are then degraded in the retinal pigment epithelium.[8] Therefore, a continuing transport of newly-synthezised outer segment components at the biosynthetic compartments of the inner segment to the outer segment has to occur. There is conciderable evidence that rhodopsin and other outer segment proteins are transported in post-Golgi membrane vesicles to the base of the connecting cilium.[9,10] Analysis of rhodopsin-bearing vesicles after subcellular fractioning of photoreceptor membranes indicates that the C-terminal domain of rhodopsin is involved in the regulation of post-Golgi sorting[11,12] and futhermore, that the directional trafficing is driven along the cellular microtubule network by cytoplasmic dynein interacting directly with the C-terminal domain of rhodopsin to the periciliary membrane of the inner segment.[13,14] However, the final step of rhodopsin delivery to the base of the outer segment has not yet been resolved and has remained an enigma for decades.[15] Although many research efforts have focused on a ciliary pathway of opsin delivery and it was commonly cited, there has been not yet clear direct evidence supporting the ciliary transport model (see review: Besharse and Horst[15]).

In the present report, we have gathered direct cytological evidence that rhodopsin is transported in the membrane of the connecting cilium from the inner segment to the outer segment of photoreceptor cells. Applying improved, highly-sensitive immunoelectron microscopical labeling with secondary antibodies conjugated to ultra-small gold in combination with silver-enhancement, prominent amounts of rhodopsin were detected by monoclonal antibodies against bovine rod opsin in the membrane of the cilium during its passage to the outer segment. In addition, our results show that rhodopsin co-localizes with myosin VIIa at the membrane of the connecting cilium of photoreceptors. Futhermore, we identified actin at the photoreceptor axoneme which most probably form the filaments supporting myosin VIIa motility at the ciliary membrane. Therefore, in our report we gathered evidence for the hypothesis on the membrane traffic through the photoreceptor cilium: The motor protein myosin VIIa linked to the plasma membrane may move along the axonemal actin filaments pulling plaques of the ciliary membrane including integral membrane proteins, e.g. rhodopsin to the base of the outer segment. Present findings also support that the symdomes of Usher syndrome 1B may be caused by defects of the function of sensory cilia.[7,16,17]

MATERIALS AND METHODS

Animals

All experiments described herein conform with the statement by the Association for Research in Vision and Ophthalmology as to care and use of animals in research. Adult albino rats (Sprangt Dawley) and C57BL76 mice were maintained on a 12/12 light/dark cycle with lights on at 7 a.m., food and water ad libitum. Bovine eyes were obtained from the local slaughter house.

Antibodies

Myosin VIIa antisera were generated in a collaborative project with Dr. D. S. Williams laboratory, La Jolla, USA, and were previously characterized.[7,17] In the present paper, we used antibodies made against a recombinant protein, corresponding to amino acids 941–1071 of mouse myosin VIIa. The antibody (known as pAb2.2 in Liu et al.[7]) was used in all the figures of the present paper. Monoclonal antibodies against bovine rod opsin were kindly provided by Dr. P. A. Hargrave (Gainesville, USA). The clones used in the present study were B6-30a1, K16-155, R2-15, and were described by Adamus et al.[18,19] A monoclonal antibody against chicken gizzard actin (clone C4) was kindly given to us by Dr. J. Lessard (Cincinetti, USA) and has been previously characterized.[20]

Western Blot Analysis

For Western blots, isolated retinae were homogenized and placed in SDS-PAGE sample buffer (62.5 mM Tris buffer, 10% glycerol, 2% SDS, 5% Mercaptoethanol, 1 mM EDTA, 0.025% Bromphenol Blue pH 6.8). Proteins were separated by SDS-PAGE, transferred electrophoretically to Immobilon-P, blocked, and probed with primary and secondary antibodies.[21] The latter was conjugated to alkaline phosphatase, so that labeling was detected by the formation of the insoluble product of 5-bromo-4-chloroindoyl phosphate hydrolysis.

Axoneme Preparation

Photoreceptor axonemes were purified from isolated bovine retinas using sucrose density gradients in combination with cytoskeleton extraction according to Horst et al.[22–24] Fractions of the gradient containing axonemes were collected and placed in SDS-PAGE sample buffer. Following transblotting, immunoreactivity to the antibodies used was determined.

Immunoelectron Microscopy

Isolated retinae were fixed in 0.1% glutaraldehyde and 4% paraformaldehyde in 0.1 M phosphate buffer (pH 7.4) for 3 hours at room temperature. Fixed tissue was dehydrated to 98% ethanol, embedded in LR White, and polymerized at 4°C under UV light for 48–60 hours.

Ultrathin sections (60–70 nm) were collected on formvar-coated nickel grids. Sections were first etched with saturated sodium periodate (Sigma, Deisenhofen, Germany) at room temperature for 3–5 minutes. The grids were preincubated with 0.1% Tween 20 in PBS, blocked with 50 mM NH_4Cl in PBS and with blocking solution (0.5% fish gelatin (Sigma) plus 0.1% ovalbumin (Sigma) in PBS. Sections were incubated with primary antibodies diluted in blocking solution at 4°C for 60 hours, washed once in PBS and twice in "IgG-gold" buffer (0.1% ovalbumin, 0.5% cold water fish gelatin, 0.01% Tween 20, 0.5 M NaCl in 10 mM phosphate buffer, pH 7.3). The sections were incubated for 2 hours with goat anti-mouse or anti-rabbit Fab's conjugated to nanogold™ (Nanoprobes, Stony Brook, N.Y., USA), diluted in "IgG-gold" buffer. After washing and post-fixation in buffered 2% glutaraldehyde the nanogold™ labeling was silver-enhanced as described by Danscher.[25] Washed grids were stained with 2% ethanolic uranyl acetate prior to examination in a Zeiss EM 912ς electron microscope.

For immunoelecton microscopic double-labelings, ultrathin LR White sections through mouse retinae were incubated first incubated with a cocktail of anti-opsin and anti-myosin VIIa antibodies in blocking buffer at least for 60 hours. To visualize antibody reaction sites, secondary antibodies conjugated to gold particles of different sizes (10 nm colloidal gold coupled goat anti-rabbit IgG and goat anti-mouse Fab′-1.4 nm nanogold™ and *vice versa* (Nanoprobes or Aurion, Wageningen, The Netherlands) were also applied mixed together in "IgG-gold" buffer. The time for silver-enhancement of double-stained (1.4 nm and 10 nm gold particles) sections was reduced to 20 minutes.

RESULTS

Western Blot Analysis of Photoreceptor Axonemes

Western blot analyses of the axonemal fraction of photoreceptor cells isolated from bovine retinae revealed that myosin VIIa and actin, as well as opsin, were present in photoreceptor axonemes (Fig. 1). In these blots, antibodies raised against a recombinantly expressed part of the myosin VIIa tail, which were previously characterized[7], react with a polypeptide at 270 kDa, the estimated and previously demonstrated molecular weight of myosin VIIa (lane 1).[7,17] The well characterized monoclonal antibody to actin (clone C4)[20] recognized a 42 kDa band the molecular weight of actins (lane 2). A cocktail of all 3 monoclonal antibodies against bovine rod opsin as well as the 3 separate monoclonal antibodies recognized two peptide bands, one of about 35 kDa, the estimated molecular weight of an opsin monomer, and the higher molecular weight dimeric opsin commonly found after SDS-PAGES (lane 3).

Myosin VIIa Localization in the Photoreceptor Cilium

To determine subcellular localization of myosin VIIa in photoreceptor cells we performed immunoelectron microscopy on ultrathin sections through LR White

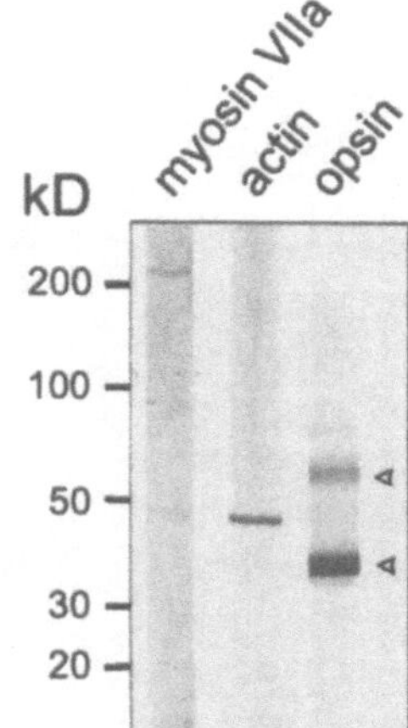

Figure 1. Immunoblot analysis of photoreceptor axonemes. Western blots were probed with antibodies against myosin VIIa (lane 1), actin (lane 2), and opsin (lane 3) after SDS-PAGE protein-separation of purified cytoskeletal fraction of photoreceptor axonemes isolated from bovine retinae. All three antibodies recognize polypeptides at expected molecular weight. Arrowheads point to the bands of opsin monomer and dimer.

embedded isolated rat and mouse retinae using previously characterized antisera against myosin VIIa.[7,17] To increase the sensitivity of the immunoreactivity on electron microscopical level, reaction sites of primary antibodies were visualized by secondary nanogold™ antibodies conjugated to ultrasmall gold particles (~1.4 nm in diameter) in combination with the silver-enhancement protocol of Danscher.[25] As expected from our previous studies[7,26], in photoreceptor cells electron microscopical immunolabel of anti-myosin VIIa was restricted to the connecting cilium (Fig. 2). At this modified cilium, prominent intense labeling of myosin VIIa was detected at the plasma membrane, while in contrast, silver-enhanced gold particles were absent from the center of the ciliary axoneme (Fig. 2).

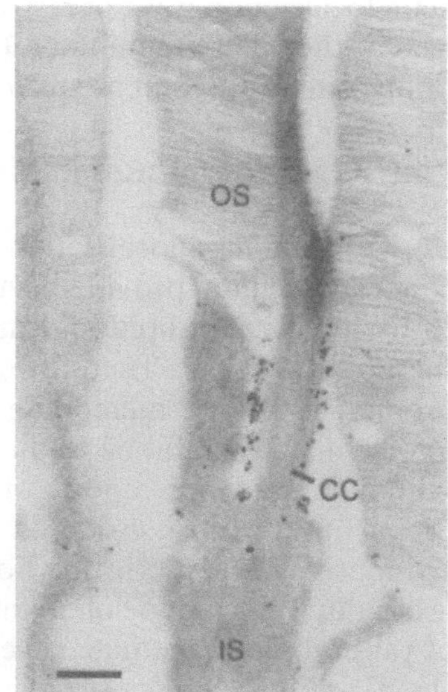

Figure 2. Localization of myosin VIIa in the photoreceptor connecting cilium (CC). Silver-enhanced immunogold labeling of myosin VIIa in a longitudinal section of the cilium of a mouse rod photoreceptor cell. The proximal connecting cilium is labeled at the plasma membrane. IS = inner segment; OS = outer segment; Scale bar = 200 nm.

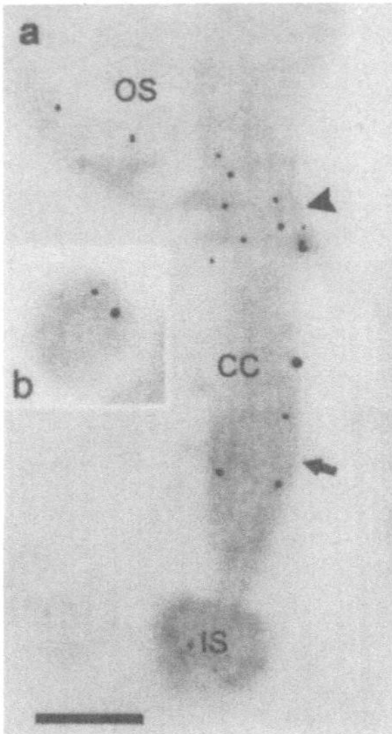

Figure 3. Localization of actin in the photoreceptor connecting cilium (CC). **a.** Silver-enhanced immunogold labeling of actin in a longitudinal section of part of a rat rod photoreceptor cell. **b.** Silver-enhanced immunogold labeling of actin in a transverse section through the connecting cilium of a rat rod cell. Actin filaments are labeled at the base of the outer segment (OS) (arrowhead) previously reported (see text). Using present sensitive immunoelectron optical method, actin is also recognized at the membrane of the connecting cilium (CC) (arrow in a). IS = inner segment; Scale bar = 280 nm.

Actin Localization at the Ciliary Membrane

We reasoned that, as in all known myosins, the myosin VIIa motile function would be based on the interaction with actin filaments. In our present study, we have already shown actin present in photoreceptor axonemes by Western blot analysis (Fig. 1, lane 2). Previous investigations by others have been demonstrated actin filaments at the base of the photoreceptor outer segment were they are supposed to be involved in disk formation interacting with conventional myosin II.[27–33] Present immunoelectron microscopy also provided first evidence of a membrane-linked actin cytoskeleton in the connecting cilia of photoreceptors. In our highly sensitive immunolabelings of sections through mammalian photoreceptor cells with anti-actin monoclonal antibodies, the actin localization was not restricted to the base of the outer segment, but was also localized at the ciliary membrane (Fig. 3) confirming our previously discussed localization of myosin VIIa (compare also Fig. 2).

Anti-Opsin Immunolabeling of Photoreceptor Cells

Our immunolocalization of myosin VIIa and actin at membrane domains of the connecting cilium provided evidence that the ciliary actin-myosin VIIa system is likely to be involved in lipid and membrane protein trafficing from the inner segment to the outer segment. Therefore, we examined the possibility that rhodopsin, the most prominent integral membrane protein in photoreceptor outer segments, was delivered in the ciliary membrane to the outer segment by the ciliary actin-myosin VIIa system. Although, previous reports of the weak opsin immunoreactivity in the ciliary membrane have been considered a major obstacle to accepting a ciliary transport of rhodopsin,[15,34] we performed our improved protocol for immunoelectron microscopy using different monoclonal antibodies against bovine rod opsin. In ultrathin sections of rat and mouse retinae silver-enhanced immunogold labeling with different monoclonal antibodies against bovine rod opsin was restricted rod photoreceptors (Fig. 4)

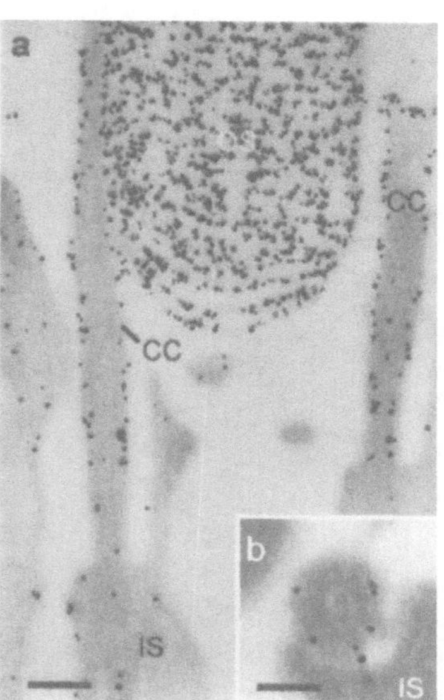

Figure 4. Localization of rhodopsin in the photoreceptor connecting cilium (CC). Silver-enhanced immunogold labeling of opsin in a longitudinal section of part of a rat rod photoreceptor cell (**a.**) in a transverse section through the connecting cilium of a rat rod cell (**b.**). As expected, monoclonal antibodies against bovine rod opsin most intensly lable the disk membranes in the rod outer segment (OS). However, rhodopsin is also recognized in the plasma membrane of photoreceptor connecting cilia in longitudinal as well as transverse ciliary sections. In the inner segment (IS), weak anti-opsin labeling is present in the cytoplasma and the plasma membrane. Scale bars: a = 250 nm, b = 150 nm.

where as cones lack any staining. In rod photoreceptor cells, most prominent labeling of opsin was observed, as expected, at disk membranes of outer segments (Figs. 4a). In rods, anti-opsin immunoreactivity was detected in the ER and the Golgi apparatus, as well as at the plasma membrane of the apical inner segment. Moreover, applying our highly sensitive detection method of silver-enhanced nanogold™ labeling, prominent amounts of rod opsin were also recognized in the membrane of the connecting cilium, contrary to previous investigations (Figs. 4a, b). In contrast, the center of the ciliary axoneme remained almost completely un-labeled. Present anti-opsin immunoreactivity at the ciliary membrane strongly indicated that rhodopsin is delivered to the outer segment of photoreceptors via the membrane of the connecting cilium.

Myosin VIIa—Rhodopsin Co-Localization at the Membrane of the Connecting Cilium Demonstrated by Electron Microscopical Double-Labelings

To prove that myosin VIIa and rhodopsin were co-localized at the membrane domain of the connecting cilium, double-labelings with purified anti-myosin VIIa antibodies raised in rabbit and anti-opsin mouse monoclonal antibodies were performed at the immunoelectron microscopic level. In these double labeling experiments, secondary antibodies conjugated to gold particles of different sizes (nanogold™ probes (1.4 nm) and 10 nm-collodial gold) were used to discriminate between the reaction sites of both antibodies. Electron microscopical analysis of double-labeled ultrathin sections through mouse photoreceptor cells convincingly demonstrated co-localization of myosin VIIa and rhodopsin at the membrane of the connecting cilium (Fig. 5). This co-loalization indicates protein-protein interaction providing considerable evidence for a myosin VIIa driven transport of rhodopsin molecules in the membrane of the photoreceptor connecting cilium.

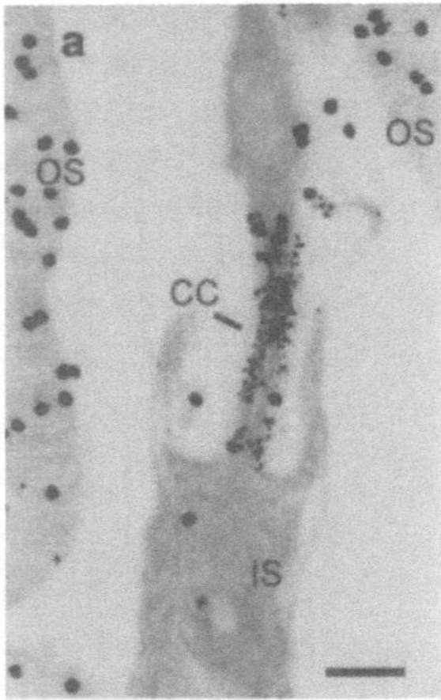

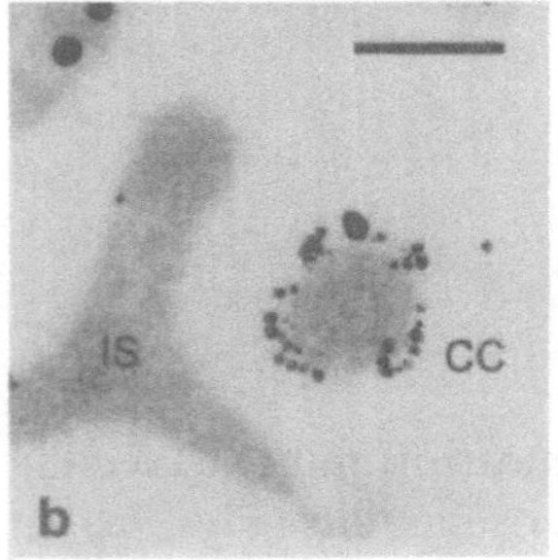

Figure 5. Co-localization of opsin and myosin VIIa in photoreceptor connecting cilium (CC). Silver-enhanced immunogold double labeling of opsin (large particles) and myosin VIIa (small particles) in a tangential section through the membrane of the connecting cilium (**a.**) and of a transverse section through the connecting cilium of mouse photoreceptor cells (**b.**). Opsin and myosin VIIa are co-localized at the connecting cilium (CC). Scale bars: a = 250 nm, b = 230 nm.

DISCUSSION

Our results indicate that, in the modified connecting cilium of mammalian photoreceptor cells, an actin-based membrane-associated cytoskeletal exists parallel to the characteristic ciliary microtubule cytoskeleton. In previous studies, actin filaments have been independently detected in the distal domain of the photoreceptor cilium by several authors.[27–33] At the base of the outer segment, actin filaments are believed to play a key role in the formation of newly generated membrane disks of the outer segment which is the basis for continuous turnover of the phototransductive membrane.[8] Further, investigations revealed that, in the distal cilium, these actin filaments are co-localized with conventional myosin II, supporting their function in membrane renewal.[32,33] In the proximal portion of the cilium, weak immunogold labeling at the ciliary membrane has been previously documented (see Fig. 8 in Arikawa and Williams),[30] but it has not been considered to be of functional significance. Our present highly-sensitive immunolabeling including silver-enhancement of nanogold™ probes confirmed the actin localization at the membrane of the connecting cilium. Our results are also in consistance with the axonemal detection of an actin-like protein in the flagellum of the green algae *Chlamydomonas*.[35] The ciliary actin may form filaments which run underneath the plasma membrane parallel to the axonemal microtubule doublets in longitudinal orientation (Fig. 6).

At the plasma membrane of the photoreceptor connecting cilium, the actin filaments are spacially associated with the unconventional myosin VIIa which we have

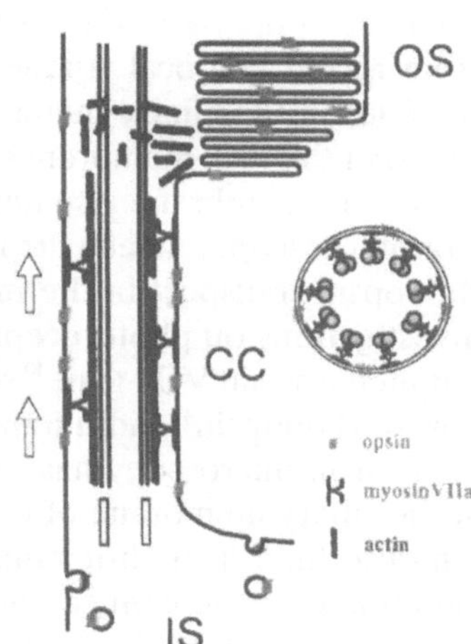

Figure 6. Diagrams of part of a rod photoreceptor cell and of an enlarged transverse section of the connecting cilium on the right illustrating the hypothesis of rhodopsin trafficing through the connecting cilium. The distribution of opsin, mysoin VIIa and actin is indicated by symbols on the lower right. Rhodopsin loaden post-Goli vesicles fuse with the plasma membrane of the apical inner segment (IS) and rhodopsin contaning membrane plaques are transported in the plasma membrane of the connecting cilium (CC) to the base of the outer segment (OS) where the membrane is incorborated into newly formed disks. This transport is probably driven by myosin VIIa moveing along axonemal actin filaments. Arrows indicate the direction of the rhodopsin transport in the plasma membrane of the connecting cilium (CC).

localized at the ciliary membrane (Fig. 2).[7,26] Myosin VIIa exhibits all the features of an actin-based mechanoenzym which forms dimers directly associated with the plasma membrane via talin-like sequences in the tail domain.[4,5] Futhermore, studies on mechanosensitive hair cells in cultured organs of Corti indicate that myosin VIIa is involved in membrane translocalization during hair cell differentiation.[36] Therefore, there is good evidence that myosin VIIa localized at the plasma membrane of the photoreceptor cilium is involved in membrane trafficing. Myosin VIIa dimers linked to the plasma membrane via their N-terminal tail may move with their actin-binding motor domain along axonemal actin filaments pulling membrane plaques inclusive with their contents (Fig. 6).

In mammalian photoreceptor cells, visual transduction occurs at the disk membranes densely packed in the outer segments. As other components of the transduction machinery of the outer segment, membrane proteins are synthesized *de novo* in the inner segment of the cell and transported in the membrane of post-Golgi vesicles to the base of the connecting cilium.[9,10] Recent studies have revealed that inner segment trafficing of rhodopsin, the major protein in outer segment disks, is regulated by its C-terminal[11,12] and that the directional rhodopsin transport is driven along the microtubule network of the inner segment by cytoplasmic dynein interacting directly with the C-terminal domain of rhodopsin.[14] There is also considerable evidence that rhodopsin-laden post-Golgi vesicles fuse with the plasma membrane at the perciliary region of the apical inner segment.[9,10] However, the final step of rhodopsin delivery to the base of the outer segment is not yet understood and has been an enigma for decades.[15] Although, a ciliary pathway of rhodopsin delivery to the outer segment base and it is commonly cited, there was not yet clear, direct evidence supporting the ciliary transport model. Many research efforts have been focused on following rhodopsin during its course through the cilium in the ciliary plasma membrane. However, weak opsin immunoreactivity at the connecting cilium has been considered as an obstacle to this model and alternative hypothesis were favored[37] (see also review: Besharse and Horst.)[15] Present results on intense nanogold™ labeling of rhodopsin in the ciliary membrane of rod photoreceptor cells provides first considerable direct electron microscopical evidence for a ciliary transport of rhodopsin from the inner segment to the base of the outer segment. Our results are als consistent with the data from recent confocal studies,[38] and freeze-fracture experiments discribing opsin-like particles in the ciliary membrane similar in shape and size to the rhodopsin particles found in the disk membranes.[39] The spacial co-localization of opsin and myosin VIIa at the ciliary membrane also demonstrated here by double labeling experiments on electron microscopic niveau strongly indicates that myosin VIIa is probably involved in rhodopsin transport in the membrane of the photoreceptor cilium. Moreover, recent investigations on photoreceptor cells in the retina of the shaker-1 mouse containing a mutated myosin VIIa gene[40] support the hypothesis of myosin VIIa driven ciliary transport of rhodopsin[41] (additional own unpublished observations). In these studies, immunoelectron microscopy has revealed that the number of labeled rhodopsin molecules in the ciliary membrane of rod photoreceptor cells is about 50% higher in the shaker-1 mouse than in its litter mates indicating disfunction of rhodopsin trafficing in the membrane of the connecting cilium. Based on our results, we postulate the working hypothesis illustrated in figure 6. Although, direct interaction between myosin VIIa and rhodopsin have not yet been demonstrated both proteins may be components of a transport-complex probably associated with additional proteins. However, it is also possible that myosin VIIa binds directly to the acid phospholipids of the ciliary membrane and transports membrane plaques including membrane proteins without any physically contact to them.

Several previous reports indicate that Usher syndrome patients have structural abnormalities in cilia present in several tissues including the neuronal retina.[16,42,43] Because all these structural changes might be caused by defects in ciliary function, it has been suggested that disfunctions of cilia might be the basis of the degenerative processes experienced by patients with Usher syndrome (e.g. Hunter et al.[43]). Our results are consistant with this hypothesis: The *retinitis pigmentosa* in Usher 1B patients carrying mutated myosin VIIa genes might result from a defective ciliary transport of outer segment components by non-functional myosin VIIa. The failure of proper rhodopsin transport probably also cause some other types of *retinitis pigmentosa.*[44,45]

In conclusion, we have localized by immunological methods a motile system based on actin and myosin VIIa at the membrane of the photoreceptor connecting cilium and suggest that rhodopsin also found in the membrane is translocated by this system in the plasma membrane of the photoreceptor cilium. The disfunction of ciliary transport in photoreceptor cells may underlie the *retinitis pigmentosa* found in Usher syndrome 1B patients. It will be interesting to determine whether myosin VIIa forms a complex with rhodopsin and additional associated proteins in the cilum, and how this suggested rhodopsin transport is regulated.

ACKNOWLEDGMENTS

Present studies are supported by Deutsche Forschungs Gemeinschaft, grant Wo 548/3-1, and FAUN-Stiftung, Nürnberg, Germany (to U.W.). The authors are most grateful to Dr. P. A. Hargrave, for kindly supplying antibodies and peptides used in present investigations and to Dr. D. S. Williams, for his collaboration in the generation of antisera against myosin VIIa. We would like to thank R. Müller, for skillful technical assistance, and C. Bode, for his assistance with immunoelectron microscopical double labelings. We also thank B. K. Huntley for helpful comments on the manuscript and attentive linguistic corrections.

REFERENCES

1. C.H. Usher, 1914, On the inheritance of Retinitis pigmentosa, with notes of cases, *R Lond Ophthalmol. Hosp. Rep.* **19**:130–236.
2. W. Kimberling and C. Möller, 1995, Clinical and molecular genetics of Usher syndrome, *J. Am. Acad. Audiol.* **6**:63–72.
3. M. Wagenaar, B. Terrahe, A. VanAaarem, P. Huygen, R. Admiraal, E. Bleeker-Wagemakers, A. Pinckers, W. Kimberling, and C. Cremers, 1995, Clinical findings in obligate carriers of type I Usher syndrome, *Am. J. Med. Genet.* **59**:375–379.
4. D. Weil, S. Blanchard, J. Kaplan, P. Guilford, F. Gibson, J. Walsh, P. Mburu, A. Varela, J. Levilliers, M.D. Weston, P.M. Kelley, W.J. Kimberling, M. Wagenaar, F. Levi-Acobas, D. Larget-Piet, A. Munnich, K.P. Steel, S.D.M. Brown, and C. Petit, 1995, Defective myosin VIIA gene responsible for Usher syndrome type 1B, *Nature* **374**:60–61.
5. Z.Y. Chen, T. Hasson, P.M. Kelley, B.J. Schwender, M.F. Schwartz, M. Ramakrishnan, W.J. Kimberling, M.S. Mooseker, and D.P. Corey, 1996, Molecular cloning and domain structure of human myosin-VIIa. The gene product defective in usher syndrome 1B, *Genomics* **36**:440–448.
6. D. Weil, G. Levy, I. Sahly, F. Levia-Cobas, S. Blanchard, A. El-Amraoui, F. Crozet, H. Philippe, M. Abitbol, and C. Petit, 1996, Human myosin VIIA responsible for the Usher 1B syndrome: A predicted membrane-associated motor protein expressed in developing sensory epithelia, *Proc. Natl. Acad. Sci. USA* **93**:3232–3237.
7. X.R. Liu, G. Vansant, I.P. Udovichenko, U. Wolfrum, and D.S. Williams, 1997, Myosin VIIa, the product of the Usher 1B syndrome gene, is concentrated in the connecting cilia of photoreceptor cells, *Cell Motil. Cytoskeleton* **37**:240–252.
8. R.W. Young, 1976, Visual cells and the concept of renewal, *Invest. Ophthalmol.* **15**:700–725.
9. D.S. Papermaster, B.G. Schneider, and J.C. Besharse, 1985, Vesicular transport of newly synthesized opsin from the Golgi apparatus toward the rod outer segment, *Invest. Ophthalmol. Visual Sci.* **26**:1386–1404.
10. D.S. Papermaster, B.G. Schneider, D. DeFoe, and J.C. Besharse, 1986, Biosynthesis and vectorial transport of opsin on vesicles in retinal rod photorecptors, *J. Histochem. Cytochem.* **34**:5–16.
11. D. Deretic and D.S. Papermaster, 1995, The role of small G-proteins in the transport of newly synthesized rhodopsin, *Prog. Retin. Eye. Res.* **14**:249–265.
12. D. Deretic, S. Schmerl, P.A. Hargrave, A. Arendt, and J.H. McDowell, 1998, Regulation of sorting and post-Golgi trafficking of rhodopsin by its C-terminal sequence QVS(A)PA, *Proc. Natl. Acad. Sci. USA* **95**:10620–10625.
13. Chuang and C.H. Sung, 1998, The cytoplasmic tail of rhodopsin acts as a novel apical sorting signal in polarized MDCK cells, *J. Cell Biol.* **142**:1245–1256.
14. A.W. Tai, J.Z. Chuang, C. Bode, U. Wolfrum, and C.H. Sung, 1998, Rhodopsin's carboxy-terminal cytoplasmic tail act as a membrane receptor for cytoplasmic dynein via ist light chain Tctex-1 (*submitted*).
15. J.C. Besharse and C.J. Horst, 1990, The photoreceptor connecting cilium—a model for the transition tone, in: *Ciliary and flagellar membranes* (R.A. Bloodgood, ed.), pp. 389–417, Plenum Publishing Coporation, New York.
16. S.D. Barrong, M. Chaitin, S. Fliesler, D. Possin, S. Jacobson, and A. Milam, 1992, Ultrastructure of connecting cilia in different forms of retinitis pigmentosa. *Arch. Ophthalmol.* **110**:706–710.
17. U. Wolfrum, X.R. Liu, A. Schmitt, I.P. Udovichenko, and D.S. Williams, 1998, Myosin VIIa as a common component of cilia and microvilli, *Cell Motil. Cytoskeleton* **40**:261–271.
18. G. Adamus, A. Arendt, Z.S. Zam, J.H. Mcdowell, and P.A. Hargrave, 1988, Use of peptides to select for anti-rhodopsin antibodies with desired amino acid sequence specificities, *Peptide Res.* **1**:42–47.
19. G. Adamus, Z.S. Zam, A. Arendt, K. Palczewski, J.H. Mcdowell, and P.A. Hargrave, 1991, Anti-rhodopsin monoclonal antibodies of defined specificity: characterization and application, *Vision Res.* **31**:17–31.
20. J.L. Lessard, 1988, Two monoclonal antibodies to actin: one muscle selective and one generally reactive, *Cell Motil. Cytoskeleton* **10**:349–362.
21. U. Wolfrum, 1995, Centrin in the photoreceptor cells of mammalian retinae, *Cell Motil. Cytoskeleton* **32**:55–64.
22. D. Fleischman and M. Denisevich, 1979, Guanylate cyclase of isolated bovine retinal rod axonemes, *Biochem.* **18**:5060–5066.

23. C.J. Horst, D.M. Forestner, and J.C. Besharse, 1987, Cytoskeletal-membrane interactions: Between cell surface glycoconjugates and doublet microtubules of the photoreceptor connecting cilium, *J. Cell Biol.* **105**:2973–2987.
24. U. Wolfrum and A. Schmitt, 1998, Rhodopsin transport in the membrane of the photorecptor cilium: myosin VIIa the product the Usher 1B gene is involved. (*submitted*).
25. G. Danscher, 1981, Localization of gold in biological tissue. A photochemical method for light and electron microscopy, *Histochem.* **71**:81–88.
26. U. Wolfrum, A. Schmitt, D.S. Williams, and R. Paulsen, 1997, Are centrin and myosin VIIa, the Usher gene product, co-localized in photoreceptors, olfactory cells, and hair cells? in: *Proc. of the 25th Göttingen Neurobiology Conference* (N. Elsner and H. Wässle, eds.), pp. 523, Thieme, Stuttgart.
27. M.H. Chaitin, B.G. Schneider, M.O. Hall, and D.S. Papermaster, 1984, Actin in the photoreceptor connecting cilium: Immunocytochemical localization to the site of outer segment disk formation, *J. Cell Biol.* **99**:239–247.
28. M.H. Chaitin and D. Bok, 1986, Immunoferritin localization of actin in retinal photoreceptors, *Invest. Ophthalmol. Visual Sci.* **27**:1764–1767.
29. D.S. Williams, K.A. Linberg, D.K. Vaughan, R.N. Fariss, and S.K. Fisher, 1988, disruption of microfilament organization and deregulation of disk membrane morphogenesis by Cytochalasin D in rod and cone photoreceptors, *J. Comp. Neurol.* **272**:161–176.
30. K. Arikawa and D.S. Williams, 1989, Organization of actin filaments and immunocolocalization of alpha-actinin in connacting cilium of rat photorecetors, *J. Comp. Neurol.* **288**:640–646.
31. M.H. Chaitin and B. Burnside, 1989, Actin filament polarity at the site of rod outer segment disk mormhogenesis, *Invest. Ophthalmol. Visual Sci.* **30**:2461–2469.
32. M.H. Chaitin and N. Coelho, 1992, Immunogold localization of myosin in the photoreceptor cilium, *Invest. Ophthalmol. Visual Sci.* **33**:3103–3108.
33. D.S. Williams, M.A. Hallett, and K. Arikawa, 1992, Association of myosin with the connecting cilium of rod photoreceptor cells, *J. Cell Sci.* **103**:183–190.
34. J.C. Besharse, 1986, Photosensitive membrane turnover: differentated membrane domains and cell-cell interaction, in: *The retina a model for biological studies. Part I* (R. Adler, D. Faber, eds.), pp. 297–352, Academic Press, Florida.
35. G. Piperno and D.J.L. Luck, 1979, An actin-like protein is a component of axonemes from *Chlamydomonas*, *J. Biol. Chem.* **254**:2187–2190.
36. G.P. Richardson, A. Forge, C.J. Kros, J. Fleming, S.D.M. Brown, and K.P. Steel, 1997, Myosin VIIA is required for aminoglycoside accumulation in cochlear hair cells, *J. Neurosci.* **17**:9506–9519.
37. J.C. Besharse and M.G. Wetzel, 1995, Immunocytochemical localization of opsin in rod photoreceptors during periods of rapid disc assembly, *J. Neurocytol.* **24**:371–388.
38. B. Matsumoto and I.L. Hale, 1993, Preparation of retinas for studying photoreceptors with confocal microscopy, in: *Methods in neurosciences Vol. 15: Photoreceptor cells.* (P.A. Hargrave, ed.), pp. 54–71, Academic Press, San Diego.
39. K. Miyaguchi and P.H. Hashimoto, 1992, Evidence for the transport of opsin in the connecting cilium and basal rod outer segment in rat retina—rapid-Freeze, deep-etch, and horseradish peroxidase labelling studies, *J. Neurocytol.* **21**:449–457.
40. F. Gibson, J. Walsh, P. Mburu, A. Varela, K.A. Brown, M. Antonio, K.W. Beisel, K.P. Steel, and S.D.M. Brown, 1995, A type VII myosin encoded by the mouse deafness gene shaker-1, *Nature* **374**:62–64.
41. D.S. Williams and X. Liu, 1998, Myosin VIIa is requiered for normal opsin transport through the connecting cilia of photoreceptors, *Mol. Biol. Cell [Supplement]* **9**:389a.
42. G.B. Arden and B. Fox, 1979, Increased incidence of abnormal nasal cilia in patients with *Retinitis pigmentosa*, *Nature* **279**:534–536.
43. D.G. Hunter, G.A. Fishman, R.S. Metha, and F.L. Kretzer, 1986, Abnormal sperm and photoreceptor axonemes in Usher's syndrome, *Arch. Ophthalmol.* **104**:385–389.
44. C.H. Sung, C. Makino, D. Baylor, and J. Nathans, 1994, A rhodopsin gene mutation responsible for autosomal dominant retinitis pigmentosa results in a protein that is defective in localization to the photoreceptor outer segment, *J. Neurosci.* **14**:5818–5833.
45. D.J. Roof, M. Adamian, and A. Hayes, 1994, Rhodopsin accumulation at abnormal sites in retinas of mice with a human P23H rhodopsin transgene, *Invest. Ophthalmol. Visual Sci.* **35**:4049–4062.

2

BLINDNESS IN USHER SYNDROME 1B

Myosin VIIa in the Retina

David S. Williams, Xinran Liu, Gordon Vansant, and Brian Ondek

Departments of Pharmacology and Neurosciences
University of California, San Diego School of Medicine
La Jolla, California 92093-0983

SUMMARY

The turnover of phototransductive membrane is an essential process for the viability of photoreceptor cells. Molecular motors are likely candidates for the trafficking of phototransductive membrane during various stages of its turnover. Our studies of molecular motors in photoreceptor cells and the RPE are aimed at understanding mechanisms involved in phototransductive membrane turnover and the cellular basis of some forms of retinal degeneration. Results described here represent initial studies on the role of myosin VIIa in the retina. Defects in the myosin VIIa gene have been found to cause forms of Usher syndrome. Myosin VIIa was localized in the apical processes of the RPE and in the connecting cilium of the photoreceptor cells. In the retinas of shaker-1 mice, which express mutant myosin VIIa, the only identifiable structural defect was the mislocalization of the melanosomes of the RPE. Disk membrane shedding did not appear to be affected. Nevertheless, the localization of myosin VIIa in the photoreceptor cilium suggests that the protein might be involved in transporting phototransductive membrane to the outer segment.

INTRODUCTION

A photoreceptor cell is the epitome of a cell adapted for a specialized function. It is extremely polarized and compartmentalized. The most specialized compartment is the one that contains the phototransductive membrane—the outer segment in vertebrate photoreceptors, and the rhabdomere in arthropod and molluscan photorecep-

Retinal Degenerative Diseases and Experimental Therapy, edited by Hollyfield *et al.*
Kluwer Academic / Plenum Publishers, New York, 1999.

tors. It has a very well-defined overall shape, with the transductive membrane packed efficiently within. Its organization provides maximal photon capture, given the constraints on its size that are imposed by the demands of visual acuity.

Many cells have short lives—that is the cells themselves are renewed. But in cells that are not replaced, such as a photoreceptor cell, the components of the cell must be turned over to prevent the build up of spontaneous macromolecular changes that might interfere with function. To turn over their phototransductive membrane, a vertebrate photoreceptor cell sheds membrane disks from the distal end of its outer segment. This process involves phagocytosis by the retinal pigment epithelium (RPE),[1] and indeed attachment of the outer segments to the RPE is required for the initiation of shedding.[2] The shed membrane disks are replaced by the formation of new ones.[3] New disk membrane is transported from its site of synthesis in the inner segment to the base of the outer segment,[4] where disk membrane morphogenesis occurs by an evagination of the plasma membrane of the connecting cilium.[5]

For a photoreceptor cell, turnover of its phototransductive membrane is a major challenge. The membrane is not turned over especially quickly. In primates and mice, it takes 8–10 days before a new disk migrates to the distal end of the outer segment and is shed.[3,6,7] The remarkable feature is that there is so much membrane and membrane protein (mainly opsin) in an outer segment, that the amount of membrane turnover is huge. The importance of phototransductive membrane turnover to the photoreceptor cell is manifest when one considers that in each human retina, some 9 billion opsin molecules are turned over on average every second*—that's one trillion per minute per person.

In recent years, it has become clearer that processes involved in phototransductive membrane turnover may be critical for photoreceptor cell viability. We had known this for some time in the case of the RCS rat, in which phagocytosis by the RPE is defective,[10] the prcd dog, in which disk membrane renewal is retarded,[11] and the rds mouse, in which disk membrane morphogenesis is blocked[12] as a result of an effective null mutation in the peripherin/rds gene.[13,14] Photoreceptor degeneration results in each case. Now, two studies on rhodopsin transgenic mice, one by Li et al.,[15] and one by Liu et al.,[16] have shown that some forms of mutant rhodopsin result in defective disk membrane morphogenesis prior to cell death. In addition, the absence of rom-1 results in the abnormal formation of disks and photoreceptor degeneration.[17] It is probably fair to assume that a major defect in some stage of the turnover of phototransductive membrane will result in photoreceptor cell death. Thus, understanding the mechanisms involved in phototransductive membrane turnover are likely to help understand the cellular basis of some forms of photoreceptor cell degeneration. Conversely, genetic defects that are responsible for photoreceptor degeneration may be linked to proteins that play essential roles in phototransductive membrane turnover. This would seem to be especially so, if the gene encodes a molecular motor. With their potential for transporting membrane vesicles and organelles, molecular motors that are critical for photoreceptor cell viability are surely very good candidates to function in phototransductive membrane turnover.

So far, two unconventional myosins—a class III myosin and a class VII myosin—have been linked to retinal degeneration. In Drosophila photoreceptors, a myosin III

* Based on a renewal rate in primate rod outer segments of 3.2 μm, or 112 membrane disks, per day,[6] an opsin concentration of $20{,}000/\mu m^2$ in the disk membrane,[8] and that each human retina contains about 100 million photoreceptor cells, of which 95% are rods.[9]

gene encodes the two ninaC proteins.[18] Mutations in the ninaC gene result in degeneration of the photoreceptor cells.[19] The ninaC proteins are rather odd myosins in that they have a kinase domain appended to the N-terminal of the myosin motor domain.[18,20] It is the myosin domain, however, that is critical for photoreceptor cell viability. Deletion of the kinase domain results in a defective electroretinogram, but not in retinal degeneration. In contrast, deletion of the myosin motor domain results in photoreceptor cell degeneration.[21]

The ninaC proteins have been shown to interact with actin filaments like true myosins,[22] but we have only sketchy information on their in vivo motor function. The two proteins have different distributions. The larger protein, p174, is found exclusively in the rhabdomere,[23,24] where it appears to link the microvillar phototransductive membrane to an axial actin cytoskeleton.[22] The smaller protein, p132, is present in the extrarhabdomeral cytoplasm.[23,24] It has been suggested that an essential role for p174 may be to transport calmodulin to the rhabdomere, or tether it there.[25] Alternatively, its placement as a link between the phototransductive membrane and actin filaments, together with the orientation of the actin filaments,[26] is consistent with a role in moving the microvillar membrane along the axial cytoskeleton.[22] The p132 protein may also function in trafficking phototransductive membrane, for in photoreceptors of flies lacking just this protein abnormal amounts of membrane accumulate in multivesicular bodies within the extrarhabdomeral cytoplasm.[22]

The second unconventional myosin to be linked to retinal degeneration is a class VII myosin, myosin VIIa. Defects in the myosin VIIa gene have been linked to Usher syndrome 1B,[27] and to atypical Usher syndrome.[28] Usher syndrome is an inherited blindness and deafness disorder. This paper concerns our initial studies aimed at elucidating the retinal function of myosin VIIa, and, in particular, determining whether this motor protein functions in the turnover of phototransductive membrane.

MATERIALS AND METHODS

Antibodies

We attempted to generate rabbit antiserum against several regions of myosin VIIa. At the time, most of the motor domain sequence (except for the N-terminal region) and a small part of the N-terminal region of the tail had been described.[27,29] A large part of the motor domain is homologous with other myosins, but we identified two separate peptide sequences that were unique to myosin VIIa, were hydrophilic, and appeared to be located on the surface of the molecule by comparison with the known structure of myosin II. These two peptides were ^{337}EARTFENLDACE and ^{475}EEYDLESID. They were synthesized with a Lys residue added to the N-terminal and conjugated to BSA with glutaraldehyde. Antisera thus generated did not give strong labeling of retinas in immunoelectron microscopy, although one lot produced in response to ^{337}EARTFENLDACE gave a distinct signal and a labeling pattern that was comparable to that of more reactive antisera obtained from a recombinant myosin VIIa protein. This recombinant protein corresponded to 131 amino acids of the tail domain of mouse myosin VIIa (Fig. 1). It contained a His_6 tag to facilitate purification and was produced by expression of a myosin VIIa RT-PCR fragment cloned into in a pET23b vector and transformed in BL21 DE3 [lys(s)] cells (see Liu et al.[30] for details).

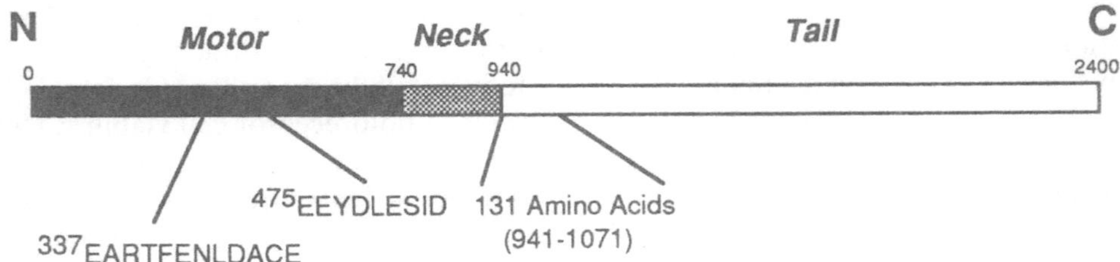

Figure 1. Diagram illustrating the motor, neck, and tail domains of myosin VIIa and the regions that were used to generate antibodies.

The antibodies were characterized by Western blot analysis.[30] Monospecific antibodies were used for immunogold labeling of retinal sections, as described.[30]

Shaker-1 Mice

Mice carrying the original shaker-1 allele (*sh1*) were purchased from Jackson Laboratories and used to start a breeding stock. They were treated according to NIH and UCSD animal care guidelines. They were maintained on a 12-hr light/12-hr dark cycle, with exposure to 10–50 lux of fluorescent lighting during the light phase. Ages of the mice used ranged from 8 weeks to 34 weeks. Microscopy of their retinas was performed by standard procedures following embedment in Epon.

The following procedure was used to determine the genotype of individual mice. PCR products of 131 bp were obtained from mouse tail genomic DNA, using primers flanking the position of the *sh1* mutation, Arg[CGG]502Pro[CCG]. PCR products from wildtype and *sh1/sh1* littermates were initially cloned and sequenced for confirmation of the mutation. Routinely, the PCR products from wildtype, *sh1/+* and *sh1/sh1* mice were dot blotted in duplicate, and hybridized with ^{32}P-labeled probes: a wildtype oligonucleotide (5U-tgccaaccggcctatga-3U) and a *sh1/sh1* mutant oligonucleotide (5U-tcataggcgggttggca-3U, corresponding to the complementary strand). Blots were washed at a stringency such that hybridizations with a single base pair mismatch were destabilized, and could be distinguished from an identical match.

RESULTS

Localization of Myosin VIIa in the Retina

An important first step in investigating the cellular function of a protein is to determine its precise subcellular localization. The first immunolocalization study of myosin VIIa in the retina was described by Hasson et al.[31] They demonstrated by immunofluorescence microscopy of thick sections of rat retina that myosin VIIa was present in the apical region of the RPE. A subsequent study, by El-Amraoui et al.,[32] reported a difference in immunofluorescence labeling of myosin VIIa between rodent and human retinas. Their labeling of rodent retinas agreed with that of Hasson et al.;[31] label was found in the apical RPE. But, in human retinas, they found diffuse labeling throughout much of the photoreceptor cells, in addition to the apical RPE. These authors argued that the expression of myosin VIIa in human photoreceptor cells, but not in mouse photoreceptor cells, may explain why myosin VIIa mutations in humans

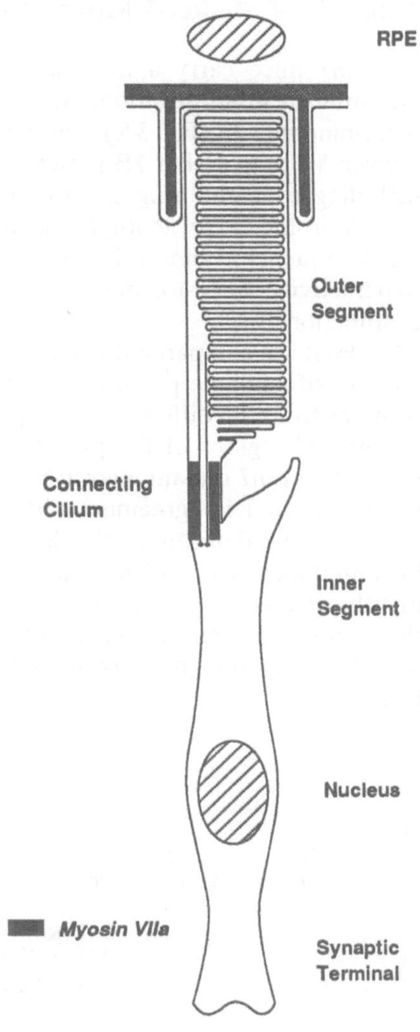

Figure 2. Diagram indicating the location of myosin VIIa in the retina. It is present in the apical RPE and the connecting cilium of each photoreceptor cell. Modified from Liu et al.[30]

result in blindness (in addition to deafness), whereas no retinal degeneration (only deafness) had been observed in mice with mutant myosin VIIa.

In immunoelectron microscopy studies of human and rodent retinas, we confirmed the presence of myosin VIIa in the apical RPE. It was located in the actin-rich apical processes that envelope the distal parts of the outer segments. In the photoreceptor cells, myosin VIIa was detected in the photoreceptor cilium. Labeling was concentrated primarily in the proximal part of the cilium; i.e. the part that links the inner and outer segments and is termed the "connecting cilium". Labeling of this very small domain is usually very difficult to detect at the level of a light microscope. We did not detect significant label elsewhere in the photoreceptor cells, and we found no difference in the distribution of label between human and rodent retinas. Nevertheless, for an unknown reason, labeling of the connecting cilium was usually more intense on sections of our human tissue than those of the mouse or rat tissue used. A summary of the distribution of myosin VIIa in the retina, as observed in our studies, is illustrated in Fig. 2. Original micrographs are shown in Liu et al.[30]

Analysis of Shaker-1 Mouse Retinas

sh1 mice carry a missense mutation in their myosin VIIa gene, resulting in an amino acid substitution, Arg502Pro.[29] This substitution occurs near the putative actin-binding site (Fig. 3A), suggesting a defect in the motor capability of the mutant myosin VIIa. In Usher 1B patients, various missense mutations in the motor domain, including a neighboring substitution, Pro503Leu, have been identified as causes of the syndrome.[33] The genotypes of littermates from *sh1/+* parents were determined by DNA sequencing and/or dot blots (Fig. 3A, B); *sh1/sh1* mice were also evident by their hyperactivity, head-tossing, and circling behavior, due to vestibular and cochlear dysfunction.[29]

First, we compared the amount of mutant myosin VIIa in *sh1/sh1* retinas with the amount of wildtype protein in retinas of control littermates. Western blots of retinal proteins from the different genotypes were labeled with myosin VIIa antibodies, which recognized regions of the protein well away from the mutated site (Fig. 1). Results showed that *sh1* mutant myosin VIIa and wildtype myosin VIIa are expressed at similar levels (Fig. 3C), in agreement with the findings of Hasson et al.[34]

Next, we determined the distribution of the mutant myosin VIIa in *sh1/sh1* retinas by immunoelectron microscopy. The mutant protein was found to have the same distribution—apical RPE (Fig. 4A, B) and the photoreceptor connecting cilium—as the wildtype protein in control mice. This finding indicates that myosin VIIa probably does not depend on its own motor capability for reaching its correct subcellular location.

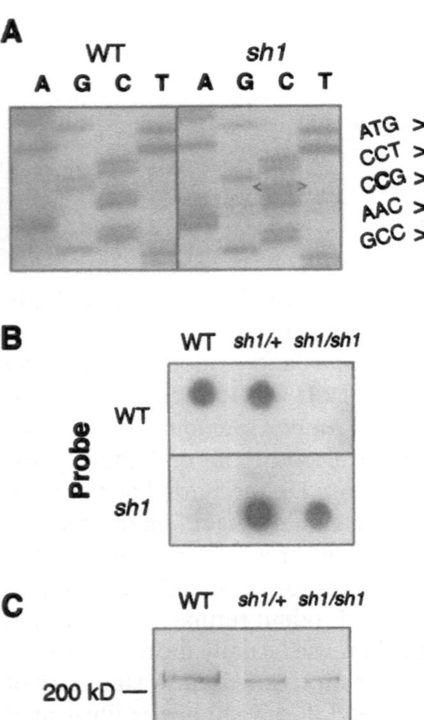

Figure 3. Genotype and retinal myosin VIIa content of *sh1* (original) mice. (A) DNA sequencing gel of wildtype myosin VIIa gene (left) and mutant myosin VIIa gene from a *sh1/sh1* mouse (right). A point mutation, G to C, in the *sh1* myosin VIIa gene results in an Arg502Pro change. (B) Example of a dot blot used to identify genotypes of mice bred from heterozygous parents. DNA was from wildtype (WT), *sh1* heterozygous (*sh1/+*), and *sh1* homozygous mutant (*sh1/sh1*) mice. In the upper panel, the probe corresponded to wild-type (WT) sequence. In the lower panel, the probe corresponded to the *sh1* sequence (*sh1*). (C) Western blot of proteins from neural retinas of normal (WT), heterozygous (*sh1/+*), and homozygous mutant (*sh1/sh1*) littermates. An equal amount of total protein was loaded in each lane, and the blot was labeled with myosin VIIa antibody.

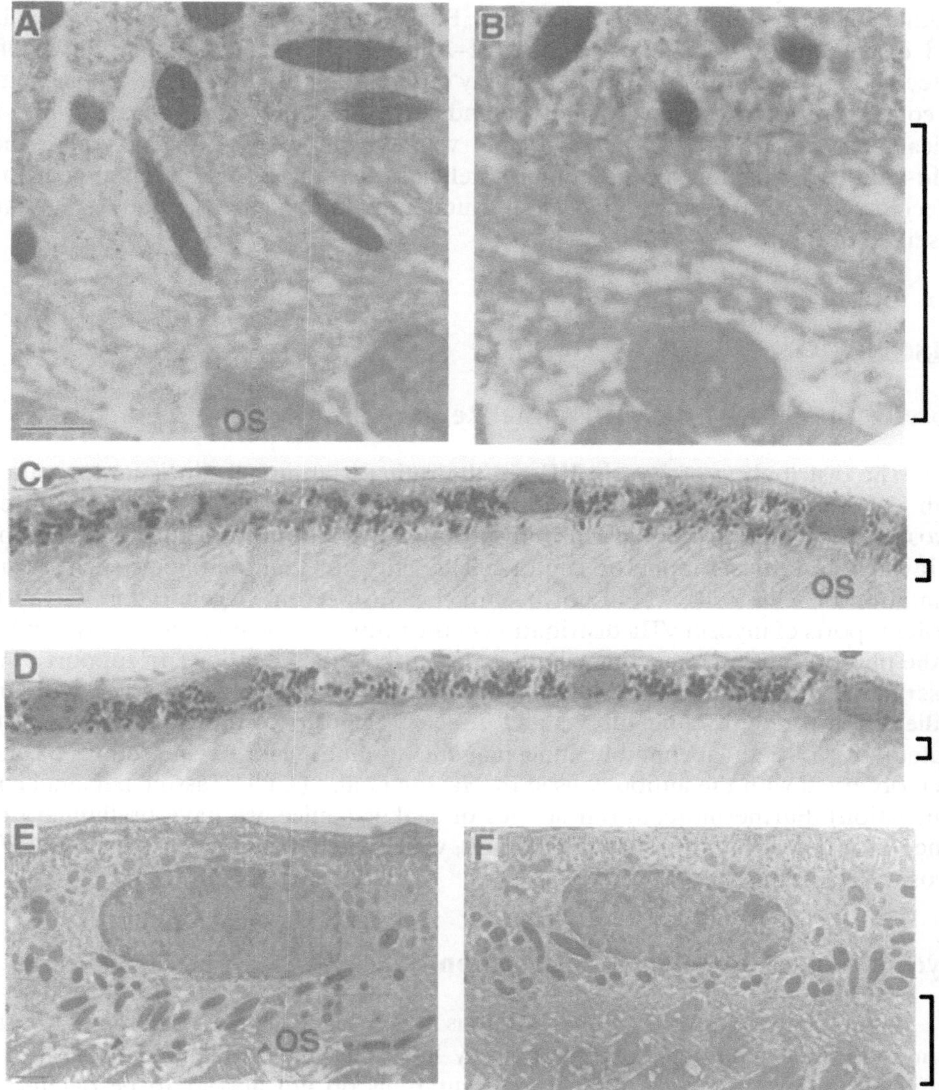

Figure 4. RPE of *shl/+* (A, C, E) and *shl/shl* (B, D, F) mice. Brackets (right) indicate apical processes. OS (arrowheads in E), photoreceptor outer segments. (A, B) Electron micrographs of apical processes immunogold-labeled with myosin VIIa antibody. Label is present in both wildtype and mutant. Scale bar = 0.5 μm. (C, D) Light micrographs of outer retina. Melanosomes are evident as black flecks. Scale bar = 5 μm. (E, F) Electron micrographs of individual RPE cells. Scale bar = 1 μm. Melanosomes are absent from the apical processes of *shl/shl* RPE. Mice for A-D were littermates; those for E and F were from another litter. From Liu et al.;[35] reproduced by permission of Nature America.

Last, we examined the retinas by microscopy. Up to the age of 34 weeks, there were no signs of retinal degeneration. The structure of the photoreceptor cells appeared normal, as did the membrane organization of the RPE apical processes. In consideration of a possible defect on the turnover of phototransductive membrane, we examined the shedding of disk membrane. Phagosomes were of normal size, number, and distribution in *sh1/sh1* RPE, fixed 1–2 hours after lights-on. The only defect detected in the *sh1/sh1* RPE was a distinctly different distribution of the melanosomes. In contrast to melanosomes in wildtype and *sh1/+* RPE, they did not extend into the apical processes (Fig. 4C–F). This defect was evident to the same extent in both light- and dark-adapted *sh1/sh1* retinas. Melanosome biogenesis did not appear to be affected, as melanosomes of all stages, including the final pigment granule stage, were observed in the cells.

DISCUSSION

Localization of Myosin VIIa in the Retina

The distribution of myosin VIIa in the retina is an important issue for patients with Usher syndrome caused by mutations in the myosin VIIa gene. The presence of myosin VIIa in photoreceptor cells indicates that a photoreceptor function of myosin VIIa might be critical for photoreceptor cell viability, and that treatments, such as transplantation of RPE cells, would not be sufficient to prevent degeneration. Given the earlier reports of myosin VIIa distribution in the retina,[31,32] the presence of myosin VIIa in the photoreceptor connecting cilium is presently still contentious. In support of our observations, we note that labeling of the connecting cilium was observed with antibodies from three different rabbits and directed against two different regions of the protein. In addition, labeling of connecting cilia in mouse and bovine retinas has now been observed with the antibody used by Hasson et al.[31] (Dr T. Hasson, personal communication). Furthermore, in our studies of shaker-1 mice, we have preliminary evidence of a defect in the connecting cilium, which is consistent with the presence of myosin VIIa in this structure.

Myosin VIIa and Retinal Degeneration

As noted in the Introduction, mutations in myosin VIIa have been found to cause forms of Usher syndrome.[27] Recently, they have also been linked to two forms of deafness without blindness,[36–38] so some mutations do not appear to be deleterious to retinal cell viability. The *sh1* mutation might fall into this class, especially since its effect on hearing is the mildest amongst all the known shaker-1 alleles.[39] However, none of the shaker-1 alleles have been found to exhibit retinal degeneration,[34] and some of these contain effective null mutations,[39] like some Usher 1B patients.[33] At present the reason why retinal degeneration resulting from mutant myosin VIIa has been observed only in humans is not clear. It is worth noting, however, that the rate of retinal degeneration typically varies greatly between man and mouse with the same genetic defect. In examples studied so far, it is more common to find degeneration occuring more quickly in the mouse retina. For example, some opsin mutations and the rd mouse (which has a mutant beta phosphodiesterase gene[40]) result in photoreceptor degeneration before the mouse photoreceptors are even fully developed.[41,42] In Usher 1B patients, retinitis pigmentosa is first detected typically at about 10 years of age.[43] If the deleterious effects of mutant myosin VIIa develop only slowly in a mouse retina, photoreceptor cell death may never become evident within a mouse's lifespan of 1–2 years.

Function of Myosin VIIa in the Retina

Analysis of shaker-1 mouse retinas indicates that the localization of melanosomes in the apical processes of the RPE requires proper myosin VIIa function. In fish and amphibian retinas, light induces melanosomes to migrate distally along the apical processes of the RPE cells in a manner that is dependent on actin filaments but not microtubules.[44–46] In contrast, the melanosomes in mouse RPE do not appear to move; comparisons of day and night adapted retinas show no difference in the position of the melanosomes (own observations). Despite this lack of adaptive movement in mammals, the present result suggests that melanosomes are conveyed at some stage along the apical processes and/or are held there by a process involving myosin VIIa. Thus, in the RPE, myosin VIIa may have a similar function to another large unconventional myosin, myosin Va,[47] which is necessary for melanosome localization in the dendrites of melanocytes.[48]

Interestingly, however, such a role seems unlikely in itself to be critical for retinal viability (retinas of albinos, although more susceptible to light damage, do not necessarily undergo degeneration). Therefore, the cause of Usher 1B blindness might lie primarily with the function of myosin VIIa in the photoreceptor cells. At present, we can only speculate on this function. A likely role, given the location of myosin VIIa in the connecting cilium, is in the transport of newly-synthesized opsin-containing membrane to the outer segment.

Outer segment proteins are synthesized in the inner segment and must therefore be routed to the outer segment. As mentioned in the Introduction, this involves large amounts of opsin, in particular. The most obvious route for the opsin is through the only link between the inner and outer segment, the connecting cilium. But, in attempts to immunodetect opsin in the connecting cilium, using a variety of procedures and antibodies, different researchers have found hardly any (e.g., refs[49,50]). This has led to the suggestion of opsin delivery to the outer segment by an extracellular route.[51] However, for the purpose of a simple calculation, let us suppose that myosin VIIa is able to move at a velocity similar to that measured *in vitro* for other myosins, such as myosin Va,[47] so that it would have the potential to transport opsin at about 500 nm/sec. Each mouse photoreceptor renews ~87 disks per day;[3,5] or, given an opsin concentration in the disk membrane of 20,000 opsins/μm^2 (ref 8), it renews an average of 72 opsins/sec. The surface area of a 1-μm length of ciliary plasma membrane is about 1 μm^2. At 500 nm/sec, opsin molecules would take 2 sec to travel this 1-μm length, so that the opsin concentration need only be 144/μm^2 to support the reported renewal rate. That is less than 1/100th of that in the disk membranes, making significant immunodetection improbable.

Hence, the presence of a motor protein, like myosin VIIa, functioning to transport opsin through the connecting cilium could account for the published images of opsin immunolabeling. Whether myosin VIIa functions in this manner remains to be determined in future studies. Moreover, the relationship of myosin VIIa's role to that of other motor proteins in the connecting cilium is unknown; in addition to myosin

VIIa, conventional myosin (myosin II),[52,53] and members of the kinesin family[54,55] have been detected in this structure.

ACKNOWLEDGMENTS

Research was supported by NIH grant EY07042 to DSW.

REFERENCES

1. Young, R.W. and Bok, D. 1969, Participation of the retinal pigment epithelium in the rod outer segment renewal process, *J. Cell Biol.* 42:392–403.
2. Williams, D.S. and Fisher, S.K. 1987, Prevention of the shedding of rod outer segment disks by detachment from the retinal pigment epithelium, *Invest. Ophthalmol. Vis. Sci.* 28:184–187.
3. Young, R.W. 1967, The renewal of photoreceptor cell outer segments, *J. Cell Biol.* 33:61–72.
4. Papermaster, D.S., Schneider, B.G., and Besharse, J.C. 1985, Vesicular transport of newly synthesized opsin from the Golgi apparatus toward the rod outer segment, *Invest. Ophthalmol. Vis. Sci.* 26:1386–1404.
5. Steinberg, R.H., Fisher, S.K., and Anderson, D.H. 1980, Disc morphogenesis in vertebrate photoreceptors, *J. Comp. Neurol.* 190:501–508.
6. Besharse, J.C. and Hollyfield, J.G. 1979, Turnover of mouse photoreceptor outer segments in constant light and darkness, *Invest. Ophthalmol. Vis. Sci.* 18:1019–1024.
7. Guerin, C.J., Lewis, G.P., Fisher, S.K., and Anderson, D.H. 1993, Recovery of photoreceptor outer segment length and analysis of membrane assembly rates in regenerating primate photoreceptor outer segment, *Invest. Ophthalmol. Vis. Sci.* 34:175–183.
8. Corless, J.M., Cobbs, W.H., Costello, M.J., and Robertson, J.D. 1976, On the asymmetry of frog retinal rod outer segment disk membranes, *Exp. Eye Res.* 23:295–324.
9. Curcio, C.A., Sloan, K.R., Kalina, R.E., and Hendrickson, A.E. 1990, Human photoreceptor topography, *J. Comp. Neurol.* 292:497–523.
10. Bok, D. and Hall, M.O. 1971, The role of the pigment epithelium in the etiology of inherited retinal dystrophy in the rat, *J. Cell Biol.* 49:664–682.
11. Aguirre, G., Alligood, J., O'Brien, P., and Buyukmihci, N. 1982, Pathogenesis of progressive rod-cone degeneration in miniature poodles, *Invest. Ophthalmol. Vis. Sci.* 23:610–630.
12. Jansen, H.G. and Sanyal, S. 1984, Development and degeneration of retina in rds mutant mice: electron microscopy, *J. Comp. Neurol.* 224:71–84.
13. Travis, G.H., Brennan, M.B., Danielson, P.E., Kozak, C.A., and Sutcliffe, J.G. 1989, Identification of a photoreceptor-specific mRNA encoded by the gene responsible for retinal degeneration slow (rds), *Nature* 338:70–73.
14. Connell, G., Bascom, R., Molday, L., Reid, D., McInnes, R.R., and Molday, R.S. 1991, Photoreceptor peripherin is the normal product of the gene responsible for retinal degeneration in the rds mouse, *Proc. Nat. Acad. Sci. (USA)* 88:723–726.
15. Li, T.S., Snyder, W.K., Olsson, J.E., and Dryja, T.P. 1996, Transgenic mice carrying the dominant rhodopsin mutation P347S—evidence for defective vectorial transport of rhodopsin to the outer segments, *Proc. Natl. Acad. Sci. USA* 93:14176–14181.
16. Liu, X., Wu, T.-H., Stowe, S., Matsushita, A., Arikawa, K., Naash, M.I., and Williams, D.S. 1997, Defective phototransductive disk membrane morphogenesis in transgenic mice expressing rhodopsin with a mutated N-terminal domain, *J. Cell Sci.* 110:2589–2597.
17. Clarke, G.A., Rossant, J., and McInnes, R.R. 1998, Rom-1 is required for outer segment morphogenesis and photoreceptor viability, *Invest. Opththamol. Vis. Sci.* 39:S962.
18. Montell, C. and Rubin, G.M. 1988, The Drosophila ninaC locus encodes two photoreceptor cell specific proteins with domains homologous to protein kinases and the myosin heavy chain head, *Cell* 52:757–772.
19. Matsumoto, H.K., Isono, K., Pye, Q., and Pak, W.L. 1987, Genes encoding cytoskeletal proteins in Drosophila rhabdomeres, *Proc. Natl. Acad. Sci. USA.* 84:985–989.

20. Ng, K.P., Kambara, T., Matsuura, M., Burke, M., and Ikebe, M. 1996, Identification of myosin III as a protein kinase, *Biochemistry* 35:9392–9399.
21. Porter, J.A. and Montell, C. 1993, Distinct roles of the Drosophila ninaC kinase and myosin domains revealed by systematic mutagenesis, *J. Cell Biol.* 122:601–612.
22. Hicks, J.L., Liu, X.R., and Williams, D.S. 1996, Role of the ninac proteins in photoreceptor cell structure: ultrastructure of ninac deletion mutants and binding to actin filaments, *Cell Motil. Cytoskel.* 35:367–379.
23. Hicks, J.L. and Williams, D.S. 1992, Distribution of the myosin I-like *ninaC* proteins in the *Drosophila* retina and ultrastructural analysis of mutant phenotypes, *J. Cell Sci.* 101:247–254.
24. Porter, J.A., Hicks, J.L., Williams, D.S., and Montell, C. 1992, Differential localizations of and requirements for the two Drosophila ninaC kinase/myosins in photoreceptor cells, *J. Cell Biol.* 116:683–693.
25. Porter, J.A., Yu, M., Doberstein, S.K., Pollard, T.D., and Montell, C. 1993, Dependence of calmodulin localization in the retina on the ninaC unconventional myosin, *Science* 262:1038–1042.
26. Arikawa, K., Hicks, J.L., and Williams, D.S. 1990, Identification of actin filaments in the rhabdomeral microvilli of Drosophila photoreceptors, *J. Cell Biol.* 110:1993–1998.
27. Weil, D., Blanchard, S., Kaplan, J., Guilford, P., Gibson, F., Walsh, J., Mburu, P., Varela, A., Levilliers, J., Weston, M.D., Kelley, P.M., Kimberling, W.J., Wagenaar, M., Levi-Acobas, F., Larget-Piet, D., Munnich, A., Steel, K.P., Brown, S.D.M., and Petit, C. 1995, Defective myosin VIIA gene responsible for Usher syndrome type 1B, *Nature* 374:60–61.
28. Liu, X.Z., Hope, C., Walsh, J., Newton, V., Mei, K.X., Liang, C., Xu, L.R., Zhou, J.M., Trump, D., Steel, K.P., Bundey, S., and Brown, S.D.M. 1998, Mutations in the myosin VIIA gene cause a wide phenotypic spectrum including atypical Usher syndrome, *Amer. J. Hum. Genet.* in press.
29. Gibson, F., Walsh, J., Mburu, P., Varela, A., Brown, K.A., Antonio, M., Beisel, K.W., Steel, K.P., and Brown, S.D.M. 1995, A type VII myosin encoded by mouse deafness gene shaker-1, *Nature* 374:62–64.
30. Liu, X., Vansant, G., Udovichenko, I.P., Wolfrum, U., and Williams, D.S. 1997, Myosin VIIa, the product of the Usher 1B syndrome gene, is concentrated in the connecting cilia of photoreceptor cells, *Cell Motil. Cytoskel.* 37:240–252.
31. Hasson, T., Heintzelman, M.B., Santos-Sacchi, J., Corey, D.P., and Mooseker, M.S. 1995, Expression in cochlea and retina of myosin VIIa, the gene product defective in Usher syndrome type 1B, *Proc. Natl. Acad. Sci. USA* 92:9815–9819.
32. El-Amraoui, A., Sahly, I., Picaud, S., Sahel, J., Abitbol, M., and Petit, C. 1996, Human Usher 1B/mouse *shaker-1*: the retinal phenotype discrepancy explained by the presence/absence of myosin VIIA in the photoreceptor cells, *Hum. Mol. Genet.* 5:1171–1178.
33. Weston, M.D., Kelley, P.M., Overbeck, L.D., Wagenaar, M., Orten, D.J., Hasson, T., Chen, Z.Y., Corey, D., Mooseker, M., Sumegi, J., Cremers, C. Moller, C., Jacobson, S.G., Gorin, M.B., and Kimberling, W.J. 1996, Myosin VIIa mutation screening in 189 Usher syndrome type 1 patients, *Amer. J. Hum. Genet.* 59:1074–1083.
34. Hasson, T., Walsh, J., Cable, J., Mooseker M.S., Brown, S.D. M., and Steel, K.P. 1997, Effects of shaker-1 mutations on myosin-VIIa protein and mRNA expression, *Cell Motil. Cytoskel.* 37:127–138.
35. Liu, X.R., Ondek, B., and Williams, D.S. 1998, Mutant myosin VIIa causes defective melanosome distribution in the RPE of shaker-1 mice, *Nature Genetics* 19:117–118.
36. Liu, X.-Z., Walsh, J., Mburu, P., KendrickJones, J., Cope, M.J. T.V., Steel, K.P., and Brown, S.D. M. 1997, Mutations in the myosin VIIA gene cause non-syndromic recessive deafness, *Nature Genetics* 16:188–190.
37. Liu, X.-Z., Walsh, J., Tamagawa, Y., Kitamura, K., Nishizawa, M., Steel, K.P., and Brown, S.D.M. 1997, Autosomal dominant non-syndromic deafness caused by a mutation in the myosin VIIA gene, *Nature Genetics* 17:268–269.
38. Weil, D., Kussel, P., Blanchard, S., Levy, G., LeviAcobas, F., Drira, M., Ayadi, H., and Petit, C. 1997, The autosomal recessive isolated deafness, DFNB2, and the Usher 1B syndrome are allelic defects of the myosin-VIIA gene, *Nature Genetics* 16:191–193.
39. Mburu, P., Liu, X.Z., Walsh, J., Saw, D., Jamie, M., Cope, T.V., Gibson, F., Kendrick-Jones, J., Steel, K.P., and Brown, S.D.M. 1997, Mutation analysis of the mouse myosin VIIA deafness gene, *Genes Funct.* 1:191–203.
40. Bowes, C., Li, T., Danciger, M., Baxter, L.C., Applebury, M.L., and Farber, D.B. 1990, Retinal degeneration in the rd mouse is caused by a defect in the beta subunit of rod cGMP-phosphodiesterase, *Nature* 347:677–680.

41. Carter-Dawson, L.D., LaVail, M.M., and Sidman, R.L. 1978, Differential effect of the rd mutation on rods and cones in the mouse retina, *Invest. Ophthalmol. Vis. Sci.* 17:489–498.
42. Olsson, J.E., Gordon, J.W., Pawlyk, B.S., Roof, D., Hayes, A., Molday, R.S., Mukai, S., Cowley, G.S., Berson, E.L., and Dryja, T.P. 1992, Transgenic mice with a rhodopsin mutation (Pro23His): a mouse model of autosomal dominant retinitis pigmentosa, *Neuron* 9:815–30.
43. Fishman, G.A., Kumar, A., Joseph, M.E., Torok, N., and Anderson, R.J. 1983, Usher's syndrome: ophthalmic and neuro-otologic findings suggesting genetic heterogeneity, *Arch. Ophthalmol.* 101: 1367–1374.
44. Murray, R.L. and Dubin, M.W. 1975, The occurrence of actinlike filaments in association with migrating pigment granules in frog retinal pigment epithelium, *J. Cell Biol.* 64:705–710.
45. Burnside, B., Adler, R., and O'Connor, P. 1983, Retinomotor pigment migration in pigment granule transport and cone movement, *Invest. Ophthalmol. Vis. Sci.* 24:1–15.
46. Troutt, L.L. and Burnside, B. 1989, Role of microtubules in pigment granule migration in teleost retinal pigment epithelial cells, *Exp. Eye Res.* 48:433–443.
47. Cheney, R.E., O'Shea, M.K., Heuser, J.E., Coelho, M.V., Wolenski, J.S., Espreafico, E.M., Forschner, P., Larson, R.E., and Mooseker, M.S. 1993, Brain myosin-V is a two-headed unconventional myosin with motor activity, *Cell* 75:13–23.
48. Provance, D.W., Wei, M., Ipe, V., and Mercer, J.A. 1996, Cultured melanocytes from dilute mutant mice exhibit dendritic morphology and altered melanosome distribution, *Proc. Natl. Acad. Sci. USA* 93:14554–14558.
49. Nir, I., Cohen, D., and Papermaster, D.S. 1984, Immunocytochemical localization of opsin in the cell membrane of developing rat retinal photoreceptors, *J. Cell Biol.* 98:1788–1795.
50. Besharse, J.C., Forestner, D.M., and Defoe, D.M. 1985, Membrane assembly in retinal photoreceptors. III. Distinct membrane domains of the connecting cilium of developing rods, *J. Neurosci.* 5:1035–1048.
51. Besharse, J.C. and Horst, C.J. 1990, The photoreceptor connecting cilium: a model for the transition zone, in: *Ciliary and flagellar membranes* (R.A. Bloodgood, eds.), pp. 409–431, Plenum Press, New York.
52. Williams, D.S., Hallett, M.A., and Arikawa, K. 1992, Association of myosin with the connecting cilium of rod photoreceptors, *J. Cell Sci.* 103:183–190.
53. Chaitin, M.H. and Coelho, N. 1992, Immunogold localization of myosin in the photoreceptor cilium, *Invest. Ophthalmol. Vis. Sci.* 33:3103–3108.
54. Beech, P.L., Pagh Roehl, K., Noda, Y., Hirokawa, N., Burnside, B., and Rosenbaum, J.L. 1996, Localization of kinesin superfamily proteins to the connecting cilium of fish photoreceptors, *J. Cell Sci.* 109:889–97.
55. Muresan, V., BendalaTufanisco, E., Hollander, B.A., and Besharse, J.C. 1997, Evidence for kinesin-related proteins associated with the axoneme of retinal photoreceptors, *Exp. Eye Res.* 64:895–903.

3

DIFFERENT MUTATIONS IN *RPE65* ARE ASSOCIATED WITH VARIABILITY IN THE SEVERITY OF RETINAL DYSTROPHY

Christian P. Hamel,[1,3]* Françoise Marlhens,[1] Jean-Michel Griffoin,[1] Corinne Bareil,[2] Mireille Claustres,[2] and Bernard Arnaud[3]

[1] INSERM U. 254
Laboratoire de Neurobiologie de l'Audition-Plasticité Synaptique
300, rue Auguste Broussonnet
34295 Montpellier cedex 5, France
[2] Laboratoire de Biochimie Génétique
Institut de Biologie
Boulevard Henri IV
34060 Montpellier
France
[3] Service d'Ophtalmologie
Hôpital Gui de Chauliac
2, avenue Bertin Sens
34295 Montpellier cedex 5
France

SUMMARY

RPE65 encodes a retinal pigment epithelium (RPE)-specific protein thought to be involved in the 11-*cis* retinoid metabolism, a key process in vision. To evaluate the visual implications of defects in this gene, we screened patients with various types of inherited retinal dystrophies. We found two nonsense mutations, 1121delA and R234X, in two sibs with Leber's congenital amaurosis. The presence of either one of the two

---

* Telephone: (33) 467 52 46 93, Fax: (33) 467 52 56 01, e-mail: biomol@mnet.fr

Retinal Degenerative Diseases and Experimental Therapy, edited by Hollyfield *et al.*
Kluwer Academic / Plenum Publishers, New York, 1999.

mutations on each allele indicates that they were compound heterozygotes. In a subsequent screening, two missense mutations, L22P and H68Y, were found in a patient with autosomal recessive retinitis pigmentosa. This individual was also a compound heterozygote. These results suggest that the level of impairment of the RPE65 protein, presumably abnormal activity in the case of amino acid substitutions and absence of protein in the case of null mutations, controls the rate at which the photoreceptors degenerate, and hence the severity of the phenotype.

1. INTRODUCTION

The retinal pigment epithelium (RPE) is a single cell-layered tissue in close contact with the photoreceptor outer segments. RPE performs many functions important for the physiology of photoreceptors,[1] some of them relying on RPE-specific proteins. These functions include the periodic phagocytosis of the outer segment tips,[2] a process whose defect leads to retinal degeneration,[3] and the metabolism of 11-*cis*-retinoids.[4] RPE65 is a 65-kDa protein specific to the RPE.[5] While the precise role of RPE65 remains unclear, its involvement in retinoid metabolism seems certain[6] and implies a functional relationship of RPE65 with photoreceptor physiology.

Recently, mutations in *RPE65* have been found to cause retinal dystrophy. They have been described in various clinical conditions including childhood-onset severe retinal dystrophy,[7] Leber's congenital amaurosis,[8,9] and retinitis pigmentosa.[9,10] We will review the mutations in *RPE65* that we have found in three patients, two of them with an apparent phenotype of Leber's congenital amaurosis (LCA) and the third one with a phenotype of retinitis pigmentosa (RP) followed by a short description of the clinical findings in these patients.

2. RESULTS

2.1. Mutations in Two Patients from One Family with LCA

The 14 exons of *RPE65* from 12 unrelated LCA patients from whom informed consent had been obtained were screened using single-strand conformation analysis (SSCA). In one family, an aberrant migration pattern was detected in exon 10 (patient III_2 in Fig. 1A). Direct sequencing of this exon revealed a deletion of nucleotide A at cDNA position 1121, in a stretch of consecutive adenine residues. This mutation, 1121delA, resulted in a frameshift and a premature termination at codon 373, 47 base pairs downstream from the deletion. This finding prompted us to further sequence the 13 other *RPE65* exons from this patient who was found to carry a second alteration in exon 7, substituting C for T at cDNA position 700 in a CpG site and resulting in a nonsense mutation, R234X. Both mutations were also found in the affected brother (patient III_3 in Fig. 1A).

The segregation of the mutations was followed in the patient's family using SSCA for 1121delA and restriction digest analysis for R234X since this latter mutation created an *Alw*N1 restriction site. This analysis revealed that 1121delA and R234X were from maternal and paternal origin, respectively, confirming that the two affected siblings III_2 and III_3 were compound heterozygotes (Fig. 1A). The heterozygous carriers of either

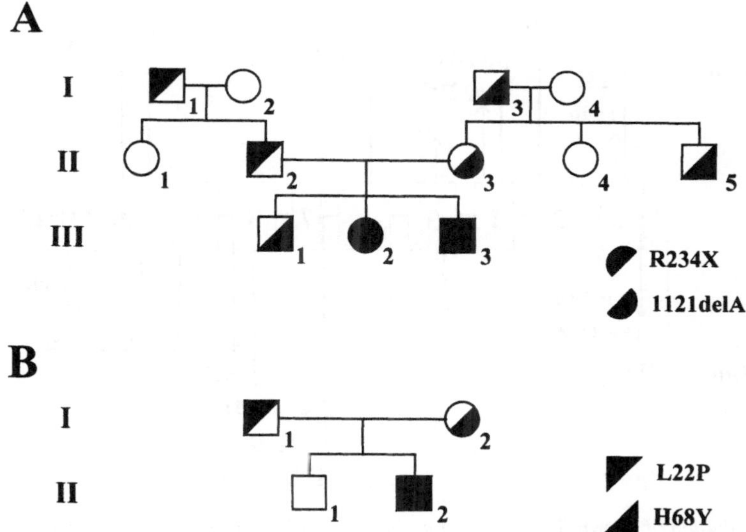

Figure 1. *Identification of mutations in RPE65*. A. Pedigree from a family with two sibs (patients III_2 and III_3) that suffered Leber's congenital amaurosis. Patients III_2 and III_3 were compound heterozygotes for translation-terminating mutations. Segregation of the paternal mutation R234X (due to a C → T transition at cDNA position 700 that changes an arginine residue for a premature stop) is depicted in upper left black symbols; that of the maternal mutation 1121 delA (resulting in a frameshift leading to a premature stop at codon 373) is shown in lower right black symbols. B. Pedigree from another family with one male (patient II_2) that suffered retinitis pigmentosa. Patient II_2 was a compound heterozygote for amino acid substitutions. Segregation of the paternal mutation L22P (due to a T → C substitution at cDNA position 65) is depicted in upper left black symbols; that of the maternal mutation H68Y (resulting from a C → T substitution at cDNA position 202) is shown in lower right black symbols.

one of the two mutations were asymptomatic. The possibility that they carry a SSCA-undetected mutation in the other *RPE65* allele with no visual impairment was excluded by sequencing all 14 *RPE65* exons in both parents of affected individuals.

2.2. Mutations in One Patient with RP

We then screened 184 unrelated patients with various types of retinal dystrophies from whom informed consent had been obtained. In one family, an aberrant migration pattern was detected in exon 3 (patient II_2 in Fig. 1B). Direct sequencing of this exon revealed a substitution of two consecutive nucleotides, changing G for T and C for T at cDNA positions 201 and 202, respectively. This substitution resulted in a missense mutation, H68Y. Following this finding, the 13 other exons from this patient were sequenced. He was found to carry a second alteration in exon 2, substituting T for C at cDNA position 65, and resulting in a second missense mutation, L22P.

The segregation of the mutations was followed in the patient's family by SSCA. This analysis revealed that H68Y and L22P were from maternal and paternal origins, respectively, confirming that the patient was a compound heterozygote (Fig. 1B).

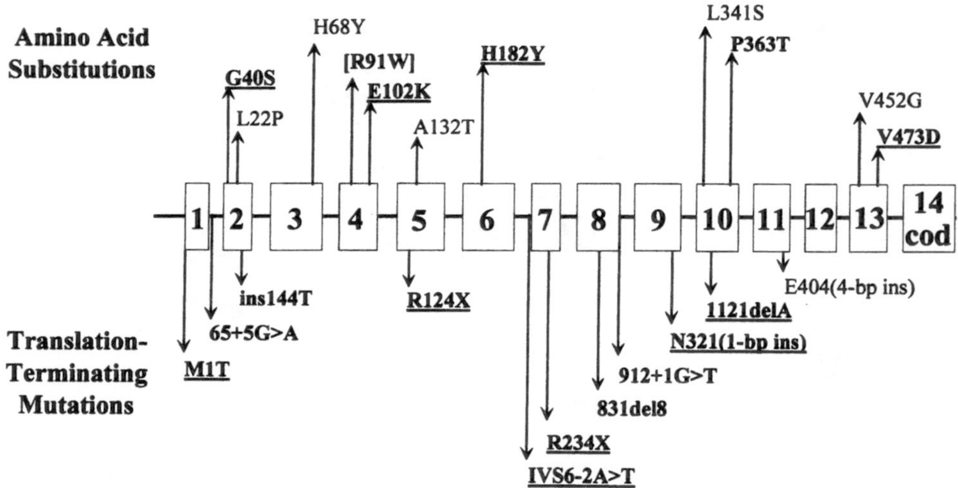

Figure 2. *Distribution of known RPE65 mutations in various phenotypes.* The 14 coding exons of *RPE65* are represented in numbered boxes. All mutations described in references 7, 8, 9, 10 fall in two parts: the amino acid substitution group (upper half of the diagram) and the translation-terminating mutation group (lower half of the diagram). Mutations found in retinitis pigmentosa (9, 10) are marked in plain front, those described in childhood-onset severe retinal dystrophy (7) are in bold front, those found in Leber's congenital amaurosis (8, 9) are underlined and in bold front, and the one present both in Leber's congenital amaurosis and in retinitis pigmentosa (9) is in brackets and in bold front.

2.3. Clinical Findings in Patients with LCA

LCA patients III_2 and III_3, aged 20 and 13 respectively, were examined. They reported nystagmus, night blindness, and inability to move since birth in the absence of systemic disease. Their medical histories and visual status were similar, although patient III_2 has ever been slightly more affected than her younger brother III_3. Patient III_2 never had color perception but patient III_3 could recognize long wave length radiations as reddish while not discriminating shorter wave length radiations (green and blue). At the time of the examination, they complained of complete night blindness, poor day vision, intense photophobia and were incapable of moving autonomously. They noticed minimal progression of their symptoms. Visual acuity was 1/50 and 1/100 in right and left eyes of patients III_2, respectively, and 1/25 in both eyes of patient III_3. Goldman perimetry was unrecordable in both patients. Fundus examination revealed numerous yellowish spots of the retina throughout the fundus. Narrowing of retinal vessels was mild and there was no bone spicule-shaped pigment deposits. In addition, macular involvement was noticeable, with some degree of atrophy and pigmentation in patient III_2. Photopic and scotopic ERGs were unrecordable in both patients. Based on these results and on the medical history the diagnosis of LCA was made.

2.4. Clinical Findings in Patient with Retinitis Pigmentosa

Fourty year-old patient II_2 reported night blindness since early childhood. As a child, in contrast to LCA patients, he had no moving difficulties in day time, and was able to read fluently until 13. He also did not mention any trouble in color discrimination. By the age of 13 however, he began to complain of decreasing visual acuity and

Table 1. Summary of the clinical features found in two patients with Leber's congenital amaurosis (1121delA + R234X) and in one patient with retinitis pigmentosa (L22P + H68Y)

Mutations	Patients (age)	Functional signs	Visual acuity OD	Visual acuity OS	Color perception	Peripheral visual field	Fundus	ERG
1121delA + R234X	III_2 (20)	Since birth, both patients have: Night blindness Nystagmus Major reading impairment Inability to move alone	1/50	1/100	No	Unrecordable	Small whitish spots Attenuated retinal vessels Macular dystrophy No bone spicule-shaped pigment deposits	Unrecordable
	III_3 (13)		1/25	1/25	Red only	Unrecordable	Small whitish spots Attenuated retinal vessels Macular involvement No bone spicule-shaped pigment deposits	Unrecordable
H68Y + L22P	II_2 (40)	Night blindness since infancy Reading and moving difficulties appeared at 13	1/33	1/25	Yes	Goldman V-4e at 12°	Attenuated retinal vessels Optic disc atrophy Macular atrophy Few bone spicule-shaped pigment deposits	Unrecordable

was placed at first row in classroom. At the time of the examination, his visual acuity had dropped to 1/33 and 1/25 in right and left eyes, respectively. Goldman perimetry found a dramatic loss of peripheral visual field with the V-4e isopter restricted to 12° at both eyes. Fundus examination revealed marked attenuated retinal vessels, waxy pallor of optic discs and rather sparse bone spicule-shaped pigment deposits in mid-peripheral retina. In addition, extended atrophy of the macular area was observable. ERG responses for photopic and scotopic systems were undetectable (<10 μV) indicating that both rods and cones were profoundly affected. Based on these findings, the diagnosis of severe retinitis pigmentosa was made.

3. DISCUSSION

Although the phenotype of the diseases caused by *RPE65* mutations are apparently different, some clinical features are shared by our three patients. These are (i) an early and severe visual impairment that seems to predominate on rod function, (ii) a prominent involvement of the macula, and (iii) the scarcity of pigmentary deposits in the retina. The severity of the disease however is variable explaining the diagnosis of LCA when nystagmus, absence of autonomous moving, poor visual acuity are observed in early childhood.

Mutated *RPE65* genes found here in LCA patients would result, if translated, in severely truncated proteins with C-terminal deletion of 33.2% and 56.3% for the 1121delA and R234X, respectively. As such, they probably do not encode functional proteins and therefore are likely to be null alleles. In the case of RP, the two missense mutations probably result in abnormal proteins since both L22 and H68 are conserved in human,[11] bovine,[12] and rat[13] RPE65. In this latter case, the RPE65 protein, in contrast to that encoded by null alleles, might exhibit some residual function in the RPE cell.

A survey of recently published mutations in *RPE65*[7,8,9,10] shows that in the case of LCA or childhood-onset severe retinal dystrophy, both translation-terminating mutations and amino acid substitutions were found. However, in the case of RP diagnosis, which is less severe than LCA and childhood-onset severe retinal dystrophy, most of the mutations found were amino acid substitutions. This suggests that the level of impairment of the RPE65 protein, presumably milder in amino acid substitutions than in translation-terminating mutations, might control the rate at which the photoreceptors degenerate. Following this hypothesis, the absence of *RPE65* products would cause a massive, early degeneration of photoreceptors whose phenotype corresponds to LCA while substitution of amino acids would lead to a slower pace of degeneration whose expected phenotype is RP. Currently, the knowledge of the function of the RPE65 protein is lacking and prevents experimental verification of this speculation.

ACKNOWLEDGMENTS

J-M G and CB are supported by SOS Rétinite, Montpellier, France. Research was sponsored in part by SOS Rétinite, IRRP and Retina France, and by the CHU de Montpellier, France. We gratefully thank Jean-Louis Pasquier for art work.

REFERENCES

1. K.M. Zinn and M.F. Marmor, 1978, *The retinal pigment epithelium*, Harvard University Press, Cambridge.
2. R.W. Young and D. Bok, 1969, Participation of the retinal pigment epithelium in the rod outer segment renewal process, *J. Cell Biol.* **42**:392–403.
3. D. Bok and M.O. Hall, 1971, The role of the pigment epithelium in the etiology of inherited retinal dystrophy in the rat, *J. Cell Biol.* **49**:664–682.
4. G. Wald, 1968, The molecular basis of visual excitation, *Nature* **219**:800–807.
5. C.P. Hamel, E. Tsilou, E. Harris, B.A. Pfeffer, J.J. Hooks, B. Detrick, and T.M. Redmond, 1993, A developmentally regulated microsomal protein specific for the pigment epithelium of the vertebrate retina, *J. Neurosci. Res.* **34**:414–425.
6. R.K. Crouch, P. Goletz, S. Yu, and T.M. Redmond, 1997, A possible role for RPE65 in retinoid processing, *Invest. Ophthalmol. Vis. Sci.* **38**:S304
7. S.-M. Gu, D.A. Thompson, C.R. Srisailapathy Srikumari, B. Lorenz, U. Finckh, A. Nicoletti, K.R. Murthy, M. Rathmann, G. Kumaramanickavel, M.J. Denton, and A. Gal, 1997, Mutations in *RPE65* cause autosomal recessive childhood-onset severe retinal dystrophy, *Nature. Genet.* **17**:194–197.
8. F. Marlhens, C. Bareil, J.M. Griffoin, E. Zrenner, P. Amalric, C. Eliaou, S.Y. Liu, E. Harris, T.M. Redmond, B. Arnaud, M. Claustres, and C.P. Hamel, 1997, Mutations in *RPE65* cause Leber's congenital amaurosis, *Nature Genet.* **17**:5–7.
9. H. Morimura, G.A. Fishman, S.A. Grover, A.B. Fulton, E.L. Berson, and T.P. Dryja, 1998, Mutations in the RPE65 gene in patients with autosomal recessive retinitis pigmentosa or Leber's congenital amaurosis, *Proc. Natl. Acad. Sci. USA* **95**:3088–3093.
10. F. Marlhens, J.-M. Griffoin, C. Bareil, B. Arnaud, M. Claustres, and C.P. Hamel, 1998, Autosomal recessive retinal dystrophy associated with two novel mutations in the *RPE65* gene, *Eur. J. Hum. Genet.* in press.
11. A. Nicoletti, D.J. Wong, K. Kawase, K.H. Gibson, T. L. Yang-Feng, J.E. Richards, and D.A. Thompson, 1995, Molecular characterization of the human gene encoding an abundant 61 kDa protein specific to the retinal pigment epithelium, *Hum. Molec. Genet.* **4**:641–649.
12. C.P. Hamel, E. Tsilou, B.A. Pfeffer, J.J. Hooks, B. Detrick, and T.M. Redmond, 1993, Molecular cloning and expression of RPE65, a novel retinal pigment epithelium-specific microsomal protein that is post-transcriptionally regulated in vitro, *J. Biol. Chem.* **268**:15751–15757.
13. G. Manès, R. Leducq, J. Kucharczak, A. Pagès, C.F. Schmitt-Bernard, and C.P. Hamel, 1998, Rat messenger RNA for the retinal pigment epithelium-specific protein RPE65 gradually accumulates in two weeks from late embryonic days, *FEBS Lett.* **423**:133–137.

4

THE *RPE65*-DEFICIENT MOUSE AS A MODEL FOR *RPE65*-ASSOCIATED LEBER'S CONGENITAL AMAUROSIS AND RELATED DISORDERS

T. Michael Redmond

Laboratory of Retinal Cell and Molecular Biology
National Eye Institute, National Institutes of Health, Bethesda
Maryland 20892

1. SUMMARY

The retinal pigment epithelium (*RPE*)-specific protein *RPE65* has been speculated to play a role in the vitamin A metabolism of the outer retina. Recently, mutations in the human *RPE65* gene have been associated with Leber's congenital amaurosis, a disorder characterized by blindness at birth, and autosomal recessive childhood-onset severe retinal dystrophies. We have developed an *RPE65*-deficient mouse and have described its phenotype. The *RPE65*-deficient mouse phenotype parallels that of the human disorders. This phenotype can be explained by a disruption of the *RPE* visual cycle causing a lack of 11-*cis*-retinal chromophore. Consequently, it is feasible to use this mouse as a model to test possible therapies for the human diseases.

2. INTRODUCTION

The retinal pigment epithelium (*RPE*) is a monolayer simple epithelium interposed between the blood vessels of the choriocapillaris and the neural retina. As such, it is involved in regulating supply of nutrients and micronutrients to the neural retina and especially the photoreceptor layer of the neural retina. Several specialized functions are performed by the *RPE*. Included among these are the formation of one side of the blood-retina barrier,[1] the periodic phagocytosis of spent outer segment discs and the degradation and recycling of these,[2,3] and the *RPE*/retina-specific vitamin A metabolism known as the visual cycle.[4] Since its discovery by Wald in 1934,[5] the elucidation

Retinal Degenerative Diseases and Experimental Therapy, edited by Hollyfield *et al.*
Kluwer Academic / Plenum Publishers, New York, 1999.

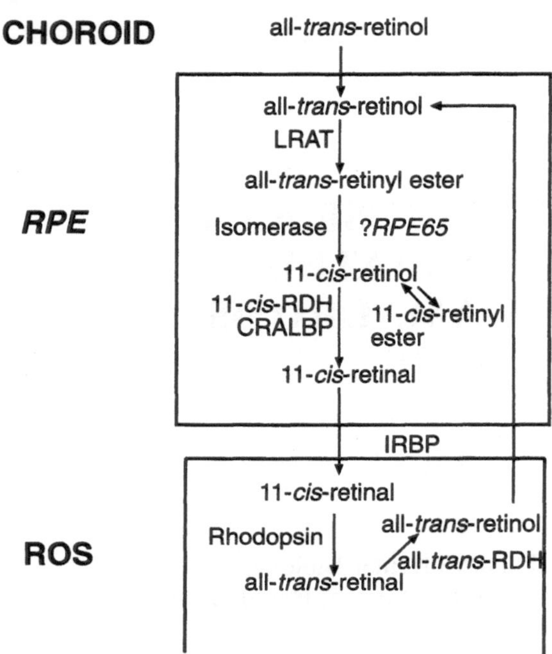

Figure 1. The *RPE*/OS visual cycle. Schematic model of the *RPE* visual cycle in the normal eye. CHOROID, choriocapillaris blood supply; *RPE*, retinal pigment epithelium; ROS, rod outer segment; LRAT, lecithin:retinol acyltransferase; 11-*cis*-RDH, 11-*cis*-retinol dehydrogenase; CRALBP, cellular retinaldehyde-binding protein; IRBP, interphotoreceptor retinoid-binding protein; all-*trans*-RDH, all-*trans*-retinol dehydrogenase.

of the steps of the visual cycle has been a major focus of research in retinal biochemistry.[4,6] In its outline, the 11-*cis*-retinal chromophore of visual pigments is photoisomerized to all-*trans*-retinal and then regenerated to the 11-*cis* isomer. This regeneration or isomerization occurs in the *RPE* of vertebrates and involves a series of enzymes and binding proteins that have been incompletely characterized.[4]

While many of the major components of the mammalian visual cycle have been identified, including rhodopsin,[7] interphotoreceptor retinoid-binding protein (IRBP),[8] ROS all-*trans*-retinol dehydrogenase,[9] lecithin:retinol acyltransferase (LRAT),[10] 11-*cis*-retinol dehydrogenase,[11] and 11-*cis* specific binding protein cellular retinaldehyde-binding protein (CRALBP),[12] the all-*trans*: 11-*cis*-retinol isomerase remains unidentified. Our understanding of the visual cycle is *RPE*resented schematically in Fig. 1. Following photoisomerization, the ROS all-*trans*-retinol dehydrogenase reduces all-*trans*-retinal to the alcohol, which is returned to the *RPE* via IRBP. LRAT catalyzes the formation of all-*trans*-retinyl esters,[10] the immediate precursor to 11-*cis*-retinol, from all-*trans*-retinol supplied from the blood or returned subsequent to photoisomerization of 11-*cis*-retinol. Enzymatic isomerization of all-*trans*-retinyl ester to 11-*cis*-retinol[13] is thought to occur with the concerted hydrolysis of the ester linkage providing the energy for isomerization.[14] Thus the enzyme catalyzing this step is thought to be an isomerohydrolase.[14] The 11-*cis*-retinol dehydrogenase oxidizes 11-*cis*-retinol to the aldehyde,[11] in cooperation with CRALBP.[12,15] A CRALBP gene mutation resulting in autosomal recessive retinitis pigmentosa is supportive of this interpretation.[16] Finally, IRBP transports the 11-*cis*-retinal across the interphotoreceptor space to the photoreceptor outer segments for incorporation into rhodopsin.[8,17,18] The possible function of *RPE65*, a conserved protein, highly and preferentially expressed in the *RPE*, in this cycle has been speculated.[19] Recent findings on the role of mutations in the *RPE65* gene in human retinal dystrophies and the generation of a mouse knockout have validated the role of *RPE65* in the visual cycle. This article reviews the identification of

RPE65, the developing understanding of its role in retinal biochemistry and the molecular genetic approach, both in human and mouse models, to elucidating its function. The similarities of the human disease and mouse model and the possible utility of the *RPE65*-deficient mouse are discussed.

3. CHARACTERIZATION OF THE *RPE*-SPECIFIC PROTEIN *RPE65*

RPE65 was first identified by the use of an *RPE*-specific monoclonal antibody[20] that recognized an *RPE*-specific protein of relative molecular weight of 65 kD on SDS-PAGE.[21] This non-glycosylated protein was found to be expressed only in *RPE* and conserved in mammals, birds, and frogs.[21] It was most effectively solubilized in the presence of detergents, suggesting an association with *RPE* cell membranes.[21] The subcellular localization of *RPE65* has been identified by biochemical fractionation as the microsomal membrane fraction,[21] and by immunoelectron microscopy as being associated with intracellular membranes.[22] *RPE65* is the major component of the microsomal membrane fraction.[21] Parenthetically, this fraction has been the primary source material for study of visual cycle enzymology.[10–15] *RPE65* is first detectable at post-natal day 4 in the rat, just before the appearance of photoreceptor outer segments, suggesting that expression of *RPE65* is coordinated with photoreceptor development.[21] Subsequent cloning and sequencing of the cDNA for *RPE65* revealed that it was a unique protein.[23] Interestingly, the deduced protein sequence lacked a signal peptide and transmembrane helices, indicating that *RPE65* was not a membrane-inserted protein.[23] This was of great significance since *RPE65* had been independently identified as the *RPE* serum retinol-binding protein (sRBP) receptor.[24–26] The secondary structure of the protein, however, belied such a function.[23] The mechanism by which *RPE65* interacts with membranes has been addressed by experiments showing that *RPE65* associates with phospholipid vesicles.[27] Cultured *RPE* cells 7 weeks in primary culture contained *RPE65* mRNA in amounts equivalent to fresh *RPE*, but no immunodetectable *RPE65* protein.[23] This was postulated to be due to post-transcriptional regulation perhaps mediated by the long AU-rich 3'-untranslated region (UTR) of the *RPE65* mRNA. In addition to the bovine cDNA,[23] the human cDNA also has a long 3'-UTR with sequence similarities to the bovine.[28] Further analysis of the bovine 3'-UTR suggests that this post-translational regulation of *RPE65* mRNA expression may be via a mechanism of translational inhibition mediated by a specific translational inhibition element.[29] In addition to the human[28] and bovine[23] cDNAs, *RPE65* has also been cloned from rat[30] and salamander.[31] Overall, *RPE65* from the various species is highly conserved. The human gene has also been cloned and sequenced,[28] and its genomic structure consists of 14 exons and intervening exons. The mouse gene has also been cloned and sequenced and its structure is similar to that of the human (Redmond et al., in p*RPE*.). The chromosomal localization of human gene has been mapped to human 1p31[28,32] and that of the mouse to distal mouse 3.[32]

4. MUTATIONS IN THE HUMAN *RPE65* GENE CAUSE RETINAL DYSTROPHIES

Given the possible importance of *RPE65* in photoreceptor related functions of the *RPE*, mutations in its gene could be construed to cause early degeneration or dystrophy of the photoreceptors. Accordingly, we searched[33] for mutations in the *RPE65* gene in patients with Leber's congenital amaurosis (LCA), a disorder characterized by blindness at birth. One patient (and, subsequently, her sibling) was found to have mutations in exon 7 and exon 10, resulting in premature termination.[33] These are likely to be null alleles. At the same time, Gu et al.[34] were investigating LCA and similar disorders, collectively termed autosomal recessive childhood-onset severe retinal dystrophies (arCSRDs), in Indian and German families. They linked these conditions to human chromosome 1p31, the locus for *RPE65*[28,32] and identified several mutations consistent with these disorders. Subsequently, mutations in *RPE65* have been found to be associated with 15% of a large group of North American LCA patients and a lower percentage of patients diagnosed with juvenile retinitis pigmentosa.[35] To date pathogenic mutations have been found in 12 out of 14 exons of *RPE65*. (The role of *RPE65* in LCA will be covered more completely in other chapters in this volume.)

5. GENERATION AND PHENOTYPE OF *RPE65* KNOCKOUT MOUSE MODEL

In view of the involvement of mutations in the human *RPE65* gene in these severe forms of retinal dystrophy, the undoubted role of *RPE65* in vitamin A metabolism, and the need for more direct clues to the precise role of *RPE65*, the necessity of obtaining an *RPE65* knockout was obvious. The method used in deriving the *RPE65* targeted disruption, the generation of the knockout and the characterization of the genotype and phenotype are described in detail elsewhere.[36] In brief, we replaced the promoter region, exons 1 to 3, and intervening introns with the neomycin resistance gene. The phenotype of this mouse confirms the crucial role of *RPE65* in *RPE* vitamin A metabolism with the finding that the *RPE65*-deficient mouse lacks rhodopsin, though it expresses opsin apoprotein in the rod OS. As a result, the rod ERG is abolished. Interestingly, however, the cone ERG remains, confirming that cone pigment regeneration can occur in the absence of the *RPE*. The biochemical basis for the phenotype, thus, can be explained by a disruption of the *RPE* visual cycle causing a lack of 11-*cis*-retinal chromophore. Morphologically, the *RPE65*-deficient mouse displays a slow retinal degeneration, evident by light microscopy. By electron microscopy, the photoreceptors of the knockout have somewhat disorganized outer segments. These, however, do have discs which appropriately express opsin apoprotein.[36] The effect of the knockout is shown schematically in Fig. 2. This absence of chromophore is accompanied by accumulation of all-*trans*-retinyl esters in the *RPE*, similar to that observed in mice injected intra-peritoneally with high doses of all-*trans*-retinyl esters.[37] Parenthetically, it can also be concluded that *RPE65* is not the *RPE* sRBP receptor[38] as has been hypothesized[24–26] since retinol uptake by the *RPE* is not impeded in the *RPE65*-deficient mouse.[36] This situation can be construed of as an 'inborn error of metabolism', restricted to the retina, with accumulation of substrate accompanied by lack of downstream product. In this instance, the defective enzyme might be concluded to be the isomerase. It is premature, however, to conclude that *RPE65* is the actual catalytic moiety. It is clear, though, that the presence of *RPE65* is required for this enzymatic activity and that the function of *RPE65* is proximate to that of the isomerase.

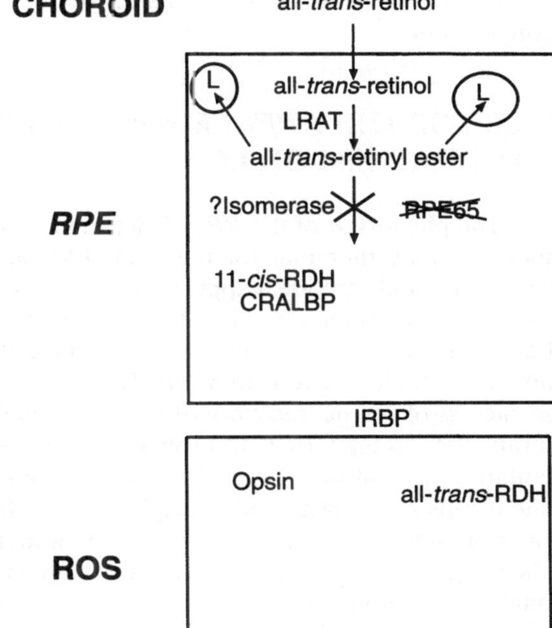

Figure 2. The effect of *RPE65* deficiency on the *RPE*/OS visual cycle. Schematic model illustrating blockage of *RPE* visual cycle by *RPE65* deficiency. CHOROID, choriocapillaris blood supply; *RPE*, retinal pigment epithelium; ROS, rod outer segment; LRAT, lecithin:retinol acyltransferase; 11-*cis*-RDH, 11-*cis*-retinol dehydrogenase; CRALBP, cellular retinaldehyde-binding protein; IRBP, interphotoreceptor retinoid-binding protein; all-*trans*-RDH, all-*trans*-retinol dehydrogenase; L, lipid droplets.

6. THE *RPE65* KNOCKOUT MOUSE AS A MODEL FOR HUMAN *RPE65* DISEASE

The differences between the specific characteristics of the knockout mouse model phenotype and the human LCA and less severe forms may be due to species-specific differences in cone expression and distribution. The *RPE65* knockout mouse retains cone function, while human LCA patients have severely disrupted cone function. It can be speculated that the loss of human cone function may be as a result of the extreme importance of foveal cones to human central vision. In this way, these highly active foveal cones might be as dependent on the *RPE* as the rod photoreceptors for supply of chromophore. Peripheral cones, like the widely scattered small cones of the mouse, might not be quite so dependent on the *RPE* for regeneration, as is the case for amphibian cones.[39,40] More precise electrophysiological and psychophysical testing of affected human and mouse cone responses appear to be warranted in deciphering this apparent difference.

In spite of these differences in photopic responses, both mouse, and human share profound effects on scotopic responses. With the recent finding that the Briard dog retinal dystrophy is due to a mutation in the canine *RPE65* gene,[41] this observation is further strengthened. The phenotype of the Briard dog has been described as a congenital stationary night blindness with variable involvement of cone cells. Canine retinal organization lies somewhere between the rodent and the human situations. Cones are much more common and prominent than in the mouse, but lack the well organized fovea of the human retina. As an interesting parallel to the mouse *RPE65* knockout phenotype, affected Briard dogs have been shown to contain numerous lipid droplets as a hallmark of the disease phenotype.[41] In view of these parallels, it will be interesting to see if human LCA due to *RPE65* mutations also accumulate retinyl esters.

7. USE OF THE *RPE65* KNOCKOUT TO EXPLORE THERAPIES FOR LCA

The phenotype of the *RPE65*-deficient mouse suggests that it may prove a useful model to study therapies for the related human LCA and arCSRD disorders. Gene transfer methods are ideal approaches since *RPE* cells quite avidly take up both adenovirus[42,43] and adeno-associated virus.[44] Both approaches can be used to see if the effects of the phenotype can be reversed. Since photoreceptor loss occurs relatively late and slowly, their rescue is more likely to occur. There are also convenient assays for the success of rescue: recovery of rhodopsin and recovery of rod ERG. As a possible therapeutic strategy, *RPE* transplantation is not as technically demanding as transplantation of photoreceptors. However the competence of the transplanted cells is an issue if cells are transitioned through culture. This is especially true as *RPE65* expression is often lost in culture.[21,23] Finally, the ultimate biochemical outcome of the *RPE65* deficiency, lack of 11-*cis* retinoids, may be amenable to pharmacological and/or nutritional intervention. The effect of retinyl ester accumulation,[36] however, must be clarified and minimized if necessary.

8. CONCLUSIONS

The *RPE65*-deficient mouse *RPE*resents a valuable model for understanding *RPE* biochemistry and the vertebrate visual cycle. While there are important differences in the details of the phenotypes of the knockout mouse and the human *RPE65*-related LCA disorders, especially regarding effects on cones, the similarities are marked. This should allow for the use of the *RPE65*-deficient mouse as a model to test possible therapies for the human diseases. It is possible that overcoming the effects on the *RPE* visual cycle may also help the cones in the human *RPE65*-related LCA disorders.

ACKNOWLEDGMENTS

I wish to acknowledge the contributions of Shirley Yu, Eric Lee, Dean Bok, Duco Hamasaki, Ning Chen, Patrice Goletz, Jian-xing Ma, Rosalie K. Crouch, and Karl Pfeiffer to the work presented at the VIII International Symposium on Retinal Degeneration and reviewed here. Additionally, I wish to thank my collaborator Christian Hamel for discussions on these topics, especially as they relate to the human diseases.

REFERENCES

1. J.G. Cunha-Vaz, 1976, The blood-retinal barriers, *Doc. Ophthalmol.* **41**:287–327.
2. R.W. Young, and D. Bok, 1969, Participation of the retinal pigment epithelium in the rod outer segment renewal process, *J. Cell Biol.* **42**:392–403.

3. R.W. Young, and D. Bok, 1970, Autoradiographic studies on the metabolism of the retinal pigment epithelium, *Invest. Ophthalmol.* 9:524–536.
4. J.C. Saari, 1994, Retinoids in photosensitive systems, in: *The Retinoids: Biology, Chemistry, and Medicine*, 2nd Ed. (Eds: Sporn, M.B., Roberts, A.B., and Goodman, D.S.) 351–385, Raven, New York.
5. G. Wald, 1968, Molecular basis of visual excitation, *Science* **162**:230–239.
6. R.J. Barry, F.J. Canada, and R.R. Rando, 1989, Solubilization and partial purification of retinyl ester synthetase and retinoid isomerase from bovine ocular pigment epithelium, *J. Biol. Chem.* **264**:9231–9238.
7. G. Travis, 1997, Insights from a lost visual pigment, *Nature Genet.* **15**:115–117.
8. Y.L. Lai, B. Wiggert, Y.P. Liu, and G.J. Chader, 1982, Interphotoreceptor retinol-binding protein: Possible transport vehicles between compartments of the retina, *Nature* **298**:848–849.
9. K. Palczewski, S. Jager, J. Buczylko, R.K. Crouch, D.L. Bredberg, K.P. Hofmann, M.A. Asson-Batres, and J.C. Saari, 1994, Rod outer segment retinol dehydrogenase: substrate specificity and role in phototransduction, *Biochemistry* **33**:13741–13750.
10. J.C. Saari, and D.L. Bredberg, 1989, Lecithin: retinol acyl transferase in retinal pigment epithelial microsomes, *J. Biol. Chem.* **264**:8636–8640.
11. A. Simon, U. Hellman, C. Wernstedt, and U. Eriksson, 1995, The retinal pigment epithelial-specific 11-cis retinol dehydrogenase belongs to the family of short chain dehydrogenases, *J. Biol. Chem.* **270**:1107–1112.
12. J.C. Saari, D.L. Bredberg, and N. Noy, 1994, Control of substrate flow at a branch in the visual cycle, *Biochemistry* **33**:3106–3112.
13. P.S. Bernstein, W.C. Law, and R.R. Rando, 1987, Biochemical characterization of the retinoid isomerase system of the eye, *J. Biol. Chem.* **262**:16848–16857.
14. P.S. Deigner, W.C. Law, F.C. Cañada, and R.R. Rando, 1989, Membranes as the energy source in the endergonic transformation of vitamin A to 11-*cis*-retinol, *Science* **244**:968–971.
15. A. Winston, and R.R. Rando, 1998, Regulation of isomerohydrolase in the visual cycle, *Biochemistry* **37**:2044–2050.
16. M.A. Maw, B. Kennedy, A. Knight, R. Bridges, K.E. Roth, E.J. Mani, J.K. Mukkadan, D. Nancarrow, J.W. Crabb, and M.J. Denton, 1997, Mutation of the gene encoding cellular retinaldehyde-binding protein in autosomal recessive retinitis pigmentosa, *Nature Genet.* **17**:198–200.
17. G.J. Jones, R.K. Crouch, B. Wiggert, M.C. Cornwall, and Chader, G.J., 1989, Retinoid requirements for recovery of sensitivity after visual-pigment bleaching in isolated photoreceptors, *Proc. Natl. Acad. Sci. USA* **86**:9606–9610.
18. G.I. Liou, Y. Fei, N.S. Peachey, S. Matragoon, S. Wei, W.S. Blaner, Y. Wang, C. Liu, M.E. Gottesman, H. Ripps, 1998, Early onset photoreceptor abnormalities induced by targeted disruption of the interphotoreceptor retinoid-binding protein gene, *J. Neurosci.* **18**:4511–4520.
19. A.F. Wright, 1997, A searchlight through the fog, *Nature Genet.* **17**:132–134.
20. J.J. Hooks, B. Detrick, C. Percopo, C. Hamel, and R. Siraganian, 1989, Development and characterization of monoclonal antibodies directed against the human retinal pigment epithelial cell, *Invest. Ophthalmol. Vis. Sci.* **30**:2106–2113.
21. C.P. Hamel, E. Tsilou, E. Harris, B.A. Pfeffer, J.J. Hooks, B. Detrick, and T.M. Redmond, 1993, A developmentally regulated microsomal protein specific for the pigment epithelium of the vertebrate retina, *J. Neurosci. Res.* **34**:414–425.
22. S.A. Vinores, W. Orman, J.J. Hooks, B. Detrick, and P.A. Campochiaro, 1993, Ultrastructural localization of *RPE*-associated epitopes recognized by monoclonal antibodies in human *RPE* and their induction in human fibroblasts by vitreous, *Graefes Arch. Clin. Exp. Ophthalmol.* **231**:395–401.
23. C.P. Hamel, E. Tsilou, B.A. Pfeffer, J.J. Hooks, B. Detrick, and T.M. Redmond, 1993, Molecular cloning and expression of *RPE65*, a novel retinal pigment epithelium-specific microsomal protein that is post-transcriptionally regulated in vitro, *J. Biol. Chem.* **268**:15751–15757.
24. C.-O. Båvik, C. Busch, and U. Eriksson, 1992, Characterization of a plasma retinol-binding protein membrane receptor expressed in the retinal pigment epithelium, *J. Biol. Chem.* **267**:23035–23042.
25. C.-O. Båvik, U. Eriksson, R.A. Allen, and P.A. Peterson, 1991, Identification and partial characterization of a retinal pigment epithelial membrane receptor for plasma retinol-binding protein, *J. Biol. Chem.* **266**:14978–14985.
26. C.-O. Båvik, F. Lévy, U. Hellman, C. Wernstedt, and U. Eriksson, 1993, The retinal pigment epithelial membrane receptor for plasma retinol-binding protein, *J. Biol. Chem.* **268**:20540–20546.
27. E. Tsilou, C.P. Hamel, S. Yu, and T.M. Redmond, 1997, *RPE65*, the major retinal pigment epithelium

microsomal membrane protein associates with phospholipid liposomes, *Arch. Biochem. Biophys.* **346**:21–27.

28. A. Nicoletti, D.J. Wong, K. Kawase, L.H. Gibson, T.L. Yang-Feng, J.E. Richards, and D.A. Thompson, 1995, Molecular characterization of the gene encoding an abundant protein specific to the retinal pigment epithelium. *Hum. Molec. Genet.* **4**:641–649.
29. S.-Y. Liu, and T.M. Redmond, 1998, Role of the 3′ untranslated region of *RPE65* mRNA in the translational regulation of the *RPE65* gene: Identification of a specific translation inhibitory element (TIE), *Arch. Biochem. Biophys.* **357**:37–44.
30. G. Manes, R. Leducq, J. Kucharczak, A. Pages, C.F. Schmitt-Bernard, and C.P. Hamel, 1998, Rat messenger RNA for the retinal pigment epithelium-specific protein *RPE65* gradually accumulates in two weeks from late embryonic days, *FEBS Lett.* **423**:133–137.
31. J.-x. Ma, L. Xu, D.K. Lockman, T.M. Redmond, and R.K. Crouch, 1998, Cloning and localization of *RPE65* mRNA in salamander cone photoreceptor cells, *Biochim. Biophys. Acta*, **1443**:255–261.
32. C.P. Hamel, N.A. Jenkins, D.J. Gilbert, N.G. Copeland, and T.M. Redmond, 1994, The gene for the retinal pigment epithelium-specific protein *RPE65* is localized to human 1p31 and mouse 3, *Genomics* **20**:509–512.
33. F. Marlhens, C. Bareil, J.-M. Griffoin, E. Zrenner, P. Amalric, C. Eliaou, S.-Y. Liu, E. Harris, T.M. Redmond, B. Arnaud, M. Claustres, and C.P. Hamel, 1997, Mutations in *RPE65* cause Leber's congenital amaurosis, *Nature Genetics* **17**:139–141.
34. S.-m. Gu, D.A. Thompson, C.R. Srisailapathy Srikumari, B. Lorenz, U. Finckh, A. Nicoletti, K.R. Murthy, M. Rathmann, G. Kumaramanickavel, M.J. Denton, and A. Gal, 1997, Mutations in *RPE65* cause autosomal recessive childhood-onset severe retinal dystrophy, *Nature Genet.* **17**:194–197.
35. H. Morimura, G.A. Fishman, S.A. Grover, A.B. Fulton, E.L. Berson, and T.P. Dryja, 1998, Mutations in the *RPE65* gene in patients with autosomal retinitis pigmentosa or Leber congenital amaurosis, *Proc. Natl. Acad. Sci. USA* **95**:3088–3093.
36. T.M. Redmond, S. Yu, E. Lee, D. Bok, D. Hamasaki, N. Chen, P. Goletz, J.-X. Ma, R.K. Crouch, and K. Pfeiffer, 1998, *RPE65* is necessary for production of 11-*cis*-Vitamin A in the retinal visual cycle, *Nature Genetics* **20**:344–351.
37. W.G. Robison, and T. Kuwabara, 1977, Vitamin A storage and peroxisomes in retinal pigment epithelium and liver, *Invest. Ophthalmol. Vis. Sci.* **16**:1110–1117.
38. J. Heller, and D. Bok, 1976, A specific receptor for retinol-binding protein as detected by the binding of human and bovine retinol-binding protein to pigment epithelial cells, *Am. J. Ophthalmol.* **81**:93–97.
39. D.C. Hood, and P.A. Hock, 1973, Recovery of cone receptor activity in the frog's isolated retina, *Vision Res.* **13**:1943–1951.
40. E.B. Goldstein, and B.M. Wolf, 1973, Regeneration of the green-rod pigment in the isolated frog retina, *Vision Res.* **13**:527–534.
41. G.D. Aguirre, V. Baldwin, S. Pearce-Kelling, K. Narfstrom, K. Ray, G.M. Acland, 1998, Congenital stationary night blindness in the dog: common mutation in the *RPE65* gene indicates founder effect, Mol. Vis. 4: 23; http://www.molvis.org/molvis/v4/p23/
42. J. Bennett, J. Wilson, D. Sun, B. Forbes, and A. Maguire, 1994, Adenovirus vector-mediated in vivo gene transfer into adult murine retina, *Invest. Ophthalmol. Vis. Sci.* **35**:2535–2542.
43. D.M. Sullivan, D.C. Chung, E. Anglade, R.M. Nussenblatt, and K.G. Csaky, 1996, Adenovirus-mediated gene transfer of ornithine aminotransferase in cultured human retinal pigment epithelium, *Invest. Ophthalmol. Vis. Sci.* **37**:766–774.
44. R.R. Ali, M.B. Reichel, A.J. Thrasher, R.J. Levinsky, C. Kinnon, N. Kanuga, D.M. Hunt, S.S. Bhattacharya, 1996, Gene transfer into the mouse retina mediated by an adeno-associated viral vector, *Hum. Mol. Genet.* **5**:591–594.

5

CRALBP AND INHERITED RETINAL DEGENERATION

Breandán N. Kennedy,[1] John C. Saari,[2] and John W. Crabb[3]

[1] Center for Transgene Research
Department of Biological Sciences
University of Notre Dame
Notre Dame, Indiana 46556
[2] Departments of Ophthalmology and Biochemistry
University of Washington
Seattle, Washington 98195
[3] Cole Eye Institute
Cleveland Clinic Foundation
Cleveland, Ohio 44195

1. SUMMARY

Missense mutations in the gene encoding cellular retinaldehyde-binding protein (CRALBP) associate with early onset autosomal recessive retinitis pigmentosa (arRP). We show here liquid chromatography electrospray mass spectrometric results from recombinant mutant CRALBP that were instrumental in establishing that the arRP modification was responsible for a loss of protein function. In addition, we present progress in the development of CRALBP knockout mice for studying *in vivo* functions and therapeutic approaches to retinal degenerations related to CRALBP.

2. INTRODUCTION

Mutations in the gene encoding cellular retinaldehyde-binding protein (CRALBP) can result in retinal disease, detectable as night blindness in children as early as ages 3 and 4, with progression to legal blindness in teenage years.[1] Specifically, an Arg to Gln mutation at residue 150 has been genetically linked with non-syndromic autosomal recessive retinitis pigmentosa (arRP).[1] Mutations in CRALBP and RPE65[2] provide the first well defined examples of arRP caused by defective genes expressed in the retinal pigment epithelium (RPE) and not in photoreceptor cells. CRALBP is abundant in RPE

Retinal Degenerative Diseases and Experimental Therapy, edited by Hollyfield *et al.*
Kluwer Academic / Plenum Publishers, New York, 1999.

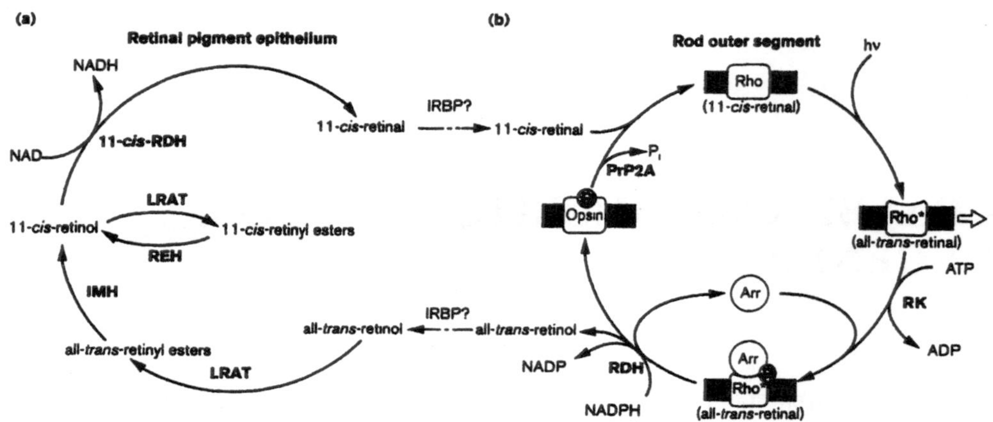

Figure 1. Mammalian visual cycle. Absorption of light (hv) isomerizes the 11-*cis*-retinal bound to rhodopsin (Rho) to all-*trans*-retinal in the rod outer segment (ROS), stimulating phototransduction. All-*trans*-retinal is reduced in the ROS by all-*trans*-retinol dehydrogenase (RDH) and all-*trans*-retinol is shuttled to the retinal pigment epithelium (RPE), probably via interphotoreceptor retinoid-binding protein (IRBP). Within the RPE, all-*trans*-retinol can be esterified by lecithin:retinol acyltransferase (LRAT) to all-*trans*-retinyl esters, which are thought to be converted to 11-*cis*-retinol by isomerohydrolase (IMH). 11-*cis*-retinol can be esterified and stored or oxidized to 11-*cis*-retinal by 11-*cis*-retinol dehydrogenase (11-*cis*-RDH) and shuttled back to the photoreceptor cells. *In vitro* CRALBP serves as a substrate carrier protein in the reaction with 11-*cis*-RDH and may also function as a regulator in the isomerization and ester hydrolase reactions. REH, retinyl ester hydrolase; Arr, arrestin; RK, rhodpsin kinase; PrP2A, protein phosphatase 2A. (Reprinted with permission from K. Palczewski and J.C. Saari, 1997, Current Opinion in Neurobiology 7: 500–504).

and Müller cells of the retina, and also expressed in lower amounts in ciliary body, cornea, pineal gland, optic nerve, brain, and transiently in iris.[3] In the RPE, CRALBP is thought to function in the mammalian visual cycle,[4] interacting with 11-*cis*-retinol dehydrogenase in the oxidation of 11-*cis*-retinol to 11-*cis*-retinal for visual pigment regeneration (Fig. 1). Recent studies demonstrate CRALBP stimulates the generation of 11-*cis*-Rol in RPE microsomal preparations,[5,6] suggesting that the protein may also serve as a regulator of the isomerization and ester hydrolase reactions (Fig. 1). How CRALBP functions in tissues not involved with visual pigment regeneration and ligands other than 11-*cis*-retinoids remain to be defined.

Gene therapy may be useful for treating CRALBP-associated retinal disease. Preliminary CRALBP gene studies[7,8] have identified two photoreceptor consensus elements (PCE-1) in the human proximal promoter and two identical sequences (GCAGGA) flanking these elements, as playing a positive role in controlling CRALBP gene expression in cultured RPE cells. The PCE-1 may have another effect in Müller cells,[9] but in ocular ciliary epithelial cell cultures,[10] CRALBP gene regulation appears similar to that observed in the RPE. The mouse CRALBP gene sequence has recently been determined, including all eight exons and about 4 kb of the promoter region.[11] The mouse and human coding sequences are about 87% identical but only ~48% identical in the 5′-flanking region. In contrast to the human gene, the mouse gene contains a consensus TATA box and one rather than two PCE-1. The human proximal promoter region contains an apparent Initiator (Inr) element, a Ying-Yang 1 (YY1) consensus site and two RPE65 related sequences.[11] Several conserved sequences have been identified in the mouse and human distal promoter region and may be associated with putative enhancer activity.[11] We document here protein structural characterization of the R150Q rCRALBP mutant previously associated with retinal degeneration and report progress toward disruption of the mouse CRALBP gene by homologous recombination.

3. METHODS

3.1. rCRALBP Mutant R150Q

Mutant R150Q recombinant human CRALBP (rCRALBP) was produced in *E coli*[1] and the protein structure verified by liquid chromatography electrospray mass spectrometry (LC ESMS) using methods documented for wildtype rCRALBP.[12,13]

3.2. Gene Targeting Vector

Plasmid Vectors pBK5′ and pBOC3′ were created by annealing complementary synthetic oligonucleotides and ligating into pGem3Zf(-) (Promega). The pGT-NT-28 vector was a generous gift from Dr. Roberto Polakiewicz, New England Biolabs, Beverly, MA. Plasmid DNA was purified using the Promega Wizard Mini-Prep System. Following restriction enzyme digestion and electrophoresis, fragment DNA was extracted from LMP agarose (FMC Bioproducts, Rockland, ME) and ligated into vectors at 16°C overnight using T4 DNA ligase (New England Biolabs). Ultracompetent cells for transformation were created according to published methods.[14] Restriction digestion, PCR mapping, and DNA sequencing were used to verify the structure of the constructs. The targeting vector was purified by the Endo-Free Plasmid Maxi Kit (Qiagen Inc., Santa Clarita, CA) and provided to Genome Systems Inc., St Louis, MO for transfection and cloning of homologous recombinants in embryonic stem cells (ES). Screening of ES clones for homologous recombinants was by Southern Blotting of *BamHI* and *KpnI* restriction digests using a 230 bp probe spanning part of mouse CRALBP exon 5 and intron 5 and located 3′ of the targeting vector (pGT-BK-KO). This probe was prepared by PCR from mouse CRALBP genomic subclone mp3.2SK[11] and purified by electrophoresis and the Promega Wizard PCR Prep System.

4. RESULTS

4.1. Mutant R150Q Recombinant CRALBP

The CRALBP mutation segregating with autosomal recessive retinitis pigmentosa results in an Arg to Gln exchange at residue 150.[1] To evaluate the effect of the mutation on biological function, a wildtype rCRALBP expression plasmid was altered by site-directed mutagenesis to express R150Q rCRALBP and the coding region sequenced to verify that no other mutations were introduced.[1] Mutant R150Q rCRALBP exhibited decreased solubility, and consequently was recovered in relatively low yield (eg, μg amounts). Inaddition, the mutant protein was unable to bind 11-*cis*-retinal.[1] To confirm the identity of the purified mutant and to exclude the presence of other possible debilitating modifications, the protein was carboxyamidomethylated, digested with trypsin and analyzed by LC ESMS. Essentially the complete primary structure of R150Q mutant rCRALBP was verified by LC ESMS (Fig. 2), including the

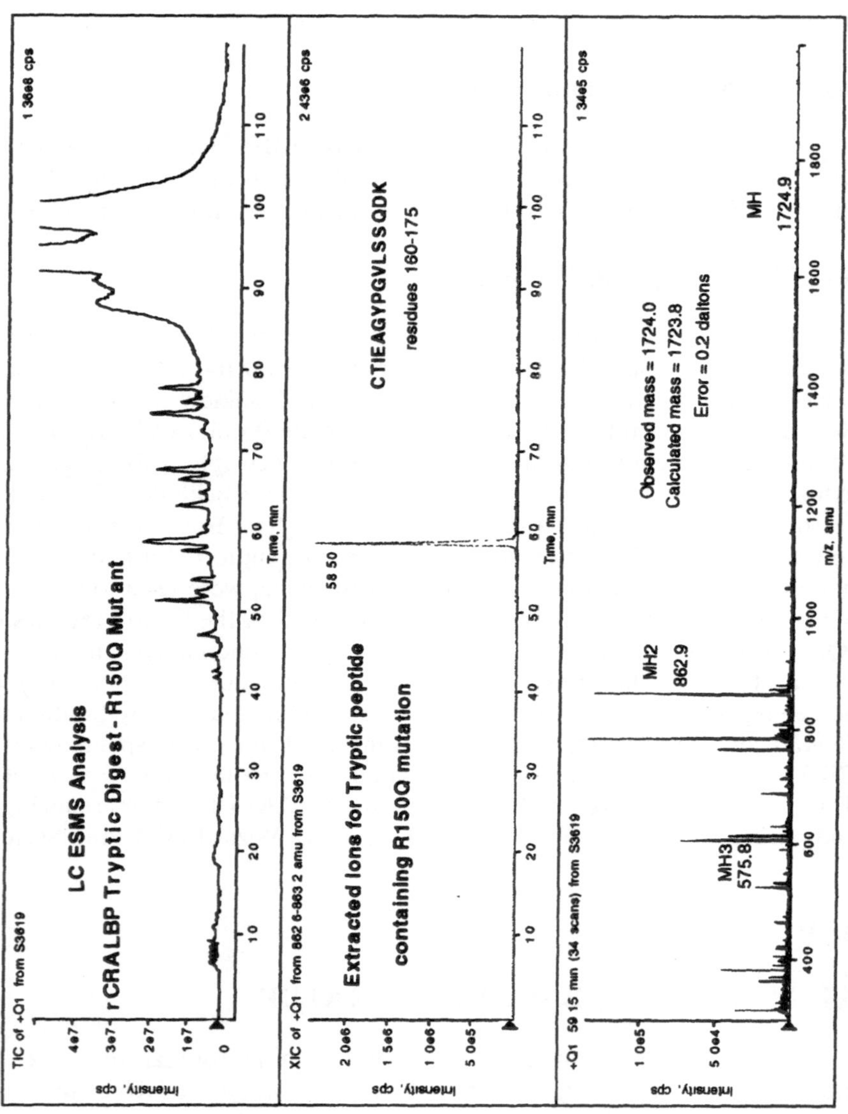
A
TIC of +Q1 from S3619
1 36e8 cps
LC ESMS Analysis
rCRALBP Tryptic Digest - R150Q Mutant
Intensity, cps
Time, min
XIC of +Q1 from 862 6-863 2 amu from S3619
2 43e6 cps
Extracted Ions for Tryptic peptide
containing R150Q mutation
58 50
CTIEAGYPGVLSSQDK
residues 160-175
+Q1 59 15 min (34 scans) from S3619
1 34e5 cps
MH3
575.8
MH2
862.9
MH
1724.9
Observed mass = 1724.0
Calculated mass = 1723.8
Error = 0.2 daltons
m/z, amu

B

```
G H H H H H H H H H   H S S G H I D D D D   K/H M S E G V G T F   R/M V P E E E Q E L   R/A Q L E Q L T T K/  50
D H G P V F G P C S   Q L P R/H T L Q K/A   K/D E L N E R/E E T   R/E E A V R/E L Q E   M V Q A Q A A S G E   100
E L A V A V A E R/V   Q E K/D S G F F L R/  F I R/A R/K/F N V G   R/A Y E L L R/G Y V   N F R/L Q Y P E L F   150
D S L S P E A V R/C   T I E A G Y P G V L   S S Q D K/Y G R/V V   M L F N I E N W Q S   Q E I T F D E I L Q   200
A Y C F I L E K/L L   E N E E T Q I N G F   C I I E N F K/G F T   M Q Q A A S L R/T S   D L R/K/M V D M L Q   250
D S F P A R/F K/A I   H F I H Q P W Y F T   T T Y N V V K P F L   K/S K/L L E R/V F V   H G D D L S G F Y Q   300
E I D E N I L P S D   F G G T L P K/Y D G   K/A V A E Q L F G P   Q A Q A E N T A F                         339
```

Figure 2. Verification of the R150Q rCRALBP Protein Structure. (A) Carboxyamidomethyl R150Q rCRALBP was digested with trypsin and analyzed by LC ESMS using a PE Sciex Model API 300 triple quadrupole electrospray mass spectrometer.[1,13] The top panel shows the total ion current (TIC) for the full scan m/z 300–2000; the middle panel shows computer extracted ions for the peptide containing the R150Q mutation and the bottom panel shows the mass spectra of this peptide. (B) Tryptic peptides identified in the R150Q rCRALBP structure from the analysis in the top panel are underlined using PeptideMap (version 2.2, PE Sciex). Slash marks (/) indicate tryptic cleavage sites.

tryptic peptide containing the engineered mutation. No evidence was found for additional mutations or structural modifications other than from carboxyamidomethylation introduced by experimental design for the analysis. The direct protein analysis corroborates the observed loss of CRALBP function resulted from the R150Q mutation.

4.2. CRALBP Gene Targeting Vector

Mouse 129/SvJ CRALBP genomic subclones mp3.6CRII, mp2.8SK, and mp3.2SK (11) and the pGT-NT28 vector (New England Biolabs Inc.) containing the Neomycin resistance (Neo^r) gene and the viral Thymidine Kinase (vTK) gene, were used to generate the CRALBP targeting vector (Fig. 3). The neomycin resistance gene was positioned within CRALBP coding exon 3, so as to interrupt CRALBP expression. The 5′ and 3′ arms were assembled in pBK5′ and pBOC3′, respectively. These vectors were designed to facilitate ligation of the 5′ arm (~4 kb to exon 3) immediately upstream of the Neo^r cassette and the 3′ arm (exon 3 to exon 5) immediately downstream of the Neo^r cassette. The final targeting vector of ~16.5 kb encompasses ~9.2 kb of the mouse CRALBP gene.

4.3. Screening Homologous Recombinants

Southern blotting was used to screen for ES clones homologously recombined with the CRALBP targeting vector. Following digestion of wildtype mouse 129/SvJ genomic DNA with *Kpn*I and *BamH*I, Southern blotting with a 230 bp 3′ flanking probe demonstrated a ~12 kb hybridizing fragment (Fig. 4). Under the same conditions, the probe hybridizes with a ~8 kb fragment from an homologously recombined allele. In conjunction with Genome Systems Inc., eight positive clones were identified by the Southern screening method outlined above (Fig. 4). Secondary Southern screening of DNA from these clones digested with *BamHI* and probed with a fragment from the Neo^r gene identified six positive clones that were injected into host blastocysts for the generation of chimeric mice. Six mice ranging from 20–80% chimeric were generated from two of these clones.

5. DISCUSSION

Structure function studies have defined CRALBP ligand stereoselectivity and photosensitivity,[15] developed topological and epitope maps,[16,17] established *in vitro* evidence for a substrate carrier function in RPE[18,19] and produced human recombinant CRALBP.[11,12] Characterization of recombinant CRALBP has localized the retinoid-binding domain to the C-terminal region,[20] and identified Gln 210 and Lys 221 as components of the retinoid-binding cavity.[17] The Arg 150 to Gln CRALBP mutation (Fig. 2) in arRP abolishes retinoid binding function and decreases protein solubility.[1] Whether Arg 150 is part of the CRALBP retinoid binding cavity is not clear;[17] however, a single Arg residue can modulate ligand specificity and affinity in the cellular retinoic acid-binding protein, intestinal fatty acid-binding protein and cellular retinol-binding protein II.

The mouse CRALBP 'knockout construct' (Fig. 3) was engineered from the mouse CRALBP gene sequence and genomic subclones[11] and has appropriate

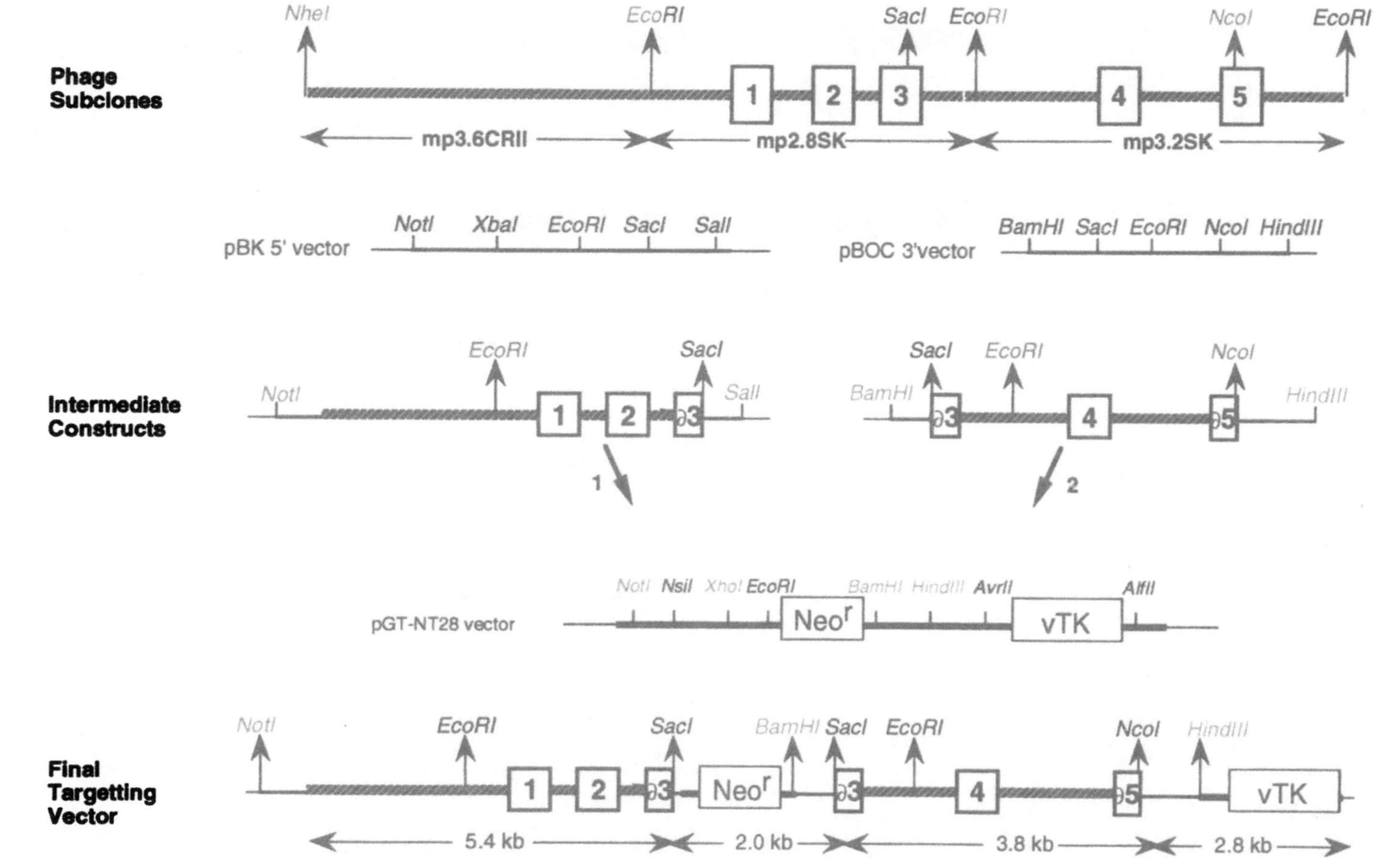

Figure 3. Construction of the CRALBP Gene Targeting Vector. The 5′ arm was engineered by ligating an *EcoRI-SacI* fragment of mp2.8SK to the compatibly digested pBK5′ vector, and subsequent ligation of an *NheI-EcoRI* fragment of mp3.6CRII using the *XbaI* and *EcoRI* sites. The assembled 5′ arm was released by *NotI* and *SalI* digestion and ligated into the pGT-NT-28 vector digested with *NotI* and *XhoI*. The 3′ arm was assembled by ligating a *SacI-EcoRI* fragment of mp2.8SK, and subsequently an *EcoRI-NcoI* fragment of mp3.2SK, into compatibly digested pBOC3′ vector. The 3′ arm was released by *BamHI* and *HindIII* digestion and ligated into the *BamHI and HindIII* sites of the pGT-NT-28 vector.

A

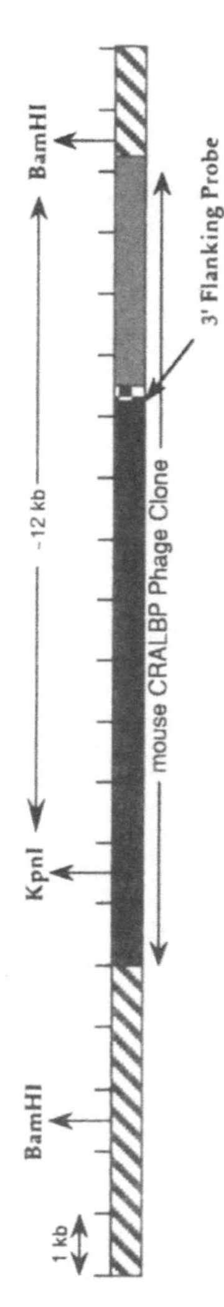

Endogenous Mouse 129/SvJ CRALBP Gene
1 kb
BamHI
KpnI
~12 kb
BamHI
mouse CRALBP Phage Clone
3' Flanking Probe

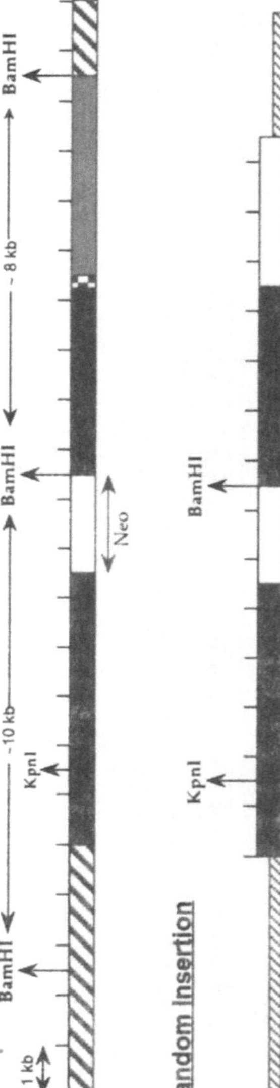

Homologous Recombination
1 kb
BamHI
KpnI
~10 kb
BamHI
Neo
~8 kb
BamHI
Random Insertion
KpnI
BamHI

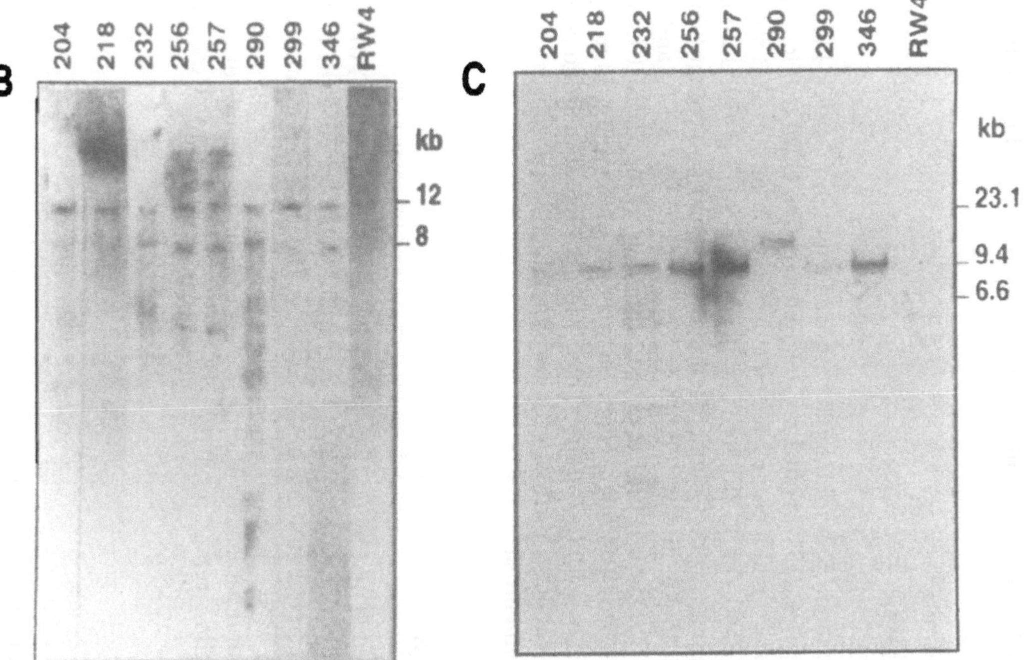

Figure 4. Identification of Homologous Recombinants. (A) Restriction maps are shown for the endogenous mouse CRALBP gene and for homologous recombination and random insertion of the transfected CRALBP targeting vector in ES cells. The region boxed in black corresponds to the CRALBP sequence in the targeting vector. (B) Southern blot analysis is shown with eight putative positive ES clones and the RW4 ES control. The 3′ flanking probe indicated in panel A was used with *BamHI* and *KpnI* digestion to yield a ~8 kb fragment for homologous recombinants and ~12 kb fragment for the endogenous gene. (C) Positive clones in panel B were rescreened by Southern analysis using a Neo^r probe and BamHI digestion. Six of the eight clones exhibit a single positive fragment of ~10 kb, as predicted for homologous recombinants.

attributes as a targeting vector for homologous recombination. Indeed, transfection of ES cells with the targeting vector allowed cloning of several homologous recombinants (Fig. 4) and the production of chimeric mice. Ongoing breeding efforts with these mice and preliminary PCR screening results suggest that homozygous CRALBP knockout mice will be available soon for phenotypic characterization in electrophysiological, morphological, and retinoid metabolism studies. We anticipate these and other transgenic mouse studies will lead to a better understanding of the *in vivo* function of CRALBP as well as to new insights regarding CRALBP-associated retinal degenerations and therapeutic strategies for treating inherited retinal diseases.

NOTE ADDED IN PROOF

Two recent reports document four additional pathogenic mutations in the human CARLBP gene, namely M225K, R233W, a one base deletion in codon Q277 that lengthens the protein from 316 to 326 residues, and a point mutation in an intron 3 splice donor site [MSI Burstedt, O Sandgren, G Holgren and K Forsman-Semb, 1999, Bothnia Dystrophy Caused by Mutations in the Cellular Retinaldhyde-Binding Protein Gene (RLBP1) on Chromosome 15q26, Invest. Ophthal. Vis Sci. 40: 995–1000; and H Morimura, EL Berson and TP Dryja, 1999, Recessive Mutations in the RLBP1 Gene Encoding Cellular Retinaldehyde-Binding Protein in a Form of Retinitis Punctata Albescens, Invest. Opthal. Vis. Sci. 40: 1000–1004]. These reports suggest that defects in the human CRALBP gene produce a form of retinitis punctata albescens distinguishable from typical retinitis pigmentosa.

ACKNOWLEDGMENTS

This study was supported in part by USPHS grants EY06603, EY02317, EY01730, and by Research to Prevent Blindness, Inc (RPB). We gratefully acknowledge the technical assistance of Robin Wesselschmidt, Genome Systems, Inc. This report includes thesis research submitted by BNK in partial fulfillment of the requirements for the doctoral degree in the Cell and Molecular Biology Program jointly administered by the University College Dublin, Ireland, and the Adirondack Biomedical Research Institute, Lake Placid, NY.

REFERENCES

1. M.A. Maw, B. Kennedy, A. Knight, R. Bridges, K.E. Roth, E.J. Mani, J.K. Mukkadan, D. Nancarrow, J.W. Crabb, and M.J. Denton, 1997, Mutation of the gene encoding cellular retinaldehyde binding protein in autosomal recessive retinitis pigmentosa, *Nature Genetics* **17**:198–200.
2. S. Gu, D.A. Thompson, C.R.S. Srikumari, B. Lorenz, U. Finckh, A. Nicoletti, K.R. Murthy, M. Rathmann, G. Kumaramanickavel, M.J. Denton, and A. Gal, 1997, Mutations in RPE65 cause autosomal recessive childhood-onset severe retinal dystrophy, *Nature Genetics* **17**:194–197.
3. J.C. Saari, J. Huang, D.E. Possin, R.N. Fariss, J. Leonard, G.G. Garwin, J.W. Crabb, and A.H. Milam, 1997, Cellular retinaldehyde-binding protein is expressed by oligodendrocytes in the optic nerve and brain, *Glia* **21**:259–268.
4. J.C. Saari, 1994, Retinoids in photosensitive systems, in: *The Retinoids* (Sporn, M.A. Roberts, A.B., and Goodman, D.S., eds) pp 351–385, Raven Press, Ltd., New York.

5. A. Winston and R.R. Rando, 1998, Regulation of isomerohydrolase activity in the visual cycle, *Biochemistry* **37**:2044–2050.
6. H. Stecher, M.H. Gelb, J.C. Saari, and K. Palczewski, 1999, Preferential release of 11-*cis*-retinol from retinal pigment epithelial cells in the presence of cellular retinaldehyde-binding protein (CRALBP). *J. Biol. Chem.* **274**:8577–8585.
7. R. Intres, S. Goldflam, J.R. Cook, and J.W. Crabb, 1994, Molecular cloning and structural analysis of the human gene encoding cellular retinaldehyde-binding protein, *J. Biol. Chem.* **269**:25411–25418.
8. B.N. Kennedy, S. Goldflam, M.A. Chang, P. Campochairo, A.A. Davis, D.J. Zack, and J.W. Crabb, 1998, Transcriptional Regulation of the Human Gene Encoding Cellular Retinaldehyde-Binding Protein; A Role for Photoreceptor Consensus Element, *J. Biol. Chem.* **273**:5591–5598.
9. S. Brodjian, B.N. Kennedy, J.W. Crabb, and V. Sarthy, 1997, Regulation of the cellular retinaldehyde-binding protein (CRALBP) gene in Müller cells, *Invest. Ophthal. Vis. Sci.* **38**:5296, S1129.
10. J. Ortega, B.N. Kennedy, J.W. Crabb, and M. Coca Prados, 1998, CRALBP gene regulation In ciliary epithelial cells, *Invest. Ophthal. Vis. Sci.* **39**:169, S39.
11. B.N. Kennedy, J. Huang, J.C. Saari, and J.W. Crabb, 1998, Characterization of the mouse gene encoding cellular retinaldehyde-binding protein, *Molecular Vision* **4**:14; http://www.molvis.org/molvis/v4/p14.
12. J.W. Crabb, Y. Chen, S. Goldflam, K. West, and J. Kapron, 1998, Methods for producing recombinant human cellular retinaldehyde-binding protein, in: *Methods In Molecular Biology*, Volume 89, *Retinoid Protocols* (C. Redfern, ed.) pp 91–104, Humana Press, Totowa, New Jersey.
13. J.W. Crabb, A. Carlson, Y. Chen, S. Goldflam, R. Intres, K.A. West, J.D. Hulmes, J.T. Kapron, L.A. Luck, J. Horwitz, and D. Bok, 1998, Structural and Functional Characterization of Recombinant Human Cellular Retinaldehyde-Binding Protein, *Protein Science* **7**:746–757.
14. D. Hanahan, 1985, Techniques for transformation of *E. Coli.* in: *DNA Cloning-A Practical Approach* (D.M. Glover, ed.) pp 109–135, IRL Press, Oxford.
15. J.C. Saari and D.L. Bredburg, 1987, Photochemistry and stereoselectivity of cellular retinaldehyde-binding protein from bovine retina. *J Biol Chem* **262**:7618-7622.
16. J.W. Crabb, V.P. Gaur, G.G. Garwin, S.V. Marx, C. Chapline, C.M. Johnson, and J.C. Saari, 1991, Topological and epitope mapping of the cellular retinaldehyde-binding protein from retina, *J. Biol. Chem.* **266**:16674–16683.
17. J.W. Crabb, Z. Nie, Y. Chen, J.D. Hulmes, K.A. West, J.T. Kapron, S.E. Ruuska, N. Noy, and J.C. Saari, 1998, CRALBP Ligand Interactions: Gln210 and Lys221 are in the Retinoid-binding Pocket, *J Biol Chem* **273**:20712–20720.
18. J.C. Saari and D.L. Bredberg, 1982, Enzymatic reduction of 11-*cis*-retinal bound to cellular retinal-binding protein, *Biochim Biophys Acta* **716**:266–272.
19. J.C. Saari, D.L. Bredburg, and N. Noy, 1994, Control of substrate flow at a branch in the visual cycle, *Biochemistry* **33**:3106–3112.
20. Y. Chen, C. Johnson, K. West, S. Goldflam, M.F. Bean, M.J. Huddleston, S.A. Carr, J.L. Gabriel, and J.W. Crabb, 1994, Mapping the retinoid-binding domain of cellular retinaldehyde-binding protein. in: *Techniques In Protein Chemistry V* (J.W. Crabb, ed.) pp 371–378, Academic Press, San Diego.

6

GUANYLYL CYCLASE GENES AND THEIR ROLE IN RETINAL DEGENERATION

Hans-Jürgen Fülle* and Rima Khankan

Mary D. Allen Laboratories for Vision Research
Doheny Eye Institute
Departments of Cell and Neurobiology and Ophthalmology
University of Southern California School of Medicine
Los Angeles, California 90033

SUMMARY

Guanylyl cyclases play a fundamental role in a wide variety of cellular processes. In response to specific regulatory signals these enzymes form the intracellular second messenger molecule cGMP. In vertebrate retinal photoreceptor cells cGMP has been identified as the primary internal messenger for visual transduction more than a decade ago. Inherited defects in cGMP metabolism have been linked to retinal degeneration and blindness even longer. Recently, specific gene defects in one of the retinal isoforms of guanylyl cyclase, *RETGC-1*, have been found to cause Leber's congenital amaurosis (LCA) and an autosomal-dominant form of cone-rod dystrophy. We investigated the functional consequences of a *RETGC-1* missense mutation (F589S) described in LCA patients. We demonstrate that this mutation markedly decreases enzyme activity when expressed *in vitro*. Function and possible linkage to eye disease of an X-chromosomal gene encoding a second retinal isoform, *RETGC-2*, are less clear. We determined the genomic structure of *RETGC-2* as a prerequisite for future mutation screening studies.

---

* To whom correspondence and reprint requests should be addressed: Dr. H.-J. Fülle, Department of Cell and Neurobiology, University of Southern California School of Medicine, 1333 San Pablo Street BMT 401, Los Angeles, CA 90033, Tel.: 323-442-6630; Fax: 323-442-6655; E-mail: fulle@hsc.usc.edu.

Retinal Degenerative Diseases and Experimental Therapy, edited by Hollyfield *et al.*
Kluwer Academic / Plenum Publishers, New York, 1999.

1. INTRODUCTION

1.1. The Guanylyl Cyclase Family

Many cellular signaling pathways involve cyclic GMP (cGMP) as an intracellular second messenger for the action of a variety of extracellular signals including peptide hormones, nitric oxide, toxins, and light.[1–3] The synthesizing enzyme guanylyl cyclase (GC) exists in both membrane and soluble fractions of most cells. The GC family of proteins is characterized by a highly conserved catalytic domain at the carboxy terminus that is also present in adenylyl cyclases. In the mammal, at least seven isoforms are found associated with the membrane, and four apparent soluble enzyme subunits were identified so far.[4] The actual number of mammalian GC is not known but more than 30 forms exist in *C. elegans*.[5] This raises the question of whether a significant number of guanylyl cyclases remain to be identified in the mammal.[4,5]

Soluble GC subunits form heme-containing heterodimeric enzymes that can be activated by nitric oxide and related nitrovasodilatory drugs.[2] Membrane GC are single membrane-spanning proteins that appear to form homodimers or higher-ordered aggregates in their basal state.[3] They can be grouped into three subfamilies according to their expression pattern and their functional ligands or activators. Phylogenetic analysis of their amino acid sequence homology and gene structure leads to a similar classification (Table 1): The natriuretic peptide receptor guanylyl cyclases—GC-A,[6] GC-B,[7] and GC-G,[8] a recently identified apparent orphan receptor GC—function as cell-surface receptors. They are ubiquitously expressed in many tissues including retina. A second subfamily is represented by GC-C, a cell surface receptor for heat-stable bacterial enterotoxins and possibly two endogenous peptides, guanylin, and uroguanylin, with a more restricted expression pattern.[9]

We have identified a third GC subfamily with an expression pattern apparently limited to sensory tissues.[10,11] GC-D is expressed in a small number of sensory neurons in the peripheral olfactory neuroepithelium. No extracellular ligands are known and it has been considered an orphan receptor.[10] Two retinal GC isoforms—GC-E and GC-F—are encoded by separate genes and were found in the eye of rat,[11] human,[12,13] and bovine.[14,15] Their human orthologues are also referred to as RetGC-1 and RetGC-2, respectively.[12,13] The two retinal GC are co-expressed in outer segments of retinal photoreceptor cells where homomers between two identical subunits are formed preferentially *in vivo*.[16] Retinal GC activity appears to be regulated through an intracellular protein domain by interaction with two homologous Ca^{2+}-binding proteins, GCAP-1, and GCAP-2.[13,17–22]

Table 1. Mammalian membrane guanylyl cyclases

Subtype (Orthologues)	Expression	Ligand (Activators)
GC-A (NPR-A)	most tissues, incl. retina	ANP, BNP
GC-B (NPR-B)	most tissues, incl. retina	CNP
GC-G	lung, intestine, skeletal muscle	?
GC-C (STaR)	intestine, lung, kidney, liver	STa, (uro-)guanylin
GC-D	subset of olfactory receptor neurons	?
GC-E (RetGC-1, ROS-GC1)	retina, pineal gland	(GCAP-1, GCAP-2)
GC-F (RetGC-2, ROS-GC2)	retina	(GCAP-2)

Abbreviations: ANP, atrial natriuretic peptide; BNP, brain natriuretic peptide, CNP, C-type natriuretic peptide; STa, *E.coli* heat-stable enterotoxin; GCAP, guanylyl cyclase activating proptein.

1.2. Cyclic GMP Function in Retinal Photoreceptor Cells

Cyclic GMP has long been known as the primary internal messenger of vertebrate phototransduction.[23,24] Changes in cGMP concentration in rod and cone outer segments control the flow of Na^+ and Ca^{2+} ions through cyclic nucleotide-gated (CNG) nonselective cation channels, and ultimately membrane potential and transmitter release of photoreceptor cells. In vertebrates, hydrolysis of cGMP, channel closure, and hyperpolarization are initiated by light-activated rhodopsin through a cascade of enzymatic reactions that lead to activation of a cGMP-specific phosphodiesterase (PDE). Resynthesis of cGMP by retinal GC is thought to be regulated by a reciprocal Ca^{2+}-sensitive feedback mechanism that allows recovery and light adaptation of photoreceptor cells.[22,25,26]

Aberrant changes in cGMP metabolism have been shown to affect function and viability of photoreceptor cells.[27] Although the precise mechanisms remain unclear, this often results in retinal degeneration and blindness. It was suggested that either the disruption or the continuous activation of the phototransduction pathway may lead to photoreceptor cell death.[28,29]

High levels of cGMP appear to be deleterious to photoreceptor cells.[30] Farber and Lolley first showed that accumulation of cGMP precedes photoreceptor cell death in a mouse mutant *retinal degeneration (rd)* strain.[31] Markedly elevated levels of retinal cGMP during postnatal development were also seen in the *rcd1* Irish setter dog with early onset rod/cone dysplasia.[32] Later it was found that mutations in the rod photoreceptor PDE β-subunit gene are responsible for production of non-functional enzyme and retinal degeneration in these animal models.[33–37]

Abnormally low levels of cGMP were measured in retinas and photoreceptor cells of the *rd* chicken before onset of degeneration. This strain is characterized by a complete absence of rod and cone phototransduction in initially normally developed retina due to a null mutation of a photoreceptor GC gene.[38]

1.3. Mutations in Human Genes Affecting Cyclic GMP Synthesis in Retinal Photoreceptor Cells

In the past few years more and more mutations causing inherited retinal degeneration and blindness in humans were identified. Many of them were found in genes that encode key proteins of the phototransduction cascade. They are thought to cause progressive dysfunction and subsequent death of retinal photoreceptor cells.[39] Examples are mutations in the α- and β-subunits of photoreceptor PDE possibly affecting cGMP hydrolysis which were linked to forms of retinitis pigmentosa.[40–42]

Retinal GC has joined this rapidly expanding list only recently. Functional defects have been implicated in the pathogenesis of at least three clinically distinct inherited retinopathies. A mutation (Y99C) in the gene encoding a GC-activating protein, GCAP-1, was linked to autosomal dominant cone dystrophy.[43] This mutation markedly alters the Ca^{2+} sensitivity of GCAP-1 and may result in constitutive activation of retinal GC with elevated cGMP synthesis even at high free Ca^{2+} concentrations as found under physiological dark conditions.[44,45]

Two other retinopathies are caused by mutations in the *RETGC-1* gene itself. Missense mutations (E837D, R838C) were identified in families with a dominant form of cone-rod dystrophy, CORD6.[46] The amino acid substitutions occur in the putative dimerization domain of RetGC-1 and may affect proper dimerization which is required for enzymatic activity.[16,46] Mutations in *RETGC-1* also cause a recessive form of Leber's congenital amaurosis (LCA1), one of the earliest and most severe of all inherited retinal dystrophies. Several frameshift mutations and one missense mutation (F589S) have been identified in LCA1 patients.[47] It has been suggested that LCA1 is caused by low levels of cGMP and permanent closure of CNG channels in photoreceptor cells of affected patients.[47] In this paper, we present evidence that the F589S missense mutation indeed causes a marked decrease in GC activity.

Previously, we mapped the *RETGC-2* gene to the long arm of human chromosome X at q22.[48] Although inherited eye diseases have not yet been linked to this locus, it is possible that a phenotype is associated with mutations in *RETGC-2*. Here, we present and compare the gene structures of *RETGC-1* and *RETGC-2* as a prerequisite for future screening studies in search for disease-causing mutations.

2. METHODS

2.1. Genomic Organization of Human *RETGC-1* and *RETGC-2* Genes

2.1.1. Amplification of Introns From Genomic DNA. Polymerase chain reaction (PCR) and exonic oligonucleotide primers were used to amplify introns with flanking exon regions of *RETGC-1* and *RETGC-2* from human genomic DNA (Promega, Madison, WI) and from human bacterial artificial chromosomes (BAC). *RETGC*-containing human BAC clones were obtained from Research Genetics (Huntsville, AL) after screening human BAC plasmid DNA pools by PCR with specific oligonucleotides. Initial primer design was based on human RetGC-1 and RetGC-2 cDNA sequences (GenBank nos. M92432 and L37378, respectively) and the exon/intron organization of the orthologue genes from mouse (GenBank nos. L41933 and L44608, respectively). Primers were usually 21-oligomers (Genosys, The Woodlands, TX) and located 50–100 nucleotides upstream or downstream of the intron positions.

2.1.2. Automated DNA Sequencing. PCR products were gel-purified (QIAEX II gel extraction kit, QIAGEN, Valencia, CA), quantified by UV determination, and directly sequenced using sequence-specific primers. Some PCR products were subcloned using a pCR-Script Amp SK(+) cloning kit (Stratagene, La Jolla, CA) and subsequently sequenced using universal (Promega) and sequence-specific primers (Genosys). In general, the nucleotide sequence of both strands was determined using an ABI Prism dRhodamine terminator cycle sequencing kit and an ABI Prism 310 Genetic Analyzer with POP-6 separation polymer according to the manufacturer's instructions (PE Applied Biosystems, Foster City, CA). Sequence data were analyzed by AutoAssembler and Sequence Navigator software (PE Applied Biosystems).

2.2. Functional Analysis of A RetGC-1 Missense Mutation (F589S)

2.2.1. Site-Directed Mutagenesis. A full-length cDNA encoding rat GC-E inserted into the mammalian expression vector pCMV5 was generously provided by Dr. David L. Garbers, UT Southwestern Medical Center, Dallas, TX.[11] A single point mutation was generated in GC-E that corresponds to a missense mutation (F589S) found in patients with LCA.[47] Codon 568 was changed from TTT (encoding phenylalanine) to TCT (encoding serine) using the QuickChange™ site-directed mutagenesis kit according the manufacturer's suggestions (Stratagene, La Jolla). The modified sequence of mutant GC-E was confirmed by DNA sequencing of both strands.

2.2.2. Transfection Expression, Western Blot Analysis and Assay of Guanylyl Cyclase Activity. A polyclonal antisera directed against the COOH-terminal amino acid sequence of rat GC-E was generously provided by Dr. David L. Garbers.[16] Wild type and mutant GC-E were transiently expressed in COS-7 cells, and determination of guanylyl cyclase activity and Western blot analysis of Triton-solubilized crude membrane fractions was performed as described previously.[11,16] The experiment was performed twice in triplicates with similar results.

3. RESULTS

3.1. Genomic Organization of Human *RETGC-1* and *RETGC-2* Genes

We determined the size and location of introns in two human genes encoding retinal guanylyl cyclase. The human *RETGC-1* gene was mapped to chromosome 17 at p13.[49] Here we report that *RETGC-1* contains 20 exons and 19 introns and confirm previous results (Fig. 1).[47] The 5′- and 3′-most exons of *RETGC-1* are non-coding. Thus, the structure of the human gene is similar to its orthologues from mouse, bovine, and canine.[48,50,51] The human *RETGC-2* gene was mapped to the X chromosome at q22,[48] but to our knowledge its genomic organization has not yet been determined. We show that the overall structure of *RETGC-2* closely corresponds to that of *RETGC-1*. It, too, is composed of 20 exons and 19 introns and corresponding amino acids are interrupted by introns (Fig. 1). A detailed description of the genomic structures will be given elsewhere (Fülle et al., in preparation).

Exon sizes range from 81 to 730 and 85 to 815 nucleotides in *RETGC-1* and *RETGC-2*, respectively. All splice junction sequences for donor and acceptor sites are in agreement with the GT/AG rule in both genes.[52] A striking difference between the two genes is the larger size of introns in *RETGC-2*. Accordingly, while *RETGC-1* spans about 17.5 kb, the *RETGC-2* gene is much larger and spans close to 50 kb (Fig. 1).

As we described earlier for the mouse orthologue genes the separation of exons corresponds to the structural and functional domains of membrane guanylyl cyclases: The extracellular domain is distributed among three exons, the single membrane-spanning domain is encoded by one exon, and the kinase-like and cyclase-catalytic domains each are encoded by seven exons.[48] The relative locations of introns within the intracellular but not the extracellular domains correspond to those found in the rat GC-A gene.[53] It was suggested that membrane guanylyl cyclases are derived from a common ancestral gene harboring kinase-like and cyclase-catalytic domains.[48] Our results on the genomic organization of the human *RETGC-1* and *RETGC-2* genes further support this hypothesis.

3.2. Functional Analysis of a Missense Mutation (F589S) Described In Families with Leber's Congenital Amaurosis

Perrault et al. first identified disease-causing mutations in the *RETGC-1* gene from four unrelated patients of North African ancestry affected with Leber's congenital amaurosis.[47] In two families, single nucleotide deletions were located in exon 2 that

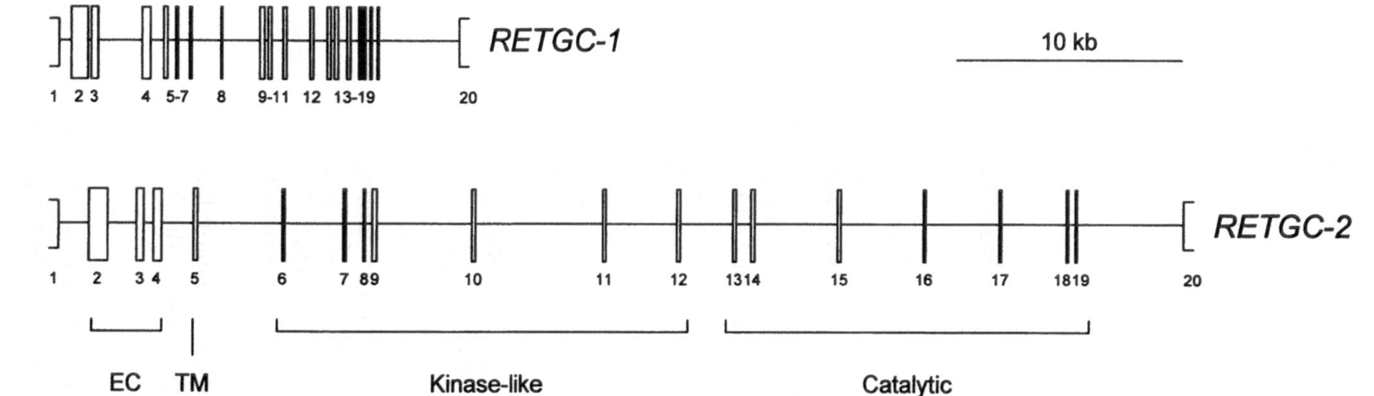

Figure 1. Exon-intron organization of the human *RETGC-1* and *RETGC-2* genes. Exons are indicated by boxes with numbers. The extracellular (EC) domains are encoded by exons 2–4, and exon 5 corresponds to the transmembrane (TM) domain. The kinase-like and cyclase catalytic domains are encoded by exons 6–12 and 13–19, respectively. Exons 1 and 20 are non-coding.

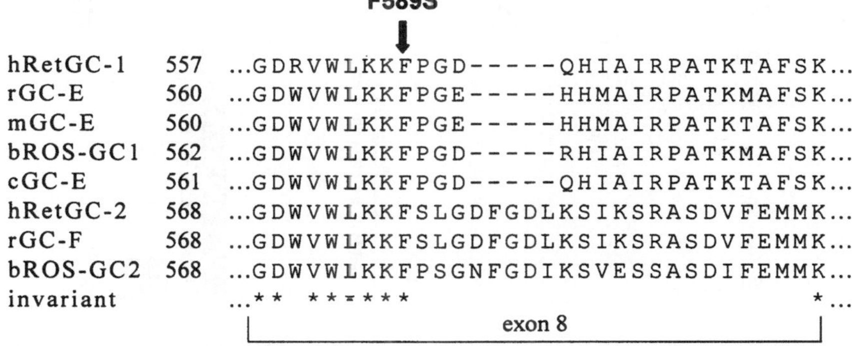

Figure 2. Amino acid sequence alignment of retinal isoforms of guanylyl cyclase from human (hRetGC-1, hRetGC-2),[12,13] rat (rGC-E, rGC-F),[11] bovine (bROS-GC1, bROS-GC-2),[14,15] and canine (cGC-E).[51] Exon 8 encodes part of a kinase like protein domain in the sensory-type subgroup of membrane guanylyl cyclases.[48] Codons are numbered from the start of translation. Invariant residues are marked by asterisks. The arrow indicates a missense mutation (F589S) identified in two families with Leber's congenital amaurosis type I (LCA1) where the phenylalanine residue is replaced by serine.[47]

result in frameshift mutations with premature stop codons. No functional RetGC-1 protein was predicted to be made in these patients. In two other families a missense mutation (F589S) was found in exon 8 changing a phenylalanine to serine.

While the nonsense mutations may be modeled by the naturally occuring gene knock-out in the *rd* chicken and targeted gene disruption of GC-E in the mouse (see chapters by S. Semple-Rowland and D. Birch in this volume), how the F589S missense mutation causes the LCA disease phenotype remained unclear. This mutation is located in a highly conserved kinase-like domain of RetGC-1. It was suggested that the amino acid change could result in severely reduced protein stability.[47]

Amino acid sequence alignment of exon 8 which encodes part of this regulatory domain shows that the phenylalanine residue is invariantly conserved in retinal GC from human, rat, bovine, and canine (Fig. 2). We modeled the F589S mutation in rat GC-E by site-directed mutagenesis of the corresponding codon. Wild type GC-E and F589S mutant GC-E were then inserted in the mammalian expression vector pCMV5 and transiently expressed *in vitro* by transfection of COS cells (Fig. 3). Western blot analysis of crude membrane fractions from transfected COS cells using a polyclonal antibody specific for the COOH-terminal amino acids of GC-E demonstrated expression of full length wild type and mutant GC proteins (Fig. 3A). Guanylyl cyclase activity was markedly reduced in the mutant as compared to wild type enzyme (Fig. 3B). In mock-transfected COS cells no protein was detected and guanylyl cyclase activity was below the detection limit of the assay.

4. DISCUSSION

Two isoforms of guanylyl cyclase are expressed in the photoreceptor layer of the mammalian retina.[11–16] Guanylyl cyclase activity is critically involved in the recovery phase of the phototransduction cascade and in maintenance of photoreceptor cell function and integrity.[22,25] While both isoforms were detected in rods and cones, *RETGC-1*

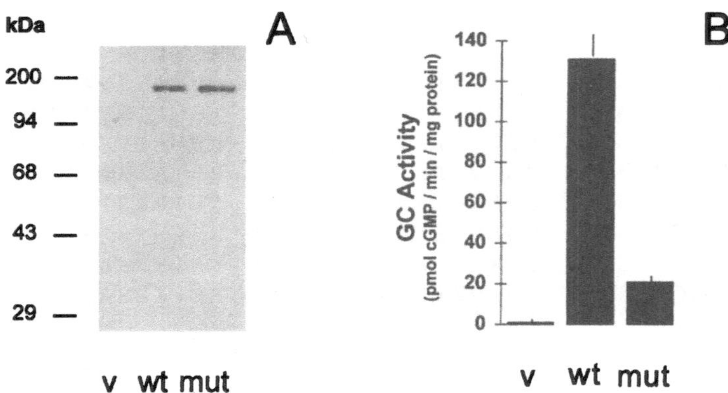

Figure 3. Analysis of *RETGC-1* missense mutation (F589S) described in families with Leber's congenital amaurosis. Triton-solubilized membranes from COS-7 cells transiently transfected with 5 µg of expression vector (*v*), wild type rat GC-E (*wt*), and mutant rat GC-E (*mut*) were analyzed as described before.[11,16] (**A**) Western blot analysis. Triton extracts (10 µg) were separated by SDS-PAGE, blotted to nitrocellulose membranes and probed with a polyclonal antibody directed against the C-terminus of rat GC-E. Positions of protein standards are shown on the left. Full-length rat GC-E is a 120 kDa polypeptide.[16] (**B**) Guanylyl cyclase activity was measured in presence of 1% Triton and Mn^{2+}.

appears to be expressed primarily in cone outer segments and to a lesser extent in rod outer segments.[12,13,17]

In humans, mutations in the *RETGC-1* gene can cause loss of vision and blindness as seen in one form of LCA. It was suggested that light-induced synthesis of cGMP is severely diminished in affected photoreceptor cells due to frameshift mutations.[47] Premature translation termination would lead to expression of truncated RetGC-1 protein lacking a functional catalytic domain. Here, we provide preliminary experimental evidence that the F589S missense mutation detected in LCA1 patients may result in reduced cGMP synthesis as well. Guanylyl cyclase activity measured in the presence of 1% Triton and Mn^{2+} is markedly decreased in the mutant compared to the wild type enzyme. Further studies will be necessary to evaluate the exact molecular mechanisms causing this decrease.

In LCA1, RetGC-2 activity seems not to be able to fully compensate for abolished RetGC1 activity and possibly reduced levels of cGMP. Although there is evidence that both isoforms are co-expressed in the same photoreceptor cells,[16] there may exist specific signaling pathways for each of the isoforms, and they may not be functionally redundant. To better understand their physiological role and their contribution to human eye disease we decided to first determine the genomic structure of their genes as prerequisite of screening studies. We found that both genes are very similar in their exon-intron organization as was expected from the structures of the orthologue genes in mouse.[48] The availability of human *RETGC-2* gene sequence information will permit a future comprehensive approach to mutation screening. As there is as yet no direct evidence for the involvement of *RETGC-2* in phototransduction, and other functions of cGMP are not well defined in photoreceptor cells, one can only speculate on possible disease phenotypes. However, knowledge of structure and location of *RETGC-2* on the long arm of the X chromosome will be helpful in the ongoing search for causes of human eye disease.

ACKNOWLEDGMENTS

We are grateful to Dr. David L. Garbers, University of Texas Southwestern Medical Center, Dallas, TX, for generously providing GC-E cDNA and specific antisera. We thank J. Koon, C. Duan, and V. Belonogoff for their expert technical assistance. R.K. was the recipient of a student research fellowship from the Fight for Sight research division of Prevent Blindness America. This work was supported by grants from the Knights Templar Eye Foundation, the James H. Zumberge Research and Innovation Fund, and the Donald E. and Delia B. Baxter Foundation. Additional support was from the Mary D. Allen Endowment, a Howard Hughes Medical Institute Resources Grant (P.I. Dr. Craft, USC Dept. of Cell & Neurobiology), and from NIH Core Vision Research Center grant EY03042 (Doheny Eye Institute).

REFERENCES

1. H.-J. Fülle and D.L. Garbers, 1994, Guanylyl cyclases: a family of receptor-linked enzymes, *Cell Biochem. Funct.* **12**(3):157–165.
2. J.G. Drewett and D.L. Garbers, 1994, The family of guanylyl cyclase receptors and their ligands, *Endocr. Rev.* **15**(2):135–162.
3. D.L. Garbers and D.G. Lowe, 1994, Guanylyl cyclase receptors, *J. Biol. Chem.* **269**(49):30741–30744.
4. B.J. Wedel and D.L. Garbers, 1997, New insights on the functions of the guanylyl cyclase receptors, *FEBS Lett.* **410**(1):29–33.
5. S. Yu, L. Avery, E. Baude, and D.L. Garbers, 1997, Guanylyl cyclase expression in specific sensory neurons: a new family of chemosensory receptors, *Proc. Natl. Acad. Sci. U.S.A.* **94**(7):3384–3387.
6. M. Chinkers, D.L. Garbers, M.S. Chang, D.G. Lowe, H.M. Chin, D.V. Goeddel, and S. Schulz, 1989, A membrane form of guanylate cyclase is an atrial natriuretic peptide receptor, *Nature* **338**(6210):78–83.
7. S. Schulz, S. Singh, R.A. Bellet, G. Singh, D.J. Tubb, H. Chin, and D.L. Garbers, 1989, The primary structure of a plasma membrane guanylate cyclase demonstrates diversity within this new receptor family, *Cell* **58**(6):1155–1162.
8. S. Schulz, B.J. Wedel, A. Matthews, and D.L. Garbers, 1998, The cloning and expression of a new guanylyl cyclase orphan receptor, *J. Biol. Chem.* **273**(2):1032–1037.
9. S. Schulz, C.K. Green, P.S. Yuen, and D.L. Garbers, 1990, Guanylyl cyclase is a heat-stable enterotoxin receptor, *Cell* **63**(5):941–948.
10. H.-J. Fülle, R. Vassar, D.C. Foster, R.-B. Yang, R. Axel, and D.L. Garbers, 1995, A receptor guanylyl cyclase expressed specifically in olfactory sensory neurons, *Proc. Natl. Acad. Sci. U.S.A.* **92**(8): 3571–3575.
11. R.-B. Yang, D.C. Foster, D.L. Garbers, and H.-J. Fülle, 1995, Two membrane forms of guanylyl cyclase found in the eye, *Proc. Natl. Acad. Sci. U.S.A.* **92**(2):602–606.
12. A.W. Shyjan, F.J. de Sauvage, N.A. Gillett, D.V. Goeddel, and D.G. Lowe, 1992, Molecular cloning of a retina-specific membrane guanylyl cyclase, *Neuron* **9**(4):727–737.
13. D.G. Lowe, A.M. Dizhoor, K. Liu, Q. Gu, M. Spencer, R. Laura, L. Lu, and J.B. Hurley, 1995, Cloning and expression of a second photoreceptor-specific membrane retina guanylyl cyclase (RetGC), RetGC-2, *Proc. Natl. Acad. Sci. U.S.A.* **92**(12):5535–5539.
14. R.M. Goraczniak, T. Duda, A. Sitaramayya, and R.K. Sharma, 1994, Structural and functional characterization of the rod outer segment membrane guanylate cyclase, *Biochem. J.* **302**(Pt 2):455–461.
15. R. Goraczniak, T. Duda, and R.K. Sharma, 1997, Structural and functional characterization of a second subfamily member of the calcium-modulated bovine rod outer segment membrane guanylate cyclase, ROS-GC2, *Biochem. Biophys. Res Commun.* **234**(3):666–670.
16. R.-B. Yang and D.L. Garbers, 1997, Two eye guanylyl cyclases are expressed in the same photoreceptor cells and form homomers in preference to heteromers, *J. Biol. Chem.* **272**(21):13738–13742.
17. A.M. Dizhoor, D.G. Lowe, E.V. Olshevskaya, R.P. Laura, and J.B. Hurley, 1994, The human photoreceptor membrane guanylyl cyclase, RetGC, is present in outer segments and is regulated by calcium and a soluble activator, *Neuron* **12**(6):1345–1352.

18. K. Palczewski, I. Subbaraya, W.A. Gorczyca, B.S. Helekar, C.C. Ruiz, H. Ohguro, J. Huang, X. Zhao, J.W. Crabb, R.S. Johnson, K.A. Walsh, M.P. Gray-Keller, P.B. Detwiler, and W. Baehr, 1994, Molecular cloning and characterization of retinal photoreceptor guanylyl cyclase-activating protein, *Neuron* **13**(2):395–404.
19. W.A. Gorczyca, A.S. Polans, I.G. Surgucheva, I. Subbaraya, W. Baehr, and K. Palczewski, 1995, Guanylyl cyclase activating protein. A calcium-sensitive regulator of phototransduction, *J. Biol. Chem.* **270**(37):22029–22036.
20. A.M. Dizhoor, E.V. Olshevskaya, W.J. Henzel, S.C. Wong, J.T. Stults, I. Ankoudinova, and J.B. Hurley, 1995, Cloning, sequencing, and expression of a 24-kDa Ca(2+)-binding protein activating photoreceptor guanylyl cyclase, *J. Biol. Chem.* **270**(42):25200–25206.
21. R.P. Laura, A.M. Dizhoor, and J.B. Hurley, 1996, The membrane guanylyl cyclase, retinal guanylyl cyclase-1, is activated through its intracellular domain, *J. Biol. Chem.* **271**(20):11646–11651.
22. E.N. Pugh, Jr., T. Duda, A. Sitaramayya, and R.K. Sharma, 1997, Photoreceptor guanylate cyclases: a review, *Biosci. Rep.* **17**(5):429–473.
23. R.N. Lolley and R.H. Lee, 1990, Cyclic GMP and photoreceptor function, *Faseb. J.* **4**(12):3001–3008.
24. E.N. Pugh, Jr. and T.D. Lamb, 1990, Cyclic GMP and calcium: the internal messengers of excitation and adaptation in vertebrate photoreceptors, *Vision Res.* **30**(12):1923–1948.
25. K.W. Yau, 1994, Phototransduction mechanism in retinal rods and cones. The Friedenwald Lecture, *Invest. Ophthalmol. Vis. Sci.* **35**(1):9–32.
26. A. Polans, W. Baehr, and K. Palczewski, 1996, Turned on by Ca^{2+}! The physiology and pathology of Ca^{2+}-binding proteins in the retina, *Trends Neurosci.* **19**(12):547–554.
27. R.N. Lolley and R.H. Lee, 1984, Phosphodiesterase dysfunction, cyclic GMP accumulation, and visual cell degeneration in early-onset inherited blindness, *Adv. Cyclic Nucleotide Protein Phosphorylation Res.* **17**:315–327.
28. P.R. Robinson, G.B. Cohen, E.A. Zhukovsky, and D.D. Oprian, 1992, Constitutively active mutants of rhodopsin, *Neuron* **9**(4):719–725.
29. G.L. Fain and J.E. Lisman, 1993, Photoreceptor degeneration in vitamin A deprivation and retinitis pigmentosa: the equivalent light hypothesis, *Exp. Eye Res.* **57**(3):335–340.
30. R.N. Lolley, D.B. Farber, M.E. Rayborn, and J.G. Hollyfield, 1977, Cyclic GMP accumulation causes degeneration of photoreceptor cells: simulation of an inherited disease, *Science* **196**(4290):664–666.
31. D.B. Farber and R.N. Lolley, 1974, Cyclic guanosine monophosphate: elevation in degenerating photoreceptor cells of the C3H mouse retina, *Science* **186**(4162):449–451.
32. G. Aquirre, D. Farber, R. Lolley, R.T. Fletcher, and G.J. Chader, 1978, Rod-cone dysplasia in Irish setters: a defect in cyclic GMP metabolism in visual cells, *Science* **201**(4361):1133–1134.
33. R.H. Lee, B.S. Lieberman, R.L. Hurwitz, and R.N. Lolley, 1985, Phosphodiesterase-probes show distinct defects in rd mice and Irish setter dog disorders, *Invest. Ophthalmol. Vis. Sci.* **26**(11):1569–1579.
34. C. Bowes, T. Li, M. Danciger, L.C. Baxter, M.L. Applebury, and D.B. Farber, 1990, Retinal degeneration in the rd mouse is caused by a defect in the beta subunit of rod cGMP-phosphodiesterase, *Nature* **347**(6294):677–680.
35. S.J. Pittler and W. Baehr, 1991, Identification of a nonsense mutation in the rod photoreceptor cGMP phosphodiesterase beta-subunit gene of the rd mouse, *Proc. Natl. Acad. Sci. U.S.A.* **88**(19):8322–8326.
36. D.B. Farber, J.S. Danciger, and G. Aguirre, 1992, The beta subunit of cyclic GMP phosphodiesterase mRNA is deficient in canine rod-cone dysplasia 1, *Neuron* **9**(2):349–356.
37. M.L. Suber, S.J. Pittler, N. Qin, G.C. Wright, V. Holcombe, R.H. Lee, C.M. Craft, R.N. Lolley, W. Baehr, and R.L. Hurwitz, 1993, Irish setter dogs affected with rod/cone dysplasia contain a nonsense mutation in the rod cGMP phosphodiesterase beta-subunit gene, *Proc. Natl. Acad. Sci. U.S.A.* **90**(9):3968–3972.
38. S.L. Semple-Rowland, N.R. Lee, J.P. Van Hooser, K. Palczewski, and W. Baehr, 1998, A null mutation in the photoreceptor guanylate cyclase gene causes the retinal degeneration chicken phenotype, *Proc. Natl. Acad. Sci. U.S.A.* **95**(3):1271–1276.
39. T.P. Dryja and T. Li, 1995, Molecular genetics of retinitis pigmentosa, *Hum. Mol. Genet.* **4**:1739–1743.
40. M.E. McLaughlin, M.A. Sandberg, E.L. Berson, and T.P. Dryja, 1993, Recessive mutations in the gene encoding the beta-subunit of rod phosphodiesterase in patients with retinitis pigmentosa, *Nat. Genet.* **4**(2):130–134.
41. S.H. Huang, S.J. Pittler, X. Huang, L. Oliveira, E.L. Berson, and T.P. Dryja, 1995, Autosomal recessive retinitis pigmentosa caused by mutations in the alpha subunit of rod cGMP phosphodiesterase, *Nat. Genet.* **11**(4):468–471.

42. T.P. Dryja, J.T. Finn, Y.W. Peng, T.L. McGee, E.L. Berson, and K.W. Yau, 1995, Mutations in the gene encoding the alpha subunit of the rod cGMP-gated channel in autosomal recessive retinitis pigmentosa, *Proc. Natl. Acad. Sci. U.S.A.* **92**(22):10177–10181.
43. A.M. Payne, S.M. Downes, D.A.R. Bessant, R. Taylor, G.E. Holder, M.J. Warren, A.C. Bird, and S.S. Bhattacharya, 1998, A mutation in guanylate cyclase activator 1A (GUCA1A) in an autosomal dominant cone dystrophy pedigree mapping to a new locus on chromosome 6p21.1, *Hum. Mol. Gen.* **7**(2):273–277.
44. I. Sokal, N. Li, I. Surgucheva, M.J. Warren, A.M. Payne, S.S. Bhattacharya, W. Baehr, and K. Palczewski, 1998, GCAP1 (Y99C) mutant is constitutively active in autosomal dominant cone dystrophy, *Mol. Cell* **2**(1):129–133.
45. A.M. Dizhoor, S.G. Boikov, and E.V. Olshevskaya, 1998, Constitutive activation of photoreceptor guanylate cyclase by Y99C mutant of GCAP-1, *J. Biol. Chem.* **273**(28):17311–17314.
46. R.E. Kelsell, K. Gregory-Evans, A.M. Payne, I. Perrault, J. Kaplan, R.-B. Yang, D.L. Garbers, A.C. Bird, A.T. Moore, and D.M. Hunt, 1998, Mutations in the retinal guanylate cyclase (*RETGC-1*) gene in dominant cone-rod dystrophy, *Hum. Mol. Gen.* **7**(7):1179–1184.
47. I. Perrault, J.M. Rozet, P. Calvas, S. Gerber, A. Camuzat, H. Dollfus, S. Chatelin, E. Souied, I. Ghazi, C. Leowski, M. Bonnemaison, D. Le Paslier, J. Frezal, J.L. Dufier, S. Pittler, A. Munnich, and J. Kaplan, 1996, Retinal-specific guanylate cyclase gene mutations in Leber's congenital amaurosis, *Nat. Genet.* **14**(4):461–464.
48. R.-B. Yang, H.-J. Fülle, and D.L. Garbers, 1996, Chromosomal localization and genomic organization of genes encoding guanylyl cyclase receptors expressed in olfactory sensory neurons and retina, *Genomics* **31**(3):367–372.
49. L. Oliveira, P. Miniou, E. Viegas-Pequignot, J.M. Rozet, H. Dollfus, and S.J. Pittler, 1994, Human retinal guanylate cyclase (GUC2D) maps to chromosome 17p13.1, *Genomics* **22**(2):478–481.
50. J.P. Johnston, F. Farhangfar, J.G. Aparicio, S.H. Nam, and M.L. Applebury, 1997, The bovine guanylate cyclase *GC-E* gene and 5′ flanking region, *Gene* **193**:219–227.
51. A. Veske, S.E. Nilsson, and A. Gal, 1998, Organization of the canine gene encoding the E isoform of retinal guanylate cyclase (cGC-E) and exclusion of its involvement in the inherited retinal dystrophy of the Swedish Briard and Briard-beagle dogs, *Biochim. Biophys. Acta* **1372**(1):69–77.
52. S.M. Mount, 1982, A catalogue of splice junction sequences, *Nucleic Acids Res.* **10**(2):459–472.
53. M. Yamaguchi, L.J. Rutledge, and D.L. Garbers, 1990, The primary structure of the rat guanylyl cyclase A/atrial natriuretic peptide receptor gene, *J. Biol. Chem.* **265**(33):20414–20420.

7

LOSS OF CONE AND REDUCTION IN ROD ERG RESPONSES IN THE GUANYLYL CYCLASE-E (GC-E) DEFICIENT MOUSE

David G. Birch,[1,2] Ruey-Bing Yang,[3] Susan W. Robinson,[3,4] and David L. Garbers[3,4]

[1] Retina Foundation of the Southwest
9900 N. Central Expressway
Suite 400, Dallas, Texas 75231-3303
[2] Department of Ophthalmology
[3] Department of Pharmacology and
[4] the Howard Hughes Medical Institute
University of Texas Southwestern Medical Center
5323 Harry Hines Boulevard
Dallas, Texas 75235-9041

ABSTRACT

Guanylyl cyclase-E (GC-E), which is specifically expressed in the retina, has been proposed as critical for normal vision. Mutations within the human homolog of the rodent GC-E gene have been reported in patients with Leber congenital amaurosis (LCA) and in patients with cone-rod dystrophy (CORD6). We have recently produced a mouse genetic model lacking the GC-E gene, and have recorded electroretinograms (ERGs) from mice at different ages. Rod ERGs to flashes over a 6 log unit range were recorded from 18 GC-E –/–, 4 GC-E –/–, and 30 GC-E +/+ mice. Derived parameters were used to assess activation and inactivation stages of transduction, inner retinal function, and cone system activity.

GC-E +/– mice were indistinguishable from normal in all measures. In GC-E –/– mice, the most striking abnormality was a severe reduction in the cone ERG by 3–5 wks of age and a complete absence of a detectable cone ERG in mice >8 wks. An early

Retinal Degenerative Diseases and Experimental Therapy, edited by Hollyfield *et al.*
Kluwer Academic / Plenum Publishers, New York, 1999.

loss of cones or a failure of cones to develop was confirmed by histological analysis. The maximum rod photoresponse amplitude (*Rm*) was 70% lower than normal at 3–5 wks despite a normal histological appearance of rods. Recovery from activation, as assessed by a double-flash technique, was more rapid in GC-E –/– than in wild-type mice. Reductions in b-wave amplitude at all ages were accompanied by a shortening of b-wave implicit time.

These findings suggest that GC-E is critical for the survival of cone, but not rod, photoreceptors.

1. INTRODUCTION

Cyclic GMP has been proposed as essential for visual signal transduction.[1] In the mammalian retina, two guanylyl cyclases, GC-E (retGC-1) and GC-F (retGC-2), appear critical for the maintenance of cGMP.[2,3] Both GC-E and GC-F are membrane bound enzymes and both closely resemble known guanylyl cyclase receptors in their domain structure.[4] In the absence of cyclases, lowered cGMP levels in the photoreceptor could lead to the permanent closure of cGMP-gated channels and subsequent failure to respond to light.

Recently, two human genetic diseases have been associated with mutations in GC-E, Leber congenital amaurosis (LCA),[5] and a form of cone-rod dystrophy (CORD6).[6] LCA can also be caused by mutations in RPE65[7] and CRX.[8]

Gene disruption represents a potentially powerful method by which to define the functions of guanylyl cyclases within the retina. The disruption of the gene for GC-E in mice causes a complete loss of detectable cone ERG responses by 8 weeks of age and a reduction of rod ERG responses throughout life.

2. METHODS

2.1. Generation of GC-E-Deficient Mice

AB1 embryonic stem (ES) cells derived from the inbred 129/SvJ mouse line, a gift from Dr. A. Bradley, were cultured on an irradiated STO-LIF fibroblast feeder layer. ES cells (1.0×107) were electroporated at 0.23 kV, 500 μFD with 25 μg of linearized targeting vector DNA. G418 (180 μg/ml) and gancyclovir (1 μM) were added to the medium for selection for 4–5 days until ES cell colonies formed. To identify clones with targeted disruption of the GC-E gene, the ES cells were expanded and genomic DNA was prepared for Southern blot analysis. Chimeric mice were generated by injecting the GC-E disrupted ES cells into C57BL/6J blastocysts that were subsequently implanted into the uteri of pseudopregnant foster females. Seven chimeras were born and mated with C57BL/6J mice to test for germline competency of the targeted clone; one of these mice transmitted the mutant allele. ES cell-derived progeny were screened for the presence of the targeted allele by Southern blot and/or PCR analyses. Heterozygous animals were inbred to generate homozygous mice lacking a functional GC-E gene (see [9]).

2.2. Histology

Eyes were removed and fixed overnight at 4°C with a solution of 10% formalin. Eyes were embedded in paraffin and 3 or 4 μm sections were cut through the optic nerve head and stained with hematoxylin and eosin.

2.3. Full-Field Electroretinography

Full-field ERGs were obtained in a ganzfeld dome from 30 GC-E +/+ mice, 4 GC-E +/– mice, and 18 GC-E –/– mice. Mice were anesthetized with 80 to 100 μl doses of a saline solution containing ketamine (40 mg/ml) and xylazine (2 mg/ml). A gold-wire coil placed on one cornea was referenced to a similar gold wire in the mouth. A needle electrode in the tail served as ground. Eyes were dilated (cyclopentolate hydrochloride) and dark adapted for at least two hours prior to testing. Signals were amplified (Tektronix AM502 differential amplifier; ×10,000; 3 dB down at 2 and 10,000 Hz), digitized (sampling rate = 1.25 to 5 kHz) and averaged on a personal computer. A Grass photostimulator provided 20 μs flashes. Cone-mediated b-waves were elicited by achromatic flashes ranging from 0.24 to 1.44 log phot td-s in 0.3 log unit steps. A rod-saturating achromatic background (40 cd/m^2) was used to isolate the cone components.

Rod-mediated a-waves were elicited by a Novatron flash unit, which produced high-intensity (up to 3.4 log scot td-s) short-wavelength (Wratten W47B: λ_{max} = 449 nm, half-bandwidth = 47 nm) flashes of 1.3 msec duration. The leading edge of the rod a-wave was fit by a computational model of the activation phase of transduction. Based on the Lamb and Pugh model of transduction, the leading edge is defined by:

$$P3(i,t) \cong \left[1-\exp\left(-i \cdot S \cdot (t-t_d)^2\right)\right] * R_m \text{ for } t > t_d \quad (1)$$

where *P3* is the sum of the responses of individual rods. The amplitude of *P3* is a function of flash intensity, *i*, and time, *t*, after flash onset. S is a sensitivity parameter that scales *i*, R_m is the maximum response, and t_d is a brief delay. As in previous studies,[10,11] t_d was held constant at 3.2 msec for all mice.

A double-flash technique was used to assess the recovery (inactivation) phase of phototransduction (11). The technique utilized two Novatron flash units within the ganzfeld dome. The first flash was an achromatic test of 1.5 log scot td-s. This was followed at various times (t) by a 3.4 log scot td-s short-wavelength (Wratten W47B: λ_{max} = 449 nm, half-bandwidth = 47 nm) probe flash, which rapidly saturated the rods. The derived amplitude at time t, of the rod response to the *test* flash should then be the difference between the probe amplitude at time t and the saturated photoresponse. The relative recovery of the test flash response is defined by:

$$R/R_m = 1 - \exp[-(t-T)/\tau_r] \quad (2)$$

where T was the critical delay before the initiation of recovery and τ_r was the recovery time constant.

Rod-mediated b-waves were elicited by short-wavelength flashes (Wratten 47A: λ_{max} = 470 nm, half-bandwidth = 55 nm) ranging from –3 to 1 log scot td-s in 0.3 log unit steps. In wild-type mice, dark-adapted responses to intensities above 0.0 log scot td-sec contained a small cone component that could be isolated by presenting the same flashes in the presence of a rod-saturating (40 cd/m^2) background. Rod-only responses were obtained by computer-subtracting this cone component from the mixed response. Using parameters obtained from fits to high-intensity a-waves, the photoreceptor component could be derived for responses of lower intensity. Oscillatory potentials were digitally-isolated with a bandpass filter 6 dB down at 54.4 and 1037 Hz. Subtracting the photoreceptor component and the oscillatory components from each response yielded derived P2, which reflects primarily the rod on-bipolar cells.[12,13] A quantitative measure of the rod-mediated P2 response was provided by fitting V—log I functions with the Michaelis-Menton relationship:

$$V = \left(\frac{I}{I+k}\right)V_{max} \tag{3}$$

to determine the parameters of sensitivity (k) and maximum P2 amplitude (V_{max}).

3. RESULTS

Figure 1 shows cone responses elicited by achromatic stimuli presented in the presence of a rod saturating (40 cd/m^2) background. GC-E +/+ mice (A) showed typical cone responses, with virtually no a-wave and a b-wave whose amplitude varied systematically with intensity while the implicit time stayed fairly constant with intensity. Over this range of intensities, the cone b-wave amplitude does not saturate, so intensity-response data can be fit with a simple linear function (see inset) to derive a measure of cone threshold (the retinal illuminance necessary for a 2.0 μV criterion amplitude). The mean (±1 SEM) cone threshold in wild-type animals was −0.42 ± 0.27 log phot td-sec, while the mean (± 1 SEM) amplitude to the highest retinal illuminance (1.44 log phot td-s) was 99.6 ± 46.5 μV. Cone b-wave responses from GC-E +/− mice

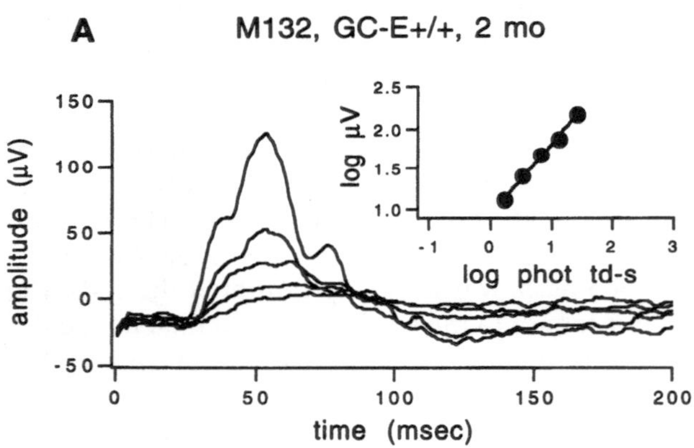

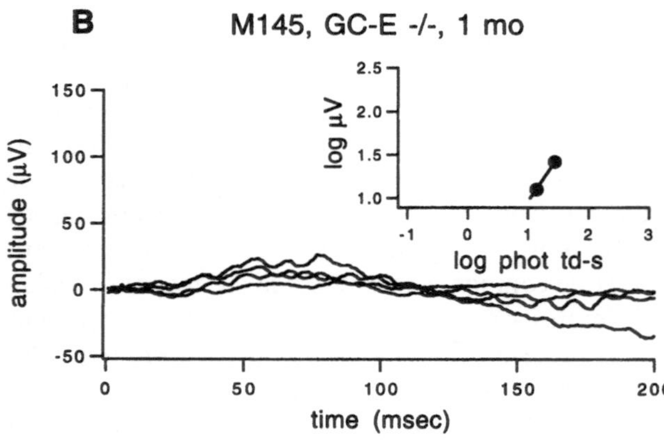

Figure 1. A. Cone responses from GC-E +/+ mouse to flashes ranging from 0.24 to 1.44 log phot td-sec in approximately 0.3 log unit steps. Responses were recorded in the presence of a rod-saturating (40 cd/m^2) background. Inset shows linear relationship between b-wave amplitude and retinal illuminance over this range. B. Corresponding data for representative 1-month-old GC-E −/− mouse.

(not shown) were indistinguishable from normal. GC-E –/– (B) mice showed a severe reduction in the cone ERG at 1 month. The maximum amplitude was reduced by a factor of 10 and the threshold for a detectable response (inset) was increased by 1 log unit. At 2 months of age and older, the cone response was non-detectable in all GC-E –/– mice.

Figure 2A shows rod responses from a wild type mouse elicited by over a 6 log unit range of retinal illuminances. At low intensities, the response consists entirely of a b-wave. At higher intensities, the a-wave grows in amplitude and oscillatory potentials become extremely prominent. Figure 2B shows the a-waves to stimuli ranging from 1.8 to 3.4 log scot td-s on an expanded scale. The dashed curves are the best-fit of the transduction model (Eq. 1) to the leading edge. The two parameters of most interest are the sensitivity parameter, S, which is related to the gain of activation stages, and R_m, which is the maximal, saturated photoresponse amplitude. For 30 wild type mice, mean (± SEM) values for log S and log R_m were 2.03 ± 0.2 $sec^{-2}[td\text{-}sec]^{-1}$ and 2.35 ± 0.16 log μV, respectively.

The top panel (A) of Fig. 3 show 200 ms of the ERG responses to stimuli ranging from –3.0 to 1.27 log scot td-s for a GC-E +/+ mouse. Responses contain at least three prominent components, photoreceptor (P3), inner retinal (P2), and oscillatory potentials, which are especially prominent in the mouse. The computational approach makes it possible to isolate these components of the rod ERG in order to assess each separately.[12] Based on the fit to the leading edge at high intensities (Fig. 2B), we can derive the underlying rod photoresponse (B) to the intensities shown in A. We can also digitally-isolate the oscillatory potentials (C). Subtracting the photoreceptor

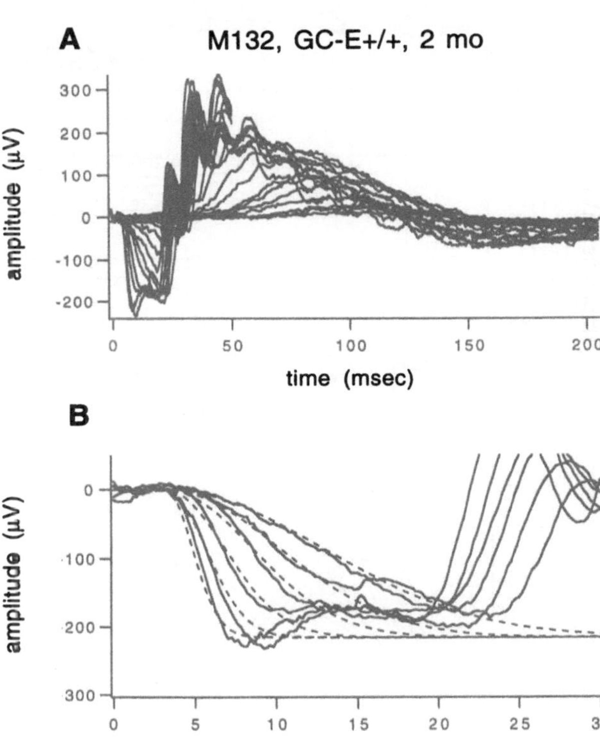

Figure 2. A. Rod responses from a wild type mouse to flashes ranging from –3 to 3.4 log scot td-sec (approximately 0.3 log unit steps). B. Responses to highest five flashes (1.82–3.4 log scot td-sec) on expanded scale. Dashed curves are simultaneous fits of Eq. 1 to all five responses, with t_d = 3.2 msec, log S = 1.9 $sec^{-2}[td\text{-}sec]^{-1}$, and log R_m = 2.33 logμV.

component and the oscillatory potentials from each response yields a corneal-positive waveform (D), derived P2, with characteristics that are remarkably similar to those of isolated rod on-bipolar cells.[14] The measured amplitudes are then well-fit by a Naka-Rushton function (inset).

Figure 4 shows rod P3 responses (top panel) and rod P2 responses (bottom panel) from a 5-month-old GC-E +/+ mouse (A), a 1-month-old GC-E –/– mouse (B), a 5-month-old GC-E –/– mouse (C), and a 5-month-old GC-E +/– mouse (D). The a-waves are clearly reduced in amplitude at 1 month and 5 months in the GC-E –/– mice, and the on-bipolar components show a comparable reduction. The responses from the GC-E +/– mouse are indistinguishable from normal.

Figure 5 summarizes rod a-wave data from all mice. Vertical bars represent the 95% confidence interval for R_m values from wild-type mice at each age. GC-E +/– mice were similar to wild-type mice. All 18 GC-E –/– mice were significantly below normal in their maximum photoresponse amplitude. There is a slight trend for amplitudes to decrease with age, but this is not statistically significant.

Figure 6 summarizes rod and cone ERG parameters. For GC-E +/– mice (white bars), and GC-E –/– (black bars), the average value of each parameter is shown rela-

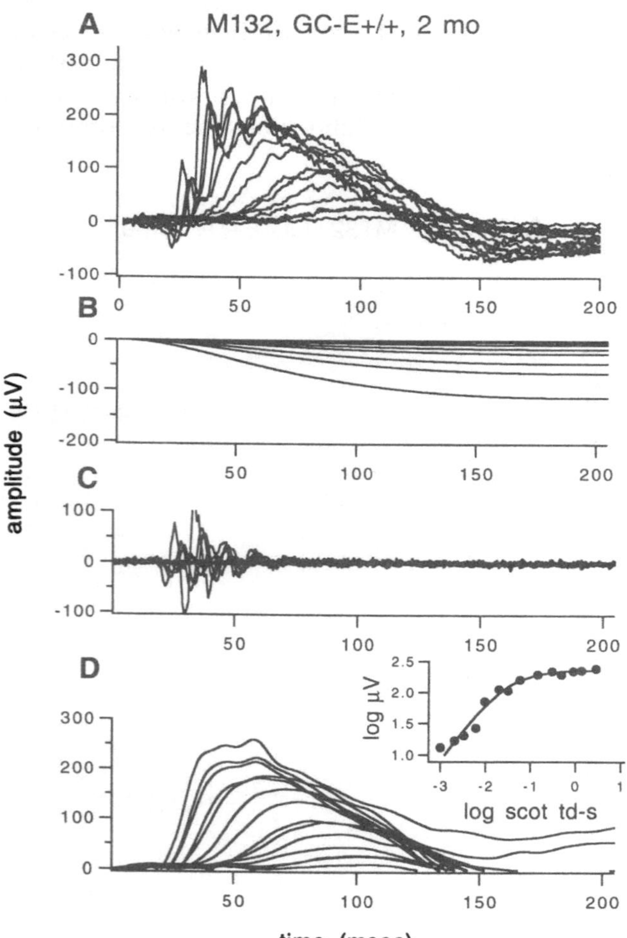

Figure 3. A. Rod ERG responses from a wild-type mouse to flashes ranging from –3 to 1.27 log scot td-sec (approx. 0.3 log unit steps). B. Photoresponse component of responses shown in A. Photoresponse components for these flashes were derived from parameters of Eq. 1. Parameters for Eq. 1 were determined from best-fits to high-intensity responses (i.e. Fig. 2B). C. Oscillatory components obtained by digitally-filtering the responses shown in A (6 dB down at 54.4 and 1037 Hz). D. Derived on-bipolar component obtained by subtracting responses in panels B and C from those shown in panel A.

tive to the mean wild-type value. For the GC-E +/– mice, no differences from normal were seen in any parameter. For the GC-E –/– mice, there was no significant change ($t = 0.7$; n.s.) in S, the gain parameter from the a-wave, suggesting that the cyclase deficiency had no effect on activation of phototransduction. Compared to wild-type mice, the saturated photoresponse amplitude, R_m, was significantly reduced ($t = 11.6$; $p < 0.001$) and there were comparable reductions in the k ($t = 9.05$; $p < 0.001$), and V_{max} ($t = 8.5$; $p < 0.001$) of the on-bipolar component. The most striking reduction was in the cone response, which was barely detectable within the first month, but non-detectable after the first month; i.e. cone threshold was elevated at least 2.5 log units.

Preliminary histological analysis of the GC-E –/– retina at the light microscopic level revealed a pronounced reduction in the number of cone nuclei by 5 weeks, and virtual absence by 6 months. Rods, on the other hand, showed no obvious degeneration at any age in the absence of GC-E. There were no apparent differences between GC-E +/+ and GC-E –/– mice in the number of rows of photoreceptor nuclei.

The timing of the on-bipolar component appeared to be faster in the GC-E –/– mice than in the wild-type mice. A somewhat extreme example of this to a moderate intensity (–0.5 log scot td-sec) flash is shown in Fig. 7. The on-bipolar response in the GC-E –/– mouse is both smaller and faster than the comparable response in the wild-type mouse. The dashed curves show the GC-E –/– responses scaled in amplitude to match the wild-type responses. Average implicit time functions for wild-type and GC-E –/– mice (Fig. 8) show that the decrease in amplitude in the GC-E –/– mice is accompanied by a significant ($t = -3.99$; $p < 0.001$) shortening of implicit time to a flash intensity of –2 log scot td-sec. Mean implicit times tend to converge at higher intensities as the rod bipolar component saturates.

In order to assess the recovery (inactivation) phase of phototransduction, a double-flash technique was used. Probe flash responses at various times following a 1.5 log scot td-sec test flash are shown in Fig. 9A. There is no detectable response to the probe for times less that 400 msec (i.e. the response to the test flash remains saturated). Beginning at 500 msec and continuing through 1500 msec, there is a progressive increase in the amplitude of the photoresponse to the probe. This recovery is plotted in Figure 9B (open symbols) along with the best fit of Eq. 2, where $T = 467$ msec and $\tau_r = 477$ msec. The closed symbols show recovery data from a 4-month-old GC-E –/– mouse, where recovery was faster than in the wild type ($T = 208$ msec, $\tau_r = 568$ msec). Summary data for 6 wild-type and 8 GC-E –/– mice are shown in Fig. 10. The period of complete suppression of the photoresponse (T) was significantly shorter in GC-E –/– mice than in GC-E +/+ mice ($t = 5.07$; $p < 0.001$).

4. DISCUSSION

The most striking abnormality in the GC-E deficient mouse was a severe reduction in the cone b-wave within the first post-natal month and the absence of a detectable cone response at 2 months of age or older. At the light microscopic level, reduced numbers of cone nuclei were evident at 5 weeks of age and few morphologically identifiable cone nuclei were evident at 6 months of age. While immunoreactive labeling with cone-specific antibodies will be necessary to fully define the time course of cone loss or failure to develop, these initial findings are at least consistent with the possibility that GC-E may be the major or only cyclase in the mouse cone photoreceptor. These findings are similar to those in the chicken *retinal degeneration* model,[15] where

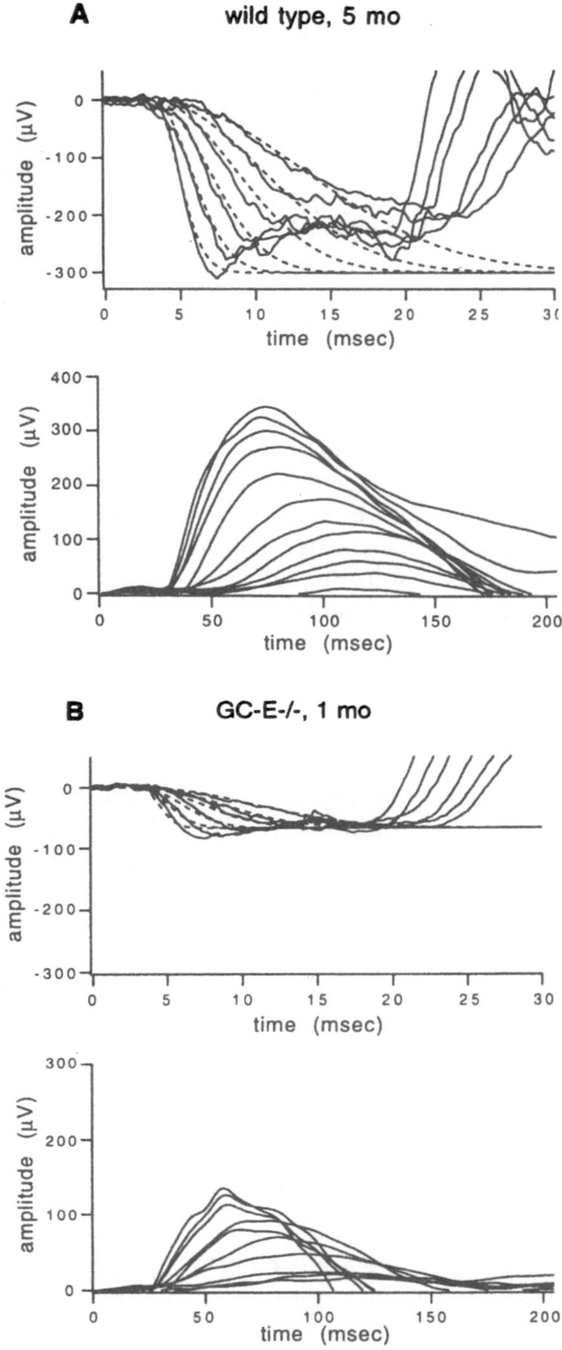

Figure 4. Photoresponses and rod on-bipolar responses derived as in Fig. 3. A. 5-month-old GC-E +/+; photoreceptor: log S = 1.9 log sec^{-2} $[td\text{-}sec]^{-1}$; log R_m = 2.48 log μV; on-bipolar: log k = −1.3 log scot td-sec; log V_{max} = 2.51 log μV. B. 1-month-old GC-E −/−; photoreceptor: log S = 2.15 log $sec^{-2}[td\text{-}sec]^{-1}$; log R_m = 1.82 log μV; on-bipolar: log k = −0.74 log scot td-sec; log V_{max} = 2.15 log μV. C. 5-month-old GC-E −/−; photoreceptor: log S = 2.34 log $sec^{-2}[td\text{-}sec]^{-1}$; log R_m = 1.79 log μV; on-bipolar: log k = −1.22 log scot td-sec; log V_{max} = 2.22 log μV. D. 2-month-old GC-E +/−; photoreceptor: log S = 1.99 log $sec^{-2}[td\text{-}sec]^{-1}$; log R_m = 2.43 log μV; on-bipolar: log k = −1.3 log scot td-sec; log V_{max} = 2.55 log μV.

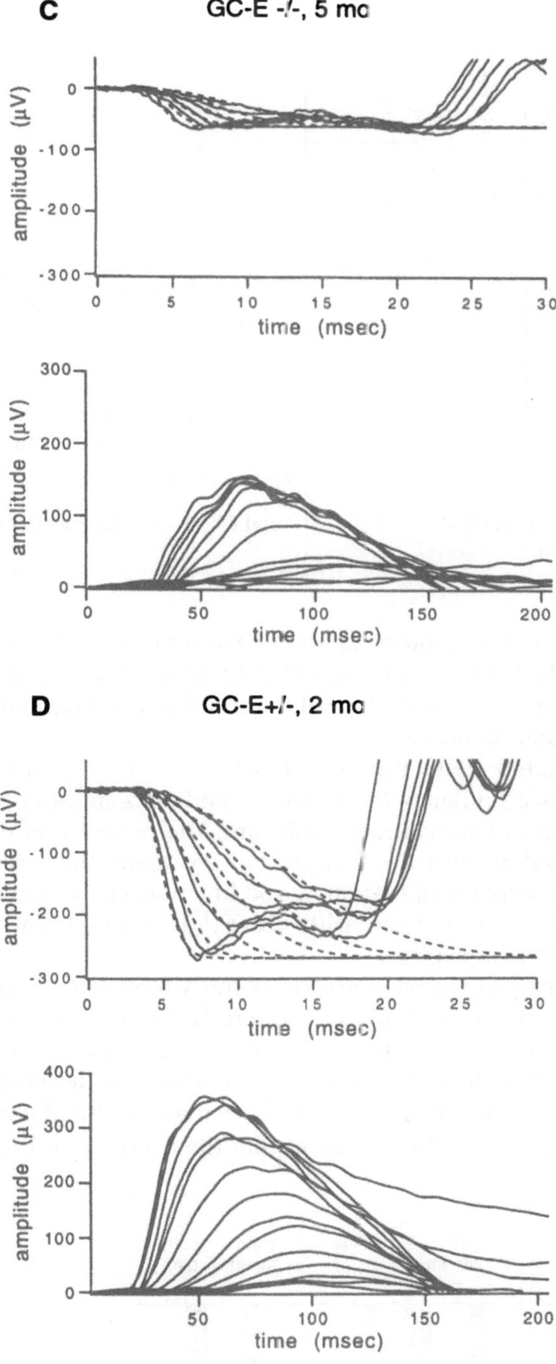

Figure 4. *Continued.*

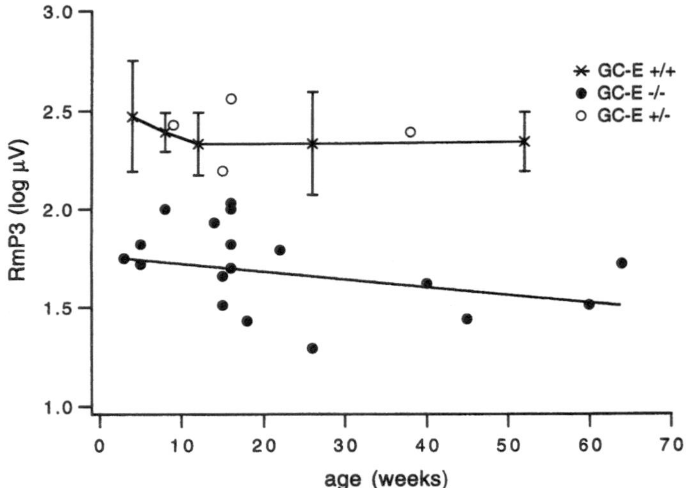

Figure 5. Log R_m values from GC-E +/+, GC-E +/–, and GC-E –/– mice. Vertical bars on wild-type means represent the 95% confidence interval.

approximately 80% of the photoreceptors in wild-type animals are cones. The absence of GC-E leads to chickens that are behaviorally blind at hatching with no measurable ERG. Photoreceptors are morphologically normal at hatching but cone degeneration becomes evident soon thereafter.

The rod-mediated ERG was detectable in GC-E –/– mice at all ages tested. Although there was a tendency for smaller amplitudes in older animals, the modest decline with age was not statistically significant. The presence of detectable rod ERG responses throughout at least the first year of life raises the possibility that another cyclase (i.e. GC-F) is capable of sustaining rod function. GC-F is expressed in rod outer segments, where it is co-localized with GC-E.[3] It is not known whether GC-F is expressed in cone photoreceptors.

Component analysis has been used previously to isolate photoreceptor and inner retinal components of the human ERG, where it provides a powerful technique for determining the site of abnormality in retinal disease.[16] The results of this paper show that it is readily applicable to the mouse ERG, where the large oscillatory potentials tend to obscure the rod b-wave. When photoreceptor, on-bipolar and oscillatory components are isolated, it is readily apparent that the decrease in ERG amplitude in the

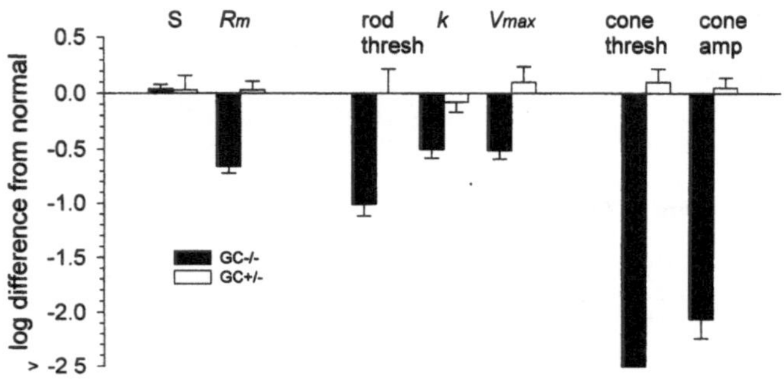

Figure 6. Summary ERG parameters from all mice. The average values are shown relative to mean wild-type values. Vertical bars represent S.E.M.

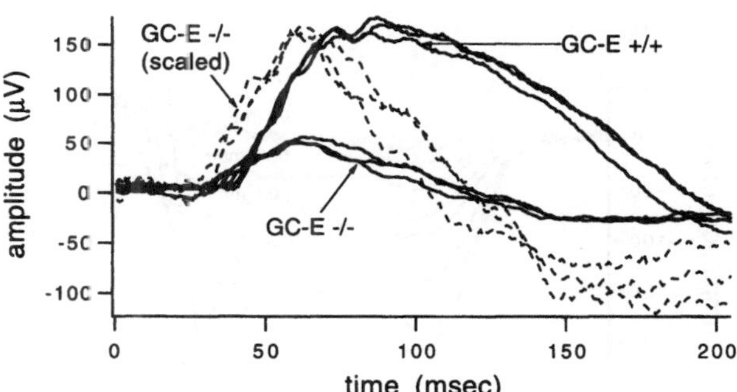

Figure 7. On-bipolar responses (three replications) to moderate intensity (–0.5 log scot td-sec) flashes in wild-type and GC-E –/– mice. Dashed curves are the GC-E –/– responses scaled in amplitude to match the wild-type responses.

GC-E deficient mouse is due to a decrease in the amplitude of the photoreceptor component. All subsequent components are decreased in proportion to the photoreceptor component. Component analysis also simplifies the measurement of implicit time. When the b-wave includes prominent oscillatory components, it is difficult to know whether the same component is being measured across intensities and between animals. Isolation of a relatively pure on-bipolar component simplifies these comparisons. Over an intermediate range of retinal illuminances, the implicit time of the response was shorter in the GC-E deficient than in wild type animals. This is extremely unusual, since in virtually every patient and animal model of retinal degeneration, the loss of amplitude is linked to an increase in b-wave implicit time, or slowing of the response (17). In fact, the decrease in implicit time in the GC-E mouse is similar to the decrease in implicit time caused by light adaptation in the normal mouse(18). Implicit time is not measurable at low flash intensities, since the reduction in amplitude from either the mutation or from light adaptation leads to non-detectable responses over this range. Similar to light-adaptation, responses are faster than those of dark-adapted wild-type mice over a middle range of intensities and converge as the on-bipolar component saturates at high intensities.

Recovery from activation was measured with a double-flash protocol similar to that used previously in humans[11,19] and mice.[20] Given the role of cyclases in the production of cGMP and restoration of the cGMP-gated channel to its open or "dark" state, the faster than normal recovery in the GC-E deficient mice was somewhat unexpected. However, a hallmark of mammalian rod light adaptation is the decrease in integration time of the rod photoresponse, and the decrease in the period of complete suppression following a saturating flash.[21,23] Thus, the faster recovery seen in the GC-

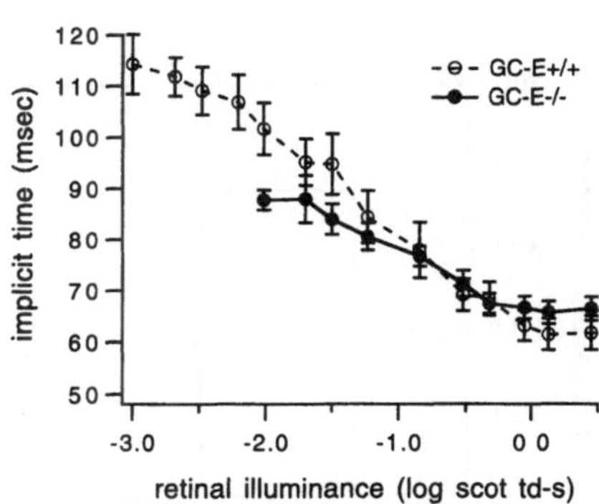

Figure 8. Mean (± S.E.M.) on-bipolar implicit time functions for wild-type and GC-E –/– mice.

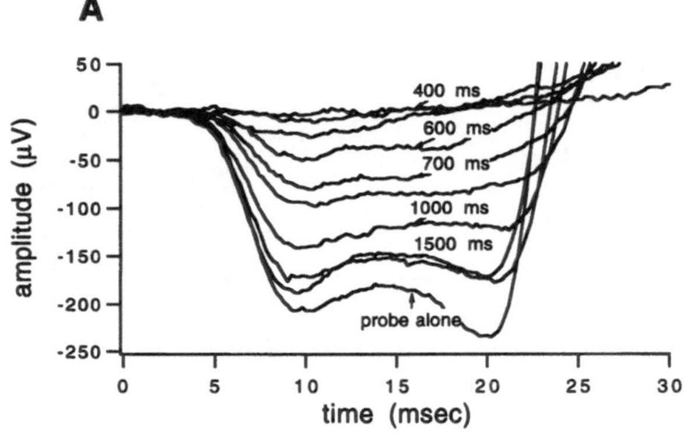

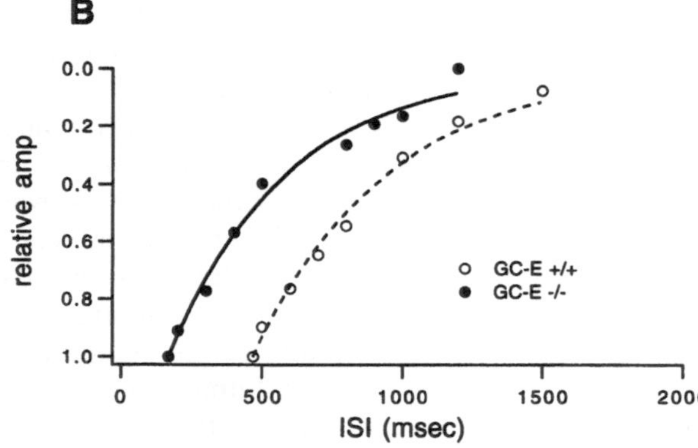

Figure 9. Photoresponse recovery derived with double-flash paradigm. A. Responses to probe flash delivered at various times after 1.5 log scot td-sec test flash. B. Relative amplitude of photoresponse to test flash (derived from Eq. 3) as a function of interval between test and probe (ISI).

E deficient mice is reminiscent of the shortening of the flash response by steady background illumination that has been observed previously in mammalian rods.

The GC-E deficient mouse phenotype differs from that associated with the rather broad phenotype of LCA (i.e.[25]). Both have abolished cone responses, but the GC-E knockout mouse has larger rod responses than patients with LCA . However, other genes are also associated with LCA and, based on recent studies, it appears that RPE65 mutations may account for 10–15%,[24] CRX mutations for 3–4%,[8] and GC-E mutations for less than 1% of the patients displaying congenital blindness. Thus, GC-E mutations appear to be extremely rare, if present at all, in North American patients with LCA. ERGs in the GC-E (retGC-1) form of LCA (LCA1) have not been reported, so it is

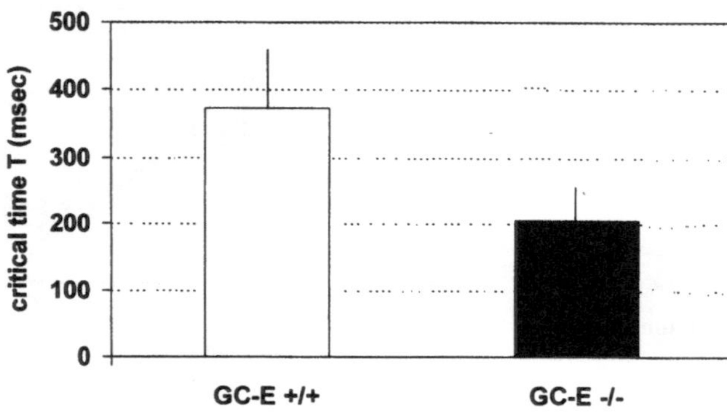

Figure 10. Summary data from double-flash paradigm in wild-type and GC-E –/– mice. The period of complete saturation (T) was derived from Eq. 3. Vertical bars represent S.E.M.

not known whether LCA1 patients retain rod ERG function. A dominant form of cone-rod dystrophy (CORD6) has been linked to the chromosome 17 locus of GC-E and associated with GC-E (retGC-1) mutations.[6] The mouse phenotype appears more consistent with cone-rod dystrophy, raising the possibility that GC-E mutations could also potentially be associated with autosomal recessive forms of this disease.

ACKNOWLEDGMENT

Supported by the Howard Hughes Medical Institute and National Institutes of Health grant EY05235 (DGB).

REFERENCES

1. Fesenko E.E., Kolesnikov S.S., Lyubarsky A.L., 1985, Induction by cyclic GMP of cationic conductance in plasma membrane of retinal rod outer segments. Nature 313:310–313.
2. Lowe D.G., Dizhoor A.M., Liu K., et al., 1995, Cloning and expression of a second photorecptor-specific membrane retina guanylate cyclase (RetGC), RetGC-2. Proceedings of the National Academy of Sciences of the United States of America 92:5535–5539.
3. Yang R.-B., Foster D.C., Garbers D.L., Fulle H.-J., 1995, Two membrane forms of guanylyl cyclase found in the eye. Proc. Natl. Acad. Sci., USA 92:602–606.
4. Yang R.-B., Fulle H.-J., Garbers D.L., 1996, Chromosomal localization and genomic organizations of genes encoding guanylyl cyclase receptors expressed in olfactory sensory neurons and retina. Genomics 31:367–372.
5. Perrault I., Rozet J.M., Calvas P., et al., 1996, Retinal-specific guanylate cyclase gene mutations in Leber's congenital amaurosis. Nature Genetics 14(4):461–464.
6. Kelsell R.E., Gregory-Evans K., Payne A.M., et al., 1998, Mutations in the retinal guanylate cyclase (RETGC-1) gene in dominant cone-rod dystrophy. Human Molecular Genetics 7(7):1179–1184.
7. Gu S.-M., Thompson D., Srikumari C., et al., 1997, Mutations in RPE65 cause autosomal recessive childhood-onset severe retinal dystrophy. Nature Genetics 17:194–197.
8. Freund C.L., Wang Q.-L., Chen S., et al., 1998, De novo mutations in the CRX homeobox gene associated with Leber congenital amaurosis. Nature genetics 18:311–312.
9. Yang R.-B., Robinson S.W., Xiong W., Yau K.-W., Birch D.G., Garbers D.L., In Press, Cone-specific dystrophy in mice lacking a retinal guanylyl cyclase.
10. Kedzierski W., Lloyd M., Birch D.G., Bok D., Travis G.H., 1997, Generation and Analysis of Transgenic Mice Expressing P216L-Substituted Rds/peripherin in Rod Photoreceptors. Invest. Ophthal. Vis. Sci. 38:498–509.
11. Birch D.G., Hood D.C., Nusinowitz S., Pepperberg D.R., 1995, Abnormal activation and inactivation mechanisms of rod transduction in patients with autosomal dominant retinitis pigmentosa and the pro-23-his mutation. Invest. Ophthalmol Vis. Sci. 36(8):1603–1614.
12. Hood D.C., Birch D.G., 1996, b- wave of the scotopic (rod) electroretinogram as a measure of the activity of human on-bipolar cells. J. Opt. Soc. Am. A. 13(3):623–633.
13. Robson J.G., Frishman L.J., 1995, Response linearity and kinetics of the cat retina: the bipolar cell component of the dark-adapted electroretinogram. Visual Neuroscience 12:837–850.
14. Ashmore J.F., Falk G., 1980, Responses of rod bipolar cells in the dark-a0dapted retina of the dogfish, *Scyliorhinus canicula*. J. Physiol. (London) 300:115–150.
15. Semple-Rowland S.L., Lee N.R., Van Hooser J.P., Palczewski K., Baehr W., 1998, A null mutation in the photoreceptor guanylate cyclase gene causes the retinal degeneration chicken phenotype. Proc. Natl. Acad. Sci., USA 95:1271–1276.
16. Hood D.C., Birch D.G., 1992, A computational model of the amplitude and implicit time of the b-wave of the human ERG. Vis. Neurosci 8(2):107–126.
17. Berson E.L., 1993, Retinitis pigmentosa. Invest Ophthalmol Vis. Sci. 34(5):1659–1676.
18. Peachey N.S., Goto Y., Al-Ubaidi M.R., Naash M.I., 1993, Properties of the mouse cone-mediated electroretinogram during light adaptation. Neuroscience Letters 162:9–11.

19. Pepperberg D.R., Birch D.G., Hood D.C., 1997, Photoresponses of human rods *in vivo* derived from paired-flash electroretinograms. Visual Neuroscience 14:73–82.
20. Lyubarsky A.L., Pugh E.N., 1996, Recovery phase of the murine rod photoresponse reconstructed from electroretinographic recordings. J. Neuroscience 16(2):563–571.
21. Tamura T., Nakatani K., Yau K.-W., 1989, Light adaptation in cat retinal rods. Science 245:755–758.
22. Nakatani K., Tamura T., Yau K.-W., 1991, Light adaptation in retinal rods of the rabbit and two other nonprimate mammals. Journal of General Physiology 97:413–435.
23. Kraft T.W., Schneeweis D.M., Schnapf J.L., 1993, Visual transduction in human rod photoreceptors. Journal of Physiology 464:747–765.
24. Morimura H., Fishman G.A., Grover S.A., Fulton A.B., Berson E.L., Dryja T.P., 1998, Mutations in the RPE65 gene in patients with recessive retinitis pigmentosa or Leber congenital amaurosis. Invest. Opthalmol. Visual Sci. 39:S293.
25. Foxman S.G., Heckenlively J.R., Bateman J.B., Wirtschafter J.D., 1985, Classification of congenital and early onset retinitis pigmentosa. Arch. Ophthalmol. 103:1502–1506.

8

GUANYLYL CYCLASE GENE IS THE DISEASE LOCUS IN THE *rd* CHICKEN

A Model for Leber Congenital Amaurosis, Type 1

Susan L. Semple-Rowland

University of Florida Brain Institute
Department of Neuroscience
100 Newell Drive
Box 100244, Rm L1-100
Gainesville, Florida 32610-0244

SUMMARY

The first description of the *retinal degeneration* (*rd*) chicken appeared in the literature in 1980.[1] From the early 1980s to 1991, Ulshafer and his colleagues carried out detailed electrophysiological, morphological, and biochemical analyses of the retinas of these animals, the results of which provided a comprehensive description of the pathology of this retinal degenerative disease, and the first clues concerning the identity of the *rd* gene. In 1990, we embarked on a series of molecular studies that would eventually lead to the identification of the *rd* gene in 1998. As a result of the tireless efforts of many individuals, the *rd* chicken is now recognized as a model for Leber congenital amaurosis, Type 1, a devastating human disease that causes blindness in newborn infants. With the identity of the *rd* gene in hand, it seems appropriate to review the nearly two decades of research that has been conducted on this model, research that has provided a firm foundation for our current efforts to restore sight in these animals using somatic gene therapy.

1. INTRODUCTION

In 1980, Cheng et al.[1] reported finding a single locus, autosomal recessive mutation in an experimental Minnesota Marker line of Rhode Island Red chickens that pro-

Retinal Degenerative Diseases and Experimental Therapy, edited by Hollyfield *et al.*
Kluwer Academic / Plenum Publishers, New York, 1999.

duced blindness in these animals at hatching. Histological examination of the retinas of adult chickens homozygous for the mutation revealed an absence of photoreceptor cells. Based on this observation, Cheng et al.[1] suggested that the rods and cones failed to develop in this model and designated the mutation, rods, and cones, with the gene symbol *rc*. Subsequent analyses of the retinas of these animals by Ulshafer and his colleagues[2] at the University of Florida revealed that the retinas of *rc* chickens appeared normal at hatching and that loss of photoreceptors was due to degeneration rather than dysplasia or lack of development. As a result of the differences between their findings and those reported by Cheng et al.,[1] Ulshafer referred to this model as the retinal degeneration (*rd*) chicken. In subsequent investigations of this model, Cheng and her colleagues and Ulshafer and his colleagues have referred to this model as either *rc* or *rd*, respectively. The recent description of the *rd* chicken model of riboflavinuria introduces yet another possible point of confusion in the literature. In this animal, the splicing defect in the gene encoding riboflavin-binding protein (ribBP) prevents the synthesis of functional ribBP, inducing death on embryonic day 13.[3] The *rd* model of riboflavinuria should not be confused with retinal degeneration chicken.

The retinal degeneration (*rd*) chicken colony housed at the University of Florida was started from a small breeding colony taken from the original *rc* colony described by Cheng in the early 1980s. Dr. Cheng currently maintains a colony of these birds at the University of British Columbia in Vancouver, British Columbia. Recent studies conducted by Cheng and her colleagues describe the presence of an additional phenotype in her colony of *rc* birds that causes abnormalities in the structure of the sarcolemmal membranes and choline transport mechanism of cardiac myocytes.[4,5] Studies to determine if this abnormality is present in the *rd* colony housed at the University of Florida have not been carried out. It is possible that this new phenotype represents a divergence of the *rc* and *rd* chicken lines that has resulted from the independent breeding programs of these birds that have been established at these two institutions. The studies described in this chapter were carried out on the *rd* chicken line that has been maintained at the University of Florida.

2. ELECTROPHYSIOLOGICAL ANALYSES

Visually guided pecking, a robust behavior exhibited by normal and heterozygous (*rd*/+) chickens within the first few days of hatching, is not observed in *rd/rd* chickens. Instead, *rd/rd* chickens, which are blind at hatch, exhibit random bobbing and nystagmiform movements of the head when standing motionless. These random head movements are manifest on the day of hatching and continue throughout the lifetime of the animal.

On the day of hatching, small, well-formed light- and dark-adapted electroretinograms (ERGs) can be recorded from normal chickens, responses that approach the amplitude of those recorded from adult animals by 7 days of age. Similar responses can be recorded from *rd*/+ chickens.[6] Attempts to record ERGs from *rd/rd* chickens under both light- and dark-adapted conditions showed that the retinas of the majority of *rd* animals do not produce responses to light stimulation.[1,2] In the few instances where ERG responses could be elicited, the amplitudes of these responses were of very small ($<25\mu V$) compared to those recorded from normal 1–3-day old chickens ($\sim400\mu V$).[6] The amplitude of the early receptor potential (ERP) when present was also abnormally small.[6] The ERP is generated by the redistribution of charge on pigment molecules as they undergo changes in conformation following absorption of light.[7] The amplitude of the ERP is linearly proportional to the number of pigment molecules activated by light,[8] and in retinas containing both rod and cone cells, the ERP is dominated by the cone cells.[9,10] The amplitudes of the R_2 components of the ERPs recorded from 1–3-day-old *rd/rd* animals were highly variable, ranging from 0–36 μV compared to 57–95 μV for normal animals.[6] At 11–15 days of age, the amplitudes of the normal R_2 waves had increased to 129–166 μV, whereas those recorded from the *rd/rd* chickens had decreased to 0–12 μV.

3. MORPHOLOGICAL ANALYSES

In light of the severe visual deficits exhibited by the *rd/rd* chickens at hatching, it was expected that the retinas of these animals would exhibit some form of retinal degeneration. It was, therefore, surprising to find that the retina, pigment epithelium, and choroid of 1-day-old *rd/rd* chickens are morphologically indistinguishable from those found in normal chickens (Fig. 1A). Examination of the development of neural connections between the retinal ganglion cells and the central nervous system of the *rd/rd* chicken revealed that the absence of retinal function in these animals does not lead to abnormal development of these projections.[11]

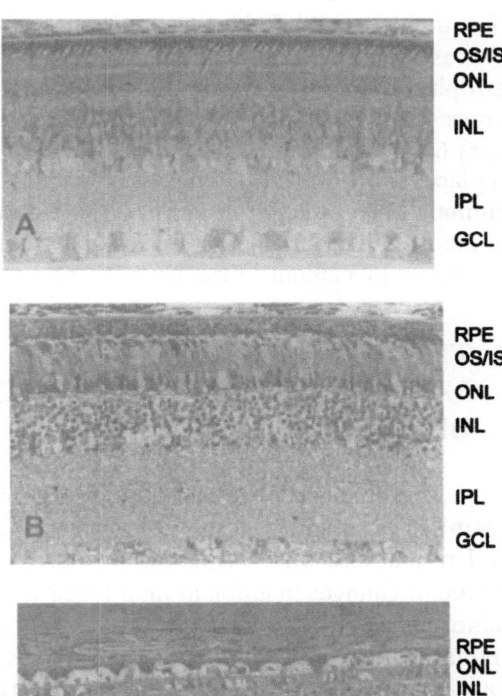

Figure 1. Light Microscopy of *rd/rd* Retina. Sections of retinas of (A) 1-day-old, (B) 21-day-old, and (C) 115-day-old *rd/rd* chickens stained with toluidine blue. RPE—retinal pigment epithelium; OS/IS—outer segments/inner segments; ONL—outer nuclear layer; INL—inner nuclear layer; IPL—inner plexiform layer; GCL—ganglion cell layer.

The sequence of degenerative changes that is observed in retinas of *rd/rd* chickens resembles that seen in many other forms of inherited retinal disease.[2,12] The first signs of retinal pathology appear 7–10 days after hatching and are localized to the photoreceptor layer in central retina (vesiculation, disorganization, and loss of the photoreceptor outer segments). With time, the degenerative changes progress toward the peripheral regions of the retina. By 21 days of age, the number of identifiable outer segments in the central retina is significantly reduced. Pyknotic nuclei are evident in the outer nuclear layer, as well as in the inner retinal layers, at this time (Fig. 1B). Large membranous inclusions resembling entire outer segments are frequently observed within the retinal pigment epithelium overlying central retina. Increased levels of glial fibrillary acid protein (GFAP), detected by immunocytochemical and western techniques, are observed in the cell bodies and processes of the Müller cells in retinas of 21- and 33-day-old *rd/rd* chickens.[13] By 60 days of age, the lesion has progressed to include mid-peripheral retina. At 115 days of age, very few photoreceptors remain in central retina, their space being filled by Müller cell processes (Fig. 1C). The photoreceptors that are still present contain clear oil droplets suggesting that they are the principal members of double cones. At this stage, gross pathological changes are also evident in the retinal pigment epithelium.[14] The overall thinning of the pigment epithelium is punctuated by the appearance of hypertrophic cells extending into the subretinal space. By 6–8 months of age, only a few photoreceptors remain in the retina and significant cell loss from the inner retina has occurred.

Intraocular ossification of the vitreous represents the end-stage of the disease in the *rd* chicken.[15,16] Evidence of vitreal ossification is first seen at 18 months of age, a process that begins after the retina and pigment epithelium have undergone complete atrophy. The deposition of calcium during the initial stages of this process occurs in areas of the vitreous body or atrophied retina lying adjacent to the choroid. Over the next 6–18 months, a matrix composed of fatty marrow and cancellous bone gradually replaces the vitreous, the degree of ossification varying between the eyes of individual animals. The osseous formation observed in the *rd* chicken is similar to phthisis bulbi in man, a condition characterized by extreme disorganization of the ocular tissues and gradual replacement of the globe by scar tissue and cancellous bone.[17]

4. THE PHYSIOLOGICAL STATE OF THE *rd* RETINA: METABOLIC, MORPHOLOGICAL, AND BIOCHEMICAL EVIDENCE

Many aspects of retinal physiology, including photoreceptor disk shedding, melatonin synthesis and release, gene expression, and visual sensitivity are regulated by light and retinal circadian oscillators. Under normal conditions, retinal function is tightly linked to changes in ambient light levels that occur over the course of a day. Relatively minor changes in the daily light regimen have been shown to be capable of inducing retinal damage and cell loss.[18,19] The retinas of *rd/rd* chickens at hatching are morphologically normal but do not produce electrical responses when stimulated by light. Is the absence of electrical activity in the *rd/rd* retina accompanied by a total loss of all light-regulated responses in this retina? To better understand the physiological state of the pre-degenerate *rd/rd* retina, glucose utilization, retinomotor function, and neurotransmitter levels were examined in the retinas of *rd* chickens as a function of light adaptation.

Measures of glucose utilization in rat[20] and amphibian[21] retinas have shown that light adaptation significantly reduces the metabolic rate of the retina relative to that observed in dark-adapted retinas. To determine if the metabolic rate of the *rd* chicken retina changes as a function of light adaptation, 1-day-old normal and *rd/rd* chickens that had been either light- or dark-adapted were given an intraperitoneal injection of the glucose analogue, ^{3}H-2-deoxy-D-glucose (2-DG).[22,23] Thirty minutes after the injection, the amount of 2-DG taken up by the retinas was measured, 80% of which had been converted into ^{3}H-2-deoxyglucose-6-phosphate. As expected from studies of other vertebrate retinas, glucose uptake by normal, light-adapted chicken retinas was approximately 30% less than that observed for dark-adapted retinas. In *rd* chickens, glucose uptake by both the light- and dark-adapted retinas was the same, the amount of 2-DG in the *rd/rd* retinas similar to that found in the normal, light-adapted retinas. This result suggested that metabolic processes in the non-functioning, pre-degenerate *rd/rd* retina are unresponsive to changes in ambient light conditions and that the retina may be arrested in a light-adapted-like state.

The retinas of lower vertebrates, including birds, undergo several morphological changes during light adaptation, the most prominent of which include elongation of the rod cells, contraction of the cone cells, and translocation of the melanin granules within the pigment epithelium into the apical processes of these cells.[24] These retinomotor movements are reversed during dark adaptation of the retina. If the *rd/rd* retina is arrested in a light-adapted-like state, then we might expect that retinomotor movements in response to changes in ambient light levels would not be present in the retinas of these animals. To examine this possibility, the relative positions of the photoreceptors and pigment epithelium in the retinas of normal and *rd/rd* chickens raised either in constant darkness or in cyclic light (12 hours light: 12 hours dark) were compared.[23] The results of these comparisons showed that the retinas of both the dark-reared and light-adapted, cyclic light-reared *rd/rd* chickens resembled the retinas of normal, light-adapted animals. In all three cases, the melanin granules located in the pigment epithelium were dispersed within the apical processes of the cells, a pattern normally associated with light adaptation. This observation was consistent with the hypothesis that the *rd/rd* retina is arrested in a light-adapted-like state.

A third piece of evidence supporting this hypothesis came from measurements of the levels of the excitatory amino acid transmitters, aspartate, and glutamate, in *rd/rd* retinas. In normal retina, light stimulation leads to hyperpolarization of the photoreceptors and a decrease in release of transmitter from these cells. If the photoreceptors in *rd/rd* retina are chronically hyperpolarized then levels of glutamate and aspartate in these cells might be expected to be higher than those observed in normal retina. HPLC and immunocytochemical analyses of the retinas of 1-, 7-, and 14-day-old *rd/rd* chickens revealed that this was the case.[25] The amounts of aspartate and glutamate in *rd/rd* retina were significantly higher than the amounts present in normal retina, the first evidence of an increase in glutamate appearing approximately 7 days after hatching. The increases in these amino acids in *rd/rd* retina were specific to the photoreceptor cell layers, the immunostaining of the inner retinal layers resembling that seen in normal chicken retina. The observed increase in glutamate and aspartate within the photoreceptor cells is consistent with the hypothesis that the *rd/rd* retina is arrested in a light adapted-like state. Since the increases in the levels of these amino acids did not appear to precede the loss of retinal function, it was concluded that abnormal excitatory amino acid metabolism was not the underlying cause of blindness in the *rd* chicken. These findings did raise the possibility that an excitotoxic mechanism[26] could play a role in triggering the degeneration of the photoreceptor cells, a possibility that has not been formally explored.

5. PHOTOTRANSDUCTION

5.1. The Vitamin A Cycle and Visual Pigments

The absence of electroretinographic responses from *rd/rd* retina suggested that the *rd* gene encoded a protein that played a key role in phototransduction. Candidate genes expressed in both rod and cone cells provided the most parsimonious explanation for the simultaneous loss of function in both of these cell types. It was also possible that a mutation in a gene expressed in either the retinal pigment epithelium, rod cells, or cone cells could produce the observed functional phenotype.

Rhodopsin and the vitamin A visual cycle were the focus of the initial biochemical analyses of the phototransduction cascade in the *rd/rd* retina. Spectroscopy, high performance liquid chromatography, and immunocytochemical techniques were used to compare the amounts of rhodopsin, interstitial retinol-binding protein (IRBP), and vitamin A compounds in dark-adapted eyes of *rd/rd* and *+/rd* chickens.[27] The results of these analyses showed that the absence of visual function in 3-day-old *rd/rd* chickens was not due to an inability to synthesize or secrete IRBP, or to form the 11-*cis* retinal isomer, both of which were present in amounts comparable to those found in the eyes of sighted 3-day-old *+/rd* chicken. Analyses of the rod visual pigment, rhodopsin, using difference spectra[27] and immunocytochemistry[28] showed that the amount and absorption characteristics of this pigment were also normal in *rd/rd* retina.

Iodopsin, the most abundant cone pigment present in chicken retina, was the focus of a second series of studies. The results of immunocytochemical and electrophysiological experiments suggested that this pigment was abnormal in predegenerate *rd/rd* retina. First, immunocytochemical studies of *rd/rd* retina using a monoclonal antibody that specifically labeled middle-to-long wavelength sensitive cones (COS-1) showed that the density of staining of cone outer segments in *rd/rd* retina was significantly less than that observed in *+/rd* retina.[29] Second, recordings of the ERP showed that this response was either absent from or severely attenuated in 1–3-day-old *rd/rd* chickens.[6] Both of these results suggested that either less iodopsin was present in *rd/rd* retina or that the structure of the iodopsin protein was altered in this mutant. To resolve this issue, biochemical and molecular analyses of iodopsin were carried out.[30] Western blot studies using a second monoclonal antibody for iodopsin, R4, showed that the amount of iodopsin in *rd/rd* retina was normal. Sequence analyses of rd/rd iodopsin cDNA clones revealed that the predicted primary structure of *rd/rd* iodopsin is normal. The epitope recognized by the COS-1 antibody is the last 6 amino acids of the C-terminus of iodopsin,[31] a region that is enriched with serine and threonine residues that has been shown to be phosphorylated by opsin kinase in a light-dependent manner.[32] Consistent with the immunocytochemical study,[29] COS-1 staining of *rd/rd* retinal proteins on western blots was found to be significantly less than that observed for normal retinal proteins. The fact that the epitopes recognized by R4 and COS-1 are different prompted the suggestion that the differences that are observed between the ability of COS-1 and R4 to stain iodopsin on western blots of *rd/rd* retinal protein could be due to abnormal phosphorylation of *rd/rd* iodopsin. This idea has not been further investigated; however, abnormalities in rhodopsin phosphorylation have been reported in two other animal models of inherited retinal disease, the *rd* mouse[33,34] and the *rcdl* Irish Setter.[35,36] In both of these models, abnormal pigment phosphorylation appears to be secondary to disruption of cGMP metabolism in the retinas of these animals.

5.2. cGMP

Disruption of cGMP metabolism in photoreceptor cells has been shown *in vitro*[37] and *in vivo*[38,39] to be detrimental to the function and ultimately, the survival of these cells. Based on this information, a series of experiments were conducted to measure levels of cGMP in the retinas of *rd/rd* chickens as a function of development and of retinal cell type.[23,40] Comparisons of levels of cGMP in *rd/rd* and *rd*/+ control retinas taken from embryonic day 12 to 35-day-old chickens revealed that levels of cGMP were nearly identical in embryonic day 12 *rd/rd* and control retinas (Fig. 2A). By embryonic day 18, the amount of cGMP present in control retina had nearly doubled while levels of cGMP in *rd/rd* retina remained near the embryonic day 12 value. The amount of cGMP present in control retina continued to increase over the next 10 days, reaching a plateau in 7-day-old animals, an increase that coincided with the development and elongation of the outer segments of the photoreceptor cells.[41] No increase was observed in *rd/rd* retina over this period of time. By 7 days post hatch, levels of cGMP in control retina were six times greater than those found in *rd/rd* retina. This difference could not be attributed to cell loss in the *rd/rd* retina since the first signs of retinal degeneration do not appear until the chickens are 7–10 days old. These data suggested that the *rd* mutation disrupts either the synthesis or hydrolysis of cGMP in the *rd/rd* chicken retina. In order to identify the retinal cell type(s) in which the cGMP abnormality is expressed, cGMP levels in each of the layers of the retina were measured. The results of these analyses showed that the amount of cGMP present in the photoreceptors of 1- to 2-day-old dark-adapted *rd/rd* retina is only 10–20% of that found in the photoreceptors of control retina (Fig. 2B). cGMP levels in the remaining layers of the control and *rd/rd* retinas were comparable to each other. Together, these observations suggested that the absence of photoreceptor function in the *rd/rd* retina was likely due to abnormal cGMP metabolism in these cells.

6. CANDIDATE GENE STUDIES

A defect in enzymes involved in photoreceptor cGMP metabolism that either disables synthesis or accelerates hydrolysis of cGMP in *rd/rd* retina would be consistent with the *rd* phenotype. Early on in our search for the *rd* gene, we examined the sequences and expression of the genes encoding visinin[42] and the alpha' subunit of cone cGMP phosphodiesterase[43] in *rd/rd* retina. Our interest in visinin stemmed from early reports that suggested that this calcium binding protein was involved in the regulation of the activity of guanylate cyclase, a suggestion that later turned out to be incorrect.[44,45] The results of our analyses of both of these candidates revealed that they were normal in the *rd* chicken.

The hunt for the *rd* gene took a decidedly positive turn with our analyses of the Ca^{2+}-sensitive guanylate cyclase activating proteins, GCAP1 and GCAP2. Both GCAP1 and GCAP2 activate photoreceptor guanylate cyclase in the presence of low concentrations of Ca^{2+}.[46–48] It was hypothesized that a mutation in either of these regulatory proteins could lead to decreased synthesis of cGMP by guanylate cyclase. Northern

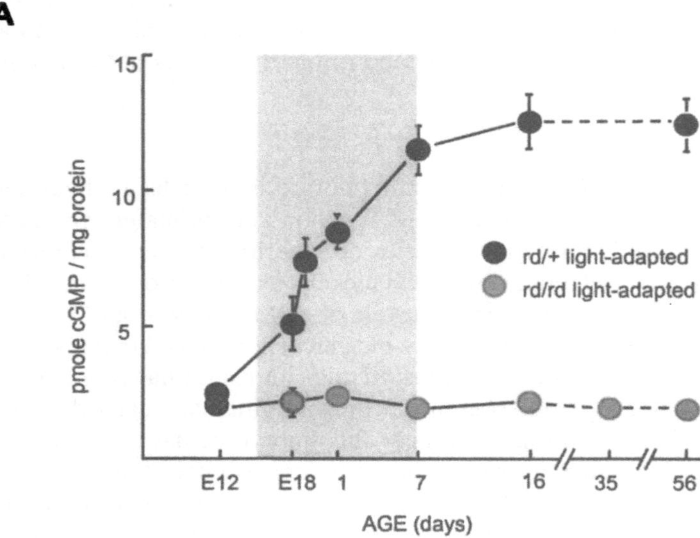
A
15
10
5
pmole cGMP / mg protein
rd/+ light-adapted
rd/rd light-adapted
E12
E18
1
7
16
35
56
AGE (days)

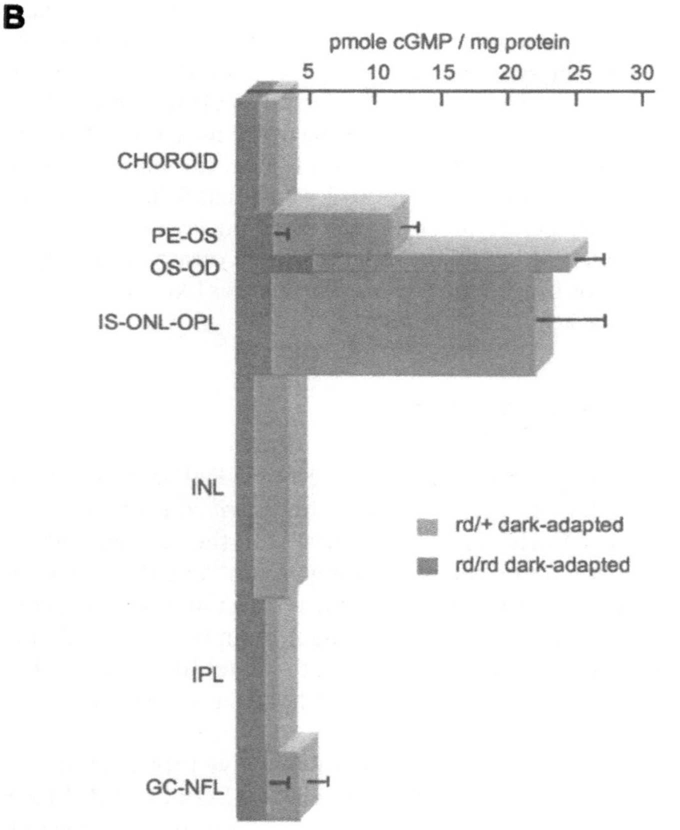
B
pmole cGMP / mg protein
5
10
15
20
25
30
CHOROID
PE-OS
OS-OD
IS-ONL-OPL
INL
IPL
GC-NFL
rd/+ dark-adapted
rd/rd dark-adapted

Figure 2. cGMP in *rd/rd* Retina. (A) Comparison of cGMP levels in *rd/+* control and *rd/rd* central retina. Animals were light adapted for 2 hr (1600 lux) before sacrifice. The shaded region corresponds to the time during which photoreceptor outer segments form and mature to their final adult dimensions. Each value is the mean ± SD of measures taken from 6–11 eyes. (B) cGMP levels in microdissected layers of 1- to 2-day-old chickens. Chickens were dark-adapted overnight prior to sacrifice. Assays were carried out on seven retina layers taken from central retina. Each value represents the mean ± SD of 3–4 eyes (10–30 sections/eye). PE-OS—pigment epithelium plus outer segments; OS-OD—outer segments plus oil droplets; IS-ONL-OPL—inner segments plus outer nuclear layer plus outer plexiform layer; INL—inner nuclear layer; IPL—inner plexiform layer; GC-NFL—ganglion cell layer plus nerve fiber layer.

blot, western blot, and sequence analyses of normal and *rd/rd* GCAP1 and GCAP2 cDNAs were carried out to determine if either of these proteins was abnormal in the mutant.[49] Northern analyses revealed that levels of both the GCAP1 and GCAP2 transcripts were slightly lower in 1–3-day-old, pre-degenerate *rd/rd* retina as compared to normal retina. This was in contrast to the western blot findings that showed that the GCAP1 protein was nearly undetectable in *rd/rd* retina while GCAP2 levels were normal. Sequence analyses of *rd/rd* GCAP1 cDNA revealed that there were no disease producing mutations in the sequence that could account for the western blot results. The absence of a disease causing mutation in GCAP1 suggested that the selective downregulation of GCAP1 protein in *rd/rd* retina is a consequence of a mutation in a gene that encodes a protein that specifically interacts with GCAP1.

7. THE *rd* GENE

The results obtained in our analyses of cGMP and GCAP in the *rd* chicken retina prompted us to begin studies of photoreceptor guanylate cyclase. At the time that we initiated these experiments, two photoreceptor-specific guanylate cyclase enzymes (GC1 and retGC2) had been cloned, both of which were found to be expressed in rod and cone photoreceptor cells.[50–54] For our analyses, we chose to focus on GC1. Western and northern blot experiments of *rd/rd* chicken retina indicated that expression of GC1 was not normal in this retina.[40] No GC1 protein was detected in the pre-degenerate retinas of 1–3-day-old *rd/rd* chickens. Northern blot analyses revealed that the GC1 transcript in *rd/rd* retina is 0.5–1.0 kb smaller than that observed in normal chicken retina (9.5 kb), and that the amount of GC1 mRNA in *rd/rd* retina is less than 10% of that found in normal retina. Sequence analyses of *rd/rd* GC1 cDNAs revealed that the sequence corresponding to exons 4–7 was replaced by an 81 bp fragment with 89% sequence identity to a portion of putative exon 9 in reverse orientation. This mutation, which is predicted to result in the deletion of the membrane spanning domain of the GC1 enzyme and the regions immediately flanking it (Fig. 3), does not disrupt the reading frame of the transcript. However, it is sufficient to destabilize the protein so that no GC1 is present in the predegenerate *rd/rd* retina.

The mutation in GC1 in the *rd* chicken is consistent with the observed cGMP phenotype. With the identification of the *rd* gene came new insights into the impact that GC1 and cGMP have on photoreceptor development and function. First, cGMP levels too low to support phototransduction do not affect the development of photoreceptor cells. This is in contrast to the arrested development of photoreceptors that is observed in the presence of abnormally high levels of cGMP.[38,39] Second, the absence of function in both cone and rod photoreceptor cells in *rd/rd* retina indicates that GC1

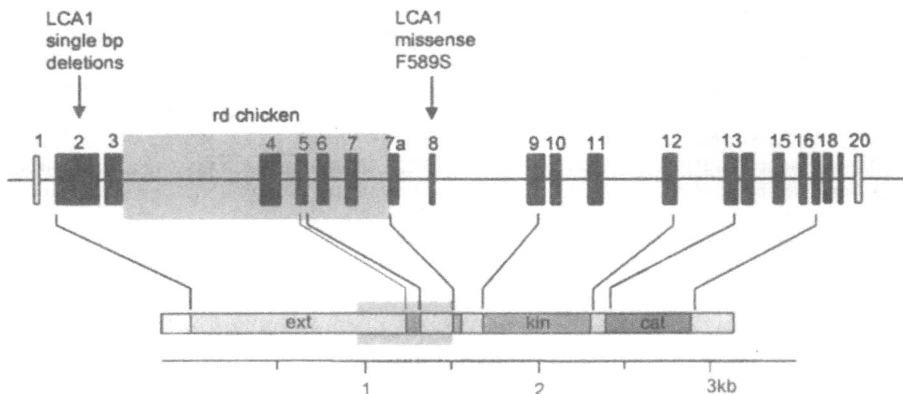

Figure 3. *rd* and LCA1 Mutations in GC1. Putative exon/intron structure of chicken GC1 based on comparisons of chicken, human and bovine GC1. The portions of the GC1 protein encoded by each exon are indicated on the map of the chicken GC1 cDNA. Exon 7A, which is depicted as a unique exon, is alternatively spliced in both normal and rd/rd GC1 transcripts.[40] Putative exons 4–7 are deleted in the *rd/rd* GC1 gene (gray shaded region). Exons 2 and 8 have been reported to contain frameshift and missense mutations in LCA1 families.[55]

plays a critical role in both cone and rod phototransduction. Finally, our analyses suggest that retGC2, a second GC enzyme present in photoreceptor cells, does not play a major role in the synthesis of cGMP necessary for phototransduction.

8. THE RD CHICKEN AND LEBER CONGENITAL AMAUROSIS, TYPE 1

Our discovery that GC1 is defective in the *rd* chicken coincided in time with the appearance of a report in the literature that showed that frameshift and missense mutations in GC1 (GUC2D) are linked to patients diagnosed with Leber congenital amaurosis, type I (LCA1)[55] The frameshift mutations in exon 2 (Fig. 3) produce premature translation termination codons downstream of the deletion. As in the *rd* chicken, these mutations would be predicted to result in the absence of GC1 in the retinas of the patients with these mutations. The missense mutation identified in exon 8 (Fig. 3) is predicted to alter the hydrophobicity of GC1, and as a result, severely compromise the stability of the enzyme.[55]

A comparison of the *rd* chicken and LCA1 phenotypes reveals that these diseases are strikingly similar (Table 1). The *rd* chicken model provides a unique system in which to test possible therapeutic strategies for the treatment of LCA1. Because the retinas of these animals are cone-dominant and can be studied throughout development, studies of the *rd* chicken model should provide new insight into cone photoreceptor disease and the developmental time frame in which treatment interventions must occur in order to achieve maximum therapeutic benefits. Our current research efforts are directed toward "rescue" of sight in the *rd* chicken using a lentiviral vector system carrying a normal copy of GC1. It is our working hypothesis that synthesis of cGMP in the rod and cone cells in this retina will restore phototransduction and prevent degeneration of these cells.

Table 1. A Comparison of *rd* Chicken and Leber Congenital Amaurosis, Type 1 Phenotypes

Leber Congenital Amaurosis	*retinal degeneration (rd)* Chicken
Greatly impaired vision/blindness at birth	Blindness at hatching
Normal fundus at birth	Retinal morphology normal at hatching
Extinguished ERGs	Extinguished ERGs
Rod and cone cells affected	Rod and cone cells affected
Autosomal recessive inheritance	Autosomal recessive inheritance
GC1 mutations linked to disease	GC1 mutation linked to disease

ACKNOWLEDGMENTS

I would like to recognize the following individuals for their research contributions that have led to our current understanding of retinal degeneration in the *rd* chicken, an animal that is now recognized as a model for Leber congenital amaurosis, Type 1: Dr. Robert Ulshafer, Dr. Nancy R. Lee, Dr. Kimberly M Cheng, Dr. Wolfgang Baehr, and Dr. Krzysztof Palczewski. Studies of the *rd* chicken in my laboratory have been supported by research grants from the National Eye Institute and the Foundation Fighting Blindness.

REFERENCES

1. K.M. Cheng, R.N. Shoffner, K.N. Gelatt, G.G. Gum, J.S. Otis, and J.J. Bitgood, 1980, An autosomal recessive blind mutant in chickens, *Poultry Sci.* **59**:2179–2182.
2. R.J. Ulshafer, C. Allen, W.W. Dawson, and E.D. Wolf, 1984, Hereditary retinal degeneration in the Rhode Island Red chicken. I. Histology and ERG. *Exp. Eye Res.* **39**:125–135.
3. I. MacLachlan, J. Nimpf, H.B. White, III, W.J. Schneider, 1993, Riboflavinuria in the rd chicken. 5′-splice site mutation in the gene for riboflavin-binding protein. *J. Biol. Chem.* **268**:23222–23226.
4. S.W. Rabkin and K.M. Cheng, 1992, A genetic abnormality of cardiac myocytes from the blind mutant (RC) chick heart: abnormalities of cardiac structure and choline transport. *Basic Res. Cardiol.* **87**:610–617.
5. S. Cerruti Sola, M. Castagnaro, and K.M. Cheng, 1997, Histological changes caused by the rc mutation in chickens. *J. Comp. Pathol.* **116**:329–338.
6. W.W. Dawson, R.J. Ulshafer, R. Parmer, and N.R. Lee, 1990, Receptor potentials in the normal and retinal degenerate (rd) chick. *Clin. Vision Sci.* **5**:285–292.
7. G.S. Brindley and A.R. Gardner-Medwin, 1966, The origin of the ERP of the retina. *J. Physiol.* **182**:185–194.
8. R.A. Cone, 1967, Early receptor potential: photoreversible charge displacement in rhodopsin. *Science*, **155**:1128–1131.
9. E.B. Goldstein and E.L. Berson, 1969, Cone dominance of the human early receptor potential. *Nature*, **222**:1272–1273.
10. E.B. Goldsterin, 1969, Contribution of cones in the early receptor potential in the rhesus monkey. *Nature*, **222**:1273–1274.
11. R.J. Ulshafer and M.B. Heaton, 1989, Axonal transport and central visual projections of ganglion cells in congenitally blind chickens. *Curr. Eye Res.* **8**:321–326.
12. R.J. Ulshafer and C.B. Allen, 1985, Hereditary retinal degeneration in the Rhode Island Red chicken: Ultrastructural analysis. *Exp. Eye Res.* **40**:865–877.
13. S.L. Semple-Rowland, 1991, Expression of glial fibrillary acidic protein by Müller cells in rd chick retina. *J. Comp. Neurol.* **305**:582–590.
14. R.J. Ulshafer and C.B. Allen, 1985, Ultrastructural changes in the retinal pigment epithelium of congenitally blind chickens. *Curr. Eye Res.* **4**:1009–1021.
15. C.B. Allen, R.J. Ulshafer, E.A. Ellis, and J.C. Woodard, 1987, Scanning electron microscopic analysis of intraocular ossification in advanced retinal disease. *Scanning Microscopy*, **1**:233–239.

16. K.C. Kelley, R.J. Ulshafer, and E.A. Ellis, 1987, Intraocular ossification in the rd chicken. *Avian Pathol.* **16**:189–197.
17. M. Yanoff and B.S. Fine, 1975, *Ocular Pathology, A Text and Atlas*, pp. 74–75, Harper and Row, Maryland.
18. J.S. Penn and T.P. Williams, 1986, Photostasis: regulation of daily photon-catch by rat retinas in response to various cyclic illuminances. *Exp. Eye Res.* **43**:915–928.
19. S.L. Semple-Rowland and W.W. Dawson, 1987, Cyclic light intensity threshold for retinal damage in albino rats raised under 6 lx. *Exp. Eye Res.* **44**:643–661.
20. B. Morjaria and M.J. Voaden, 1979, The uptake of [3H]2-deoxy glucose by light- and dark-adapted rat retinas in vivo. *J. Neurochem.* **32**:1881–1883.
21. S.S. Goldman and P. Witkovsky, 1986, Glycogen metabolism in an amphibian retina. *Exp. Eye Res.* **43**:267–272.
22. N.J. Ruth, R.J. Ulshafer, and K.C. Kelly, 1985, Depressed glucose utilization in the rd chicken retina. *Invest. Ophthalmol. Vis. Sci. (Suppl)* **26**:62.
23. N.J. Ruth, 1991, Cyclic Nucleotides in the (rd) Retinal Degenerate Chicken Retina. Ph.D. dissertation, University of Florida, Gainesville, FL.
24. B. Burnside and B. Nagle, 1983, Retinomoter movements of photoreceptors and retinal pigment epithelium: mechanisms and regulation. *Prog. Retinal Res.* **2**:67–109.
25. R.J. Ulshafer, D.M. Sherry, R. Dawson, Jr., and D.R. Wallace, 1990, Excitatory amino acid involvement in retinal degeneration. *Brain Res.* **531**:350–354.
26. J. Sattayasai, J. Zappia, and D. Ehrlich, 1989, Differential effects of excitatory amino acids on photoreceptors of the chick retina: an electron-microscopical study using the zinc-iodide-osmium technique. *Vis. Neurosci.* **2**:237–45.
27. C.D.B. Bridges, R.A. Alvarez, S-L. Fong, G.I. Liou, and R.J. Ulshafer, 1987, Rhodopsin, vitamin A, and interstitial retinol-binding protein in the rd chicken. *Invest. Ophthalmol. Vis. Sci.* **28**:613–617.
28. R.J. Ulshafer, G. Adamus, E. Clausnitzer, and P.A. Hargrave, 1990, Rhodopsin in normal and rd chick retinas. *Invest. Ophthalmol. Vis. Sci.* (suppl) **31**:546.
29. R.J. Ulshafer, E.L. Clausnitzer, D.M. Sherry, A. Szel, and P. Rohlich, 1990, Immunocytochemical identification of outer segment proteins in the rd chicken. *Exp. Eye Res.* **51**:209–216.
30. S.L. Semple-Rowland and D.A. Green, 1994, Molecular and biochemical analyses of iodopsin in rd chick retina. *Invest. Ophthalmol. Vis. Sci.* **35**:2550–2557. (also see erratum in *Invest. Ophthalmol. Vis. Sci.* **35**:3126).
31. P. Rohlich and A. Szel, 1993, Binding sites of photoreceptor-specific antibodies COS-1, OS-2, and AD. *Curr. Eye Res.* **12**:935–944.
32. Y. Fukada, K. Kokame, T. Okano, Y. Shichida, T. Yoshizawa, J.H. McDowell, P.A. Hargrave, and K. Palczewski, 1990, Phosphorylation of iodopsin, chicken red-sensitive cone visual pigment. *Biochemistry*, **29**:10102–10106.
33. T.A. Shuster and D.B. Farber, 1986, Rhodopsin phosphorylation in developing normal and degenerative mouse retinas. *Invest Ophthalmol. Vis. Sci.* **27**:264–268.
34. K. Palczewski, D.B. Farber, and P.A. Hargrave, 1991, Elevated level of protein phosphatase 2A activity in retinas of rd mice. *Exp. Eye Res.* **53**:101–105.
35. S.Y. Schmidt, U.P. Andley, C.A. Heth, and J. Miller, 1986, Deficiency in light-dependent opsin phosphorylation in Irish setters with rod-cone dysplasia. *Invest. Ophthalmol. Vis. Sci.* **27**:1551–1559.
36. D.J. Takemoto, J. Cunnick, and L.J. Takemoto, 1986, Reduced rhodopsin phosphorylation during retinal dystrophy. *Biochem. Biophys. Res. Comm.* **135**:1022–1028.
37. R.J. Ulshafer, C.A. Garcia, and J.G. Hollyfield, 1980, Sensitivity of photoreceptors to elevated levels of cGMP in the human retina. *Invest. Ophthalmol. Vis. Sci.* **19**:1236–41.
38. D.B. Farber and R.N. Lolley, 1974, Cyclic guanosine monophosphate: elevation in degenerating photoreceptor cells of the C3H mouse retina. *Science*, **186**:449–451.
39. G. Aguirre, D. Farber, R. Lolley, R.T. Fletcher, and G.J. Chader, 1978, Rod-cone dysplasia in Irish setters: a defect in cyclic GMP metabolism in visual cells. *Science*, **201**:1133–1134.
40. S.L. Semple-Rowland, N.R. Lee, J.P. Van Hooser, K. Palczewski, and W. Baehr, 1998, A null mutation in the photoreceptor guanylate cyclase gene causes the retinal degeneration chicken phenotype. *Proc. Natl. Acad. Sci. USA*, **95**:1271–1276.
41. K. Meller and W. Tetzlaff, 1976, Scanning electron microscopic studies on the development of the chick retina. *Cell Tiss. Res.* **170**:145–159.
42. S.L. Semple-Rowland and H. van der Wel, 1992, Visinin: biochemical and molecular comparisons in normal and rd chick retina. *Biochem. Biophys. Res. Commun.* **183**:456–461.

43. S.L. Semple-Rowland and D.A. Green, 1994, Molecular characterization of the a'-subunit of cone photoreceptor cGMP phosphodiesterase in normal and rd chicken. *Exp. Eye Res.* **59**:365–372.
44. S. Kawamura, O. Kuwata, M. Yamada, S. Matsuda, O. Hisatomi, and F. Tokunaga, 1996, Photoreceptor protein s26, a cone homologue of S-modulin in frog retina. *J. Biol. Chem.* **271**:21359–21364.
45. A. Polans, W. Baehr, and K. Palczewski, 1996, Turned on by Ca^{2+}! The physiology and pathology of Ca^{2+}-binding proteins in the retina. *Trends Neurosci.*, **19**:547–554.
46. W.A. Gorczyca, A.S. Polans, I. Surgucheva, I. Subbaraya, W. Baehr, and K. Palczewski, 1995, Guanylyl cyclase activating protein. A calcium-sensitive regulator of phototransduction. *J. Biol. Chem.* **270**:22029–22036.
47. A.M. Dizhoor, E.V. Olshevskaya, W.J. Henzel, S.C. Wong, J.T. Stults, I. Ankoudinova, and J.B. Hurley, 1995, Cloning, sequencing, and expression of a 24-kDa Ca(2+)-binding protein activating photoreceptor guanylyl cyclase. *J. Biol. Chem.* **270**:25200–25206.
48. K. Palczewski, I. Subbaraya, W.A. Gorczyca, B.S. Helekar, C.C. Ruiz, H. Ohguro, J. Huang, X. Zhao, J.W. Crabb, R.S. Johnson, K.A. Walsh, M.P. Gray-Keller, P.B. Detwiler, and W. Baehr, 1994, Molecular cloning and characterization of retinal photoreceptor guanylyl cyclase-activating protein. *Neuron* **13**:395–404.
49. S.L. Semple-Rowland, W.A. Gorczyca, J. Buczylko, B.S. Helekar, C.C. Ruiz, I. Subbaraya, K. Palczewski, and W. Baehr, 1996, Expression of GCAP1 and GCAP2 in the retinal degeneration (rd) mutant chicken retina. *FEBS Lett* **385**:47–52.
50. A.W. Shyjan, F.J. de Sauvage, N.A. Gillett, D.V. Goeddel, and D.G. Lowe, 1992, Molecular cloning of a retina-specific membrane guanylyl cyclase. *Neuron* **9**:727–737.
51. D.G. Lowe, A.M. Dizhoor, K. Liu, Q. Gu, M. Spencer, R. Laura, L. Lu, and J.B. Hurley, 1995, Cloning and expression of a second photoreceptor-specific membrane retina guanylyl cyclase (RetGC), RetGC-2. *Proc. Natl. Acad. Sci. USA*, **92**:5535–5539.
52. R.M. Goraczniak, T. Duda, A. Sitaramayya, and R.K. Sharma, 1994, Structural and functional characterization of the rod outer segment membrane guanylate cyclase. *Biochem. J.* **302**:455–461.
53. R.-B. Yang, D.C. Foster, D.L. Garbers, and H.-J. Fulle, 1995, Two membrane forms of guanylyl cyclase found in the eye. *Proc. Natl. Acad. Sci. USA*, **92**:602–606.
54. R.-B. Yang, and D.L. Garbers, 1997, Two eye guanylyl cyclases are expressed in the same photoreceptor cells and form homomers in preference to heteromers. *J. Biol. Chem.* **272**:13738–13742.
55. I. Perrault, J.M. Rozet, P. Calvas, S. Gerber, A. Camuzat, H. Dollfus, S. Chatelin, E. Souied, I. Ghazi, C. Leowski, M. Bonnemaison, D. Le Paslier, J. Frezal, J.-L. Dufier, S. Pittler, A. Munnich, and J. Kaplan, 1996, Retinal-specific guanylate cyclase gene mutations in Leber's congenital amaurosis. *Nat. Genet.* **14**:461–464.

LEBER'S CONGENITAL AMAUROSIS

Clinical Heterogeneity and Electroretinography in 27 Patients

Matthias C. Grieshaber,[1] Eugen Boltshauser,[2] and Günter Niemeyer[1]

[1] Clinical ERG-Laborator
Department of Ophthalmology
[2] University Children's Hospital
Zürich, Switzerland

1. SUMMARY

Purpose: Clinical evaluation of 42 patients with primarily suspected diagnosis of Leber's congenital amaurosis (LCA).

Methods: We used the electroretinogram (ERG), visual acuity, fundus appearance, age of onset, and systemic findings to confirm or revise the diagnosis of LCA.

Results: Twenty-four patients had a non-recordable ERG, and 3 showed small, residual, mainly cone-driven b-waves. Profound visual loss ranged from no light perception to 20/200. Visual acuity of light perception was frequent at <1 year of age. Three patients never had light perception. Most patients had a stable visual acuity despite progressive retinal pigmentary changes. A normal fundus was seen predominantly in patients younger than 1 year of age, but could also be found in patients of all age groups. The first, and frequently the only pigmentary changes were vascular attenuation and rarefaction, later in life heterogeneous retinal findings appeared. None of the 9 reexamined patients showed a normal fundus. Nystagmus and strabismus were principal presenting symptoms. Eleven patients presented with the oculodigital sign, three of them were enophthalmic, and 12 patients showed roving eye movements. Patients with additional (nonocular) findings such as mental retardation (n = 12), renal (n = 3) and skeletal (n = 4) abnormalities revealed no different ERG- or

Address for correspondence: G. Niemeyer, M.D., Department of Ophthalmology, University Hospital, 8091 Zürich, Switzerland, Fax 411 255 4472, Telephone 411 255 4982, e-mail niemeyer@opht.unizh.ch

Retinal Degenerative Diseases and Experimental Therapy, edited by Hollyfield *et al.*
Kluwer Academic / Plenum Publishers, New York, 1999.

ophthalmoscopic findings than the remaining patients. Twelve patients revealed abnormal psychomotor development.

From the initial cohort of 42 patients with suspected LCA, the diagnosis had to be revised in 15 cases as follows: juvenile RP (n = 8), infantile Refsum syndrome (n = 1), and 6 patients could not yet be classified with certainty.

Conclusion: Bilateral visual impairment in infants should be assessed clinically and electroretinographically within the first year. Later in life, the diagnosis of LCA can be a challenge due to its phenotypical heterogeneity. Systemic disorders are independent of ocular findings. In patients older than two years with non-recordable ERG, visual acuity better than 20/200, progressive visual loss, normal pupillary responses, absent oculodigital sign and absence of systemic abnormalities, juvenile retinitis pigmentosa should be considered.

2. INTRODUCTION

Leber's congenital amaurosis (LCA) is the earliest and most severe form of inherited retinopathy as described by Theodor Leber.[1] The autosomal recessive disease[2,3] is characterized by profound visual loss or total blindness, searching nystagmus, hyperopia, and eye poking (oculodigital sign), usually recognized within the first months of life. The highly reduced or non-recordable electroretinogram (ERG) and exclusion of metabolic disorders corroborate the diagnosis.

We present 27 patients with the diagnosis of LCA, that evolved from revising the diagnoses of 42 cases with primarily suspected congenital amaurosis. Clinical and electroretinographic results were often obtained under general anesthesia and reevaluated at times by repeat examination or assessment of the patients' history.

3. PATIENTS AND METHODS

We reviewed 42 patients retrospectively in whom LCA was suspected between 1968 and 1996 at the Department of Ophthalmology, University Hospital of Zürich. The initial diagnosis was based on an absent or highly reduced ERG in infancy. Out of these 42 patients, 27 met the following inclusion criteria for LCA: (1) markedly reduced or non-recordable ERG, (2) profound visual loss documented during the first year, (3) no evidence for a metabolic disorder.

A thorough history was taken and a complete ocular examination was performed in most patients, assessing visual acuity, testing of the pupillary responses to light if possible, eye movements, slit-lamp biomicroscopy, retinoscopy in cycloplegia, direct, and indirect ophthalmoscopy and fundus photography. We revised the diagnosis in 15 cases out of the total of 42 patients with suspected LCA.

3.1. Electroretinography

Complementing the clinical observations electrophysiology is used to objectively assess or exclude retinal degeneration in infants. The electrophysiological technique, recording with Henkes contact lens electrodes, was full-field, single flash ERG with selective stimulation of the rod-system following 20 min. of dark adaptation and subsequently using a rod-desensitizing white background to assess the function of the cone-system.[4,5]

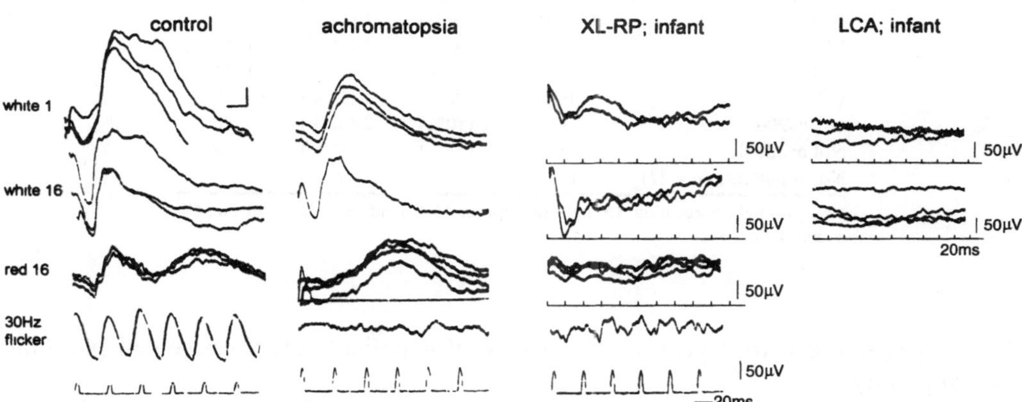

Figure 1. Electroretinographic traces recorded under general anesthesia from a healthy infant (left column) and from patients with achromatopsia, with X-linked retinitis pigmentosa and with LCA. The stimuli (see text for details) are marked as abbreviations on the left margin. Calibration: 50 µV vertical and 20 msec horizontal. Note that the "red 16" stimulus separates two b-wave components in the control (left), and that the early cone-driven component is missing in the corresponding trace in the second column, indicating missing cone-activity in achromatopsia. The infant affected by LCA presented without measurable ERG responses in the first year of life.

We recorded ERGs in infants in general anesthesia (Fig. 1) after brief dark adaptation when poor vision or nystagmus are reported, when malformations or syndromes with retinal involvement are suspected, or in older patients with mental retardation.[6,7] Briefly, Sevofluran inhalation for 20 to 30 min was useful to perform white and "red" stimulation as well as 30 Hz flicker ERG, followed by ophthalmoscopy and fundus photography. The long wavelengths ("red") filter in front of the white flash unit allows to separate the rod-driven from cone-driven b-waves even after only several min of dark adaptation (Fig. 1, left column). With inherent limitations, the test can be performed using skin electrodes placed on the lower lid, but the amplitudes of b-waves are about 1/8 of those recorded by contact lens electrodes.[8]

3.2. Neuropediatric Examination

A neuropediatric examination was performed in the University Children's Hospital Zürich on patients suspected of having psychomotor retardation. The developmental evaluation included assessment of cognitive, language (receptive and expressive), fine, and gross motor skills. Computed axial tomography, magnet resonance imaging (MRI), and metabolic examinations were performed if indicated. We reexamined nine patients and repeated the ERG recording in three cases.

4. RESULTS

The age of the 27 patients with LCA (14 males and 13 females) ranged from 4 months and 6 $^{9}/_{12}$ years of age (mean 2 $^{3}/_{12}$ years) at the first electroretinographic examination. The patients were divided in four age groups (younger than 1 year, 1 to 2 years,

Table 1. Visual acuity and age in LCA

Vision*	<1 year	1–2 years	3–5 years	6–7 years
NLP	1 (8%)	1 (14%)	1 (16%)	0
LP	9 (75%)	0	1 (16%)	1 (50%)
HM	0	1 (14%)	0	1 (50%)
CF	0	0	1 (16%)	0
</= 20/200	0	1 (10%)	2 (32%)	0
Not available	2 (16%)	4 (57%)	1 (16%)	0
No. of patients (n = 27)	12	7	6	2

*NLP no light perception, LP light perception, HM hand movements in 1 m, CF counting fingers.

3 to 5 years, and 6 to 7 years). The history of 4 patients (15%) revealed parental consanguinity.

Twenty-four patients had a non-recordable ERG (89%), and three showed minimal ERG potentials with residual cone-driven responses. On review, we recorded similar ERG as during the first ERG assessment in two cases, and in one patient the traces were not clearly distinguishable from eye movements. Figure 1 illustrates the electroretinographic data of three infants with achromatopsia, with retinal degeneration such as XL-RP, and with LCA compared with a normal ERG from an infant (control).

Absent visual contact was noted frequently by the parents and by the examiners. Profound visual loss ranged from no light perception to 20/200 (Table 1). In most patients younger than 1 year of age light perception was present. Only 3 patients had no light perception, as assessed at age 5 months, 18 months, and 3 years. Their anterior segments and fundus appearance were normal. Visual acuity of 20/200 was seen in three patients between 1 and 4 years of age. In 7 children, 4 of which exhibited mental retardation, no reliable assessment of visual function could be obtained. The pupillary response to light was found to be normal in three, abnormal in 18, and not documented in 6 patients.

The majority had stable visual acuity despite progressive retinal pigmentary and vascular changes with increasing age. In three cases visual loss was described as slowly increasing. Only one patient had a significant decline in vision from 20/200 to hand movements in 1 m over 11 years. He presented with marked nystagmus and the fundi showed severe atrophy of the pigment epithelium. Moreover, the retinal vessels *disappeared completely* between age 1 and 11 years (Fig. 2). We have never observed such dramatic change in retinal degenerations so far, but it can occur in cats affected by inherited retinal degenerations (Prof. G. Aguirre, personal communication, and photographic documentation). Another patient with continuing visual deterioration developed prominent pigmentary changes and white flecks in the macula. None of these patients developed keratoconus or cataracts.

4.1. Ophthalmoscopic Observations

The ophthalmoscopic findings are summarized in Table 2. A normal fundus was seen predominantly in patients younger than 1 year, but could also be found in all age groups. Two patients of the youngest group developed subtle pigmentary mottling within a year. Attenuation of retinal arterioles was seen in 50% of patients in the

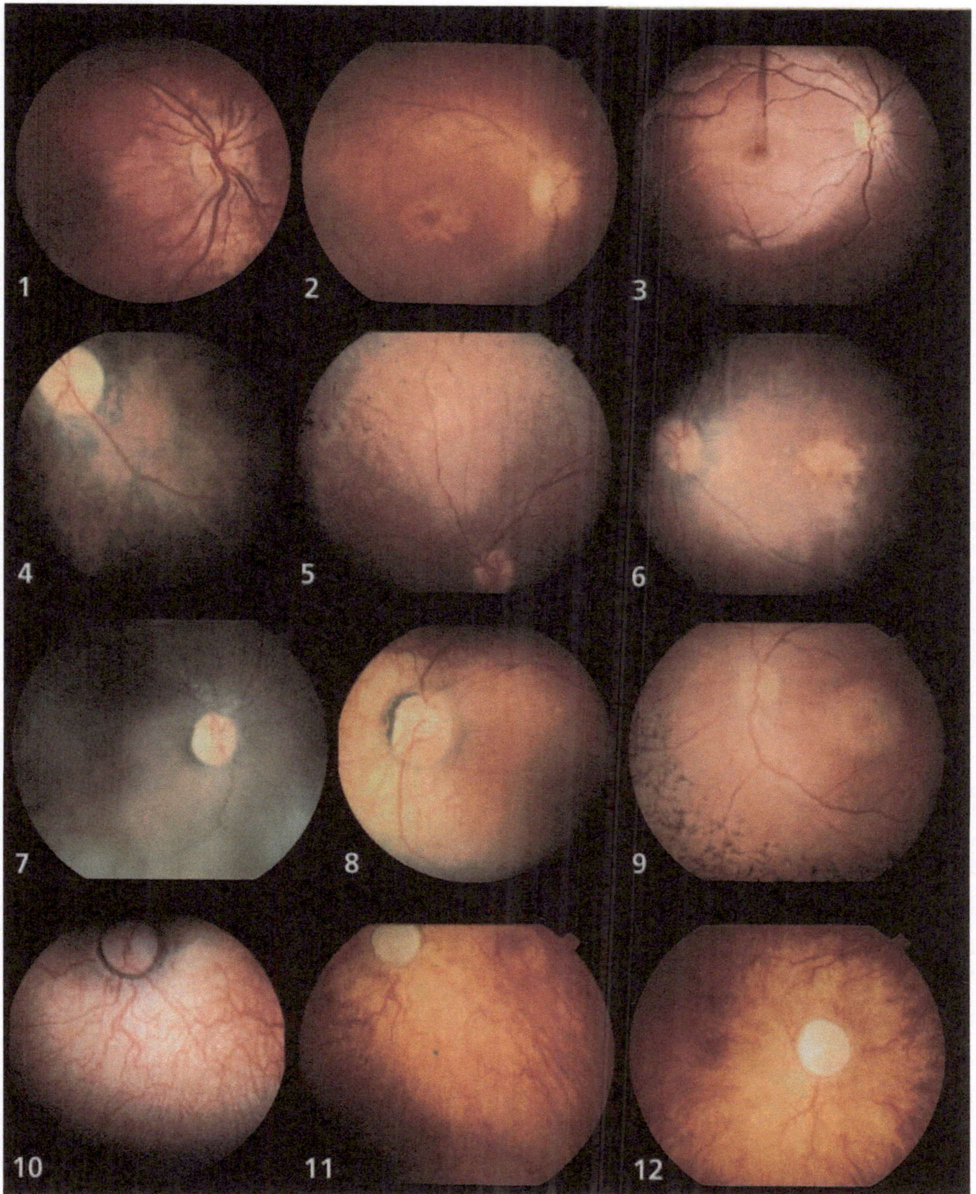

Figure 2. Fundus photographs of 10 patients affected by LCA. (1) Three year-old boy with nearly normal appearance of the fundus and non-recordable ERG. (2) Twenty-nine year-old female patient with rarefaction and marked narrowing of the retinal vessels, degenerative changes of the macula, and non-recordable ERG. (3) Three year-old girl presenting with fine salt-and-pepper pigmentary changes in the periphery, slight narrowing of the retinal vessels and highly reduced ERG amplitudes. The fixation marker appears in nearly central position despite a visual acuity of 0.1. (4) Six year-old girl with geographic irregularity of the RPE-pigment, narrowed retinal vessels, and non-recordable ERG. (5) Twenty year-old female patient who developed marked mid-peripheral pigmentary changes as well as narrowing of the retinal vessels, small optic disc in presence of non-recordable ERG. (6) Six year-old girl exhibiting irregular hypopigmentary fundus changes and, mostly peripherally white intraretinal flecks, slightly narrowed retinal vessels, and non-recordable ERG. (7) Thirteen year-old girl with highly reduced and narrowed retinal vessels. discrete degenerative macular changes and a non-recordable ERG. (8) Seven month-old boy with marked hypopigmentation of the mid-peripheral and peripheral fundus, marked narrowing of the retinal vessels and non-recordable ERG. (9) Fifteen year-old boy with markedly narrowed retinal vessels, degenerative macular changes and with a network of bone spicules-like hyperpigmentations in the midperiphery and a non-recordable ERG. (10) Left fundus of a boy with nystagmus. Photograph taken at age one, revealing marked lack of pigment in the RPE. At 12 years of age pictures 11 (left) and 12 (right) could be taken despite a continuous nystagmus. The photographs reveal that the retinal vessels have apparently *disappeared completely*, a development that can also be observed in cats with retinal degeneration.

Table 2. Fundus findings in relation to age in LCA

Age[1]	number of patients n = 27	normal fundus	optic nerve pallor	vascular attenuation	macular pigment	salt and pepper	bone spicules pgm.	RPE-atrophy[2]	marbleization
<1 yr.	12	4 (33%)	—	6 (50%)	3 (25%)	5 (42%)	—	1 (8%)	1 (8%)
1–2 yrs.	7[3]	1% (14%)	—	3 (42%)	3 (42%)	3 (42%)	—	1 (14%)	—
3–5 yrs.	6	1 (16%)	1 (16%)	4 (66%)	1 (16%)	1 (16%)	—	2 (32%)	1 (16%)
6–7 yrs.	2	1 (50%)	—	1 (50%)	1 (50%)	—	1 (50%)	1 (50%)	—

[1]Age at time of first electroretinal examination.
[2]Retinal pigment epithelial atrophy.
[3]One patient: no data.

youngest age group, but was more pronounced in older patients. In patients of 1 to 2 years of age, similar fundus appearances were found as described for patients <1 year of age. The optic discs were not recognized to be atrophic until after the second year. By age 3 to 5 years, pigmentary changes were prominent and tended to be more extensive in the peripheral retina, and highly attenuated retinal vessels were seen in two thirds of these patients. In the oldest age group, one patient presented with normal fundus appearance, whereas the other patient revealed pronounced chorioretinal atrophy. Of the 9 reexamined patients none showed a normal fundus.

4.2. Associated Ocular Features

Nystagmus was seen in 16, and strabismus was found in 11 out of the total of 27 patients.

Refractive errors were ascertained in 14 patients and ranged from +8.00 to −7.50 diopters. From 12 patients who were hyperopic, six had +4.00 diopters or more. Two patients were myopic with −5.75 to −7.50 diopters. Posterior subcapsular cataracts and vitreal pigment dusting were observed in two patients. One patient presented with keratoglobus at 29 years of age. Eleven patients presented with the oculodigital sign, and three of them were enophthalmic. Photophobia was reported in six patients. Twelve patients showed roving eye movements.

In 15 patients out of the initial total of 42 infants and children we had to revise the diagnosis as follows: juvenile retinitis pigmentosa (RP; n = 8), infantile Refsum syndrome (n = 1),[9] and 6 patients could not be definitely classified (see Table 3).

Table 3. Summary of major findings in LCA and differential diagnosis

		absent ERG	visual acuity*	eye poking	nystamus
total patients	n = 42				
confirmed LCA	n = 27	n = 24 (89%)	MLP to 20/200	n = 11	n = 27
juvenile RP	n = 8	n = 6 (75%)	HM to 20/70	n = 0	n = 3
not classifiable	n = 6	n = 2 (33%)	CF to 20/100	n = 0	n = 3
infantile Refsum	n = 1	n = 1	NLP	n = 0	n = 1

*NLP: no light perception, HM: hand movements, CF: counting fingers in 0.5 m.

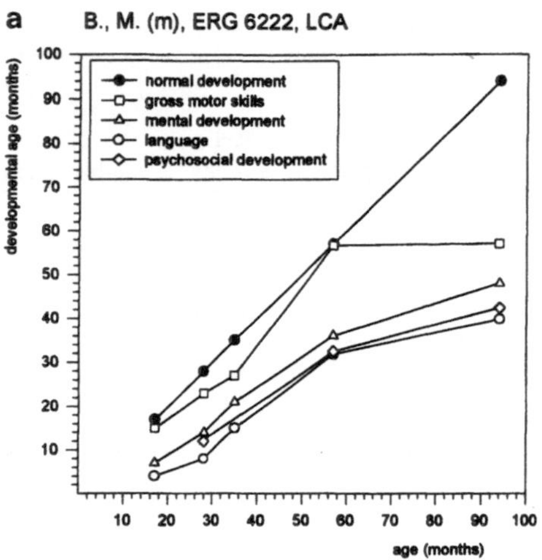

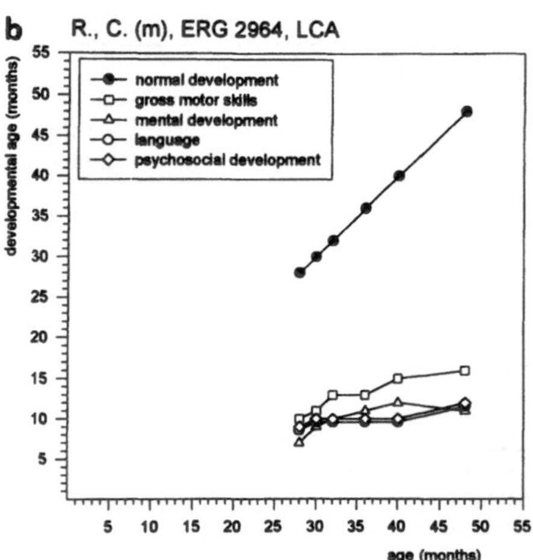

Figure 3. Delayed psychomotor development in 2 patients with LCA. a) developmental data of a 7 year old boy. b) developmental data of a 4 year old boy with LCA.

4.3. Neurodevelopment

The majority of children (n = 15) revealed a normal neurodevelopment. Twelve patients showed psychomotor retardation, 4 of them were severely retarded (Fig. 3). An abnormal gross motor skill was often the first sign of retardation. Even mild retardation (n = 8) in fine and gross motor skills became evident in the first few months of life. Hypotonia was an additional finding in two patients, one of which showed hypoplasia of the cerebellar vermis in MRI. Normally, mildly delayed patients caught up in the first decade. Interestingly, the neurological development remained stable or improved without changes in visual acuity. On the other hand, infants with severe psychomotor deficiency showed no development at all. The best results were achieved in major motor skill. All patients who were severely retarded had a pronounced hypotonia. Skeletal anomalies were found in three, renal dysgenesis in two patients.

5. DISCUSSION

The incidence of LCA at the University Hospital of Zürich is about 1.9% of all retinal dystrophies[10] and thereby less than up to 5%, as reported earlier.[11,12]

5.1. Electroretinography

An *abolished/non-recordable* or *highly reduced ERG* is typical, but not specific for LCA. On review, the majority of patients with juvenile RP and infantile Refsum syndrome had an extinguished ERG, as also described for the Joubert[13,14] and Zellweger syndromes.[15] ERG with minimal residual potentials show functional remains of the retina not excluding the diagnosis of LCA. Of interest is the greater affection of the rod-system, that correlates with histopathological description.[16]

In view of the retinal development of a normal newborn, the age of the patient at the time of the first ERG is important. Retinal sensitivity reaches adult values by about 6 months, and maximal amplitudes and latencies were similar to adult by age 12 months.[17] ERG of infants younger than one year should be repeated after one year to exclude maturation-related electrophysiological improvement, which would argue against LCA.[18] We suggest to perform ERG testing within the first two years to distinguish LCA from early manifested RP (juvenile or syndromic) or acquired retinopathy after rubeola, rubella and syphilis infections[19] especially if ophthalmoscopic changes are missing.

We favour the procedure of ERG recording and photographic documentation in brief anesthesia in the light of diagnostic challenges in infants such as distinction between malfunction of the rod- from that of the cone-system. A broad band, long wavelengths filter is convenient to separate reliably the cone-driven from the rod-driven b-wave, even after only a few minutes of dark adaptation, and the 30 Hz flicker stimulus is useful to rapidly assess the cone-function selectively. An advantage of ERG recording in anesthesia is the opportunity for retinoscopy and fundus photography. This entire effort appears justified by providing the basis for clearcut information to the parents, and of the geneticist, as well as a basis for specific educational planning for the child with visual dysfunction. It appears important to detect retinal dysfunction in combination with hearing loss (Usher syndrome) as early as possible. Two examples illustrate this point: the diagnosis of rod monochromatism or of congenital stationary night blindness based on an examination in anesthesia enables us to orient the parents about the nonprogressive character of these retinal malfunction entities.

5.2. Ocular Features

Light perception does not exclude LCA. The majority of our patients had light perception as visual acuity. The high percentage of patients younger than one year (75%) with only light perception can be influenced by inadequate cooperation of infants. On the other side, an earlier assignment to ERG-testing for patients with uncertain light perception is possible. Patients with extraocular findings revealed no different visual acuity compared to the other patients, if assessment was possible.

We took 20/200 as the upper limit for visual acuity as mentioned previously.[15,20] The majority had a stable visual acuity despite progressive retinal pigmentary changes with increasing age.

A normal ophthalmoscopic aspect was often observed in patients younger than one year, but retinal pigmentary changes usually evolve during the first decade. A variety of aspects are described. The first and frequently only retinal changes were vascular attenuation and rarefaction with the arterioles being more affected than venules. Later in life very heterogeneous pigmentary changes characterize LCA, including salt-and-pepper, bone spicules pigmentation, as well as widespread retinal pigment epithelium- (RPE) and chorioretinal atrophy. A less common feature consists of yellowish-white flecks in the midperipheral retina as seen in two patients. Pallor of the optic disc was not seen until after the second year, unlike the description by Heher et al.,[21] and a macular coloboma was not found in our patients.

Hyperopia (n = 12) is typical for LCA and more frequent than myopia (n = 2). The degree of hyperopia does not correlate with any other ocular or nonocular symptoms. The association of an uncomplicated form of LCA with high hyperopia without systemic findings has been described by Foxman et al.[11] and by Wagner et al.[22]

Nystagmus (n = 16) and strabismus (n = 11) were principal symptoms and causes for the first ophthalmological consultations.

Bilateral visual impairment in infants should be assessed clinically and electroretinographically within the first year. Later in life, Leber's congenital amaurosis can be hardly diagnosed due to the heterogeneous ocular findings. In patients older than two years with non-recordable ERG, but visual acuity better than 20/200, rapid progressive visual loss, normal pupillary responses, absent oculodigital sign and no systemic abnormalities juvenile RP should be considered.[23]

5.3. Non-Ocular Changes and Syndromes with LCA-Like Ocular Phenotype

Further, we propose to differentiate LCA from metabolic syndromes particularly from neuronal ceroid lipofuscinoses, infantile Refsum, and Bassen-Kornzweig syndrome the latter being treatable. Therefore biochemical investigations such as phytanic acid, very long chain fatty acids, cholesterol, and vitamin E are essentiel for a complete assessment to exclude LCA.

5.4. Genetics

Although Th. Leber described the clinical phenomenology of LCA over one hundred years ago, little was known about the pathophysiology until recently. LCA, autosomal-recessively inherited, is usually regarded as the consequence of either impaired development of photoreceptors or extremely early degeneration of cells that have developed normally.[16] In 1995, for the first time a gene for LCA was mapped to chromosome 17p13.1 (LCA1).[24] Later, missense- and frameshift-mutations of the retinal guanylate cyclase (retGC) gene were found in patients with LCA and may account for the impaired production of cGMP in the retina.[25] Recently, a histological study of a human fetus (sixth month) with retGC gene defects confirmed the hypothesis of a severe impairment of the photoreceptors during prenatal development in LCA.[26] An additional locus for LCA of a highly inbred family was mapped to chromosome 6, an area where other retinal dystrophies have been mapped previously such as Stargardts disease, North Carolina macular dystrophy, and progressive chorioretinal dystrophy.[27]

RPE65, an abundant 61 kDa protein,[28] is expressed specifically by the RPE. Although the function of RPE65 is unknown, it may play an important role in the RPE/photoreceptor vitamin-A metabolism. It is in association with the serum retinol-binding protein[29,30] and with the RPE-specific 11-cis retinol dehydrogenase, an enzyme active in the synthesis of the visual pigment chromophore 11-cis retinal.[31]

Mutations in RPE65 cause autosomal recessive, childhood-onset, and severe retinal dystrophy (arCSRD).[32] Phenotypic variation associated with different RPE65 gene mutations are suggested.[33] Another study showed mutations in the RPE65 gene responsible for about 2% of recessive RP and 15% of LCA.[34] However, it is not definitely clarified if RPE65 mutations cause LCA.[35] At the occasion of reexamination we obtained blood samples from patients for future genetic analysis.

5.5. Differential Diagnosis of Impaired Vision in Infants

As discussed in previous articles[5,36] retinal and cortical electrophysiology should be applied to differentiate the following conditions causing visual impairment in infants: RP with very early onset, achromatopsia, congenital stationary night blindness, delayed visual maturation, and cortical blindness. Other congenital conditions including cataract, optic nerve hypoplasia, congenital optic atrophy, severe microphthalmus, and coloboma are clinically obvious and usually do not require electrophysiologic testing. Skeletal and renal anomalies are rare. Systemic disorders are independent of ocular findings and do not predict outcome.

ACKNOWLEDGMENTS

We thank Mrs. A. Schaefer for excellent assistance in the Clinical ERG-Laboratory, Dr. B. Grbovic (Department of Anesthesiology—University Hospital Zürich) for continuous collaboration, Dr. M. Steinlin (University Childrens' Hospital Zürich) for examining several patients, and Dr. K. Landau for constructive comments on the manuscript.

REFERENCES

1. T. Leber, 1869, Ueber Retinitis pigmentosa und angeborene Amaurose, *Graefes Arch. Clin. Exp. Ophthalmol.* **15**:1–25.
2. C.H. Alström and O.A. Olson, 1957, Heredo-retinopathia congenitalis, Monohybrida recessiva autosomalis, *Hereditas* **43**:1–77.
3. S.R. Lambert, S. Sherman, D. Taylor, A. Kriss, R. Coffey, and M. Pembrey, 1993, Concordance and recessive inheritance of Leber congenital amaurosis, *Am. J. Med. Genet.* **46**:275–277.
4. P. Gouras, 1970, Electroretinography: Some basic principles. *Investig. Ophthalmol.* **9**:557–569.
5. G. Niemeyer, 1998, Electroretinography: An useful tool to assess retinal function in clinical neurophysiology. In: *9th european congress of clinical neurophysiology*, edited by De Weerd Stalberg E.V. and Zidar J. Al W., Ljubljana, Slovenia: Monduzzi Editore, pp. 201–208.
6. G. Niemeyer, B. Grbovic, and B. Gloor, 1993, Elektroretinographie-Diagnostik beim Kleinkind. Electroretinographic diagnosis in infants, *Klin. Monatsbl. Augenheilkd.* **202**:417–421.
7. G. Niemeyer, 1999, Vom Symptom zur Elektroretinographie-Diagnose (From symptom to ERG-diagnoses), *Klin. Monatsbl. Augenheilkd.* (in press).
8. A. Kriss and I. Russell-Eggitt, 1992, Electrophysiological assessment of visual pathway function in infants, *Eye* **6**:145–153.
9. R.J.A. Wanders, E. Boltshauser, B. Steinmann, M.A. Spycher, R.B.H. Schutgens, H. van den Bosch, and J.M. Tager, 1990, Infantile phytanic acid storage disease, a disorder of peroxisome biogenesis: a case report, *J. Neurol. Sci.* **98**:1–11.
10. G. Niemeyer and P. Stähli, 1996, ERG-Diagnose und Differentialdiagnose: Untersuchungsergebnisse über 6 Jahre. Electroretinographic diagnoses and differential diagnoses: Results over 6 years, *Klin. Monatsbl. Augenheilkd.* **208**:306–310.
11. S.G. Foxman, J.R. Heckenlively, J.B. Bateman, and J.D. Wirtschafter, 1985, Classification of congenital and early onset retinitis pigmentosa, *Arch. Ophthalmol.* **103**:1502–1506.
12. J. Kaplan, D. Bonneau, J. Frézal, A. Munnich, and J.L. Dufier, 1990, Clinical and genetic heterogeneity in retinitis pigmentosa, *Hum. Genet.* **85**:635–642.
13. M. Steinlin, M. Schmid, K. Landau, and E. Boltshauser, 1997, Follow-up in children with Joubert syndrome, *Neuropediatrics* **28**:204–211.
14. B. Kendall, D. Kingsley, S.R. Lambert, D. Taylor, and P. Finn, 1990, Joubert syndrome: a clinico-radiological study, *Neuroradiology* **31**:502–506.
15. S.R. Lambert, A. Kriss, D. Taylor, R. Coffey, and B. Soc, M. Pembrey, 1989, Leber's congenital amaurosis: A follow-up and diagnostic reappraisal of 75 patients, *Am. J. Ophthalmol.* **107**:624–631.
16. K. Mizuno, Y. Takei, M.L. Sears, W.S. Peterson, R.E. Carr, and L.M. Jampol, 1977, Leber's congenital amaurosis, *Am. J. Ophthalmol.* **83**:34–42.
17. A.B. Fulton and R.M. Hansen, 1985, Electroretinography: Application to clinical studies of infants, *J. Pediat. Ophthalmol. & Strabismus* **22**:251–255.

18. I. Casteels, W. Spileers, Ph. Demaerel, P. Casaer, P. De Cock, L. Dralands, and L. Missotten, 1996, Leber congenital amaurosis—differential diagnosis, ophthalmological, and neuroradiological report of 18 patients, *Neuropediatrics* **27**:189–193.
19. G. Niemeyer, 1989, Indikationen zu elektrophysiologischen Untersuchungen des Auges, *Klin. Monatsbl. Augenheilkd.* **194**:333–336.
20. R. Schroeder, M.B. Mets, and I.H. Maumenee, 1987, Leber's congenital amaurosis. Retrospective review of 43 cases and a new fundus finding in two cases, *Arch. Ophthalmol.* **105**:356–359.
21. K.L. Heher, E.L. Traboulski, and I.H. Maumenee, 1992, The natural history of Leber's congenital amaurosis, age-related findings in 35 patients, *Ophthalmology* **99**:241–245.
22. R.S. Wagner, A.R. Caputo, L. Nelson, and D. Zanoni, 1985, High hyperopia in Leber's congenital amaurosis, *Arch. Ophthalmol.* **103**:1507–1509.
23. M.C. Grieshaber, and G. Niemeyer, 1998, Kongenitale Amaurose Leber: Diagnose, Verlauf und Differentialdiagnose. Leber's congenital amaurosis: Diagnosis, course, and differential diagnosis, *Klin. Monatsbl. Augenheilkd.* **212**:309–310.
24. A. Camuzat, H. Dollfuss, J.M. Rozet, S. Gerber, D. Bonneau, M. Bonnemaison, M.-C. Briard, J.-L. Dufier, I. Ghazi, C. Leowski, J. Weissenbach, J. Frézal, A. Munnich, and J. Kaplan, 1995, A gene for Leber's congenital amaurosis maps to chromosome 17p, *Hum. Mol. Genet.* **4**:1447–1452.
25. I. Perrault, J.M. Rozet, P. Calvas, S. Gerber, A. Camuzat, H. Dollfus, S. Châtelin, E. Souied, I. Ghazi, C. Leowski, M. Bonnemaison, D. Le Paslier, J. Frézal, J.-L. Dufier, S. Pittler, A. Munnich, and J. Kaplan, 1996, Retinal-specific guanylate cyclase gene mutations in Leber's congenital amaurosis, *Nature genetics* **14**:461–464.
26. I. Perrault, J.M. Rozet, A.L. Delezoide, S. Gerber, E. Souied, A. Cabot, A. Munnich, and J. Kaplan, 1998, Histological study and in situ hybridization of the retGC gene in a human fetus affected with Leber's congenital amaurosis (LCA1), *Invest. Ophthalmol. Vis. Sci.* **39**:1042.
27. S. Dharmaraj, J.M. Robitaille, D. Zhu, Y.Y. Li, and I.H. Maumenee, 1998, An additional locus for Leber's congenital amaurosis maps to chromosome 6, *Invest. Ophthalmol. Vis. Sci.* **39**:296.
28. A. Nicoletti, D.J. Wong, K. Kawase, L.H. Gibson, T.L. Yang-Feng, J.E. Richards, and D.A. Thompson, 1995, Molecular characterization of the human gene encoding an abundant 61 kDa protein specific to the retinal pigment epithelium, *Hum. Mol. Gen.* **4**:641–649.
29. C.-O. Bavik, U. Eriksson, R.A. Allen, and P.A. Peterson, 1991, Identification and partial characterization of a retinal pigment epithelial membrane receptor for plasma retinol-binding protein, *J. Biol. Chem.* **266**:14978–14985.
30. C.-O. Bavik, C. Busch, and U. Eriksson, 1992, Characterization of a plasma retinol-binding protein membrane receptor expressed in the retinal pigment epithelium, *J. Biol. Chem.* **267**:23035–23042.
31. A. Simon, U. Hellman, C. Wenstedt, and U. Eriksson, 1995, The retinal pigment epithelial-specific11-cis retinal dehydrogenase belongs to the family of short chain alcohol dehydrogenases, *J. Biol. Chem.* **270**:1107–1112.
32. S. Gu, D.A. Thompson, C.R. Srisailapathy Srikumari, B. Lorenz, U. Finckh, A. Nicoletti, K.R. Murthy, M. Rathmann, G. Kumaramanickavel, M.J. Denton, and A. Gal, 1997, Mutations in RPE65 cause autosomal recessive childhood-onset severe retinal dystrophy, *Nature Genetics* **17**:194–197.
33. E. Apfelstedt-Sylla, S. Gu, A. Gal, and E. Zrenner, 1998, Clinical findings in patients with retinal dystrophies caused by mutations in RPE65, *Invest. Ophthalmol. Vis. Sci.* **39**:967.
34. H. Morimura, G.A. Fishman, S.A. Grover, A.B. Fulton, E.L. Berson, and T.P. Dryja, 1998, Mutations in the RPE65 gene in patients with recessive retinitis pigmentosa or Leber's congenital amaurosis, *Invest. Ophthalmol. Vis. Sci.* **39**:293.
35. J.M. Rozet, I. Perrault, I. Ghazi, S. Gerber, E. Souied, A. Cabot, J.L. Dufier, A. Munnich, and J. Kaplan, 1998, Exclusion of the RPE65 gene in 26 families affected with Leber's congenital amaurosis (LCA), *Invest. Ophthalmol. Vis. Sci.* **39**:297.
36. S.R. Lambert, D. Taylor, and A. Kriss, 1989, The infant with nystagmus, normal appearing fundi, but an abnormal ERG, *Survey of Ophthalmol.* **34**:173–186.

10

RETINAL DEGENERATION IS ACCELERATED WHEN A MUTANT RHODOPSIN TRANSGENE IS EXPRESSED ON A HAPLOID OR NULL RHODOPSIN BACKGROUND

Jeanne Frederick,[1*] Nataliia Krasnoperova,[2] Kirstin Hoffmann,[3] Wolfgang Baehr,[1] Janis Lem,[2,4] and Klaus Rüther[3]

[1] Moran Eye Center
University of Utah Health Science Center
Salt Lake City, Utah 84132
[2] Department of Ophthalmology
New England Medical Center
Boston, Massachusetts 02111
[3] Charité-Virchow Augenklinik
Humboldt University
13353 Berlin, Germany
[4] Tufts University School of Medicine
Boston, Massachusetts 02111

1. SUMMARY

The rod opsin Pro23His mutation accounts for approximately 12% of autosomal dominant retinitis pigmentosa (adRP) cases in the United States. To study mechanisms leading to photoreceptor degeneration, we investigated the effects of mutant and wild-type opsin stoichiometry on retinal morphology and function. We crossbred a transgenic mouse line expressing the triple mutant, V20G, P23H, and P27L (GHL), with rhodopsin knockout mice. Retinal morphology of 30-day old GHL^{+} mice with two functional copies of the rod opsin gene (GHL^{+}, rho$^{+/+}$), one functional copy (GHL^{+}, rho$^{+/-}$)

* Corresponding author; mailing address: John A. Moran Eye Center, 50 N. Medical Dr., Salt Lake City, UT 84132. Ph. 801-585-3573; fax 801-585-1515.

Retinal Degenerative Diseases and Experimental Therapy, edited by Hollyfield *et al.*
Kluwer Academic / Plenum Publishers, New York, 1999.

or no functional copies of the rod opsin gene (GHL$^+$, rho$^{-/-}$) was examined. Although mice of all three genotypes underwent retinal degeneration, the severity of the retinopathy correlated inversely with the number of wild-type opsin genes present. Mice with no functional wild-type opsin gene were most severely affected, while those with two functional copies were the least affected. Mice with a single functional wild-type gene were intermediate in the degenerative phenotype. Correspondingly, changes in fundus morphology and ERG function were most prominent in GHL$^+$, rho$^{-/-}$ mice, whereas GHL$^+$, rho$^{+/+}$ mice were similar to wild-type mice.

2. INTRODUCTION

Transgenic mice expressing mutant rhodopsin transgenes[1] are a powerful tool with which to investigate the etiology of human retinitis pigmentosa (RP)[2]. More than 70 mutations have been identified in the human rhodopsin gene, and most of them are thought to cause dominant RP.[3] We had generated previously a mouse line that expresses a mutant rhodopsin transgene[4]. The transgene consisted of the complete mouse opsin gene with 3 amino acid substitutions in exon 1 (see Fig. 1). One mutation

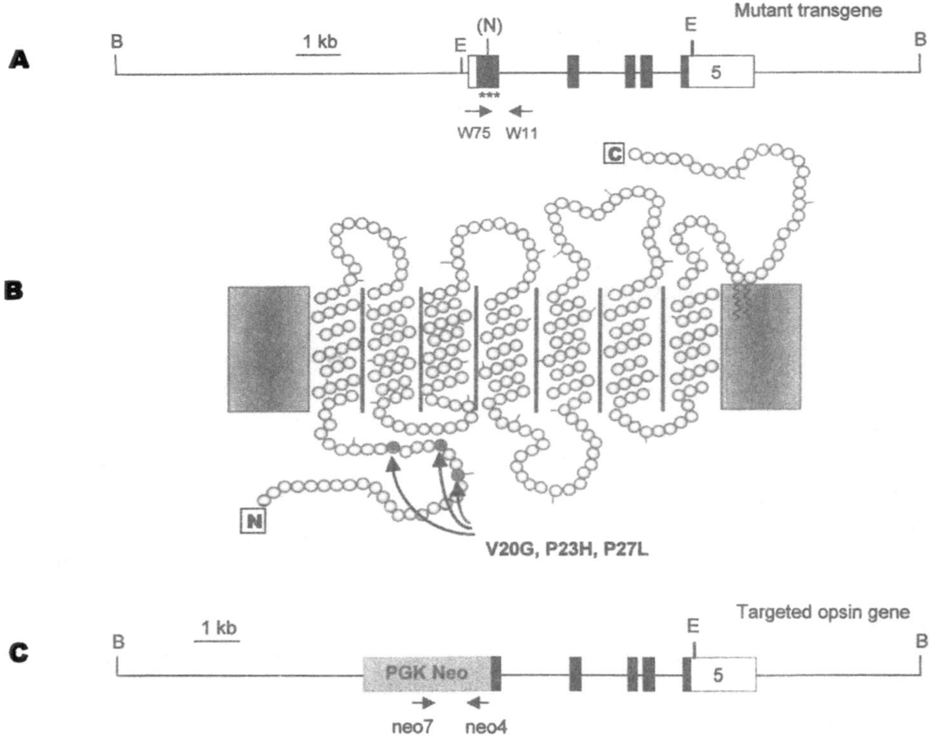

Figure 1. Transgene, mutant rhodopsin, and targeted opsin gene. A) The mouse opsin transgene containing mutations in exon 1 as described previously[4]; (N) denotes a deleted *NcoI* site. Asterisks symbolize the three amino acid substitutions illustrated in Fig. 1B. B) Two-dimensional model of mouse rhodopsin. The three amino acids mutated in the transgene are identified near the N-terminus in the extracellular (intradiscal) space. C) The targeted opsin gene. The first 111 codons of exon 1, including the translational start, were replaced by a neomycin gene.

was P23H, the most frequent rhodopsin mutation in human autosomal dominant RP in the United States. The phenotype of the resulting mouse line was a slowly progressing photoreceptor degeneration, preceded by a significant shortening of rod outer segments.[4] Subsequent studies showed that basal discs were disorganized and that disk shedding was normal[5]. In an independent model carrying a P23H human transgene, the mutant rhodopsin was found to accumulate in the outer plexiform layer, indicating a misrouting of mutant rhodopsin after biosynthesis.[6] Studies with recombinant P23H rhodopsin showed that some of the protein misfolded, and that the misfolded protein was susceptible to trypsin digestion.[7]

Knockout of the rhodopsin gene by gene targeting had differential effects on photoreceptor structure and function. Photoreceptors of mice heterozygous for the knockout gene showed misorientation of some disk membranes, but outer segment lengths in adult mice were only slightly shortened and the retinas underwent a very slow degeneration. Homozygous knockouts did not develop outer segments at all and underwent a rapid degeneration.[8,9] In addition to its role in photoreception, rhodopsin has a structural role, similar to that of rds/peripherin and rom1 that are thought to be exclusively structural disk proteins (for review, see[10]). Here, addition of the GHL transgene on the $rho^{-/-}$ background had dramatic effects on the rate of photoreceptor degeneration.

3. METHODS

3.1. Mouse Breeding

GHL^{+}, $rho^{+/-} \times rho^{+/-}$ parental mice were bred to so that littermate mice could be analyzed for histopathology and ERG function. The six possible genotypes arising from this mating were: GHL^{-} in a $rho^{-/-}$, $rho^{+/-}$ or $rho^{+/+}$ genetic background or GHL^{+} in a $rho^{-/-}$, $rho^{+/-}$ or $rho^{+/+}$ genetic background.

3.2. Genotyping

Genomic DNA was extracted from tail biopsies for use as templates in the polymerase chain reaction (PCR). Genotypes of mice were determined using 2 primer pairs. The transgene and wild-type gene were distinguished by differential restriction digestion of the 1.3 kb amplification product produced by the primers W75 (5′-AGA CTG ACA TGG GGA GGA ATT C) and W11 (5′-GTG CCT GTG ATC ACA GCA CTT). Digestion with *Nco I* produced 689, 431, and 197 bp fragments from the endogenous wild-type opsin gene, or 886 and 431 bp fragments from the GHL transgene, owing to deletion of one *Nco I* site in the construct (Fig. 1). Presence of the targeted opsin gene was detected by PCR amplification of a 498 bp fragment from the neomycin (neo) gene using the primers Neo4 (5′-CGG GAG CGG CGA TAC CGT AAA GC) and Neo7 (5′-GAA GCG GGA AGG GAC TGG CTG CTA). Genotypes were assigned as described in Figure 2 legend.

3.3. Microscopy

Right eyes were oriented by a cautery mark to the superior pole, enucleated and immersed immediately in a fixative of 4% paraformaldehyde/0.1% glutaraldehyde in 0.1 M phosphate buffer, pH 7.4. Following fixation overnight at 4°C, the anterior segment of each globe was removed and the eyecup was bisected along the vertical meridian. The half eyecup containing the optic nerve was postfixed in 1% osmium tetroxide in phosphate buffer, dehydrated through a series of graded ethanols and embedded in Spurr's resin. Sections 0.5–1 μm thick were stained with 1% methylene blue/azure II in sodium borate solution, and photographed in superior regions equidistant from the optic nerve using a Zeiss 40X oil objective lens.

3.4. Electroretinograms (ERGs)

Mice were dark-adapted at least 2 hours prior to recording ERGs. Under dim red light, pupils were dilated using 1% atropine/0.5% tropicamide and animals were anesthetized with 20 mg xylazine and 40 mg ketamine per kilogram of body weight, administered subcutaneously. Body temperature was maintained with a heating pad during testing. ERGs were recorded using a monopolar corneal contact lens electrode (Medical Workshop, Groningen, Netherlands). Silver reference and neutral electrodes were placed subcutaneously 3 mm below the eye and in the forehead, respectively. Mice were placed into a Ganzfeld bowl (Toennies Multiliner Vision, Höchberg, Germany) for recordings. The signal was amplified 10,000-fold with a 1 to 300 Hz bandpass filter. Noise level was 1.0 μV v_{eff} from 0 to 10 kHz. Signals were digitized at a rate of 1.7 kHz.

The Ganzfeld stimulus was a white flash with a color temperature of 6,000 K and 50 μsec duration. The maximal stimulus intensity of 15 $cdsm^{-2}$ could be attenuated by neutral density filters (Schott) mounted in three filter wheels. Fine regulation was achieved by voltage adjustment. For the scotopic ERG, 8 flash intensities from 4×10^{-5} $cdsm^{-2}$ to 1 $cdsm^{-2}$ were used. At the four lower intensities, 5 responses were averaged, with an interstimulus interval (ISI) of 2 seconds. At higher intensities, only 2 responses were recorded (ISI = 5 seconds). At even higher stimulus intensities ($>4 \times 10^{-3}$ $cdsm^{-2}$), marked oscillatory potentials (OPs) became apparent, making the peak of the b-wave invisible. Therefore, the curve was smoothed by a low pass filter to obtain the b-wave minus OPs. The amplitude was measured from the trough of the a-wave to the peak of the b-wave. The OPs were then recorded in a separate step using a 100 to 1,000 Hz bandpass filter.

For the photopic ERG, a 30 $cdsm^{-2}$ background light was used. A single flash response to a 15 $cdsm^{-2}$ light flash was recorded (digitizing rate 2.56 kHz, gain 25.000, average of 16). Scotopic and photopic ERGs were repeated three times during each session to confirm the stability of recordings. Final amplitude values were the mean of three determinations.

3.5. Fundus Photography

Fundus photographs were obtained using a Rodenstock Scanning Laser Ophthalmoscope (Rodenstock, München, Germany) fitted with a 90 diopter lens.

4. RESULTS

4.1. Genotype Frequencies

Each N_2 parent was $rho^{+/-}$ and, in addition, the mother carried the mutant rhodopsin transgene. Given these parents, all possible genotypes may be generated with

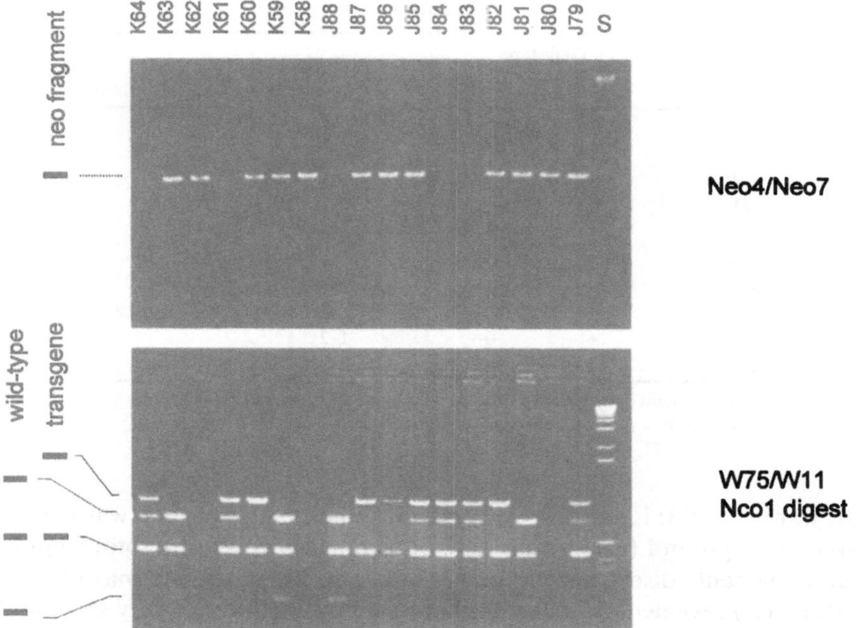

Figure 2. Genotypic analysis. This figure shows agarose gels identifying the genotypes of mice in this report. Genotypes were assigned using 2 primer pairs run separately. Above, the neomycin gene was amplified by primers Neo4/Neo7 in twelve of seventeen mice (71%), indicating disruption of one or both opsin alleles. Presence of a 498 bp band was diagnostic for the null opsin gene. Below, the transgene GHL^{+} was transmitted in ten mice (59%) as determined from the 886 bp fragment produced following digestion with the restriction enzyme, *Nco I*. A 689 bp band was diagnostic for the endogenous wild-type opsin gene. Mice with both the 689 bp band (lower panel) and 498 bp band (upper panel) were rho$^{+/-}$, while mice without the 498 bp band but retaining the 689 bp band were rho$^{+/+}$. Conversely, mice without the 669 bp band but retaining the 498 bp band were rho$^{-/-}$.

frequencies predicted as follows: GHL^{-}, rho$^{+/-}$ (25%); GHL^{+}, rho$^{+/-}$ (25%); GHL^{-}, rho$^{+/+}$ (12.5%); GHL^{+}, rho$^{+/+}$ (12.5%); GHL^{-}, rho$^{-/-}$ (12.5%); and GHL^{+}, rho$^{-/-}$ (12.5%). One litter of ten was used for histopathology (Figure 2, J79-J88) and contained all six possible genotypes. A second litter of seven was used for assessment of ERG function (Fig. 2, K58-K64) and contained 4 of the six possible genotypes.

4.2. Histopathology

Retinal morphology of littermate mice was determined at 30 postnatal days. The rate of degeneration was compared among the different genotypes by examination of: i) thickness of the outer nuclear layer (ONL), which provides an estimate of the number of surviving photoreceptor cells, and ii) ratio of the photoreceptor cell outer segment (OS) to inner segment (IS) lengths (Table 1).

Retinal morphologies corresponding to each genotype are shown in Fig. 3. Sections of rho$^{+/+}$, rho$^{+/-}$, and rho$^{-/-}$ mouse retinas are shown for reference (upper row), and recapitulate earlier findings.[9] Among these, the number of rows of photoreceptor nuclei were maintained; however, relative to wild-type (A), rod outer segments were slightly shorter and disorganized in rho$^{-/-}$ (B) and absent in rho$^{-/-}$ (C) mice. The OS:

Table 1.

Mouse #	Genotype GHL	rho	ONL Thickness (rows)	OS:IS ratio
J88	–	+/+	10	2:1
J83	+	+/+	9	1.6:1
J84	+	+/+	10	
J81	–	+/–	10	1.8:1
J79	+	+/–	5	0.2:1
J85	+	+/–	4–5	
J80	–	–/–	10	0:1
J82	+	–/–	2–3	0:0.2
J86	+	–/–	1–2	
J87	+	–/–	2–3	

GHL = mutant opsin transgene; rho = endogenous opsin gene; ONL = outer nuclear layer; OS:IS = ratio of outer segment length to inner segment length.

IS ratios were 2:1, 1.8:1, and 0:1, respectively. When the transgene was expressed on a wild-type background (D), retinas had, on average, 9 rows of photoreceptor nuclei with outer segments disorganized and shorter than the norm. Retinas of littermates GHL$^+$, rho$^{+/-}$ (E) revealed 4–5 rows of photoreceptor nuclei with very short and disorganized outer segment material whereas mice GHL$^+$, rho$^{-/-}$ (F) had 1–3 rows of photoreceptor nuclei with no outer segments. The OS:IS ratios were 1.6:1, 0.2:1, and 0:0.2, respectively. These data show that photoreceptor cell degeneration became more severe with the loss of each functional wild-type opsin allele.

4.3. Assessment of Visual Function by Electroretinography

Scotopic ERG amplitudes of the GHL$^+$, rho$^{-/-}$ mouse were already very low at postnatal day 30 (Fig. 4). Amplitudes of the highest intensity of the scotopic recording (1 cdsm^{-2}), the second oscillatory potential (usually the highest), and the cone responses were plotted in Fig. 5. Measurements from individual rho$^{-/-}$ and rho$^{+/-}$ mice, as well as GHL$^+$, rho$^{-/-}$, and GHL$^+$, rho$^{+/+}$ mice, were compared to earlier measurements of five mice from the original transgenic strain (C57BL/6), that is, with and without the transgene on wild-type background. The data showed that the transgene combined with a homologous rhodopsin knockout leads to a severe loss of rod function. There was no detectable rod function at all. The same was true for the scotopic oscillatory potentials. In contrast, the cone response was well detectable although attenuated. Consistent with published findings,[8] the rho$^{-/-}$ mice also showed a severe attenuation of the scotopic response at PN 30; the clearly detectable response may be a contribution from cone potentials. Photopic function appeared unaffected. At this early time point there appeared no appreciable difference in retinal function among the heterozygous rhodopsin knockout mouse, GHL$^+$, rho$^{+/+}$ mouse, and wild-type mouse, probably due to the slow rate of the degeneration.

4.4. Fundus Observations

Fundus morphology was examined using the same litter of mice used to study ERGs. At postnatal day 30, the fundus of each of seven littermates was visualized by conventional ophthalmoscopy using a 90 diopter lens. No visible changes were observed

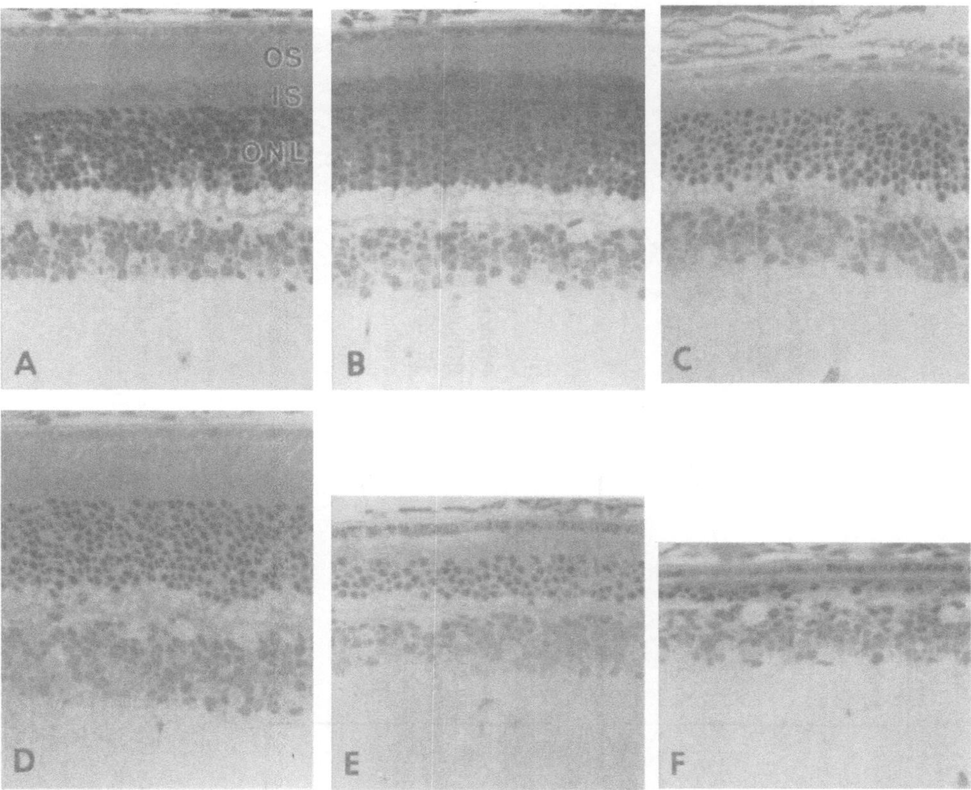

Figure 3. Retinal histology of A) $rho^{+/+}$, B) $rho^{+/-}$, and C) $rho^{-/-}$ mice at PN 30 (top row). Retinas of littermates carrying the transgene (bottom row): D) GHL^{+}, $rho^{+/+}$, E) GHL^{+}, $rho^{+/-}$, and F) GHL^{+}, $rho^{-/-}$. In each vertical pairing, the presence of the GHL transgene accelerated the rate of degeneration. Abbreviations: outer segment layer, OS; inner segment layer, IS; outer nuclear layer, ONL.

among the four genotypes represented. However, the fundus pictures differed upon reexamination at PN55. In the GHL^{+}, $rho^{-/-}$ mouse, the pigment epithelium had become mottled, especially in the periphery, with fine and coarse pigment irregularities (see Fig. 6, middle). Changes were undetectable in the $rho^{-/-}$ or GHL^{+}, $rho^{+/+}$ mice. Thus, incremental loss of functional wild-type opsin genes in the presence of this mutant transgene appears correlated with more extensive photoreceptor degeneration, attenuation of photoreceptor function and pigmentary alteration.

5. DISCUSSION

Parameters that define the rate at which a retinal degeneration progresses are poorly understood. The little that is known derives from transgenic mouse studies. One important parameter is the total level of opsin that is present. It has been shown previously that even over-expression of wild-type opsin in a transgenic mouse can cause a retinal degeneration.[11] Another parameter is the amount of mutant opsin protein expressed, which can be affected by transgene copy number and integration site. Dif-

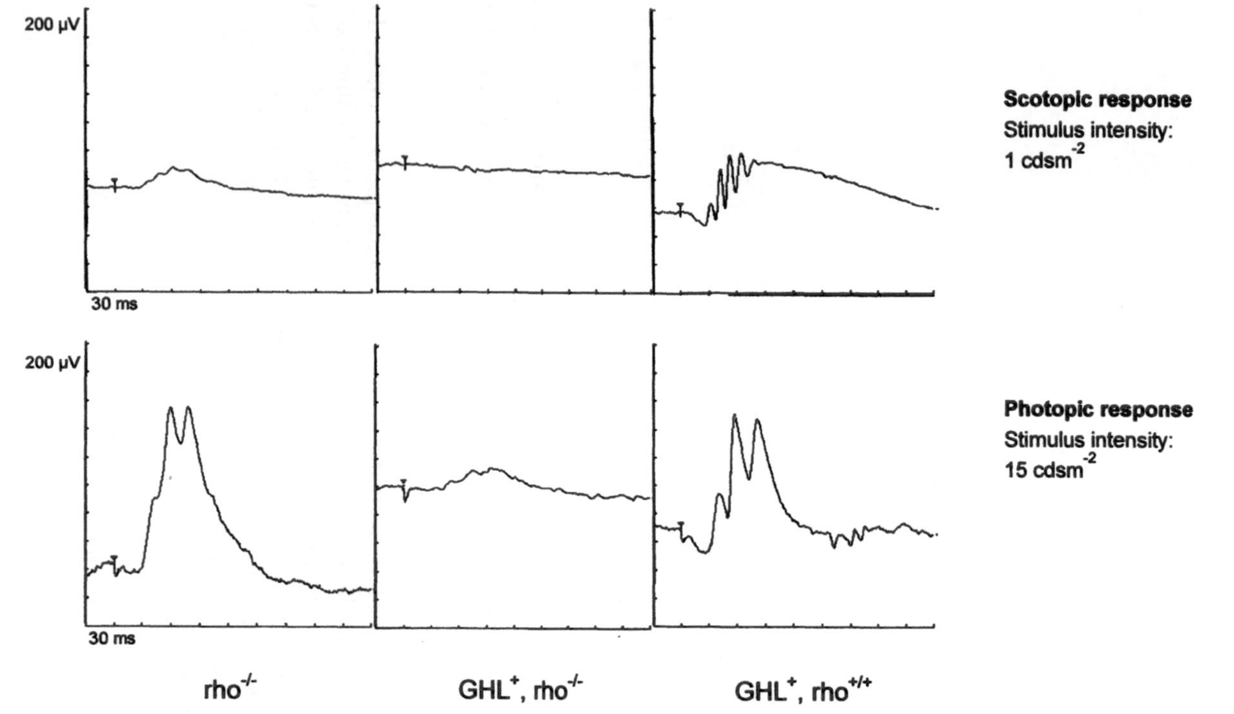

Figure 4. Original ERG recordings. Top row: Scotopic response at 1 $cdsm^{-2}$ of the GHL^+, $rho^{+/+}$ eye does not differ from a recording of a normal mouse eye. The dominant oscillatory potentials are typical for the mouse ERG. Bottom row: Photopic response with single flash stimulation (15 $cdsm^{-2}$). Recordings of $rho^{-/-}$ mice and GHL^+, $rho^{+/+}$ mice do not differ from the norm.

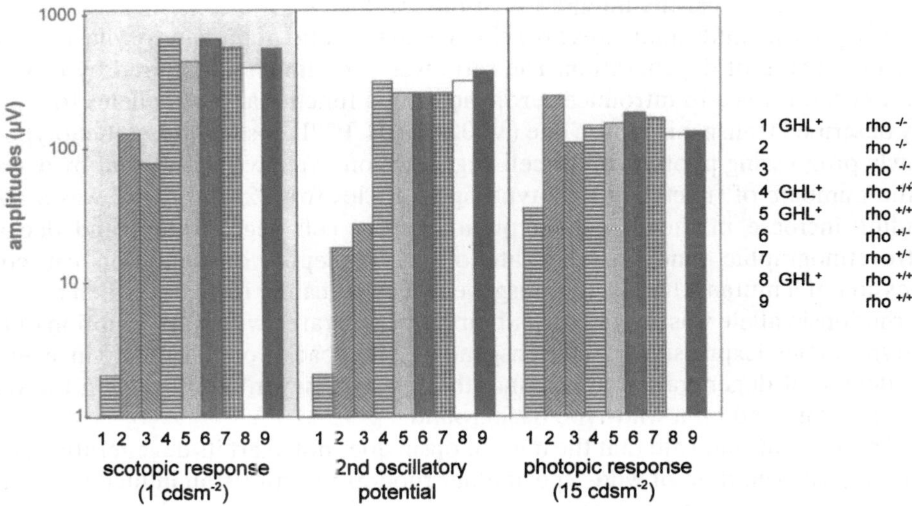

Figure 5. Amplitude plots of three different ERG recordings. Data of lanes 1–7 were from individual littermate mice while, for comparison, the GHL$^+$, rho$^{+/+}$(lane 8), and wild-type (lane 9) data were each the average of 5 age-matched mice of the original C57BL/6 strain (standard deviation in GHL$^+$, rho$^{+/+}$: 75, 57, and 28 μV respectively; in wild-type, 64, 79, and 24 μV respectively).

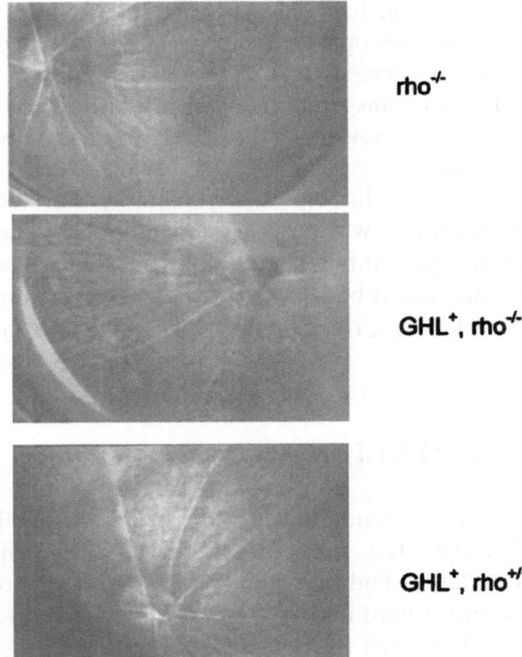

Figure 6. Scanning laser ophthalmoscope pictures of the mouse fundus at PN55. Picture quality is compromised by the small dimension of the mouse eye. The GHL$^+$, rho$^{-/-}$ fundus shows mottling of the pigment epithelium while those of rho $^{-/-}$ mice and GHL$^+$. rho$^{+/+}$ mice appear normal.

ferent transgenic mouse lines expressing the same transgene exhibit different rates and degrees of degeneration, which is attributed to differences in levels of transgene expression arising from transgene integration at different sites.

The present study addresses how changes in the ratio of mutant to wild-type opsin affect the degree of degeneration. The ratio was experimentally altered by using rod opsin knockout mice to introduce zero, one or two functional opsin alleles to a previously described transgenic mouse line (V20G, P23H, P27L opsin) whose phenotype was a slowly progressing photoreceptor cell degeneration. We have shown that by decreasing the numbers of functional wild-type opsin alleles from 2 to 0, there was a corresponding increase in the degree of photoreceptor cell degeneration and decay of electroretinographic function. The rate of photoreceptor degeneration caused by expression of a mutant rhodopsin transgene was significantly increased when one wild-type rhodopsin allele was disrupted, and further accelerated with the disruption of both wild-type alleles. Expression of the transgene on a null background resulted in an equivalent degree of degeneration at one month as seen at seven months when the transgene was expressed on a wild-type background.

These results indicate that the mutant opsin does not exert its degenerative effects autonomously. The ratio of wild-type: mutant rhodopsin expression influences the time course of photoreceptor cell degeneration. Acceleration of degeneration occurring with the disruption of one wild-type opsin allele suggests that the mutant rhodopsin may interfere with the normal transport or folding of wild-type rhodopsin. Alternatively, the differential rates of degeneration may reflect a mass-action effect. In the presence of higher levels of wild-type opsin, the wild-type opsin may be preferentially incorporated into disk membranes, thereby producing membranes more closely approximating the composition of normal membranes.

That changes in the ratio of mutant to wild-type opsin affect the rate of degeneration have also been demonstrated in transgenic rats expressing a mutant P23H mouse opsin[12]. In those studies, ribozymes targeted specifically to the mutant P23H messenger RNA were injected via an adeno-associated virus vector into retinas of transgenic rats. Degeneration in treated transgenic rats was decreased, presumably by lowering the level of mutant opsin mRNA and its associated mutant protein.

Our studies raise the possibility that decreasing the ratio of mutant to wild-type opsin may slow the rate of degeneration. This could be done by enhancing wild-type opsin expression in degenerating retinas. The advantage of this approach is that a single therapy could be used across a wide spectrum of mutant opsin degenerations, instead of designing a ribozyme specific for each opsin mutant.

ACKNOWLEDGMENTS

These studies were supported by the National Eye Institute (EY08123, WB; EY12008, JL), the Foundation Fighting Blindness (JL), a center grant from the Foundation Fighting Blindness (WB), Research to Prevent Blindness (Career Development Award and James S. Adams Scholars Award, JL), Fight-for-Sight (JL), awards from Research to Prevent Blindness, Inc. (Department of Ophthalmology, University of Utah; and Senior Investigator Award, WB) and the Deutsche Forschungsgemeinschaft (KR).

REFERENCES

1. T.P. Dryja and T. Li, 1995, Molecular genetics of retinitis pigmentosa, *Hum. Mol. Genet.* **4**:1739–1743.
2. J. Lem and C.L. Makino, 1996, Phototransduction in transgenic mice, *Curr. Opin. Neurobiol.* **6**:453–458.
3. K. Gregory-Evans and S.S. Bhattacharya, 1998, Genetic blindness: current concepts in the pathogenesis of human outer retinal dystrophies, *Trends Genet* **14**:103–108.
4. M.I. Naash, J.G. Hollyfield, M.R. Al-Ubaidi, and W. Baehr, 1993, Simulation of human autosomal dominant retinitis pigmentosa in transgenic mice expressing a mutated murine opsin gene, *Proc. Natl. Acad. Sci. U.S.A.* **90**:5499–5504.
5. X.R. Liu, T.H. Wu, S. Stowe, A. Matsushita, K. Arikawa, M.I. Naash, and D.S. Williams, 1997, Defective phototransductive disk membrane morphogenesis in transgenic mice expressing opsin with a mutated N-terminal domain, *J. Cell Sci.* **110**:2589–2597.
6. D.J. Roof, M. Adamian, and A. Hayes, 1994, Rhodopsin accumulation at abnormal sites in retinas of mice with a human P23H rhodopsin transgene, *Invest. Ophthalmol. Vis. Sci.* **35**:4049–4062.
7. X. Liu, P. Garriga, and H.G. Khorana, 1996, Structure and function in rhodopsin: correct folding and misfolding in two point mutants in the intradiscal domain of rhodopsin identified in retinitis pigmentosa, *Proc. Natl. Acad. Sci. U.S.A.* **93**:4554–4559.
8. M.M. Humphries, D. Rancourt, G.J. Farrar, P. Kenna, M. Hazel, R.A. Bush, P.A. Sieving, D.M. Sheils, N. McNally, P. Creighton, A. Erven, A. Boros, K. Gulya, M.R. Capecchi, and P. Humphries, 1997, Retinopathy induced in mice by targeted disruption of the rhodopsin gene, *Nature Genet.* **15**:216–219.
9. J. Lem, P.D. Calvert, B. Kosaras, D.A. Cameron, M. Nicolo, C. Makino, and R.L. Sidman, 1999, Morphological, physiological and biochemical changes in rhodopsin knockout mice, *Proc. Natl. Acad. Sci. U.S.A.*, submitted.
10. G.H. Travis, 1997, Insights from a lost visual pigment [news; comment], *Nat. Genet* **15**:115–117.
11. J.E. Olsson, J.W. Gordon, B.S. Pawlyk, D. Roof, A. Hayes, R.S. Molday, S. Mukai, G.S. Cowley, E.L. Berson, and T.P. Dryja, 1992, Transgenic mice with a rhodopsin mutation (Pro23His): A mouse model of autosomal dominant retinitis pigmentosa, *Neuron* **9**:815–830.
12. A.S. Lewin, K.A. Drenser, W.W. Hauswirth, S. Nishikawa, D. Yasumura, J.G. Flannery, and M.M. LaVail, 1998, Ribozyme rescue of photoreceptor cells in a transgenic rat model of autosomal dominant retinitis pigmentosa, *Nat. Med* **4**:967–971.

11

RHODOPSIN C-TERMINAL SEQUENCE QVS(A)PA DIRECTS ITS SORTING TO THE ROS IN RETINAL PHOTORECEPTORS

Dusanka Deretic,[1,2] Sonia Schmerl,[1] Paul A. Hargrave,[3] Anatol Arendt,[3] and J. Hugh McDowell[3]

[1] Department of Ophthalmology
[2] Anatomy and Cell Biology
University of Michigan
Ann Arbor, Michigan 48105
[3] Department of Ophthalmology
University of Florida
Gainesville, Florida 32610

ABSTRACT

Several mutations that cause severe forms of autosomal dominant retinitis pigmentosa (ADRP) cluster in the C-terminal region of rhodopsin. Recent studies have implicated the C-terminal domain of rhodopsin in its trafficking on specialized post-Golgi membranes to the rod outer segment (ROS) of the photoreceptor cell. In addition, aberrant subcellular localization of rhodopsin has been observed in transgenic animals carrying C-terminal mutations. To test if sequence within the C-terminal domain regulates rhodopsin sorting to the ROS, we tested the effects of synthetic peptides that mimic this domain on intracellular trafficking of rhodopsin reconstituted in the frog retinal cell free system. The rhodopsin C-terminal sequence QVS(A)PA is highly conserved among different species. Synthetic peptides corresponding to the C-terminal of frog (AA 330–354) and bovine (AA 324–348) rhodopsin inhibited post-Golgi trafficking by 60% and 50%, respectively, and arrested newly synthesized rhodopsin in the trans-Golgi network (TGN). Peptides corresponding to the cytoplasmic loops and several control peptides had no effect. To assess the role of the last 5 amino acids QVS(A)PA in rhodopsin trafficking, and to model three naturally occurring mutations: Q344ter (lacking the last 5 amino acids QVAPA), V345M and P347S, we introduced equivalent substitutions into the frog C-terminal peptide. Each of these substitutions completely abolished the inhibitory activity of the peptides. Our data could help to explain detrimental effect of mutant rhodopsin in patients with ADRP. We propose that amino acids QVS(A)PA comprise a sorting signal that is recognized by specific factor(s) in the trans-Golgi network. A lack of recognition of this sequence, due to mutations in the last five amino acids causing ADRP most likely results in abnormal post-Golgi membrane formation and in an aberrant subcellular localization of rhodopsin.

INTRODUCTION

In the past several years a number of research groups has gathered evidence for the crucial role of aberrant rhodopsin trafficking in some forms of retinal degenerations[1–5] and for an absolute requirement for rhodopsin in ROS morphogenesis.[6–8] Mutations found in the rhodopsin gene account for 10% of all cases and represent the most common known cause of retinal degenerations classified as retinitis pigmentosa.[9] More than 70 mutations cause the autosomal dominant form of the disease (ADRP).[2] This indicates that the presence of the mutant protein initiates intracellular interactions undesirable for the photoreceptor cells. The ultimate response of the injured cell is the initiation of an apoptotic cascade which leads to cell death and retinal degeneration.[10]

Two converging approaches have led to our present understanding of the role of rhodopsin in the regulation of its sorting in photoreceptor cells. The research in this direction was driven by the need to explain the detrimental effects of rhodopsin mutations in patients with ADRP. Mutant rhodopsins responsible for ADRP were transfected into an embryonic human kidney cell line, and some of the mutants were transported inefficiently to the plasma membrane.[11] These findings raised the possibility that defective transport may be partially responsible for the retinal injury in some patients with ADRP. Mutations clustered at the C-terminal of rhodopsin appear to cause particularly severe forms of the disease.[12,13] One "hot spot" involves the codon for the next to the last amino acid (proline) where every possible mutation that changes the amino acid at that position (a total of six) leads to retinal degeneration and blindness.[14,15] Mutations in V345[16] and truncations of the last five amino acids Q344ter[17,18] all lead to retinal degenerations, and yet these amino acids are not implicated in any light-dependent function.[3,19] Modeling of C-terminal mutations in transgenic animals has indicated that mutant rhodopsin is defective in localization to the rod outer segments.[2,3,20] Mice expressing Q344ter transgene (missing amino acids QVAPA) accumulated truncated rhodopsin in the plasma membrane of photoreceptor inner segments,[3] while mice carrying the P347S mutation accumulated extracellular vesicles containing rhodopsin.[2] Abnormal rhodopsin localization was also reported in transgenic pigs expressing P347L mutation.[20] Studies in transgenic animals suggest that rhodopsin transport may be defective but have not defined the specific step affected by the mutations.

A different approach that we designed to study intracellular regulators of rhodopsin trafficking has, surprisingly, identified a crucial role of the C-terminal of rhodopsin in its post Golgi transport.[21] We have developed a frog retinal cell-free system derived from photoreceptor-enriched post-nuclear supernatant (PNS) that supports formation of rhodopsin-bearing post-Golgi membranes in an ATP, GTP, and cytosol-dependent manner. Using this cell free assay we found that anti-rhodopsin C-terminal mAb 11D5 inhibits post-Golgi trafficking and arrests rhodopsin in the trans-Golgi network (TGN).[21] TGN is the tubular network in close proximity to the trans-Golgi cisternae and represents the central sorting station of the cell where proteins and lipids destined for different subcellular domains are segregated from each other and sorted into post-Golgi carriers.[22] Sorting of membrane proteins destined for different subcellular compartments of polarized epithelial cells is based on the presence of specific, dominant determinants in their cytoplasmic domains that interact with regulatory proteins,[23,24] or on the interaction with specific lipids either by direct partitioning or by binding of their oligosacharides to lipid-associated lectins.[25,26] Homologous sorting pathways also exist in neurons.[27] Therefore, arrest of trafficking in the photoreceptor TGN in the presence of anti-rhodopsin C-terminal antibody suggests that antibody bound to rhodopsin interferes with its interaction with the specific recognition factor.[21]

To delineate the potential regulatory sequence within the C-terminus of rhodopsin and model the effects of severe retinitis pigmentosa alleles on rhodopsin sorting we next used synthetic peptides as competitive inhibitors of rhodopsin trafficking in the frog retinal cell-free system.[28] Here we summarize that study and report that the amino acids QVS(A)PA, the five C-terminal residues, and valine (V) and proline (P) in particular, play a critical role in regulating rhodopsin sorting into specific post-Golgi membranes that bud from the TGN. These observations indicate that the defective recognition of mutant rhodopsin in patients with ADRP may result specifically in abnormal post-Golgi membrane formation, or mistargeting of mutant rhodopsin.

MATERIALS AND METHODS

Southern leopard frogs, *Rana berlandieri*, were purchased from Rana Co. (Brownsville, TX). [^{35}S]-Express protein labeling mixture (1000 Ci/mmol), was from DuPont NEN, ATP, creatine phosphate, creatine phosphokinase (800 U/mg) and hexokinase (450 U/mg) were from Boehringer-Mannheim.

Peptide Synthesis

Peptides corresponding to the frog[29] and bovine rhodopsin sequence[30,31] were synthesized by solid phase technique using Fmoc derivatives of amino acids as described.[28] The following peptides were tested: bovine AA 3–14, 63–75, 141–153, 234–245, 231–252, 310–321, 324–348, 330–348, 331–342, and frog AA 330–354, 330–349, 330–354 V351M, and 330–354 P353S.

Pulse-Labeling of Frog Retinas, Preparation of Photoreceptor-Enriched Postnuclear Supernatant (PNS), and in Vitro Incubation of PNS

Pulse-labeling, retinal fractionation and preparation of retinal PNS were performed as described.[21] Radiolabeled PNS was preincubated with synthetic peptides (dissolved in 50 μl of 10 mM Hepes-KOH, pH 7.0) for 30′ on ice. The standard assay for cell-free post-Golgi membrane formation was as previously described.[21,28] Gel electrophoresis and quantitative analysis of radiolabeled rhodopsin in retinal subcellular fractions was performed as described.[21,28]

RESULTS

Rhodopsin Trafficking in the Retinal Cell-Free System

In this study we employed frog retinal cell-free system derived from photoreceptor-enriched post-nuclear supernatant (PNS).[21] The kinetics of intracellular trafficking of rhodopsin in this cell-free system parallels that in vivo.[21] After one hour of pulse labeling in vivo, movement of radiolabeled rhodopsin through the subcellular compartments during in vitro chase is monitored by the separation of biosynthetic membranes on linear sucrose density gradients. We have previously determined the identity and the distribution of photoreceptor subcellular compartments in these gradients by their buoyant density, morphology, immunoreactivity with antibodies against compartment specific markers, sensitivity to brefeldin A (BFA), by their galactosyltransferase and sialyltransferase activity, by their lipid composition and by their profile of membrane-associated small GTPases of the rab family.[21,32–38] Newly synthesized rhodopsin is predominantly found in the Golgi after one hour of pulse labeling in vivo, with only a small fraction remaining in the endoplasmic reticulum (ER).[21,32] The kinetics of the transfer of radiolabeled rhodopsin from the ER to the Golgi complex and its export from the Golgi indicates that the rate of transport through the Golgi is significantly slower than the ER-to-Golgi transport.[32] Therefore, the majority of radiolabeled protein is localized in the Golgi at the onset of a typical in vitro chase experiment and any subsequent redistribution of rhodopsin predominantly corresponds to its Golgi-to-post-Golgi trafficking.

Figure 1 illustrates the ATP-dependent transfer of radiolabeled rhodopsin to the post-Golgi fraction during incubation in the cell-free system. In this assay, isolated frog retinas are first pulse-labeled for one hour, so that the bulk of the newly synthesized rhodopsin is localized in the Golgi complex identified by the galactosyltransferase activity of the trans-Golgi cisternae, as previously described.[32] Following two hours of chase in vitro, in the presence of an ATP-depleting system, radiolabeled rhodopsin remained in the Golgi, indicating a complete arrest of trafficking in the absence of ATP as previously reported[21] and shown in Figure 1. In the presence of ATP, and an ATP-regenerating system, intracellular trafficking continued in vitro, the Golgi and ER content of newly synthesized rhodopsin was significantly reduced, TGN membranes became enriched in radiolabeled rhodopsin and post-Golgi membranes became highly labeled as shown in Figure 1, consistent with the previously reported findings.[21]

Synthetic Peptides Corresponding to the C-Terminal of Rhodopsin Inhibit Its Post-Golgi Trafficking

Post-Golgi trafficking in the frog retinal cell-free system is inhibited by anti-rhodopsin mAb 11D5, and its Fab fragments, and rhodopsin is arrested in the TGN.[21] The antigenic epitope of inhibitory mAb 11D5 has been mapped within the nine amino acids comprising the C-terminal of rhodopsin.[32] Figure 2 schematically represents the transmembrane model of rhodopsin showing the mAb 11D5 antigenic site as well as the locations of mutations in the C-terminal domain identified in patients with ADRP.

To study the role of the C-terminal domain we synthesized peptides corresponding to the cytoplasmic or intradiscal domains of rhodopsin (as shown in Fig. 2) and tested their effect on rhodopsin trafficking in vitro. Progression of radiolabeled

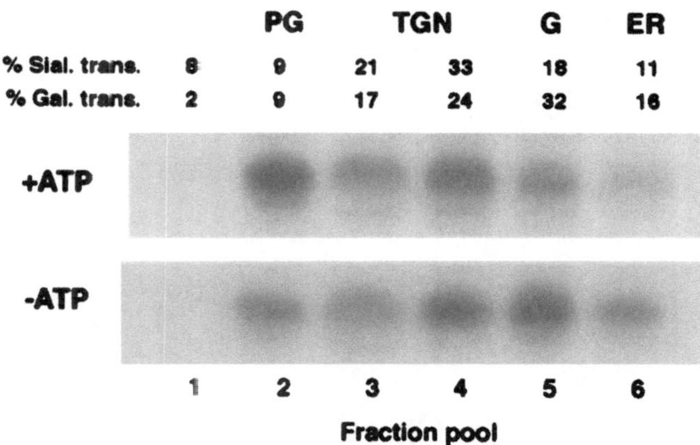

Figure 1. Rhodopsin trafficking to the post-Golgi membranes in the retinal cell-free system. Isolated frog retinas were pulse-labeled for 60 min. and radiolabeled postnuclear supernatant was incubated for 2 hours in the presence or absence of ATP. Following sucrose gradient separation subcellular fractions were pooled according to the kinetics of their acquisition of newly synthesized rhodopsin as previously described.[21] Aliquots corresponding to radiolabeled membrane proteins from two retinas were separated by SDS-PAGE and autoradiographed. In the absence of ATP (−ATP), rhodopsin remains in the Golgi (G) during in vitro chase indicating that the assay is ATP-dependent.[21] Upon addition of ATP (+ATP), radiolabeled rhodopsin exits the Golgi and appears in the TGN and in low density post-Golgi fractions (PG). The distribution of Golgi and TGN membranes has been determined by their galactosyltransferase or sialyltransferase activities, respectively.[21,32,33] Activities of pooled membrane fractions are indicated. ER- endoplasmic reticulum. Reproduced from Deretic et al., 1998.[28]

rhodopsin through the biosynthetic compartment was significantly affected by the addition of the 50 μM frog C-terminal peptide (AA 330–354) (Fig. 3). Newly synthesized rhodopsin was predominantly associated with the membranes identified by their sialyltransferase activity as the TGN, and its appearance in the post-Golgi membranes was reduced by 60%. The distribution of radiolabeled rhodopsin closely resembled its previously described distribution in the presence of mAb 11D5,[21] suggesting a common mechanism for inhibition at the level of the TGN. Unlike the inhibition caused by the absence of ATP that reflects an early and complete block of trafficking in all compartments during in vitro chase, in the presence of rhodopsin C-terminal peptide some rhodopsin trafficking initially occurred (Fig. 3). Approximately 25% of the radiolabeled rhodopsin redistributed from the ER and the Golgi at the end of the chase period. The rate of clearance from these compartments was comparable to the control with ATP. This suggests that rhodopsin trafficking at the level of the ER and the Golgi is unaffected by the C-terminal peptide and that its inhibitory effect is exhibited once rhodopsin reaches the TGN.

Since the C-terminal domain is highly conserved among rhodopsin species[39] we tested the effect of bovine peptides (AA 324–348) and (AA 330–348) on rhodopsin trafficking in the frog retinal system. At the same concentration (50 μM) these peptides inhibited post-Golgi trafficking by ~50%[28] (data not shown). The primary sequences of frog and bovine rhodopsin in the C-terminal domain are highly homologous, but frog rhodopsin contains two inserts, one of one and one of five amino acids[29] (see Fig. 2). The three dimensional structure of the bovine C-terminal domain has been determined using NMR.[40,41] The comparison of frog and bovine sequence suggests that, if the frog peptide assumes a similar conformation, the conserved amino acids cluster on one face of this domain.

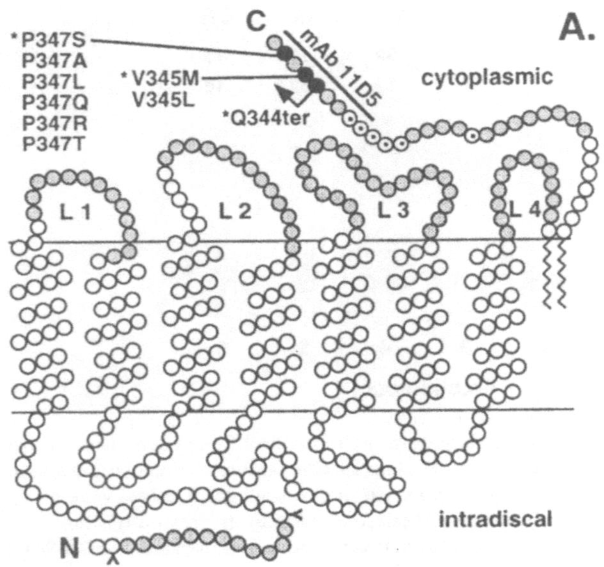

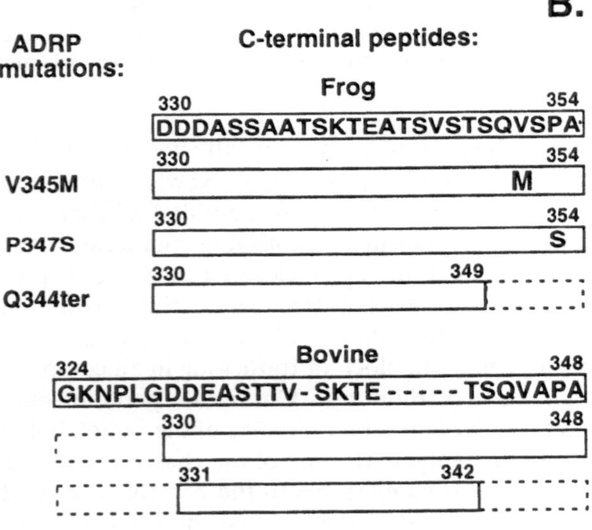

Figure 2. Model of rhodopsin showing locations of ADRP mutations in the C-terminal domain, the mAb 11D5 antigenic site, locations and sequences of the peptides used in this study. A. Peptides synthesized for this study are highlighted in gray in the transmembrane model of rhodopsin. L1-L4 are cytoplasmic loops. C-terminal amino acids mutated in retinitis pigmentosa are shown in black. Mutations modeled in this study are indicated with an asterisk. B. Sequences of the C-terminal peptides from frog and bovine rhodopsin used in this study. Since frog rhodopsin contains two (one and five amino acids) insertions in this region these are represented by dashes in the bovine peptide sequence and with the dotted open circles in the model. Reproduced from Deretic et al., 1998.[28]

Peptides Corresponding to the Cytoplasmic Loops of Rhodopsin Have No Effect on Its Trafficking

Since the C-terminal comprises only a part of the cytoplasmic surface of rhodopsin we tested if neighboring cytoplasmic loops contribute to the sorting signal. Peptides corresponding to the third cytoplasmic loop (AA 234–245) and (AA 231–252), known to contain sites of interaction with transducin[42] arrestin[43] and rhodopsin kinase,[44] the second cytoplasmic loop (AA 141–153) and a control NH_2 terminal (intradiscal) domain (AA 3–14) had no effect on rhodopsin trafficking (data not shown). Peptides corresponding to the first (AA 63–75) and the fourth loop (AA 310–321) showed minimal inhibition (~20%). Since these loops are in close proximity to the C-terminal domain in the three dimensional structure[45] it is possible that they make some contribution to the recognition site.

ADRP Mutations Q344ter (Lacking Amino Acids QVAPA), V345M, and P347S Introduced into the Frog C-Terminal Peptide Eliminate Its Inhibitory Effect on Rhodopsin Trafficking

The five amino acids QVS(A)PA are the most conserved in the C-terminal domain. To test their importance we modeled a naturally occurring stop codon mutation that removes the last five amino acids of rhodopsin Q344ter.[17,18] We synthesized a truncated frog C-terminal peptide (AA 330–349) and followed its effect on post-Golgi trafficking of rhodopsin. As shown in Fig. 3, deletion of QVSPA from the inhibitory C-terminal peptide abolished its effect on rhodopsin trafficking. To further define amino acids critical for the recognition of the C-terminal sequence of rhodopsin, we synthesized two peptides corresponding to amino acids 330–354 of frog rhodopsin with single amino acid substitutions of V351M and P353S. These substitutions correspond to mutations V345M and P347S found in patients with ADRP.[15,16,46] Addition of these peptides to the cell-free system had no effect on rhodopsin trafficking while in the same series of experiments the parent peptide 330–354 reduced its exit from the TGN by 60% (Fig. 3). The subcellular distribution of radiolabeled rhodopsin paralleled its distribution in the control. This suggests that C-terminal peptides containing substitutions in amino acids QVS(A)PA equivalent to ADRP mutations can not successfully compete with rhodopsin for the binding to the specific recognition factor(s) and therefore can not affect its intracellular trafficking.

DISCUSSION

We defined the C-terminal of rhodopsin as the domain that regulates its post-Golgi trafficking. Specifically, the rhodopsin C-terminal sequence QVAPA (QVSPA in frog) that is highly conserved among different species, is absolutely critical for the regulatory function of this domain. Our data suggest that the last five amino acids of rhodopsin contain a signal that is specifically recognized in the TGN. Amino acids valine (–3 from the C terminal) and proline (–1) are absolutely required for signal recognition, while serine (–2) can be substituted by alanine as evidenced by a successful competition of the bovine peptide with frog rhodopsin.

In the present study we employed the frog retinal cell-free system and mapped the domain that directs intracellular trafficking using synthetic peptides from the C-terminal of frog and bovine rhodopsin. The majority of the peptides that we have tested in this assay have been successfully used in other studies delineating functional domains of rhodopsin.[42,47] Moreover, the NMR analysis of the bovine C-terminal peptide revealed a compact globular structure that could form a structural domain of rhodopsin.[40,41] These peptides arrested rhodopsin in the TGN in a manner indistinguishable from that of the inhibitory mAb 11D5. This suggests that the recognition of the C-terminal sequence occurs in the TGN, prior to post-Golgi membrane formation. The addition of the C-terminal peptides interferes with these processes most likely by competing for the sorting factors and affecting the rate of rhodopsin exit from the TGN. Peptides lacking the QVS(A)PA sequence, or containing single amino acid substitutions mimicking ADRP mutations within this sequence, do not inhibit post-Golgi

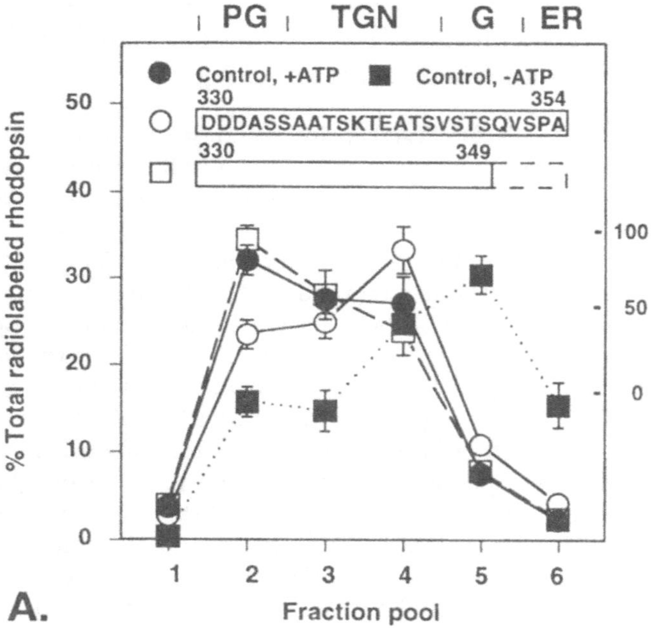
PG
TGN
G
ER
Control, +ATP
Control, -ATP
330
354
DDDASSAATSKTEATSVSTSQVSPA
330
349
% Total radiolabeled rhodopsin
Fraction pool
A.

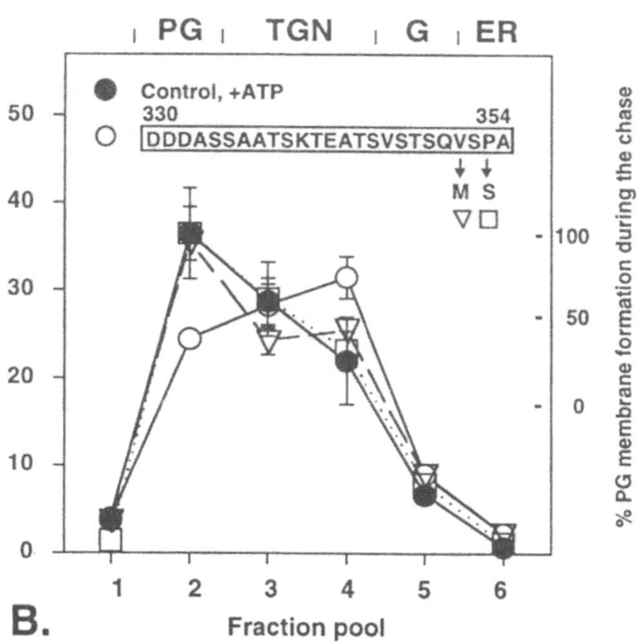
PG
TGN
G
ER
Control, +ATP
330
354
DDDASSAATSKTEATSVSTSQVSPA
M
S
% PG membrane formation during the chase
Fraction pool
B.

trafficking of rhodopsin. This indicates that the recognition factor(s) that presumably binds to rhodopsin in the TGN either binds directly to this sequence, or that this sequence is critical for maintaining the proper three dimensional conformation of the remainder of the C-terminal domain.

The consequence of the lack of, or the "attenuation" of the putative sorting signal at the C-terminal of rhodopsin, as seen in patients with ADRP, could be delivery of rhodopsin to a different subcellular compartment. It is possible that the lack of the recognition of the sorting signal causes entry into a different population of post-Golgi carriers destined for the inner segment plasma membrane or the synapse. While the existence of these pathways in the photoreceptor cell is purely speculative, evidence from transgenic animals suggests that mutant rhodopsin is indeed found in these domains.[3,20] Moreover, loss of cell polarity manifested in neurite sprouting in patients with ADRP, and rhodopsin was also present in these newly formed structures.[1] These abnormal photoreceptor processes found in the inner retina in patients with rhodopsin mutations are thought to interfere with signal transmission and cause ERG defects.[48] It is possible that in conjunction with heterotrimeric GTPases, not normally found in the same membrane, rhodopsin can modulate the cAMP cascade which is the backbone of intracellular signaling.[19,49] This could possibly be the path that leads to the initiation of apoptotic pathways since high levels of cAMP have been found in photoreceptors of transgenic mice carrying rhodopsin P347S mutation and undergoing apoptosis.[19] Longer transit time and exposure to glycosyl transferases in the Golgi, may result in aberrant oligosacharide processing. These oligosacharides may be recognized by the lectins present in the TGN and used as "tags" for sorting into a different population of post-Golgi membranes.

Based on the data presented here, we propose that the polarized sorting of rhodopsin at the exit from the TGN is mediated by the interaction of the C-terminal sequence QVS(A)PA with a recognition domain of a regulatory protein that allows entrance of rhodopsin molecules into the specific post-Golgi carriers targeted to the ROS. Our data suggest that the defective recognition of mutant rhodopsin in patients with ADRP interferes with proper sorting of rhodopsin and may result in abnormal post-Golgi membrane formation, longer transit time through the Golgi, or mistargeting of mutant rhodopsin to the inner segment and the synapse which may destabilize the structure and, consequently, affect the function of photoreceptor cells.

Figure 3. Inhibition of post-Golgi rhodopsin trafficking with C-terminal synthetic peptides in the frog retinal cell-free system. Deletion of QVSPA and substitutions that mimic ADRP mutations V345M and P347S eliminate the inhibitory activity of the C-terminal peptide. A. Radiolabeled frog retinal PNS was preincubated with the frog rhodopsin C-terminal peptides prior to cell-free chase in the presence of ATP. The distribution of radiolabeled rhodopsin was measured in a PhosphorImager in three separate experiments. The data are presented as the means ± S.E. In the control sample without peptide added (control, +ATP), in addition to the post-Golgi membranes already formed in the isolated retinas during the pulse (~15%)[21] an additional ~16% of total radiolabeled rhodopsin accumulates in the post-Golgi membrane fraction (fraction pool 2) during the cell-free chase. In the absence of ATP (−ATP) further trafficking does not occur. ATP-dependent appearance of rhodopsin in post-Golgi membranes during the chase is inhibited by 60% in the presence of the 50 μM C-terminal peptide 330–354 and rhodopsin is arrested in the TGN. Truncation of the last five amino acids eliminates the inhibitory effect of the C-terminal peptide. Inhibition of rhodopsin trafficking is assessed over the −ATP background; the ordinate on the right refers only to the ATP-dependent appearance of rhodopsin in the post-Golgi membrane fraction, when this background is subtracted. B. The distribution of radiolabeled rhodopsin was measured in four separate experiments. Substitution of the amino acid V351M (−3 from the C-terminal) or P353S (−1) eliminates the inhibitory effect of the C-terminal peptide. Modified from Deretic et al., 1998.[28]

ACKNOWLEDGMENTS

We are grateful to Dr. David Papermaster for his support and helpful discussions. Supported by NIH grant EY-6891 and the Core Grant for Vision Research EY-07003 (D. D.) and EY-6226 and an unrestricted departmental award from Research to Prevent Blindness (P. A. H.).

REFERENCES

1. Li Z.Y., Kljavin I.J., Milam A.H. 1995. Rod photoreceptor neurite sprouting in retinitis pigmentosa. *J Neurosci.* **15**:5429–5438.
2. Li T., Snyder W.K., Olsson J.E., Dryja T.P. 1996. Transgenic mice carrying the dominant rhodopsin mutation P347S: evidence for defective vectorial transport of rhodopsin to the outer segments. *Proc Natl Acad Sci USA*. **93**:14176–14181.
3. Sung C.H., Makino C., Baylor D., Nathans J. 1994. A rhodopsin gene mutation responsible for autosomal dominant retinitis pigmentosa results in a protein that is defective in localization to the photoreceptor outer segment. *J Neurosci.* **14**:5818–5833.
4. Colley N.J., Cassill J.A., Baker E.K., Zuker C.S. 1995. Defective intracellular transport is the molecular basis of rhodopsin-dependent dominant retinal degeneration. *Proc Natl Acad Sci USA*. **92**:3070–3074.
5. Roof D.J., Adamian M., Hayes A. 1994. Rhodopsin accumulation at abnormal sites in retinas of mice with a human P23H rhodopsin transgene. *Invest Ophthalmol Vis Sci.* **35**:4049–4062.
6. Rosenfeld P.J., Cowley G.S., McGee T.L., Sandberg M.A., Berson E.L., Dryja T.P. 1992. A null mutation in the rhodopsin gene causes rod photoreceptor dysfunction and autosomal recessive retinitis pigmentosa. *Nat Genet.* **1**:209–213.
7. Humphries M.M., Rancourt D., Farrar G.J., Kenna P., Hazel M., Bush R.A., Sieving P.A., Sheils D.M., McNally N., Creighton P., Erven A., Boros A., Gulya K., Capecchi M.R., Humphries P. 1997. Retinopathy induced in mice by targeted disruption of the rhodopsin gene. *Nat Genet.* **15**:216–219.
8. Kumar J.P., Bowman J., O'Tousa J.E., Ready D.F. 1997. Rhodopsin replacement rescues photoreceptor structure during a critical developmental window. *Dev Biol.* **188**:43–47.
9. Berson E.L. 1996. Retinitis pigmentosa: unfolding its mystery. *Proc Natl Acad Sci USA*. **93**: 4526–4528.
10. Papermaster D.S., Windle J. 1995. Death at an early age. Apoptosis in inherited retinal degenerations. *Invest Ophthalmol Vis Sci.* **36**:977–983.
11. Sung C.H., Schneider B.G., Agarwal N., Papermaster D.S., Nathans J. 1991. Functional heterogeneity of mutant rhodopsins responsible for autosomal dominant retinitis pigmentosa. *Proc Natl Acad Sci USA*. **88**:8840–8844.
12. Berson E.L., Rosner B., Sandberg M.A., Weigel DiFranco C., Dryja T.P. 1991. Ocular findings in patients with autosomal dominant retinitis pigmentosa and rhodopsin, proline-347-leucine. *Am J Ophthalmol.* **111**:614–623.
13. Sandberg M.A., Weigel DiFranco C., Dryja T.P., Berson E.L. 1995. Clinical expression correlates with location of rhodopsin mutation in dominant retinitis pigmentosa. *Invest Ophthalmol Vis Sci.* **36**:1934–1942.
14. Sullivan L.S., Daiger S.P. 1996. Inherited retinal degeneration: exceptional genetic and clinical heterogeneity. *Mol Med Today*. **2**:380–386.
15. Dryja T.P., McGee T.L., Hahn L.B., Cowley G.S., Olsson J.E., Reichel E., Sandberg M.A., Berson E.L. 1990. Mutations within the rhodopsin gene in patients with autosomal dominant retinitis pigmentosa. *N Engl J Med.* **323**:1302–1307.
16. Berson E.L., Sandberg M.A., Dryja T.P. 1991. Autosomal dominant retinitis pigmentosa with rhodopsin, valine-345-methionine. *Trans Am Ophthalmol Soc.* **89**:117–130.
17. Sung C.H., Davenport C.M., Hennessey J.C., Maumenee I.H., Jacobson S.G., Heckenlively J.R., Nowakowski R., Fishman G., Gouras P., Nathans J. 1991. Rhodopsin mutations in autosomal dominant retinitis pigmentosa. *Proc Natl Acad Sci USA*. **88**:6481–6485.
18. Jacobson S.G., Kemp C.M., Sung C.H., Nathans J. 1991. Retinal function and rhodopsin levels in autosomal dominant retinitis pigmentosa with rhodopsin mutations. *Am J Ophthalmol.* **112**:256–271.

19. Weiss E.R., Hao Y., Dickerson C.D., Osawa S., Shi W., Zhang L., Wong F. 1995. Altered cAMP levels in retinas from transgenic mice expressing a rhodopsin mutant. *Biochem Biophys Res Commun.* **216**:755–761.
20. Li Z.Y., Wong F., Chang J.H., Possin D.E., Hao Y., Petters R.M., Milam A.H. 1998. Rhodopsin transgenic pigs as a model for human retinitis pigmentosa. *Investigative Ophthalmology & Visual Science.* **39**:808–819.
21. Deretic D., Puleo Scheppke B., Trippe C. 1996. Cytoplasmic domain of rhodopsin is essential for post-Golgi vesicle formation in a retinal cell-free system. *J Biol Chem.* **271**:2279–2286.
22. Griffiths G., Simons K. 1986. The trans Golgi network: sorting at the exit site of the Golgi complex. *Science.* **234**:438–443.
23. Matter K., Mellman I. 1994. Mechanisms of cell polarity: sorting and transport in epithelial cells. *Curr Opin Cell Biol.* **6**:545–554.
24. Schekman R., Orci L. 1996. Coat proteins and vesicle budding. *Science.* **271**:1526–1533.
25. Simons K., Ikonen E. 1997. Functional rafts in cell membranes. *Nature.* **387**:569–572.
26. Scheiffele P., Peränen J, Simons K. 1995. N-glycans as apical sorting signals in epithelial cells. *Nature.* **378**:96–98.
27. Dotti C.G., Simons K. 1990. Polarized sorting of viral glycoproteins to the axon and dendrites of hippocampal neurons in culture. *Cell.* **62**:63–72.
28. Deretic D., Schmerl S., Hargrave P.A., Arendt A., McDowell J.H. 1998. Regulation of sorting and post-Golgi trafficking of rhodopsin by its C-terminal sequence QVS(A)PA. *Proc Natl Acad Sci USA.* **95**:in press.
29. Pittler S.J., Fliesler S.J., Baehr W. 1992. Primary structure of frog rhodopsin. *FEBS Lett.* **313**:103–108.
30. Hargrave P.A., McDowell J.H., Curtis D.R., Wang J.K., Juszczak E., Fong S.L., Rao J.K., Argos P. 1983. The structure of bovine rhodopsin. *Biophys Struct Mech.* **9**:235–244.
31. Nathans J., Hogness D.S. 1983. Isolation, sequence analysis, and intron-exon arrangement of the gene encoding bovine rhodopsin. *Cell.* **34**:807–814.
32. Deretic D., Papermaster D.S. 1991. Polarized sorting of rhodopsin on post-Golgi membranes in frog retinal photoreceptor cells. *J Cell Biol.* **113**:1281–1293.
33. Deretic D., Papermaster D.S. 1993. Rab6 is associated with a compartment that transports rhodopsin from the trans-Golgi to the site of rod outer segment disk formation in frog retinal photoreceptors. *J Cell Sci.* **106**:803–813.
34. Deretic D., Papermaster D.S. 1993. Isolation of post-Golgi membranes transporting newly synthesized rhodopsin. In: Hargrave P.A., ed. Methods for the Study of Photoreceptor Cells. New York: Rockefeller University Press, 108–120. (Conn PM, ed. Methods in Neurosciences; vol 15).
35. Deretic D., Huber L.A., Ransom N., Mancini M., Simons K., Papermaster D.S. 1995. rab8 in retinal photoreceptors may participate in rhodopsin transport and in rod outer segment disk morphogenesis. *J Cell Sci.* **108**:215–224.
36. Deretic D., Papermaster D.S. 1995. The Role of Small G-Proteins in the Transport of Newly Synthesized Rhodopsin. In: Osborne N.N., Chader G.J., eds. Progress in Retinal and Eye Research. New York: Pergamon Press, 249–265. vol 14).
37. Rodriguez de Turco EB, Deretic D, Bazan NG, Papermaster DS. 1997. Post-Golgi vesicles cotransport docosahexaenoyl-phospholipids and rhodopsin during frog photoreceptor membrane biogenesis. *J Biol Chem.* **272**:10491–10497.
38. Deretic D. 1997. Rab proteins and post-Golgi trafficking of rhodopsin in photoreceptor cells. *Electrophoresis.* **18**:2537–2541.
39. Hargrave P.A., McDowell J.H. 1992. Rhodopsin and phototransduction. *Int Rev Cytol.* **137b**:49–97.
40. Yeagle P.L., Alderfer J.L., Albert A.D. 1995. Structure of the carboxy-terminal domain of bovine rhodopsin [letter]. *Nat Struct Biol.* **2**:332–834.
41. Yeagle P.L., Alderfer J.L., Albert A.D. 1996. Structure determination of the fourth cytoplasmic loop and carboxyl terminal domain of bovine rhodopsin. *Mol. Vis.* **2**:12.
42. Konig B., Arendt A., McDowell J.H., Kahlert M., Hargrave P.A., Hofmann K.P. 1989. Three cytoplasmic loops of rhodopsin interact with transducin. *Proc Natl Acad Sci USA.* **86**:6878–6882.
43. Krupnick J.G., Gurevich V.V., Schepers T., Hamm H.E., Benovic J.L. 1994. Arrestin-rhodopsin interaction. Multi-site binding delineated by peptide inhibition. *J Biol Chem.* **269**:3226–3232.
44. Thurmond R.L., Creuzenet C., Reeves P., Khorana H.G. 1997. Structure and function of rhodopsin: Peptide sequences in the cytoplasmic loops of rhodopsin are intimately involved in the interaction with rhodopsin kinase. *Proc. Natl. Acad. Sci. USA.* **94**:1715–1720.
45. Unger V.M., Hargrave P.A., Baldwin J.M., Schertler G.F. 1997. Arrangement of rhodopsin transmembrane alpha-helices. *Nature.* **389**:203–206.

46. Dryja T.P., Hahn L.B., Cowley G.S., McGee T.L., Berson E.L. 1991. Mutation spectrum of the rhodopsin gene among patients with autosomal dominant retinitis pigmentosa. *Proc Natl Acad Sci USA*. **88**:9370–9374.
47. Palczewski K., Arendt A., McDowell J.H., Hargrave P.A. 1989. Substrate recognition determinants for rhodopsin kinase: studies with synthetic peptides, polyanions, and polycations. *Biochemistry*. **28**:8764–8770.
48. Milam A.H., Li Z.Y., Cideciyan A.V., Jacobson S.G. 1996. Clinicopathologic effects of the Q64ter rhodopsin mutation in retinitis pigmentosa. *Invest Ophthalmol Vis Sci*. **37**:753–765.
49. Dickerson C.D., Weiss E.R. 1995. The coupling of pertussis toxin-sensitive G proteins to phospholipase A2 and adenylyl cyclase in CHO cells expressing bovine rhodopsin. *Exp Cell Res*. **216**:46–50.

12

THE CYTOPLASMIC TAIL OF RHODOPSIN ACTS AS A SORTING SIGNAL IN POLARIZED PHOTORECEPTORS AND MDCK CELLS

Ching-Hwa Sung[1,2] and Jen-Zen Chuang[1]

[1] Department of Ophthalmology
[2] Department of Cell Biology and Anatomy
The Margaret M. Dyson Vision Research Institute
Joan and Sanford I. Weill Medical College of Cornell University
1300 York Avenue, New York, New York 10021

SUMMARY

In this report, we demonstrate that wild-type rhodopsin is targeted to the apical plasma membrane via the *trans*-Golgi network (TGN) upon expression in polarized epithelial Madin-Darby canine kidney (MDCK) cells. Truncated rhodopsin with a deletion of 32 C-terminal residues shows a non-polar steady-state distribution. Addition of the C-terminal 39 residues of rhodopsin redirects the basolateral membrane protein CD7 to the apical membrane. Fusion of rhodopsin's cytoplasmic tail to a cytosolic protein glutathione *S*-transferase (GST) also targets this fusion protein (GST-Rho39Tr) to the apical membrane. We conclude that the carboxy-terminal tail of rhodopsin contains a novel cytoplasmic apical sorting determinant. This result is in agreement with previous studies showing that the cytoplasmic tail of rhodopsin mediates its vectorial transport from its site of synthesis in the rod photoreceptor cell body to the rod outer segment, where phototransduction occurs. Several mutant rhodopsins, found in patients with autosomal dominant retinitis pigmentosa (ADRP), have an amino acid change at the carboxy-terminus and these mutants are defective in their outer segment localization. It will be of interest to examine the relationship between the defects in rhodopsin's targeting/transport and the pathogenesis of retinal degeneration.

INTRODUCTION

Rhodopsin, a seven transmembrane G-protein coupled receptor, is the light absorbing protein in rod photoreceptors. In the photoreceptor, rhodopsin is synthesized

Retinal Degenerative Diseases and Experimental Therapy, edited by Hollyfield *et al.*
Kluwer Academic / Plenum Publishers, New York, 1999.

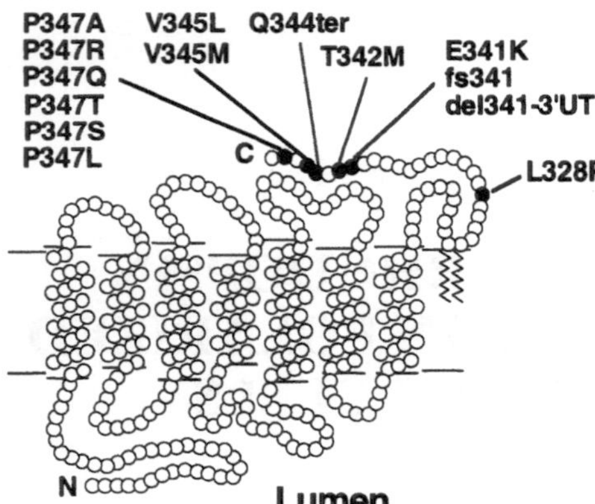

Figure 1. Transmembrane model of human rhodopsin showing the locations of mutations near the carboxy-terminus responsible for ADRP. Fourteen mutations are shown: L328P;[45] E341K;[46] *del* 341-3'UT,[47] a 42 base-pair deletion that removes the last eight codons; *fs* 341,[48] an 8 base pair deletion beginning in codon 341; T342M;[46] Q344ter;[5] V345M;[49] V345L;[50] P347A;[51] P347R;[52] P347Q;[50] P347T;[45] P347S;[53] and P347L.[5,53] N and C mark the amino- and carboxy-termini, respectively. The carboxy-terminus faces the cytosol. *fs*, frameshift; *del*, deletion.

and processed through the endoplasmic reticulum-Golgi pathway[1] before it is sorted at the TGN.[2] Post-Golgi rhodopsin-bearing vesicles are then vectorially transported to a specialized, cilium-derived compartment—the outer segment—in which phototransduction occurs. Rhodopsin is targeted to the outer segment with nearly perfect fidelity;[3] more than 90% of rhodopsin in the photoreceptor is accumulated in this region at steady-state.[4] Nevertheless, the mechanism underlying the sorting and targeting of rhodopsin in rod cells remains unclear. Several pieces of evidence, taken together, suggest that the cytosolic carboxy-terminal sequences of rhodopsin are involved in its polarized targeting and localization. First, the physiological importance of the C-terminal region of rhodopsin was suggested by the high conservation of these amino acid sequences among different species. A second clue was the non-random clustering of mutations associated with ADRP. RP is a hereditary degenerative disease of photoreceptor cells which ultimately leads to blindness. More than a dozen of the known autosomal dominant RP mutations are found near the carboxy-terminus of rhodopsin (Fig. 1). Third, the importance of the C-terminus in sorting was shown by transgenic mice expressing a naturally occurring rhodopsin mutant, Q344ter,[5] in which the carboxy-terminal 5 amino acids are missing. Whereas the endogenous wild-type rhodopsin was properly localized in the transgenic rod outer segment, the Q344ter mutant was found to be distributed throughout the entire rod cell plasma membrane.[4] More recently, other ADRP-related rhodopsin C-terminal transgenes—P347S and P347L—have also been found to be similarly impaired in their outer segment targeting.[6,7] Despite their defective polarized expression in photoreceptors, all these rhodopsin C-terminal mutants fold correctly and exhibit normal plasma membrane localization when expressed in non-polarized tissue culture cells such as 293S and COS-1.[8–10] Finally, in a cell-free system, Deretic et al.[11] have shown that an antibody against rhodopsin's C-terminus can arrest the exit of newly synthesized rhodopsin-bearing vesicles from TGN isolated from frog photoreceptors.

Proper sorting and targeting of membrane proteins and lipids are critical in generating and maintaining the plasma membrane asymmetry of polarized neuronal and epithelial cells. Renal epithelial MDCK cells, which form polarized monolayers with distinct apical and basolateral surfaces, provide a useful model for studying vectorial membrane trafficking. Proteins in the secretory pathway are sorted into distinct transport vesicles during transit through the TGN before they are recruited to either the apical or basolateral plasma membrane in MDCK cells.[12] The fidelity of this intracellular sorting in MDCK cells is very high, suggesting that it is tightly regulated via recognition of specific sorting signals. All basolateral sorting signals described to date reside in the cytoplasmic domain of the sorted proteins,[13–16] whereas apical targeting motifs have been found in lumenal ectodomains (see review[17]). Lumenal modifications—glycosyl phosphatidylinositol (GPI) membrane anchors[18,19] or glycans[20,21]—have also been suggested to play a role in apical protein sorting.

In the present report, we demonstrate that newly synthesized rhodopsin is predominantly delivered directly from the TGN to the apical plasma membrane in MDCK cells. Removal of 32 amino acids from the cytoplasmic tail of rhodopsin results in a randomization of its membrane localization. Moreover, the addition of rhodopsin's cytoplasmic tail can redirect the sorting of other proteins to the apical side of polarized MDCK cells. We propose that the cytoplasmic amino acid sequences of rhodopsin act as an apical sorting signal in MDCK cells.

METHODS

Constructs

Stop codons or amino acid changes were introduced into human rhodopsin cDNA by site-directed mutagenesis[8] using the oligos 5′-GGCCACCTAGCTCGTCT-3′ (Δ5), 5′-TCACCCAGTTAGTTCTTGCC-3′ (Δ22), 5′-GTGGTGAGCTAGCAGTTCCG-3′ (Δ32), 5′-AGTGGGTTCTTGCCGGAGGAGATGGTGGTGAGCA-3′ ($Cys^{322}Cys^{323} \rightarrow Ser^{322}Ser^{323}$) priming on single-stranded template DNA. In each case, the mutated region was recloned into pCB6, a CMV-driven expression vector that had not undergone mutagenesis, and the entire inserted coding region was sequenced.

For the CD7-Rho39 construct, the C-terminal 39 amino acids of human rhodopsin (aa^{310}-aa^{348}) was PCR amplified using the forward primer 5′-CGAC*CTCGAG*AACAAGCAGTTCCGGAACTGCATGC and the reverse primer 5′-ATGC*TCTAGA*AGTCCTAGGCAGGTCTTAGGC. Plasmid CD7BB is a CMV expression vector (pCDM8, Invitrogen Co.) that contains a CD7 cDNA plus a XhoI site immediately 3′ of the stop codon. For the GST-Rho39Tr fusion construct, a yeast expression plasmid pDB-Rho39Tr[23] containing the coding sequence of a triple repeat of the terminal 39 residues of rhodopsin was digested with XbaI and NdeI, and inserted into Xba1- and NdeI- digested pBC vector. pBC is a eukaryotic GST fusion vector containing a SV40 enhancer/promoter followed by the open reading frame of GST.[22] The resulting fusion contains 50 irrelevant amino acids between the GST and the first rhodopsin sequence and 28 amino acids in between each tandem repeat of the rhodopsin C-terminal sequence. In the GST-Rho39pal⁻Tr construct, the rhodopsin sequences in the GST-Rho39Tr construct were replaced by PCR fragments generated from the $Cys^{322}Cys^{323} \rightarrow Ser^{322}Ser^{323}$ mutant.

Cell Culture and Transfection

MDCK cells (type II) were maintained in DMEM supplemented with 5% fetal bovine serum, penicillin (50 U/ml) and streptomycin (50 mg/ml). Transfection of MDCK cells was performed using lipofection following the manufacturer's instructions (Life Technologies). Stable cell lines were obtained by G418 (500 μg/ml) selection and positive clones were screened by immunofluorescent staining. MDCK stable transfectants were seeded at high density (1.5×10^6 cells/24mm filter or equivalent density) in Costar™ polyester clear Transwells for 5–7 days to allow the development of a tight monolayer. For each construct, at least two independent clonal lines were selected and used for the experiments.

Immunofluorescent Staining

For surface immunofluorescent staining, filter-grown MDCK monolayers were washed with ice-cold phosphate buffered saline containing 2mM $MgCl_2$ and 0.2mM $CaCl_2$ (PBS-C/M), and incubated with monoclonal antibody (mAb) B6-30 (anti-rhodopsin N-terminus)[24] or T3–3A1 (anti-CD7 N-terminus, ATCC) from both sides for 30min at 4°C. After extensive washing with cold PBS-C/M, cells were fixed with 2% paraformaldehyde for 30min, quenched with 50mM ammonium chloride for 10min, rinsed, then incubated with biotinylated goat anti-mouse antibody (1:200; Vector Laboratory) and FITC-streptavidin (1:25, Amersham). To immunolabel the permeabilized cell, MDCK monolayers were fixed with paraformaldehyde, quenched, and permeabilized with 0.075% saponin, 0.5% BSA in PBS-C/M for 30min. The cells were then incubated with anti-GST antibody (1:1000 dilution, Sigma Co.) followed by biotinylated goat anti-rabbit antibody and FITC-streptavidin. For nuclear staining, cells were treated with DNase-free RNase (1:100, 5′ prime—3′ Prime Inc. Boulder, CO) during secondary antibody incubation and then incubated with propidium iodide (10µg/ml) for an additional 15min. The immunostaining was examined using a laser scanning confocal microscope (Molecular Dynamics, Sunnyvale, CA).

Domain Selective Biotinylation/Membrane Targeting Assay

To quantitate the steady-state protein distribution on the apical and basolateral sides of MDCK cells, a domain selective biotinylation/immunoprecipitation assay was performed as described.[25] Briefly, filter-grown MDCK cells expressing rhodopsin were surface labeled either apically or basolaterally using biotin-LC-hydrazide (Pierce Chemical Co.). After biotinylation, the filters were washed with cold PBS-C/M and excised. Cells were then solubilized in lysis buffer (1% Triton X-100, 50mM Tris, pH 7.5, 150mM NaCl, and 2mM EDTA) containing protease inhibitor cocktail (1µm PMSF, 2µg/mL aprotinin, 2µg/mL leupeptin, and 0.7µg/mL pepstatin) at 4°C for 1hr and the cell lysates were centrifuged at 13krpm for 20min. Our preliminary results have shown that rhodopsin was almost completely solubilized in Triton X-100 under these extraction conditions. The supernatant was immunoprecipitated with protein-A Sepharose (Pharmacia LKB) precoated with rabbit anti-mouse antibody plus mAb B6-30 or mAb T3-3A1. The eluted immunocomplexes were electrophoresed on sodium dodecyl sulfate-polyacrylamide gels (SDS-PAGE), transferred to nitrocellulose, and blotted with ^{125}I-streptavidin (Amersham) as described. Quantitation of biotinylated rhodopsin recovered from the apical and basolateral domains was determined by a phosphorimager (Molecular Dynamics).

A membrane targeting assay[26] was carried out to follow the delivery of newly synthesized rhodopsin, CD7, or CD7-Rho39 to the membrane. MDCK monolayers were pulse labeled with EXPRE35S35S protein labeling mix (DuPont NEN) from the basolateral side for 20min. At varying time points during the chase, cells were chilled to 4°C and biotinylated from the apical or basolateral surface using biotin-LC-hydrazide. Following biotinylation, cells were lysed and immunoprecipitated as described above. The immunoprecipitates were dissociated from the Sepharose beads by boiling the samples in 5% SDS for 5 min, and then diluting them with lysis buffer for a subsequent streptavidin agarose precipitation. This immunoprecipitate was separated by SDS-PAGE and visualized by fluorography.

Subcellular Fractionation

Fractionation of cytosol (S100) and total membrane (P100) was carried out according to a protocol described previously[27] with minor modifications. Briefly, cells were washed and scraped from Transwell filters in ice-cold buffer (50 mM Tris, pH 7.5, 150 mM NaCl, and 2 mM EDTA) and spun for 5 min at 2000 rpm. Cells were resuspended in isotonic buffer (10 mM Tris, pH 7.2, 0.2 mM $MgCl_2$, 250 mM sucrose, 1 mM EDTA plus protease inhibitors), sheared by 22 gauge needles, and homogenized with a Balch homogenizer. Post-nuclear supernatants, prepared by spinning the cell homogenates at 3000 g for 10 min, were then centrifuged for 45 min at 100,000 g (Beckman TLA 100.3 rotor). The resulting supernatant (S100) was supplemented with of 5X concentrated lysis buffer whereas the pellet (P100) was resuspended in 1X lysis buffer (50 mM Tris, pH 8.0, 150 mM NaCl, 2 mM EDTA, 1% NP-40, 0.5% deoxycholate plus protease inhibitors). Samples were then lysed on ice for 15 min and cleared by centrifugation at 100,000 g, 15 min. Equal volumes of P100 and S100 were subjected to SDS-PAGE followed by immunoblotting analysis.

RESULTS

Rhodopsin's Carboxy-Terminus Is Necessary for Its Apical Sorting in MDCK Cells

The membrane distribution of native rhodopsin in polarized MDCK cells was first determined by surface immunofluorescent staining. Monolayers of MDCK cells stably expressing wild-type rhodopsin were incubated from both sides with mAb B6–30, which recognizes the extracellular N-terminus of rhodopsin at 4°C for 30 min. The cells were then washed, fixed, and rhodopsin was visualized by incubating with biotinylated anti-mouse IgG followed by FITC conjugated-streptavidin. *En face* images of confocal microscopic analysis revealed a characteristic tile-like, apical labeling of rhodopsin (Fig. 2A, *upper panel*). The vertical scan confirmed that rhodopsin is almost exclusively localized on the apical membrane (Fig. 2A, *lower panel*). To determine the relative distribution of rhodopsin quantitatively, we used a domain-selective biotinylation/immunoprecipitation assay in which MDCK monolayers were selectively biotinylated from either surface. The cell lysates were immunoprecipitated by mAb B6-30, electrophoresed, and transferred onto nitrocellulose membrane. The biotinylated rhodopsin was then detected and quantified by the amount of bound ^{125}I-streptavidin. In agreement with the immunofluorescent labeling, the majority of the wild-type rhodopsin (~84%) was found to be expressed apically (Fig. 2B).

To test whether the cytoplasmic tail of rhodopsin contains apical sorting information, a nested set of C-terminal deleted mutant rhodopsins was constructed and these rhodopsins were stably expressed in MDCK cells (Fig. 3, *upper panel*). Using both surface immunofluorescent labeling (Fig. 3, *middle panel*) and the domain selective

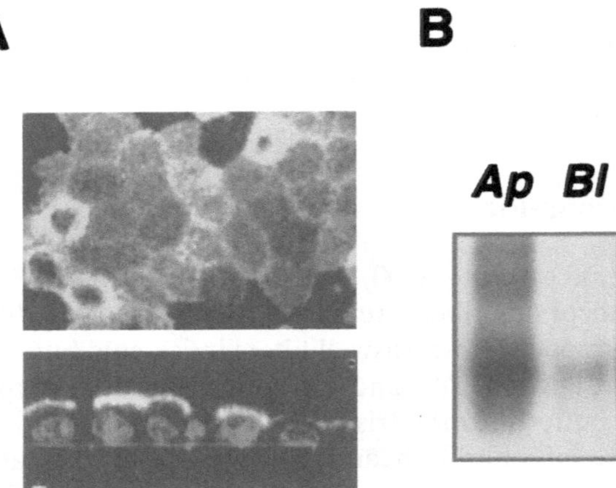

Figure 2. The steady-state surface distribution of wild-type rhodopsins expressed in stably transfected MDCK cells. A. Rhodopsin was detected by surface-immunofluorescent labeling using mAb B6-30 and the staining was analyzed by laser scanning confocal microscopy. Optical sections horizontal (*top*) and perpendicular (*bottom*) to the monolayer are shown. Propidium iodide was included in the staining to show the nuclei in the Z section (shown in *gray*). Bar = 5 μm. B. The immunoprecipitated, biotinylated rhodopsin proteins were electrophoresed, transferred to nitrocellulose, and probed with ^{125}I-streptavidin as described in *Methods*. *Ap*: apically biotinylated; *Bl*: basolaterally biotinylated.

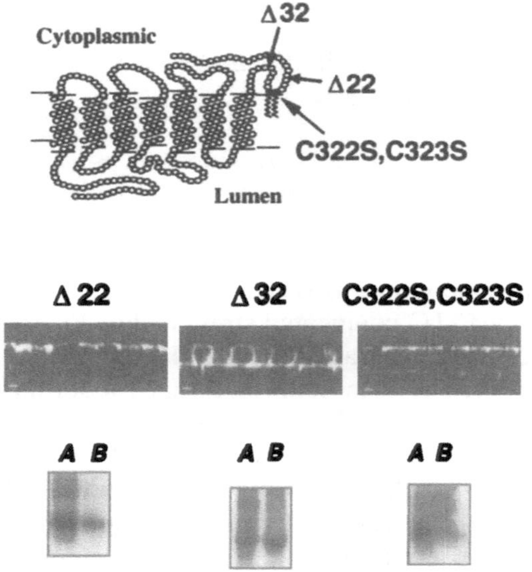

Figure 3. Surface distributions of mutant rhodopsins in polarized MDCK cells. (*Top panel*) Rhodopsin topology showing carboxy-terminal mutant constructs used to characterize the apical sorting domains. The zigzags indicate the palmitoyl membrane anchors. Δ22, and Δ32 are carboxy-terminal deletion mutants in which the last 22, and 32 amino acids are missing, respectively. In the $Cys^{322}Cys^{323} \rightarrow Ser^{322}Ser^{323}$ (C322S, C323S) mutant, the cysteines at positions 322 and 323 were replaced with serines. (*Middle panel*) Surface immunofluorescent staining was carried out as described above and optical sections perpendicular to the monolayer are shown. Bar = 5 μm. (*Bottom panel*) Domain selective biotinylation/immunoprecipitation assay shows the steady-state distribution of each mutant rhodopsin on the apical (*Ap*) and the basolateral (*Bl*) surfaces.

biotinylation/immunoprecipitation assay (Fig. 3, *bottom panel*), we found that a deletion of the terminal 5 residues (Δ5, ~80% apical, data not shown) or 22 residues (Δ22, ~81% apical) did not affect the apical sorting pattern of rhodopsin. However, a deletion of 32 amino acids resulted in the loss of apical polarity, with the truncated mutant rhodopsin (Δ32) expressed on both the apical and basolateral surfaces at similar levels. These results suggest that the terminal 32 residues of rhodopsin's cytoplasmic tail contain the necessary information required for its apical targeting.

Rhodopsin's C-Terminus Is Sufficient to Redirect CD7 to the Apical Surface of MDCK Cells

To address whether the cytoplasmic tail of rhodopsin is not only necessary but also sufficient for apical sorting, we examined the membrane targeting of a fusion protein, CD7-Rho39, produced by fusion of the CD7 gene to the sequence of the terminal 39 amino acids of rhodopsin. CD7 is a type I integral membrane protein which has been reported to be a "bulk-flow" membrane marker in MDCK cells, and it appears on both the apical and the basolateral surfaces at roughly equivalent levels.[28] Nevertheless, to our surprise, in three independent MDCK stable clones we examined, CD7 was detected basolaterally (~82%) in the biotinylation/immunoprecipitation assay (Fig. 4B). Indirect immunofluorescence and laser-scanning confocal microscopy of cells stained with antibodies against CD7 showed that immunofluorescent CD7 signal was observed predominantly on the basolateral surface although some the signal can also be detected on the apical surface at variable levels (Fig. 4A). We later confirmed that CD7 is primarily sorted to the basolateral surface of MDCK cells through a direct pathway from the TGN (see below). This discrepancy regarding the targeting of CD7 likely reflects variability in the sorting patterns of MDCK sublines; this phenomenon has also been observed for other proteins.[29,30] In contrast to CD7, we found that the fusion protein CD7-Rho39, produced by fusion of the entire rhodopsin's cytoplasmic tail 39 amino acids to the carboxy-terminus of CD7, was apically sorted in MDCK cells assayed by surface immunofluorescent labeling using anti-CD7 antibody (Fig. 4A). Little or no specific basolateral staining was detected in these cells. In agreement with the immunostaining results, ~89% of CD7-Rho39 fusion proteins were detected on the apical surface biochemically (Fig. 4B). The apical localization of the CD7-Rho39 fusion demonstrates that rhodopsin's C-terminal sequence is able to provide sufficient

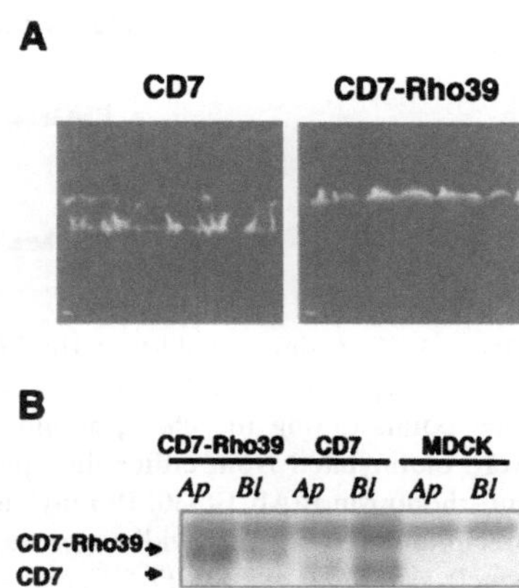

Figure 4. Apical targeting of CD7 by rhodopsin's carboxy-terminus. A. Polarized MDCK monolayers stably expressing CD7 and CD7-Rho39 were surface immunofluorescent labeled with mAb T3-3A1 and vertical optical sections obtained from confocal microscopic analysis are shown. Bar = 5 μm. B. The levels of steady-state surface distribution of CD7 (~40 kD) and CD7-Rho39 (~44 kD) were determined by the domain-selective biotinylation/immunoprecipitation assay. As was shown previously,[28] an endogenous MDCK protein which migrates slightly slower than CD7-Rho39 on SDS-PAGE is also precipitated by mAb T3-3A1.

apical-sorting information to function independently on another membrane protein. Finally, it is unlikely that the apical targeting of CD7-Rho39 is due to inactivation of a putative basolateral targeting signal of CD7 because that CD7-Rho7—a truncated form of CD7-Rho39 in which the distal 32 residues of rhodopsin's cytoplasmic tail were deleted—was predominantly sorted basolaterally (data not shown), arguing that the addition of extra nonspecific sequences to CD7's carboxy-terminus has little or no effect on its membrane localization.

Rhodopsin and CD7-Rho39 Are Directly Targeted from the TGN to the Apical Surface

To investigate the intracellular routing of wild-type rhodopsin in MDCK cells, we traced the cell-surface delivery of newly synthesized rhodopsin using a pulse-chase/membrane targeting assay. Confluent MDCK monolayers were pulse labeled with ^{35}S-methionine/cysteine, chased for different times, biotinylated from either the apical or basolateral surface, lysed, and immunoprecipitated by mAb B6-30. The precipitated rhodopsin molecules were subsequently precipitated by streptavidin-agarose, so that only the radiolabeled rhodopsin that had reached the cell surface by the time of biotinylation was detected by autoradiography. After 1 hr of chase, rhodopsin begins to appear on the apical surface as indicated by a smear of bands upon SDS-PAGE analysis (Fig. 5, *top panel*). This appearance is typical for proteins that have undergone heterogeneous Golgi sugar processing; in this state, they are characterized by their resistance to endo H cleavage (data not shown). These smear bands can be trimmed to a core polypeptide of ~38 kD by treating the immunoprecipitates with PNGase F (Fig. 5, *bottom panel*). It is evident that newly synthesized rhodopsin is delivered predominantly to the apical plasma membrane at all time points tested, suggesting that rhodopsin follows a direct route from the TGN to the apical membrane.

The intracellular transport of CD7-Rho39 fusion protein was also examined using the same membrane targeting assays. At all chase time points, CD7-Rho39 fusion proteins were detected predominantly on the apical surface (Fig. 6, *top panel*), whereas CD7 alone was predominantly targeted to the basolateral side (Fig. 6, *bottom panel*). These results are consistent with the steady-state surface distribution of these two

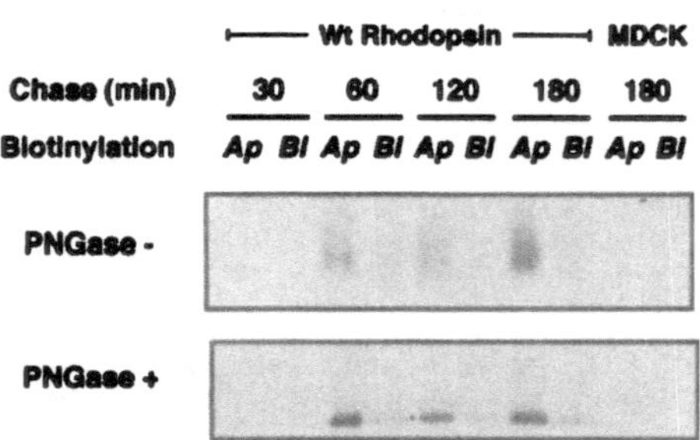

Figure 5. Targeting of wild-type rhodopsin from the TGN to the apical surface. Vectorial delivery of wild-type rhodopsin to the cell surface was analyzed by a membrane targeting assay (see *Methods*). At various time points during the chase, as indicated, ^{35}S-metabolically labeled, rhodopsin-expressing MDCK cells were biotinylated from either the apical (*Ap*) or basolateral (*Bl*) surface, and immunoprecipitated with anti-rhodopsin mAb B6-30. Biotinylated, radiolabeled immunoprecipitates obtained by a subsequent re-precipitation with streptavidin agarose, were either treated without (*top panel*) or with PNGase F (*bottom panel*), separated by SDS-PAGE, and visualized by autoradiography.

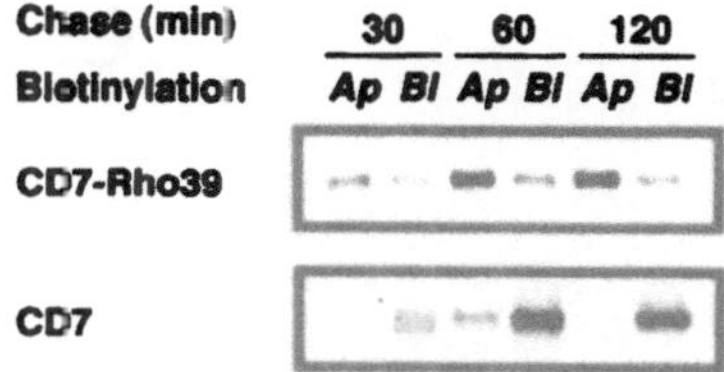

Figure 6. Targeting of CD7-Rho39 directly from the TGN to the apical surface. Targeting of CD7-Rho39 (*top panel*) and CD7 (*bottom panel*) was analyzed by the membrane targeting assay described in Fig 4. In this experiment, the CD7-Rho39 fusion protein was precipitated by anti-rhodopsin C-terminus mAb 1D4 and CD7 was precipitated by anti-CD7 mAb T3–3A1. *Ap*: apically biotinylated; *Bl*: basolaterally biotinylated.

proteins. The direct apical targeting of CD7-Rho39 fusion protein suggests that similar mechanisms may be used for the sorting/targeting of both CD7-Rho39 and wild-type rhodopsin in MDCK cells.

Palmitoylation Is Not Required for the Apical Targeting of Rhodopsin

Fatty acylation, such as palmitoylation, has been suggested not only provides membrane anchoring for these otherwise cytosolic proteins but also mediates their proper cellular targeting.[31–34] Two highly conserved cysteine residues, Cys^{322} and Cys^{323}, located in the cytoplasmic tail of rhodopsin, are known to be palmitoylated and membrane anchored.[35,36] So far, no signal transduction function has been assigned to this palmitoyl modification of rhodopsin.[37,38] These palmitoylation signals are deleted in the Δ32 mutant rhodopsin, as described above, which has altered domain expression in MDCK cells. We therefore examined whether the palmitoyl moiety of rhodopsin plays a role in its sorting. A non-palmitoylated, site-directed mutant rhodopsin $Cys^{322}Cys^{323} \rightarrow Ser^{322}Ser^{323}$ was stably transfected into MDCK cells and its expression was examined by both immunostaining (Fig. 3, *middle panel*) and domain selective membrane targeting assay (Fig. 3, *bottom panel*). We concluded that the $Cys^{322}Cys^{323} \rightarrow Ser^{322}Ser^{323}$ mutant exhibits apical expression in MDCK cells (~82% apical), indistinguishable from the wild-type rhodopsin. This result suggests that neither the $Cys^{322}Cys^{323}$ residues nor their resulting palmitoylation is essential for rhodopsin's apical sorting.

The Palmitoylated Rhodopsin Cytoplasmic Tail Enables a Cytosolic Protein to Be Delivered to the Apical Surface

To confirm that there is no recessive information present in the lumenal or transmembrane domains of rhodopsin or CD7 which is necessary for the apical targeting activity, we tested whether the cytoplasmic tail of rhodopsin can be recognized in the absence of any transmembrane or lumenal domains. We made a fusion protein GST-Rho39Tr, in which the cytosolic protein GST was fused to a triple-repeat of the C-terminal 39 residues of rhodopsin (Fig. 7A). A triple repeat was chosen because the placement of the putative signal at varying distances from GST minimizes any steric hindrance of the signal by GST. We reasoned that the GST-Rho39Tr would likely be membrane anchored via the palmitoylation signal provided by the rhodopsin sequences. A non-palmitoylated GST fusion (GST-Rho39pal^{-}Tr) was also generated by

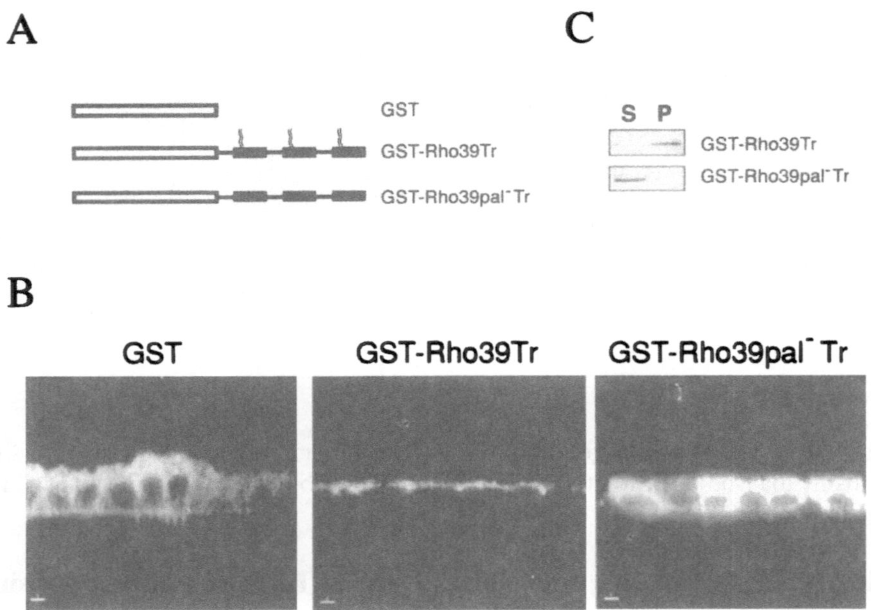

Figure 7. The apical targeting of GST by addition of rhodopsin's carboxy-terminus. A. Schematic diagrams of GST, GST-Rho39Tr, and GST-Rho39pal$^-$Tr. GST-Rho39Tr is a fusion protein in which a triple-repeat of the terminal 39 amino acids of rhodopsin (filled boxes) was fused to the carboxy-terminus of GST (open box). GST-Rho39pal$^-$Tr is a similar fusion protein except that the 39 rhodopsin residues were derived from the Cys^{322}Cys^{323}$\rightarrow$Ser^{322}Ser^{323} mutant so that the palmitoylation signal (zigzag) was removed. B. Immunolocalization of GST, GST-Rho39Tr, and GST-Rho39pal$^-$Tr in MDCK cells. Permeabilized MDCK monolayers were labeled with anti-GST antibody followed by biotinylated anti-rabbit antibody and FITC-streptavidin. Optical sections perpendicular to the monolayer are shown. Both GST and GST-Rho39pal$^-$Tr show diffuse, nucleus-excluded cytoplasmic staining. Bar = 5 µm. C. Subcellular distributions of GST-Rho39Tr and GST-Rho39pal$^-$Tr in MDCK cells. MDCK cell homogenates were fractionated into cytosolic (S100) and total cellular membrane (P100) fractions by centrifugation at 100,000 g for 45 min. Equal fractions of S100 and P100 were electrophoresed on SDS-PAGE and immunoblotted with mAb 1D4 followed by alkaline phosphatase anti-mouse IgG. NBT/BCIP substrates were used for color development.

replacing the wild-type rhodopsin sequence with the Cys^{322}Cys^{323}$\rightarrow$Ser^{322}Ser^{323} mutant sequence for comparison (Fig. 7A). MDCK cells stably expressing control GST, GST-Rho39Tr, or GST-Rho39pal$^-$Tr were paraformaldehyde fixed, permeabilized, and immunolabeled with anti-GST antibody. We found that the GST-Rho39Tr fusion protein was almost exclusively accumulated on the apical plasma membrane (Fig. 7B, *middle panel*). Likewise, the subcellular fractionation assay showed that the GST-Rho39Tr fusion protein was predominantly localized in the membrane fraction (P100) (Fig. 7C). Conversely, the GST-Rho39pal$^-$Tr protein was associated with the cytosolic (S100) fraction (Fig. 7C) and exhibited a uniform, diffuse cytoplasmic staining (Fig. 7B, *right panel*). This staining was indistinguishable from that of the control GST protein (Fig. 7 B, *left panel*). Since neither the cysteine residues nor the palmitoylation signal is essential for rhodopsin's transport, we predict that the palmitoylation itself of GST-Rho39Tr does not encode specific apical targeting information. However, this result suggests to us that recognition of the apical sorting signal contained in the rhodopsin terminal 39 residues requires membrane anchorage, which can be provided by palmitoylation and/or transmembrane domains.

DISCUSSION

In the present report, we demonstrate that wild type rhodopsin is predominantly targeted to the apical surface in polarized MDCK cells. Removal of the terminal 32 amino acids from the cytoplasmic tail of rhodopsin abolished this vectorial targeting. The truncated protein was sent to both the apical and the basolateral membrane at roughly equivalent levels, presumably due to the lack of any sorting information. To confirm that the sorting determinants are present in the cytoplasmic tail of rhodopsin, we show that the C-terminal 39 residues of rhodopsin can redirect apical targeting of two heterologous proteins—CD7 and GST. To our knowledge, this is the first time that sorting of an apical protein was found to be mediated by cytoplasmic amino acids.

The Physiological Relevance of the Apical Targeting of Rhodopsin in MDCK Cells and in Photoreceptors

The topographic correlation between the apical membrane localization of rhodopsin in polarized MDCK cells and the initial "apical" inner segment plasma membrane targeting of rhodopsin is intriguing. Coincidentally, Na^{+}/K^{+} ATPase, a molecule which accumulates basolaterally in most epithelia,[39,40] appears to be exclusively expressed in the "lateral plasma membrane" of the photoreceptor inner segment.[41] Such correlation suggest that polarized MDCK cells may provide a useful model for studying photoreceptor protein transport.

Common mechanisms have suggested to be used in sorting proteins to the apical/basolateral surface of MDCK cells and the axonal/dendritic region of neuronal cells.[42] However, a photoreceptor is both more complicated in morphology and more compartmentalized than a conventional neuron. It is not difficult to imagine that rhodopsin might have to make multiple sorting decisions as it travels from the TGN to the outer segment. In fact, a great deal of immunoelectron microscopic analyses have revealed that this might be the case: rhodopsin synthesized near the nucleus is transported as small membrane vesicles along the inner segment cytoplasm until they reach and fuse with the apical extremity of the inner segment plasma membrane.[3,43,44] By a mechanism still under debate, either through the connecting cilium or through the extracellular space, rhodopsin is further transported to the outer segment and eventually incorporated into the outer segment disc membranes. Bearing this in mind, we are thus not too surprised by the discrepancy observed in photoreceptor and MDCK cells: a short C-terminal deletion of 5 amino acids is sufficient to alter rhodopsin's outer segment localization in transgenic mice, yet a longer deletion is necessary to abolish its apical sorting in MDCK cells.

We speculate that a hierarchy of sorting mechanisms is utilized in rhodopsin trafficking in photoreceptors. The first step, which may be analogous to apical sorting in MDCK cells, targets rhodopsin from the TGN to the inner segment plasma membrane near the base of the outer segment. The second step, which may require a signal at the carboxy terminus; transfers rhodopsin from the plasma membrane onto newly forming disc membranes. A third mechanism to prevent its backward diffusion away from the outer segment may also be involved. In this model, a 5 amino-acid truncation does not affect the ability of rhodopsin to be targeted to the apical inner segment plasma membrane but drastically reduces its efficiency to be further transferred to the outer segment and thus rhodopsin subsequently diffuses to the lateral plasma membrane. However, it is also possible that two or more independent sorting motifs exist in rhodopsin's cytoplasmic region and these motifs are used differently in MDCK cells and photoreceptors. It is conceivable that rhodopsin contains two distinct apical targeting signals—one at the distal and one at the proximal part of rhodopsin's cytoplasmic tail. The machinery recognizing the distal apical sorting signal may be present in photoreceptors but absent in MDCK cells. Alternatively, either the proximal or distal signal is individually sufficient in MDCK cells whereas only the distal sorting signal can be used in photoreceptors. Further characterization of the required targeting signals in rod photoreceptors and in MDCK cells and the identification of the cellular components interacting with rhodopsin's C-terminus will be needed to distinguish between these models.

Possible Mechanisms of Apical Targeting of Rhodopsin

MDCK cells have been proved to be a useful model in studying intracellular trafficking of many exogenously introduced proteins. In this report, we demonstrate that the cytoplasmic amino acids of rhodopsin are able to act as an apical sorting signal in polarized MDCK cells. We surmise that the mechanism for the sorting of rhodopsin must be distinct from that used for the recognition of previously described apical sorting signals such as lumenal ectodomains, GPI anchors, or glycans. It is unclear how these sequences direct its apical transport but two models are proposed. First, the cytoplasmic determinant may be recognized by a cytosolic factor(s) which is responsible for forming apically-targeted transport vesicles. This scenario is known to be the mechanism underlying the sorting of several basolateral proteins. Second, the cytoplasmic tail of rhodopsin could be transported as a result of interacting with other apically targeted proteins. Further analysis is required to address this issue and distinguish these possibilities.

ACKNOWLEDGMENTS

We thank Dr. R. Molday for mAb 1D4, Dr. P. Hargrave for mAb B6-30, Dr. B. Seed for CD7 cDNA, Drs. Y.-T. Chen, and W.J. Nelson for a modified CD7BB plasmid, Drs. D.L. Price and M.F. Stinski for pCB6 vector, and Dr. B. Chatton for pBC vector. We are grateful to Ms. Maria Eugene-Alfonzo and Ramee Lee for technical assistance. This work was supported by grants to C.-H. Sung from the NIH (EY11307), Research Preventing Blindness, and Foundation Fighting Blindness, and to J.-Z. Chuang from the NIH Tri-Institutional Training Program in Vision (EY07138).

REFERENCES

1. D. Bok, M.O. Hall, and P. O'Brien, 1977, International Cell Biology, (B.R. Brinkley and K.R. Porter, eds.), p. 608, Rockefeller University Press, New York.
2. R. Schmied, and E. Holtzman, 1989, Involvement of the Golgi apparatus in sorting of materials to opposite ends of frog rod retinal photoreceptors, *J. Neurobiol.* **20**:115–138.
3. D.S. Papermaster, B.G. Schneider, D. Defoe, and J.C. Besharse, 1986, Biosynthesis and vectorial transport of opsin on vesicles in retinal rod photoreceptors, *J. Histochem. Cytochem.* **34**:5–16.

4. C.-H. Sung, C. Makino, D. Baylor, and J. Nathans, 1994, A rhodopsin gene mutation responsible for autosomal dominant retinitis pigmentosa results in a protein that is defective in localization to the photoreceptor outer segment, *J. Neurosci.* **14**:5818–5833.
5. C.-H. Sung, C.M. Davenport, J.C. Hennessey, I.H. Maumenee, S.G. Jacobson, J.R. Heckenlively, R. Nowakowski, G. Fishman, P. Gouras, and J. Nathans, 1991, Rhodopsin mutations in autosomal dominant retinitis pigmentosa, *Proc. Natl. Acad. Sci. USA* **88**:6481–6485.
6. T. Li, W.K. Snyder, J.E. Olsson, and T.P. Dryja, 1996, Transgenic mice carrying the dominant rhodopsin mutation P347S: evidence for defective vectorial transport of rhodopsin to the outer segments, *Proc. Natl. Acad. Sci. USA* **93**:14176–14181.
7. Z.Y. Li, F. Wong, J.H. Chang, D.E. Possin, Y. Hao, R.M. Petters, and A.H. Milam, 1998, Rhodopsin transgenic pigs as a model for human retinitis pigmentosa. *Invest. Ophthalmol. Vis. Sci.* **39**:808–819.
8. C.-H. Sung, B.G. Schneider, N. Agarwal, D.S. Papermaster, and J. Nathans, 1991, Functional heterogeneity of mutant rhodopsins responsible for autosomal dominant retinitis pigmentosa, *Proc. Natl. Acad. Sci. USA* **88**:8840–8844.
9. C.-H. Sung, C.M. Davenport, and J. Nathans, 1993, Rhodopsin mutations responsible for autosomal dominant retinitis pigmentosa. Clustering of functional classes along the polypeptide chain, *J Biol. Chem.* **268**:26645–26649.
10. S. Kaushal, and H. Khorana, 1994, Structure and function in rhodopsin. 7. Point mutations associated with autosomal dominant retinitis pigmentosa, *Biochemistry* **33**:6121–6128.
11. D. Deretic, L. Huber, N. Ransom, M. Mancini, K. Simons, and D.S. Papermaster, 1995, rab8 in retinal photoreceptors may participate in rhodopsin transport and in rod outer segment disk morphogenesis, *J Cell Sci.* **108**:215–224.
12. A. Wandinger-Ness, M.K. Bennett, C. Antony, and K. Simons, 1990, Distinct transport vesicles mediate the delivery of plasma membrane proteins to the apical and basolateral domains of MDCK cells, *J. Cell Biol.* **111**:987–1000.
13. W. Hunziker, C. Harter, K. Matter, and I. Mellman, 1991, Basolateral sorting in MDCK cells requires a distinct cytoplasmic domain determinant, *Cell* **66**:907–920.
14. J.E. Casanova, G. Apodaca, and K.E. Mostov, 1991, An autonomous signal for basolateral sorting in the cytoplasmic domain of the polymeric immunoglobulin receptor, *Cell* **66**:65–75.
15. A. Le Bivic, Y. Sambuy, A. Patzak, N. Patil, M. Chao, and E. Rodriguez-Boulan, 1991, An internal deletion in the cytoplasmic tail reverses the apical localization of human NGF receptor in transfected MDCK cells, *J. Cell Biol.* **115**:607–618.
16. K. Matter, W. Hunziker, and I. Mellman, 1992, Basolateral sorting of LDL receptor in MDCK cells: the cytoplasmic domain contains two tyrosine-dependent targeting determents, *Cell* **71**:741–753.
17. W.J. Nelson, 1992, Regulation of cell surface polarity from bacteria to mammals, *Science* **258**:948–955.
18. D.A. Brown, B. Crise, and J.K. Rose, 1989, Mechanism of membrane anchoring affects polarized expression of two proteins in MDCK cells, *Science* **245**:1499–1501.
19. M.P. Lisanti, I.W. Caras, M.A. Davitz, and E. Rodriguez-Boulan, 1989, A glycophospholipid membrane anchor acts as an apical targeting signal in polarized epithelial cells, *J. Cell Biol.* **109**:2145–2156.
20. L. Fiedler and K. Simons, 1995, The role of N-glycans in the secretory pathway, *Cell* **81**:309–312.
21. C. Yeaman, A.H. Le Gall, A.N. Baldwin, L. Monlauzeur, A. Le Bivic, and E. Rodriguez-Boulan, 1997, The O-glycosylated "stalk" domain is required for apical sorting of neurotrophin receptors in polarized MDCK cells, *J. Cell Biol.* **139**:929–940.
22. B. Chatton, A. Bahr, J. Acker, and C. Kedinger, 1995, Eukaryotic GST fusion vector for the study of protein-protein associations *in vivo*: application to interaction of ATFa with Jun and Fos, *BioTechniques* **18**:142–145.
23. J.-Z. Chuang and C.-H. Sung, 1998, The carboxy-terminus of rhodopsin functions as an apical sorting signal in MDCK cells, *J. Cell Biol.* **142**:1–12.
24. G. Adamus, A. Arendt, Z.S. Zam, J.H. McDowell, and P.A. Hargrave, 1988, Use of peptides to select for anti-rhodopsin antibodies with desired amino acid sequence specificities, *Pept. Res.* **1**:42–47.
25. M.P. Lisanti, A.L. Bivic, M. Sargiacomo, and E. Rodriguez-Boulan, 1989, Steady-state distribution and biogenesis of endogenous Madin-Darby canine kidney glycoproteins: evidence for intracellular sorting and polarized cell surface delivery, *J. Cell Biol.* **109**:2117–2127.
26. A. Le Bivic, F.S. Real, and E. Rodriguez-Boulan, 1989, Vectorial targeting of apical and basolateral plasma membrane proteins in a human adenocarcinoma epithelial cell line, *Proc. Natl. Acad. Sci. USA* **86**:9313–9317.

27. C.O. Van Hooff, J.C. Holthuis, A.B. Oestreicher, J. Boonstra, P.N. De Graan, and W.H. Gispen, 1989, Nerve growth factor-induced changes in the intracellular localization of the protein kinase C substrate B-50 in pheochromocytoma PC12 cells, *J. Cell Biol.* **108**:1115–1125.
28. C. Haller and S.L. Alper, 1993, Nonpolarized surface distribution and delivery of human CD7 in polarized MDCK cells, *Am. J. Physiol.* **265**:1069–1079.
29. G.K. Ojakian, R.E. Romain, and R.E. Herz, 1987, A distal nephron glycoprotein that has different cell surface distributions on MDCK cell sublines, *Am. J. Physiol.* **253**:C433-C443.
30. R. Mays, K.A. Siemers, B.A. Fritz, A.W. Lowe, G. van Meer, and W.J. Nelson, 1995, Hierarchy of mechanisms involved in generating Na/K-ATPase polarity in MDCK epithelial cells, *J. Cell Biol.* **130**:1105–1115.
31. Y. Liu, D.A. Fisher, and D.R. Storm, 1994, Intracellular sorting of neuromodulin (GAP-43) mutants modified in the membrane targeting domain, *J. Neurosci.* **14**:5807–5817.
32. A.M. Shenoy-Scaria, D.J. Dietzen, J. Kwong, D.C. Link, and D.M. Lublin, 1994, Cysteine 3 of Src family protein tyrosine kinase determines palmitoylation and localization in caveolae, *J. Cell Biol.* **126**:353–363.
33. G. Garcia-Cardena, P. Oh, J. Liu, J.E. Schnitzer, and W.C. Sessa, 1996, Targeting of nitric oxide synthase to endothelial cell caveolae via palmitoylation: implications for nitric oxide signaling, *Proc. Natl. Acad. Sci. USA* **93**:6448–6453.
34. P.W. Shaul, E.J. Smart, L.J. Robinson, A. German, I.S. Yuhanna, Y. Ying, R.G. Anderson, and T. Michel, 1996, Acylation targets endothelial nitric-oxide synthase to plasmalemmal caveolae, *J. Biol. Chem.* **271**:6518–6522.
35. M.L. Applebury and P.A. Hargrave, 1986, Molecular biology of the visual pigments. *Vision Res.* **26**:1881–1895.
36. D.I. Papac, K.R. Thornburg, E.E. Bullesbach, R.K. Crouch, and D.R. Knapp, 1992, Palmitylation of a G-protein coupled receptor. Direct analysis by tandem mass spectrometry, *J. Biol. Chem.* **267**:16889–16894.
37. S.S. Karnik, T.P. Sakmar, H.-B. Chen, and H.G. Khorana, 1988, Cysteine residues 110 and 187 are essential for the formation of correct structure in bovine rhodopsin, *Proc. Natl. Acad. Sci. USA* **85**:8459–8463.
38. S. Osawa and E.R. Weiss, 1994, The carboxy-terminus of bovine rhodopsin is not required for G protein activation, *Mol. Pharmacol.* **46**:1036–1040.
39. R.W. Hammerton, K.A. Kizeminski, R.W. Mays, D.A. Wollner, and W.J. Nelson, 1991, Mechanism for regulating cell surface distribution of Na^+, K^+-ATPase in polarized epithelial cells, *Science* **254**:847–850.
40. K.A. Siemers, R. Wilson, R.W. Mays, R.A. Tyan, D.A. Wollner, and W.J. Nelon, 1993, Delivery of Na^+, K^+-ATPase in polarized epithelial cells, *Science* **260**:554–556.
41. B.G. Schneider and E. Kraig, 1990, Na, K-ATPase of the photoreceptor: selective expression of $\alpha3$ and $\beta2$ isoforms, *Exp. Eye Res.* **51**:553–564.
42. C.G. Dotti and K. Simons, 1990, Polarized sorting of viral glycoproteins to the axon and dendrites of hippocampal neurons in culture, *Cell* **62**:63–72.
43. D.S. Papermaster and B.G. Schneider, 1982, Biosynthesis and morphogenesis of outer segment membranes in vertebrate photoreceptor cells, in *Cell Biology of the Eye*, (D.S. McDevitt, editor), pp. 477–531, Academic Press, New York.
44. D.M. Defoe and J.C. Besharse, 1985, Membrane assembly in retinal photoreceptors. II. Immunocytochemical analysis of freeze-fractured rod photoreceptor membranes using anti-opsin antibodies, *J. Neurosci.* **5**:1023–1034.
45. J.A. Rodriguez, C.A. Herrera, D.G. Birch, J.R. Heckenlively, and S.P. Daiger, 1993, Rhodopsin mutations in patients with retinitis pigmentosa, *Am. J. Hum. Genet.* **53**:1224.
46. K. Scott, P.A. Sieving, E. Bingham, V.J. Bhagat, J. Sullivan, M. Alpern, and J.E. Richards, 1993, Rhodopsin mutations associated with autosomal dominant retinitis pigmentosa, *Am. J. Hum. Genet.* **53**:147.
47. G. Restagno, M. Maghtheh, S. Bhattacharya, M. Ferrone, S. Garnerone, R. Samuelly, and A. Carbonara, 1993, A large deletion at the 3′ end of the rhodopsin gene in an Italian family with a diffuse form of autosomal dominant retinitis pigmentosa, *Hum. Mol. Genet.* **2**:207–208.
48. M. Horn, P. Humphries, M. Kunisch, C. Marchese, E. Apfelstedt-Sylla, L. Fugi, E. Zrenner, P. Kenna, A. Gal, and J. Farrar, 1992, Deletions in exon 5 of the human rhodopsin gene causing a shift in the reading frame and autosomal dominant retinitis pigmentosa, *Hum. Genet.* **90**:255–257.
49. T.P. Dryja, L.B. Hahn, G.S. Cowley, T.L. McGee, and E.L. Berson, 1991, Mutation spectrum of the rhodopsin gene among patients with autosomal dominant retinitis pigmentosa, *Proc. Natl. Acad. Sci. USA* **88**:9370–9374.

50. R. Vaithinathan, E.L. Berson, and T.P. Dryja, 1994, Further screening of the rhodopsin gene in patients with autosomal dominant retinitis pigmentosa, *Genomics* **21**:461–463.
51. E.M. Stone, K. Vandenburgh, A.E. Kimura, B.L. Lam, G.A. Fishman, J.R. Heckenlively, R.A. Castillo, and V. Sheffield, 1993, Novel mutations in the peripherin (RDS) and rhodopsin genes associated with autosomal dominant retinitis pigmentosa (ADRP). *Invest. Ophthalmol. Vis. Res.* **34**:1149.
52. A. Gal, A. Artlich, M. Ludwig, G. Niemeyer, K. Olek, E. Schwinger, and A. Schinzel, 1991, Pro-347-Arg mutation of the rhodopsin gene in autosomal dominant retinitis pigmentosa, *Genomics* **11**:468–470.
53. T.P. Dryja, T.L. McGee, L.B. Hahn, G.S. Cowley, J.E. Olsson, E. Reichel, M.A. Sandberg, and E.L. Berson, Mutations within the rhodopsin gene in patients with autosomal dominant retinitis pigmentosa, *New Engl.J. Med.* **323**:1302–1307.

13

A HOMOZYGOUS 1-BASE PAIR DELETION (1147delA) IN THE ARRESTIN GENE IN AUTOSOMAL RECESSIVE RETINITIS PIGMENTOSA

Mitsuru Nakazawa, Yuko Wada, and Makoto Tamai

Department of Ophthalmology
Tohoku University School of Medicine
Sendai, 980-8574, Japan

1. SUMMARY

Retinitis pigmentosa (RP) is a genetically heterogeneous group of diseases that show progressive degeneration of photoreceptors. Its incidence is about one in 3000 to 10,000 people in many countries, and it is a leading cause of blindness in the adult population. Several kinds of mutations have been reported in patients with autosomal recessive RP (arRP) in genes that are normally expressed in rod photoreceptors and that function in the process of phototransduction. Arrestin, one of the photoreceptor-specific proteins, plays a role in the cycle of rhodopsin regeneration.[1] A homozygous 1-base pair deletion mutation in the arrestin gene, designated the arrestin 1147delA, has been known as a frequent cause of Oguchi disease, a form of congenital stationary night blindness, in Japanese patients.[2] Interestingly, molecular genetic screening in patients with arRP, including a sibling of a patient with Oguchi disease, to search for mutations in the arrestin gene revealed that the same 1147delA mutation is also associated with arRP and is cosegregated with the disease. We report on 3 patients with arRP associated with the homozygous 1147delA mutation in the arrestin gene as evidence of variable expressivity of the mutation in the arrestin gene.

2. INTRODUCTION

Arrestin is a photoreceptor-specific soluble protein that normally plays an important role in quenching the phototrasduction cascade by inactivating phosphorylated-

Retinal Degenerative Diseases and Experimental Therapy, edited by Hollyfield *et al.*
Kluwer Academic / Plenum Publishers, New York, 1999.

activated rhodopsin.[1] Because of its important role in conjunction with phototransduction cascade, the arrestin gene has been thought to be one of candidate genes that may cause some forms of hereditary retinal degenerations. Oguchi disease is a form of congenital stationary night blindness that is characterized as a typical golden-yellow fundus reflex and extremely retarded dark adaptation. It is inherited as an autosomal recessive trait, and most Japanese patients with Oguchi disease have had homozygous 1-base pair deletion of the arrestin gene designated as 1147delA mutation,[2] which is located in exon 11 of the gene. Because the 1147delA mutation is found quite frequently in Japanese patients with Oguchi disease and it has not yet found in other patient popullation, this mutation can be a kind of a founder effect occurred in Japanese. Although Oguchi desease is categorized in the group of stationary night blindness, this disease has also been reported in other members of the same family with retinitis pigmentosa (RP)[3] which is progressive form of retinal degeneration, and partial chorioretinal degeneration is sometimes associated with Oguchi desease.[4] These findings have suggested the possibility that mutations in the arrestin gene are related not only with Oguchi disease but also other types of hereditary retinal degeneration such as RP. This possibility accelated us to perform mutation screening on patients with autosomal recessive RP (arRP) to search for mutations in exon 11 of the arrestin gene and we detected the 1147delA mutation in 3 patients. We report, here, molecular genetic and clinical characteristics of these three cases with arRP associated with a homozygous 1147delA mutation in the arrestin gene.

3. MATERIALS AND METHODS

We first screened exon 11 of the arrestin gene to seach for a mutation in a patient of arRP (case 1) whose younger sister was a patient with Oguchi disease associated with 1147delA mutation in the arrestin gene. Next, we molecular genetically examined a patient with arRP which also showed golden-yellow fundus reflex in the peripheral retina (case 2). Thirdly, we found a patient with arRP (case 3) whose electrortinograms were resemble to those of case 1, and subsequently screened exon 11 of the arrestin gene in this patient. Family trees of these 3 patients are shown in Fig. 1. Then we extended the molecular genetic examination to other 117 Japanese patients with arRP

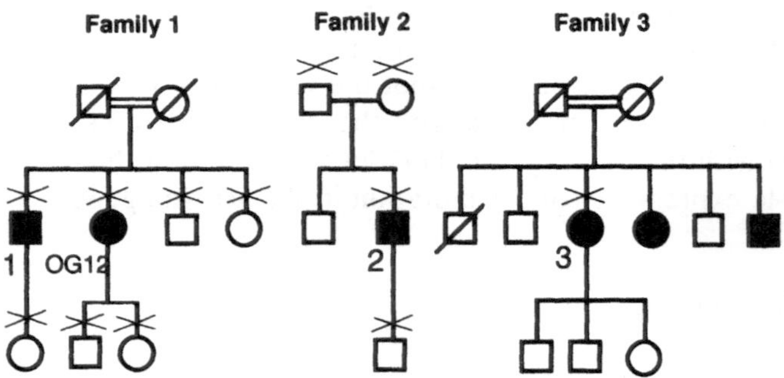

Figure 1. Pedigrees showing affected (solid symbols) and unaffected (open symbols) members with retinitis pigmentosa. Numbers 1, 2, and 3 indicate cases 1, 2, and 3, respectively. Hatched symbol indicates a patient with Oguchi disease (OG 12). Squares indicate male members; circles, females; X, examined; slash, deceased; =, consanguinous marriage.

to search for the 1147delA mutation in the arrestin gene. The mutation screening was performed by polymerase chain reaction (PCR) followed by single strand conformation polymorphism (SSCP) analysis previously described.[4] The abnormally migrated DNA bands on SSCP analysis were subsequently sequenced. The PCR product was subcloned into a plasmid vector (pGEM, Promega). The nucleotide sequencing analysis of the subcloned genomic DNA was performed by dideoxynucleotide chain termination protocol using ALF automatic sequencer (Pharmacia, Sweeden). At least 5 subclones were sequenced to determine a mutation for each patient.

4. RESULTS

We found a homozygous 1147delA mutation in all of cases 1, 2, and 3 (Fig. 2). As far as we examined, the mutation was cosegregated with the disease (Fig. 3). However, no mutation was found in exon 11 in other 117 patients with arRP. Also, we did not find this mutation in 100 chromosomes of 50 normal control subjects.

Although clinical features of cases 1, 2, and 3 have been well described in our previous report,[5] we briefly report here their main clinical characteristics in the next chapter.

5. CASE REPORTS

5.1. Case 1

A 58-year-old man (Fig. 1) had initially noticed night blindness during his early teens, when he had been diagnosed as having RP by an ophthalmologist. His parents were first cousins. His 55-year-old sister (OG 12) had Oguchi disease associated with the arrestin 1147delA mutation.[4] Fundus findings showed atrophic macular lesions in both eyes associated with pigmentary retinal degeneration in midperipheral and

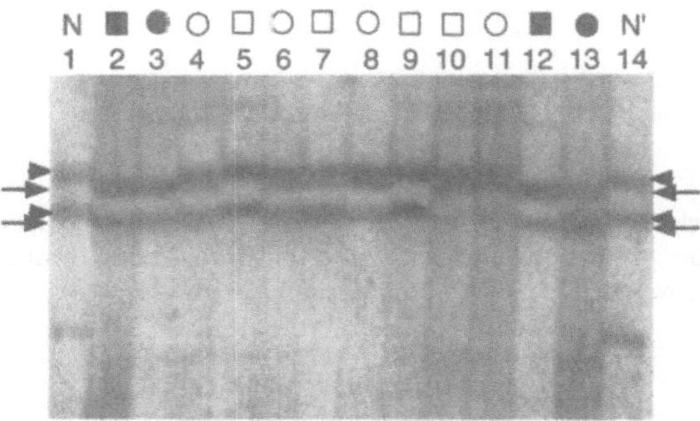

Figure 2. Results of SSCP analysis of amplified DNA fragments encoding exon 11 of the arrestin gene from affected (solid symbols) and non-affected (open symbols) members. A member with a hatched symbol indicates a patient with Oguchi disease. Squares indicate male members and circles indicate female members. Normal controls are designated N and N'. Lanes are 2, case 1; 3, OG 12; 4 and 5; siblings of case 1; 6, a daughter of case 1; 7 and 8, children of OG 12; 9, a grandchild of OG 12; 10 and 11, parents of case 2; 12, case 2; and 13, case 3. Arrowheads indicate the position of the normal allele, and arrows show the position of the mutated allele.

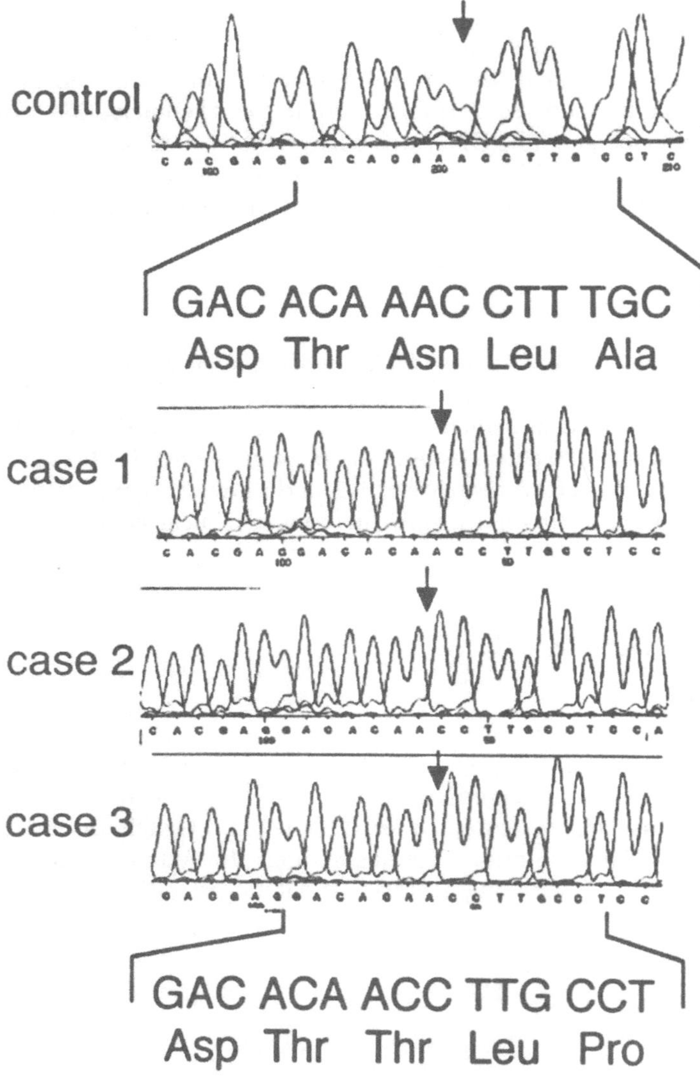

Figure 3. Results of nucleotide sequencing analysis of cases 1, 2, and 3. In the upper most sequence, a normal allele from a normal subject shows adenine nucleotide (A) at the nucleotide position 1147 (arrow). Sequences from cases 1, 2, and 3 show a deletion of A at this position.

peripheral retina in both eyes.[5] Fluorescein angiography revealed hypofluorescent areas indicating chorioretinal atrophy in the macular and pericentral regions and hyperfluorescent areas in the central and midperipheral regions in both eyes.[5] The mixed rod and cone electroretinogram (ERG) disclosed decreased amplitudes of both a-waves and extinguished amplitudes of b-waves.[5] The amplitudes of 30-Hz flicker ERG were reduced in both eyes.[5] Scotopic ERG was nonrecordable.

5.2. Case 2

A 35-year-old man (Fig. 1) had had impaired visual acuity in his teens. In addition to a gradual progression of visual impairment, he also noticed disturbance of night vision. Fundus findings demonstrated bilateral pigmentary retinal degeneration in the posterior portion that extended from the macular area to the midperipheral retina, and abnormal golden-yellow fundus reflex in the peripheral area.[5] Fluorescein angiography disclosed a diffuse hyperfluorescent area in the posterior portion associated with hypofluorescent areas in the pericentral region in both eyes.[5] The mixed rod and cone ERG showed severely reduced amplitudes of a-waves and extinguished responses of b-waves.[5] The amplitudes of 30-Hz flicker ERG were reduced in both eyes.[5]

5.3. Case 3

A 72-year-old woman (Fig. 1) was referred to our clinic because of RP. She had experienced night blindness and constriction of visual field for 20 years. Her parents were first cousins (Fig. 1). Fundus findings showed diffuse pigmentary retinal degeneration in the midperipheral retina associated with attenuated retinal vessels.[5] Fluorescein angiography disclosed diffuse hyperfluorescence in the posterior and midperipheral regions associated with patchy hypofluorescent areas along the vascular arcade.[5] The mixed rod and cone ERG disclosed reduced amplitudes of a-waves and extinguished amplitudes of b-waves bilaterally.[5] The amplitudes of 30-Hz flicker ERG were reduced bilaterally.[5]

6. DISCUSSION

The 1147delA mutation in the arrestin gene has been found as a frequent cause of Oguchi disease in the Japanese population. In the present study, we also found the same mutation in patients with arRP. The previous mutation screening revealed that no other possible disease-causing mutation had been found in the rhodopsin, peripherin/*RDS*, and *ROM-1* genes in our present cases. The results suggest that the mutation in the arrestin gene can cause RP as well as Oguchi disease and that variable expressivity is produced by the mutation in the arrestin gene.

In our previous study on phenotypic characteristics of patients with Oguchi disease, we also found slight variability in clinical features in patients with the arrestin 1147delA mutation.[4] For instance, partial chorioretinal degeneration was seen in a patient with Oguchi disease. And this finding led us to speculate that the arrestin gene might also be a candidate gene of RP. We therefore extended our study and identified that a sibling (case 1) of the patient with Oguchi disease (OG 12) had RP that also was associated with the same mutation as Oguchi disease. Further study revealed the homozygous 1147delA mutation in the arrestin gene in two additional patients with RP. Phenotypic characteristics of these three patients with the 1147delA mutation were also variable.[5] Together with our previous studies,[4,5] clinical expressions associated with the arrestin 1147delA mutation were distributed in Oguchi disease, Oguchi disease with partial chorioretinal degeneration, and RP with or without the golden-yellow fundus reflex. It has been suggested that this variability may be a spectrum of phenotypes caused by a mutation in the arrestin gene. Also, this speculation can explain the coexistence of patients with Oguchi disease and RP in the same family,[3] although the exact mechanism of such a wide range of clinical features is unknown.

It should be noted that the frequency of the 1147delA mutation in arRP may not be so high, because we found only 3 patients with arRP having the 1147delA mutation among 120 patients with arRP. However, we can not estimate the exact frequency of mutations in the arrestin gene until all entire coding region of the gene is screened. Further molecular genetic examination would be needed to clarify this point.

ACKOWLEDGMENT

This study was supported in part by a grant from the Research Committee on Chorioretinal Degenerations and Optic Atrophy, the Ministry of Health and Welfare of the Japanese Government (Dr. Nakazawa), Tokyo, Japan, and a grant-in-aid for scientific research from the Ministry of Education, Science, and Culture of the Japanese Government (Dr. Nakazawa, C-2-09671782), Tokyo, Japan.

REFERENCES

1. H. Kühn, S.W. Hall, and U. Wilden, 1984, Light induced binding of 48-kDa protein to photoreceptor membranes is highly enhanced by phosphorylation of rhodopsin, *FEBS Lett.* **176**:473–478.
2. S. Fuchs, M. Nakazawa, M. Maw, M. Tamai, Y. Oguchi, and A. Gal, 1995, A homozygous 1-base pair deletion in the arrestin gene is a frequent cause of Oguchi disease in Japanese, *Nat. Genet.* **10**:360–362.
3. M. Yamanaka, 1969, Histologic study of Oguchi's disease, *Am. J. Ophthalmol.* **68**:19–26.
4. M. Nakazawa, Y. Wada, S. Fuchs, A. Gal, and M. Tamai, 1997, Oguchi disease: Phenotypic characteristics in patients associated with the frequent 1147delA mutation in the arrestin gene, *Retina.* **17**:17–22.
5. M. Nakazawa, Y. Wada, and M. Tamai, 1998, Arrestin gene mutations in autosomal recessive retinitis pigmentosa. *Arch. Ophthalmol.* **116**:498–501.

14

mRNA ANALYSIS OF OGUCHI PATIENTS WITH THE FREQUENT 1147delA MUTATION IN THE ARRESTIN GENE

Yuko Wada,[1]* Mitsuru Nakazawa,[2] Toshiaki Abe,[1] and Makoto Tamai[1]

[1] Department of Ophthalmology
Tohoku University School of Medicine
[2] Department of Ophthalmology
Hirosaki University School of Medicine

I. ABSTRACT

Arrestin is a 45 KDa protein and is mainly located in the membrane of photoreceptor rod cells and pineal gland. It is well known that 1147delA nutation in the arrestin gene causes Oguchi disease and autosomal recessive retinitis pigmentosa in Japanese patients. We examined expression of arrestin in mRNA isolated from whole blood cells of patients with Oguchi disease, who had the 1147delA mutation in the arrestin gene. Arrestin was expressed at the level of mRNA in all of three patients with Oguchi disease. mRNA from Oguchi patients did not have the same 1147delA mutation as was seen in the arrestin gene but had the normal sequence. The result indicated the possibility of RNA editing in blood cells in our patients.

II. INTRODUCTION

Arrestin is a soluble 45 KDa protein and expressed mainly in rod photoreceptor cells and pineal gland at high level.[1–3] Recent reports have shown that in human arrestin is expressed not only in photoreceptor cells and pineal gland but also in iris, anterior capsule of lens, choroid, and brain.[4–8]

*Corresponding Author: Yuko Wada, MD, Department of Ophthalmology, Tohoku University School of Medicine, 1-1 Seiryo-machi, Aoba-ku, Sendai, Miyagi 980-8574, Japan, tel: 022-717-7294; fax: 022-717-7298; e-mail: YW@oph.med.tohoku.ac.jp

Retinal Degenerative Diseases and Experimental Therapy, edited by Hollyfield *et al.*
Kluwer Academic / Plenum Publishers, New York, 1999.

The functions of arrestin are believed to inhibit the phototransduction cascade by binding to the photo-activated phosphorylated rhodopsin by directly interacting with a-transducin.[1,2,9] Arrestin is also known to induce experimental autoimmune uveitis (EAU).[2]

Recently, linkage analysis on an Indian family showed that the locus of Oguchi disease was on chromosome 2q, in which arrestin gene is located.[10] After this report, mutation screening was performed with Japanese patients with Oguchi disease to search for mutations in the arrestin gene and it was disclosed that a homozygous 1147delA mutation in the arrestin gene was frequent mutation in the Japanese patients with Oguchi disease.[11] We also examined mRNA from Oguchi patients associated with the 1147delA mutation in the arrestin gene to examine the presence of abnormal mRNA in blood cells.

III. MATERIALS AND METHODS

Patients and Materials

Peripheral blood was taken from three patients with Oguchi disease associated with a homozygous 1147delA mutation in the arrestin gene[12] and one normal control. Informed consent was obtained for all subjects.

Extraction of mRNA and Synthesis of cDNA

mRNA were isolated from 400 of fresh peripheral blood by using oligo-dT cellulose (Pharmacia Biothecn, Uppsala, Sweden). These cells were pelleted and suspended in extraction buffer (4 M guanidium thiocynanate and 0.5% N-lauroyl sarcosine). Cleared cell homogenate by centrifugation was mixed with oligo-dT cellulose. This mixture was washed by high salt buffer (10 mM Tris-HCl, pH 7.5, 1 mM EDTA, 0.5 M NaCl) for five times followed by low salt buffer (10 mM Tris-HCl, pH 7.5, 1 mM EDTA, 0.1 M NaCl) for three times. Subsequently, mRNA was eluted by prewarmed elution buffer (10 mM Tris-HCl, pH 7.5, 1 mM EDTA). First strand cDNA was generated by random hexadeoxynucleotides at 0.2 in each reaction, which was catalysed by Moloney Murine Leukemia Virus reverse transcriptase (Phamacia Biotech).

Reverse Transcriptase-Polymerase Chain Reaction (RT-PCR)

We set primers to amplify 175 bp between exons 11 and 12 in the arrestin gene (foward primer: 5′-GCACTTTGACCAAGACGTTG-3′, reverse primer: 5′-TTGATCTGGTAAGACACCAG-3′) to avoid amplification. The RT-PCR reactions were carried out in a thermocycler (Perkin Elmer, Norwalk, CT, U.S.A.). Briefly, the PCR reaction mixture (50) contained 1.01 of cDNA solution, 20 of each primer, 200 each of dATP, dCTP, dGTP, and TTP, 50 mM of KCl, 10 mM of Tris-HCl (pH 8.3), 1.5 mM of MgCl and 0.001% gelatin, and 2.5 units of Taq polymerase for 45 cycles. Following denaturation of the reaction mixture at 94C for 2 minutes, PCR was performed as follows: 45 cycles consisting of 94C for 1 minute for denaturation, 45C for 2 minutes for anealing, and 72C for 2 minutes for polymerization. Each PCR product was analysed in 1.5% agarose gel (Seakem, FMC Bioproducts, Rockland, ME, U.S.A.) containing 0.005 mg/ml of ethisium bromide. DNA was detected by using an ultraviolet transilluminator. As a negative control, distilled water without containing mRNA sample was used for mRNA solution at the same time of each RT-PCR procedure All other solutions and procedures were completely the same in each RT-PCR.

Polymerase Chain Reaction (PCR) of Genomic DNA Samples

Genomic DNA samples were isolated from the same blood samples of patients with Oguchi disease using QIAmp Blood Kit? @(Qiagen, Hilden, Germany). The exon 11 of the arrestin gene was amplified from the genomic DNA samples by the method previously described.[11] The amplified DNA fragments were subsequently processed for nucleotide sequencing analysis.

Sequence Analysis

For sequence analysis analysis, the amplified PCR products was subcloned into a plasmid vector (pGEM: Promega, Madison, WI, U.S.A.) and sequenced with an automatic DNA sequencer (A.L.F. DNA Sequencer, Pharmacia, Sweden) using a dideoxy-chain termination protocol.[13]

IV. RESULTS

After we obtained samples, mRNA was immediately isolated and cDNA was subsequently synthesized. The RT-PCR was carried out with our primers, which were set to amplify exons 11 and 12 in the arrestin gene. The results of RT-PCR are shown in Fig. 1. The nucleotide sequencing analyses of these RT-PCR products revealed that a-arrestin is expressed in all our samples.

Subsequently, the same RT-PCR analysis was performed in mRNA samples independently purified from three patients with Oguchi disease, who had the homozygous 1147delA mutation in the arrestin gene. The result disclosed that the RT-PCR products were the same size as this from retina. The nucleotide sequencing analysis revealed that the RT-PCR products from all of three Oguchi patients did not have the same 1147delA mutation as had been seen in genomic DNA but had normal sequence (Fig. 2).

V. DISCUSSION

Recently, by using RT-PCR technique, ectopic transcription has been reported. Furthermore this technique is applied to the diagnosis of some kinds of hereditary metabolic disorders without performing liver biopsy.

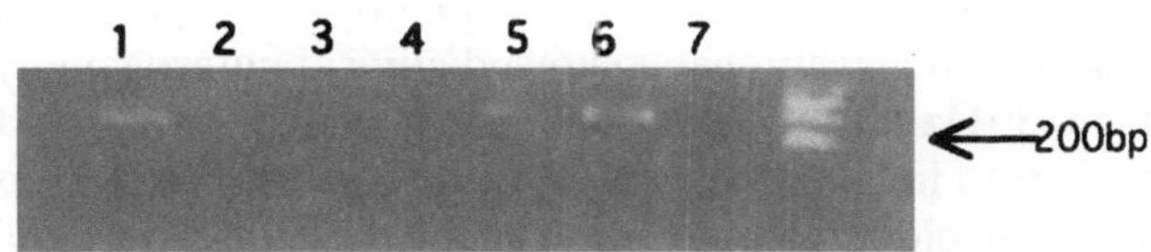

Figure 1. Result of RT-PCR of the arrestin gene. 1, retina; 2–4, blood from Oguchi patients; 5,6, blood from a normal human control; 7, negative control.

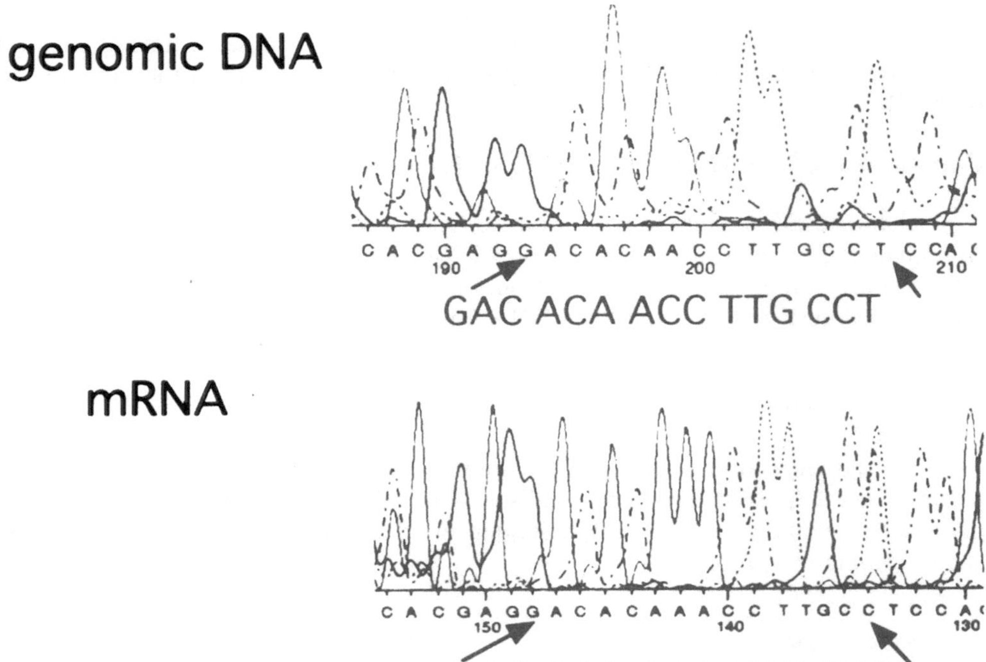

Figure 2. The results of sequence analysis of mRNA and genomic DNA of a Oguchi patient associated with 1147delA mutation in the arrestin gene. This sequence showed a part of exon 11. The nucleotide sequence derived from mRNA did not have a 1147delA mutation, whereas genomic DNA had the mutation. The same pattern was also seen in the other two Oguchi patients.

Until now it has been reported expression of the arrestin was in the retina and pineal gland at high level and in the iris, choroid, lens epithelial cells with anterior capsule attached and brain at low level.[1–3,8] In this study, we examined ectopic transcription of the arrestin gene in human blood cells. By RT-PCR, it was disclosed that arrestin was expressed in our all samples.

Arrestin has high homology with b-arrestin, so we had to prove that these RT-PCR products came from a-arrestin. To eliminate this possibility, we performed nucleotide sequencing analyses and the results disclosed that a-arrestin was expressed in all our samples. We discovered that human a-arrestin gene was ectopically expressed in human blood cells other than retina, pineal gland, although biological significance is still unknown.

In 1995, it was reported that 1147delA mutation in the arrestin gene caused Oguchi disease in Japanese patients.[10,11] Subsequently, we reported that patients with 1147delA mutation in the arrestin gene showed clinical heterogeneity, for example, in fundus appearance, 30 Hz-flicker ERG and visual acuity.[12] This study disclosed that arrestin is also expressed in human blood, therefore, as the second step, we investigated the mRNA from fresh blood samples of Oguchi patients associated with 1147delA mutation in the arrestin gene. The sequence analysis showed that mRNA from blood had the normal sequence. We carefully eliminated the possibility that this finding was due to a result of contamination. For instance, we used a negative control in which no mRNA sample was contained at the same time when RNA samples were processed. And we did not detect any amplified product in the negative control. In addition, each Oguchi patient's RT-PCR was independently carried out at defferent time point to rule out a possibility of contamination.

Recently, discrepancy between genomic DNA and mRNA has been found and this phenomenon is called RNA editing. The errors in the sequence are corrected by the process of transcription. Fred et aL reported the possibility that RNA editing was a mean of producing phenotypic variability.[14] Therefore, possible explanation of the discrepancy between sequences of mRNA and genomic DNA is RNA editing. Because the genomic DNA samples from all three Oguchi patients showed homozygous 1147delA and they did not show any heterozygosity as far as we examined, the mechanism by which the 1147delA was corrected is largely unknown. In this study, we could not investigate if RNA editing occurred in the photoreceptor cells of patients with Oguchi disease, because we could not get mRNA from retina of these patients at the present time. But because each patient with Oguchi disease we reported showed complete cosegregation between 1147delA mutation and clinical findings,[11,12] it is reasonable that Oguchi disease is caused by the 1147delA mutation in the arrestin gene. Further investigation using some other tissues from Oguchi patients will help us speculate the presence of RNA editing in these patients. We also speculate that the ratio of RNA editing in the retina with Oguchi disease may vary with their clinical severity. Further investigation using some other tissues from Oguchi patients will help us speculate the presence of RNA editing in these patients. Furthermore, recently we reported three independent patients with progressive retinal degeneration had the same 1147delA mutation in the arrestin gene.[15,16] We hypothesize that these patients almost do not have RNA editing in their retinas. These speculations are reasonable to explain that 1147delA mutation in the arrestin gene causes not only Oguchi disease but also progressive retinal degeneration. Further investigation is necessary to clarify the relationship between mRNA editing and clinical severity with Oguchi disease.

ACKNOWLEDGMENT

This study was supported in part by a grant from the Research Committee on Chorioretinal Degenerations and Optoc Atrophy, the Ministry of Health and Welfare of the Japanese Government (Dr. Nakazawa), Tokyo, Japan and a grant-in-aid for scientific research from the Ministry of Education, Science and Culture of the Japanese Government (Dr. Nakazawa, C-2-09671782), Tokyo, Japan.

A part of this paper was accepted in Japanese Journal Ophthalmology in 1998.10.2

REFERENCES

1. Wacker W.B., Donoso L.A., Kalsow C.M., et al. 1977, Experimental allergic uveitis. Isolation, characterization, and localization of a soluble uveitopathogenic antigen from bovine retina, J Immunol 119:1948–1958.
2. Gery I., Mochizuki M., and Nussenblatt R.B., 1986, Retinal specific antigens and immunopathogenic processed they provoke, Prog Retinal Res. 5:75–109.
3. Shinohara T., Donosol L., Tsuda M., Yamaki K., and Singh V.K., 1989, S-Antigen. Structure, function, and experimental autoimmune uveitis (EAU), Prog Retinal Res. 8:51–66.
4. Zaal J., Doekes G., Breebaart A.C., and Kijlstar A., 1986, Quantitative determination of S-antigen in human ocular tissues, aqueous humor, and serum, Curr Eye Res. 5:763–775.

5. Mirshahi M., Borgese F., Razaghi A., Scheuing U., et al. 1989, Immunological detection of arrestin, a phototransduction regulatory protein, in the cytosol of nucleated erythrocytes, FEBS Lett. 258: 240–243.
6. Faure J.-P., and Mirshahi M., S-antigen in nonocular tissues, 1990, Curr Eye Res. 9:163–7
7. Breitman M.L., Tsuda M., Usukura J., et al. 1991, Expression of S-antigen in retina, pineal gland, lens, and brain is derected by 5-flanking sequences, J Bio Chem. 226:15505–15510.
8. Qian L., Abe T., Kikuchi T., et al. 1994, Corticosteroids enhance S-antigen expression in ocular tissues of rats with experimental autoimmune uveitis, Exp Mol Pathol. 60:27–38.
9. Yamaki K., Tsuda M., and Shinohara T., 1988, The sequence of human retinal S-antigen reveals similarities with a-transducin, FEBS Lett. 234:39–43.
10. Maw M.A., John S., Jablonka S., et al. 1995, Oguchi disease: Suggestion of linkage to markers on chromosome 2q, J Med Genet. 32:396–398.
11. Fuchs S., Nakazawa M., Maw M.A., et al. 1995, A homozygous 1-base pair deletion in the arrestin gene is a frequent cause of Oguchi disease in Japan, Nat Genet. 10:360–362.
12. Nakazawa M., Wada Y., Fuchs S., Gal A., and Tamai M., 1997, Oguchi disease. Phenotypic characteristics of patients with the frequent 1147delA mutation in the arrestin gene, Retina. 17:17–22.
13. Sanger F., Donelson J.E., Coulson A.R., Kossel H., and Fisher D., 1973, Use of DNA polymerase I primed by a synthetic oligonucleotide to determine a nucleotide sequence in phage f1 DNA, Proc Natl Acad Sci USA. 70:1209–1213.
14. Fred W., Dominique P., Helema H., et al. 1998, Frameshift mutants of b-amyloid precursor protein and ubiquitin-B in Alzheimer's and Down patients, Science. 279:242–246.
15. Nakazawa M., Wada Y., and Tamai M., 1998, Arrestin gene mutations in autosomal recessive retinitis pigmentosa, Arch Ophthalmol. 116:498–501.
16. Wada Y., Nakazawa M., and Tamai M. A patient with progressive retinal degeneration associated with homozygous 1147delA mutation in the arrestin gene, 1997, In: LaVail M.M., Hollyfield J.G., Anderson R.E., ed. Degenerative Retinal Diseases. New York: Plenum Press. 319–322.

MUTATION SPECTRUM IN THE *REP-1* GENE OF JAPANESE CHOROIDEREMIA PATIENTS

Yoshihiro Hotta,[1]* Keiko Fujiki,[1] Mutsuko Hayakawa,[1] Misako Takeda,[1] Nao Kohno,[1] Atsushi Kanai,[1] Yukihiko Mashima,[2] Mikirou Mori,[3] Osamu Okajima,[3] Akira Murakami,[4] Masaru Yoshii,[4] Masayuki Matsumoto,[5] Seiji Hayasaka,[5] Nobuko Tagami,[6] Yasushi Isashiki,[7] and Norio Ohba[7]

[1] Department of Ophthalmology
Juntendo University School of Medicine, Tokyo, Japan
[2] Department of Ophthalmology
School of Medicine, Keio University, Tokyo, Japan
[3] Department of Ophthalmology
The University of Tokyo School of Medicine, Tokyo, Japan
[4] Department of Ophthalmology
National Defense Medical College, Tokorozawa, Japan
[5] Department of Ophthalmology
Faculty of Medicine, Toyama Medical, and Pharmaceutical University
Toyama, Japan
[6] Department of Ophthalmology
Kansai Medical University, Moriguchi, Japan
[7] Department of Ophthalmology
Kagoshima University Faculty of Medicine, Kagoshima, Japan

ABSTRACT

Background: Although the mutations of the *REP-1* gene were reported in the Caucasian patients with choroideremia, no report was found in non-Caucasian patient except our previous case report. We performed multi-institutional study of the *REP-1* gene in Japanese patients with choroideremia.

* To whom correspondence should be addressed:
Department of Ophthalmology, Juntendo University School of Medicine, 3-1-3 Hongo, Bunkyo-ku, Tokyo 113, Japan Tel. +81-3-3813-3111, Fax. +81-3-3817-0260

Retinal Degenerative Diseases and Experimental Therapy, edited by Hollyfield *et al.*
Kluwer Academic / Plenum Publishers, New York, 1999.

Methods: Twenty four unrelated Japanese families with choroideremia were analyzed. It included 24 patients and a carrier. The method for screening was a non-radioisotopic modification of single strand conformation polymorphism (SSCP). The PCR products from the patients and the carrier were screened and directly sequenced using an automated DNA sequencer.

Results: SSCP analysis showed abnormal band shifts in the patients and the carrier. Direct sequence analysis showed 15 kinds of mutations in 19 families. They included 8 nonsense mutations, 7 frameshift mutations, and 4 point mutations in the splice site.

Conclusion: The mutations in a racial background different from families reported before confirm that the *REP-1* gene defect causes choroideremia. Most of Japanese patients with choroideremia are caused by the *REP-1* gene mutations. No missense mutation has been found in Japanese patients either. Although large deletion was sometimes recognized in European patients, we could not find it in this series.

INTRODUCTION

Choroideremia is an X-linked chorioretinal dystrophy in which affected males suffer from progressive night blindness and visual field constriction leading often total blindness. Female carrier is usually asymptomatic, however can be diagnosed because of characteristic pigment abnormalities of the ocular fundus. Analyses of microscopic deletions in patients with choroideremia have assigned the choroideremia gene to small interval of Xq21. Human choroideremia, *REP-1* gene was successfully isolated by the positional cloning procedure and the genomic structure of the *REP-1* gene, which consists of 15 exons, was determined.[1–2] We previously reported a Japanese family with choroideremia with *REP-1* gene mutation.[3]

The proband is 31-year-old male who was diagnosed as the pinealoma in his 12-year-old. His asymptomatic mother is a 57-year-old female and is diagnosed as a carrier showing diffuse spotty areas within the pigment epithelium at the mid-peripheral area of the ocular fundus. A hemizygous A to CC mutation was detected in nucleotide position 1608 of the *REP-1* gene in the patient, resulting in the absence or truncation of the predicted REP-1 protein. The heterozygous mutation was detected in his carrier mother.

Although about 50 mutations of the *REP-1* gene were reported in the North American and European patients with choroideremia,[4–10] no report was found in non-Caucasian patients except our previous report.[3] We performed multi- institutional study of the *REP-1* gene in Japanese patients with choroideremia.

MATERIAL AND METHODS

Twenty four individual Japanese families with choroideremia were analyzed, including 24 patients (a brother) and a carrier; in one family sample were not available from a patient but from an obligate carrier. Genomic DNAs were extracted from leukocytes of peripheral blood collected from 24 patients with choroideremia and a carrier. Exons 1 to 15 were amplified by polymerase chain reaction (PCR) using the genomic DNAs and primers described before.[4] The PCR was carried out in a volume of 100 µl in a DNA Thermal Cycler (Perkin Elmer Corporation, USA). PCR condition is shown in Table 1.

Table 1. Annealing temperature of the PCR amplification*

Exon	Van Bokhoven[6] (°C)	This study (°C)
1	59	60
2	51	54
3	54	54
4	61	60
5A, 5B	53	57
6	51	54
7	51	52
8	53	52
9	51	50
10	61	60
11–15	55	54

*Denature, 94°C; primer extention, 72°C.

The method for mutation screening was a nonradioisotopic modification of single-strand conformation polymorphism (SSCP). Aliquots (10 µl) of PCR products were subsequently diluted with an equal volume of 95% formamide and denatured at 95°C for 40 min. and then underwent electrophoresis in nondenaturing 12.5% polyacrylamide gel. Electrophoresis was performed at 20°C with the Genephor system. After electrophoresis, DNA bands were visualized by silver staining. If the genetic defect was suspected in the SSCP analysis, the PCR amplified DNA cut out from 1.4% low melting agarose gels were directly sequenced. Direct sequence was performed using Taq Dye Deoxy™ Terminator Cycle Sequencing Kit by an automated DNA sequencer, Model 373A (Applied Biosystems, USA).

RESULTS

The results are summarized as shown in Table 2. We found 15 kinds of mutations in 19 families. They included 8 nonsense mutations, 7 frameshift mutations, 4 point mutations in splice sites. No missense mutation was observed.

Figure 1 showed the examples of the SSCP analysis. SSCP analysis showed abnormal band shift in the patients. Fig. 2 showed the examples of the direct sequence analysis. Direct sequence analysis showed various genetic defect in the *REP-1* gene in the patients. The sequence result of the carrier is slightly complicated after the position of the mutation suggesting existence of both wild and mutant alleles.

The mutations were confirmed using both sense and antisense primers.

DISCUSSION

The genetic defects of *rhodopsin* and *peripherin/RDS* gene causing the retinal degeneration was well investigated and reported in North America, Europe, and Japan.[11–14] However, the genetic defect of the *REP-1* gene causing choroideremia was reported only in Caucasian patients except our previous report.[3–10] The mutation was found in a racial background different from families reported before confirming that the *REP-1* gene defect causes choroideremia.

We analyzed 24 patients and a carrier from 24 families and found 15 kinds of mutations in 19 patients and a carrier from 19 families. We employed in this study the

Table 2. REP-1 gene mutations in Japanese patients with choroideremia

Patient	Exon	Position	Mutation	Result
K5(23)	intron 1	80 (−1)	AG→AC	splice error
T4(31)	intron 1	80 (−1)	AG→AC	splice error
T6(32)	intron 1	80 (−1)	AG→AC	splice error
HT(17)	intron 2	146 (+1)	GT→AT	splice error
KM(wr)*	3	170	TGG→TAG	Try 47 stop
EH(10)	5	345–348	deletion (TCAG)	frameshift
SH(11)	5	345–348	deletion (TCAG)	frameshift
T2(30)	5	594–595	deletion (GC)	frameshift
KY(1)	5	655	CAA→TAA	Gln 209 stop
DT(8)**	6	787	CGA→TGA	Arg 253 stop
T(9)**	6	787	CGA→TGA	Arg 253 stop
TU(6)	6	829	CGA→TGA	Arg 267 stop
FI(5)	6	838	CGA→TGA	Arg 270 stop
K4(22)	6	838	CGA→TGA	Arg 270 stop
NE(7)	8	1064	TCA→TGA	Ser 345 stop
MM(13)	9	1238	TCA→TGA	Ser 403 stop
MK(3)	10	1313–1314	deletion (TC)	frameshift
HK(4)	10	1374	insertion (C)	frameshift
HM⁺	13	1608	A→CC	frameshift
TM(25)	14	1789	deletion (G)	frameshift

*Carrier only.
**Brothers.
⁺A case who was previously reported[3].

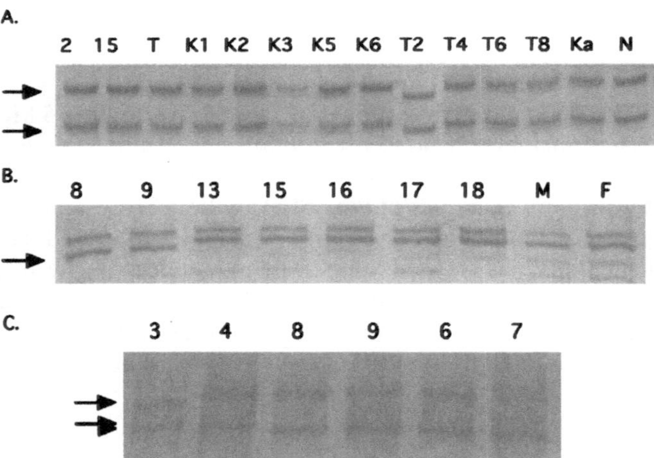

Figure 1. Results of the SSCP analysis of the *REP-1* gene. A, exon 5B; N, normal control; 2, 15, T, T2, T4, T6, T8, T6, T8, K1, K2, K3, K5, K6, Ka, choroideremia patients; the identical abnormal band shifts (arrows) are observed in patient T2. B, exon 6; M, F, normal controls; 8, 9, 13, 15, 16, 17, 18, choroideremia patients; the identical abnormal band shift (arrow) is observed in patients 8 and 9 (brothers). C, exon 10; 3, 4, 6, 7, 8, 9, choroideremia patients; the identical large abnormal band shifts are observed in patients 3 (arrows); the small band shift is also observed in patient 4 (arrow head).

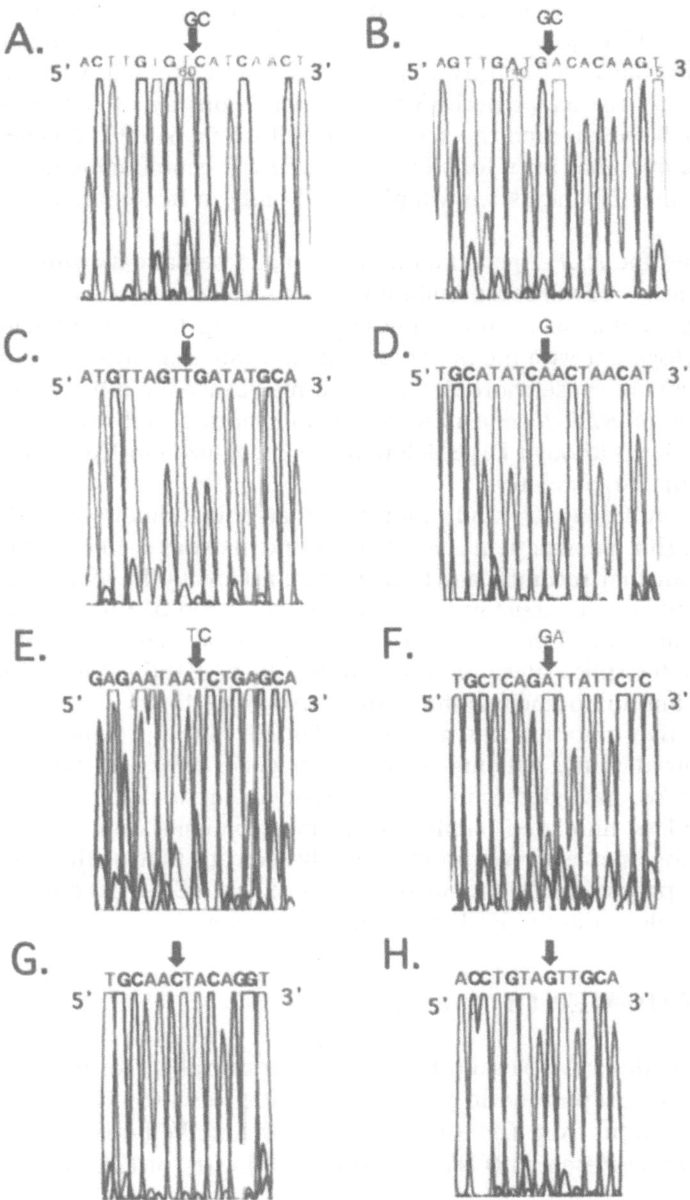

Figure 2. Nucleotide sequences of the *REP-1* gene determined by the automated DNA sequencer using sense (A, C, E, G) and antisense (B, D, F, H) primers. PCR product was amplified and directly sequenced by dideoxy terminator method as described in the text. Arrows indicate the position of the mutation. The hemizygous mutations are detected in the patients. A, B, a deletion (GC) of nucleotide positions (nts) 594 and 595 in patient T2; C, D, a nonsense mutation of codon 253 (CGA to TGA) in patient 8; E, F, a deletion (TC) of nts 1313 and 1314 in patient 3; G, H, an insertion (C) between nts 1374 and 1375 in patient 4.

nonradioisotopic SSCP. Although the nonradioisotopic SSCP required more amount of DNA than the radioisotopic SSCP, the nonradioisotopic SSCP is safer and easier, is adequate in the clinical laboratory. Since no band shift of SSCP was recognized in 3 out of the other 5 patients, we performed the direct sequence of all the exons of *REP-1* gene. Although we checked the sequence of all exons of *REP-1* gene in the three patients, no mutation has been found yet. Since *REP-1* gene fragments was not amplified well in another 2 patients, we will plan the Southern blot analysis using the REP-1 cDNA probe.

We showed the *REP-1* gene mutations in Japanese patients with choroideremia. The defects consists of nonsense mutations, deletions, an insertion, and mutations in splice junctions. No missense mutation has been found in Japanese patients either. The present results together with the previous study indicate that missense mutation of the *REP-1* gene may not cause choroideremia and may cause the other retinal dystrophy. Further analysis of *REP-1* gene is required both normal individuals and the other retinal dystrophies. Although large deletion was recognized in European patients, we could not find the large deletion.

More recent European study indicates that point mutations and small insertion/deletion in European patients have been found in exons 5–14 of the *REP-1* gene.[10] They pointed out that recombinant REP-1 protein lacking the 49 amino acids in the C-terminus is still able to assist in the geranylgeranylation of Rab proteins, indicating that mutations in exon 15 may not cause choroideremia.[10] The mutations in Japanese patients include the splice sites mutations around exons 1 and 2, and a nonsense mutation of exon 3, but do not include mutation in exon 4 or 15.

The causative mutation has not been identified in approximately one-third of European Choroideremia patients screened for gene defects in the REP-1 protein coding region.[10] We have determined the causative genetic defects of *REP-1* gene in 19 from 24 (79.1%) unrelated families in Japanese patients with choroideremia. The present result together with the previous study indicate that further investigation is required in the promoter and the intron region of the *REP-1* gene in choroideremia patients without defect in the REP-1 protein coding region.

ACKNOWLEDGMENTS

The authors thank Ms Misako Takeda, Ms Naoko Yahagi for technical assistance.

This work was supported in part by grants for Research Committee on Chorioretinal Degenerations from The Ministry of Health and Welfare of Japan, and Grant-in-Aid for Scientific Research (C05807166) from Ministry of Education, Science, and Culture.

REFERENCES

1. F.P.M. Cremers, D.J.R. van de Pol, L.P.M. van Kerkhoff, B. Wieringa, and H.-H. Ropers, 1990, Cloning of a gene that is rearranged in patients with choroideremia. *Nature* **347**:674–677.
2. H. van Bokhoven, J.A.J.M. van den Hurk, L. Bogerd, C. Philippe, S. Gilgenkrantz, P. de Jong, H.-H. Ropers, and F.P.M. Cremers, 1994, Cloning and characterization of the human choroideremia gene. *Hum Mol Genet* **3**:1041–1046.
3. Y. Hotta, K. Fujiki, M. Hayakawa, N. Kohno, H. Kitagawa, R. Doi, and A. Kanai, 1997, A hemizygous A to CC base change of the *CHM* gene causing choroideremia associated with pinealoma. *Graefe's Arch Clin Exp Ophthalmol* **235**:653–655.

4. F.P.M. Cremers, E.-M. Sankila, F. Brunsmann, M. Jay, B. Jay, A. Wright, A.J.L.G. Pinckers, M. Schwartz, T.J.R. van de Pol, B. Wieringa, A. de la Chapelle, I. Pawlowitzki, and H.-H. Ropers, 1990, Deletions in patients with classical choroideremia vary in size from 45 kb to several megabases. *Am J Hum Genet* **47**:622–628.
5. E-M, Sankila, R. Tolvanen, J.A.J.M. van den Hurk, F.P.M. Cremers, and A. de la Chapelle, 1992, Aberrant splicing of the CHM gene is a significant cause of choroideremia. *Nature Genet* **1**:109–113.
6. H. Van Bokhoven, M. Schwartz, S. Andréasson, J.A.J.M. van den Hurk, L. Bogerd, M. Jay, K. Rüther, B. Jay, I.H. Pawlowitzki, E.-M. Sankila, A. Wright, H.-H. Ropers, T. Rosenberg, and F.P.M. Cremers, 1994, Mutation spectrum in the *CHM* gene of Danish and Swedish choroideremia patients. *Hum Mol Genet* **3**:1047–1051.
7. N. Nesslinger, G. Mitchell, P. Strasberg, and I.M. MacDonald, 1996, Mutation analysis in Canadian families with choroideremia. *Ophthalmic Genetics* **17**:47–52.
8. P. Forsythe, A. Maguire, R. Fujita, C. Moen, A Swaroop, and J. Bennett, 1997, A carboxy-terminal truncation of 99 amino acids resulting from a novel mutation (Arg555—stop) in the CHM gene leads to choroideremia. *Exp Eye Res* **64**:487–490.
9. L. Beaufrère, S. Tuffery, C. Hamel, C. Bareil, B. Arnaud, J. Demaille, and M. laustres, 1997, The protein truncation test (PTT) as a method of detection for choroideremia mutations. *Exp Eye Res* **65**:849–854.
10. J.A.J.M. van den Hurk, M. Schwartz, H. van Bokhoven, T.J.R. van de Pol, L. Bogerd, A.J.L.G. Pinckers, E.M. Bleeker-Wagemakers, I.H. Pawlowitzki, K. Rüther, H.-H. Ropers, and F.P.M. Cremers, 1997, Molecular basis of choroideremia (CHM): Mutations involving the Rab escort protein-1 (*REP-1*) gene. *Hum Mutat* **9**:110–117.
11. T.P. Dryja, 1992, Doyne lecture, Rhodopsin, and autosomal dominant retinitis pigmentosa. *Eye* **6**: 1–10.
12. B.S. Shastry, 1994, Retinitis pigmentosa and related disorders, Phenotypes of rhodopsin and peripherin/RDS mutations. *Am J Med Genet* **52**:467–474.
13. A.C. Bird, 1995, Retinal photoreceptor dystrophies Ll. Edward Jackson memorial lecture. *Am J Ophthalmol* **119**:543–562.
14. Y. Hotta, K. Fujiki, M. Hayakawa, H. Sakuma, H. Kawano, A. Kanai, A. Murakami, M. Yoshii, K. Akeo, S. Okisaka, M. Matsumoto, S. Hayasaka, Y. Isashiki, N. Ohba, T. Shiono, and M. Tamai, 1997. Screening of candidate genes on Japanese retinal dystrophies, in: *Degenerative Retinal Diseases*, Volume 28 (M.M. LaVail, J.E. Hollyfield, and R.E. Anderson eds), pp. 255–261, Plenum Press, New York.

16

PHENOTYPE-GENOTYPE CORRELATIONS IN RETINAL DEGENERATIONS CAUSED BY ABCR GENE MUTATIONS

Jean-Michel Rozet,[1]* Sylvie Gerber,[1]* Imad Ghazi,[2] Isabelle Perrault,[1] Eric Souied,[1] Dominique Ducroq,[1] Annick Cabot,[1] Jean-Louis Dufier,[2] Gabriel Coscas,[3] Gisèle Soubrane,[3] Arnold Munnich,[1] and Josseline Kaplan[1]

[1] Unité de Recherches sur les Handicaps Génétiques de l'Enfant
INSERM U-393
Hôpital des Enfants Malades, 149 rue de Sèvres
75743 Paris Cedex 15, France
[2] Service d'Ophthalmologie, Hôpital Necker
Paris, France
[3] Clinique Ophthalmologique,
Hôpital Intercommunal de Créteil
Créteil, France

ABSTRACT

Stargardt disease (STGD) and late-onset fundus flavimaculatus (FFM) are autosomal recessive conditions leading to macular degenerations in childhood and adulthood, respectively. Recently, mutations of the photoreceptor cell-specific ATP binding transporter gene (ABCR) have been reported in STGD on one hand, and in some age related macular dystrophies (AMD) on the other hand. In addition, autosomal recessive retinitis pigmentosa (RP19) has recently been ascribed to mutations in the same gene. To provide the first genotype-phenotype correlations in ABCR gene mutations, the screening of the 50 exons encoding ABCR have been performed in a large series of FFM and STGD families as well as in familial cases of AMD and in pedigrees segregating both STGD and RP19. Giving consideration to the results of this study, we

* Equally contributed.

Retinal Degenerative Diseases and Experimental Therapy, edited by Hollyfield *et al.*
Kluwer Academic / Plenum Publishers, New York, 1999.

suggest that homozygosity for truncating ABCR gene mutations result in severe RP while compound heterozygosity for frameshift and missense mutations at this locus result in a severe macular disorder, STGD. Conversely, all mutations identified in milder macular disorder, FFM, were missense mutations. Finally, some AMD patients were found to carry heterozygote ABCR gene mutations, suggesting that these mutations could represent in some cases a susceptibility factor for the disease.

INTRODUCTION

Stargardt disease (STGD), originally described in 1909, is an autosomal recessive condition of childhood, characterized by a bilateral loss of central vision over a period of several months.[1] This frequent cause of macular degeneration in children accounts for about 7% of all retinal dystrophies.[2] It has an early onset (between 7 and 12 years of age), a rapidly progressive course and a poor final visual outcome. The final visual acuity ranges from 1/60 to 3/60 and the peripheral visual field remains normal throughout life.

The STGD gene has been mapped to the short arm of chromosome 1[3] in a narrow genetic interval, subsequently assigned to band p22.1.[4] In addition, we have provided evidence of genetic homogeneity of the disease.[3] On the other hand, in 1965, Franceschetti described another form of fleck fundus disease termed late-onset *fundus flavimaculatus* (FFM, ref 5) with a Stargardt-like phenotype, but differing by a later onset, a more slowly progressive course and a better prognosis. When visual acuity loss begins in the first two decades, the disease is termed STGD while the term "FFM", also called late STGD, is favored when the disease begins later in life and has a slowly progressive course.[6] Yet, the two conditions have long been regarded as the variable manifestations of the same disease. Indeed, the FFM gene has been mapped to the STGD locus, supporting the view that the two conditions are allelic disorders.[7] Moreover, owing to the broad spectrum of age of onset in FFM (17 to 63 years), this genetic mapping addressed the intriguing question of the relevance of this locus in age-related macular degenerations (AMD) and, more generally, of a possible continuum of macular disorders at this locus from early childhood to the elderly.[7]

Mutations in the photoreceptor cell-specific ATP-binding transporter (ABCR) gene have been recently reported in STGD,[8] in some AMD[9] and more interestingly in retinitis pigmentosa (RP19, ref 10). The ATP-Binding Cassette (ABC) family, also called "traffic ATPases", belongs to a superfamily of membrane proteins involved in the energy-dependent transport of a wide variety of substrates across membranes.[8,11] Here, we report on the screening of the 50 ABCR exons[12] in 55 unrelated STGD and FFM families as well as in f in pedigrees segregating STGD and AMD or STGD and RP19.

PATIENTS AND METHODS

Patients

A total of 58 patients belonging to 55 unrelated families were split into two groups based on their age at onset of the disease (STGD before 12 years, FFM after 17 years). Among them, 40 families had STGD (15 multiplex families) and 15 families had late-

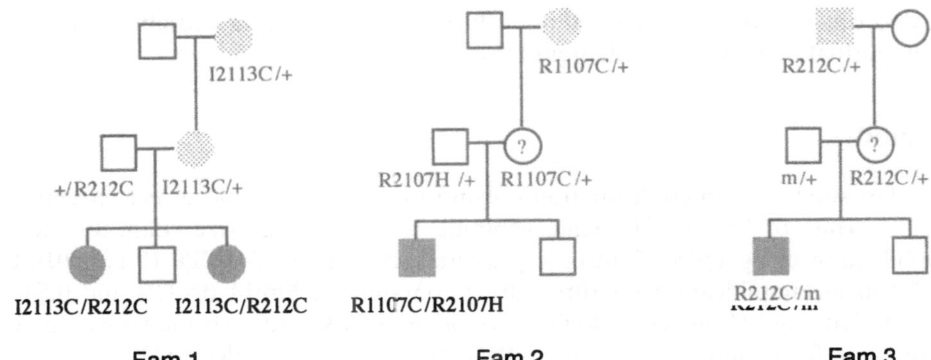

Figure 1. Pedigrees and segregating of ABCR gene mutations in STGD/AMD families. + = wild type allele, dark shading = STGD, light shading = AMD.

onset FFM (4 multiplex families). Ophthalmological data were available for each affected individual. The time course of the disease over approximately ten years (1–40 years) was obtained by interviewing the patients or their parents and a pedigree was established. The minimal criteria for diagnosis of STGD and FFM have been previously described.[3,7] Only 34/110 alive parents were older than 60 years and 49/220 alive grand-parents were older than 70 years. Among them, 1/34 parent and 3/49 grand-parents were known to be affected with AMD (Fig. 1).

Finally, we have ascertained 5 unrelated families segregating both STGD and RP (Fig. 2). In these families, RP had an early onset (before the end of the first decade) and was characterized by both rod and cone impairment. The fundus examination

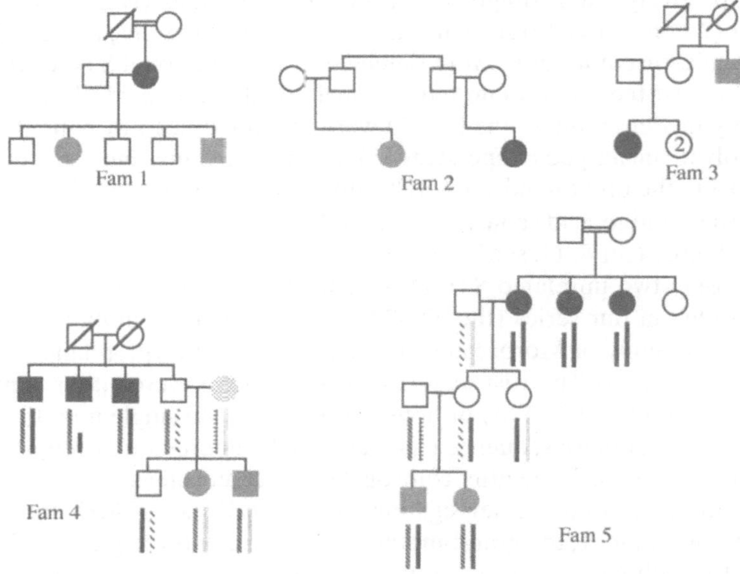

Figure 2. Pedigrees and haplotype analysis in families segregating both STGD and RP. The haplotypes are only given for the two largest pedigrees. Dark shading = STGD, black = RP, light shading = AMD.

showed bone corpuscular pigmentation throughout the retina and abnormal atrophy of the retinal pigment epithelium and the choriocapillaris.

Methods

For single strand conformation polymorphism analysis and direct sequencing of the 50 exons of the ABCR gene, genomic DNA (200ng) was submitted to PCR amplification using 1 µM of intronic primers[12] and alpha ^{33}P dCTP (0.1 µl, 10 mCi/ml NEN) in an amplification mixture (20 µl) containing 200 mM dNTPs and 0.5 U Taq DNA polymerase (Life Technologies). Amplified DNA (2 µl) was mixed with an equal volume of formamide loading dye (95% formamide, 20 mM EDTA, 0.05% bromophenol blue, 0.05% xylene cyanol), denatured for 10 min at 95°C, loaded onto a polyacrylamide gel (20 × 45 × 0.04 cm, MDE) and electrophoresed at room temperature for 14 h at 3 W. Gels were transferred onto 3MM Whatman paper, dried and autoradiographied with Kodak X-OMAT films for 16–48 h.

For direct sequencing, amplification products were loaded onto a 0.7% agarose gel, purified by phenol-chloroform extraction and recovered by ethanol precipitation. Purified fragments were directly sequenced using the same specific intronic primers and the PRISM™ Ready Reaction Sequencing Kit (Perkin Elmer Cetus), on an automatic fluorometric DNA sequencer (Applied Biosystems).

For genotyping, hypervariable microsatellites flanking the STGD locus on chromosome 1p13–p22 were used.[3,5]

RESULTS

A total of 25 different mutations were identified in 21/40 STGD and 5/15 FFM patients, respectively. Interestingly, the truncating mutations (9/25) were identified in STGD only (2 non sense, 2 frameshift, and 5 splicing mutations, n = 10). In addition, 10 different missense mutations were detected in ten additional STGD cases. Most of them (5/10) converted an arginine into an uncharged amino acid (Table 1, Fig. 3). By contrast, only missense mutations (n = 7) were identified in FFM patients (n = 5). All of them involved uncharged amino acids (4 leucines, 1 alanine, and 2 glycines, see Table 1 and Fig. 3). On the other hand, all ABCR missense mutations identified in STGD and FFM concerned amino acids conserved in RmP[13] contrasting with the other members of the ABC family (ref 8, 11, see Table 1).

So far, only two unrelated STGD patients were found to be homozygotes for a mutant genotype in our series (the 571-2A ➡ G splice site mutation in intron 5 in a consanguineous family of Moroccan origin, an arginine to tryptophane substitution at codon 1640 in exon 35). The vast majority of our patients were either compound heterozygotes (5 STGD and 2 FFM) or heterozygotes for one single mutation (14 STGD and 3 FFM). None of these sequence changes have been previously reported[8] nor were they identified in any of 50 healthy controls (100 chromosomes).

Concerning the families segregating both STGD and AMD, the mother and grand-mother in family 1, the grand-mother in family 2 and the grand father in family 3 were affected with AMD and were found to carry one of the two different mutations identified in their relative affected with Stargardt disease (I2113C, family 1, JEG; R1107C, family 2, JUL; R212C, family 3, GEN; see Table 1 and Fig. 1).

Table 1. Mutations in the ABCR gene in STGD and FFM families.

Nucleotide change	Amino acid change	Domain	Conserved aa in: ABCs	Conserved aa in: RmP	Phenotype	# families	Comment
(571–2)A→G	**splicing mutation**				**STGD**	**1**	**HAD[1]**
(1938–2)A→G	**splicing mutation**				**STGD**	**1**	
(4668+2)T→C	**splicing mutation**				**STGD**	**1**	
(4735+2)T→A	**splicing mutation**				**STGD**	**1**	
del(5196+1–5196+6)	**splicing mutation**				**STGD**	**1**	**LOZ[2]**
2570 delT	**frameshift mutation**				**STGD**	**1**	
3209insGT	**frameshift mutation**				**STGD**	**2**	**CHE[2]**
G3754T	**E1252X**				**STGD**	**1**	
C3994T	**Q1332X**				**STGD**	**1**	
C52T	R18W	IC	–	+	STGD	1	**JEG[2]**
C634T	R212C	EC	–	+	STGD	5	
G1908T	Q636H	IC	–	+	STGD	1	GEN[2], JEG[2]
C3056T	T1019M	IC	–	+	STGD	1	LOZ[2]
C3322T	R1107C	IC	–	+	STGD	1	
C4916T	R1640W	IC	+	+	STGD	2	JUL[2]
G5929A	G1977S	ATP2	+	+	STGD	1	MAR[1]
C6320A	R2107H	IC	+	+	STGD	1	JUL[2]
G6337G	I2113C	IC	–	+	STGD	1	GEN[2]
C3114T	A1038V	IC	–	+	STGD	2	CHE[2]
					+FFM	+1	VIL[2]
T1622C	L541P	EC	–	+	FFM	1	VIL[2]
T31C	L11P	IC	+	+	FFM	1	
G3272A	G1090E	IC	+	+	FFM	1	
G4522T	G1508C	IC	+	+	FFM	1	
C5908T	L1970F	IC	+	+	FFM	1	GON[2]
T5912G	L1971R	IC	+	+	FFM	1	GON[2]

Mutations refer to the Standard nomenclature. Truncating mutations are indicated by bold letters. EC: extracellular domain, IC: intracytoplasmic domain; ATP2: second ATP-binding site; RmP: mouse Rim protein, ABCs: other members of the ABC subfamily (mouse ABC1, mouse ABC2, human ABCC), +: conserved amino acid, –: non-conserved amino acid. The comment field indicates [1]homozygous families and [2]compound heterozygous families in which two independent altered alleles were identified.

On the other hand, under the hypothesis of one single disease causing gene, the linkage analyses performed in the two largest pedigrees segregating STGD and RP19 showed consistent linkage with 1p22 markers with no recombination event (Fig. 2). In addition, sequence analyses of the 50 exons and intron-exon boundaries of the ABCR gene performed in all five families segregating the two phenotypes revealed mutations in one of them (family 2, see Fig. 4A). Indeed, we have identified an apparently heterozygote mutation in the acceptor splice site of intron 13 (1938-1G ➡ A) in the STGD patient (II1) of this family. This mutation created a Mse1 restriction site and is expected to result in a frameshift. Restriction analysis of the entire family revealed that the patient and her healthy father (I2) were both heterozygote for the splice mutation (Fig. 4A). In the patient with RP (II2), apparent homozygosity for the same mutation was observed. Subsequent restriction analysis revealed that her father (I2) was also

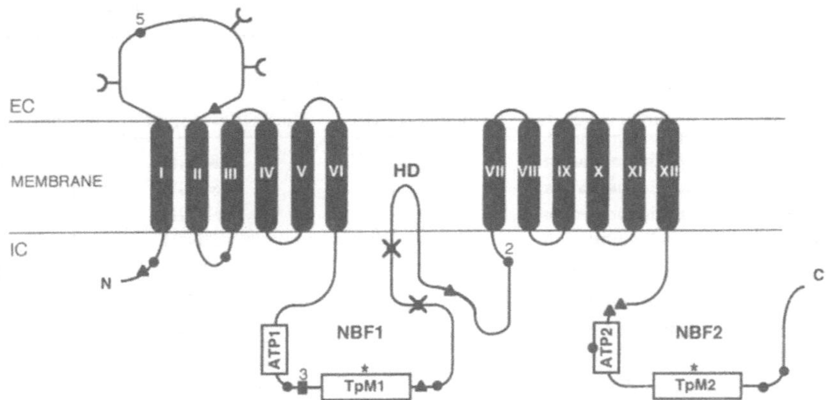

Figure 3. Predicted topology of ABCR and position of point mutations associated with STGD and FFM. EC: extracellular domain and IC: cytoplasmic domain. The 12 predicted membrane-spanning α-helical segments are indicated with roman numbers. Y predicted N-glycosylation site. ATP1: first ATP-binding site; ATP2: second ATP-binding site; TpM_1: first transporter signature-motif; TpM_2: second transporter signature-motif; NBF1: first nucleotide binding fold; NBF2: second nucleotide binding fold; HD: proline rich hydrophobic domain. *PKA-phosphorylation sites. The Approximate positions of missense mutations identified in STGD, FFM and in both phenotypes are indicated by dots, triangles, and a dark square, respectively. A crossed-out dot shows the approximate positions of nonsense mutations and a dot with a number above it indicates that the mutation was found more than once (in n families).

heterozygote for the 1938-1G ➡ A splice mutation. Most interestingly, her mother (I4) was apparently homozygote for the normal allele providing strong evidence for maternal non-contribution at the ABCR locus (Fig. 4 A and B). Accordingly, unambiguous maternal contribution was observed at the other polymorphic loci tested (including flanking intragenic polymorphisms 1268 A/C and 2828 G/A, see Fig. 4B). Although no direct evidence for deletion of the ABCR gene could be given, the apparent segregation of a null allele through the pedigree is consistent with deletion of the genomic region encompassing the acceptor splice site in intron 13.

DISCUSSION

The ABCR protein is a member of the adenosine triphosphate (ATP) binding cassette transporter superfamily[9] involved in transport of a wide variety of substrates across cellular membranes. Recently, ABCR gene mutations were identified in STGD[8] and in some AMD patients[9] as well as in a particular form of RP (RP19) previously mapped to chromosome 1p22.[10,14] The purpose of the present study was to investigate the allelic heterogeneity at the ABCR locus and to look for possible genotype-phenotype correlations in a large series of STGD and FFM patients, on one hand, and in some pedigrees segregating both STGD and AMD or both STGD and RP, on the other hand.

Among the 25 hitherto unreported ABCR gene mutations described here, 10/25 were truncating mutations. Interestingly, mutations truncating the ABCR gene product were found in STGD patients only. In addition, all of them, except one (HAD family), were either heterozygote for a frameshift mutation or compound heterozygote for

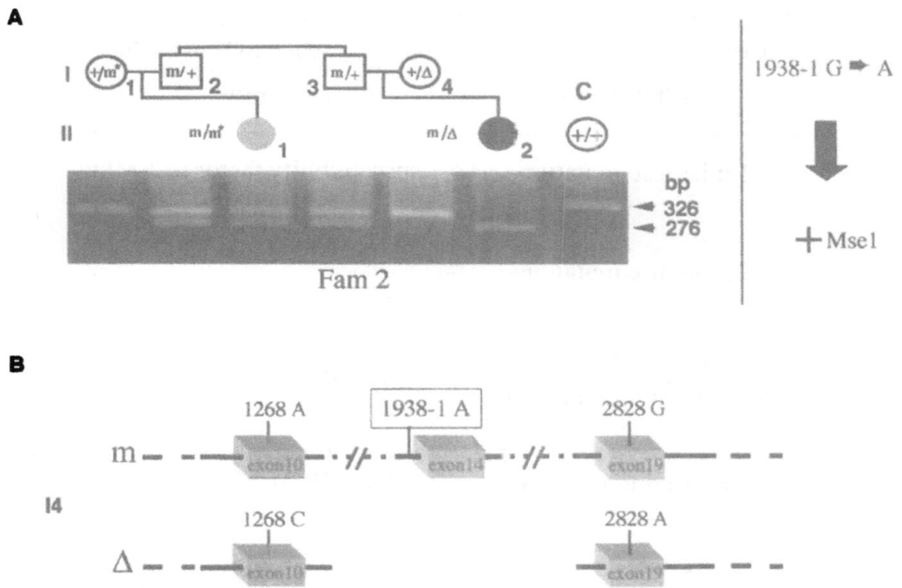

Figure 4. A. Pedigree of the family and restriction analysis of the 1938-1 G → A mutation. STGD, RP. The 1938-1 G → A mutation created a Msel site. PCR-amplified genomic DNA was digested using Msel. Heterozygote carriers of the 1938-1 G → A (I2, I3, II1) display three fragments of 326bp (undigested), 276bp and 50bp (not visible), respectively. There is a maternal non-contribution at the ABCR locus, as the RP patient (II1) is hemizygote for the ABCR gene mutation while her mother (I4) displays the 276bp fragment only. B. Schematic representation of the two ABCR alleles of I4. m: allele harbouring the 1938-1 G → A mutation, Δ: allele carrying a large deletion of unknow size, encompassing exon 14 and flanked by exons 10 and 19 as shown by two heterozygote polymorphisms 1268A/C and 2828 G/A, respectively.

frameshift and missense mutations. Most of the missense mutations detected in STGD (5/9) converted charged into uncharged amino acids. By contrast, all mutations identified in FFM patients were missense mutations involving uncharged amino acids and none of them (but the A1038V mutation) were detected in STGD. Interestingly, the A1038V mutation was found in compound heterozygotes (one STGD, family CHE, and one FFM, family VIL), the second mutation being another missense mutation in the FFM family (L541P) but a frameshift mutation in STGD (Table 2). Furthermore, all ABCR missense mutations detected in STGD and FFM patients involved amino acids conserved in the mouse orthologue (ref 13, see Table 1). Nevertheless, no correlation between the severity of the phenotype and the conservation of the mutated amino acid in other members of the ABC subfamily could be established (ref 8, 11, see Table 1).

On the other hand, all patients 'relatives known to be affected with AMD, in our series, were found to carry an ABCR mutation identified in a STGD proband. It seems particularly important to look for macular degeneration in parents and grand parents of STGD and FFM patients so as to decide whether the incidence of AMD in this subgroup is indeed higher than in control population.

Moreover, under the hypothesis of one single disease causing gene, the linkage analyses performed in pedigrees segregating STGD and RP (2/5 only were large

Table 2. Genotype-phenotype correlations in retinal degenerations caused by ABCR gene mutations. CRD = cone-rod dystrophy

GENOTYPE	PHENOTYPE
1 missense mutation	➡ susceptibility factor to AMD
2 missense mutations	↗ early STGD ↘ late FFM
1 missense mutation + 1 truncating mutation	➡ early STGD
2 truncating mutations	➡ RP (or CRD)

enough) showed consistent linkage with 1p22 markers with no recombination event, strongly suggesting that despite striking clinical differences, the two phenotypes were allelic. In addition, we have identified a splicing mutation of the ABCR gene in two first cousin patients suffering RP and STGD, respectively. It is important to note that these two first cousins were of the same age and displayed unambiguously different retinal disorders. Indeed, STGD in the first patient (II1, Fig. 4A) started at the end of the first decade and only affected the macular area while RP in the second one (II2, Fig. 4A) had an early onset (before the age of 7) and involved the whole retina. In this family, heterozygosity for a splice mutation in exon 13 of the ABCR gene resulted in STGD while compound heterozygosity for the splice mutation and large intragenic deletion flanked by exons 10 and 19, resulted in severe retinitis pigmentosa (Fig. 4B). This observation is consistent with the report of RP19 in a Spanish family harbouring a homozygote frameshift deletion of the ABCR gene.[10]

Based on these results, we suggest that homozygosity for truncating ABCR gene mutations results in RP phenotype, while compound heterozygosity for frameshift and missense mutations or two missense mutations at this locus result in STGD (Table 2). Conversely, FFM phenotype is always the result of two missense mutations (Tables 1 and 2). Concerning the consanguineous STGD family with an homozygous splice site mutation (HAD, see Table 1), it is most likely that this mutation will not affect the reading frame.

On the other hand, it is worth noting that no mutation has been hitherto identified in 4/5 families segregating both STGD and RP and in 29/55 families with either STGD or FFM. In addition, 17/26 STGD or FFM patients were found to be heterozygotes for one single mutation although 7/17 cases were multiplex, strongly suggesting an autosomal recessive inheritance. This point could be partly explained by the defective sensitivity of the PCR-SSCP method to detect mutations (about 80%, ref 15). Nevertheless, it is also possible that some of the allelic mutations lie in a not screened region of the gene, such as the introns or the promotor region. Considering the detection of a large heterozygote deletion in one RP patient and her mother, it is likely that a major part of the RP and STGD patients with no mutation detected the ABCR coding sequence may harbour large heterozygote deletions undetectable by PCR as the normal allele is normally amplified blinding the deleted allele. This hypothesis has been already raised and confirmed for the factor eight gene in 50% of severe form of hemophilia.[16]

Clinical heterogeneity at a single locus has been previously reported in retinal dystrophies. For example, mutations of the peripherin gene have been associated with autosomal dominant RP, *retinitis punctata albescens* and several forms of macular dystrophies.[17–19] Similarly, mutations in the bPDE and rhodopsin genes have resulted in either RP or congenital stationary night blindness.[20–23] Considering the variable clinical expression of ABCR gene mutations and their wide range of age of onset (including age-related macular dystrophies, 9), the present study addresses the intriguing question of the relevance of the gene in other retinal dystrophies, especially as ABCR is expressed in rods[8] and is a major gene in the ATP-dependent transport of a wide variety of substrates and cellular debris across the cellular membranes of the retinal pigment epithelium. More generally, a variety of ABCR gene mutations could possibly account for a continuum of macular and/or retinal degenerations that had an onset from early childhood to late adulthood.

ACKNOWLEDGMENTS

We were grateful to Monique Dailhat for her help in preparing this manuscript. This research was supported by the Association Retina France and Association Française contre les Myopathies.

REFERENCES

1. Stargardt K.: Uber familiäre, progressive degeneration under makulagegend des augen. Albrecht. von. Graefes. Arch. Ophthalmol. 1909;71:534–550.
2. Kaplan J., Bonneau D., Frézal J., and Munnich A., and Dufier J.L.: Clinical and genetic heterogeneity in retinitis pigmentosa. Hum. Genet. 1990;85:635–642.
3. Kaplan J., Gerber S., Larget-Piet D., Rozet J.M., Dollfus H., Dufier J.L., Odent S., Postel-Vinay A., Janin N., Briard M.L., Frézal J., Munnich A.: A gene for Stargardt's disease (*fundus flavimaculatus*) maps to the short arm of chromosome 1. Nature. Genet. 1993;5:308–311.
4. Rozet J.M., Gerber S., Perrault I., Camuzat A., Calvas P., Viegas-Pequignot E., MOlina-Gomez D., Le Paslier D., Chumakov I., Munnich A., and Kaplan J.: Structure and physical mapping of DR1, a TATA-binding protein-associated phosphoprotein gene, to chromosome 1p22.1 ans its exclusion in Stargardt disease (STGD). Genomics. 1996;36:554–556.
5. Franceschetti A. François J.: *Fundus flavimaculatus* Arch. Ophthalmol. 1965;25:505–530.
6. Welever RG.: Stargardt's macular dystrophy. Arch. Ophthalmol. 1994;112:752–754.
7. Gerber S., Rozet J.M., Bonneau D., Souïed E., Camuzat A., Dufier J.L., Amalric P., Weissenbech J., Munnich A., and Kaplan J.: A gene for late-onset *fundus flavimaculatus* with macular dystrophy maps to chromosome 1p13. Am. J. Hum. Genet. 1995;56:396–399.
8. Allikmets R., Singh N., Sun H., Shroyer N.F., Hutchinson A., Chidambaram A., Gerrard B., Baird L., Stauffer D., Peiffer A., Rattner A., Smallwood P., Li Y., Anderson K.L., Lewin R.A., Nathan J., Leppert M., Dean M., and Lupski J.R.: A photoreceptor cell-specific ATP-binding transporter gene (*ABCR*) is mutated in recessive Stargardt macular dystrophy. Nature. Genet. 1997;15:236–246.
9. Allikmets R., Shroyer N.F., Singh N., Seddon J.M., Lewis RA, Bernstein P.S., Peiffer A., Zabriskie A., Li Y., Hutchinson A., Dean M., Lupski J.R., and Leppert M.: Mutation of the Stargardt disease gene (*ABCR*) in age-related macular degeneration. Science. 1997;277:1805–1807.
10. Martinez-Mir, Paloma E., Allikmets R., Ayuso C., del Rio T., Dean M., Vilageliu L., Gonzales-Duarte R., and Balcells S.: Retinitis pigmentosa caused by a homozygous mutation in the Stargardt disease gene ABCR. Nature. Genet. 1998;18:11–12.

11. Allikmets R., Gerrard B., Hutchinson A., and Dean M.: Characterization of the human ABC superfamily: Isolation and mapping of 21 new genes using the expressed sequence tags database. Hum. Mol. Genet. 1996;5:1649–1655.
12. Gerber S., Rozet J.M., van de Pol T.J.R., Hoyng C.B., Munnich A., Blankenagel A., Kaplan J., and Cremers F.P.M.: Complete exon-intron structure of the retina specific ATP binding transporter gene (*ABCR*) allows the identification of novel mutations underlying Stargardt disease. Genomics. 1998;48:139–142.
13. Arazian S.M., and Travis G.H.: The photoreceptor rim protein is an ABC transporter encoded by the gene for recessive Stargardt's disease (*ABCR*). FEBS Letters 1997;409:247–252.
14. Martinez-Mir A., Bayes M., Vilageliu L., Grinberg D., Ayuso C., del Rio T., Garcia-Sandoval B., Bussaglia E., Baiget M., Gonzales-Duarte R., and Balcells S.: A new locus for autosomal recessive retinitis pigmentosa (RP19) maps to 1p31-p21. Genomics. 1997;40:142–146.
15. Hayashi K., and Yandell D.W.: How sensitive is PCR-SSCP? Hum. Mut. 1993;2:338–346.
16. Maugard C., Tuffery S., Aguilar-Martinez P., Schved J.F., Gris J.C., Demaille J., and Claustres M.: Protein truncation test: detection of severe haemophilia A mutation analysis of factor VIII transcripts. Hum. Mut. 1998;11:18–22
17. Nichols B.E., Sheffiield V.C., Vandenburgh K., et al. Butterfly-shaped pigment dystrophy of the fovea caused by a point mutation in codon 167 of the RDS gene. Nature. Genet. 1993;3:202–207.
18. Kajiwara K., Sandberg M.A., Berson E.L., and Dryja T.P.: A null mutation in the human peripherin/RDS gene in a family with autosomal dominant retinitis punctata albescens. Nature. Genet. 1993;3:208–212.
19. Wells J., Wroblewski J., Keen J., et al. Mutations in the human retinal degeneration slow (RDS) gene can cause either retinitis pigmentosa or macular dystrophy. Nature. Genet. 1993;3:213–218.
20. Rosenfeld P.J., Cowley G.S., McGee T.L., et al. A *null* mutation in the rhodopsin gene causes rod photoreceptor dysfunction and autosomal recessive retinitis pigmentosa. Nature. Genet. 1992;1:208–213.
21. Dryja T.P., Berson E.L., Rao V.R., and Oprian D.D.: Heterozygous missense mutation in the rhodopsin gene as a cause of congenital stationary night blindness. Nature. Genet. 1993;4:280–283.
22. McLaughlin M.E., Sandberg M.A., Berson E.L., and Dryja T.P.: Recessive mutations in the gene encoding the beta-subunit of rod phosphodiesterase in patients with retinitis pigmentosa. Nature. Genet. 1993;4:130–134.
23. Gal A., Orth U., Baehr W., Schwinger E., and Rosenberg P.J.: Heterozygous missense mutation in the rod cGMP phosphodiesterase b-subunit gene in autosomal dominant stationary night blindness. Nature. Genet. 1994;7:64–67.

17

AUTOSOMAL DOMINANT MACULAR DEGENERATION LOCALIZED TO CHROMOSOME 6q BY LINKAGE ANALYSIS

I. B. Griesinger, P. A. Sieving, and R. Ayyagari

W.K. Kellogg Eye Center, University of Michigan
1000 Wall Street
Ann Arbor, Michigan 48105

SUMMARY

We are studying a large autosomal dominant family with a broad clinical spectum of macular degeneration. By analyzing microsatellite markers on chromosome 6, significant positive lod sores were obtained (D6S455: Z = 4.49 at theta = 0). Linkage analysis localizes the disease-causing gene to a 22cM interval on chromosome 6q11–13 and excludes the loci for North Carolina Macular Dystrophy and Progressive Bifocal Chorioretinal Atrophy. Three other retinal/macular dystrophies map to overlapping intervals and could be allelic forms of the same gene or from a cluster of genes causing retinal/macular degeneration.

1. INTRODUCTION

Macular dystrophies represent a group of disorders that are characterized by degeneration of retinal, pigment epithelial and/or choroidal tissue, most severely affecting the macula.[1] Several macular disorders have been described, ranging from the most common and complex genetic condition of age related macular degeneration (AMD) to a group of dystrophies due to single-gene involvement with autosomal or X-linked Mendelian inheritance.[2] More than 20 loci have been identified indicating the genetic heterogeneity of these disorders (MIM). In addition there is considerable allelic heterogeneity. Mutations in RDS/Peripherin are responsible for causing macular degeneration and butterfly macular degeneration and also result in widespread retinal degeneration or retinitis pigmentosa (RP).[3-5] The ATP binding transporter gene (ABCR) was shown to be associated with Stargardt's macular degeneration, with

Retinal Degenerative Diseases and Experimental Therapy, edited by Hollyfield *et al.*
Kluwer Academic / Plenum Publishers, New York, 1999.

recessive RP, and possibly with an AMD susceptibility locus.[6,7] The genes that cause Best's macular degeneration[8] and Sorsby's fundus dystrophy[9] have also been cloned.

We are studying a large, five-generation pedigree with autosomal dominant macular degeneration. The severity and the age of onset of the disease are highly variable. Here we report linkage of the disease locus to chromosome 6q. Four other macular degenerative diseases have been localized previously to 6q: Stargardt's-like dominant macular degeneration,[10] North Carolina Macular Degeneration (MCDR1),[11,12] Progressive Bifocal Chorioretinal Atrophy (PBCRA),[13] and Cone-Rod Dystrophy 7 (CORD7).[14] We exclude linkage with NCMD and PBCRD and have localized the disease-causing gene to an interval that overlaps the critical region for CORD7 and Stargardt's-like dominant macular degeneration.

2. METHODS

Clinical Methods: Ophthalmic examination included measurement of visual acuity, fundoscopy, fluorescein angiography, and color vision testing with Fransworth D15 test and Ishihara plates. Electroretinography (ERG) was performed with a single flash under light- and dark-adapted condition or a 30Hz flicker stimulus under light-adapted condition, using standard protocols and with normal values reported previously.[15]

Genotyping and Linkage analysis: Blood samples were obtained from 25 family members and 8 spouses. The DNA was extracted as described by Ayyagari et al.[16] Genotyping was performed with microsatellite loci as described in Padma et al.[17] The PCR products were separated by denaturing gel electrophoresis and visualized by autoradiography.

Linkage Analysis: Two point linkage analysis was performed using the MLINK program (version 5.1) of the LINKAGE package.[18] Allele frequencies given in the genome data base were used. The phenotype in this family was analysed as an autosomal dominant trait with a frequency of 0.0001 for the affected allele. LOD scores were calculated using complete penetrance, 90% penetrance and affecteds-only.

3. RESULTS

3.1. Clinical Description

A large, five generation pedigree was ascertained (Fig. 1). The primary pathology clinically involves RPE atrophy in the fovea and parafovea and is progressive over the course of many years. Peripapillary RPE changes were observed in most of the affected individuals (Fig. 2). Severity of the disease and the age of onset are highly variable in this family making the definitive clinical diagnosis of "normal individuals" difficult in younger family members. Despite the progressive acuity deficit, the color vision was excellent in all affected individuals except a 55 year old severely affected woman who had a large central scotoma.

ERG findings did not indicate widespread rod or cone involvement (Fig. 3). Dark-adapted rod and photopic cone b-wave amplitudes were normal for all affected individuals. The flicker ERG amplitudes were normal in 6 of 8 affected individuals, but

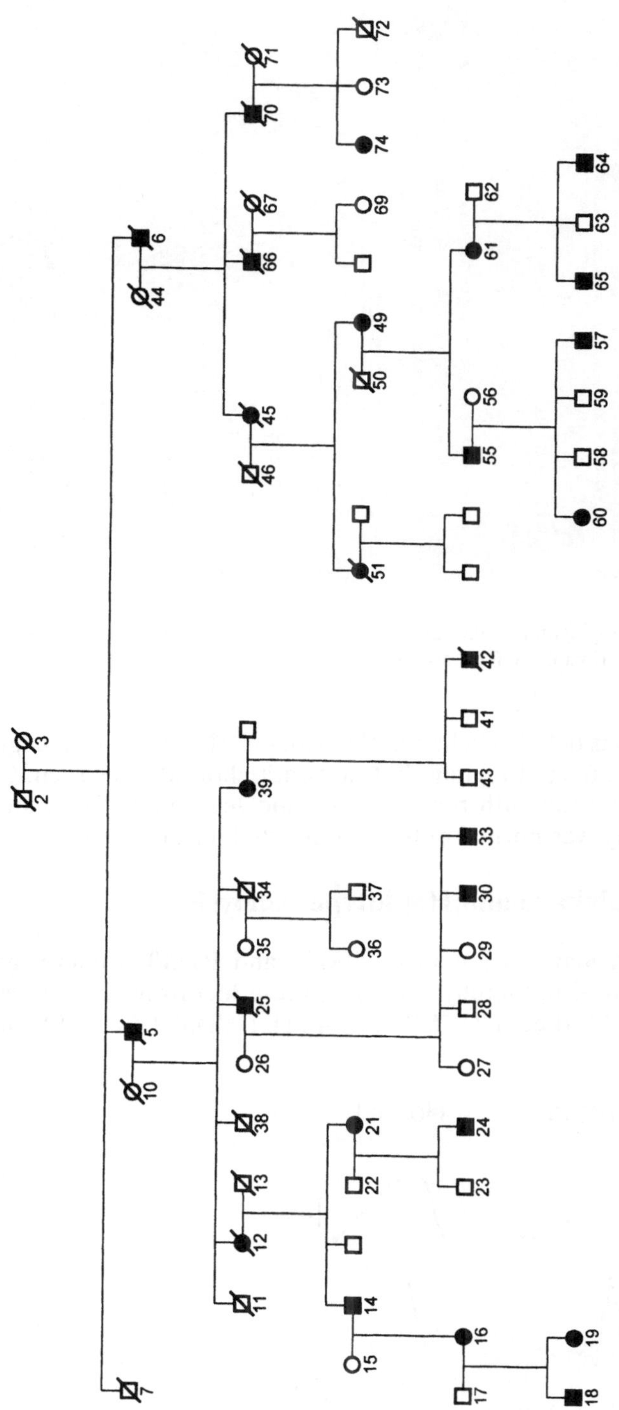

Figure 1. Pedigree of family H785 with autosomal dominant macular degeneration.

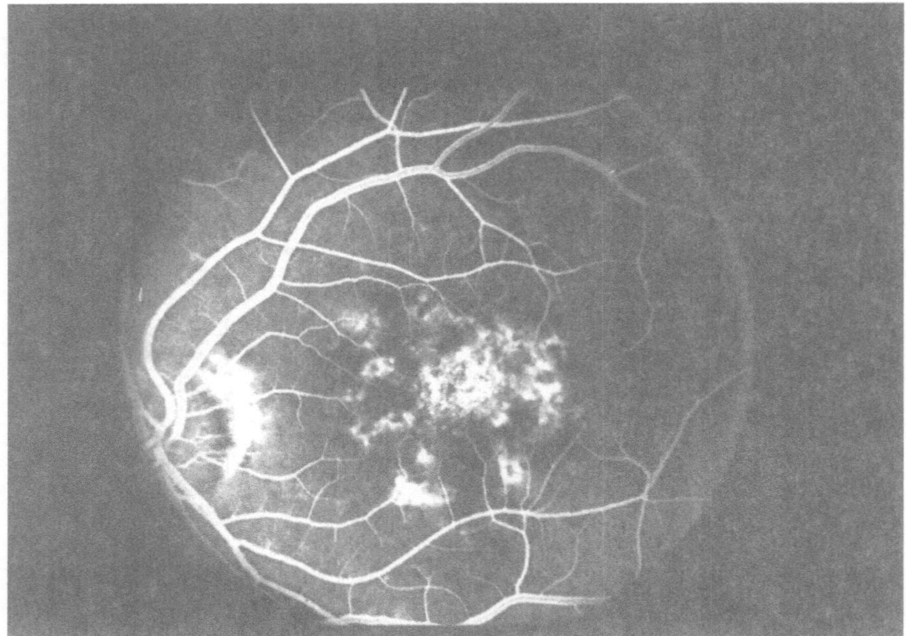

Figure 2. Fluorescein angiogram of the left eye of a 47 y/o affected male with a visual acuity of 20/200 showing macular and peripapillary RPE atrophy.

the flicker timing was delayed in half of the patients. The fluorescein angiogram showed window defects in affected regions, but no "dark choroid" was seen. Goldman visual fields were generally full with both the V4e and I4e targets. Dark-adapted absolute threshold sensitivity was normal outside of affected macular regions.

3.2. Results of Linkage and Haplotype Analysis

Analysis with markers linked to TIMP3 and RDS/Peripherin gave lod scores more negative than –2.0, thereby excluding these loci from linkage with the disease gene in our family. Markers D6S1557 (Z = 4.31), D6S455 (Z = 4.49), and D6S1609 (Z

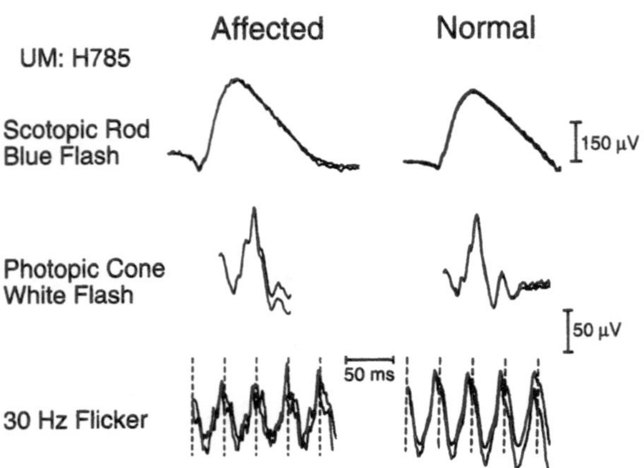

Figure 3. Electroretinogram of the same affected male as in figure 2. Rod and cone ERG amplitudes are normal but the 30 Hz flicker timing is delayed.

Table 1. Pairwise linkage data using affecteds-only

	Two point lod scores at theta =							
Markers	0.00	0.05	0.10	0.20	0.30	0.40	Z max	Theta
D6S271	−6.12	−0.13	0.24	0.40	0.32	0.17	−2.00	0.005
D6S294	−6.30	−0.01	0.29	0.33	0.20	0.07	−2.00	0.004
D6S402	−1.94	0.69	0.69	0.44	0.21	0.09	0.71	0.070
D6S1557	4.31	3.94	3.54	2.66	1.70	0.75	4.31	0.000
D6S455	4.49	4.08	3.66	2.77	1.82	0.86	4.49	0.000
D6S280	3.75	3.37	2.98	2.18	1.37	0.60	3.75	0.000
D6S460	4.36	3.94	3.51	2.61	1.67	0.76	4.36	0.000
D6S1609	5.97	5.48	4.97	3.87	2.65	1.31	5.97	0.000
D6S1644	3.45	3.10	2.75	2.02	1.29	0.60	3.45	0.000
D6S1613	3.65	3.32	2.98	2.21	1.39	0.59	3.65	0.000
D6S1631	3.22	2.86	2.49	1.72	0.97	0.36	3.22	0.000
D6S275	−1.13	2.21	2.15	1.69	1.10	0.50	2.21	0.070
D6S300	−4.36	0.25	0.56	0.59	0.39	0.14	−2.00	0.001

= 5.97) gave significant positive lod scores at zero recombination, thus localizing the disease causing gene in this family to 6q. Because of the variable penetrance of the phenotype, linkage analysis was performed using a range of penetrance values and also using only clinically affected individuals. The results did not change significantly neither for different penetrance values nor for analysis with affecteds-only. The results of linkage analysis with only the 16 affected individuals are presented in table 1. Analysis with markers D6S300 gave lod scores more negative than −2, thereby excluding MCDR1[11] and PBCRA[13] loci from linkage to the disease locus in our family (Table 1).

Utilizing the genotype data of the markers across the linked region, haplotypes were constructed keeping the number of recombinations to the minimum. Recombination events observed in affected individuals place the disease causing gene in this family to a 22cM interval between markers D6S402 and D6S275 (data not shown). Two additional individuals have been identified with key recombinations that are likely to reduce the interval further to a 3cM region, but the clinical examination is pending to determine the disease status.

4. DISCUSSION

The macular degeneration gene in a five-generation family has been localized to 6q13 by linkage and haplotype analysis. The clinical phenotype in this family is quite variable with regard to severity of the disease and the age of onset. TIMP3 and RDS/Peripherin genes were excluded from linkage with the disease locus in this family. Analysis of markers on chromosome 6 localizes the disease gene to the q arm of chromosome 6. Five retinal diseases forming two different groups of loci have been mapped to the q arm of chromosome 6 (Fig. 4): they include MCDR1,[11] PBCRA,[13] CORD7,[14] Stargardt's like macular degeneration,[10] and a recessive form of RP.[19]

Linkage results and haplotype analysis exclude MCDR1[11,12] and PBCRA[13] loci from being involved in this family. The 22cM critical gene interval in this family overlaps the intervals for CORD7, Stargardt's-like macular degeneration, and arRP have been localized (Fig. 4). The phenotype of the family is distinctly different from arRP and from CORD7. For example, in the CORD7 the scotopic rod responses were barely

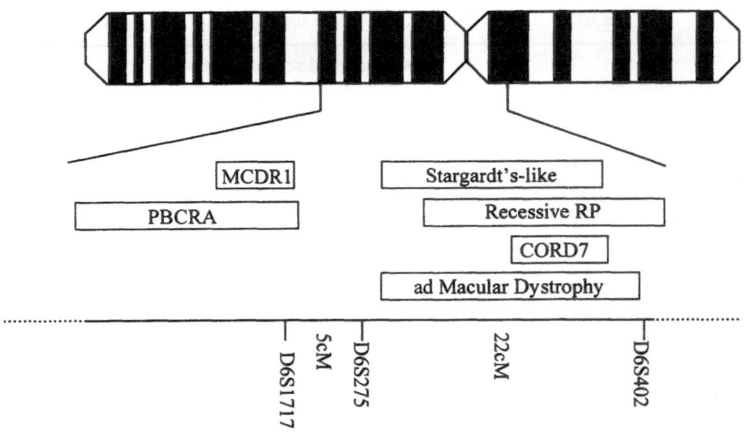

Figure 4. Retinal and macular dystrophies mapped to the centromeric region of human chromosome 6. The interval of Stargardt's-like autosomal dominant macular degeneration, CORD7, and a recessive form of RP overlap with the interval of the gene causing macular degeneration in this family. MCDR1 and PBCRA have been excluded.

detectable, and all cone responses were severely attenuated; however, results from only one individual are reported.[14] Affected individuals of families with arRP localized to 6q have non-detectable rod-driven ERG.[19] In our family rod ERGs of the affected members were normal, and the cone ERGs showed only minimal changes. The clinical symptoms of the affected individuals in our family are most similar to Stargardt's-like macular degeneration.[10] The visual acuities reported in Stargardt's-like dominant macular degeneration were 20/200 or worse in all individuals over age 36; this is somewhat more severe than we observed in our family with the variable severity and age of onset. In conclusion, all these three diseases localized to an overlapping critical region could be caused by allelic variants of the same gene or by the involvement of a cluster of different genes.

ACKNOWLEDGMENTS

This work was supported by the Foundation Fighting Blindness, Hunt Valley, MD. We wish to thank the family members for participating in this study.

REFERENCES

1. P.A. Sieving, 1995, Diagnostic issues with inherited retinal and macular dystrophies, *Seminars in Ophthal.* **10**:279–294.
2. C.F. Inglehearn, 1998, Molecular genetics of human retinal dystrophies. *Eye* **12**:571–579.
3. J. Wells, J. Wroblewski, J. Keen, C. Inglehearn, C. Jubb, A. Eckstein, M. Jay, G. Arden, S. Bhattacharya, and F. Fitzke, 1993, Mutations in the human retinal degeneration slow (RDS) gene can cause either retinitis pigmentosa or macular dystrophy [see comments], *Nat. Genet.* **3**:213–218.
4. K. Kajiwara, M.A. Sandberg, E.L. Berson, and T.P. Dryja, 1993, A null mutation in the human peripherin/RDS gene in a family with autosomal dominant retinitis punctata albescens [see comments], *Nat. Genet.* **3**:208–212.

5. B.E. Nichols, A.V. Drack, K. Vandenburgh, A.E. Kimura, V.C. Sheffield, and E.M. Stone, 1993, A 2 base pair deletion in the RDS gene associated with butterfly-shaped pigment dystrophy of the fovea, *Hum. Mol. Genet.* **2**:601–603.
6. R. Allikmets, N. Singh, H. Sun, N.F. Shroyer, A. Hutchinson, A. Chidambaram, B. Gerrard, L. Baird, D. Stauffer, and A. Peiffer, et al. 1997, A photoreceptor cell-specific ATP-binding transporter gene (ABCR) is mutated in recessive Stargardt macular dystrophy [see comments], *Nat. Genet.* **15**:236–246.
7. R. Allikmets, N.F. Shroyer, N. Singh, J.M. Seddon, R.A. Lewis, P.S. Bernstein, A. Peiffer, N.A. Zabriskie, Y. Li, and A. Hutchinson, et al. 1997, Mutation of the Stargardt disease gene (ABCR) in age-related macular degeneration [see comments], *Science* **277**:1805–1807.
8. K. Petrukhin, M.J. Kiosti, B. Bakall, W. Li, G. Xie, T. Marknell, O. Sandgren, K. Forsman, G. Holmgren, and S. Anderson, et al. 1998, Identification of the gene responsible for Best macular dystrophy. *Nature Genet.* **19**:241–247.
9. B.H. Weber, G. Vogt, R.C. Pruett, H. Stohr, and U. Felbor, 1994, Mutations in the tissue inhibitor of metalloproteinases-3 (TIMP3) in patients with Sorsby's fundus dystrophy, *Nat. Genet.* **8**:352–356.
10. E.M. Stone, B.E. Nichols, A.E. Kimura, T.A. Weingeist, A. Drack, and V.C. Sheffield, 1994, Clinical features of a Stargardt-like dominant progressive macular dystrophy with genetic linkage to chromosome 6q [see comments], *Arch. Ophthalmol.* **112**:765–772.
11. K.W. Small, J.L. Weber, A. Roses, F. Lennon, J.M. Vance, and M.A. Pericak-Vance, 1992, North Carolina macular dystrophy is assigned to chromosome 6, *Genomics* **13**:681–685.
12. C.G. Sauer, H.D. Schowrm, M. Ulbig, A. Blankenagel, K. Rohrschneider, D. Pauleikhoff, T. Grimm, and B.H.F. Weber, 1997, An ancesteral core haplotype defines the critical region harbouring the North Carolina macular dystrophy gene (MCDR1), *J. Med. Genet.* **34**:961–966.
13. R.E. Kelsell, B.F. Godley, K. Evans, P.A. Tiffin, C.Y. Gregory, C. Plant, A.T. Moore, A.C. Bird, and D.M. Hunt, 1995, Localization of the gene for progressive bifocal chorioretinal atrophy (PBCRA) to chromosome 6q, *Hum. Mol. Genet.* **4**:1653–1656.
14. R.E. Kelsell, K. Gregory-Evans, C.Y. Gregory, Holder G.E., Jay M.R., Weber B.H.F., A.T. Moore, A. Bird, and D.M. Hunt, 1998, Localization of a gene (CORD7) for a dominant Cone-Rod Dystrophy to chromosome 6q. *Am. J. Hum. Genet.* **63**:274–279.
15. J.E. Richards, C. Kuo, M. Boehnke, and P.A. Sieving, 1991, Rhodopsin Thr58Arg mutation in a family with autosomal dominant Retinitis Pigmentosa, *Opthalmol.* **98**:1797–1805.
16. R. Ayyagari, Y. Li, R.J.H. Smith, M.Z. Pelias, and J.F. Hejtmancik, 1995, Fine mapping of the Usher Syndrome Type 1C to chromosome 11p14 and idnetification of flanking markers by haplotype analysis, *Mol. Vis. (http://www. emory. edu/MOLECULAR_VISION/v1/ayyagari.* **1**:••.
17. T. Padma, R. Ayyagari, J.S. Murty, S. Basti, T. Fletcher, G.N. Rao, M. Kaiser-Kupfer, and J.F. Hejtmancik, 1995, Autosomal dominant zonular cataract with sutural opacities localized to chromosome 17q11–12, *Am. J. Hum. Genet.* **57**:840–845.
18. G.M. Lathrop, and J.M. Lalouel, 1984, Easy calculations of lod scores and genetic risks on small computers, *Am. J. Hum. Genet.* **36**:460–465.
19. A. Ruiz, S. Borrego, I. Marcos, and G. Antinolo, 1998, A major locus for autosomal recessive retinitis pigmentosa on 6q, determined by homozygosity mapping of chromosomal regions that contain gamma-aminobutyric acid-receptor clusters. *Am. J. Hum. Genet.* **62**:••.

18

THE GENE PRODUCT OF IMPG1 IS THE GLYCOPROTEIN SPACR, NOT AN IPM PROTEOGLYCAN

Shreeta Acharya,[1] Ignacio R. Rodriguez,[2] and Joe G. Hollyfield[1]

[1] Cole Eye Institute and The Lerner Research Institute
The Cleveland Clinic Foundation
Cleveland, Ohio 44195
[2] The National Eye Institute
National Institutes of Health
Bethesda, Maryland 20892

1. SUMMARY

From affinity purified SPACR we obtained seven peptides whose sequences reveal 100% identity to the deduced sequences in IMPG1 (accession number AF047492).[1] The gene product of IMPG1 is described as a chondroitin-6-sulfate proteoglycan. We report here that SPACR is the most prominent 147–150 kD molecule present in the IPM. SPACR is not a chondroitin sulfate proteoglycan since it is not a product of chondroitinase ABC digestion and does not react with a highly specific chondroitin-6-sulfate antibody. Additionally, the deduced amino acid sequence of IMPG1 reveals no established glycosaminoglycan attachment site. We conclude that the glycoprotein SPACR is the gene product of IMPG1. SPACR is a heavily sialated glycoprotein containing both N- and O-linked glyconjugates. It is not the core protein of an IPM specific chondroitin sulfate proteoglycan.

2. INTRODUCTION

Photoreceptors project from the outer retinal surface into a carbohydrate-rich interphotoreceptor matrix (IPM).[2] Unique glycoconjugates, some of which are organized into a relatively aqueous insoluble complex through interaction with hyaluronan[1,3] are present in the IPM. Wheat germ agglutinin (WGA) strongly decorates the

Retinal Degenerative Diseases and Experimental Therapy, edited by Hollyfield *et al.*
Kluwer Academic / Plenum Publishers, New York, 1999.

domains surrounding rods and weakly decorates the domains surrounding cones.[4] In contrast, peanut agglutinin (PNA), almost exclusively decorates[5,6,7] the cone matrix, and a small population of rod photoreceptors.[8] We have characterized the major WGA-binding glycoprotein in the human IPM, which we refer to as SPACR (*S*ialo*P*rotein *A*ssociated with *C*ones and *R*ods). SPACR, which has a molecular weight of around 150 kD, was isolated and purified from the IPM by lectin affinity chromatography. A polyclonal antibody prepared against SPACR intensely labels the rod-associated matrix with weaker labeling of the matrix surrounding cones.[9] This pattern and intensity of immunoreactivity mimics the differential distribution of WGA binding to the rod and cone matrix domains. Using a variety of lectins in conjunction with neuraminidase and glycosidase digestions, we established that (a) SPACR is a glycoprotein, (b) the molecule is heavily sialylated, (c) both N- and O-linked glycoconjugates are present in the molecule, and (d) these glycoconjugates account for approximately 30% of the molecular mass.[9]

Following enzymatic digestion of purified SPACR, we isolated seven peptides for sequencing which would permit cloning of the SPACR gene. The sequences obtained were identical to predicted peptides encoded by the cDNA of IMPG1 (accession number AF047492). The gene product of IMPG1 is reported to be a chondroitin sulfate proteoglycan core protein present in the IPM.[10,11,12,13] Although both IMPG1 and SPACR are located in the IPM, our protein and carbohydrate analysis clearly indicates that SPACR is a glycoprotein and not a proteoglycan.

IMPG1 was previously referred to as IPM150. Although a full description of human IPM150 has not been published in the peer reviewed literature, several abstracts describe IPM150 as a PNA-binding molecule, which can also be recognized by a chondroitin-6-sulfate antibody following chondroitinase digestion.[11,12,13] Could SPACR and IMPG1 be products of the same gene, differing only in post-translational modifications? We demonstrate here that the gene product of IMPG1 is not a proteoglycan, but instead is the IPM glycoprotein SPACR.[1]

3. RESULTS AND DISCUSSION

The controversy between the claim that IMPG1 is a proteoglycan and our observations that SPACR is a glycoprotein, could be reconciled if SPACR and IMPG1 share the same polypeptide but contain different associated glycoconjugates. If IMPG1 is the core protein of a chondroitin-6-sulfate proteoglycan, it should not enter a 7.5% SDS/PAGE unless the chondroitin sulfate chains are degraded with chondroitinase. This rationale was used in the analysis of the prominent 147–150 kD band in IPM samples before and after chondroitinase digestion. No substantial differences in the amount of protein; the amount of immunoreactivity to an anti-SPACR antibody; the amount of WGA binding; or the amount of PNA binding were evident in comparisons between control and chondroitinase ABC digested IPM samples (Fig. 1). These data indicate that no proteoglycan core protein with a molecular weight of 147–150 kD is released following chondroitinase digestion.

We also evaluated the distribution of chondroitin-6-sulfate immunoreactivity in IPM samples digested with chondroitinase ABC using 3B3, a monoclonal antibody, which is highly specific for chondroitin-6-sulfate Δdisaccharide (ΔDi6S).[14] In dot blots and in immunoblots, no differences were found between the digested and undigested

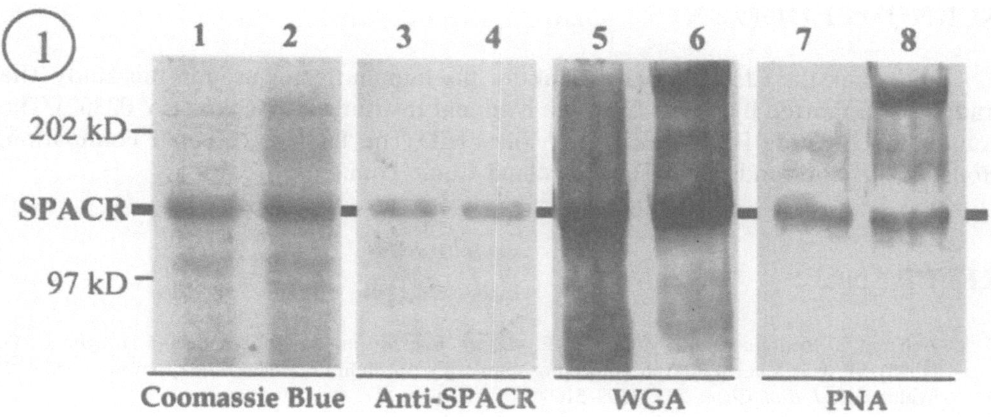

Figure 1. Isolated IPM samples were incubated in the absence of enzyme (odd numbered lanes) or were digested with chondroitinase ABC (even numbered lanes), before separation with SDS/PAGE (lanes 1–2) and transferred to membranes for Western (lanes 3–4) or lectin blotting (lanes 5–8). Note the presence of the SPACR band (at the level of the thick background line) in control and chondroitinase treated samples. There is a slightly darker appearance of the SPACR band in lanes 2, 6, and 8, but this is due to an elevation of Coomassie Blue and lectin staining throughout the lane following enzyme treatment. When the densities are quantitated following adjustment for differences in background, no change in the density of the SPACR band is evident following enzyme treatment [1].

IPM samples, indicating that the gene product of IMPG1, is not a chondroitin-6-sulfated proteoglycan core protein (data not shown).

Although our direct biochemical analysis failed to demonstrate a proteoglycan core protein with a molecular weight of 147–150 kD, is there any sequence information in IMPG1 which would indicate that this gene may code for a proteoglycan core protein? If the gene product of IMPG1 is the core protein of a proteoglycan, as has been stated in earlier reports,[11,12,13] the predicted polypeptide must contain a consensus sequence required for GAG attachment. In proteoglycans, GAGs are linked to serine residues in the core protein by way of a trisaccharide bridge, consisting of a xylose residue attached to serine, followed by two galactose residues.[15] Serine residues that function as GAG attachment sites require the consensus sequence S-G-X, where X represents either glutamic acid or aspartic acid.[16,17] Of the 701 amino acids present in the predicted polypeptide of IMPG1, 77 serine residues are present, but none of these residues are part of a consensus sequence for GAG attachment. Since no other consensus site for GAG attachment has ever been documented in a mammalian proteoglycan core protein, there is no sequence data to support the claim that IMPG1 codes for an IPM proteoglycan.

The IMPG1/SPACR gene has been mapped to chromosome 6q13–q15.[16] Linkage analysis has placed Stargardt's disease and North Carolina macular dystrophy near this location.[19,20,21] Although no specific disease causing mutations have yet been identified,[18,19] the gene for this abundant IPM protein is a potential candidate locus for inherited retinal disorders. If this gene is involved in any form of inherited retinal degeneration, as experiments are devised to probe the causal events which lead to degenerative changes, it will be important to recognize that the product of this gene is the glycoprotein SPACR, not an IPM chondroitin sulfate proteoglycan.

ACKNOWLEDGMENTS

We thank the Cleveland Eye Bank for the human tissues used in this study. The study was supported by grants from the National Institutes of Health (EY 02362), The Foundation Fighting Blindness, Hunt Valley, MD, The Retina Research Foundation, Houston, TX and funds from The Cleveland Clinic Foundation.

REFERENCES

1. Acharya, S., Rodriguez, I.R., Moreira, E.F., Midura, R.J., Misono, K., Todres, E., and Hollyfield, J.G. (1998). SPACR: a novel interphotoreceptor matrix glycoprotein in human retina that interacts with hyaluronan. *J. Biol. Chem.* **273**:31599–31606.
2. Röhlich, P. (1970). The interphotoreceptor matrix: electron microscopic and histochemical observations on the vertebrate retina. *Exp. Eye Res.* **10**:80–96.
3. Hollyfield, J., Rayborn, M., Tammi, M., and Tammi, R. (1998). Hyaluronan in the interphotoreceptor matrix of the eye: Species differences in content, distribution, ligand binding and degradation. *Exp. Eye Res.* **66**:241–248.
4. Hollyfield, J.G., Rayborn, M.E., Landers, R.A., and Myers, K.A. (1990). Insoluble interphotoreceptor domains surround rod photoreceptors in the human retina. *Exp. Eye Res.* **50**:107–110.
5. Hageman, G.S. and Johnson, L.V. (1987). Chondroitin 6-sulfate glycosaminoglycan is a major constituent of primate cone photoreceptor matrix sheaths. *Curr. Eye Res.* **6**:639–646.
6. Johnson, L.V. and Hageman, G.S. (1987). Enzymatic characterization of peanut agglutinin-binding components in the retinal interphotoreceptor matrix. *Exp. Eye Res.* **44**:553–556.
7. Blanks, J.C., Hageman, G.S., Johnson, L.V., and Spee, C. (1988). Ultrastructural visualization of primate cone photoreceptor matrix sheaths. *J. Comp. Neurol.* **270**:288–300.
8. Iwasaki, M., Meyers, K.M., Rayborn, M.E., and Hollyfield, J.G. (1992). Interphotoreceptor matrix in the human retina. Cone-like domains surround a small population of rod photoreceptors. *J. Comp. Neurol.* **319**:277–284.
9. Acharya, S., Rayborn, M.E., and Hollyfield, J.G. (1998). Characterization of SPACR, a sialoprotein associated with cones and rods present in the interphotoreceptor matrix of the human retina: immunological and lectin binding analysis. *Glycobiology* **8**:997–1006.
10. Kirchoff, M.A., Anderson, K., Johnson, L.V., and Hageman, G.S. (1990). Composition and distribution of insoluble interphotoreceptor matrix constituents. *Invest. Ophthalmol. Vis. Sci. (Supplement)* **31**:153 (abstract).
11. Kuehn, M.H., Mullins, R.F., and Hageman, G.S. (1993). Retinal interphotoreceptor matrix proteoglycan core protein sequences are unique and highly conserved. *Invest. Ophthal. Vis. Sci. (Supplement)* **34**:2443 (abstract).
12. Kuehn, M.H. and Hageman, G.S. (1995). Characterization of a cDNA encoding IPM150, a novel human interphotoreceptor matrix chondroitin 6-sulfate proteoglycan. *Invest. Ophth. Vis. Sci. (Supplement)* **36**:3264 (abstract).
13. Kuehn, M.H., Stone, E.M., and Hageman, G.S. (1997). Molecular analysis of IPM 150, a photoreceptor cell-specific proteoglycan. *Invest. Ophthalmol. Vis. Sci. (Supplement)* **38**:2788 (abstract).
14. Plaas, A., Wong-Palms, S., Roughley, P., Midura, R., and Hascall, V. (1997). Chemical and immunological assay of the nonreducing terminal residues of chondroitin sulfate from human aggrecan. *J. Biol. Chem.* **272**:20603–20610.
15. Roden, L. (1980). In *The biochemistry of glycoproteins and proteoglycans*. (Ed W. Lennarz). Pp. 267–371. Plenum Publishing Co. NY:
16. Chopra, R.K., Pearson, C.H., Pringle, G.A., Fackre, D.S., and Scott, P.G. (1985). Dermatan sulphate is located on serine-4 of bovine skin proteodermatan sulphate: Demonstration that most molecules possess only one glycosaminoglycan chain and comparison of amino acid sequences around glydosylation sites in different proteoglycans. *Biochem. J.* **232**:277–279.
17. Bourdon, M.A., Oldberg, A., Pierschbacher, M., and Ruoslahti, E. (1985). Molecular cloning and sequence analysis of a chondroitin sulfate proteoglycan cDNA. *Proc. Nat. Acad. Sci. USA* **82**:1321–1325.
18. Gehrig, A., Felbor, U., Kelsell, R., Hunt, D., Maumenee, I., and Weber, B. (1998). Assessment of the

interphotoreceptor matrix proteoglycan-1 (IMPG1) gene localised to 6q13–q15 in autosomal dominant Stargardt-like disease (ADSTGD), progressive bifocal chorioretinal atrophy (PBCRA), and North Carolina macular dystrophy (MCDR1). *J. Med. Genet.* **35**:641–645.
19. Kelsell, R., Gregory-Evans, K., Gregory-Evans, C., Holder, G., Jay, M., Weber, B., Moore, A., Bird, A., and Hunt, D. (1998). Localization of a gene (CORD7) for a dominant cone-rod dystrophy to chromosome 6q. *Am. J. Hum. Genet.* **63**:274–279.
20. Sauer, C., Schworm, H., Ulbig, M., Blakenagel, A., Rohrschneider, K., Pauleikhoff, D., Grimm, T., and Weber, B. (1997). An ancestral core haplotype defined the crital region harboring the north carolina macular dystrophy gene (MCDR1). *J. Med. Genet.* **34**:961–966.
21. Stone, E., Nichols, B., Kimura, A., Weigeist, T., Drack, A., and Sheffield, V. (1994). Clinical features of a Stargardt-like dominant progressive macular dystrophy with genetic linkage to chromosome 6q. *Arch. Ophthalmology* **112**:765–772.

19

BARDET-BIEDL SYNDROME

Phenotypic Characteristics Associated with the BBS4 Locus

Alessandro Iannaccone,[1]* Benedetto Falsini,[2] Neena Haider,[3] Giuseppe Del Porto,[4] Edwin M. Stone,[5] and Val C. Sheffield[3]

[1] Laboratory for Visual Electrophysiology
Department of Ophthalmology
University of Tennessee
Memphis, Tennessee
[2] Laboratory for Visual Electrophysiology
Institute of Ophthalmology
Catholic University
Rome, Italy
[3] Howard Hughes Medical Institute and Department of Pediatrics
University of Iowa, Iowa City, Iowa
[4] Department of Experimental Medicine
Section of Medical Genetics
University La Sapienza, Rome
Italy
[5] Department of Ophthalmology
University of Iowa, Iowa City, Iowa

SUMMARY

Herein reported are the systemic and ophthalmologic features of four patients with Bardet-Biedl syndrome (BBS) from a consanguineous pedigree. Molecular genetic analyses showed linkage with the BBS4 locus (chromosome 15), and excluded linkage to the other known loci on chromosome 3 (BBS3), 11 (BBS1), and 16 (BBS2), respectively. Patients were 15 to 26 years old (M = 3; F = 1) at last examination, and follow-

* To whom all correspondence should be addressed: 956 Court Avenue, Suite D-228, Memphis, TN 38163, USA; phone: (901) 448-1376; fax: (901) 448-7193; e-mail: iannacca@mail.eye.utmem.edu

Retinal Degenerative Diseases and Experimental Therapy, edited by Hollyfield *et al.*
Kluwer Academic / Plenum Publishers, New York, 1999.

up was up to 10 years. An array of clinical and functional tests were performed to characterize the ocular phenotype of this pedigree.

Mild obesity and dystrophic extremities were observed in all four subjects, and mild to moderate mental retardation was noted in three. All three male subjects had also cryptorchidism. Two siblings also had dental abnormalities. Renal function investigations were unrevealing.

Night blindness was the presenting symptom in all patients (onset: birth to 4 years of age). Visual acuity and color vision were severely affected early on in three subjects, moderately in the fourth. Further deterioration was documented in the follow-up. On ophthalmoscopy, retinal vessels were severely attenuated in all cases. Retinal pigmentary changes were mild and sectoral in all subjects in the earlier phases of the disease. Optic discs were pale (partial to complete atrophy) in all patients. Increasing retinal pigmentation and disc pallor was documented over the years. Maculas appeared initially healthy in all subjects, but mild abnormalities became ophthalmoscopically apparent over time. Goldmann visual fields were variably constricted, with good preservation of the inferior field in one subject as late as 26 years of age. Maximal electroretinogram responses were recordable only in two subjects at first examination, and suggested possible partial preservation of rod function in the early phases of the retinopathy. Later in the course of the disease, only cone-mediated function could be identified. Despite the healthy appearance on ophthalmoscopy, macular cone function was profoundly altered in all subjects.

In summary, the ophthalmologic manifestations of this BBS4 pedigree do not appear to be distinct from those of other reported BBS phenotypes. Systemic manifestations were only in part matching those described for other BBS4 kindreds, indicating substantial interfamilial heterogeneity. Particularly, obesity was not nearly as prominent a feature as reported for Arab-Bedouin BBS4 large consanguineous pedigrees. Intrafamilial variability was also noted, adding further to the complexity of this syndrome.

1. INTRODUCTION

Following the original descriptions in the late 1800s and early 1900s,[1–3] several reports described in detail the ophthalmologic and systemic features associated with Bardet-Biedl syndrome [(BBS) see Iannaccone et al.[4] for a review). The classic cardinal features of this autosomal recessive syndrome are: (1) pigmentary retinopathy, (2) dystrophic extremities, (3) obesity, (4) hypogenitalism, and (5) mental retardation. Evidence has accumulated over the past two decades to indicate that at least two additional features must be held as typical of BBS patients, i.e. (6) renal involvement[5–7] and (7) dental abnormalities.[8,9] With the exception of retinopathy, the other six features are not invariably present in each affected subject, and incomplete phenotypes are particularly frequent.

In the past several years, significant progress has been made in identifying several distinct loci associated with BBS pedigrees.[10–15] This has demonstrated the genetic heterogeneity associated with this condition, and provided in part a rationale for the observed phenotypic variability. Reports on pedigrees with multiple affected individuals are few, and so are those describing the phenotypes associated with these recently characterized distinct loci.[16–19]

We herein report the systemic and ophthalmologic features of four BBS patients from a single consanguineous pedigree that were genetically characterized and shown to be linked to the BBS4 locus on chromosome 15.

2. SUBJECTS AND METHODS

The four patients (M = 3; F = 1) examined in this study were 15 to 26 years old at the date of last exam. All patients originated from small neighboring communities in the southern outskirts of Rome. Initially believed to be unrelated and mere cases of independent isolate consanguinity, the four nuclear families where the affected individuals were identified (see Fig. 1, pedigree) were found to have a common ancestor (case I-2) who married twice, giving origin to the two branches of this pedigree. To perform linkage analysis, blood samples were obtained from 13 subjects, the four patients, and nine unaffected relatives.

Ophthalmologic examination included best corrected Snellen visual acuity; color vision confrontation tests (Panel D-15 or Roth 28 Hue); slit lamp anterior segment exam and posterior segment high-magnification (+78D Volk lens, Mentor, OH) biomicroscopy and indirect binocular ophthalmoscopy; Goldmann kinetic visual fields to I-4e, III-4e, and V-4e targets (expressed as percent of normal). Standard Ganzfeld flash electroretinography[20] (maximal ERG responses in all subjects; whenever possible, also dark-adapted rod ERGs and light-adapted cone ERGs, to either flicker or transient stimuli). Contact lens or palpebral skin electrodes were used. Macular cone function was assessed by means of a previously described method.[21] Light-adapted focal ERGs (FERGs) were obtained in response to red flickering stimuli (9 × 9 deg) generated by an array of LEDs modulated sinusoidally. When possible, responses were recorded over a range of temporal frequencies (3.7–52 Hz). The fundamental response (F1) at 41 Hz and the second harmonic (F2) at 8 Hz were held as primary indicators of outer and inner retinal function, respectively. Both the amplitude (in μV) and the phase (in degrees) of the responses were determined, the latter representing an estimate of the implicit time of the response.

History was aimed at determining type (e.g., night blindness, light aversion, etc.) and age of onset of visual disturbances. To characterize the array of systemic abnormalities, a standardized questionnaire was administered by one of us (A.I.) asking about the presence of (1) pre-existing obesity in infancy and/or current weight control therapy (if no overt obesity was observed); (2) pre-existing dystrophic extremities (polydactyly, syndactyly, and brachydactyly) and/or history of surgical removal of extra digits or other hand or foot appendices; (3) learning disabilities (school performance); (4) secondary sexual development anomalies (+ virilism, dysmenorrhea or other menstrual abnormalities in females, cryptorchidism in males); (5) urinary tract abnormalities and/or kidney dysfunction (BUN, creatinine); (6) documented elevated blood pressure; (7) dental alterations (microdontia, missing teeth, or other related problems); (8) cardiovascular abnormalities or ECG evidence of conduction defects; (9) liver function problems and/or documented liver enzyme elevation; (10) coexisting endocrinologic abnormalities; (11) gait alterations; (12) bone anomalies; and (13) hearing loss.

No formal physical exam was performed, except for visual and palpatory exam of both hands and feet, measurement of arterial blood pressure and gait examination. Whenever indicated, ECGs and dental arcade X-rays were also obtained. Renal func-

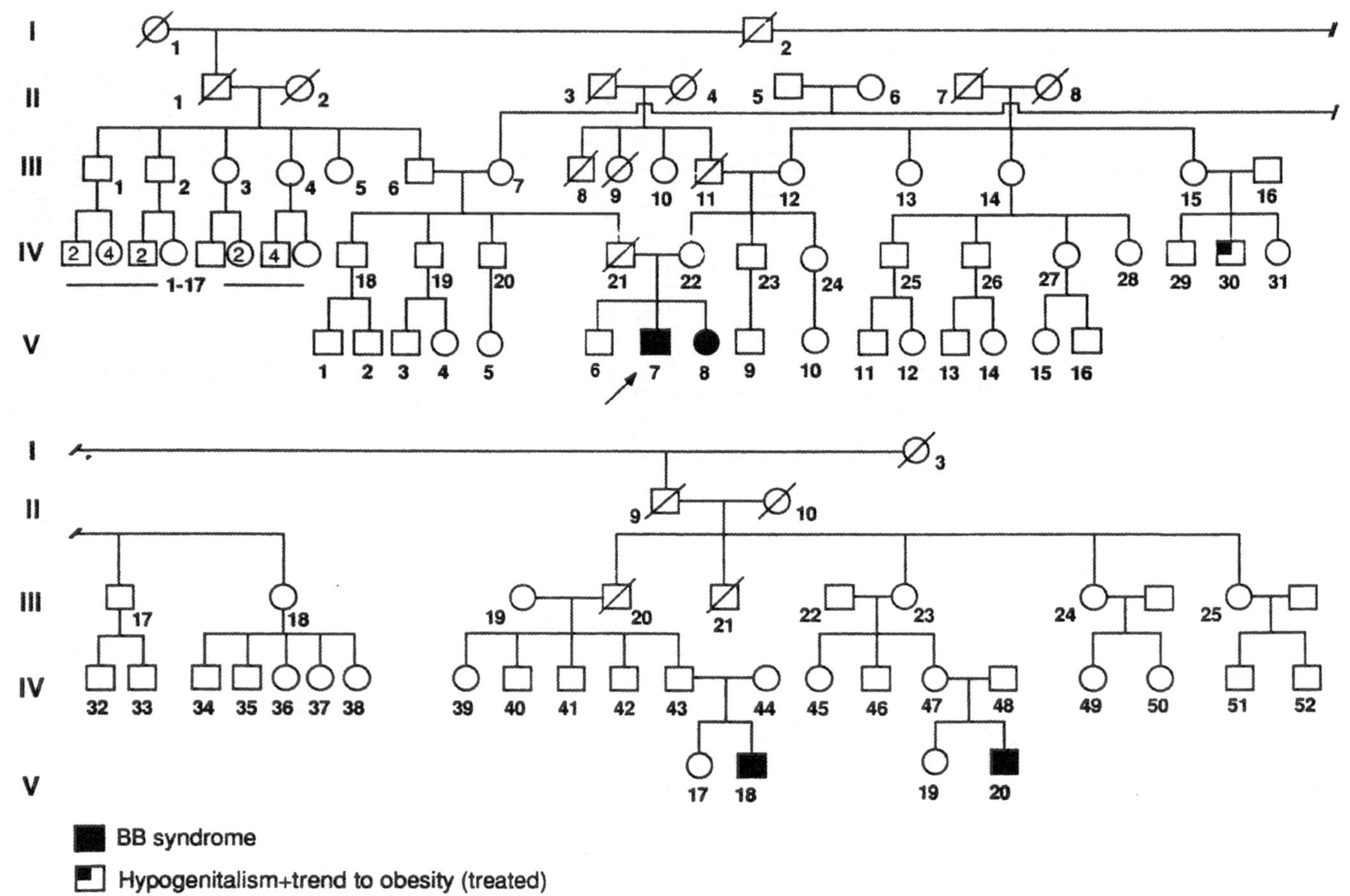

Figure 1. BBS4 (chromosome 15-linked) pedigree.

tion profiles were obtained in three subjects, kidney ultrasound in two, and urography was performed in one.

3. RESULTS

3.1. Molecular Genetic Analysis

Genotyping of DNA samples with polymorphic markers flanking and within the four known BBS loci excluded linkage to BBS1 (chromosome 11), BBS2 (chromosome 16), and BBS 3 (chromosome 3), based on recombination between the markers and the disease phenotype. Linkage to the BBS4 locus on chromosome 15 was demonstrated based on sharing of a homozygous haplotype within the disease interval by all four affected individuals. Homozygosity for the shared haplotype was not found in unaffected individuals in the pedigree.

3.2. Systemic Findings (Table 1)

Dystrophic extremities (a combination of polydactyly, syndactyly, and brachydactyly of the hands) were observed in all four subjects. Mild obesity was also reported since birth in all subjects. Dietary measures had been enforced prior to our examination in all cases, making obesity virtually unnoticeable, except for a gynecoid fat distribution in the male subjects (i.e., primarily at the hips and buttocks). Mild to moderate mental retardation was noticeable in three subjects. Case 4 (V-20) did not show any appreciable cognitive impairment, as he was able to complete successfully graduate studies in natural sciences. Hypogenitalism was reported by two subjects, and all three male subjects had also cryptorchidism. Renal ultrasound in two subjects and urography in one were unrevealing. Two siblings also had dental abnormalities, currently under investigation.

3.3. Ophthalmologic Findings (Table 2)

The main clinical and functional findings are summarized in Table 2. All patients had signs and symptoms of retinopathy. Night blindness was the presenting symptom

Table 1. Summary of systemic findings (cardinal features)

Patient (Pedigree Position)	Case 1 (V-7)	Case 2 (V-8)	Case 3 (V-18)	Case (V-20)
Age at last exam/Sex	24/M	15/F	21/M	26/M
Obesity	Mild	Mild	Mild	Mild
Dstrophic Extremities	•	•	•	•
Mental Retardation	Mild	Moderate	Mild	—
Hypogenitalism	—	•	•	—
Cryptorchidism	•	N/A	•	•
Renal disease	—	?	?	—
Dental Abnormalities	•	•	?	—

Legend: • = present; – = absent; ?: unknown or currently under investigation; N/A = not applicable.

Table 2. Clinical and functional ophthalmologic characteristics

Patient (Pedigree Position)	Case 1 (V-7)	Case 2 (V-8)	Case 3 (V-18)	Case 4 (V-20)
Age at last exam/Sex	24/M	15/F	21/M	26/M
Visual Acuity (best eye)	20/300	20/100	20/200	20/200
Refraction (type)	myopic	myopic	hyperopic	myopic
Astigmatism (diopters, avg.)	3.25	4.50	2.50	2.25
Nystagmus	•	—	•	—
Macula	Thinned	Normal	RPE Mottling	Cellophane
Pigmentary Retinopathy	•	•	•	•
Pigment Extension (degs., avg.)	360	360	90 (+360 S&P)	195
Grading of Findings				
Pigmentary Grading	3–4	2	1	1
Optic Disc	4	3	3	3
Retinal Vessels	3	4	4	3
Color Vision	GDc	GDc	Ac	GDc
Visual Field (% of normal)				
I-4e (interocular avg.)	—	N/A	—	—
V-4e (interocular avg.)	3%	N/A	25%	60%
ERG responses (avg. amplitude and timing, μV/msec)				
Rod	N/R	N/R	N/R	N/R
Maximal	N/R	13/118	N/R	N/R
Cone	N/R	N/A	1.3/49	N/R

Legend:
Pigmentary grading: 0 = absent; 1 = sparse and thin; 2 = sparse and thick; 3 = moderate and thick; 4 = diffuse and thick; 5 = confluent.
Optic disc grading: 0 = normal; 1 = mildly pale; 2 = waxy deerskin-like discoloration; 3 = temporal atrophy; 4 = diffuse atrophy.
Retinal vessels grading: 0 = normal; 1 = mildly attenuated; 2 = moderately attenuated; 3 = severely attenuated; 4 = thread-like appearance.
• = present; – = absent; S&P = salt and pepper fundus appearance; GDc = generalized dyschromatopsia; Ac = achromatopsia (unable to distinguish any color); N/A = not available; N/R = non recordable; * = 30-Hz flicker response.

in all patients (onset: between birth and 4 years of age). Visual acuity and color vision were also severely compromised early on in three subjects, with further progression (Fig. 2) documented in one (case 1, V-7). The course of foveal compromise in case 4 (V-20) was different. He had only moderate visual acuity compromise in the earliest stages of the disease, but experienced a sharp loss in foveal function in the second decade of disease duration (mid 20s). However, color vision was profoundly abnormal since his first examination.

On ophthalmoscopy, retinal vessels were severely attenuated in all cases. Retinal pigmentary changes were mild and sectoral in all subjects in the earlier phases of the disease, and as late as 21 years of age. Optic discs were pale (partial to complete atrophy) in all patients. Increasing retinal pigmentation and disc pallor was documented over the years. Maculas appeared initially healthy in 3 out of 4 subjects, except case 3 who had mild atrophy of the RPE at his first and only examination. Mild abnormalities (cellophane maculopathy) became ophthalmoscopically apparent also in case 4 in his 20s, as visual acuity began to drop more rapidly, and in case 1 (macular retinal thinning). On clinical grounds, however, the more evident changes in ophthalmoscopic appearance occurred at the optic disc level: pallor progressed from stage 2 to 4 in case 1 and from stage 2 to 3 in case 4, the latter also associated with parapapillary exudative manifestations.

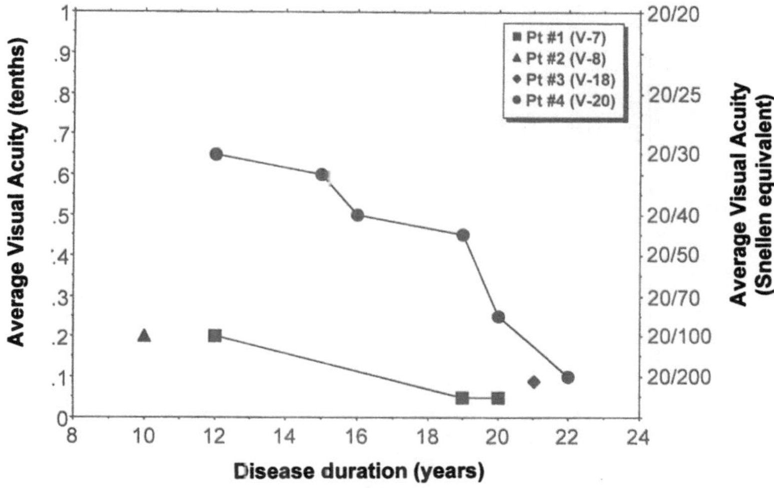

Figure 2. Visual acuity in the BBS4 phenotype: longitudinal observations.

Goldmann visual fields were severely constricted in case 1–3, but not in case 4, who had good preservation of the inferior field to the V-4e target (Fig. 3). Mild progression of the mid-peripheral scotomas was subsequently documented. Of note, no case had reliable detection of the standard I-4e target, including case 4, suggestive of a significant loss of central retinal sensitivity.

Despite the healthy or minimally affected appearance on ophthalmoscopy, macular cone function as assessed by FERG was profoundly altered in the two tested subjects. Plots of amplitude vs. phase of the responses (Fig. 4) demonstrated severe alteration of cone-mediated macular function in all instances. In relative terms, there were greater amplitude losses than phase delays, and phase delays tended to be greater for photoreceptor responses (F1) than for postreceptoral ones (F2).

Maximal ERG responses recorded with corneal electrodes (data not shown) were recordable only in case 1 at the age of 14 (1988), moderately delayed and larger in

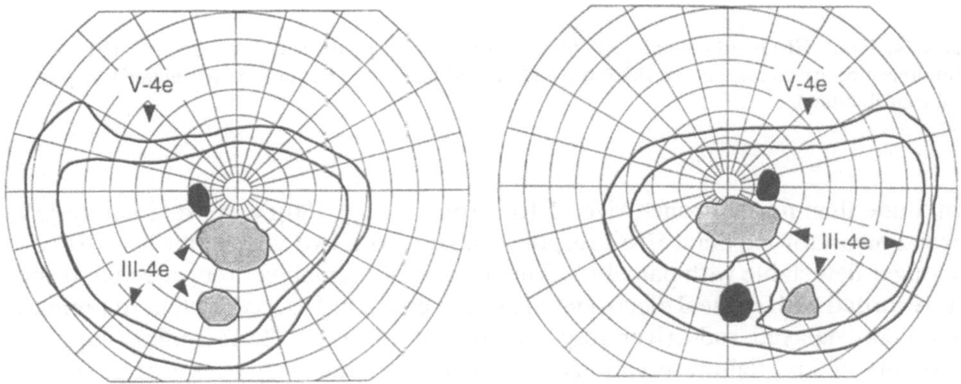

Figure 3. Visual fields of case 4 (V-20) show inferior preservation with the V-4e target in both eyes.

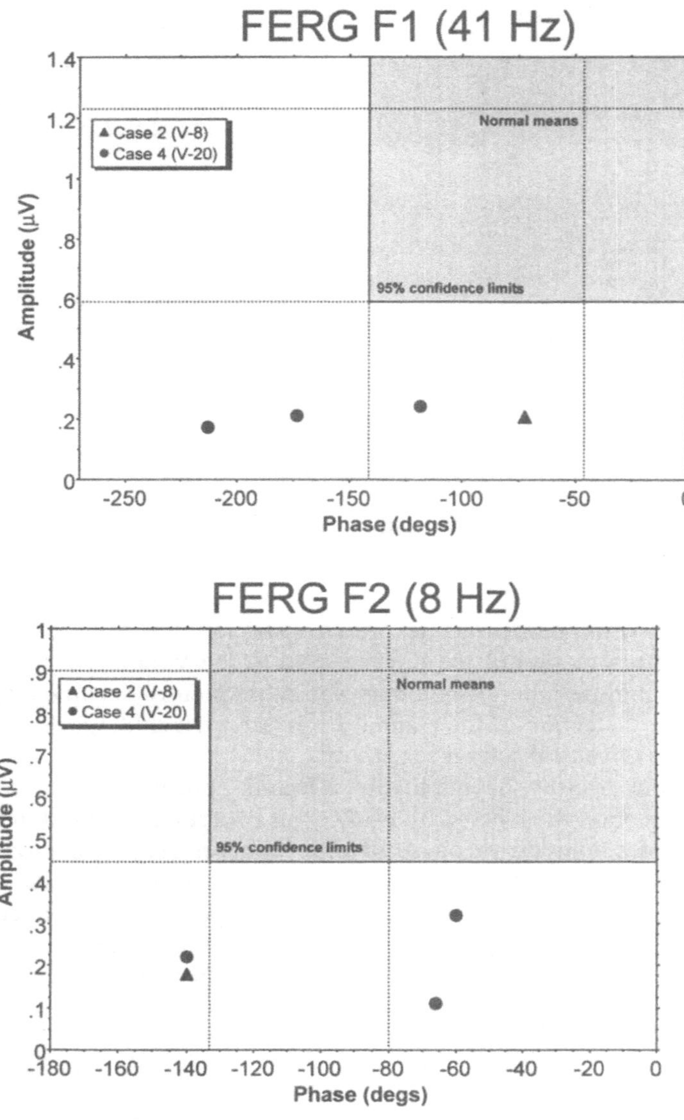

Figure 4. Focal ERGs (amplitude vs. phase). Top panel, fundamental component (F1) at 41 Hz (photoreceptor-driven responses). Bottom panel, second harmonic (F2) at 8 Hz (dominated by postreceptoral contributions).

amplitude than normal cone-driven ERGs (on average, 180 μV, approximately 35% of normal mean for maximal ERG responses). However, two years later his ERGs were no longer detectable with palpebral skin electrodes. A small yet reproducible response was recorded with corneal electrodes also from case 2 in 1996, consistent in timing with a markedly delayed, rod-driven response (data not shown). These findings support the possibility of a partial preservation of rod function in the early phases of the retinopathy in the BBS4 phenotype, followed by a rapid decline. Although a large portion of the inferior visual field was maintained in case 4, no ERG response could be detected at any time, despite extensive signal averaging. This marked mismatch has been noted already in other cases of BBS,[22] apparently occurring in this condition more frequently than in non-syndromic retinitis pigmentosa.[23]

4. DISCUSSION

Investigation of this pedigree showed an advanced form of retinal degeneration in all patients. Onset of symptoms took place consistently within the first four years of life. Profound cone and rod function abnormalities were detected in all cases. Accordingly, evidence of both central and peripheral visual function compromise was found in all subjects. Severe partial or diffuse atrophic optic nerve changes were observed in all cases, likely contributing to the severity of the phenotype.

Marked obesity was not a prominent feature in the pedigree herein described. Although dietary restrictions has been applied prior to our examination, obesity was reportedly mild in all patients since birth. This is different from previous reports on Arab-Bedouin BBS4 kindreds,[17] yet similar to a recently described Scandinavian family.[19] Of note, in the Arab-Bedouin patients, obesity was very prominent, more than in any other genetically distinct form of BBS.[17] This discrepancy may be accounted for by allelic heterogeneity, or by differences in the genetic background. However, dystrophic upper extremities were observed in all subjects. This has been consistently reported in both Arab-Bedouin and Scandinavian pedigrees.[17,19] Also, consistent with previous reports on BBS4 patients,[16–19] dental abnormalities were frequently noted, kidney compromise appeared to be infrequent, and evidence of cardiovascular compromise was lacking. Anomalies of secondary sexual development were present in all subjects. All males had cryptorchidism, the only female had delayed menarche and hypogenitalism (the latter found also in one male). Mental retardation was widely variable, ranging from absent to moderate.

Case 4 (V-20) demonstrated a somewhat milder phenotype than the other three subjects, with sectoral pigmentary retinopathy, partial inferior visual field preservation, and good visual acuity for a significantly longer length of time. However, color vision was abnormal early on in the course of his disease, overall suggesting early dysfunction of foveal cones also in this case. He also had no overt mental retardation, as he completed successfully graduate studies.

ERG measurements raise the possibility that, although night blindness was the presenting symptom in all patients, rod function may have been partially preserved in the earlier stages of the disease at least in the first two patients (V-7 and V-8, nuclear family #1). In the later stages, however, all patients have virtually non-recordable ERG signals, and only cone function appears to be retained. FERGs demonstrate persistent, although profoundly impaired, macular cone function.

Longitudinal measurements show that, within approximately 20 years of disease duration, all patients have visual acuities ≤20/100. Due to the early onset of symptoms in all cases, this predicts these levels of central compromise to be attained by the mid 20s in BBS4 patients. These findings confirm and expand previous observations on Scandinavian BBS4 kindreds.[17]

In summary, the ophthalmologic manifestations of this BBS4 pedigree do not appear to be significantly different from those of other reported BBS phenotypes, although some findings suggest the possibility of a partial preservation of rod function in the earlier stages of the disease. Intrafamilial variability in both ophthalmologic and systemic manifestations was documented. Comparison to other reported kindreds also indicates interfamilial variability. To date, the reasons for this marked phenotypical heterogeneity remain unknown. Several speculative mechanisms may be advocated. It is possible that the phenotypic effects of the BBS4 causal gene (and perhaps all other BBS genes) may be modulated by other, currently unknown, alleles. Also, variability may be accounted for by putative compound heterozygosity for a widely expressed product, yielding phenotypes of different severity. Similar to what has been observed for members of the tubby gene family,[24] alterations affecting the alternate 5′ splicing patterns of the causal genes may also modulate different phenotypical manifestations. Alternatively, similar to the OTX-like homeobox gene *CRX*,[25] regulatory genes affecting downstream the expression of several other gene products may be involved in BBS. Identification of the BBS4 gene and of the causal mutations will shed light on these aspects.

ACKNOWLEDGMENTS

The authors are grateful to Barbara Grammatico, B.S., and Carmelilia De Bernardo, B.S., for assistance with blood sample collection, and to Mr. W. Brad Rouse for helping with data analysis and technical assistance. This study was supported by NIH grant EY11298 (to VCS and EMS) and by an unrestricted grant from Research to Prevent Blindness, Inc. (UT Memphis Dept. of Ophthalmology).

REFERENCES

1. J.Z. Laurence and R.C. Moon, 1866, Four cases of retinitis pigmentosa occurring in the same family, and accompanied by general imperfections of development, *Ophthalmol Rev.* **2**:32–41.
2. G. Bardet, 1920, Sur un syndrome d'obésité congénitale avec polydactilie et rétinite pigmentaire (contribution à l'étude des formes cliniques de l'obésité hypophysaire), *Thèse de Paris (Le Grand)* **47**:9–107.
3. A. Biedl, 1922, Ein Geschwisterpaar mit adiposo-genitaler Dystrophie, *Dtsch. Med. Wochenschr.* **48**:1630.
4. A. Iannaccone, G. De Propris, S. Roncati, E. Rispoli, G. Del Porto, and M.R. Pannarale, 1997, The ocular phenotype of the Bardet-Biedl syndrome: Comparison to non-syndromic retinitis pigmentosa, *Ophthalmic Genetics* **18**:13–26.
5. D.N. Churchill, P. McManamon, and R.M. Hurley, 1981, Renal disease—a sixth cardinal feature of the Laurence-Moon-Biedl syndrome, *Clinical Nephrology* **16**:151–154.
6. J.S. Green, P.S. Parfrey, J.D. Harnett, N.R. Farid, B.C. Cramer, G. Johnson, O. Heath, P.J. McManamon, E. O'Leary, and W. Prise-Phillips, 1989, The cardinal manifestations of the Bardet-Biedl Syndrome, a form of Laurence-Moon-Biedl Syndrome, *N. Engl. J. Med.* **321**:1002–1009.
7. D. O'Dea, P.S. Parfrey, J.D. Harnett, D. Hefferton, B.C. Cramer, and J. Green, 1996, The importance of renal impairment in the natural history of Bardet-Biedl syndrome, *Am. J. Kidney Dis.* **27**:776–783.
8. J.L. Kobrin, C.L. Ternand, W.H. Knobloch, and D.D. Johnson, 1990, Dental abnormalities as a component of the Laurence-Moon-Bardet-Biedl syndrome, *Ophthalmic Paediatr. Genet.* **11**:299–303.
9. M.K. Borgström, R. Riise, K. Tornqvist, and L. Granath, 1996, Anomalies in the permanent dentition and other oral findings in 29 individuals with Laurence-Moon-Bardet-Biedl syndrome, *J. Oral Pathol. Med.* **25**:86–89.
10. A.E. Kwitek-Black, R. Carmi, G.M. Duyk, K.H. Buetow, K. Elbedour, R. Parvari, C.N. Yandava, E.M. Stone, and V.C. Sheffield, 1993, Linkage of Bardet-Biedl syndrome to chromosome 16q and evidence for non-allelic heterogeneity, *Nature Genet.* **5**:392–396.
11. M. Leppert, L. Baird, K.L. Anderson, B. Otterud, J.R. Lupski, and R.A. Lewis, 1994, Bardet-Biedl

syndrome is linked to DNA markers on chromosome 11q and is genetically heterogeneous, *Nature Genet.* **7**:108–112.

12. V.C. Sheffield, R. Carmi, A.E. Kwitek-Black, T. Rokhlina, D. Nishimura, G.M. Duyk, K. Elbedour, S.L. Sunden, and E.M. Stone, 1994, Identification of a Bardet-Biedl syndrome locus on chromosome 3 and evaluation of an efficient approach to homozygosity mapping, *Hum. Mol. Genet.* **3**:1331–1335.
13. R. Carmi, T. Rokhlina, A.E. Kwitek-Black, K. Elbedour, D. Nishimura, E. Stone, and V.C. Sheffield, 1995, Use of a DNA pooling strategy to identify a human obesity syndrome locus on chromosome 15, *Hum. Mol. Genet.* **4**:9–13.
14. N. Haider, A.E. Kwitek-Black, A. Iannaccone, R. Carmi, K. Elbedour, D.Y. Nishimura, E.M. Stone, and V.C. Sheffield, 1996, Genetic and physical fine mapping of the chromosome 15 Bardet-Biedl syndrome (BBS4) locus, *Am. J. Hum. Genet.* **59**:A220
15. E.A. Bruford, R. Riise, P.W. Teague, K. Porter, K.L. Thomson, A.T. Moore, M. Jay, M. Warburg, A. Schinzel, N. Tommerup, K. Tornqvist, T. Rosenberg, M. Patton, D.C. Mansfield, and A.F. Wright, 1997, Linkage mapping in 29 Bardet-Biedl syndrome families confirms loci in chromosomal regions 11q13, 15q22.3-q23, and 16q21, *Genomics* **41**:93–99.
16. K. Elbedour, N. Zucker, E. Zalzstein, Y. Barki, and R. Carmi, 1994, Cardiac abnormalities in the Bardet-Biedl syndrome: Echocardiographic studies of 22 patients, *Am. J. Med. Genet.* **52**:164–169.
17. R. Carmi, K. Elbedour, E.M. Stone, and V.C. Sheffield, 1995, Phenotypic differences among patients with Bardet-Biedl syndrome linked to three different chromosome loci, *Am. J. Med. Genet.* **59**:199–203.
18. P.L. Beales, A.M. Warner, G.A. Hitman, R. Thakker, and F.A. Flinter, 1997, Bardet-Biedl syndrome: a molecular and phenotypic study of 18 families, *J. Med. Genet.* **34**:92–98.
19. R. Riise, S. Andreasson, M.K. Borgström, A.F. Wright, N. Tommerup, T. Rosenberg, and K. Tornqvist, 1997, Intrafamilial variation of the phenotype in Bardet-Biedl syndrome, *Br. J. Ophthalmol.* **81**:378–385.
20. M.F. Marmor and E. Zrenner, 1995, Standard for clinical electroretinography (1994 update), *Doc. Ophthalmol.* **89**:199–210.
21. B. Falsini, G. Iarossi, V. Porciatti, E. Merendino, A. Fadda, S. Cermola, and L. Buzzonetti, 1994, Postreceptoral contribution to macular dysfunction in retinitis pigmentosa, *Invest. Ophthalmol. Vis. Sci.* **35**:4282–4290.
22. A. Iannaccone, E.M. Vingolo, E. Rispoli, G. De Propris, P. Tanzilli, and M.R. Pannarale, 1996, Electroretinographic alterations in the Laurence-Moon-Bardet-Biedl phenotype, *Acta Ophthalmol. Scand.* **74**:8–13.
23. A. Iannaccone, E. Rispoli, E.M. Vingolo, P. Onori, K. Steindl, D. Rispoli, and M.R. Pannarale, 1995, Correlation between Goldmann perimetry and maximal electroretinogram response in retinitis pigmentosa, *Doc. Ophthalmol.* **90**:129–142
24. M.A. North, J.K. Naggert, Y. Yan, K. Noben-Trauth, and P.M. Nishina, 1997, Molecular characterization of TUB, TULP1, and TULP2, members of the novel tubby gene family and their possible relation to ocular diseases, *Proc. Natl. Acad. Sci. USA* **94**:3128–3133.
25. T. Furukawa, E.M. Morrow, and C.L. Cepko, 1997, Crx, a novel otx-like homeobox gene, shows photoreceptor-specific expression and regulates photoreceptor differentiation, *Cell* **91**:531–541.

20

GENETIC MAPPING AND EVALUATION OF *PDE6A* IN 49 SPANISH FAMILIES WITH AUTOSOMAL RECESSIVE RETINITIS PIGMENTOSA

A. Martínez-Mir,[1] E. Paloma,[1] S. Balcells,[1] L. Vilageliu,[1] S. J. Pittler,[2] and R. Gonzàlez-Duarte[1]

[1]Departament de Genètica, Facultat de Biologia
Universitat de Barcelona
Avda. Diagonal, 645. 08071. Barcelona
Spain
[2]Departments of Biochemistry and Molecular Biology, Ophthalmology and the Center for Eye Research
University of South Alabama College of Medicine
Mobile, Alabama 36688-0002

SUMMARY

Autosomal recessive retinitis pigmentosa (arRP) is a degenerative disease of photoreceptors in which defects in the genes encoding rhodopsin (*RHO*), α and β subunits of the cGMP phosphodiesterase (*PDE6A* and *PDE6B*, respectively), α subunit of the cGMP gated channel (*CNCG*), *RPE65* and retinaldehyde-binding protein 1 (*RLBP1*) have been reported. Additionally, linkage analyses have defined four arRP loci on 1q31–q32 (*RP12*), 6p21.3 (*RP14*; mutations in the *TULP1* gene), 1p13–p21 (*RP19*; mutation in the *ABCR* gene), and 2q31–33. However, the molecular basis of the disease has not been ascertained in over 50% of cases, yet.

The *PDE6A* gene encodes the rod specific form of the α subunit of cGMP phosphodiesterase (PDE6). Together, the α and β subunits of cGMP PDE6 catalyze the breakdown of cGMP in response to light, a critical step in the light transduction pathway. The *PDE6B* gene has been analyzed in several arRP populations and it appears to account for less than 5% of all arRP cases. In contrast, fewer mutations have been reported in the *PDE6A* gene in arRP patients. Here we describe the genetic mapping of the *PDE6A* gene and the analysis of its involvement in arRP in 49 Spanish

Retinal Degenerative Diseases and Experimental Therapy, edited by Hollyfield *et al.*
Kluwer Academic / Plenum Publishers, New York, 1999.

families. Linkage analysis performed in the CEPH panel of families allowed us to localize the gene at 0 cM from D5S434 (Z = 10.5). Cosegregation and homozygosity studies with an intragenic polymorphism, a very close marker (D5S434) and two flanking markers (D5S2013 and D5S436) ruled out *PDE6A* as the cause of arRP in 38 pedigrees. In the eleven remaining families, SSCP analysis of the *PDE6A*-coding region did not detect any disease causing mutation in the affected members. These results strongly suggest that mutations in the *PDE6A* gene are not responsible for arRP in these families.

1. INTRODUCTION

Retinitis pigmentosa (RP) is a hereditary eye disease characterized by slow outer-retinal degeneration. Its prevalence has been estimated to be 1:3000–1:5000.[1,2] Primary clinical features include night blindness and constriction of visual fields, leading to severe visual loss.[3] RP can be inherited as an autosomal dominant, an autosomal recessive or an X-linked disorder.[1,4,5] In addition, both allelic and non-allelic heterogeneity has been documented within each hereditary pattern.

To undertake the molecular analysis of arRP two main approaches have been used, namely, genomic search of loci by linkage in extended pedigrees and mutational analysis in candidate genes.[6] To date, linkage analyses have defined four arRP loci, on 1q31–q32 (*RP12*),[7] 6p21.3 (*RP14*)[8] (mutations in the *TULP1* gene[9,10]), 1p13–p21 (*RP19*)[11] (mutation in the *ABCR* gene[12]), and 2q31–33,[13] respectively. On the other hand, the candidate gene approach has revealed that arRP can be caused by mutations in the genes encoding: rhodopsin (*RHO*),[14] α,[15,16] and β[17–21] subunits of the cGMP phosphodiesterase (*PDE6A* and *PDE6B*, respectively), α subunit of the cGMP gated channel (*CNCG*),[22] *RPE65*,[23] and retinaldehyde-binding protein 1 (*RLBP1*).[24] However, despite extensive research, the molecular basis of arRP is still poorly understood. Most of the arRP loci and genes explain a few cases each, and *PDE6B* stands as the only exception, accounting for approximately 5% of all arRP cases.

We have previously analyzed a total of 10 RP genes and candidate genes: *RHO*, *PDE6B*, peripherin/*RDS*, *ROM1* (rod outer segment membrane protein), *GCAP* (guanylate cyclase activating protein), *RCV* (recoverin), *SAG* (S-antigen), *PDEG* (γ subunit of cGMP PDE6), *NRL* (neural retina-specific leucine zipper), and *IRBP* (interphotoreceptor retinoid-binding protein) in a panel of arRP families from Spain.[25,26]

We have now focused on the *PDE6A* gene, which encodes the α subunit of rod cGMP phosphodiesterase.[27] Although so far few mutations have been described in arRP patients,[15,16] its role in the phototransduction cascade and its high homology to *PDE6B* make this gene an interesting choice. Human *PDE6A* is a single-copy gene spanning 45–50 kb and consisting of 22 exons (unpublished data). It was physically mapped to 5q31.2–q34, distal to the *SPARC* locus.[27,28] The purpose of this study has been to evaluate the involvement of *PDE6A* in 49 arRP Spanish families. Due to the large number of exons to be analyzed, direct mutation analysis would be both expensive and tedious. Thus, we have first undertaken a genetic mapping of *PDE6A*. Several polymorphic markers were identified in the close vicinity of the gene and they were later used in cosegregation and homozygosity analyses in the arRP families. Only those patients belonging to families that were not excluded or non informative were further analyzed for mutations in the *PDE6A* coding region by SSCP and sequencing.

2. METHODS

2.1. Families

49 arRP families, comprising 29 consanguineous and 20 non-consanguineous pedigrees, were used to analyze *PDE6A* involvement. To localize *PDE6A* genetically, families 102, 884, 1331, 1332, 1347, 1362, 1413, and 1416 from the CEPH reference panel were studied.

2.2. Clinical and Electrophysiologic Studies

Most of the families were examined ophthalmologically at the Hospital de la Santa Creu i Sant Pau in Barcelona or at the Fundación Jiménez Díaz in Madrid. All patients were assessed with best corrected visual acuity, slit-lamp biomicroscopy, colour vision by the Farnsworth-Munsell test, ocular tension by air tonometer, perimetry, using the Octopus 500 automated perimeter and electroretinography using the Sereme Spectral Ophthalmologie. In addition, both direct and indirect ophthalmoscopy were performed.

Electroretinography was performed using needle electrodes, after pharmacologic (tropicamida 0.5%) dilation of the pupils and adaptation to the dark for 40 minutes. Electronic flash stimuli of 1500–3000 lux were used in all cases. Photo-scotopic ERG was performed under white light, scotopic ERG under blue light (480 nm) and photopic ERG under red light (600 nm).

2.3. Analysis of DNA Polymorphisms

Genomic DNA was obtained as described by Miller et al. with minor modifications.[29]

The 210A → G polymorphism at the 5′ end of intron 5 of *PDE6A* (GenBank Acc. #U39155) was assessed by a mismatched PCR amplification (using primers: F 5′ GAGAGAAAAAAAGTAAAAAAAAA*C*C 3′ and R 5′ GTTACTGTGAAACCC-CAATAAG 3′) followed by *Msp*I digestion (the less frequent allele creates a target for this enzyme). The digestion products were resolved on an 8% acrylamide gel.

The 409A → C polymorphism in exon 1 of *PDE6A*[16] was PCR amplified according to Huang et al.[15] The predicted length of the amplified sequence is 563 bp. The amplification product was digested with *Msp*I (since the polymorphism creates a target for this enzyme) and resolved on an 8% acrylamide gel.

All the other markers used in this study were from the MapPairs set (Research Genetics, Huntsville, AL) and were analyzed according to the manufacturer's indications.

2.4. Linkage and Homozygosity Analyses

Two-point and multipoint linkage analyses were performed using the MLINK and CMAP programs from the LINKAGE package, version 5.1.[30]

The intragenic polymorphism 409A → C, and the markers D5S434, D5S2013, and D5S436 were used (in this order) to evaluate cosegregation between *PDE6A* and arRP in the families. Homozygosity at either of these markers in the affected members of consanguineous families was also considered a criterion to assess *PDE6A* involvement.[31]

2.5. PCR-SSCP Analysis of the *PDE6A* Gene

Oligonucleotide primers to amplify the 22 exons of the *PDE6A* gene are described in reference 15. Exon 1 was analyzed in two overlapping fragments because of its large size. For this purpose, two additional primers internal to exon 1 were designed: exon 1A reverse 5′TGCAGGAGGAAGCACAGCTT 3′, exon 1B forward 5′ACAGAGAAAT-GCATCTTCAATG 3′. PCR conditions were as follows: 2 min at 94°C; 35 to 40 cycles of (40 s at 94°C followed by 30 s at an annealing/extension temperature that ranged from 47°C to 60°C depending on the fragment); 5 min at 72°C.

For SSCP analyses, 1 µl of the PCR product was mixed with 6 µl of 95% formamide, 0.3% xylene cyanol, 0.3% bromophenol blue, 10 mM EDTA solution. The samples were then denatured by incubation at 80°C for three minutes and placed on ice. Electrophoresis was carried out using a Mini PROTEAN II (Bio-Rad) non-denaturing polyacrylamide gel. Every sample was analyzed in three different conditions combining different polyacrylamide concentrations (8 or 12%), different glycerol concentrations in the gel (0% or 5%) and two temperatures (room temperature or 4°C). After electrophoresis the gels were silver stained as follows. The gel was first incubated with a 10% ethanol solution for 5 min. After incubation for 3 min with a 1% HNO_3 solution it was washed in deionized water, and then incubated for 20 min with a 12 mM $AgNO_3$ solution. The gel was washed again and incubated (up to 5 min) with a freshly prepared mixture of 280 mM Na_2CO_3 and 0.02% formaldehyde. The gel was finally soaked in a 10% acetic acid solution and dried.

Fragments that showed aberrant patterns compared to normal in the SCCP test were amplified, purified (Wizard PCR Preps, Promega, Madison, WI) and directly sequenced by fluorescent dideoxy cycle sequencing (ABI 373A Fluorescent DNA Sequencer, Perkin Elmer Cetus, Norwalk, CT).

3. RESULTS

3.1. Genetic Mapping of the *PDE6A* Gene

In order to characterize a new polymorphism in intron 5, PCR primers were designed to amplify its 5′ end. After examining 58 chromosomes from unrelated individuals, a polymorphic site was detected: 210A → G. To assess the frequency of the two alleles, a mismatched primer was designed that creates an *Msp*I site. Five out of the 58 chromosomes analyzed, carried the G allele. Given its low heterozygosity value (0.157), the 210A → G polymorphism was not used for the genetic mapping of the *PDE6A* gene.

Instead, the *PDE6A* intragenic polymorphism 409A → C (heterozygosity = 0.263) was used to localize the gene with respect to markers from a high-density map.[32] Four of the eight CEPH families used in this study (1331, 1332, 1347, and 1416) were informative for this polymorphism with a total of 52 informative meioses. An array of 31 markers from chromosome 5 was used for the two-point analysis. Several recombination events placed the *PDE6A* gene to a 5.1 cM interval between markers D5S436 and D5S2013. Six markers within this region cosegregate with *PDE6A*, being the most informative D5S2033 ($Z_{max} = 9.626$) and D5S434 ($Z_{max} = 10.536$) (Table 1). To confirm this result, multipoint analysis was performed between the *PDE6A* gene and five selected markers. A value of $Z_{max} = 14.2$ at $\upsilon = 0$ was obtained for marker D5S434 (Fig. 1).

Table 1. Two-point lod scores between PDEA and markers on Chromosome 5 q

	Lod score (Z) at recombination fraction (θ)								
Marker	0	0.01	0.05	0.10	0.20	0.30	0.40	θ_{max}	Z_{max}
D5S436	$-\infty$	7.489	7.599	7.134	5.789	4.145	2.233	0.03	7.674
D5S2033	9.626	9.489	8.922	8.176	6.543	4.685	2.538	0.00	9.626
D5S434	10.536	10.383	9.756	8.935	7.144	5.114	2.771	0.00	10.536
D5S2013	$-\infty$	8.087	8.178	7.685	6.267	4.519	2.454	0.03	8.263
D5S673	$-\infty$	11.049	10.944	10.197	8.238	5.898	3.184	0.02	11.152

3.2. Cosegregation and Homozygosity Analyses in arRP Families

The *PDE6A* intragenic polymorphism 409A → C was initially used to assess cosegregation to the arRP locus in 49 Spanish families. In the 29 consanguineous pedigrees, homozygosity of the *PDE6A* marker in arRP patients was also tested (see Fig. 2). Thirty five families in which *PDE6A* could not be excluded were further typed with marker D5S434, and if necessary, with the flanking markers D5S2013 and D5S436. In total, the involvement of the *PDE6A* gene was ruled out in 38 families.

3.3. Molecular Analysis of the *PDE6A* Gene

Cosegregation and homozygosity analyses did not rule out the *PDE6A* gene as the cause of arRP in 11 families. Therefore, one affected member of each of these families was screened for mutations in the *PDE6A* coding region. Exons 2 to 22 were

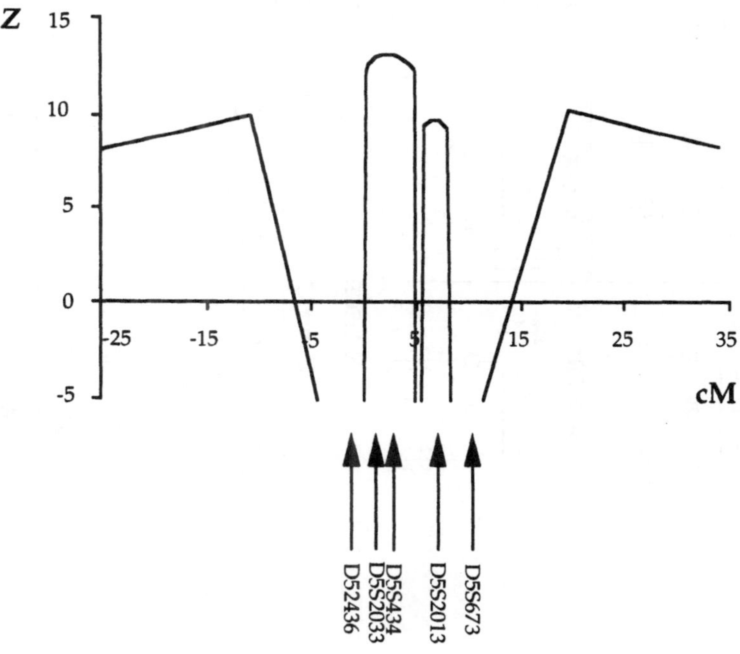

Figure 1. Multipoint linkage analysis showing the most probable localization for the *PDE6A* gene over marker D5S434.

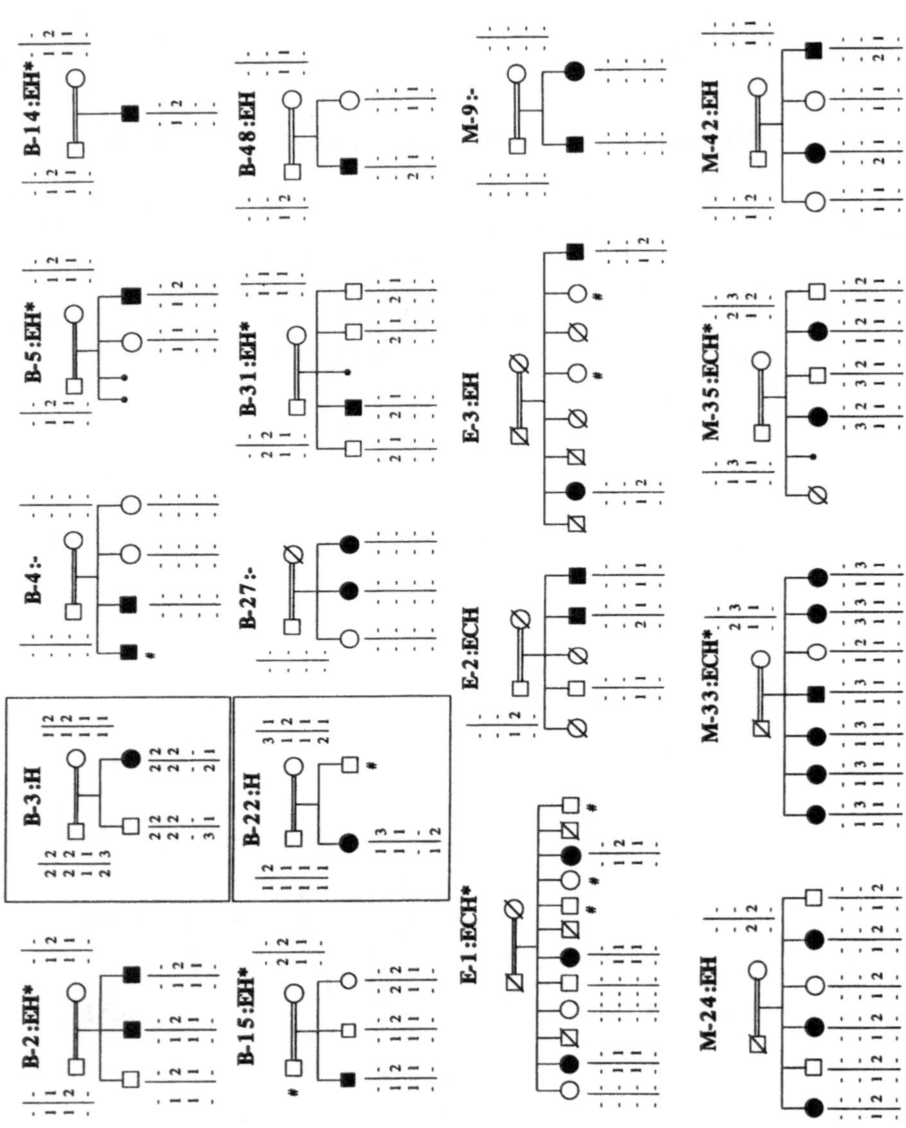
A
B-2:EH*
B-3:H
B-4:-
B-5:EH*
B-14:EH*
B-15:EH*
B-22:H
B-27:-
B-31:EH*
B-48:EH
E-1:ECH*
E-2:ECH
E-3:EH
M-9:-
M-24:EH
M-33:ECH*
M-35:ECH*
M-42:EH

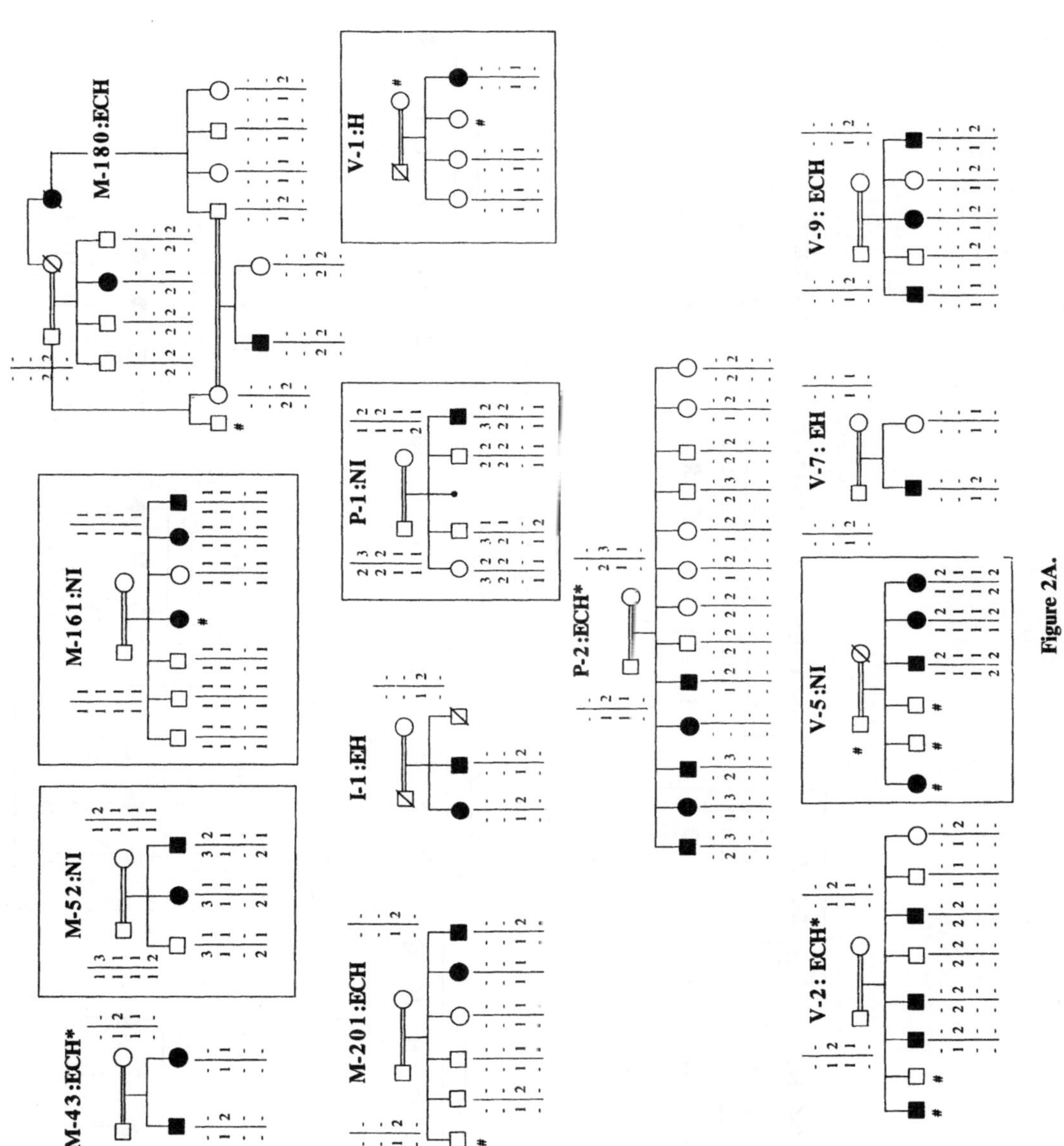
M-180:ECH
M-161:NI
M-52:NI
M-43:ECH*
V-1:H
P-1:NI
I-1:EH
M-201:ECH
P-2:ECH*
V-9: ECH
V-7: EH
V-5:NI
V-2: ECH*

Figure 2A.

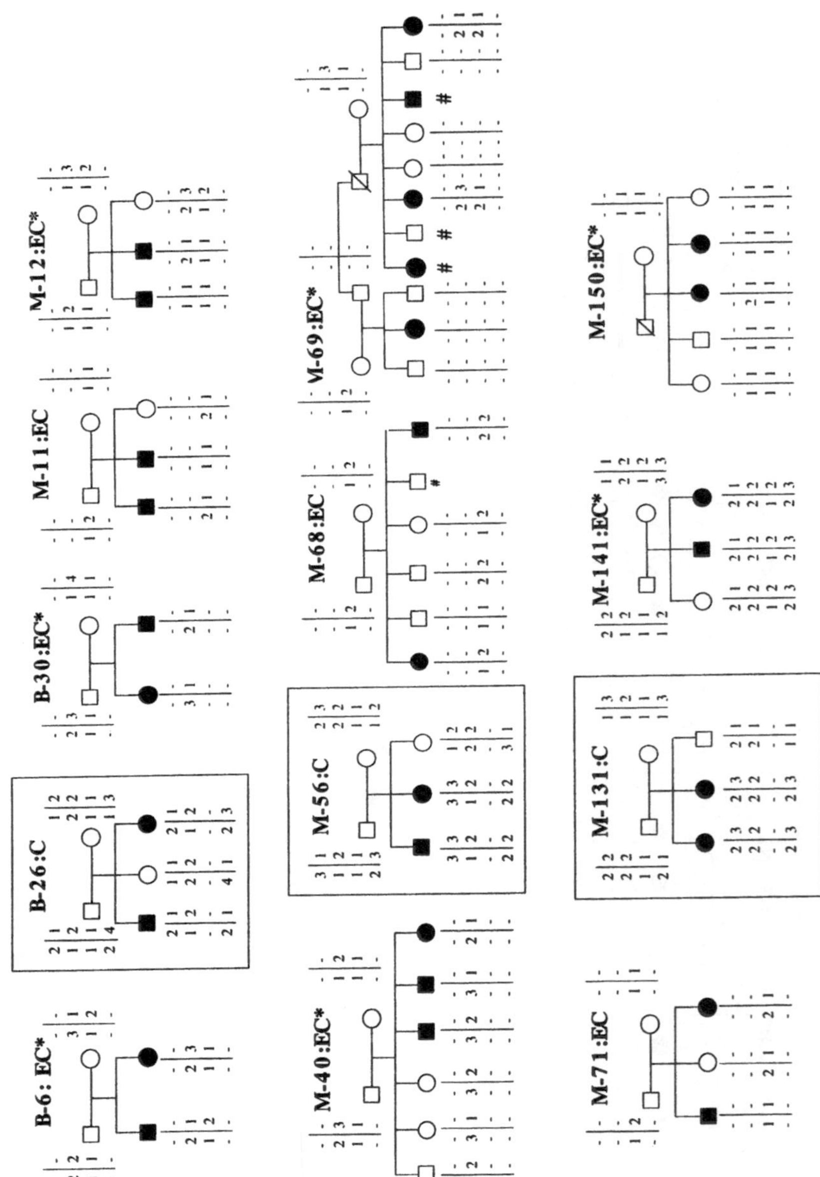
B
B-6: EC*
B-26:C
B-30:EC*
M-11:EC
M-12:EC*
M-40:EC*
M-56:C
M-68:EC
M-69:EC*
M-71:EC
M-131:C
M-141:EC*
M-150:EC*

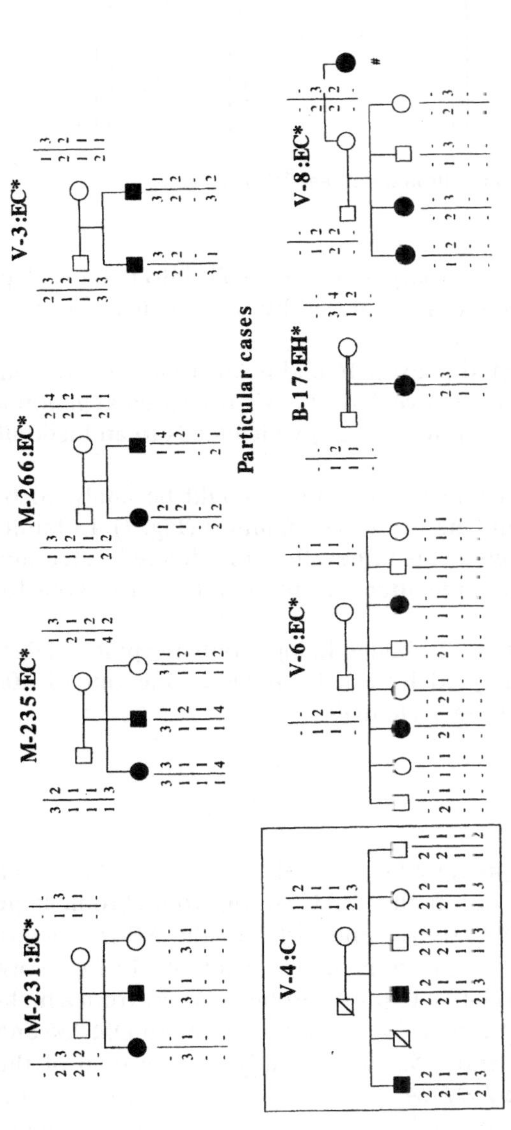

Figure 2. Cosegregation and homozygosity analyses in the panel of 49 Spanish arRP families. Haplotypes for markers D5S436, D5S434, *PDE6A* 409A → C polymorphism, and D5S2013 are shown. The families which have not been excluded are boxed. C: cosegregation between the markers and the disease locus; H: homozygosity for the marker allele; EC/EH: exclusion by lack of cosegregation/homozygosity; ECH: exclusion by lack of cosegregation and homozygosity; *: data from external markers.

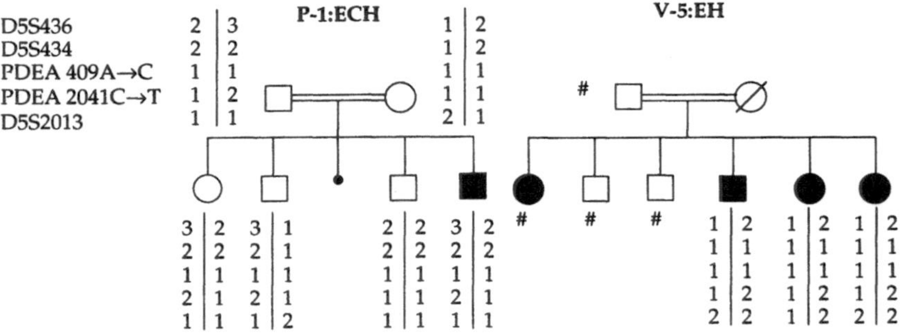

Figure 3. Haplotype and mutation results in the P1 and V5 pedigrees.

analyzed individually whereas exon 1, due to its larger size, was split into two overlapping fragments. PCR amplification products were analyzed by SSCP gels using three different conditions (see Materials and Methods).

An aberrant SSCP pattern was detected in exon 16 and it was found in two consanguineous families (P-1 and V-5). Sequencing revealed a C → T transition at genomic nucleotide 2041 (nucleotide numbering from reference 27) which leads to an histidine to tyrosine substitution at residue 655.

Because this change affects an *Nla*III restriction site, it could be analyzed by restriction analysis in all the available members of the two families (Fig. 3). In family P-1, the affected sib is heterozygous, as well as his father, the non affected sister, and one non affected brother. In family V-5 the three affected sibs are all heterozygous for the change.

To assess the possibility of the 2041C → T change being a polymorphism, DNA from 52 control individuals was also analyzed by SSCP analysis. One out of 104 chromosomes analyzed did show the 2041C → T change.

4. DISCUSSION

The *PDE6A* gene was originally mapped by Pittler et al.[27] at 5q31.2–q34 using a panel of somatic cell hybrids. Later, Li et al.[28] built a YAC contig from chromosome 5q31–q33, refining the localization of the *PDE6A* gene distal to the *SPARC* locus. However, no data on the genetic localization of the gene was available. In this work we report the genetic mapping of *PDE6A* relative to polymorphic markers from a high-density map.[32] Our data suggest that it is placed in a 5 cM interval between D5S436 and D5S2013, and closely linked to D5S2033 and D5S434. Recently, a gene map of the human genome was published[33] where the *PDE6A* gene is reported to map in a radiation hybrid containing markers D5S434 and D5S2013. These data are in agreement with our genetic mapping. Moreover, if we consider both sets of data a further refinement of the *PDE6A* location (between D5S2033 and D5S2013) is strongly suggested.

A remarkable number of loci and genes are known to be involved in arRP. We have previously analyzed a total of 10 RP genes and candidate genes and found the cause of the disease in a very small number of families.[25,26] In this paper we analyze the possible role of the *PDE6A* gene in a panel of 49 arRP Spanish families. A combined strategy based on cosegregation and homozygosity studies followed by mutation screening has been used. The genetic mapping of the gene and the cosegregation and homozygosity studies with the intragenic polymorphism 409A → C and the close markers D5S434, D5S2013, and D5S436 let us to exclude this gene as the cause of arRP in 38 out of 49 families (77.6%).

In the remaining 11 families SSCP analysis was undertaken in the whole coding region of the gene. As a result of this search for mutations only one aberrant banding pattern was detected, corresponding to the DNA fragment of exon 16. Sequencing revealed the change 2041C → T which leads to the amino acid substitution H655Y. This change was found in two different consanguineous families (P-1 and V-5) as well as in 1 out of 104 control chromosomes, what represents a frequency of 2/22 in RP chromosomes and 1/104 in control chromosomes, respectively. Both arRP families had been scored as non informative in the cosegregation and homozygosity analysis (Fig. 2). However, after analyzing the segregation of the 2041C → T variant (Fig. 3) we might exclude the *PDE6A* gene as the cause of the disease in families P-1 and V-5 by lack of cosegregation and/or lack of homozygosity. This variant was previously reported in heterozygosity by Meins et al.[16] in an arRP family for which the mutation in the second allele remained unidentified. Those authors stated that they might have missed the second mutation or that they could be facing a hypothetical digenic inheritance case with *PDE6A* and another unknown gene. In their sample, the frequency of the variant is of 1/188 arRP chromosomes and it was not detected in 60 control chromosomes.

The residue His655 is enclosed in the putative catalytic domain of the PDEA protein and it is conserved in the bovine and human PDE6A. This arguments strongly support the possibility of the change H655Y being a disease-causing mutation. However, if this is the case, the cosegregation and the mutation results in families P-1 and V-5 only fit in a putative digenic pattern with another unknown gene.

Alternatively, the variant 2041C → T could be a non-pathogenic polymorphism present in the general population in a very low frequency. This would explain why we found it in 1 out of 104 control chromosomes. In this regard, it should be noted that it was not detected in a sample of 167 adRP patients, 173 arRP patients and 95 control individuals studied by Huang et al..[15] However, the ethnic origin of this last sample (North American) is different from that studied in our work (Spanish) and that reported by Meins et al. (German and Indian).

In summary, the data reported in this study suggest that it is unlikely that mutations in the *PDE6A* gene are responsible for arRP in this panel of 49 Spanish families. This is in agreement with the two previous analyses by Huang et al.[15] and Meins et al.[16] where *PDE6A* is shown to be involved in around 1% of arRP cases. The reason for the difference with *PDE6B*, explaining around 5% of arRP, remains still unexplained. However, it suggests that the two subunits may play different roles, *PDE6B* being more critical for the correct function of the enzyme.

ACKNOWLEDGMENTS

The authors thank the families for their generous cooperation and Drs. C. Ayuso, M. Beneyto, M.A. Ramos-Arroyo, and I. Tejada who kindly provided DNA samples of some of their patients. We also thank Drs. P. Chivelet, T. Solans, B. García-Sandoval, T. del Río, B. Goldaracena, and C. Vilela for clinical assessment, L. Oliveira for technical assistance and Robin Rycroft for revising the English. The DNA from the CEPH families was kindly provided by Drs. S. Povey and F. Moreno.

This work was supported by Spanish CICYT (SAF93-0479-062-01 and SAF96-0329), the Federación de Asociaciones de Afectados de Retinosis Pigmentaria del Estado Español (FAARPE) to RG-D, and NIH/NEI grant EY09924 to SJP.

REFERENCES

1. P. Humphries, J.G. Farrar, P. Keena, and P. McWilliam, 1990, Retinitis pigmentosa: genetic mapping in X-linked and autosomal forms of the disease, *Clin Genet* **38**:1–13.
2. E.L. Berson, 1993, Retinitis-pigmentosa—The Friedenwald lecture, *Invest Ophthalmol Visual Sci* **34**:1659–1676.
3. J.R. Heckenlively, S.L. Yoser, L.H. Friedman, and J.J. Oversier, 1988b, Clinical findings and common symptoms in retinitis pigmentosa, *Am J Ophtalmol* **105**:504–511.
4. J.A. Boughman, P.M. Conneally, and W.E. Nance, 1980, Population genetic studies of retinitis pigmentosa, *Am J Hum Genet* **32**:223–235.
5. J.R. Heckenlively, 1988a, Autosomal dominant retinitis pigmentosa, in: *Retinitis pigmentosa*, (H. Ewan, eds), pp. 125–149, Philadelphia: Lippincott.
6. T.P. Dryja, 1990, Deficiencies in sight with the candidate gene approach, *Nature* **347**:614.
7. S. van Soest, L.I. Vandenborn, A. Gal, G.J. Farrar, L.M. Bleeker-Wagemakers, A. Westerveld, P. Humphries, L.A. Sandkuijl, and A.A.B. Bergen, 1994, Assignment of a gene for autosomal recessive retinitis pigmentosa (RP12) to chromosome 1q31–q32.1 in an inbred and genetically heterogeneous disease population, *Genomics* **22**:499–504.
8. J.A. Knowles, Y. Shugart, P. Banerjee, T.C. Gilliam, C.A. Lewis, S.G. Jacobson, and J. Ott, 1994, Identification of a locus, distinct from RDS-peripherin, for autosomal recessive retinitis pigmentosa on chromosome 6p, *Hum Mol Genet* **3**:1401–1403.
9. S.A. Hagstrom, M.A. North, P.M. Nishina, E.B. Berson, and T.P. Dryja, 1998, Recessive mutations in the gene encoding the tubby-like protein TULP1 in patients with retinitis pigmentosa, *Nat Genet* **18**:174–176.
10. P. Banerjee, P.W. Kleyn, J.A. Knowles, C.A. Lewis, B.M. Ross, E. Parano, S.G. Kovats, J.J. Lee, G.K. Penchaszadeh, J. Ott, S.G. Jacobson, and T.C. Gillian, 1998, *TULP1* mutation in two extended Dominican kindreds with autosomal recessive retinitis pigmentosa, *Nat Genet* **18**:177–179.
11. A. Martínez-Mir, M. Bayés, L. Vilageliu, D. Grinberg, C. Ayuso, T. del Río, B. García-Sandoval, E. Bussaglia, M. Baiget, R. Gonzàlez-Duarte, and S. Balcells, 1997, A new locus for autosomal recessive retinitis pigmentosa (RP19) maps to 1p13 1p21, *Genomics* **40**:142–146.
12. A. Martínez-Mir, E. Paloma, R. Allikmets, C. Ayuso, T. del Río, M. Dean, Ll. Vilageliu, R. Gonzàlez-Duarte, and S. Balcells, 1998, Retinitis pigmentosa caused by a homozygous inactivating mutation in the Stargardt disease gene *ABCR*, *Nat Genet* **18**:11–12.
13. M. Bayés, B. Goldaracena, A. Martínez-Mir, M.A. Iragui-Madoz, T. Solans, P. Chivelet, E. Bussaglia, M.A. Ramos-Arroyo, M. Baiget, L. Vilageliu, S. Balcells, R. Gonzàlez-Duarte, and D. Grinberg, 1998, A new autosomal recessive retinitis pigmentosa locus maps on chromosome 2q31–q33, *J Med Genet* **35**:141–145.
14. P.J. Rosenfeld, G.S. Cowley, T.L. McGee, M.A. Sandberg, E.L. Berson, and T.P. Dryja, 1992, A null mutation in the rhodopsin gene causes rod photoreceptor dysfunction and autosomal recessive retinitis pigmentosa, *Nat Genet* **1**:209–213.
15. S.H. Huang, S.J. Pittler, X.H. Huang, L. Oliveira, E.L. Berson, and T.P. Dryja, 1995, Autosomal recessive retinitis pigmentosa caused by mutations in the alpha subunit of rod cGMP phosphodiesterase, *Nat Genet* **11**:468–471.
16. M. Meins, A. Janecke, C. Marschke, M.J. Denton, K.G. Kumaramanickavel, S. Pittler, and A. Gal, 1997, Mutations in PDE6A, the gene encoding the α subunit of rod photoreceptor cGMP-specific phosphodiesterase, are rare in autosomal recessive retinitis pigmentosa, in: *Degenerative Retinal Disease*, (M.M. La Vail, J.G. Hollyfield, R.E. Anderson, eds), pp. 237–244, Plenum Publishing Corporation, New York.
17. M.E. McLaughlin, M.A. Sandberg, E.L. Berson, and T.P. Dryja, 1993, Recessive mutations in the gene encoding the b-subunit of rod phosphodiesterase in patients with retinitis pigmentosa, *Nat Genet* **4**:130–133

18. M. Bayés, M. Giordano, S. Balcells, D. Grinberg, L. Vilageliu, I. Martínez, C. Ayuso, J. Benítez, M.A. Ramos-Arroyo, P. Chivelet, T. Solans, D. Valverde, S. Amselem, M. Goossens, M. Baiget, R. Gonzàlez-Duarte, and C. Besmond, 1995, Homozygous tandem duplication within the gene encoding the β-subunit of rod phosphodiesterase as a cause for autosomal recessive retinitis pigmentosa, *Hum Mutat* **5**:228–234.
19. D. Valverde, T. Solans, D. Grinberg, S. Balcells, L.L. Vilageliu, M. Bayés, P. Chivelet, C. Besmond, M. Goossens, R. Gonzàlez-Duarte, and M. Baiget, 1996b, A novel mutation in exon 17 of the beta-subunit of rod phosphodiesterase in two RP sisters of a consanguineous family, *Hum Genet* **97**:35–38.
20. D. Valverde, M. Baiget, R. Seminago, E. del Rio, B. García-Sandoval, T. del Río, M. Bayés, S. Balcells, A. Martínez, D. Grinberg, and C. Ayuso, 1996a, Identification of a novel ARG552GLN in exon 13 of the β-subunit of rod phosphodiesterase gene in a Spanish family with autosomal recessive retinitis pigmentosa, *Hum Mut* **8**:393–394.
21. M. Danciger, J. Blaney, Y.Q. Gao, D.Y. Zhao, J.R. Heckenlively, S.G. Jacobson, and D.B. Farber, 1995, Mutations in the PDE6B gene in autosomal recessive retinitis pigmentosa, *Genomics* **30**:1–7.
22. T.P. Dryja, J.T. Finn, Y.W. Peng, T.L. Mcgee, E.L. Berson, and K.W. Yau, 1995, Mutations in the gene encoding the alpha subunit of the rod cGMP-gated channel in autosomal recessive retinitis pigmentosa, *Proc Natl Acad Sci USA* **92**:10177–10181.
23. S. Gu, D. Thompson, C.R.S. Srikumari, B. Lorenz, U. Finckh, A. Nicoletti, K.R. Murthy, M. Rathmann, G. Kumaramanickavel, M.D. Denton, and A. Gal, 1997, Mutations in RPE65 cause autosomal recessive childhood-onset severe retinal dystrophy, *Nat Genet* **17**:194–197.
24. M.A. Maw, B. Kennedy, A. Knight, R. Bridges, K.E. Roth, E.J. Mani, J.K. Mukkadan, D. Nancarrow, J.W. Crabb, and M.J. Denton, 1997, Mutation of the gene encoding cellular retinaldehyde-binding protein in autosomal recessive retinitis pigmentosa, *Nat Genet* **17**:198–200.
25. M. Bayés, A. Martínez-Mir, D. Valverde, E. del Río, L. Vilageliu, D. Grinberg, S. Balcells, C. Ayuso, M. Baiget, and R. Gonzàlez-Duarte, 1996, Autosomal recessive retinitis pigmentosa in Spain: evaluation of four genes and two loci involved in the disease, *Clin Genet* **50**:380–387.
26. R. Gonzàlez-Duarte, M. Bayés, A. Martínez-Mir, D. Valverde, S. Balcells, M. Baiget, L. Vilageliu, and D. Grinberg, 1997, Strategies for the genetic analysis of autosomal recessive retinitis pigmentosa in Spanish families, in: *Degenerative Retinal Diseases*, (M.M. LaVail, J.G. Hollyfield, R.E. Anderson, eds), pp. 263–275, Plenum Publishing Corporation, New York.
27. S.J. Pittler, W. Baehr, J.J. Wasmuth, D.G. McConnell, M.S. Champagne, P. van Tuinen, D. Ledbetter, and R.L. Davis, 1990, Molecular characterization of human and bovine rod photoreceptor cGMP phosphodiesterase α-subunit and chromosomal localization of the human gene, *Genomics* **6**:272–283.
28. X. Li, C.A. Wise, D. Le Paslier, A.L. Hawkins, C.A. Griffin, S.J. Pittler, M. Lovett, and E.W. Jabs, 1994, A YAC contig of approximately 3 Mb from human chromosome 5q31 → q33, *Genomics* **19**:470–477.
29. S.A. Miller, D.D. Dyke, and H.F. Polesky, 1988, A simple salting out procedure for extracting DNA from human nucleated cells, *Nucl Acids Res* **16**:1215.
30. G.M. Lathrop and J.M. Lalouel, 1984, Easy calculations of Lod scores and genetic risks on small computers, *Am J Hum Genet* **36**:460–465.
31. E.S. Lander and D. Botstein, 1987, Homozygosity mapping: a way to map human recessive traits with the DNA of inbred children, *Science* **236**:1567–1568.
32. C. Dib, S. Faure, C. Fizames, D. Samson, N. Drouot, A. Vignal, P. Millasseu, S. Marc, J. Hazan, E. Seboun, M. Lathrop, G. Gyapay, J. Morissette, and J. Weissenbach, 1996, A comprehensive genetic map of the human genome based on 5264 microsatellites, *Nature* **380**:152–154.
33. G.D. Schuler, M.S. Boguski, E.A. Stewart, L.D. Stein, G. Gyapay, K. Rice, R.E. White, P. Rodríguez-Tome, A. Aggarwal, E. Bajorek, S. Bentolila, B.B. Birren, A. Butler, A.B. Castle, N. Chiannilkulchai, A. Chu, C. Clee, S. Cowles, P.J. Day, T. Dibling, N. Drouot, I. Dunham, S. Duprat, C. East, and T.J. Hudson, 1996, The human transcript map, *Science* **274**:547–562.

GRADING DEVICE FOR LIGHT PERCEPTION WITH RETINITIS PIGMENTOSA

Makoto Tamai,* Hiroshi Kunikata, and Masahiro Tsunoda

Department of Ophthalmology
Tohoku University School of Medicine
Sendai Japan

SUMMARY

Purpose

The goal of this study is to innovate a device for grading very low visual function expressed as light perception in patients such as retinitis pigmentosa (RP) and makes it possible to evaluate effects for various therapeutic modalities or their changes following the progress of the disease.

Methods

The device was composed of a pair of goggles with a white light emission diode, 5 mm in diameter, an earphone for each ear, a control cabinet which emits light and a clicking stimuli, a hand-held grip with a button, and a recorder for calculating correct answer for stimuli. The light intensity and duration were changed in three levels, 10cd/m2(C), 100C, 1000C, and its duration, 0.1second(S), 0.3S, 1S independently. As a result, they were graded into nine different levels of intensities and delivered in a random sequence after a click sound of 0.7 sec prior to the stimulation. Patients were asked to push a button when they recognized these stimuli. Each eye was separately stimulated 5 times in each intensity and duration, 45 times per eye, therefore, a total of 90 times. Ten RP patients with light perception were examined and their visual ability

*Correspondence: Makoto Tamai, M.D. Department of Ophthalmology Tohoku University School of Medicine 1-1 Seiryoumachi, Aoba-ku, Sendai 980-8574, Japan

Retinal Degenerative Diseases and Experimental Therapy, edited by Hollyfield *et al.*
Kluwer Academic / Plenum Publishers, New York, 1999.

was graded. The FP was examined before and after cataract extraction, stellate ganglion block and lipo-prostaglandin E1 was administered in two cases with RP and we studied the possibility of differentiating their functions and also reproducibility with this device.

Results

We could differentiate changes of LP with this device and observe changes of very low grade vision with treatment.

Conclusion

This device is valuable for evaluating low visual function such as LP of RP patients. It will be valuable for following their vision and also discriminating very small changes of visual function with medical and surgical treatment in the future.

INTRODUCTION

Visual functions lower than expressing the ability to analyze an image (the minimum separable) is designated by the sensitivity of the retina in receiving the stimulus (the minimum visible) as light perception(LP) and non-light perception (NLP) when patients cannot discriminate between on and off settings of a test light[1]. At present, LP is difficult to grade. If a reliable measurement of the function with perimetry can be established, the level of sensation can be expressed by sensitivity of rod or cone system[2]. For patients who cannot detect visual stimuli, it has been impossible to accurately measure their visual impairment. In other words, we could only ask our patients whether the sensation of stimulated light was brighter or dimmer than the previous one. So, in daily practice, many patients with retinitis pigmentosa (RP) are diagnosed as "LP" by their physicians. Until recently we have had no problem with these methods even though patients wanted to know their residual functions more precisely. Naturally, it takes a long period for patients' visual function to decrease from LP to NLP. They are very eager to know how much and how rapidly their natural visual abilities are decreasing or increasing with their doctor's prescription or with their own health care.[3] For eye doctors who have been trying to treat RP with new methods, it has been urgent to find a new evaluating method for measuring low vision ability, such as LP or NLP with RP. Because it is very difficult or perhaps impossible to evaluate such effects with conventional electrophysiological methods or even other psychological ones. In this study, we tried to innovate a device for grading the LP in RP patients and evaluate the effects of treatment such as cataract extraction or with drugs.

METHODS

From September 1997 to August 1998, we examined 10 patients whose visual function was LP due to RP in the Outpatient Clinic in the University Hospital. A Low Vision Evaluator, LoVE in short, was composed of goggles with a light emission diode(LED) and its generator, a set of earphones to hear a click before stimulation, a hand grip with a button, and a printer for recording. The LED, 5 mm in diameter, was set separately at the center of each goggle for both eyes. The goggles were affixed snugly to the skin with a sponge liner and was painted black to shield any inward scattering light. The distance between each goggle with LED could be adjusted depending on the patients' pupillary distance and to fit each orbital margin in order to block any scattered light from the other side of the stimulation. The stimulus light intensity was changed in three levels: 10cd/m2(C), 100C, 1000C, and their duration in 0.1second(S), 0.3S, 1S, respectively. Each eye was illuminated separately, so the patient was stimulated with 9 different kinds of stimulus intensity. Each stimulation was preceded 0.7S by a click of 0.3S sent through the earphones to alert the patient of the forthcoming light stimulation. In one series of examinations, each eye was randomly stimulated 5 times with one intensity and duration, so 45 times per eye with a total of 90 times for both eyes. In another series of examinations, the click was delivered 10 times without light stimuli The purpose of these trials was to draw patients' attention to the examination and also to discriminate false positive responses. In total, 100 clicks with 90 light stimuli were delivered in one series. The number of incorrect responses (pushing the button when there was a click but no light stimulation) was printed out as X(score)/100. These scores were evaluated to assess the reliability of the data for each patient. In cases of over 5/100, we didn't tell the patients because we considered the data to be unreliable. Such cases, however, were very rare. So in once series of tests, 100 clicks with 90 light stimuli were delivered. Two RP patients, a 67-year-old man and a 57-year-old female with LP, had their cataracts extracted and the effect of surgery to their visual function was evaluated. They were administered lipoprostaglandin E1 for 10 days following surgery since this drug was reported to be effective to improve visual function in RP.[4] Also cervical stellate ganglion block and hyperbaric oxygen chamber were tried for the same purpose.[5] Before this series of tests, the purpose of this study was fully explained to each patient and they gave their informed consent in writing.

RESULTS

The sequential changes of the level of light perception of an RP patient are shown in Figure 1. The right eye of a 67-year-old man operated on. Fig. 1a shows the result of the examination the day prior to the cataract extraction, 1b to 1e show the results from the 1st through 4th day after surgery. The level of light perception once decreased but gradually recovered to the pre-operative level at the 3rd day dramatically improved on the 4th day of surgery. With the conventional method using the light source of an indirect ophthalmoscope, his results showed the same LP before and after the operation.

With the administration of lipo-prostaglandin E1 which was started two weeks after the cataract extraction, his LP dramatically improved as shown in Figures 2a to 2d. Figure 2a and b are those of the right and left eye respectively before this treatment. After administration for 10 days, his LP improved, especially in the left eye. He could respond much better to the stimulation to 100C and also some to 10C. The improved function was not kept so long and in the evening of the same day, his LP returned to the level before injection.

Three days after finishing the administration, he also got the stellate ganglion block following hyperbaric oxygen chamber. Figure 3a and b are the right and left eye before treatment and 3c and d, those after, respectively. He showed no improvement in the left and decreased in the right as shown in Fig. 3d and 3c, so the trial was stopped.

We repeated these trials with the other patient, a 57-year-old female with RP.

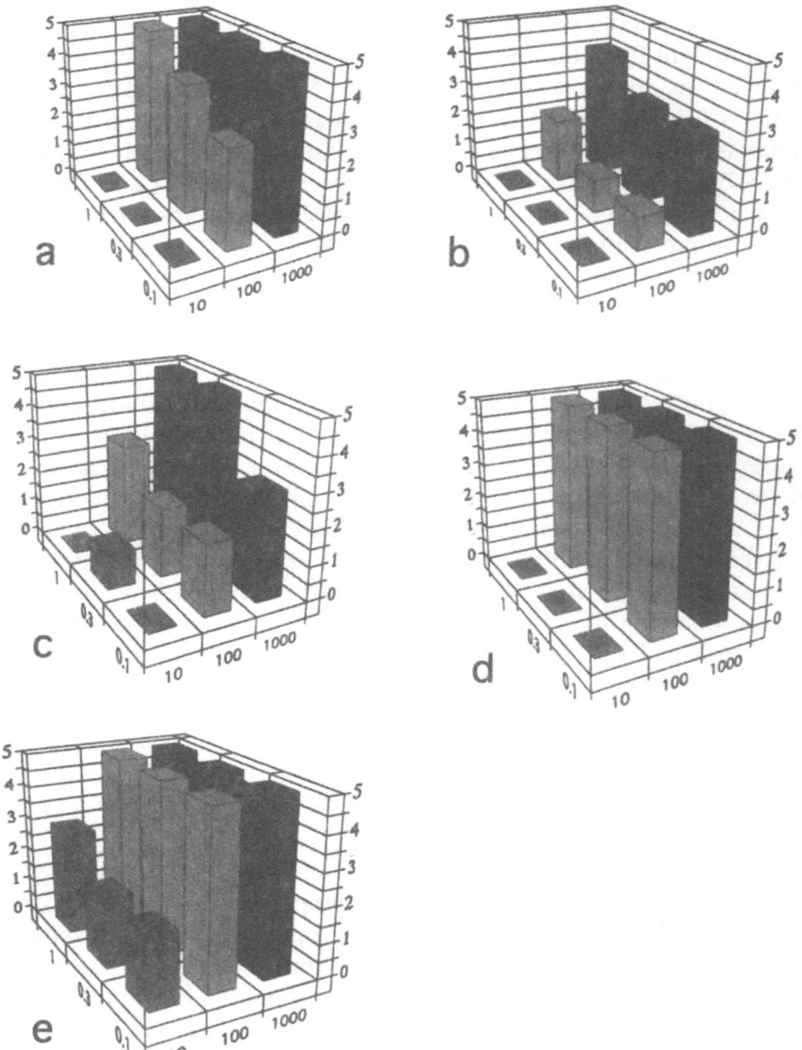

Figure 1. The profile of light sensation with the LoVE device in the RP patient before (a) and after (b to e) cataract extraction. He is a 65-year-old male and the visual function before surgery was light perception. The visual field could not be detected and electroretinogram was non-recordable but he could correctly discriminate the light projection with conventional method which was performed with the bright light of an indirect ophthalmoscope. Before surgery, he could respond to the light stimulation with 1000C in any duration but partly failed with 100C and none with 10C. His error score was 1 in 100 trials which showed these results were very reliable. After surgery, his LP decreased once as shown in Figs. 1b and 1c (2 days following surgery) but recovered to the preoperative level at 3 (Fig. 1d) and better at 4 (Fig. 1e). His error score was 3/100, 0/100, 0/100, and 1/100 respectively. In this and the following figures, x-axis means the intensity of light , the y-axis, the duration of the stimulation and z-axis, the total number of correct answers, 45 in total trials to one eye.

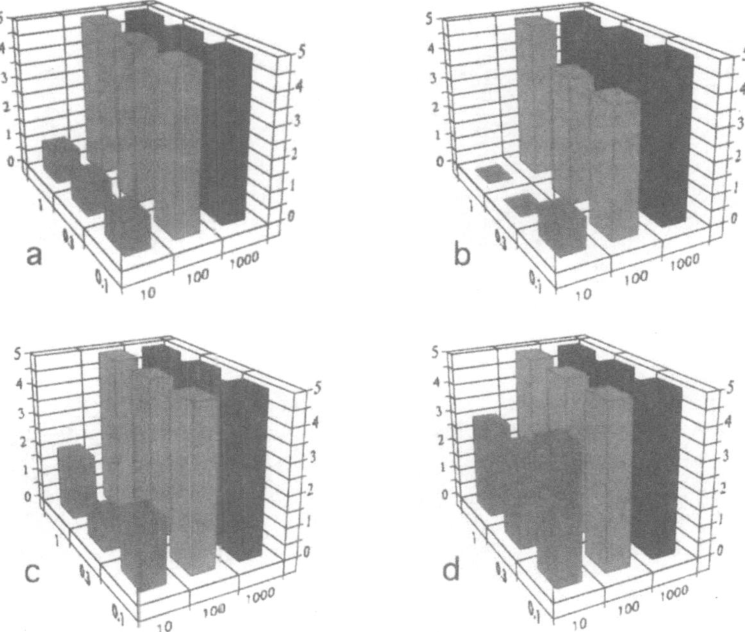

Figure 2. The profile of light sensation with the LoVE device in the RP patient before (a and b) and after the treatment with lipoprostaglandin E1(c and d). The right eye shows Fig. 2a and 2c and the left, 2b and 2d. The effect of the drug was not clear in the right, but remarkable in the left eye. The error score was 1/100, 0/100, 0/100, and 1/100, respectively.

Figure 4 shows the total number of correct answers in each trial before and after cataract extraction and administration of lipo-prostaglandin E1. Before surgery, her LP level 10 to 15 correct answers out of 45 trials in the left eye and none in the right. Cataract extraction was performed on her left eye on August 4th of this year as marked 1) in the figure. The level of LP improved the day following the surgery and remained so. On August 13th, she began the treatment of lipoprostaglandin E1 as shown 2) in the figure and continued for 10 days. The improvement was not so dramatic in the left eye but the LP in the right eye appeared in the second half of the treatment. The error score was 0–3/100 throughout these examinations.

DISCUSSION

To our knowledge, residual visual functions of RP patients have been recorded with automated light- and dark-adapted perimetery or threshold response of cone sensitivity with Humphrey perimetry 2. But in cases with very low vision such as light perception, it has been impossible to obtain reliable data with high reproducibility. With this device, it is possible to obtain a reproducible pattern to some extent of light perception of RP. For example, Fig. 1e, and 2a are the results of the same right eye, of the first case, without any treatment but two weeks interval. The number of correct answers was 37 out of 45 in Fig. 1e and 33 in Fig. 2a. In the former, he could answer at the lowest

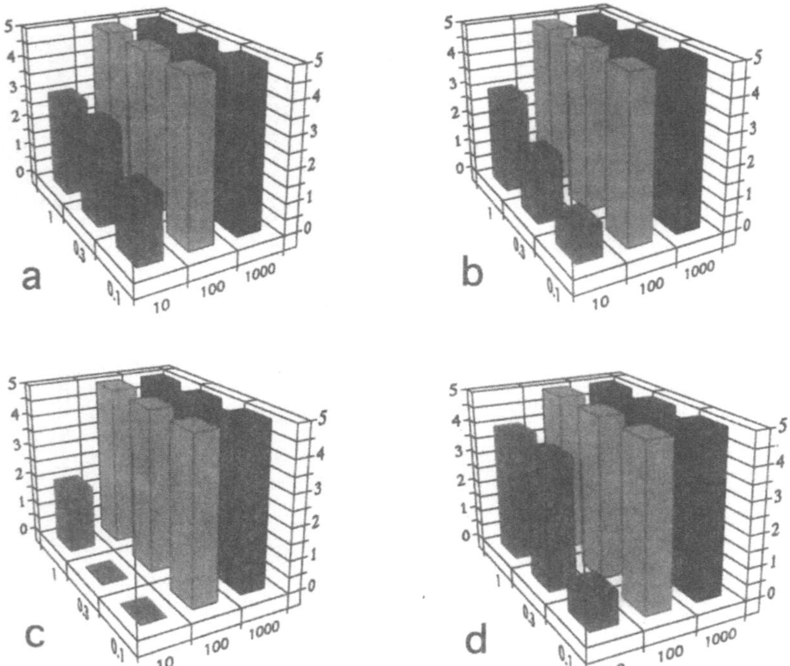

Figure 3. The effect of cervical stellate ganglion block and hyperbaric oxygen chamber to the function of the RP patient. The level of LP was all correct with 1000C and 100C but difficult with 10C as shown in a (right) and b(left eye). After the block and oxygen treatment, the level of the LP was worse in the right and no remarkable difference with 10C, so the treatment was stopped.

intensity of light, 10C more correctly than the latter. But to stimulations with 1000C and 100C, he could answer perfectly and his left eye maintained the level of LP at least to 100C, and partly to 10C. Figure 2c and 3a are also the same eye but after a 3-day interval. The same consistent pattern of correctness can be seen.

This device has several advantages for recording low visual functions. They are:

1. There is no physical harm such as insertion of contact lens electrodes on the cornea or use of adhesive paste for decreasing resistance to the electrodes.
2. The goggles block any light from outside, so we can obtain a dark adapted condition after a certain period by wearing them. In other words, we don't need a special room to control the adaptation level.
3. Earphones can block most noise and the click draws the patient's attention to the test. Also, patients can concentrate during this examination even with a certain level of environmental noise, which occurs in most offices.
4. The trials with clicks without light stimulation made it possible to estimate the reliability of this examination in relation to the responses given for light stimulation. The level of error was relatively low throughout the test for each patient.
5. The stimulus intensity of 1000C was needed for grading the low visual function such as LP in RP patients. Since each stimulus was a flash of light, the adaptation level of the photoreceptors was not changed.

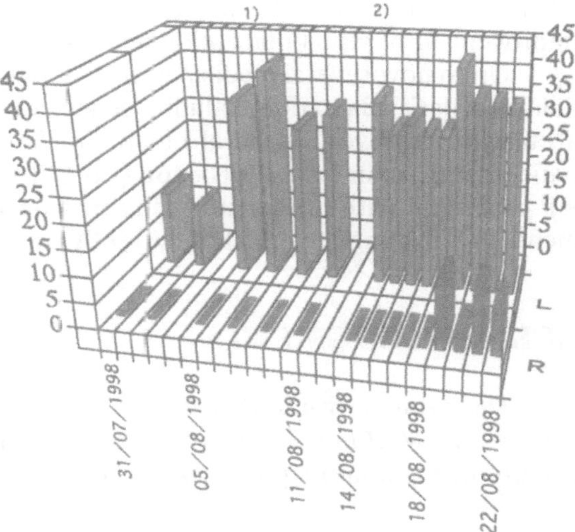

Figure 4. The effect of cataract extraction and lipoprostaglandin E1 to another patient. Before surgery, the patient could not answer correctly to any intensity of stimulation given to the right eye. In the left eye, he could answer 10 to 15 stimuli out of 45 trials. After surgery to the left eye, on August 4th shown 1) in the figure, his level had dramatically improved. On August 13th, the treatment of lipoprostaglandin E1 was started as marked 2) in the figure. The left eye did not show as much improvement, but his right eye could respond to light stimuli with 1000C in the 6th and the following days.

6. All stimuli and recordings are programmed in the control box and delivered automatically. It takes less than 7 minutes and anyone can perform this test, even RP patients by themselves.

In patients with RP, non-recordable electroretinograms have been emphasized as one of the characteristics in this clinical entity, even if they still have quite good vision. So it is difficult to differentiate their visual function, especially in such low vision with the amplitude of electrical responses. When we try to find or evaluate effects of some medical or surgical treatment for RP patients, it becomes a significant problem. The present results show this device can tell us the possibility of quantitative expression of the LP level, whether increased, the same or decreased by the pattern of each record.

The reason why the level of LP decreased after surgery in the first case is due to the postoperative inflammation and slight corneal edema (Fig. 1). In the second case, we paid more attention to the duration of ultrasound application time for cataract extraction and tried to reduce the postinflammatory reaction with subconjunctival steroid administration. As a result, the day following the surgery, the LP improved much more than before (Fig. 4). In reality, the second patient showed much less cell inflammation and flare in the anterior chamber and almost no corneal edema.

The stellate ganglion block followed by the hyperbaric oxygen treatment has been reported to have some effect for improving visual function of RP patients.[5] But this treatment is still controversial and finding no any positive effect, in fact, only a negative effect,[6] the trial was suspended.

In both cases, the patients' visual function clearly improved comparing the before and after treatment with lipoprostaglandin E1 as shown in Figs. 2 and 4. Surprisingly, the patients did not feel that their light sensation had improved even when they were show the data which showed a and definite improvement. These results suggest the level of retinal perception (the minimum visible) and recognition with their visual cortex (the minimum cognizable) is not parallel. In the future, careful attention must be paid when evaluating the effects of new treatments, such discrepancies between real improvement of visual perception and the level of recognition.

Up to now, several medicines, such as helenine, have been used to improve visual field and dark adaptation, kallidinogenase to increase retinal blood flow, tocopherol acetate to improve peripheral blood circulation and inosine to activate cells in RP patients, at least in Japan. Idebenon (Co-Q10) and vitamin A are also administered to RP patients.[7,8] With further study, this new device could prove to be a valuable instrument in evaluating the effect of currently prescribed medicine and/or vitamins.

ACKNOWLEDGMENTS

The authors wish to thank the cooperation of Mr. Yoshikawa, M. Mayo Co. Ltd, Nagoya, and Mr. Ohta, T. Japan Medical Instrument Center Co. Ltd. for taking our idea and producing it into a prototype machine.

REFERENCES

1. Duke-Elder, Sir S. The examination of the visual functions, pp. 366–380. Chapter XIV The foundations of ophthalmology, System of Ophthalmology Vol. VII Henry Kimpton, London, 1962.
2. Jacobson S.G., Walter J.V. et al. Automated light- and dark-adapted perimetery for evaluating retinitis pigmentosa. Ophthalmology, 93:1604–1611, 1986.
3. Discussion with Ms Cristina Fasser, President of International Retinitis Pigmentosa Association at the office of IRPA in Zurich, Switzerland on July 25, 1998.
4. Horiguchi M., Miyake Y., Tomita N., and Suzuki S. A therapeutic trial of intraveous lipoprostaglandine E1 (Lipo-PGE1) for type 2 (Massof) retinitis pigmentosa. Report of Research Committee on Chorioretinal Degenerations, The Ministry of Health and Welfare of Japan, 130–131, 1994.
5. Atsumi K., Ogino N., and Honda Y. Hyperbaric oxygen therapy for retinitis pigmentosa. Report of Research Committee on Chorioretinal Degenerations, The Ministry of Health and Welfare of Japan, 215–218, 1992.
6. Shiono T., Miura T., Chida Y., Nakazawa M., and Tamai M. Is hyperbaric oxygen effective on retinitis pigmentosa? Report of Research Committee on Chorioretinal Degenerations, The Ministry of Health and Welfare of Japan, 203–204, 1993.
7. Ooba N., Isashiki Y., Uto M., Adachi E. et al. Systemic administration of idebenon to patients with retinitis pigmentosa-A multicenter study. Report of Research Committee on Chorioretinal Degenerations, The Ministry of Health and Welfare of Japan. 225–227, 1992.
8. Berson E.L., Rosner B. et al. A randamized trial of vitamin A and vitamin E supplementation for retinitis pigmentosa. Arch Ophthalmol., 111:761–772, 1993.

22

BLUE CONE MONOCHROMACY

Macular Degeneration in Individuals with Cone Specific Gene Loss

Radha Ayyagari, Laura E. Kakuk, Yumiko Toda, Caraline L. Coats, Eve L. Bingham, Janet J. Szczesny, Joost Felius, and Paul A. Sieving

Department of Ophthalmology
W.K. Kellogg Eye Center
University of Michigan
1000 Wall Street, Ann Arbor, Michigan 48105

1. INTRODUCTION

Blue cone monochromacy (BCM) is an X-linked ocular disorder in which affected males have normal short-wavelength-sensitive (blue) cone and rod function but lack medium-(green) and long-wavelength-sensitive (red) cone function. Affected males characteristically have visual acuities of 20/100 to 20/200, myopia, nystagmus, and minimal foveal granularity with otherwise normal fundus findings.[1]

Alterations in color genes were shown to be associated with BCM. Vollrath et al., showed that the color genes are arranged in a tandem array on the X-chromosome with the red pigment gene 5′ to the green pigment gene(s).[2] Expression of the red and green genes is regulated by a 600 bp locus control region (LCR) present in the upstream region of the red gene.[3] The sequence between the red and green genes and 3′ to the green gene was identified to code for the Tex 28 gene the function of which is unknown.[4] BCM was shown to result from homologous recombination or gene conversion between the red and green pigment genes.[5]

During the course of studying 15 families with BCM we have noticed unusual macular changes co-segregating with BCM in 3 families. Here we describe phenotypic characteristics observed in these families and the structure of the upstream locus control region (LCR), red, green, and Tex 28 genes.

Retinal Degenerative Diseases and Experimental Therapy, edited by Hollyfield *et al.*
Kluwer Academic / Plenum Publishers, New York, 1999.

2. METHODS

2.1. Clinical Analysis

The pedigrees of the 3 BCM families exhibiting unusual macular changes are given in Fig 1. Individuals available for clinical examination are indicated in each pedigree with an asterisk.

Affected individuals underwent detailed clinical examination including fundoscopy and fluorescein angiography. Psychophysical dark-adapted thresholds were evaluated foveally and at 5 points from 60° temporal to 20° nasal along the horizontal meridian using a Goldmann-Weekers Dark Adaptometer. A standard clinical electroretinogram was recorded (details in Fig. 2). Color vision was evaluated using a combination of the Ishihara test, AOC pseudo-isochromatic plates, Farnsworth Dichotomous D-15 panel test, and Berson's BCM plates,[6] all performed under CIE Standard Illuminant C from a MacBeth Easel lamp. Additionally, in some subjects 2° Rayleigh matches were determined on a Nagel anomaloscope. The BCM plates have been designed to distinguish blue-cone monochromats from complete achromats. The Rayleigh match evaluates possible residual medium (green) and long-wavelength-sensitive (red) cone function.

2.2. Molecular Analysis

DNA was isolated from venous blood,[7] and the structure of the red and green genes, as well as the 20 kb region 5′ upstream to the red gene transcription initiation site, and the Tex 28 gene (Fig. 3a) were analyzed.

2.2.1. Southern Blot Analysis. Presence of possible gross structural changes in the red and green genes was tested using probes described by Nathans et al.[5] Probes Br, Cr, Dr, Bg, Cg, and Dg detect changes in the red and green genes (Fig. 3b). The upstream region of the red gene is analyzed with probe Zr against Hind III digests of genomic DNA; probe Zr is an 800 bp fragment located 8.2 kb upstream from the red gene transcription start site (Fig. 3).

2.2.2. PCR Amplification. Exons of the red and green genes, the 5′ upstream red gene region and the Tex 28 gene were amplified using PCR. Some of the primers designed to amplify red and green genes and the upstream red gene region were derived from Gene Bank sequences S44029, Z47066, Z68193, Z46936, M13300, M13301, M13304, M13305, K03491, K03492, U45954. Additional primers used are reported elsewhere.[8–13] Primers to analyze the exons of the Tex28 gene(s) were derived from the 3′ sequence of the red and green genes (Gene bank # Z46936, Z49258). All unpublished primers are given in Table 1. Most of the PCR reactions were carried out at 94°C for 30 seconds, 55°C for 30 seconds, and 72°C for 40 seconds for 35 cycles. PCR products were analyzed on agarose or polyacrylamide gel to check for deletions.

2.2.3. Sequencing. Exons of the green gene and the 600 bp LCR sequence in family LA793 were amplified by PCR using primers given in Table 1. Sequencing the amplified products was carried out using the Amersham cycle sequencing kit.

Table 1. Primers used in the analysis of color genes and Tex 28

Primer	Gene Specificity	Location*	Accession	Reference	Sequence 5′-3′
1	upstream red	842-861	Z47066		TCTAGCGGGACTCATCACCT
2	upstream red	1176-1157	Z47066		CTCTCTCTTCAGGGCAGTGG
3	upstream red	8127-8146	Z47066		CTTGAGACCAGGAGGCAGAG
4	upstream red	8444-8425	Z47066		TCCTGCTAGGCCACTTGTTT
5	upstream red	10696-10715	Z47066		CGTCAGAGTGGGGACAAGTT
6	upstream red	10887-10868	Z47066		TTTCCTGTCATTCAGGGACC
7	upstream red	1-20	S44029		CGAATTCGGACCAGCCTAGC
8	upstream red	461-443	S44029		TCACAGATCCTGACTAATG
9	upstream red	443-461	S44029		CATTAGTCAGGATCTGTGA
10	upstream red	971-951	S44029		CAGAGATCCAGGAGGGCATCA
11	upstream red	951-971	S44029		TGATGCCCTCCTGGATCTCTG
12	upstream red	1481-1461	S44029		CAGCTCAGCCTCATAATCTCA
13	upstream red	1461-1481	S44029		TGAGATTATGAGGCTGAGCTG
14	upstream red	1964-1945	S44029		GCCATTTGCCTTTGACACTT
15	upstream red	1945-1964	S44029		AAGTGTCAAAGGCAAATGGC
16	upstream red	2373-2353	S44029		GGTCCCACACTCTTGCAGAAG
17	upstream red	2353-2373	S44029		CTTCTGCAAGAGTGTGGGAC
18	upstream red	3012-2995	S44029		ATCCAAGAATGTGAGACC
19	upstream red	2995-3012	S44029		GGTCTCACATTCTTGGAT
20	upstream red	3397-3380	S44029		CAAGTGCCTTTCTTCCTCT
21	upstream red	3380-3397	S44029		AGAGGAAGAAAGGCACTTG
22	upstream red	4297-4279	S44029		GACTTCTTAATGGGTACAG
23	upstream red	4279-4297	S44029		CTGTACCCATTAAGAAGTC
24	upstream red	4830-4811	S44029		GGCTACCCATAGCAAGAATG
25	upstream red	4811-4830	S44029		CATTCTTGCTATGGGTAGCC
26	upstream red	5430-5410	S44029		CGCACATCCTGCTGTTGGCTT
27	upstream red	5410-5430	S44029		AAGCCAACAGCAGGATGTGCG
28	upstream red	6018-6000	S44029		CATGGCTATGGAAAGCCCT
29	upstream red	5869-5889	S44029		GCCCAATTAAGAGATCAGGTA
30	not specific; red/green	608-588	M13300		CTGGCTCACGGAGTTGCTGTT
31	not specific; red/green	252-236	M13301		GCCAGAGACCTGGTTCACAAT
32	red	121-141	M13304		GGTGGTGATGATCTTTGCGTA
33	not specific; red/green	266-246	M13305		GCAGTGAAAGCCTCTGTGACT
34	not specific; red/green	1479-1499	U45954		ACAACTCCCTATGCCTGGGTC
35	green	83-100	K03491		GAGAGATGGATGGTGGTC
36	green	179-162	K03492		GGCTTACGCTGAGTGGGG
37	green	106-88	K03491		CTTGCAGACCACCATCCAT
38	green	243-263	M13306		AGTAGCTGGGATTACAGGTTT
39	upstream red	2104-2125	S44029		AGTCCAACATCTAATCTTCCAC
40	upstream red	2437-2418	S44029		CTAGAAAGTCACTCAGCACC
41	green	181-160	K03492		GGGGCTTACGCTGAGTGGGGTG
42	red	83-100	M13302		GAGAGGTGGCTGGTGGTG
43	tex 28	20494-20513	Z68193		GAAAGGATCCTGGCAAAACA
44	tex 28	21188-21148	Z68193		CTTTCCCTGTTGCTGGAGAG
45	tex 28	6191-6210	Z46936		TTGGTTTGGGGGTATCAGAA
46	tex 28	6715-6696	Z46936		CGGAACTGAAGAGCACACAC
47	tex 28	22158-22177	Z46936		CCTTTCACCCTCCCATAGGT
48	tex 28	23011-23030	Z46936		AACGTGTCCAATTCTCCCAC
49	tex 28	26537-26556	Z46936		CCTGACCTCAAATGATTCCG
50	tex 28	27179-27160	Z46936		GGCCAGCACTTCATTGTCTT
51	tex 28	1392-1411	Z49258		CCCTTTAACCCAGCAGAACA
52	tex 28	1761-1742	Z49258		CGTGTGATTGTTGAGGAGGA

3. RESULTS

3.1. Clinical Findings

The disease inheritance in all of these families is X-linked recessive. Affected males in all families have poor visual acuity and minimal color discrimination from the earliest age of report.

Family UM:M196 presented the most remarkable example of macular degeneration cosegregating with BCM. Two brothers (individuals 18 and 19) and a male cousin (individual 31) were examined (Fig. 1), and all had bilateral macular atrophy. The age of onset was documented in one to be by 32 years of age (individual 18), when he was examined at the University of Michigan with acuities of 20/100 OU. That examination documented bilateral macular changes consisting of pigmentary clumping and discoloration of the RPE. This had been mistaken as a case of Stargardt's macular degeneration at that time. When he was next examined at the University of Michigan at age 59, he was pseudophakic and had best corrected acuities of 20/200 bilaterally. Fundus examination showed extensive, circumscribed RPE atrophy and loss of the choriocapillaris across a 3 disc-diameter region in the center of both maculae (Fig. 4b). The appearance of the peripheral fundus was unremarkable. He failed both the Ishihara

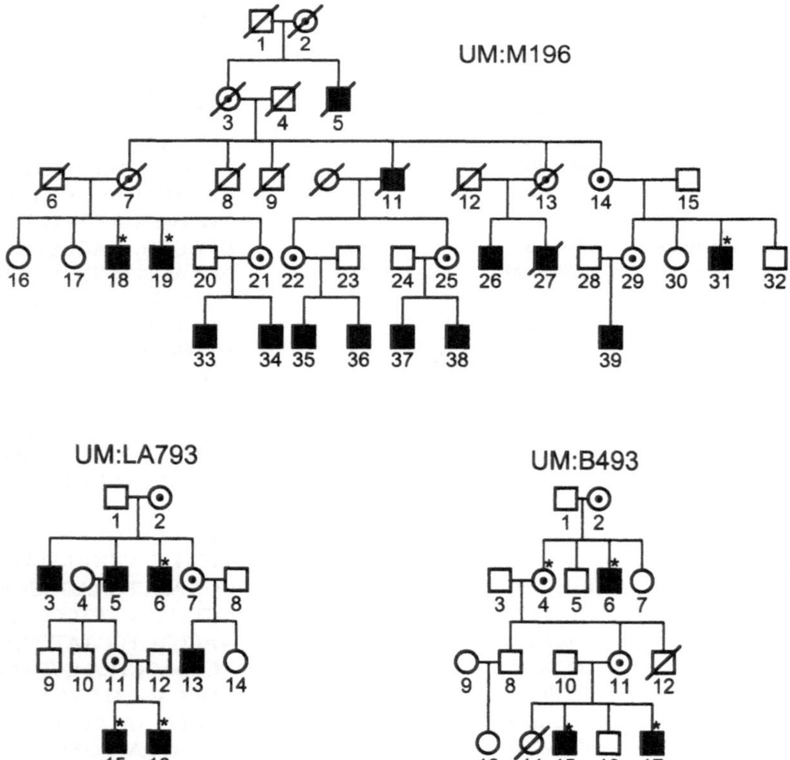

Figure 1. Pedigrees of families UM:M0196, UM:B493, and UM:LA793. Individuals available for clinical examination were denoted with "*".

and D-15 tests. He also had difficulty identifying the BCM plates presumably due to the large region of macular atrophy. On Nagel anomaloscopy, his results were not always consistent, but he made the Rayleigh match across the entire intensity range, indicating the absence of long or middle-wavelength sensitive cones. His visual fields were full on Goldmann perimetry with the V4e target. His electroretinogram (ERG) had normal scotopic b-wave amplitudes, but the photopic responses were reduced to 3 μV b-wave (Fig. 2).

His younger brother had 20/200 bilateral acuity at age 51 (Fig. 1: UM:M196 individual 19). Fundus examination showed circumscribed reddish lesions from RPE thinning across the central 1 disc-diameter in both maculae. A fluorescein angiogram documented an early window, consistent with RPE thinning. He was unable to identify any of the Ishihara plates and made numerous random cross-over errors on D-15 testing. However, he correctly identified all of the BCM plates, indicating the presence of functioning short-wavelength sensitive (blue) cones. His ERG was normal for the dark-adapted rod b-wave, whereas the cone-driven photopic b-wave was nearly absent (Fig. 2). A 45-year-old male cousin (Fig. 1, UM:M196 individual 31) who had BCM was also found to have bilateral macular atrophy (Fig. 4a) and 20/400 acuities. His color vision discrimination was negligible.

This family documents severe macular atrophy cosegregating with BCM and occurring at ages too young to ascribe to an "age-related" process. We have also found atrophic maculopathy in older males in two other BCM families. In family UM: LA793 (Fig. 1), an older male (individual 6) reported further reduction of visual acuity beginning at about age 60. Records from the University of Michigan documented poor color discrimination at a young age and acuities of 20/80 and 20/70 at age 42. This had dropped to 20/100 and 20/80 at age 61, at which time he had normal scotopic rod b-wave amplitudes but photopic cone responses were nearly absent (Fig. 2). He was unable to correctly identify any Ishihara plates and made numerous random crossing errors on the D-15 test, but he passed the BCM plate test. When we re-examined him at age 67, his acuities remained 20/80 OU, best corrected. Both maculae showed discolored central lesions and an angiogram documented early, circumscribed, parafoveal window defects consistent with RPE thinning (Fig. 4c). At this point his color vision, as well as that of 2 nephews (individuals 15 and 16; Fig. 1), was typical for BCM: They all completely failed the Ishihara test, and made many crossover errors on the D-15. Their Rayleigh matches spanned the entire intensity range, indicating the absence of functional long and middle-wavelength sensitive cones in the central 2° of the maculae. They made no errors on the BCM Plate test, indicating the presence of functional short-wavelength sensitive (blue) cones.

Family UM:B493 was identified with BCM by vision testing of younger family members (individuals 15 and 17; Fig. 1). The electroretinogram was remarkable in this family for a small but recordable photopic b-wave of 30 μV (Fig. 2). The waveform was anomalous in that it lacked an a-wave and slowed only the slow b-wave as is reported for the photopic ERG elicited from blue cones.[14] The 30Hz flicker response was absent, as would be expected for blue cone activity. The scotopic rod-driven b-wave had normal amplitude. Their color vision was typical for BCM: they completely failed the AOC Pseudo-isochromatic Plates, and made many crossover errors on the D-15. Their Rayleigh matches spanned the entire spectral range, indicating the absence of functional long and middle-wavelength sensitive cones in the maculae. They made no errors on the BCM Plate test, indicating the presence of functional short-wavelength sensitive cones.

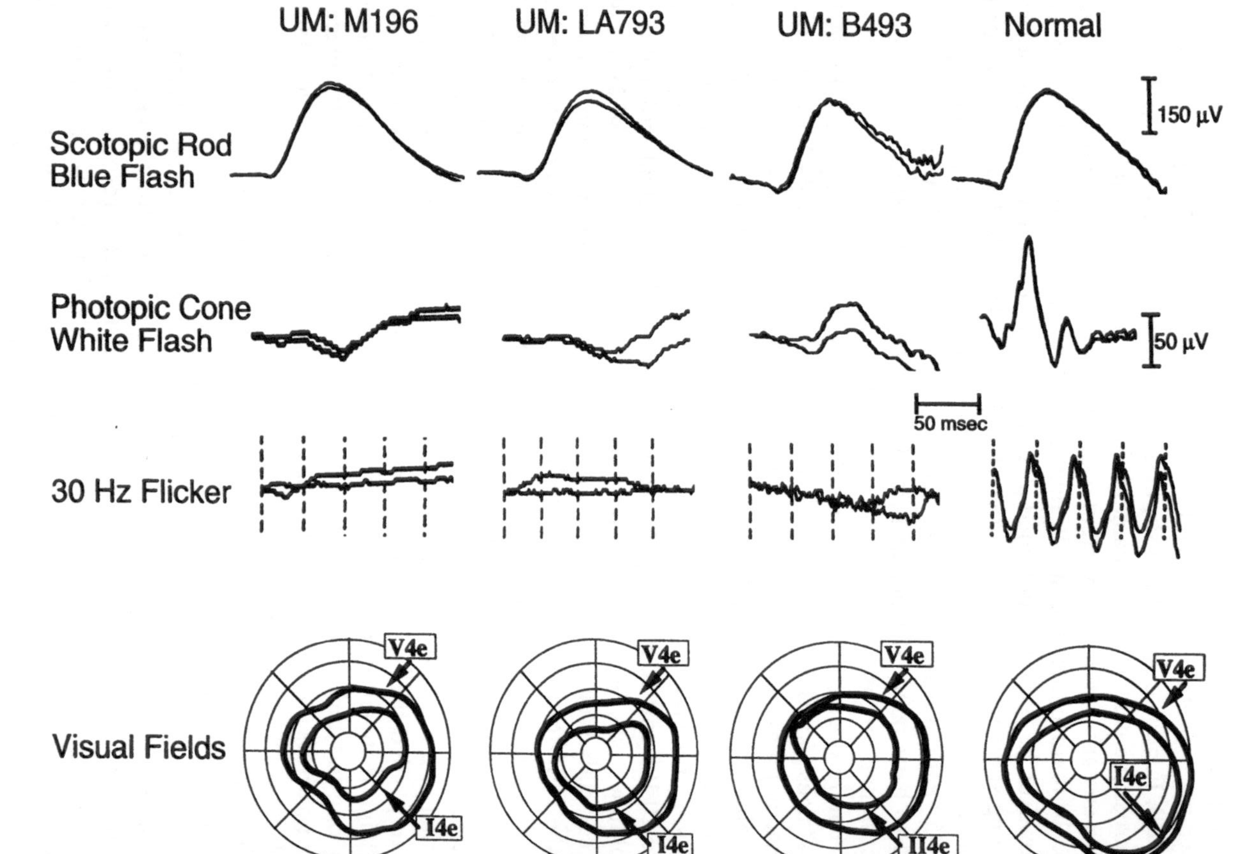

Figure 2. ERGs and Goldmann visual fields of affected individuals from families UM:M196 (individual 18), UM:LA793 (individual 6), UM:B493 (individual 15), and a normal subject.

Two older relatives were examined, a 70-year-old female carrier (UM:B493, individual 4) and a 56-year-old BCM-affected brother (individual 6) (Fig. 1). Both individuals showed extensive RPE thinning along the macular arcade vessels and coarse RPE granularity. The carrier had acuities of 20/30 OU best corrected. She was examined in the field and only limited testing was performed. On the AOC pseudo-isochromatic plates, she consistently missed 4 plates for each eye separately. Her brother had 20/200 acuity bilaterally and was a 6 diopter myope, which accounted for the mild peripapillary RPE degeneration but would not explain the perimacular RPE granularity and early atrophy. He completely failed the AOC pseudo-isochromatic plate test.

3.2. Analysis of the LCR and Upstream Red Gene Region

Family UM:M196: Analysis with the Nathans' Zr probe to screen the upstream region in individuals IV-3 and IV-13 gave a novel band. The size of this band is greater than the 7.2 kb expected band (Fig. 3). This indicates that a deletion or rearrangement in the upstream region had altered the fragment size of the Hind III restriction digestion product. We further analyzed this 10 kb upstream region by PCR using primers given in table 1. Primers located 0–8.0 kb upstream from the initiating methionine of exon 1 of the red gene did not amplify any product, whereas the primers located 8.0–10.0 kb upstream gave the expected size products. These results indicate that the 8.0 kb region that includes the LCR upstream from the red gene has been deleted (Fig. 3c).

The 3′ end of the deletion was explored with two primer pairs in the region immediately around the beginning of exon 1. Primer pair 27 & 28 lie immediately upstream of exon 1, and these failed to amplify any product (expected to be 608 bp); the 5′ end of reverse primer 28 is complementary to and terminates with the ATG methionine that begins the coding sequence of exon 1. Primer pair 29 & 30 demonstrated that the deletion did not extend into exon 1; primer 29 lies 149 bp upstream of exon 1 in the 5′ untranslated region, and primer 30 overlaps the 3′ end of exon 1; together primers 29 & 30 amplified the 262 bp fragment that includes all of exon 1 and the transcription initiation site that lies 43 bp upstream of exon 1. Consequently the deletion terminates within a 459 bp interval between primers 27 and 29 in the upstream region before the beginning of exon 1.

Family UM:B493: Hybridization of Hind III digested genomic DNA blots with probe Zr did not detect any band suggesting the absence of the region corresponding to the probe. Further analysis with PCR primers designed to analyze the upstream red gene region up to 20 kb (Table 1) indicated the presence of a deletion extending 18 kb from the red gene transcription initiation start site (Fig. 3c).

Family UM:LA793: Analysis of Hind III digested genomic DNA blots with Zr probe shows the presence of a 7.2 kb fragment suggesting the lack of gross alteration in the upstream red gene region detected by the Zr probe. Amplification of the upstream red gene region by PCR also did not detect any alteration or deletion. No microdeletions or point mutations were identified by the sequence analysis of 600 bp LCR region (Fig. 3c).

3.3. Analysis of Color Genes and Tex 28

Family UM:M196: Southern blot analysis of the red and green gene regions gave the expected fragment sizes when hybridized with Nathans'[5] probes Br, Cr, Dr, and Bg,

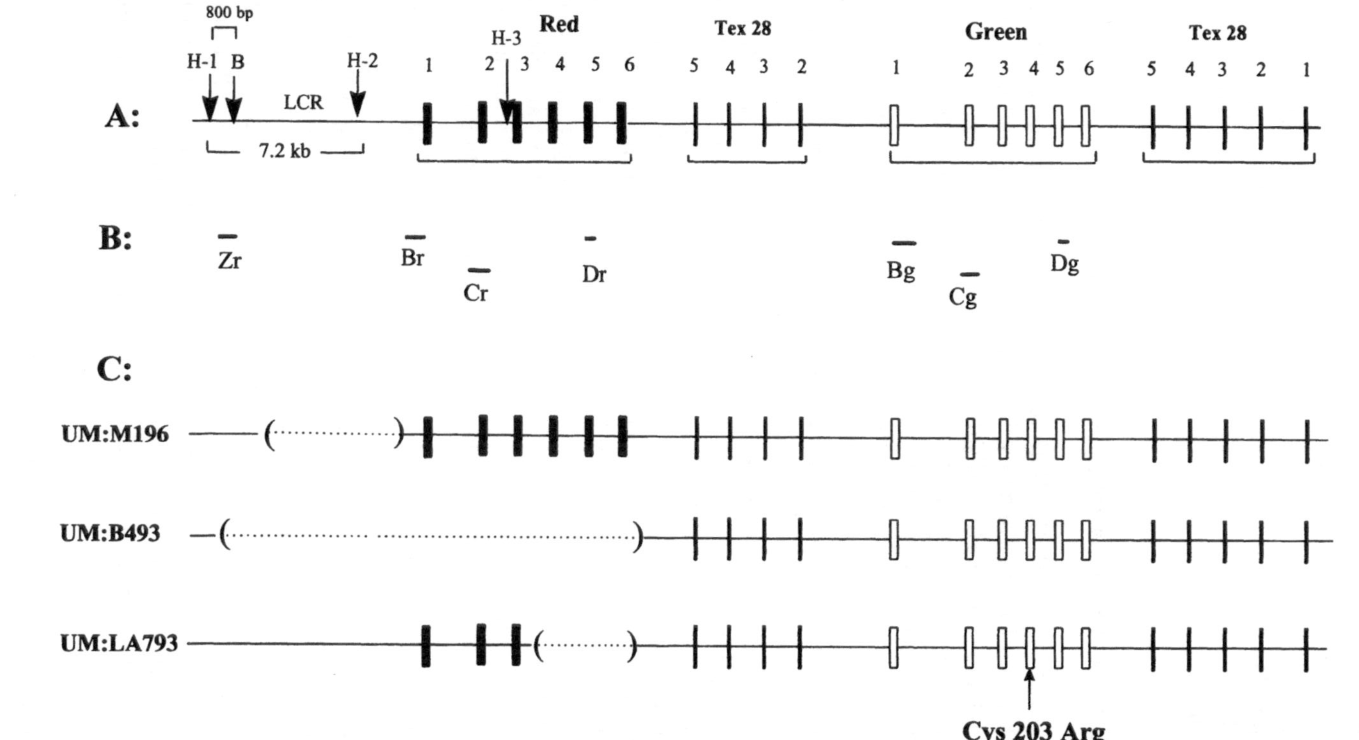

Figure 3. (A) Organization of the red, green, and Tex 28 genes in normal individuals. Boxes indicate the position of exons and numbers above the boxes indicate the number of the exons. Hind III sites in the upstream red gene region and in intron 2 of the red gene are labeled as H-1, H-2, and H-3. "B" represents a Bam HI site. (B) Position of the probes used for Southern blot analysis. Zr is a 800 bp probe corresponding to the sequence located between the Hind III and Bam HI sites in the upstream red gene region. Br and Bg hybridize to exon 1 of the red and green genes respectively. Cr and Cg correspond to exon 2 of the red and green genes and Dr and Dg represent exon 5 of the red and green genes respectively. (C) Structure of the red, green, and Tex28 genes in families UM:M0196, UM:B493, and UM:LA793. Solid lines indicate the region that is present. Dotted lines in brackets indicate deleted region. The position of the sequence shown in figure 3C corresponds to the structure of red, green and Tex28 genes shown in figure 3A. Cys 203 Arg mutation in the green gene in family UM:LA793 was also shown.

Cg, Dg (Fig 3b), suggesting that there were no gross rearrangements. Further analysis was done with PCR primers that specifically amplify the exons of the red gene in segments ranging from approximately 200 bp–8 kb (Table 1), and no abnormality was indicated. Primers are available for the green gene (Table 1) but were not used because the multiple copies would mask any deletion in a single green gene. Five primer pairs were used to amplify exons of the Tex28 gene in segments of approximately 350–950 bp, and the appropriate size bands were obtained. The exons of Tex28 are identically duplicated in the multiple copies that are interspersed between the red and green genes, and this analysis can not distinguish between them. However, all results from probing Southern blots and from PCR were consistant with the expectation that the red and green genes and all the five exons of the Tex28 gene are intact (Fig. 3c).

Family UM:B493: Analysis with probes Br, Cr, Dr, Bg, Cg, and Dg showed the absence of exons 1, 2, and 5 of red genes corresponding to Br, Cr, and Dr respectively and the presence of exons 1,2, and 5 of green gene. Further analysis with primers that specifically amplify the red gene revealed the absence of the entire red gene (Fig. 3c).

Family UM:LA793: Southern blot analysis revealed the presence of exons 1 and 2 of the red and green genes. A probe that hybridizes to exon 5 of red and green genes indicated the absence of exon 5 of the red gene while identifying the fragment corresponding to the green gene exon 5. Further analysis of all the exons of the red gene by PCR indicated the loss of exons 4, 5, and 6 of the red gene (Fig. 3c).

Sequencing analysis of all the exons of the green genes in this family revealed the presence of Cys 203 Arg mutation which was reported to alter the spectral properties of the green gene by disrupting a disulfide bond[10] (Fig. 3c).

All the exons of Tex 28 gene were found to be present in all the three families tested.

4. DISCUSSION

Bilateral progressive macular atrophy is unusual in BCM. The strongest case for this being distinct from aging maculopathy in our families is in UM:M0196, in which 3 affected males had macular atrophy diagnosed between the ages of 32–51 y/o. This age is remarkably young for ascribing it to an aging process. The segregation of maculopathy with BCM in cousins suggests that this results from the molecular genetic alterations responsible for BCM and is not due to some unknown third factor. In family UM:LA793, clear early parafoveal RPE atrophy is seen on the fluorescein angiogram (Fig. 4), but this was first documented at age 67 which makes the connection with BCM less certain. In family UM:B493, both the older female carrier and the 56-year-old affected male show clear RPE pathology surrounding the macula but not at the center directly. Younger BCM affected individuals of ages 5–40 years examined in families UM:B493 and UM:LA793 (Fig. 1) did not show signs of macular atrophy.

Fleischman and O'Donnell[15] noted a BCM family with macular atrophy and noted the development of pigmentary changes in the fovea over a period of twelve years. A deletion of 41 kb which includes the LCR and part of the red pigment gene was identified in this family (HS129).[8] One additional BCM affected male with progressive bilateral retinal degeneration was reported with a deletion of a 2.5 kb region located upstream of the red gene (E:HS106).[5] This deletion includes the 600 bp locus control region. The deletion in both these cases includes the locus control region located 3.1 to 3.7 kb upstream of the red gene.

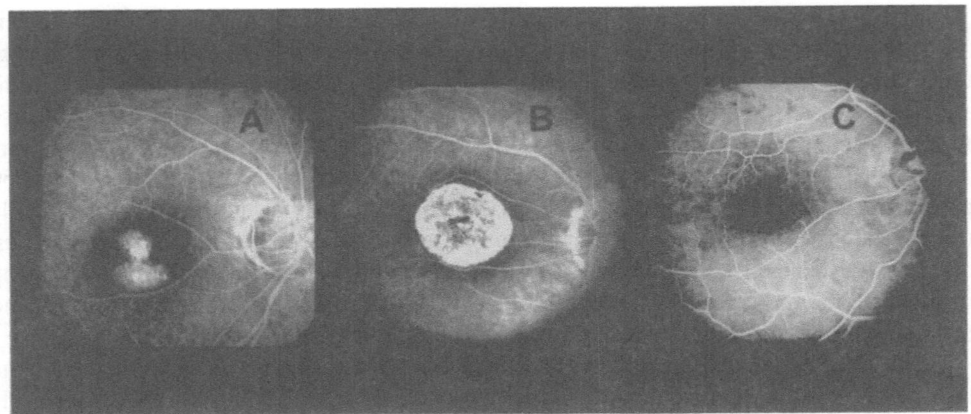

Figure 4. Fluorescine angiograms of affected individuals. A: Family UM:M196 individual 31 (age 45). B: Family UM:M196 individual 19 (age 59). C: Family UM:LA793 individual 6 (age 67).

The deletion in two out of three families presented here with BCM and macular atrophy also include the LCR. In family UM:M196 the deletion extends across an 8 kb interval in the upstream region, but the 3′ end of the deletion stops before the red gene transcription initiation site and does not extend into the red gene. The red, green, and Tex28 genes appear not to be involved. In the other family UM:B493 the deletion includes the LCR extending 18 kb upstream from the red gene transcription initiation start site besides the loss of the entire red gene. All exons of the Tex 28 and the green genes are present in this family. However, in family UM:LA793 the upstream red gene region was found to be intact without any point mutations or deletions in the LCR region. Loss of red and green gene function in this family can be explained by the loss of exons 4, 5, and 6 of the red gene and the presence of a Cys 203 Arg mutation in the green gene respectively. Though there is wide variation in the alterations observed in red/green color genes in these three BCM families with macular changes, these alterations lead to loss of functional red and green genes.

Expression of red and green opsins is controlled by the 600 bp sequence LCR.[3,16] These are the major structural proteins present in the cone photoreceptor outer segments.[17] Absence of rhodopsin expression results in lack of formation of rod photoreceptor outer segments and leads to degeneration of rods and subsequently cones.[18] In parallel, one can speculate that the absence of cone outer segments in BCM affected individuals may result in macular atrophy. However, macular atrophy is not common in individuals affected with BCM. We also have several additional BCM families with deletions of the entire LCR region that do not show macular degeneration.[19] Similarly, Nathans et al.[8] described five different BCM families with deletions in the LCR region and did not report macular atrophy. Thus, deletion of the LCR apparently does not always cause macular degeneration in BCM families. In addition, family UM:LA793 presented here with BCM and macular atrophy shows the presence of an intact upstream red gene region.

As the macular changes were observed in older individuals of BCM families we are currently evaluating other older affected males in our 15 families to understand whether a putative connection exists between macular atrophy and BCM in aging males. Because of the limited number of individuals available to examine this uncommon entity, it will be difficult to establish a tight correlation with the BCM genetics. Nevertheless, this may alert other investigators to watch for maculopathy in this cone-specific genetic dystrophy.

ACKNOWLEDGMENTS

FFB Molecular Genetics Grant (Ayyagari) and FFB Center Grant, both from The Foundation Fighting Blindness, Hunt Valley, MD; R01-EY-6094 (PAS); NEI Vision Core Grant EY0-7003. The authors thank Dr. Jeremy Nathans for providing his BCM probes, Dr. Atanacio Tiongson for some of the clinical information on IV-13, Bradley Nelson for preparation of the ocular fundus photographs and Jennifer Kemp for the ERG and visual fields drawings.

REFERENCES

1. E.L. Berson, M.A. Sandberg, A. Maguire, W.C. Bromley, and T.H. Roderick, 1986, Electroretinograms in carriers of Blue Cone Monochromatism, *Am. J. Ophthalmol.* **102**:254–261.
2. D. Vollrath, J. Nathans, and R.W. Davis, 1988, Tandem array of human visual pigment genes at Xq28, *Science* **240**:1669–1672.
3. Y. Wang, J.P. Macke, S.L. Merbs, D.J. Zack, B. Klaunberg, J. Bennett, J. Gearhart, and J. Nathans, 1992, A locus control region adjacent to the human red and green visual pigment genes, *Neuron* **9**:429–440.
4. M.C. Hanna, J.T. Platts, and E.F. Kirkness, 1997, Identification of a gene within the tandem array of Red and Green color pigment genes. *Genomics* **43**:384–386.
5. J. Nathans, C.M. Davenport, I.H. Maumenee, R.A. Lewis, J.F. Hejtmancik, M. Litt, E. Lovrien, R. Weleber, B. Bachynski, and F. Zwas, 1989, Molecular genetics of human blue cone monochromacy, *Science* **245**:831–838.
6. E.L. Berson, M.A. Sandberg, B. Rosner, and P.L. Sullivan, 1983, Color plates to help identify patients with blue cone monochromatism. *Am. J. Ophthalmol.* **95**:741–747.
7. G.I. Bell, J.H. Karam, and W.J. Rutter. 1981, Polymorphic DNA region adjacent to the 5′ end of the human insulin gene. Proc Natl Acad Sci **78**:5759–63.
8. J. Nathans, I.H. Maumenee, E. Zrenner, B. Sadowski, L.T. Sharpe, R.A. Lewis, E. Hansen, T. Rosenberg, M. Schwartz, and J.R. Heckenlively, 1993, Genetic heterogeneity among blue-cone monochromats, *Am. J. Hum. Genet.* **53**:987–1000.
9. S.S. Deeb, D.T. Lindsey, Y. Hibiya, E. Sanocki, J. Winderickx, D.Y. Teller, and A.G. Motulsky, 1992, Genotype-phenotype relationships in human red/green color-vision defects: molecular and psychophysical studies, *Am. J. Hum. Genet.* **51**:687–700.
10. J. Winderickx, E. Sanocki, D.T. Lindsey, D.Y. Teller, A.G. Motulsky, and S.S. Deeb, 1992, Defective colour vision associated with a missense mutation in the human green visual pigment gene, *Nat. Genet.* **1**:251–256.
11. E. Reyniers, M.N. Van Thienen, F. Meire, K. De Boulle, K. Devries, P. Kestelijn, and P.J. Willems, 1995, Gene conversion between red and defective green opsin gene in blue cone monochromacy, *Genomics* **29**:323–328.
12. J. Neitz, M. Neitz, and P.M. Kainz, 1996, Visual pigment gene structure and the severity of color vision defects, *Science* **274**:801–804.
13. M. Neitz, J. Neitz, and G.H. Jacobs, 1995, Genetic basis of photopigment variations in human dichromats, *Vision Res.* **35**:2095–2103.
14. P. Gouras and C.J. MacKay, 1990, Electroretinographic responses of the short-wavelength-sensitive cones, *Invest. Ophthalmol. Vis. Sci.* **31**:1203–1209.
15. J.A. Fleischman and F.E.J. O'Donnell, 1981, Congenital X-linked Incomplete Achromatopsia: Evidence for slow progression, carrier fundus findings, and possible genetic linkage with Glucose-6-Phosphate Dehydrogenase Locus. *Arch. Ophthalmol.* **99**:468–472.
16. S.A. Shaaban and S.S. Deeb, 1998, Functional analysis of the promoters of the human Red and Green visual pigment genes. *Invest. Ophthalmol. Visual. Sci.* **39**:885–896.

17. J. Nathans, D. Thomas, and D.S. Hogness, 1986, Molecular Genetics of Human Color Vision: The Genes Encoding Blue, Green, and Red Pigments, *Science* **232**:193–202.
18. M.M. Humphries, D. Rancourt, G.J. Farrar, P. Kenna, M. Hazel, R.A. Bush, P.A. Sieving, D.M. Sheils, N. McNally, P. Creighton, et al. 1997, Retinopathy induced in mice by targeted disruption of the rhodopsin gene [see comments], *Nat. Genet.* **15**:216–219.
19. Ayyagari R, Bingham E, Toda Y, et al. Blue Cone Monochromacy: Molecular rearrangements and genotype-phenotype correlations. [Abstract] Invest Ophthalmol Vis Sci, 1997;38:(4)3179.

23

THE PERIPHERIN/*rds* GENE

Structural and Functional Analyses

Tong Cheng[1] and Muna I. Naash[1,2]*

[1] Department of Ophthalmology and Visual Sciences
[2] Department of Molecular Genetics
University of Illinois at Chicago
Chicago, Illinois

SUMMARY

Mutations in the gene encoding the disk rim specific protein, peripherin/*rds*, have been implicated in the pathogenesis of both autosomal dominant retinitis pigmentosa (ADRP) and various forms of macular dystrophy (MD). Peripherin/*rds* plays an important role in the morphogenesis and maintenance of the disk rim structure. This role is supported by interactions with other proteins. To better understand how these mutations lead to the development of disease, knowledge of the structure-function relationships of the peripherin/*rds* gene is needed. To facilitate these studies, genomic clones encoding the murine gene were isolated using bovine cDNA sequences as probes. Sequence analysis of clones that contained the entire coding sequence for mouse peripherin/*rds* revealed the exon-intron organization of the gene. The gene is composed of three exons (581, 247, and 213 bp) and two introns. The first and second introns are 8.6 kb and 3.7 kb in size, respectively. Two major (1.6 and 2.7 kb) and three minor (4.0, 5.5, 6.5 kb) transcripts were detected on RNA blots. Several transcription start sites were mapped within 26

*Address correspondence to: Dr. Muna I. Naash, Department of Ophthalmology and Visual Sciences, University of Illinois College of Medicine, 1855 West Taylor Street (M/C 648), Chicago, IL 60612, Tel: (312) 413-1164, Fax: (312) 413-9763
E-mail: munanaas@uic.edu.

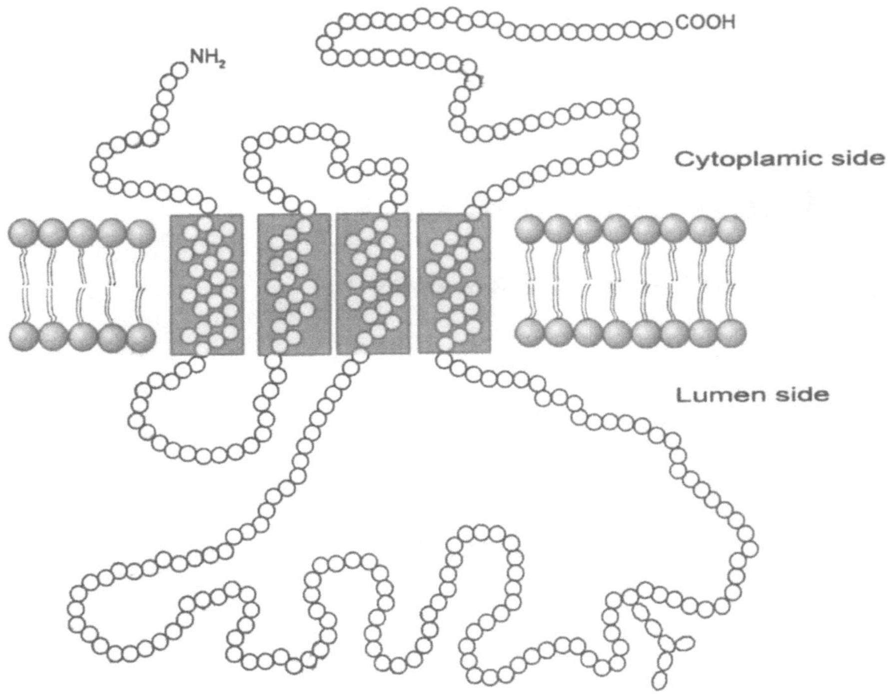

Figure 1. Molecular organization of the peripherin/*rds* protein. Schematic representation of the structural organization of peripherin/*rds* within the disk membrane. Amino acids are indicated by open circles. The disk membrane is represented by filled rectangles. The lumen and cytoplasmic sides are indicated.

nucleotides located approximately 200 bp upstream from the translation initiation site. However, the lengths of the 3′untranslated portion of the transcripts varied because of utilization of different polyadenylation signals.

INTRODUCTION

The disks of the outer segments (OS) of both rods and cones are comprised of two distinct domains, the lamellae and the rim regions. Peripherin/*rds* is restricted to the rim and the basal regions adjacent to the cilia of rod and cone OS, where disc morphogenesis occurs.[1–4] Mammalian peripherin/*rds* encodes a 346-amino acid polypeptide chain containing four possible transmembrane regions and three potential N-linked glycosylation sites.[5] The portions of peripherin/*rds* exposed on the cytoplasmic side of the disk membrane include a conserved and positively charged N-terminal region of 17 amino acids, a conserved loop of 22 amino acids and a highly charged C-terminal segment of 63 amino acids[3,6] (Fig. 1). The lumenal portion of peripherin/*rds* contains one small intradiscal hydrophilic loop of 21 amino acids and one large hydrophilic loop of 142 amino acids. While the exact function of peripherin/*rds* is not known, it may be involved in the morphogenesis and maintenance of the disk rim structure. This is suggested by the presence of peripherin/*rds* at the rim region of the disk membrane as well as by the phenotype of *retinal degeneration slow* (*rds*) mutants. Peripherin/*rds* may function by anchoring cytoskeletal proteins that link the rim region of the disks with the plasma membrane.[7] Alternatively, the portion of peripherin/*rds* exposed to the intradiscal space may interact with other peripherin/*rds* molecules or other proteins to form and maintain the rim domain.

In *rds*/+ mice, the phenotype is a semi-dominant trait characterized by haploinsufficiency of peripherin/*rds* arising from insertion of a 9.2-kilobase (kb) repetitive genomic element within exon 2 of the gene.[8,9] The corresponding mRNA molecule never escapes the nucleus and is not translated. *Rds*/+ mice are considered a good animal model for null mutations in the peripherin/*rds* gene. OS membrane biogenesis occurs in the *rds/rds* mutant but the membrane is unable to fold into the proper disk structure. We have carried out studies on *rds*/+ mice to compare rod and cone involvement at the structural and functional level.[10] The data indicate that the *rds* defect is more deleterious to rods than cones. This may reflect a different role of peripherin/*rds* in maintaining the structure of the rod OS. The phenotype in *rds*/+ mice was rescued by expressing wild-type peripherin/*rds* in the rod photoreceptor cells.[11] Since rescue occurred despite the presence of the mutant transcription product, Travis and Bok concluded that the putative *rds* protein does not have a dominant effect.[11,12] Hence, the phenotype in *rds*/+ is probably due to haploinsufficiency of a null allele.[12]

Interest in peripherin/*rds* has increased following the identification of mutations that are associated with autosomal dominant retinitis pigmentosa (ADRP)[13–20] and different forms of macular dystrophies.[15–20,21,22] As shown in table I, these mutations result either from single nucleotide substitutions,[16,19,21,22] or a small deletion.[17] While it is not known how mutations in this molecule lead to such clinically different disease phenotypes, a possible explanation involves a different role for peripherin/*rds* in rod and cone OS morphogenesis. Transgenic mice expressing different mutations in peripherin/*rds* may provide model systems that can be used to increase our understanding of how these mutations lead to blindness.[23] To ensure the proper regulation of peripherin/*rds* transgenes in these models, the native genomic regulatory sequences must first be identified and included in the transgene. In the present study, we describe the isolation and characterization of the mouse peripherin/*rds* gene. The coding region of this gene spans approximately 12.5 kb of genomic sequences with three exons interrupted by two introns. Multiple transcription initiation sites were found within a 26 bp region. Several transcripts, arising from utilization of different poly A signals, were identified by Northern blot analysis.

MATERIALS AND METHODS

Isolation and Characterization of Peripherin/*rds* Genomic Clones

Phage DNA was prepared by a modification of the previously described method.[24] DNA from the λ6-1-1 clone, which contains the entire coding region of the mouse peripherin/*rds* gene, was subjected to SalI/EcoRI digestion and the resulting fragments were subcloned into pKS+ Bluescript plasmid (Stratagene, La Jolla, CA). Exons, large portions of introns, and the entire 5′ and 3′-franking regions were sequenced using the dideoxy mediated chain termination method with Sequenase$^{2.0}$ (U.S. Biochemical Corp, Cleveland, OH) and oligonucleotide primers synthesized in a sequential overlapping manner. Final sequence was determined from both strands. Reaction products were resolved on 5% denaturing polyacrylamide gels using standard techniques. Sequence comparisons were made using the PC/GENE program (IntelliGenetics, Inc. Mountain View, CA).

RNA Blot Analysis

Mouse retinas were collected from C57BL/6 mice for RNA preparation. The mice were reared under cyclic light (12h in dark and 12h in light) with 5–10ft-cd illumination during the light cycle. Total RNA was extracted from these tissues using TRIzol™ Reagent (GIBCO-BRL, Gaithersburg, MD), electrophoretically separated on a 1% agarose gel containing 18% formaldehyde, and then transferred to nitrocellulose membrane and hybridized with exon-specific (exon 1–3) or intron-specific (intron 1–2) probes that were generated by PCR. The hybridization condition was at 42°C in 50% formamide/5x SSC/1x Denhardt's/0.5% SDS/0.5mg/ml ssDNA. The filters were washed using conditions described previously.[24]

Southern Blot Analysis

A total of 20μg of genomic DNA or 5μg of purified lambda clone was digested with different restriction enzymes and electrophoretically separated on a 0.8% TAE gel. DNA fragments were then transferred to Nytra Plus nylon membrane (Scheicher & Schuell, Keene, NH) and subjected to prehybridization and hybridization under the conditions used for screening of the genomic library.[24] Blots were probed with a ^{32}P-labeled clones containing either the entire mouse peripherin/*rds* coding sequences or its extreme 5′ or 3′ sequences.

Isolation of the 5′ Flanking Region of the Peripherin/*rds* Gene

The 5′ flanking region of the peripherin/*rds* gene was isolated using the PromotFinder DNA Walking kit (Clontech Laboratories, Inc., Palo Alto, CA). Each kit contains five libraries of restriction enzyme digested (EcoRV, ScaI, PvuII, SspI, and DraI), uncloned, adapter-ligated genomic DNA from BALB/c mice. The primary PCR reaction was performed with the gene specific primer and adapter primer provided in the kit). The PCR product was diluted and used as a template for the second PCR reaction using nested adapter primer provided in the kit and a nested gene-specific primer. A 1.76kb PCR product was obtained, subcloned and sequenced.

Mapping the 3′ Polyadenylation Sites

The 3′ polyadenylation sites were determined as described previously (25). 5μg total retinal RNA and an oligo (dT) mcs primer (5′-GGCCACGCGTCGACTAGTACGACT$_{17}$-3′) were used to produce the first strand of retinal cDNA. This oligo (dT) primer consisted of an oligo (dT$_{17}$) domain and a 29-nucleotide multiple cloning site (mcs). Amplification with this primer introduced a nucleotide sequence (mcs 5′-GGCCACGCGTCGACTAGTACGAC-3′) into the synthesized cDNA. This cDNA was then used in a second PCR with peripherin/*rds*-specific primers. The PCR conditions were set as described previously.[25] The size of the amplified DNA was determined after electrophoresis on a 1% agarose gel. The PCR fragments were subcloned into T-vector and completely sequenced to identify the polyadenylation sites.

A. Bovine Peripherin/*rds* cDNA

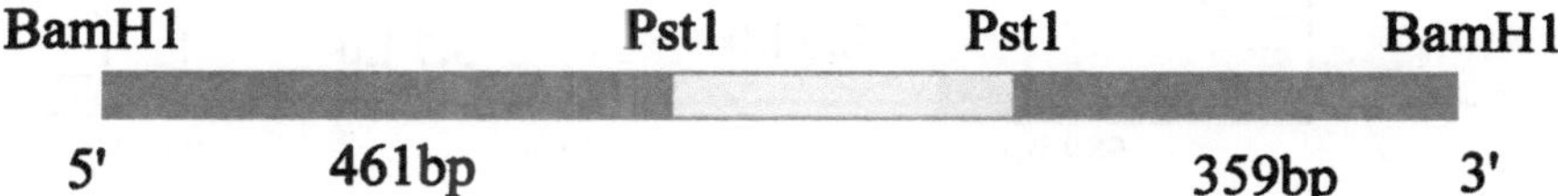

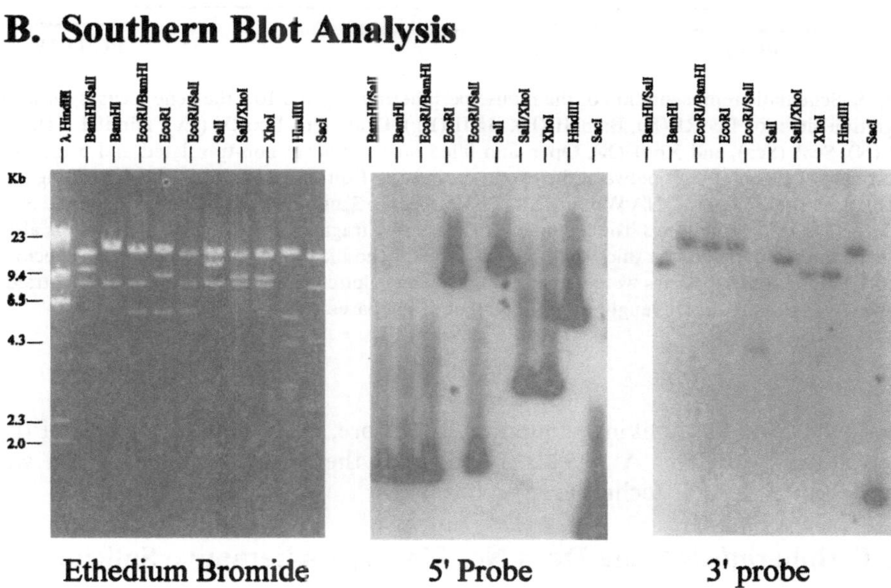

Figure 2. DNA blot analysis of the λ6-1-1 clone. DNA extracted from the λ6-1-1 clone was digested with several restriction enzymes. The digested DNA was electrophoretically fragmented on a 1% TAE agarose gel, stained with ethidium bromide (shown on the left panel) and transferred to a nitrocellulose membrane. The blot was first probed with the 5′ region of bovine peripherin/*rds* cDNA (middle panel), stripped, and then reprobed with the 3′ region of bovine peripherin/*rds* cDNA (right panel).

RESULTS

Structure and Organization of the Mouse Peripherin/*rds* Gene

A ^{32}P-labeled full-length bovine peripherin/*rds* cDNA[1] was used as a probe to screen the mouse EMBL genomic library (Clontech). Three different λclones (λ6-1-1, 6-1-2, 6-1-3) were isolated from ~10^6 plaque forming units (PFUs) and analyzed by means of restriction enzyme mapping and Southern blotting. The λ 6-1-1 clone had the longest insertion and was used for further study. This clone hybridized to the 5′ (461 bp, BamHI/PstI fragment) and 3′ (350 bp, BamHI/PstI fragment) ends of the bovine peripherin/*rds* cDNA (Fig. 2), indicating that it contained the entire mouse peripherin/*rds* cDNA sequence (1). Sequence analysis of the λ6-1-1 clone revealed that the mouse peripherin/*rds* gene contains three exons interrupted by two introns (Fig. 3). The sequences of the exon/intron boundaries correspond to the consensus sequences reported previously for other genes.[26] The λ6-1-1 clone was found to be

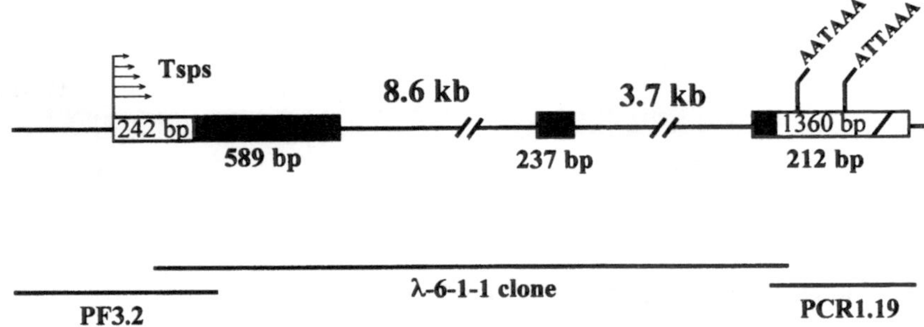

Figure 3. Schematic representation of the mouse peripherin/*rds* gene. Top, the gene locus is shown with the recognition sites for EcoRI (E), BamHI (B), BglII (Bg), DraI (Dr), EcoRV (Ev), HindIII (H), KpnI (K), NcoI (N), ScaI (Sca), and XhoI (X). Open and filled boxes indicate non-translated and translated exons, respectively. Horizontal lines between the exons are introns. Bottom shows the 3.2 kb 5′ flanking region isolated by PromoterFinder™ DNA Walking Kit (PF3.2) and the 3′ untranslated fragment identified by 3′RACE (PCR 1.19) as thick, solid lines. The alignment of these two fragments with the λ6-1-1 clone is also shown. Thin solid lines underneath the thick solid lines represent genomic fragments that have been subcloned from the λ6-1-1 clone. The subclones were named using the single letter symbols of their flanking restriction sites, followed by the approximate length of the fragment in kilobases.

missing the 5′ and 3′ flanking sequences. Therefore, the 5′ flanking region of the gene was isolated by PCR.[24] A 1.19 kb fragment of the 3′ untranslated region was also identified by 3′ RACE technique.[26]

The Peripherin/*rds* Gene Does Not Undergo Alternative Splicing

RNA blot analysis revealed that multiple RNA species of the mouse peripherin/*rds* gene are expressed.[8,24] To determine whether each of the peripherin/*rds* mRNA transcripts contains all three exons, a set of exon-specific (exon 1–3) and intron-specific (intron 1–2) probes were generated by PCR. These probes were hybridized with mouse retinal total RNA (Fig. 4). As shown in Fig. 4, when exon-specific probes were used, two major (1.6 & 2.7 kb) and several minor bands (5.5, 6.5, & 8.0 kb) were consistently observed. These results show that all exon sequences were present in each of the different RNA transcripts, implying that the peripherin/*rds* gene does not undergo alternative splicing. When the intron-I specific probe was used (lane 2), an 8.0 kb band was seen. This band may be due to the presence of incompletely spliced intron I within one of the major transcripts. Alternatively, intron I may contain another gene. Two bands (5.5 and 6.5 kb) were detected when the intron-II specific probe (lane 4) was used. The sizes of these two products correspond exactly to the expected sizes of the two major transcripts (1.6 & 2.7 kb) plus intron II, suggesting that they might be unprocessed heterogeneous RNA (hnRNA).

One Peripherin/*rds* Gene Is Expressed in Both Rod and Cone Photoreceptors

It has been suggested that a single peripherin/*rds* gene may be expressed in both rod and cone photoreceptor cells. Evidence for this came from immunocytochemical

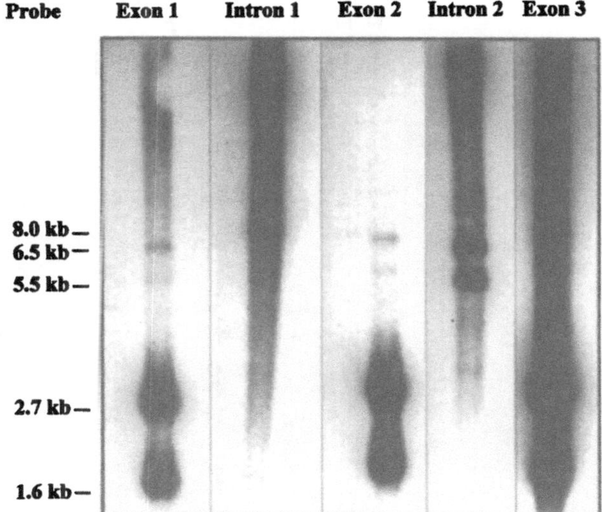

Figure 4. Absence of alternative splicing of the mouse peripherin/*rds* gene. Total RNA was extracted from normal mouse retinas and hybridized with α^{32}P-labeled DNA exon and intron specific probes. The probe used in each lane is marked accordingly and the sizes of the transcripts are also indicated.

studies.[27] The outer segment of both rod dominant (rat) and cone dominant (ground squirrel) retinas were strongly labeled with anti-peripherin/*rds* antibody. Further evidence was provided by the failure of development of cone as well as rod outer segments in the *rds* mouse.[110,28] To obtain further molecular evidence for expression of a single peripherin/*rds* gene in both rods and cones, we performed two experiments. First, the complexity of the mouse peripherin/*rds* gene was determined by Southern blot analysis. Genomic DNA isolated from a Balb/c mouse was digested with different restriction enzymes and hybridized with the mouse peripherin/*rds* cDNA probe. As demonstrated in Fig. 5, a simple genomic pattern was observed, suggesting the presence of only a single gene that codes for mouse peripherin/*rds*. The size of this gene is estimated to be ~24 kb.

Second, we also evaluated peripherin/*rds* expression in *rd/rd* mice at ages representative of the degenerative process.[10] At early ages (postnatal days P5 and P15), the degeneration is incomplete and the expression of both peripherin/*rds* and rod opsin can be detected. However, by P70 rod degeneration is complete and rod opsin expression cannot be detected above the background level.[29] At the same age, it was possible to detect a low level of expression of peripherin/*rds* in *rd/rd* mice (Fig. 1A from ref[29]). Since the *rd* defect is rod-specific, and affects rods more than cones, the residual peripherin/*rds* signal must reflect the small population of cone photoreceptors that survive the early cascade of photoreceptor degeneration. Overall, these results indicate that the same peripherin/*rds* gene may be expressed in both rod and cone photoreceptor cells.

The 5′ Flanking Region of the Mouse Peripherin/*rds* Gene

A PCR based technique was used to isolate the 5′ flanking region of the mouse peripherin/*rds* gene. A 1.7 kb fragment was cloned. Continuity of this fragment with the coding region of the gene was confirmed by both PCR and Southern blot (paper in

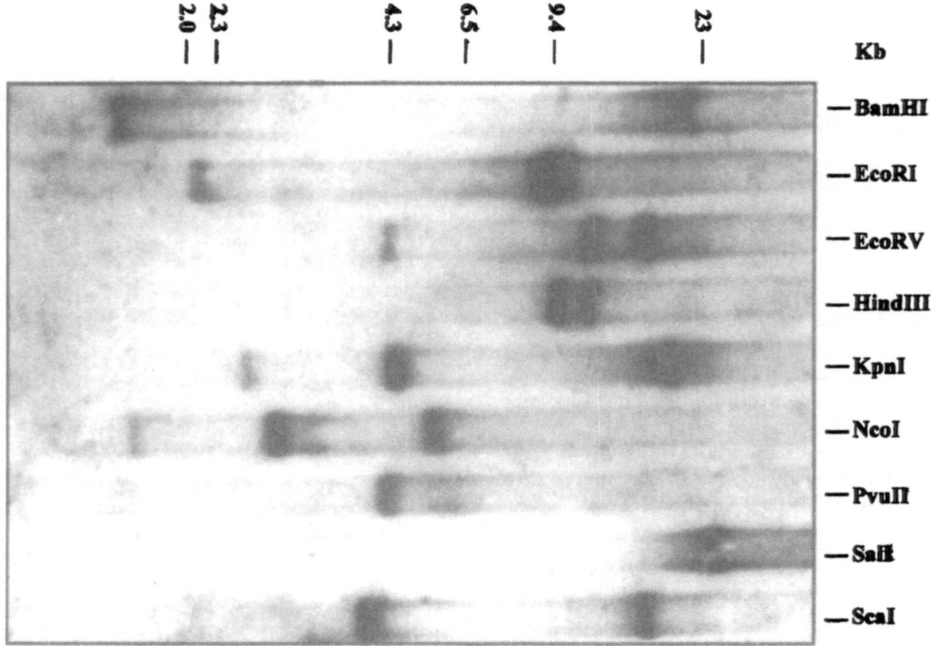

Figure 5. Southern blot analysis of the mouse peripherin/*rds* gene: 20 μg of genomic DNA from a Balb/c mouse was digested with restriction enzyme. The digested DNA was electrophoresed on a 0.8% TAE gel and transferred to nylon membrane. The blot was then hybridized with the mouse peripherin/*rds* full length cDNA using standard conditions (see experimental method). The result was visualized by autoradiography.

preparation). Sequence analysis of this clone revealed several unique features, including the presence of B1 and B2 repetitive sequences. Progressive deletions within this region were generated and used to direct expression of the Luciferase gene in both Y-79 and non-photoreceptor cells. We are in the process of using green fluorescent protein (*gfp*) as a reporter gene to determine the promoter activity and cell type specificity of the 5′ proximal region of this gene in transgenic mice.

Sequence Comparison of the 3′ Untranslated Region

Conservation of the immediate 3′ untranslated region of the peripherin/*rds* gene across species was also investigated. Sequence comparison of mouse, rat, bovine, human, and cat peripherin/*rds* revealed the presence of two highly homologous regions located within the first 100 bp downstream of the TGA stop codon (Fig. 6). The mouse sequence corresponding to the first of these regions, located between 2 to 12 bp downstream of the stop codon, shows 90% homology with the rat sequence, 73% with bovine, 73% with human, and 82% with cat. The mouse sequence corresponding to the second area, located between 31 to 47 bp downstream of the stop codon shows 100% homology with the rat sequence, 94% with bovine, 81% with human, and 75% with cat. Comparison of mouse and human sequences further downstream of the termination codon revealed the presence of several other domains with high degrees of homology. Only areas with homology greater than 80% are indicated in Fig. 6.

Table 1. Pathogenic and polymorphic changes in the human peripherin/*rds* gene

	Mutations	Domain	Phenotype	References
Retinitis Pigmentosa	Arg46stop	D1	RP like	(Meins et al. 1993)
	Cys118del	M3	RP	(Farrar et al. 1991)
	Leu126Arg	D2	RP	(Kajiwara et al. 1994)
	Lys153Arg	D2	RP	(Jacobson et al. 1994)
	Cys165Tyr	D2	RP	(Souied et al. 1995)
	Asp173Val	D2	RP	(Gruning et al. 1994)
	206-209del	D2	RP	(Gannon et al. 1993)
	Pro210Ser	D2	RP	(Kemp et al. 1994)
	Phe211Leu	D2	RP	(Souied et al. 1995)
	Ser212Gly	D2	RP	(Farrar et al. 1992)
	Cys214Ser	D2	RP	(Saga et al. 1993)
	Pro216Ser	D2	RP	(Fishman et al. 1994)
	Pro216Leu	D2	RP	(Kajiwara et al. 1991)
	Pro219del	D2	RP	(Kajiwara et al. 1991)
	Asn244Lys	D2	RP	(Nakazawa et al. 1994)
	Gly266Asp	M4	RP	(Kajiwara et al. 1992)
	304-307del	C-tem	RP	(Heinzmann et al. 1995)
	Arg13Trp	N-term	RP digenic + ROM1 Arg16His	(Jacobson et al. 1995)
	Leu45Phe	D1	RP digenic + ROM1 1 bp del	(Jacobson et al. 1995)
	Leu185Pro	D2	RP digenic + ROM1	(Kajiwara et al. 1994)
Macular dystrophy/ pattern dystrophy	Ser27Phe	M1	Cone dystrophy	(Fishman et al. 1997)
	Ile32Val	M1	ARMD	(Kajiwara et al. 1994)
	Arg142Trp	D2	CACD	(Hoyng et al. 1995)
	Arg172Trp	D2	Macular dystrophy	(Wells et al. 1993)
	Arg172Gln	D2	Macular dystrophy	(Wells et al. 1993)
	Tyr184Ser	D2	Cone rod dystrophy	(Wada et al. 1995)
	Asn244His	D2	Cone-rod dystrophy	(Nakazawa et al. 1994)
	Trp25stop	M1	Punctata albescans	(Kajiwara et al. 1993)
	Met67Del	M2	Pattern/cone-rod dystrophy	(Jacobson et al. 1994)
	Gly68Arg	M2	Pattern/cone-rod dystrophy	(Cideciyan et al. 1995)
	Phe140ins	D2	Pattern dystrophy	(Keen et al. 1994)
	Asp157Asn	D2	Pattern dystrophy	(Cideciyan et al. 1995)
	Gly167Asp	D2	Butterfly dystrophy	(Nichols et al. 1993)
	Arg172Gly	D2	Butterfly dystrophy	(Nichols et al. 1993)
	Lys193del	D2	Pattern/cone-rod dystrophy	(Jacobson et al. 1994)
	Pro210Arg	D2	Variable pattern dystrophy	(Jackson et al. 1993)
	Arg220Gln	D2	Pattern dystrophy	(Cideciyan et al. 1995)
	Tyr258Stop	D2	Adult vitelliform	(Wells et al. 1993)
	Pro299delTG	C-term	Butterfly dystrophy	(Nichols et al. 1993)
	Leu307delT	C-term	Pattern type dystrophy	(Gruning et al. 1994)
Variable	Ile32ins	M1	Variable	(Jacobson et al. 1996)
	67-69del	M2	Variable	(Jacobson et al. 1996)
	Lys153del	D2	Variable	(Weleber et al. 1993)
Polymorphism	Ser72Cys	M1	Polymorphism	(Fishman et al. 1997)
	Glu304Gln	C-term	Polymorphism	(Jordan et al. 1992)
	Lys310Arg	C-term	Polymorphism	(Jordan et al. 1992)
	Gly338Asp	C-term	Polymorphism	(Jordan et al. 1992)

Point mutations that alter or delete single amino acids are identified by their codon number, whereas insertions or deletions that change the reading frame are identified by the first nucleotide deleted or that preceding an insertion (all numbers relate to the original cDNA sequence published by Travis et al., 1991). RP = retinitis pigmentosa, ARMD = age-related macular dystrophy, CACD = central areolar choroidal dystrophy. D = intradiscal loop, M = transmembrane.

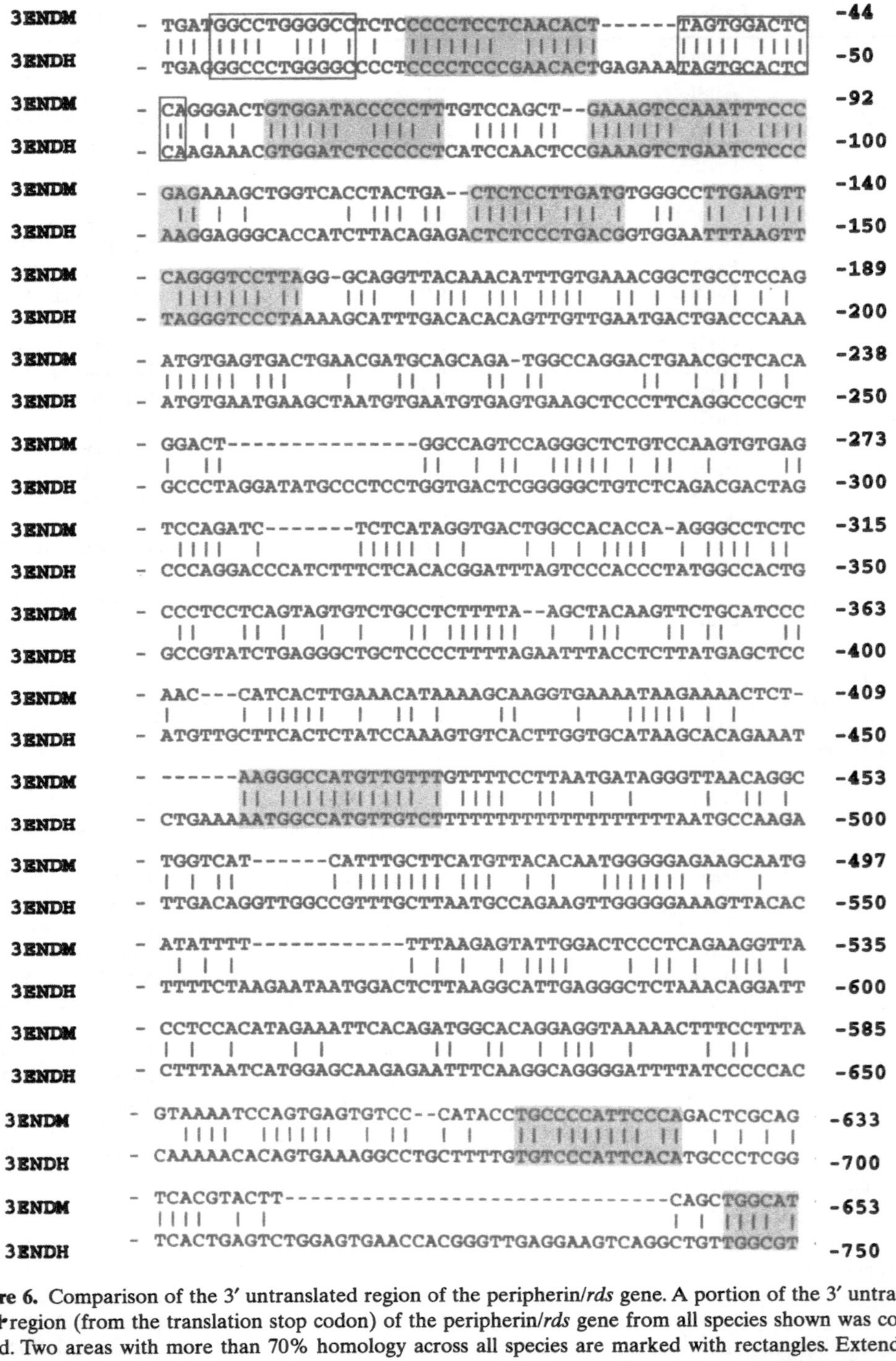

```
3ENDM  - TGATGGCCTGGGGCCTCTCCCCCTCCTCAACACT------TAGTGGACTC  -44
3ENDH  - TGAGGGCCCTGGGGCCCCTCCCCTCCCGAACACTGAGAAATAGTGCACTC  -50

3ENDM  - CAGGGACTGTGGATACCCCCTTTGTCCAGCT--GAAAGTCCAAATTTCCC  -92
3ENDH  - CAAGAAACGTGGATCTCCCCCTCATCCAACTCCGAAAGTCTGAATCTCCC  -100

3ENDM  - GAGAAAGCTGGTCACCTACTGA--CTCTCCTTGATGTGGGCCTTGAAGTT  -140
3ENDH  - AAGGAGGGCACCATCTTACAGAGACTCTCCCTGACGGTGGAATTTAAGTT  -150

3ENDM  - CAGGGTCCTTAGG-GCAGGTTACAAACATTTGTGAAACGGCTGCCTCCAG  -189
3ENDH  - TAGGGTCCCTAAAAGCATTTGACACACAGTTGTTGAATGACTGACCCAAA  -200

3ENDM  - ATGTGAGTGACTGAACGATGCAGCAGA-TGGCCAGGACTGAACGCTCACA  -238
3ENDH  - ATGTGAATGAAGCTAATGTGAATGTGAGTGAAGCTCCCTTCAGGCCCGCT  -250

3ENDM  - GGACT---------------GGCCAGTCCAGGGCTCTGTCCAAGTGTGAG  -273
3ENDH  - GCCCTAGGATATGCCCTCCTGGTGACTCGGGGGCTGTCTCAGACGACTAG  -300

3ENDM  - TCCAGATC-------TCTCATAGGTGACTGGCCACACCA-AGGGCCTCTC  -315
3ENDH  - CCCAGGACCCATCTTTCTCACACGGATTTAGTCCCACCCTATGGCCACTG  -350

3ENDM  - CCCTCCTCAGTAGTGTCTGCCTCTTTTA--AGCTACAAGTTCTGCATCCC  -363
3ENDH  - GCCGTATCTGAGGGCTGCTCCCCTTTTAGAATTTACCTCTTATGAGCTCC  -400

3ENDM  - AAC---CATCACTTGAAACATAAAAGCAAGGTGAAAATAAGAAAACTCT-  -409
3ENDH  - ATGTTGCTTCACTCTATCCAAAGTGTCACTTGGTGCATAAGCACAGAAAT  -450

3ENDM  - ------AAGGGCCATGTTGTTTGTTTTCCTTAATGATAGGGTTAACAGGC  -453
3ENDH  - CTGAAAAATGGCCATGTTGTCTTTTTTTTTTTTTTTTTTAATGCCAAGA  -500

3ENDM  - TGGTCAT------CATTTGCTTCATGTTACACAATGGGGGAGAAGCAATG  -497
3ENDH  - TTGACAGGTTGGCCGTTTGCTTAATGCCAGAAGTTGGGGGAAAGTTACAC  -550

3ENDM  - ATATTTT------------TTTAAGAGTATTGGACTCCCTCAGAAGGTTA  -535
3ENDH  - TTTTCTAAGAATAATGGACTCTTAAGGCATTGAGGGCTCTAAACAGGATT  -600

3ENDM  - CCTCCACATAGAAATTCACAGATGGCACAGGAGGTAAAAACTTTCCTTTA  -585
3ENDH  - CTTTAATCATGGAGCAAGAGAATTTCAAGGCAGGGGATTTTATCCCCCAC  -650

3ENDM  - GTAAAATCCAGTGAGTGTCC--CATACCTGCCCCATTCCCAGACTCGCAG  -633
3ENDH  - CAAAAACACAGTGAAAGGCCTGCTTTTGTGTCCCATTCACATGCCCTCGG  -700

3ENDM  - TCACGTACTT----------------------------CAGCTGGCAT  -653
3ENDH  - TCACTGAGTCTGGAGTGAACCACGGGTTGAGGAAGTCAGGCTGTTGGCGT  -750
```

Figure 6. Comparison of the 3′ untranslated region of the peripherin/*rds* gene. A portion of the 3′ untranslated region (from the translation stop codon) of the peripherin/*rds* gene from all species shown was compared. Two areas with more than 70% homology across all species are marked with rectangles. Extended comparison was performed between mouse and human and the areas with more than 80% homology are indicated as shaded rectangles. 3ENDM, 3′ end of the mouse peripherin/*rds* gene; 3ENDH, 3′ end of the human peripherin/*rds* gene.

```
3ENDM  - GTCCCAGCATC-CTCCACTCACCAAGTGCTGGGTGAGCCACTGC------  -696
3ENDH  - GTCCCAGCACCACACCATCCCTAAAGTGCCAGGTGATCTCCTGTGGCTCA  -800
3ENDM  - -CTGAGAAGGCCCCGGGTGAGTTTGCTGTCCTT-TGTGGGAGAGATGTGG  -744
3ENDH  - TCGGTGGAAGCAGTGGGGTAGGCTGCTGCCCTGCTGTGGAAGAGGAGCAA  -850
3ENDM  - TGATCAATTAGGAATCTGCCACTTGGGGACCAAGCCCTGCCTCCTGGTGG  -794
3ENDH  - CAATCAGACATGAGTCCACCCTTTGGAGACCAAGCCTCAGCTCTTGGTGG  -900
3ENDM  - TCT--------------CAGATGGCCACAACAGTCTTTGTGGACTGCACT  -830
3ENDH  - GCCAAGGGACACCCACACAGGTGGCCATCACAGCCCCATGGACAACACTA  -950
3ENDM  - GTCCAGGAGGGCAAAGA---------CCCCTGGGAGCTTCTGTTCTCCCT  -871
3ENDH  - ATTGTCCACAGCAAAGGGCAAGGAATCCTCTGGGAGCTTCTTCCGTTTCT  -1000
3ENDM  - CCCACAAAG-----------------CACTATG---GGAATGCTTCCGTA  -901
3ENDH  - TCCCCCAAGATACCCATCTTGAAAAACACTATTTCTGGAATGCTTCTGCA  -1050
3ENDM  - TCCAGGAAGATTCTG-GAGATAGGTTATCTTTCCGAGCTAGCGTGTGTGG  -950
3ENDH  - TCAAAGGAGATTCTTTGAGATAGCCCATCTTCCTGAGCTAGCAAATACAG  -1100
3ENDM  - AAAGTTGCTGTTTCTTCTGGGAACGTCAGGCAAACTCTCAGGGGAAGCAA  -1000
3ENDH  - GAGTTTTCACTTTCTTTAGGAAAGAGAAGCTTTCAGGGGAAGGAGAGAAT  -1150
3ENDM  - AAATATGATT----------------------------------------  -1010
3ENDH  - GATTTTGCTGACTTCCCAAGCCCTGGTGACCAGACCAAGGCAGGGCCCAG  -1200
3ENDM  - --------------TTGGTGGCACCTT---GTGGGATGGCACCCACAAGG  -1043
3ENDH  - CATAATTCCTCCAGTTGGATGAACATTCAAGAGAGCTCGTTCCTACCTGG  -1250
3ENDM  - ACTGTTGTGGCCATCTGAGACCACCAGGAGGCAGGGCTTGGTCCCCAAGG  -1093
3ENDH  - CTGGAGACCGAGGCCAGAAGGCAAAAACCAGAAAGGGAACAGTCCATAAC  -1300
3ENDM  - GCTCAGCTGCTGCTGGCCTGTGCTGTGTGGTTCTA---------------  -1128
3ENDH  - TTACCTCTGCTTCTGACCGATGGTGTTTGGGAATAGGTTACTTTGGACTG  -1350
3ENDM  - --------------------------------------------------  -1128
3ENDH  - AGTTTGGGTTCTTTGCTGTCCTAAGAACTTTAGTGTAGAGAAAATAAGAC  -1400
3ENDM  - ---------------------------------ACTCTCTCAGGGGGAAGA  -1146
3ENDH  - TTCTGGTGCTGCTGGGGTATGTTCTGGGCTTAATTCCCCCAAGCAGAAGA  -1450
3ENDM  - CCAGATTAGAGATGCTGGCACACCCAACGA-ACTATGGCCCCAATGTTGG  -1195
3ENDH  - CCAGATCCAAGATGTTTGGACACCCTGTCAGACGTTGGTCCCAAGTTTAA  -1500
3ENDM  - TTAGTATTCTGAGTCTCTCAGAG--CAAGAAATGGCCCAGA---------  -1234
3ENDH  - TTAGATTTCTGAATCTCGTTGAGGCCAAGGAATGATCCATACTGAAAAAA  -1550
3ENDM  - TGCTGAGCCAGGCATCTGTGGCGGGGACCCTCGGGGTCCTGCCGTCACCT  -1284
3ENDH  - TGCTGAGCCAGCCATCTTTGGCAAAGGTCCCTGAGCTCTTGCTATCTCTC  -1600
```

Figure 6. *Continued.*

```
3ENDM   - CTCAGGAGGGCTGAGACTATGGTGAAAGATACTCTAGACCCATTTCTAAC -1334
               ||    |   |        |  |  | | |||||| ||||     |
3ENDH   - AAGAGTGCTGAGAACCACGGTGAAAGTGCTGCTCTAGGCCCACAAGTGTA -1650

3ENDM   - TATGCTGTCAACCATCGTATTAAGAGATATTAAAAACT-ATATTGTATG -1382
              |    |         | | || ||||| | |||| | ||| ||||||
3ENDH   - ACTATGCTGTTAACAGCTGTCAATAGATAATTAAAATTCATACTGTATG -1699
```

Identity: 801 (57.96%)

Figure 6. *Continued.*

DISCUSSION

This study was undertaken to clone and characterize the mouse peripherin/*rds* gene and its 5′ flanking region. The cDNA sequences of bovine,[1] human[30], mouse,[8] rat[31], canine,[32] cat[33], and *Xenopus laevis*[34] peripherin/*rds* have been previously determined. This protein is highly conserved across these species, with 92.5% amino acid sequence identity shared by mouse and bovine, 91.3% by mouse and human, 97.1% by mouse and rat,[2] 94% by mouse and canine, and 70% by mouse and *Xenopus*. We recently cloned and sequenced the skate homologue.[35] Skate peripherin/*rds* shares approximately 70% identity with that from dog, cat, human, bovine, mouse, and rat.

Analysis of the gene structure of mouse and human peripherin/*rds* shows the presence of three exons and two introns.[9–24] Although the sizes of these introns are different, the locations of introns I and II are conserved in both the mouse and human.[9–24] Intron I spans 8.6 kb and interrupts the large intradiscal loop and intron II spans 3.7 kb and interrupts the fourth transmembrane domain.[24] While our estimated size for intron I agrees with previously reported findings,[9] our size for intron II differs significantly. We determined the locations of these introns by sequence divergence from the cDNA. The length was estimated by combining the sizes of different contiguous subcloned fragments that did not contain any cDNA sequences. In addition, the sizes of introns I and II were confirmed by PCR amplification using genomic DNA from Balb/c and C57BL/6 mice as templates and primers corresponding to the coding regions flanking the introns. Since genomic DNA from both Balb/c and C57BL/6 mice was used, the observed difference in the size of intron II is not the result of strain differences.

Analyses of the 5′ flanking region showed that the peripherin/*rds* gene initiates transcription at seven different sites within 26 nucleotides.[24] Of all these transcription initiation sites, only one, T at position –220, has been reported previously[9].

Increased knowledge of the structural and functional relationships of the peripherin/*rds* gene will have great impact on understanding its role in human retinal diseases. Availability of the 5′ flanking sequence is of particular importance, as this will allow patients to be screened for mutations in this region of the gene that could result in reductions in expression levels. Alterations in the promoter sequences of peripherin/*rds* have not yet been shown to associate with any retinal diseases. However, it is possible that some forms of retinal diseases are caused by mutations in the promoter elements that regulate expression of this gene. The presence of repeat elements in the promoter region may make this area a hot spot for recombination events that could lead to one or more forms of retinal abnormalities. Of note is the presence of several short repeat sequences in intron I and II of the gene,[24] including twelve repeats of triplets (aat) in intron I. The same length of repeats has also been found in other genes.[36] Whether the presence or the length of these repeats plays a role in regulating the function of the mouse peripherin/*rds* gene remains to be determined.

Conservation of portions of the 3′ untranslated region of the peripherin/*rds* gene in mouse and human may also be significant. Although no pathogenic mutations have been identified to date in these areas, there is increasing evidence that the 3′ untranslated region of eukaryotic genes contains sites for important signals that determine message stability,[37] translational initiation,[38] and intracellular mRNA localization.[39]

ACKNOWLEDGMENTS

We thank Dr. Robert Molday for providing the bovine cDNA clone. This work was supported by grants from the National Eye Institute (EY-10609 and EY-01792); The Foundation Fighting Blindness; Inc., Baltimore, MD; and the Knights Templar Eye Foundation, Springfield, IL.

REFERENCES

1. Connell, G.J. and Molday, R.S. (1990) Molecular cloning, primary structure, and orientation of the vertebrate photoreceptor cell protein peripherin in the rod outer segment disc membrane. *Biochem.* 29, 4691–4698.
2. Connell, G., Molday, L.L., Reid, D., and Molday, R.S. (1991) Molecular structure and properties of peripherin/rds the normal product of the gene responsible for retinal degeneration in the rds mouse. In Retinal Degenerations. R. Anderson, J. Hollyfield, and M. LaVail, editors. CRC Press, Boca Raton. 467–477.
3. Travis, G.H., Sutcliffe, J.G., and Bok, D. (1991) The *retinal degeneration slow* (*rds*) gene product is a photoreceptor disc membrane-associated glycoprotein. *Neuron* 6:61–70.
4. Arikawa, K., Molday, L.L., Molday, R.S., and Williams, D.S. (1992) Localization of peripherin/*rds* in the disk membranes of cone and rod photoreceptors: relationship to disk membrane morphogenesis and retinal degeneration. *J. Cell Biol* 116, 659–667.
5. Connell, G., Bascom, R.A., Molday, L., Reid, D., McInnes R.R., and Molday, R.S. (1991) Photoreceptor peripherin is the normal product of the gene responsible for retinal degeneration in the *rds* mouse. *Proc. Natl. Acad. Sci. USA* 88, 723–726.
6. Molday, R.S., Hicks, D., and Molday, L. (1987) Peripherin. A rim-specific membrane protein of rod outer segment discs. *Invest. Ophthalmol. Vis. Sci* 28:50–61.
7. Roof, D.J. and Heuser, J.E. (1982) Surfaces of rod photoreceptor disk membranes: integral membrane components. *J. Cell Biol.* 95:487–500.
8. Travis, G.H., Brennan, M.B., Daneilson, P.E., Kozak, C.A., and Sutcliffe, J.G. (1989) Identification of a photoreceptor-specific mRNA encoded by the gene responsible for retinal degeneration slow (*rds*). *Nature* 338:70–73.
9. Ma, J., Norton, J.C., Allen, A.C., Burns, J.B., Hasel, K.W., Burns, J.L., Sutcliffe, J.G., and Travis, G.H. (1995) *Retinal degeneration slow* (*rds*) in mouse results from simple insertion of a t haplotype-specific element into protein-coding exon II. *Genome* 28:212–219.
10. Cheng, T., Peachey, N.S., Li, S., Goto, Y., Cao, Y., and Naash, M.I. (1997) The effect of peripherin/rds haploinsufficiency on rod and cone photoreceptors. *J. Neurosci* 17, 1–11.
11. Travis, G.H., Groshan, K.R., Lloyd, M., and Bok, D. (1992) Complete rescue of photoreceptor dysplasia and degeneration in transgenic *retinal degeneration slow* (*rds*) mice. *Neuron* 9:113–119.

12. Travis, G.H. and Bok, D. (1993) A molecular characterization of the *retinal degeneration slow* (*rds*) mouse mutation. In Retinal Degeneration: Clinical and Laboratory Applications. J.G. Hollyfield, R.E. Anderson, and M.M. LaVail, editors. Plenum Press, New York & London. 219–230.
13. Farrar, G.J., Jordan, S.A., Kenna, P., Humphries, M.M., Kumar-Singh, R., McWilliam, P., Allamand, V., Sharp, E., and Humphries, P. (1991) Autosomal dominant retinitis pigmentosa: localization of a disease gene (RP6) to the short arm of chromosome 6. *Genomics* 11:870–874.
14. Kajiwara, K., Hahn, L.B., Mukai, S., Travis, G.H., Berson, E.L., and Dryja, T.P. (1991) Mutations in the human retinal degeneration slow gene in autosomal dominant retinitis pigmentosa. *Nature* 354: 480–483.
15. Kajiwara, K., Sandberg, M.A., Berson, E.L., and Dryja, T.P. (1993) A null mutation in the human peripherin/RDS gene in a family with autosomal dominant retinitis punctata albescens. *Nature Genet.* 3:208–212.
16. Wells, J., Wroblewski, J., Keen, J., Inglehearn, C., Jubb, C., Eckstein, A., Jay, M., Arden, G.B., Battacharya, S.S., Fitzke, F., and Bird, A.C. (1993). Mutations in the human retinal degeneration slow (RDS) gene can cause either retinitis pigmentosa or macular dystrophy. *Nature Genet.* 3:213–218.
17. Weleber, R., Carr, R.E., Murphey, W., Sheffield, V., and Stone, E. (1993). Phenotypic variation including retinitis pigmentosa, pattern dystrophy, and fundus flavimaculatus in a single family with a deletion of codon 153 or 154 of the peripherin/RDS gene. *Arch. Ophthalmol.* 111:1531–1542.
18. Kemp, C.M., Jacobson, S.G., Cideciyan, A.V., Kimura, A.E., Sheffield, V.C., and Stone, E.M. (1994) RDS gene mutations causing retinitis pigmentosa or macular degeneration lead to the same abnormality in photoreceptor function. *Invest. Ophthalmol. Vis. Sci.* 35:3154–3162.
19. Kikawa, E., Nakazawa, M., Chida, Y., Shiono, T., and Tamai, M. (1994) A novel mutation (asn244lys) in the peripherin/rds gene causing autosomal dominant retinitis pigmentosa associated with bull's-eye maculopathy detected by nonradioisotopic sscp. *Genomics* 20:137–139.
20. Molday, R.S. (1994) Peripherin/rds and rom-1: molecular properties and role in photoreceptor cell degeneration. In Progress in retinal and eye research. N.N. Osborne and G.J. Chader, editors. Pergamon Press Ltd., Great Britain. 271–299.
21. Nichols, B.E., Sheffield, V.C., Vandenburgh, K., Drack, A.V., Kimura, A.E., and Stone, E.M. (1993) . Butterfly-shaped pigment dystrophy of the fovea caused by a point mutation in codon 167 of the RDS gene. *Nature Genet.* 3:202–207.
22. Nakazawa, M., Kikawa, E., Chida, Y., and Tamai, M. (1994) 1994. Asn244His mutation of the peripherin/rds gene causing autosomal dominant cone-rod degeneration. *Hum. Mol. Genet.* 3:1195–1196.
23. Kedzierski, W., Lloyd, M., Birch, D.G., Bok, D., and Travis, G.H. (1997) Generation and analysis of transgenic mice expressing P216L-substituted rds/peripherin in rod photoreceptors. *Invest. Ophthalmol. Vis. Sci.* 38, 498–509.
24. Cheng, T., Al-Ubaidi, M.R., and Naash, M.I. (1997) Structural and developmental analysis of the mouse peripherin/rds gene. *Som. Cell. Mol. Genet.* 23, 165–183.
25. Al-Ubaidi, M.R., Pittler, S.J., Champagne, M.S., Triantafyllos, J.T., McGinnis, J.F., and Baehr, W. (1990) Mouse opsin: gene structure and molecular basis of multiple transcripts. *J. Biol. Chem.* 265: 20563–20569.
26. Al-Ubaidi, M.R., Ramamurthy, V., Maa, M.-C., Ingolia, D.E., Chinsky, J.M., Martin, B.D., and Kellems, R.E. (1990) Structural and functional analysis of the murine adenosine deaminase gene. *Genomics* 7:476–485.
27. Hargrave, P.A. and McDowell, J.H. (1992) Rhodopsin and phototransduction: a model system for G protein-linked receptors. *FASEB J.* 6:2323–2331.
28. Hawkins, R.K., Jansen, H.G., and Sanyal, S. (1985) Development and degeneration of retina in *rds* mutant mice: photoreceptor abnormalities in the heterozygotes. *Exp. Eye Res.* 41:701–720.
29. Carter-Dawson, L.D., LaVail, M.M., and Sidman, R.L. (1978) Differential effect of the *rd* mutation on rods and cones in the mouse retina. *Invest. Ophthalmol. Vis. Sci.* 17:489–498.
30. Travis, G.H., Christerson, L., Danielson, P.E., Klisak, I., Sparkes, R.S., Hahn, L.B., Dryja, T.P., and Sutcliffe, J.G. (1991) The human retinal degeneration slow (RDS) gene: chromosome assignment and structure of the mRNA. *Genomics* 10:733–739.
31. Begy, C. and Bridges, C.D.B. (1990) Nucleotide and predicted protein sequence of rat retinal degeneration slow (rds). *Nucl. Acids Res* 18:3058–3058.
32. Ray, K., Acland, G.M., and Aguirre, G.D. (1996) Nonallelism of *erd* and *prcd* and exclusion of the canine rds/peripherin gene as a candidate for both retinal degeneration loci. *Invest. Ophthalmol. Vis. Sci.* 37:783–794.

33. Gorin, M.B., Snyder, S., To, A., Narfstrom, K., and Curtis, R. (1993) The cat RDS transcript: candidate gene analysis and phylogenetic sequence analysis. *Mamm. Genome* 4:544–548.
34. Kedzierski, W., Moghrabi, W.N., Allen, A.C., Jablonski-Stiemke, M.M., Azarian, S.M., Bok, D., and Travis, G.H. (1996) Three homologs of rds/peripherin in *Xenopus leavis* photoreceptors that exhibit covalent and non-covalent interactions. *J. Cell Sci.* 109:2551–2560.
35. Li, C., Al-Ubaidi, M.R., and Naash, M.I. (1997) Isolation and characterization of the skate peripherin/rds gene. *Invest. Ophthalmol. Vis. Sci.* 38:S219
36. Murphy, S.P., Garbern, J., Odenwald, W.F., and Lazzarini, R.A. (1988) Differential expression of the homeobox gene *Hox-1.3* in F9 embryonal carcinoma cells. *Proc. Natl. Acad. Sci. USA* 85:5587–5591.
37. Amara, F.M., Chen, F.Y., and Wright, J.A. (1994) Phorbol ester modulation of a novel cytoplasmic protein binding activity at the 3′-untranslated region of mammalian ribonucleotide reductase R2 mRNA and role in message stability. *J. Biol. Chem.* 269, 6709–6715.
38. Svitkin, Y.V., Meerovitch, K., Lee, H.S., Dholakia, J.N., Kenan, D.J., Agol, V.I., and Sononberg, N. (1994) Internal translation initiation on poliovirus RNA: Further characterization of La function in poliovirus translation in vitro. *J. Virol.* 68:1544–1550.
39. Lantz, V. and Schedl, P. (1994) Multiple *cis*-acting targeting sequences are required for *orb* mRNA localization during *Drosophila* oogenesis. *Mol. Cell. Biol.* 14:2235–2242.

AUTOSOMAL DOMINANT RETINITIS PIGMENTOSA AND HYPEROPIA

E. M. Vingolo,[1] R. Allingham,[2] C. Palitto,[1] L. Scipioni,[1] and L. Iacobelli[1]

[1] University "La Sapienza" of Rome
Institute of Ophthalmology
Medical Retina Section
[2] Duke University
Glaucoma Unit
Durham, North Carolina

Several studies demonstrated that genetically factors seem to influence the onset and progression of juvenile glaucoma as showed by different studies which identified linkage with chromosome #3q[1,2,3] and #1q21–q31.[4] Even if clearly is possible identify inherited factors in open angle chronic glaucoma (OACG) until now no candidate genes have been proposed for this disease.

Hyperopia, between several factors involved, frequently results associated in OACG patients, even if up today is not possible indicate a candidate gene for hyperopia.

Frequently myopic refractive shift is associated in rod-cone tapetoretinal degeneration,[5] and very rarely hyperopia is reported in RP families and only one other family with autosomal dominant retinitis pigmentosa (ADRP) associated with hyperopic refraction.[5]

In our flies one three generation family in which ADRP and Hyperopia are associated in nine affected individuals even with different degree starting from +2 to +7 diopters. From a preliminary study there is a statistically significative difference between refractive error in RP patients and non-affected members in the family.

Starting from this point we postulate that distance between RP gene is not so great with at least one gene of hyperopia in this family; the goal of this study is to evaluate the family with linkage analysis to try gene localization, for RP and Hyperopia.

Linkage analysis on the affected and non-affected individuals could be able to indicate position of Hyperopia gene and consequently could be useful for a screening in OACG patients. This will allow individuating high-risk patients, which can be included in reinforced surveillance program.

Retinal Degenerative Diseases and Experimental Therapy, edited by Hollyfield *et al.*
Kluwer Academic / Plenum Publishers, New York, 1999.

Several different rhodopsin gene mutations have been identified in the last years in pedigrees with ADRP. In view of the differences in the molecular nature and location of these mutations, defining the phenotype has become increasingly important in order to identify the clinical counterpart to the different functional abnormalities of the photopigment molecule.[1–11] This is particularly true if one considers the notable clinical and functional heterogeneity associated to the different genotypes.

In the last few years in vitro expression studies on different mutant rhodopsins have allowed a better classification of rhodopsin gene defects on biochemical grounds, revealing the existence of two major classes of mutations[11] C class I, resembling the wild type rhodopsin in their accumulation to high levels in the plasma membrane and their ability to join in vitro to 11-cis retinal to form a photolabile visual pigment: these mutations tend to cluster near the carboxyl terminus.

Phenotypes associated with this class of mutations tend to be consistently severe. Class II mutations accumulate to much lower levels, are inefficiently transported from the endoplasmic reticulum to the plasma membrane, and produce little or no photolabile pigment upon incubation in vitro with 11-cis retinal: this class of rhodopsin mutants is likely to be defective in protein folding and/or stability, and has been associated to various clinical pictures, ranging from mild (sector) to severe from of RP.

The severity of RP due to class II rhodopsin mutants appears to depend mostly on the involvėd domain of the molecule. Those affecting the intradiscal domain seem to yield the better prognosis, while transmembrane mutations are associated with more variable clinical pictures, closely depending on the biological relevance of the involved residue. Here we describe the phenotype of a large adRP pedigree from Italy in which hyperopic shift has been demonstrated. The clinical picture associated with this mutation was one association between these two diseases in which hyperopia seems to be related to a shorter axial length. Data are compared to findings from report describing the phenotype yielded by mutations at the same locus, and to those of another large pedigree previously described. Thus is associated with hyperopic refraction in all affected patients.

Hyperopia is uncommon in patients with retinitis pigmentosa, which is usually more closely associated with myopia. Sieving and Fishman found that myopia occured in 201 of 268 eyes with retinitis pigmentosa (75%) as compared to 12% in a normal population, and the refractive errors ranged from +450 to −1625 diopters. Vingolo et al. found myopia in 48.4% of 471 eyes in a population of 244 pure RP patients. Myopia in RP patients represents only a simple refractive error and typical scleral and chorioretional myopic complications seem not influence the evolution of RP.

It is possible that the axial hyperopia may have represented a genetically associated trait or that degeneration of the retina and retinal pigment epithelium may have influenced the developmental growth of the globe in some way.

It is noteworthy that the association between retinitis pigmentosa and hyperopia could be useful to identify patients with glaucoma, which is a disease in most cases combined with hyperopia.

The studied pedigree (Fig. 1) was from central Italy in origin, and was elaborated from a propositus affected with adRP. Nine further families between 8 and 61 yrs of age have been investigated according to previously published criteria that are summarized in the following paragraph. Clinical and functional ophthalmologic procedures: Available patients were investigated as follows:

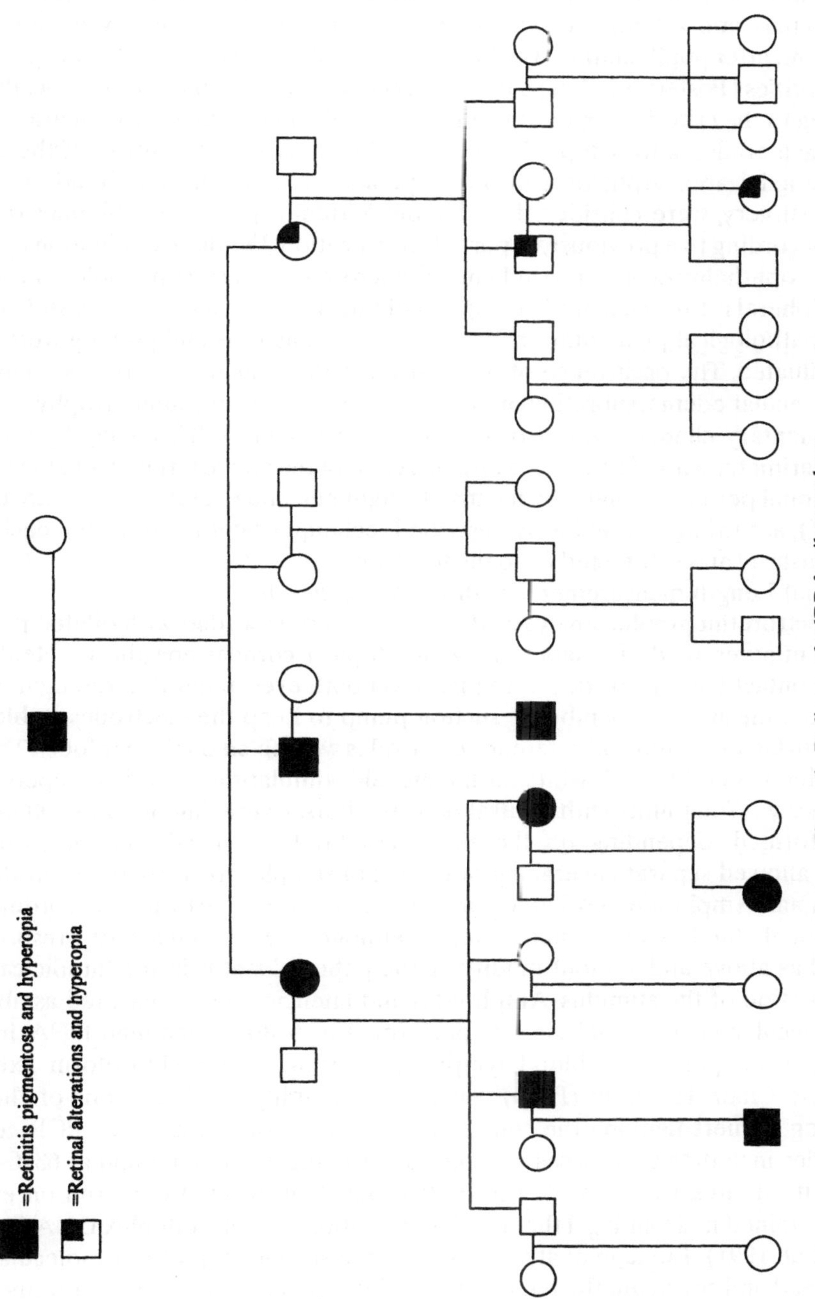

Figure 1. Pedigree of adRP family members.

Detailed medical history to define the age of onset of symptoms of the disease, with particular interest for night vision impairment, difficulty with side vision, and progressive loss of visual acuity (VA);

Ophthalmological evaluation: Best corrected distant and near VA with a standard Snellen chart and astigmometric. Anterior and posterior segment were investigated (before and after pupil dilatation) with particular reference to the following parameters. Lens opacities (Posterior Subcapsular Cataracts, PSCs) were biomicroscopically graded according to the criteria proposed by Fishman et al and modified by Pannarale et al. Six progressive grades were adopted evaluating the concentric extension of the opacities from the axial area. Aphakic and pseudophakic eyes, which had already undergone cataract surgery, were considered separately. Vitreal aspects were biomicroscopically graded according to a previously reported classification. Fundus examination by indirect binocular ophthalmoscopy and slit lamp biomicroscopy, to assess possible central and/or mid-peripheral retinal changes. The aspects of the optic disc and retinal vessels, the presence of pathological pigmentation, its topographical aspects and grading were particularly evaluated. The occurrence of macular alterations, such as epiretinal membranes, cystoid macular edema, subretinal neovascular membranes, macular atrophy or scarring and pigmentary changes was also recorded. Visual Field (VF) testing by Goldmann kinetic perimetry with I4e and III4e targets. Areas of the above targets were measured on conventional perimetry charts by means of a light pen and expressed in square centimeters (cm^2), according to previously described technique. Scotomata within each isopter were measured and subtracted from the total area.

Axial Length measurements with a-scan ecography.

Electroretinographic analyses: Rod ERGs were recorded with dilated pupils and after 30 minutes of dark-adaptation. After topical corneal anesthesia, Henkes-type corneal contact lens electrodes were placed on both eyes under dim red light and connected to a mechanical membrane suction pump to keep the electrodes stable on the corneal surface. Ground and reference electrodes were fixed on the ear lobes. The signal was elicited with a $2\,cd{\cdot}m^2/s$ white flash Ganzfeld stimulation at a 0.5 Hz repetition rate attenuated y 2.7 log units with neutral density filters. Averaging of 80 to 100 iterations was performed, depending on the signal amplitude. Sophisticated post-processing analyses allowed separate averaging of groups of samples to verify the reproducibility of the signal. Amplitude and latency measurements were performed according to the international standards for clinical electroretinography. Maximal ERG response was recorded as above and without modifying the patients' status in the faradic cage, with no attenuation of the stimulus. Amplitudes and latencies were measured as above.

Molecular genetic analysis was performed as follows: genomic DNA has been extracted from peripheral blood lymphocytes using phenol-chloroform extraction. Polimerase Chain Reaction (PCR) was used to amplify the five exons of rhodopsin gene using primers flanking the coding regions. The program used for PCR consisted of 30 cycles including two successive steps at 94°C for one minute and at 63°C for two and a half minutes, respectively. The PCR products generated were run on ethidium bromide-stained hydrolink gel that resolves mutations as heteroduplex DNA fragments (Keen et al. 1991). Linkage analysis has been the second step of our molecular study. Ten affected and seven unaffected members of this pedigree were typed for microsatellite polymorphisms.

The seven affected members had loss of visual acuity, five of them presented cataract in both eyes. Fundus changes characteristics of retinitis pigmentosa were noted in all the affected members of the family (Fig. 2).

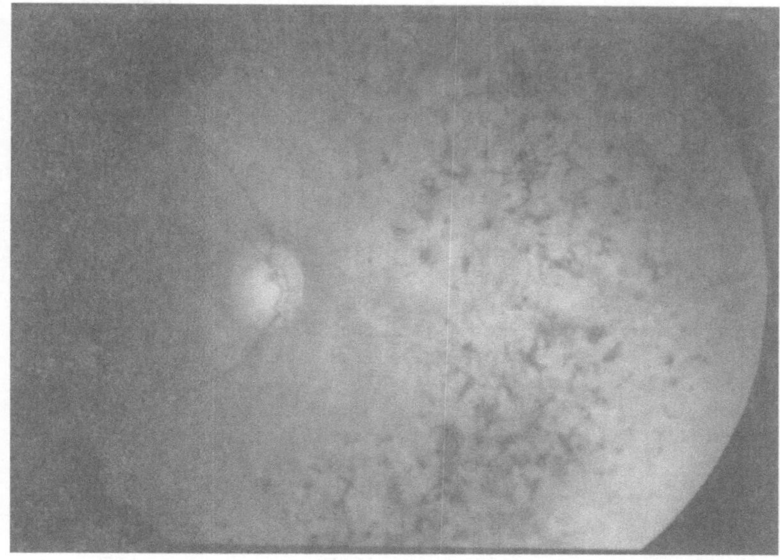

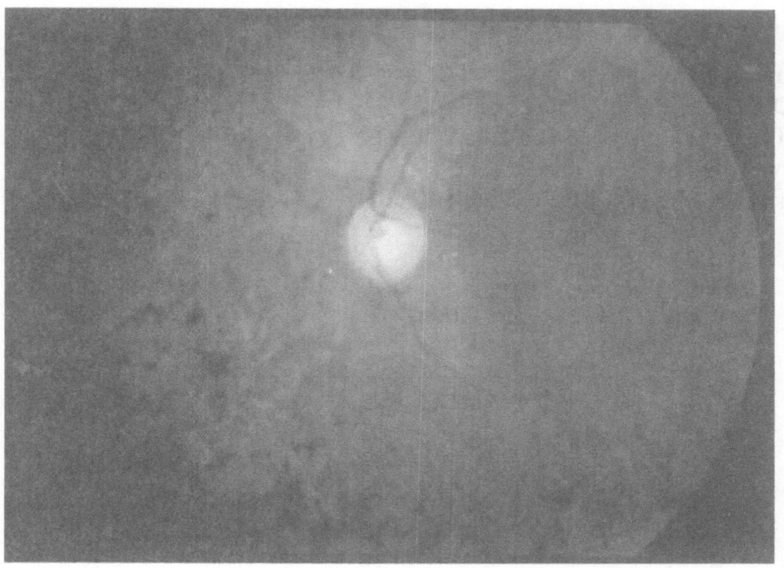

Figure 2. Retinal changes of adRP family affected members.

Table 1. Mean refractive error for RP and normal fundus family members

Refractive error			
RP Patients Descriptive Statistics		*NON RP* Patients Descriptive Statistics	
	Refraction		Refraction
Mean	3,214	Mean	–2,150
Std. Dev.	1,980	Std. Dev.	2,461
Std. Error	529	Std. Error	778
Count	14	Count	10
Minimum	1,000	Minimum	–5,750
Maximum	7,000	Maximum	250
# Missing	0	# Missing	0
Variance	3,922	Variance	6,058
Coef. Var.	616	Coef. Var.	–1,145
Range	6,000	Range	6,000
Sum	45,000	Sum	–21,500
Sum Squares	195,625	Sum Squares	100,750
Geom. Mean	2,697	Geom. Mean	•
Harm. Mean	2,273	Harm. Mean	•
Skewness	836	Skewness	–,432
Kurtosis	–,412	Kurtosis	–1,501
Median	3,000	Median	–1,250
IQR	2,750	IQR	4,750
Mode	1,750	Mode	0
10% Tr. Mean	3,083	10% Tr. Mean	–2,000
MAD	1,375	MAD	1,375

The mean spherical equivalent in the affected members of the family was +3214 diopters (DS ± 198 diopters) as compared to –2150 diopters (SD ± 2461 diopters) for the non affected members difference between two groups was highly statistically significantive ($p = 0.0012$) (Table 1).

The axial eye lengths of the affected members were 21.529 ± 1.164 mm all of them

Table 2. Axial length mean and standard deviation for RP and normal fundus family members

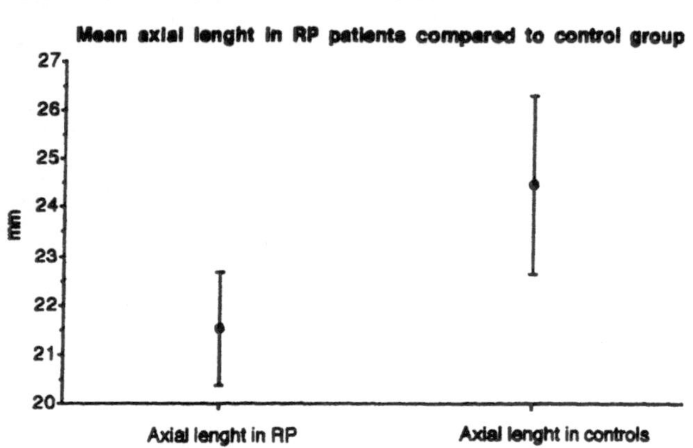

presented values less than 22 mm which suggest that much of hyperopia was probably axial (Table 2).

Non Affected members presented a mean axial length of 24.485 ± 1.843 mm, this difference was highly statistically significative ($p = 0.0012$) (Table 2).

Data shown clearly evidentiate a difference either in refractive error and in axial length in both groups, statistical analysis points out that this difference could be related to a different inherited pattern because hyperopic shift clearly cosegregate with RP.

Remains to discuss if these cosegregation is due to an unique gene defect that encode a metabolic problem for retinal photoreceptors and development of the eye or to two different gene that are located very near on the same chromosome.

Unfortunately the molecolar genetics analysis are still in progress and we do not have at the moment any data about the involved gene.

Even if myopic refractive shift is frequently associated in rod-cone tapeto-retinal degeneration, and very rarely hyperopia is reported in RP families we have reported another family autosomal dominant retinitis pigmentosa (ADRP) associated with hyperopic refraction.

REFERENCES

1. Mackey D.A. et al. Glaucoma inheritance study in tasmania (GIST II autosomal dominant optic atrophy (chr #3q) resembling low tension glaucoma in some family members. Inv. Ophthalmol. 37: (3) suppl. 146. (1996).
2. Petruhkin K. et al. Mapping the juvenile glaucoma gene. Inv. Ophthalmol. 37: (3) suppl. 149. (1996).
3. Wiggs J.L., Allingham R. et al. The juvenile glaucoma gene on 1q21–q31 is not associated with primary open angle glaucoma. Inv. Ophthalmol. 37: (3) suppl. 2071. (1996).
4. Raymond V. et al. Variable espressivity of autosomal dominant open angle glaucoma at the GLC1A locus. Inv. Ophthalmol. 37: (3) suppl. 2079. (1996).
5. Forte R., Jannaccone A., et al. Myopic shift of refractive error in retinitis pigmentosa. Inv. Ophthalmol. 37: (3) suppl. 2283. (1996).
6. Berson E.L., Rosner B., Sandberg M.A., Dryja T.P. Ocular findings in patients with Autosomal Dominant Retinitis Pigmentosa and a Rhodopsin gene defect (Pro-23-His). Arch Ophthalmol 1991;109:93–101.
7. Fishman G.A., Stone E.M., Gilbert L.D., et al. Ocular findings associated with a rhodopsin gene codon 58 transversion mutation in autosomal dominant retinitis pigmentosa. Arch Ophthalmol 1991;109:1387–1393.
8. Heckenlively J.R., Rodriguez J.A., Daiger S.P. Autosomal Dominant Sectoral Retinitis Pigmentosa. Two families with transversion mutation in codon 23 of rhodopsin. Arch Ophthalmol 1991;109:84–91.
9. Jacobson S.G., Kemp C.M., Sung C.H., Nathans J. Retinal function and rhodopsin levels in autosomal dominant retinitis pigmentosa with rhodopsin mutations. Am J Ophthalmol 1991; 112: 256–271.
10. Stone E.M., Kimura A.E., Nichols B.E., et al. Regional distribution of retinal degeneration in patients with the Proline to Histidine mutation in codon 23 of the Rhodopsin gene. Ophthalmology 1991;98:1806–1813.
11. Andreasson S., Ehinger B., Abrahamson M., Fex G. A six generation family with autosomal dominant retinitis pigmentosa and a rhodopsin gene mutation (Arg135Leu). Ophthalmic Paediatr Genet 1992;13:145–153.
12. Fishman G.A., Stone E.M., Gilbert L.D., Sheffield V.C. Ocular findings associated with a rhodopsin gene codon 106 mutation. Glycine-to-Arginine change in autosomal dominant retinitis pigmentosa. Arch Ophthalmol 1992;110:646–653.
13. Fishman G.A., Vandenburgh K., Stone E.M., et al. Ocular findings associated with rhodopsin gene codon 267 and codon 190 mutations in dominant retinitis pigmentosa. Arch Ophthalmol 1992;110:1582–1588.
14. Niemeyer G., Trüb P., Schinzel A., Gal A. Clinical and ERG data in a family with autosomal dominant RP and Pro-347-Arg mutation in the rhodopsin gene. Doc Ophthalmol 1992;79:303–311.

15. Del Porto G., Vingolo E.M., David D., et al. Clinical features of autosomal dominant retinitis pigmentosa associated with the Gly-188-Arg mutation of the rhodopsin gene. In: Hollyfield J.G., La Vail M.M., Anderson R.E., eds. Retinal Degeneration: Clinical and Laboratory Applications. New York: Plenum Publishing Corporation, 1993;91–101.
16. Kim R.Y., Al-Maghtheh M., Fitzke F.W., et al. Dominant retinitis pigmentosa associated with two rhodopsin gene mutations. Leu-40-Arg and an insertion disrupting the 5′-splice junction of exon 5. Arch Ophthalmol 1993;111:1518–1524.
17. Apfelstedt-Sylla E., Kunisch M., Horn M., et al. Ocular findings in a family with autosomal retintis pigmentosa and a frameshift mutation altering the carboxyl terminal sequence of rhodopsin. Brit J Ophtalmol 1993;77:495–501.
18. Owens S.L., Fitzke F.W., Inglehearn C.F., et al. Ocular manifestations in autosomal dominant retinitis pigmentosa with a Lys-296-Glu rhodopsin mutation at the retinal binding site. Brit J Ophthalmol 1994;78:353–358.
19. Shastry B.S. Retinitis pigmentosa and related disorders: phenotypes of rhodopsin and peripherin/RDS mutations. Am J Med Genet 1994;52:467–474.
20. Pannarale M.R. La determinazione della correzione astigmatica. Indicazioni e tecniche d'uso di quadranti astigmometrici personali. Boll Ocul 1977;56:71–77.
21. Fishman G.A., Anderson R.J., Lourenco P. Prevalence of posterior subcapsular lens opacities in patients with retinitis pigmentosa. Br J Ophthalmol 1985;69:263–266.
22. Onori P., Vingolo E.M., Forte R., et al. Vitreous changes in patients affected with retinitis pigmentosa. ARVO abstracts. Invest Ophthalmol Vis Sci 1993;34 (Suppl.):1074.
23. Iannaccone A., Rispoli E., Vingolo E.M., et al. Correlation between Goldmann perimetry and maximal ERG response in retinitis pigmentosa. Doc Ophthalmol 1995 (in press).
24. Rispoli E., Iannaccone A., Vingolo E.M. ERG in retinitis pigmentosa: low-noise recording techniques. Doc. Ophthalmol. 1994;88:27–37.
25. Marmor M.F., Arden G.B., Nilsson S.E., Zrenner E. Standard for Clinical Electroretiniography. Arch Ophthalmol 1989;107:816–819.

25

INTRAOCULAR PRESSURE IN TAPETORETINAL DEGENERATIONS

E. M. Vingolo, C. Palitto, L. Scipioni, and L. Iacobelli

University "La Sapienza" of Rome
Institute of Ophthalmology
Medical Retina Section

1. INTRODUCTION

Several Authors developed an evaluation of intraocular pressure (IOP) in normal population and in different groups of individuals, always measured with applanation tonometry that seems to be the most affordable and repeatible system.

Also in patients affected by Retinitis Pigmentosa, since its discovery, were reported elevated values of intraocular pressure, with different rate of incidence.

Neverthless in older data not very sensitive techniques may be indicated as cause of misdiagnosis of elevated IOP in RP patients either with qualitative (Schmidt, 1949) and quantitative with Schiotz Tonometer (Panepinto and Lo Cascio, 1951—Rizzoli, 1953). In more recent studies with applanation tonometry (Ruiz, 1963—Kobge, 1975—Ammannati, 1984 and ourselves, 1993) were reported discordant data in which low range IOP measurement were accompained to data with high IOP levels.

Unfortunately in RP patients other common diagnostic tools and useful techniques in Primary Open Angle Glaucoma (POAG) as Visual field testing, Nerve Fiber Analysis and cup/disk ratio are unusable for evaluating the patients.

This is particularly evident because in tapetoretinal degenerations, Visual Field is early constricted, optic disk is waxy or subatrophic even in young age and frequently there is a nerve fiber layer damage due to degenerative progress.

More helpful may be gonioscopy and some provocative tests like darkness test or ibopamine test, but since these are based on evaluation of IOP they lack in efficacy.

In this view seems to be a very crucial point to know which is the mean value and the normal range of IOP in RP patients, in order to evaluate every possible pathologic deviations from this for start adequate therapy to prevent every further damage due to a glaucomatous stage.

Retinal Degenerative Diseases and Experimental Therapy, edited by Hollyfield *et al.*
Kluwer Academic / Plenum Publishers, New York, 1999.

On these basis we evaluated two different groups of patients: one affected by RP and the other of patients selected as representative of normal population, age, and sex matched.

2. MATERIALS AND METHODS

In all RP patients diagnosis was obtained following the examinations demonstrating: bilateral involvement, rod function damage, and progressive evolution of retinal damage in previous three years.

From the Medical Retina Section of our Institute of Ophthalmology were selected 711 patients affected by different forms of tapetoretinal degeneration (342 male and 369 female) with mean age of 39,21 ± 14,13 years (range 11–76).

As a control group from a medical screening of workers in a main company were selected 689 subjects for a total amount of 1378 eyes (435 male and 254 female) with mean age of 42,02 ± 11,33 years (range 21–65), were excluded patients with medical history of glaucoma or high IOP.

All RP Patients were evaluated according a standard protocol as reported in previous study including Best Corrected Visual Acuity measurement, Anterior and Posterior segment examination and for RP group ERG recording according the ISCEV standard procedure.

After a wash-out period of two weeks, IOP was evaluated random in different time of the day with the following procedure: Use of a Goldmann applanation tonometer on Haag-Streit Slit Lamp, after instillation of topical anesthesia (oxybuprocain 0.4%) and fluorescein stripes application, mean of two reads from each eye.

All RP patients were divided in two groups: typical RP (tRP) in which were included patients with isolated ocular involvement, and syndromic RP in which were included patients with Usher's (U's) syndrome (77 patients—11,18%) and Laurence Moon Bardet Biedl's (LMBB's) Syndrome (13 patients—1,60%).

Data Analysis was performed with Statview 4.0 data pack for Macintosh Computers calculating for each group main descriptive statistics and data frequency distribution.

3. RESULTS

In RP patients mean IOP value was 14,317 ± 2,011 (range 10–22) mmHg while in Normal sample was 13,970 ± 2,131 (9–20) mmHg (Fig. 1).

Only 27 patients (3.91%) in RP group presented IOP values of 20–22 mmHg that we considered borderline for glaucoma, of these only 3 patients (0,44%) were found affected by POAG after provocative tests (Fig. 2).

Value of IOP presented normal distribution both in control and in RP group. Unpaired two tails Student T test did not showed significative differences between two groups ($p = 0.135$).

Analysis of three RP population subgroups (Fig. 3) showed a mean value of IOP for tRP of 14.255 ± 2,014, for U's of 14.102 ± 1,991, and finally for LMBB's of 14.367 ± 2,007.

Also in this case no statistically significative difference was found between these three subgroups and with control group.

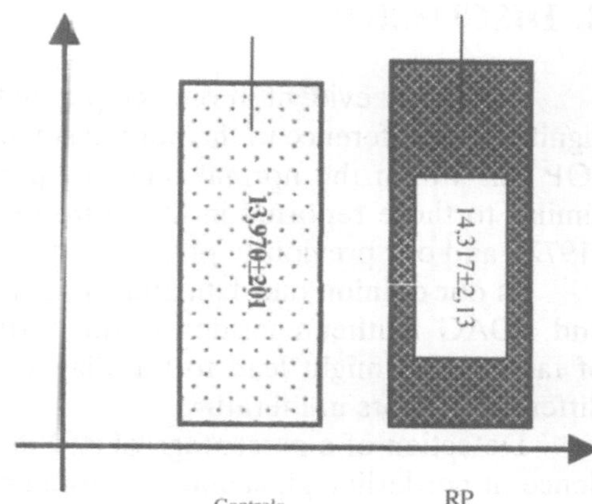

Figure 1. IOP in Normal and in RP patients.

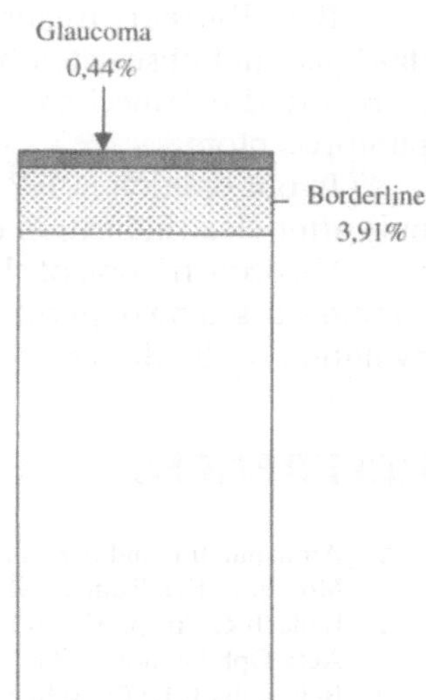

Figure 2. Rate of glaucoma in RP.

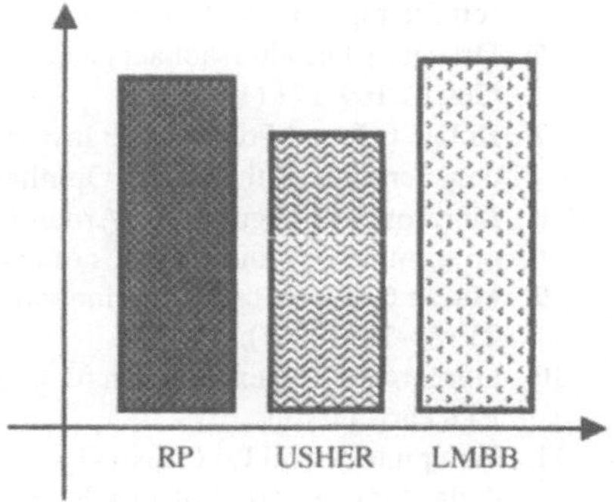

Figure 3. IOP in RP, Usher and LMBB.

4. DISCUSSION

As clearly evident in our RP population there was a mean value of IOP without significative difference to the normal subjects. Moreover in the majority of our patients IOP was within the normal range of population. Considering actual data these are similar to those reported in literature by Draeger (1959), Ourgard (1959), Bengston (1972), and our previously published data (1993) on a lower sample of RP patients.

In our opinion our data allow to conclude that no correlation exists between RP and POAG neither considering the IOP value nor evaluating the whole complex of factors that might lead to the diagnosis of glaucoma as previously postulated by different Authors in literature.

Detection of a percentage of 3.91% of borderline IOP is within the rate of incidence of borderline glaucoma in normal population (2–4%), also the increase of IOP in clearly glaucomatous patients in 0.4% of cases is due probably to a casualty.

Finally we think that is important to take a consideration about the role of antiglaucomatous drugs as beta-blockers that are widely used in therapy for high IOP levels.

Beta Blockers treatment on eyes of RP patients, in which choroidal vessels are damaged and obstructed by the degenerative process, may produce vasoconstriction, as reported by Bucci and coll. (1990), in the choriocapillaris with further damage in photoreceptors.

In our opinion in RP patients a treatment for high IOP values need to be started only after clear indication of relevant POAG diagnosed by provocative tests.

Vasoconstriction of choroidal microcirculation on a degenerated retina may cause a rapid worsening of performances of visual function in RP patients resulting in a faster evolution of the disease.

REFERENCES

1. Ammannati P. and Porciatti V.: La fluoroangiografia nella retinite pigmentosa. LXIV SOI pp. 281–312 Monduzzi ED. Roma. (1984).
2. Badeeb O., Trope G., and Musarella M.: Primary angle closure glaucoma and retinitis pigmentosa. Acta Ophthalmol. **71**(6):727–732 (1993).
3. Bellarminoff L.: Ein seltener fall von retinitis pigmentosa complicirt durch glaukom. Arch. J. Augenh. **27**:53–61 (1983).
4. Bengston B.O.: Some factors affecting the distribution of intraocular pressure in a population. Acta Ophthalmol. **50**:33–46 (1972).
5. Bucci M.G., Pescosolindo N., Mariotti S.P., and Lentini F.P.: Comportamento dell'ampiezza del polso oculare dopo l'instillazione di••-bloccanti. Boll. Ciotola G.: Sull'associazione del glaucoma con la retinite pigmentosa Boll. Ocul. **29**:489–495 (1950).
6. Draeger j: Die altersabhaengigen Anderungen des normalen augennendruckes. Ber. Dtsch. Ophthalm. Ges. **62**:169–173 (1959).
7. Kobge O.I. and Follmann P.: Investigation into the acqueous humor dynamics in primary pigmentary degeneration of the retina. Ophthalmologica **171**:169–175 (1975).
8. Marmor M.F., Aguirre G., Arden G., Berson E.L, and Birch D.G.: Retinitis Pigmentosa Symposium on terminology and methods of examination Ophthalmology **90**:126–131 (1983)
9. Mazza C.: Su un caso di retinopatia pigmentosa associata a sordomutismo e glaucoma. Ann. Ottalm. **93**:180–187 (1967).
10. Ourgard A.G.: L'exploration functionnelle de l'oeil glaucomateaux. Soc. Fr. Ophtalm. 229–244 Masson ED. Paris (1961).
11. Panepinto V. and Lo Cascio G.jr.: Sull'Associazione della degenerazione pigmentaria della retina con il glaucoma. Giorn. Ital. Oftalmol. **4**:209–219 (1951).

12. Rizzoli E.: Alcuni rilievi gonioscopici in casi di retinosi pigmentosa. Rass. Ital. Ottalmol. **22**:168–175 (1953).
13. Ruiz Barranco F.: Consideraciones sobre el problema de las degeneraciones tapetoretinianas. Arch. Soc. Ophthalmol. Hisp.-Amer. **23**:193–203 (1963).
14. Schmidt-Rimpler: Retinitis Pigmentosa Associated with glaucoma Am. J. Ophthalm. **32**:1337–1350 (1949).
15. Vingolo E.M., Pescosolido N., and Iannaccone A.: Rilievo epidemiologico della pressione endoocu-lare in un gruppo di pazienti affetti da Retinite Pigmentosa. Boll. Ocul. **72**: (Suppl. 1) 123–134 (1993).
16. Ocul. **69**:285–297 (1990).

Section II

GENE THERAPY FOR PHOTORECEPTOR RESCUE

26

RIBOZYMES DIRECTED AGAINST MESSENGER RNAs ASSOCIATED WITH AUTOSOMAL DOMINANT RETINITIS PIGMENTOSA

Lynn C. Shaw, Patrick O. Whalen, Kimberly A. Drenser, Wei-MingYan, William W. Hauswirth, and Alfred S. Lewin

Department of Molecular Genetics and Microbiology
and Gene Therapy Center
Box 100266
University of Florida
Gainesville, Florida, 32605

SUMMARY

Ribozymes are RNA enzymes that can be modified to cleave almost any target RNA. Because of this versatility, they are useful in digesting transcripts of dominant mutant genes leading to retinal disease, including ADRP. One approach to gene therapy for this disease would be to reduce the expression of the mutant opsin that interferes with the accumulation of active rhodopsin and is responsible for a substantial fraction of ADRP. We have developed ribozymes designed to cleave mutant mRNA molecules leading to several opsin missense and nonsense mutations that cause retinal degeneration in human and in animal models. These include the P23H, G90D, S334 termination, and the P347S mutations. Both hairpin and hammerhead ribozymes have been tested. The efficiency of cleavage depends on the nucleotide triplet at the cleavage site and its context in the mRNA. All ribozymes tested are capable of cleaving short synthetic targets, longer cloned targets and full length opsin mRNA containing the mutant target sequence. In no case has cleavage of the wild-type opsin been detected. For these reasons, ribozymes hold promise as sequence-specific tools for gene therapy for ADRP. Animal tests of these ribozymes are underway using Adeno-Associated Virus vectors to deliver DNA copies of the ribozymes to affected transgenic animals.

Retinal Degenerative Diseases and Experimental Therapy, edited by Hollyfield *et al.*
Kluwer Academic / Plenum Publishers, New York, 1999.

INTRODUCTION

The prospect of gene therapy for retinitis pigmentosa (RP) is a consequence of the rapid identification of mutations responsible for many forms of retinal degeneration during the past decade.[1–4] According to Daiger and colleagues (*http://www.sph.-uth.tmc.edu/Retnet/*), 58 genetic loci are associated with RP. Mutations in 18 of these loci cause the autosomal dominant form of the disease (ADRP) which represents approximately one quarter of all RP cases. Five of the disease genes have been cloned, and each encodes a protein of the visual cycle or of the rod disc membrane. Mutations in rhodopsin and the structural protein peripherin/rds are responsible for more than one third of ADRP, and more than 80 human mutations have been mapped in the gene for rod opsin alone.

Clearly, therapy will require more than the identification of disease genes. Accurate *animal models* of the human disease are needed to understand the pathogenic process and to permit preclinical testing of the therapy. Fortunately, in the case of RP, a variety of naturally occurring and transgene-induced disease models exist in rodents, dogs and, in pigs.[5–10] Some of these models recapitulate the pathogenic mechanism of RP over an accelerated period, making them ideal for testing therapies. In addition to a model system, however, there must be a *delivery vector* for introducing a therapeutic gene. The delivery system might include everything from naked DNA to artificial chromosomes, but two viral vectors, derived from Adeno-Associated Virus (AAV) and Lentivirus have shown considerable promise in terms of their ability to transduce photoreceptor and ganglion cells without inducing inflamation or a severe immune response.[11,12] While developers of the HIV-1-based lentiviral vectors have safety issues to address, AAV vectors are based on a non-pathogenic agent and are currently in use for human clinical trials.

Gene therapy for a loss-of-function mutation usually consists of replacing the defective gene with a functional copy of the gene. Treatment of a dominant genetic disease like ADRP is conceptually more difficult, since allele-specific ablation of a mutant gene or its product appear to be required. *Ribozymes*, which are enzymes comprised of RNA, are well-suited to this purpose. Naturally occurring ribozymes include RNase P, self-splicing introns, and small self-cleaving virus-associated sequences. Members of all three types can be harnessed to cleave or to repair mutant mRNA, but the small virus-derived activities, the so-called hammerhead and hairpin ribozymes, are the easiest to design for the purpose of gene therapy. These ribozyme contain a highly-structured catalytic domain linked to substrate binding "arms" that can be modified to bind almost any mRNA.[13,14] The basic strategy, therefore, is to adjust the binding domain of a ribozyme so that it base-pairs with the mutated mRNA and not the wild-type transcript. This will lead to cleavage of that mRNA, leaving the wild-type message intact (Fig. 1). Following internal cleavage by the ribozyme, the mutant mRNA will be degraded rapidly by cellular enzymes. This approach requires that the amount of normal protein produced from the wild-type gene is sufficient to preserve photoreceptor function. In the case of rhodopsin, at least, this appears to be true.

For such a strategy to succeed, target sites for ribozyme cleavage must either be created by the mutation in the disease gene or must occur in close proximity to the disease causing mutation. The minimal target for a hairpin ribozyme consists of the 4 nucleotide sequence $^{5'}$BNCUG$^{3'}$ where B represents any nucleotide but adenosine and N represents any of the four ribonucleotides. This sequence occurs frequently enough to appear several times in most messages, but not enough that it will necessarily

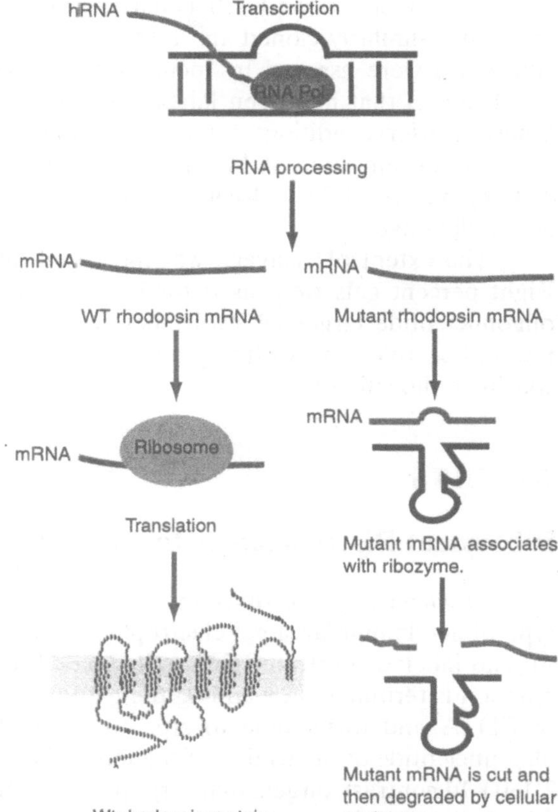

Figure 1. Ribozymes block translation of mutant rhodopsin mRNA. Following RNA splicing and nuclear transport, wild-type opsin mRNA is translated in the cytoplasm of rod cells and inserted into membranes. Hammerhead and hairpin ribozymes recognize mutant forms of opsin mRNA and cleave it. This internal cleavage leads to rapid degradation of the mutant RNA by cellular nucleases.

coincide with a disease causing mutation. The hammerhead target is usually represented as $^{5'}NUX^{3'}$ in which N can be any nucleotide and X can be anything but guanosine. Some variation of this base triplet occurs through mRNA sequences, so that it is often possible to locate a cleavage site near a missense mutation associated with disease. Even when this is not the case, it may be possible to use the techniques of *in vitro* selection to isolated ribozymes that cleave with the desired specificity.[15,16]

We have modified hammerhead and hairpin ribozymes so that they are able to cleave the mutant opsin mRNA present in several animals models of ADRP. In this paper we discuss issues concerning their analysis *in vitro*, including the choice of target site, assay conditions, and the value of these tests for predicting behavior *in vivo*.

METHODS

The methods employed in these experiments have been described elsewhere in detail.[17] Briefly, DNA sequences encoding ribozymes were prepared by chemical synthesis and were cloned in a plasmid vector (pHC) in association with a promoter for T7 RNA polymerase. These plasmids were linearized by treatment with restriction endonuclease MluI and were transcribed *in vitro*. The resultant ribozyme transcripts were purified by phenol/chloroform extraction in the presence of 0.5% SDS and gel filtration on Sephadex G-50 (Pharmacia). They were incubated with target RNA molecules similarly cloned and transcribed in the presence of ^{32}P-UTP. Alternatively, ribozymes were assayed by incubation with chemically synthesized ribonucleotides (14–17 mers) that had been labeled at the 5′ end using ^{32}P-ATP and polynucleotide kinase. Buffer conditions for cleavage reactions included 40 mM Tris HCl, pH 7.5 and varying amounts of $MgCl_2$. Reactions were routinely in volumes of 10 μl and at temperature of 37°C. Reactions were stopped by adding 8 M urea, 25 mM EDTA gel-loading dye.

The extent of cleavage was measured on acrylamide gels containing 8M urea. Eight percent gels were used for long targets (>40 nt) and 15% gels were used for oligonucleotide targets. A Molecular Dynamics Phosphorimager was used to quantitate radioactivity in substrate and product bands, using a calibration curve of known specific radioactivity.

RESULTS

Ribozymes Discriminate between Mutant and Wild-Type Opsin Targets

In vitro assays for ribozyme activity are conveniently performed using the same type of acrylamide/urea gel electrophoresis as is used for DNA sequencing. Typically, ^{32}P-end labeled substrate RNA is incubated for varying intervals with ribozyme. The reaction is terminated by adding the gel loading buffer, which contains large amounts of EDTA and formamide to stop the reaction. If such reactions contain a target oligonucleotide of the wild-type sequence, then, typically, no digestion is observed (Fig. 2, left). In contrast, targets that form perfect helices with ribozymes are readily cleaved (Fig. 2, right). This reaction shows the activity of a hammerhead ribozyme designed to cleave the mRNA present in a mouse model of retinitis pigmentosa containing the G90D rhodopsin mutation. This target contains 4 nucleotide changes relative to the

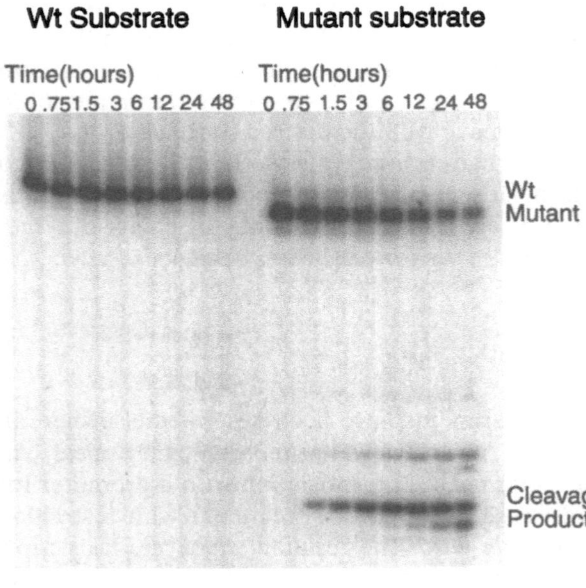

Figure 2. A hairpin ribozyme cleave the P23H opsin target but not the wild-type target. End-labeled 17-nucleotide RNA molecules were incubated with a hairpin ribozyme designed to cleave the VPP target. These targets were based on the sequence of wild-type mouse opsin (left) or the transgene present in the VPP mouse model of ADRP. Incubations were at 37°C and contained 20 mM $MgCl_2$. Samples were withdrawn at the times indicated, and all samples were separated on a 15% acrylamide/8 M urea gel.

normal mouse opsin sequence, two in codon 90 and two silent mutations in codons 91 and 92. While cleavage by this ribozyme was slow (70% cleavage in 24 hours), it was completely specific. No cleavage of wild-type target was detected in the same time course.

While in this instance, wild-type target was not cleaved, another hammerhead ribozyme designed for a G90D target cleaved mutant and wild-type RNA with nearly equal efficiency (*data not shown*). This ribozyme targeted the sequence $^{5'}$GGTC*TTC*G-GAG**AT**$^{3'}$. The underlined TTC represents the NUX triplet preceding the cleavage site, and the 3′ terminal AT in bold are the only differences between the wild-type and mutant sequence (wild-type = GA). Since the dinucleotide change was at the end of the target sequence and the four bases on either side of the cut site were the same in mutant and wild-type targets, no selectivity was obtained with this hammerhead. It is important, therefore, for target selectivity, to design ribozyme arms containing mismatches with the wild-type message as near as possible to the cleavage site.

Secondary Structure Affects the Rate of Cleavage in an RNA Target

We have assayed ribozymes using long (34–90 nt) RNA targets produced enzymatically from DNA templates or using short (14–17 nt) chemically synthesized RNA oligonucleotides. The latter always cleave more rapidly than the former (Fig. 3). In this experiment a hammerhead ribozyme designed to cleave the mutated mRNA present in the P347S mutant opsin gene was incubated with a cloned target (34 nt) or an oligo target (17 nt). While digestion of the longer target exhibited a one hour lag, cleavage of the synthetic target was robust and immediate. Nearly 90% of the target was digested within 10 minutes. Since both targets have the same cleavage site and flanking sequences, the difference in cleavage rates is probably attributable to the secondary structure of the longer molecule. Extensive internal structure can limit accessibility of the ribozyme to the cleavage site, and this issue remains a concern for the utilization

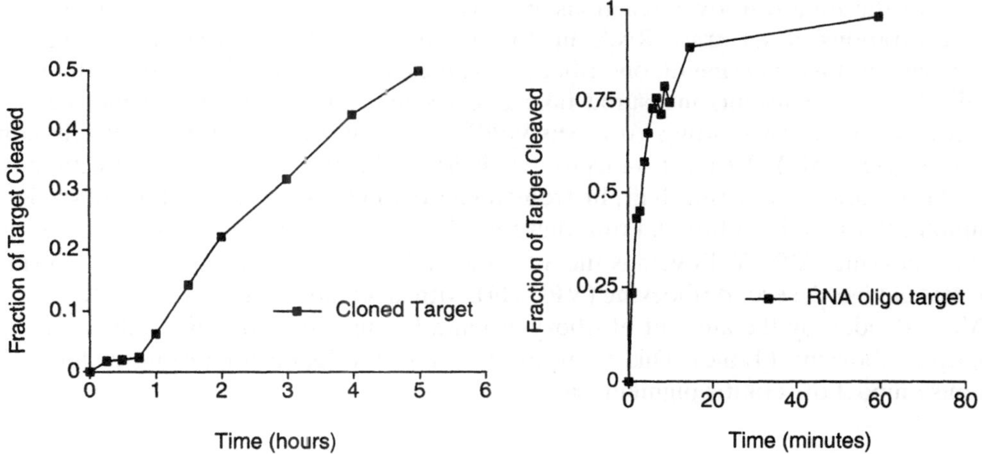

Figure 3. Ribozymes cleave oligonucleotide targets better than longer targets. A hammerhead ribozyme designed to cleave the transgene present in the P347S pig, was incubated with a 34 nucleotide cloned target that was transcribed *in vitro* (A) or with a 17 nucleotide synthetic RNA (B) for the times indicated. Incubations were at 37°C and contained 20 mM $MgCl_2$. The fraction of target cleaved was determined following separation of the products on a denaturing acrylamide gel.

Figure 4. Reaction pathway for cleavage by hammerhead ribozymes.

of ribozymes *in vivo*.[18,19] Nevertheless, the oligonucleotides provide expedient targets for testing ribozyme kinetics, especially in determining the cleavage rates independently of on- and off- rates for longer RNA molecules.

Kinetic Analysis of Ribozymes

The overall rate of ribozymes is described by several microscopic rate constants[20,21] that characterize the binding and dissociation of substrate (k_1 and k_{-1}), the cleavage and ligation steps (k_2 and k_{-2}), and the dissociation of one or both cleavage products (Fig. 4). The hammerhead ribozyme does not re-ligate cleaved products, though the hairpin ribozyme can ligate products with the same or greater rate than it cleaves.[16] We have used two methods to analyze rates of ribozyme reaction: single turnover kinetics and multi-turnover kinetics. In single turnover kinetics, the ribozyme is in excess of substrate: We used 10 nM target oligonucleotide and 100 to 700 nM ribozyme. Under these circumstances, all of the substrate is bound by ribozyme, and each ribozyme will cleave at most one substrate molecule. By plotting the initial velocity (Vo) of the reaction versus Vo/[R], we can obtain an estimate of the forward rate constant for cleavage (k_2) (Fig. 5A).

In the multi-turnover reactions, ribozyme at 5 nM is incubated with increasing concentrations of substrate RNA, up to 300 nM. Since the substrate is in excess of ribozyme in this experiment, one ribozyme can cleave more than one substrate molecule. The rate constants measured take into account all of the steps in the reaction pathway (Fig. 4). By plotting Vo versus Vo/[S] we obtain Vmax and the Michaelis constant K_M. (Fig. 5B.) These constants for two hairpin ribozymes are compared with those of the naturally occurring hairpin from tobacco ringsport virus satellite RNA in a bimolecular reaction (Table). From these results, we opted to discontinue study of the one ribozyme (VPP-WT) versus the VPP target, because this ribozyme was too slow to test *in vivo*. A second ribozyme (VPP-H4), with an extended helix 4, exhibited a k_{CAT} (V_{MAX} divided by the amount of ribozyme) and K_M similar to the naturally occurring hairpin ribozyme (Table). This ribozyme is being developed for testing in the VPP mouse model of retinitis pigmentosa.[22]

Table 1

Ribozyme	k_{cat} (min^{-1})	K_M (nM)
α-VPP Hairpin	0.09	258
α-VPP-H4 Hairpin	0.18	45
TRSV Hairpin	2.0	34

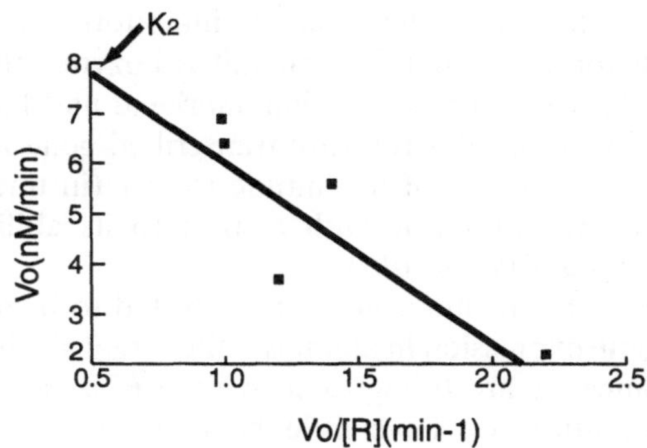

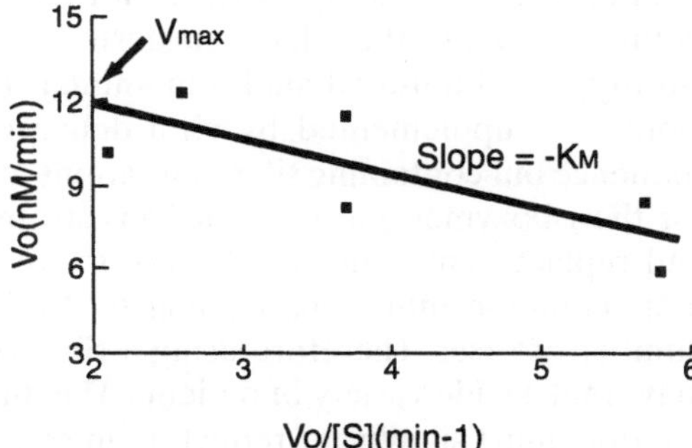

Figure 5. Eadie-Hofstee plots for the kinetic analysis of single turnover (top) and multiple turnover (bottom) of a hairpin ribozyme.

Testing on Full-Length Opsin mRNA

Before animal testing, it is desirable to test ribozymes for the ability to cleave the target sequence in the context of full-length mRNA and in an extract containing the full complement of retinal mRNA species. We have done this by extracting total RNA from retinas of animals that bear the mutant transgene under study. These animals have been developed and characterized in the laboratories of Dr. Muna Naash, Dr. Fulton Wong, and Dr. Matthew LaVail. Retinal RNA is incubated with active ribozyme or with a control, inactive ribozyme under cleavage conditions. Thereafter the relative amounts of wild-type and mutant mRNA are quantitated using an RT-PCR assay.[17] In every case, we have found that ribozymes active against synthetic targets are able to cleave the full length message containing the cognate target (*data not shown*). As expected from the experiment described in Fig. 2, cleavage of the full length message is much slower than that of oligos or cloned targets.

DISCUSSION

We have performed these studies in preparation for testing the ribozymes in animal models of RP using viral vectors to deliver copies of ribozyme coding sequence. Our initial results using recombinant AAV delivery vectors and the P23H rat model of La Vail and co-workers is described elsewhere in this volume by Flannery et al. The kinetic characterization of ribozymes may not be a completely accurate predictor of whether they will succeed as inhibitors of gene expression *in vivo*, but it is a good predictor of whether they will fail. Ribozymes that perform poorly in multi-turnover assays relative to the natural hammerhead and hairpin ribozymes are unlikely to block the expression of a robustly transcribed gene like opsin. In our hands, a ribozyme more than an order of magnitude slower (in terms of k_{cat}) showed no difference from an inactive ribozyme with respect to its ability to block retinal degeneration *in vivo* (*unpublished results*).

Even ribozymes demonstrated to be very active in these assays may not inhibit gene expression in animals if the target site is inaccessible within the full length mRNA. Since we are trying to cleave the RNA at specific locations, i.e. at or near the site of mutation, we have little flexibility in choosing targets sites. Fortunately, there are methods available to test accessibility of particular sites in mRNA *in vitro*,[18] and these should be used prior to costly animal or cell culture tests.

Our approach of using AAV-delivered ribozymes to reduce the expression of dominant genes leading to RP is not the only possible application of this technology. Humphries and others have suggested using ribozymes to block the expression both wild-type and mutated alleles in animals (and patients) with ADRP. These ribozymes would be supplemented by viral delivery of an allele encoding the normal protein sequence but containing silent mutations at the nucleotide level so that the target site(s) for the ribozyme(s) are absent. This strategy has the appeal that one set of ribozyme and replacement gene could serve to treat a spectrum of opsin mutations. Even the most common mutations leading to ADRP (e.g. P23H) represent a relatively small number of cases. Therefore using a ribozyme in conjunction with a replacement gene may treat a wide variety of patients. The difficulties with this approach are also obvious: not only must significant reduction in gene expression be achieved with the ribozyme, but the replacement gene must be adequately regulated and expressed. In any case, it is clear that careful biochemical characterization of the ribozymes is a prerequisite to their use for gene therapy.

Kinetic constants for hairpin ribozymes in bi-molecular reactions. K_{cat} is the $V_{MAX}/[R]$. K_M is the Michaelis constant and reflects the substrate concentration at ½ V_{MAX}. The VPP hairpin contains a helix 4 of 3 nucleotide pairs. In VPP-H4, this helix has been extended to 6 base pairs and the 3 base loop has been converted to a tetraloop. TRSV hairpin is the naturally occurring hairpin ribozyme derived from the satellite RNA of tobacco ringspot virus.

ACKNOWLEDGMENTS

This work was support by grants from the March of Dimes Foundation, the Foundation Fighting Blindness and the National Institutes of Health. We wish to thank Mr. James Thomas Jr. for expert technical assistance; Dr. Muna Naash and Dr. Fulton Wong for providing sequences and retinas from transgenic animals.

REFERENCES

1. Daiger, S.P., L.S. Sullivan, and J.A. Rodriguez, 1995, Correlation of phenotype and genotype in inherited retinal degeneration. *Behav. Brain Sci.* 18:452–467.
2. Heckenlively, J.R. 1988. Retinitis Pigmentosa. J.B. Lippencott, Philadelphia.
3. Sung, C.H., C.M. Davenport, J.C. Hennessey, I.H. Maumensee, S.G. Jacobson, J.R. Heckenlively, R.

Nowakowski, R. Fishman, P. Gouras, and J. Nathans, 1991, Rhodopsin mutations in autosomal dominant retinitis pigmentosa. *Proc. Natl. Acad. Sci.* 88:6481–6485.

4. Berson, E, 1993, Retinitis pigmentosa. *Inv. Ophth. Vis. Sci.* 34:1659–1676.
5. Naash, M.I., J.G. Hollyfield, M.R. Al-Ubaidi, and W. Baehr, 1993, Simulation of human autosomal dominant retinitis pigmentosa in transgenic mice expressing a mutated murine opsin gene. *Proc. Natl. Acad. Sci.* 90:5499–5503.
6. Steinberg, R.H., J.G. Flannery, M.I. Naash, P. Oh, M.T. Matthes, D. Yasumura, C. Lau-Villacorta, J. Chen, and M.M. LaVail, 1996, Transgenic rat models of inherited retinal degeneration caused by mutant opsin genes. *Inv. Ophth. Vis. Sci.* 37:S698.
7. Li, Z.Y., F. Wong, J.H. Chang, D.E. Possin, Y. Hao, R.M. Petters, and A.H. Milam, 1998, Rhodopsin transgenic pigs as a model for human retinitis pigmentosa. *Invest. Ophthalmol. Vis. Sci.* 39:808–819.
8. Petters, R.M., C.A. Alexander, K.D. Wells, E.B. Collins, J.R. Sommer, M.R. Blanton, G. Rojas, Y. Hao, W.L. Flowers, E. Banin, A.V. Cideciyan, S.G. Jacobson, and F. Wong, 1997, Genetically engineered large animal model for studying cone photoreceptor survival and degeneration in retinitis pigmentosa [see comments]. *Nat. Biotechnol.* 15:965–970.
9. Akhmedov, N.B., N.I. Piriev, S. Pearce-Kelling, G.M. Acland, G.D. Aguirre, and D.B. Farber, 1998, Canine cone transducin-gamma gene and cone degeneration in the cd dog [In Process Citation]. *Invest. Ophthalmol. Vis. Sci.* 39:1775–1781.
10. Zhang, Q., G.M. Acland, C.J. Parshall, J. Haskell, K. Ray, and G.D. Aguirre, 1998, Characterization of canine photoreceptor phosducin cDNA and identification of a sequence variant in dogs with photoreceptor dysplasia. *Gene* 215:231–239.
11. Flannery, J.G., S. Zolotukhin, M I. Vaquero, M.M. LaVail, N. Muzyczka, and W.W. Hauswirth, 1997, Efficient photoreceptor-targeted gene expression in vivo by recombinant adeno-associated virus. *Proc. Natl. Acad. Sci. USA* 94:6916–6921.
12. Miyoshi, H., M. Takahashi, F.H. Gage, and I.M. Verma, 1997, Stable and efficient gene transfer into retina using an HIV-based lentiviral vector. *Proc. Natl. Acad. Sci.* 94:10319–10323.
13. Birikh, K.R., P.A. Heaton, and F. Eckstein, 1997 The structure, function, and application of the hammerhead ribozyme. *Eur. J. Biochem.* 245:1–16.
14. Butcher, S.E., J.E. Heckman, and J.M. Burke, 1995, Reconstitution of hairpin ribozyme activity following separation of functional domains. *J. Biol. Chem.* 270:29648–29651.
15. Vaish, N.K., P.A. Heaton, O. Fedorova, and F. Eckstein, 1998, In vitro selection of a purine nucleotide-specific hammerhead-like ribozyme. *Proceedings Of The National Academy Of Sciences Of The United States Of America* 95:2158–2162.
16. Berzal-Herranz, A., S. Joseph, and J.M. Burke, 1992, In vitro selection of active hairpin ribozymes by sequential RNA-catalyzed cleavage and ligation reactions. *Genes. Dev.* 6:129–134.
17. Drenser, K.A., A.M. Timmers, W.W. Hauswirth, and A.S. Lewin, 1998, Ribozyme-targeted destruction of RNAs associated with ADRP. *Inv. Ophth. Vis. Sci.* 39:in press.
18. Birikh, K.R., Y.A. Berlin, H. Soreq, and F. Eckstein, 1997, Probing accessible sites for ribozymes on human acetylcholinesterase RNA. *RNA.* 3:429–437.
19. Heidenreich, O., S.H. Kang, D.A. Brown, X. Xu, P. Swiderski, J.J. Rossi, F. Eckstein, and M. Nerenberg, 1995. Ribozyme-mediated RNA degradation in nuclei suspension. *Nucleic. Acids. Res.* 23:2223–2228.
20. Fedor, M.J. and O.C. Uhlenbeck, 1992, Kinetics of intermolecular cleavage by hammerhead ribozymes. *Biochemistry.* 31:12042–12054.
21. Nesbitt, S., L.A. Hegg, and M.J. Fedor, 1997, An unusual pH-independent and metal-ion-independent mechanism for hairpin ribozyme catalysis. *Chemistry & Biology* 4:619–630.
22. Goto, Y., N.S. Peachey, H. Ripps, and M.I. Naash, 1995, Functional abnormalities in transgenic mice expressing a mutant rhodopsin gene. *Invest. Ophthalmol. Vis. Sci.* 36:62–70.

RIBOZYME-MEDIATED GENE THERAPY FOR AUTOSOMAL DOMINANT RETINAL DEGENERATION

John G. Flannery,[1] Alfred S. Lewin,[2,4] Kimberly A. Drenser,[2] Shimpei Nishikawa,[5] Douglas Yasumura,[5] Matthew M. LaVail,[5] and William W. Hauswirth[2,3,4]

[1] Departments of Vision Science, Neuroscience, and Biophysics
University of California, Berkeley, California
[2] Department of Molecular Genetics and Microbiology
[3] Department of Ophthalmology
[4] Gene Therapy Center
University of Florida College of Medicine, Gainesville, Florida
[5] Departments of Anatomy and Ophthalmology
Beckman Vision Center
University of California School of Medicine, San Francisco, California

1. SUMMARY

Ribozymes targeted to a point mutation (histidine for proline substitution at position 23) in rhodopsin were evaluated as therapeutic agents for photoreceptor degeneration in an animal model of autosomal dominant retinitis pigmentosa (ADRP). Recombinant adeno-associated virus (AAV) vectors incorporating a proximal bovine rod opsin promoter were used to transfer either hairpin or hammerhead ribozyme genes to photoreceptors. AAV was administered by subretinal injection at postnatal day (P) 15 to rats expressing a rod opsin transgene with a proline-to-histidine substitution in exon 1 at codon 23 of the rhodopsin gene (P23H). *In vivo* levels of normal and mutant RNAs were measured by allele-specific RT-PCR at P60. Eyes were examined by histopathology and morphometric analysis, as well as by ERG at approximately P60, P75 or P90. Expression of either a hammerhead or hairpin ribozyme markedly slowed the rate of photoreceptor degeneration for at least 3 months as determined by outer nuclear layer thickness as well as by inner and outer segment length. Catalytically inactive control ribozymes had a significantly smaller effect on the retinal degeneration. Minimal rescue effects were seen in retinas injected with either AAV

Retinal Degenerative Diseases and Experimental Therapy, edited by Hollyfield *et al.*
Kluwer Academic / Plenum Publishers, New York, 1999.

containing a gene for green fluorescent protein rather than the ribozyme gene or buffer only. The level of mutant opsin RNA relative to wild-type RNA was also reproducibly lower in ribozyme-treated retina. Finally, ERG comparisons of ribozyme-treated versus control eyes demonstrated functional rescue. We conclude that ribozyme-directed cleavage of mutant mRNAs appears to be a potentially effective therapy for autosomal dominant retinal degeneration.

2. INTRODUCTION

The major challenge in developing a successful therapy for dominant disease is to deliver an agent that will suppress the action of the dominant allele without adversely affecting the normal physiology of the photoreceptor. Suppression could occur at the protein level, through use of mutant-specific antibodies, or at the DNA level, by correction of the mutated allele. Intervention at the level of RNA, however, is more practical than direct gene correction, based on current technology, and can block expression earlier than antibodies. While antisense oligonucleotides have been successfully employed to inhibit the expression of disease genes, catalytic RNA molecules, ribozymes, are more likely to be allele-specific inhibitors, because of their target binding domains can be short (<12 nucleotides). In addition, ribozymes have the advantage of catalytic turnover–multiple target RNA molecules can be cleaved by a single ribozyme. Ribozymes have been shown to work in vivo in a variety of model systems,[1–8] and we reasoned that they could be used to block the synthesis of mutated proteins that lead to retinal degeneration. Our approach has been to deliver DNA copies of ribozymes specific for the mutant allele using recombinant AAV and the proximal rhodopsin promoter to direct synthesis in rod photoreceptor cells.[9] Here we demonstrate the feasibility of this approach using a rat model of ADRP caused by a P23H rhodopsin transgene.

3. METHODS

3.1. rAAV-Ribozyme Constructs

Recombinant AAV constructs were based on the pTR-UF2 vector[10] that we previously used to direct GFP expression specifically to rat photoreceptors.[9] We replaced the 472-bp murine rod opsin promoter with a 691 bp fragment of the proximal bovine rod opsin promoter. In addition, the gfp reporter gene was replaced with the ribozyme genes. The bovine promoter fragment contains three proximal promoter elements and the endogenous transcriptional start site at its 3′ end.[11] In preliminary experiments, we established that this promoter element would support high efficiency, rat photoreceptor-specific expression *in vivo* (unpublished results). Active and inactive ribozymes were designed, tested, and cloned as described by Drenser et al.[12] Each ribozyme gene was followed by an internally cleaving hairpin ribozyme derived from plasmid pHC[13] resulting in ribozyme cassettes of 140–152 bp. Self-cleavage at the internal cutting site in the primary ribozyme RNA leaves identical 3′ ends on each mature ribozyme. The ribozyme cassette was preceded by an intron derived from SV40 and followed by a polyadenylation signal in order to promote nuclear export and stability of the ribozyme, which may be essential for efficacy.[14] Plasmid DNA constructs were packaged into AAV particles by standard procedures.[10] Recombinant AAV titers were determined using both an infectious center assay[15] and a DNase resistant physical particle assay employing a quantitative, competitive PCR of the *neo*^{r} gene contained within all rAAV-ribozyme particles. Each of the four rAAV-ribozyme virus preparations contained 10^{10} to 10^{11} DNase resistant particles per ml and 10^8 to 10^9 infectious center units per ml. Contaminating helper adenovirus and wild-type AAV, assayed by serial dilution cytopathic effect or infectious center assay respectively[64] were less than five orders of magnitude lower than rAAV.

3.2. Subretinal Injection of rAAV

Line three albino transgenic rats (P23H-3) on an albino Sprague-Dawley background[16] (produced by Chrysalis DNX Transgenic Sciences, Princeton, NJ) were injected at the ages of P14 or P15. Animals were anesthetized by ketamine/xylazine injection, and a local anesthetic (proparacain HCl) was applied topically to the cornea. An aperture was made through the inferior cornea of the eye with a 28-gauge needle. Subretinal injections of 2 µl were then made by inserting a blunt 32 gauge needle through the opening and delivering the rAAV suspension into the subretinal space in the posterior retina. The intent was to inject into the subretinal space of the posterior superior hemisphere, but we sometimes found histologically that the injection site was located just inferior to the optic nerve head. The opposite eye was either uninjected, injected subretinally with PBS, with control rAAV containing no ribozyme or inactive forms of the ribozymes; in some cases PBS or inactive ribozymes were compared to uninjected eyes in the same animals. Injections were performed under direct observation with an operating microscope, and the subretinal location of the injection was visualized; however, the precise extent of the retinal detachment induced by the injection usually could not be resolved optically. This anterior approach of subretinal injection results in a 20–40% incidence of cataracts, regardless of the agent injected, including PBS, and such animals were not used for ERG analysis.

3.3. Electroretinography

Rats, dark adapted overnight and then in dim red light, were anesthetized with intramuscular injections of xylazine (13 mg/kg) and ketamine (87 mg/kg). Full-field scotopic ERGs were elicited with 10-µsec flashes of white light and responses were recorded using a UTAS-E 2000 Visual Electrodiagnostic System (LKC Technologies, Inc., Gaithersburg, MD). The corneas of the rats were anesthetized with a drop of 0.5% proparacaine hydrochloride, and the pupils were dilated with 1% atropine and 2.5% phenylephrine hydrochloride. Small contact lenses with gold wire loops were placed on both corneas with a drop of 2.5% methylcellulose to maintain corneal hydration. A silver wire reference electrode was placed subcutaneously between the eyes and a ground electrode was placed subcutaneously in the hind leg. Stimuli were presented at intensities of −1.1, 0.9, and 1.9 log cd m^{-2} at 10-second, 30-second, and 1-minute intervals, respectively. Responses were amplified at a gain of 4,000, filtered between 0.3 to 500 Hz and digitized at a rate of 2,000 Hz on two channels. Three responses were averaged at each intensity. The a-waves were measured from the baseline to the peak in the cornea-negative direction, and b-waves were measured from the cornea-negative peak to the major cornea-positive peak. For quantitative comparison of differences between the two eyes of rats, the values from all the stimulus intensities were averaged for a given animal.

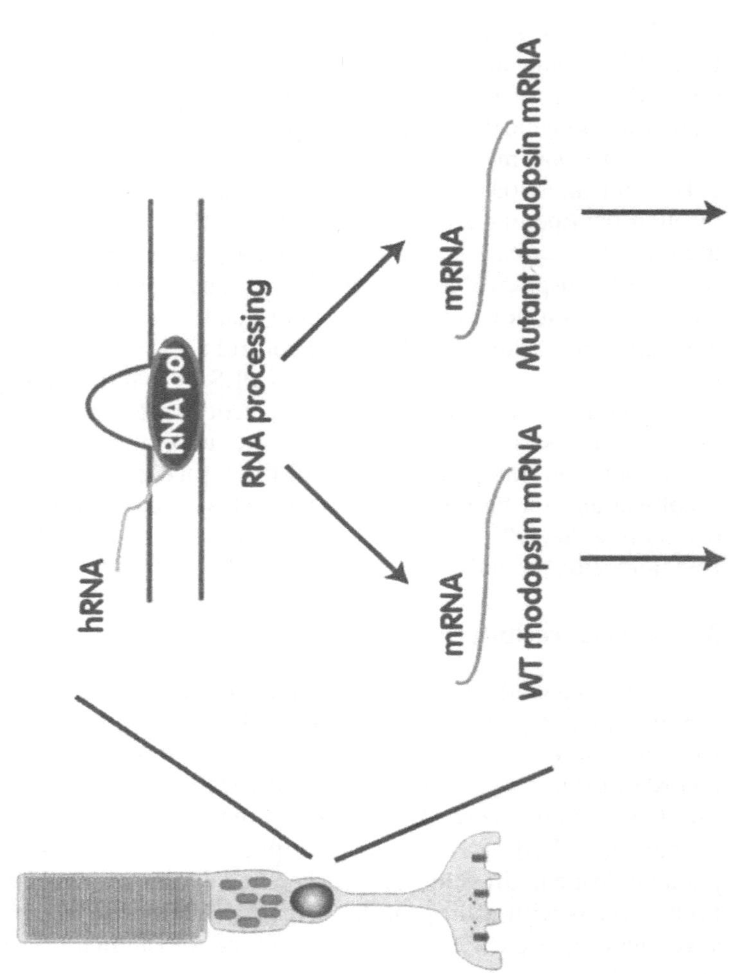
hRNA
RNA pol
RNA processing
mRNA
WT rhodopsin mRNA
mRNA
Mutant rhodopsin mRNA

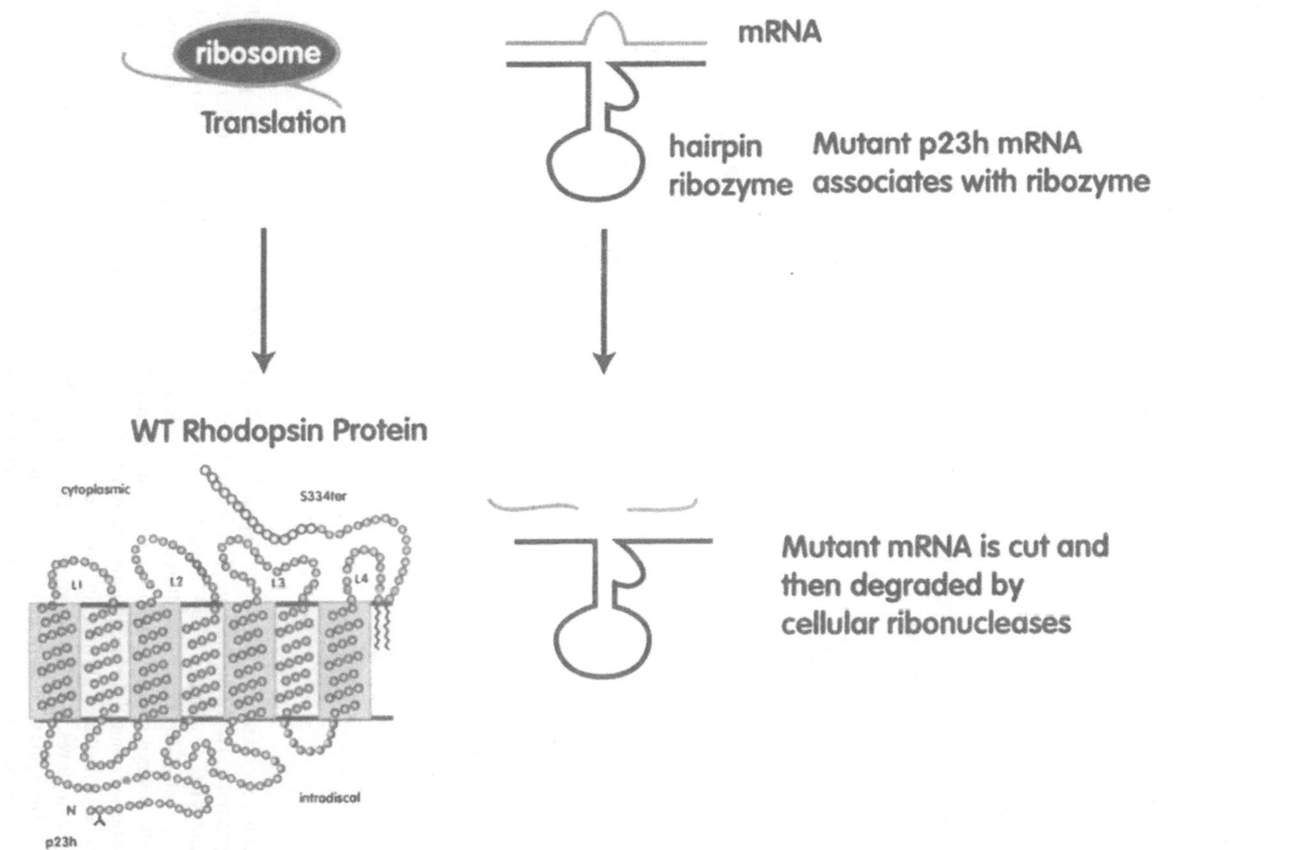

Figure 1. Ribozyme action in vivo. The AAV vector transduces the photoreceptor, transferring the passenger gene to the chromosomal DNA of the rod. The opsin promoter drives expression of the ribozyme gene in the rod, creating P23H hairpin or hammerhead ribozymes within the cell. These small catalytic antisense RNA molecules hybridize and process the complementary P23H RNA target. After processing the target, the ribozymes are released intact and are able to hybridize and process other mutant RNA molecules. Cellular ribonucleases recognize the cleaved mRNA by its loss of an initiation or termination codon, and degrade it. A reduction in the quantity of P23H mRNA generates a proportional reduction in the amount of mutant P23H opsin expressed.

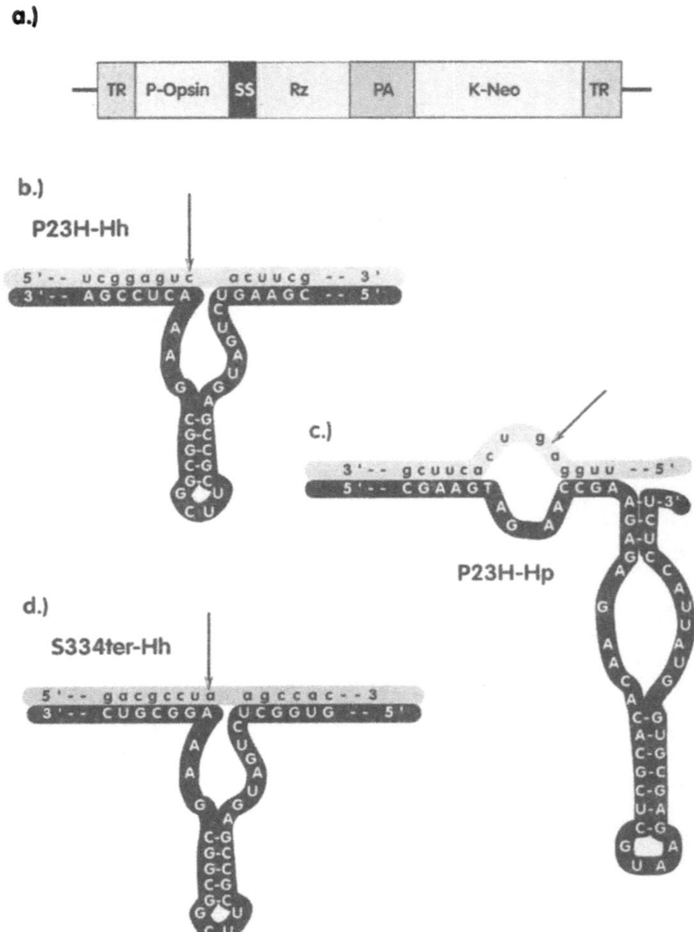

Figure 2. AAV ribozyme cassette. a. Ribozyme expression is driven by 691-bp fragment of the proximal bovine rod opsin promoter. The ribozyme gene (Rz) was followed by an internally cleaving hairpin ribozyme derived from plasmid pHC resulting in a 140–152 bp ribozyme cassettes. The ribozyme cassette was preceded by a splice donor/acceptor site (SS) and followed by a polyadenylation signal (PA) in order to promote nuclear export of the ribozyme, which may be essential for efficacy. T he *neo*^{r} gene (K-Neo) and the thymidine kinase promoter is included for quantitation of AAV particle titers. Chromosomal integration of the AAV passenger gene is mediated through its inverted terminal repeat (TR). Secondary structure of the ribozyme-substrate complexes: b. P23H hammerhead ribozyme. c. P23H hairpin ribozyme. d. S334ter hammerhead ribozyme. Ribozymes are shown with the complementary sequences of the mutant opsin mRNA. The arrow marks the cleavage site, and dashes indicate Watson-Crick base pairs.

3.4. Histology and Morphometry

The rats were euthanized by overdose of carbon dioxide inhalation and immediately perfused intracardially with a mixture of mixed aldehydes (2% formaldehyde and 2.5% glutaraldehyde). Eyes were removed and embedded in epoxy resin, and one μm thick histological sections were made along the vertical meridian.[17] Tissue sections were aligned so that the ROS and Müller cell processes crossing the inner plexiform layer were continuous throughout the plane of section. This assured that the sections were not oblique, and the thickness of the ONL and lengths of RIS and ROS were measured as described elsewhere.[18] Briefly, 54 measurements of each layer or structure were made at set points around the entire retinal section. These data were either averaged to provide a single value for the retina, or plotted as a distribution of thickness or length across the retina. We also compared the greatest three contiguous values for ONL thickness in each retina, to determine if any region of retina (e.g., nearest the injection site) showed proportionally greater rescue; although most of these values were slightly greater than the overall mean of all 54 values, they were no different from control values than the overall mean. Thus, the overall mean was used in the data cited, since it was based on a much larger number of measurements.

3.5. RT-PCR

For quantification of opsin mRNA retina from ribozyme injected or control eyes, retina were isolated without fixation and total RNA immediately extracted using the Rneasy Minikit (Qiagen, Santa Clarita, CA). RT-PCR was performed using the Pharmacia First-Strand cDNA synthesis kit employing oligo dT as the primer. Wild-type and transgene opsin cDNAs were amplified using a three primer system described by Drenser et al.[12] Primers specific for β-actin cDNA[19] were included in each reaction for internal standardization.

4. RESULTS

4.1. In Vitro Evaluation of Ribozymes

Ribozymes were designed to recognize and cleave the unique transcript produced by the P23H transgene. The mutant target sequence "5′-UCGG*A*G**U***C*U**A**CUUCG-3′" contains two differences from the wild-type mRNA (indicated in bold). The hairpin ribozyme (Hp11) cleaved 3′ to the first adenosine residue (underlined) and the hammerhead ribozyme (Hh13) cleaved 3′ to the central cytosine residue (underlined). Control ribozymes (Hp11i and Hh13i, respectively) retained the targeting domains but contained fatal flaws in their catalytic domains. *In vitro*, the active hammerhead ribozyme (Hh13) was able to cleave 20% of the P23H target within 10 min. of incubation and by 5 hours greater than 80% was converted to the expected products. The specificity and detailed kinetic properties of these ribozymes *in vitro* have been described elsewhere.[12] In multiturnover experiments, both ribozymes exhibited kinetic constants (K_m and k_{cat}) similar to those of naturally occurring ribozymes. The two active ribozymes produced negligible cleavage of the wild-type transcript even in the presence of high $MgCl_2$ concentrations. Control ribozymes (Hp11i and Hh13i) containing inactivating mutations in their catalytic domains were without measurable activity on any substrate. Using total RNA derived from retinas of P23H rats on P62, both the hairpin and the hammerhead ribozymes were able to cleave the mRNA product of the mutant transgene selectively.

4.2. In Vivo Studies

Expression of the mutated opsin transgene begins at about postnatal day (P) 5 in rats, leading to a gradual death of photoreceptor cells. The (P23H)3 rats develop an apparently normal retina up to P15, although there are somewhat more pyknotic photoreceptor nuclei in the outer nuclear layer (ONL) than in non-transgenic control rats. Thereafter, death of photoreceptor cells is almost linear until about P60, resulting in loss of about 40% of the photoreceptors (Fig. 3 lower panel). After P60, the rate of cell loss decreases, until by one year the retinas have less than a single row of photoreceptor nuclei. We injected the rAAV-ribozyme vector into the interphotoreceptor space between the photoreceptors and the adjacent retinal pigment epithelium at P14 or P15. Rats were sacrificed and eyes examined at three time points between P60–P90 (Table 1). At these ages in uninjected control eyes of P23H-3 rats, the ONL thickness, which is an index of photoreceptor cell number,[20] was reduced to about 60% of normal.

Ribozyme-injected eyes showed a modest but significant decrease in the accumulation of transcript derived from the P23H transgene. Control eyes (pooled data from two uninjected and six PBS-injected eyes) exhibited little variation in the level of transgene opsin mRNA (average transgene level was 25.5 ± 3.1% of the wild-type level). Eyes injected with either active ribozyme uniformly exhibited lowered transgene mRNA levels relative to total opsin mRNA in the same eye. Retinas receiving the hairpin ribozyme Hp11 showed a 15.3 ± 3.3% decrease in transgene expression, and those with the hammerhead ribozyme Hh13 showed a decrease of 11.1 ± 5.1% decrease. Such reproducible selective *in vivo* cleavage of mutant mRNA leads to the possibility of a related attenuation in the mutant phenotype.

4.3. Anatomic Rescue

Histologically, eyes injected with the ribozymes retained significantly more photoreceptors at P60, P75, and P90 than uninjected contralateral control eyes. Retinas receiving a subretinal injection of Hh13 at P14–15 retained 88% of the normal ONL

Table 1. Number of eyes measured histologically in ribozyme experiments

Age of Rats	P60	P75	P90
Treatment:			
Normal (non-transgenic)	3	—	3
Hp11 (hairpin)	4	10	4
Hh13 (hammerhead)	4	—	4
Hp11i ("inactive" hammerhead)	6	1	1
Hh13i ("inactive" hairpin)			
PBS	3	3	—
Bovine Opsin Promoter (BOPS)	3	—	—
Uninjected transgenic	16	8	8

P, postnatal day. Animals were killed at these ages ± 3 days.

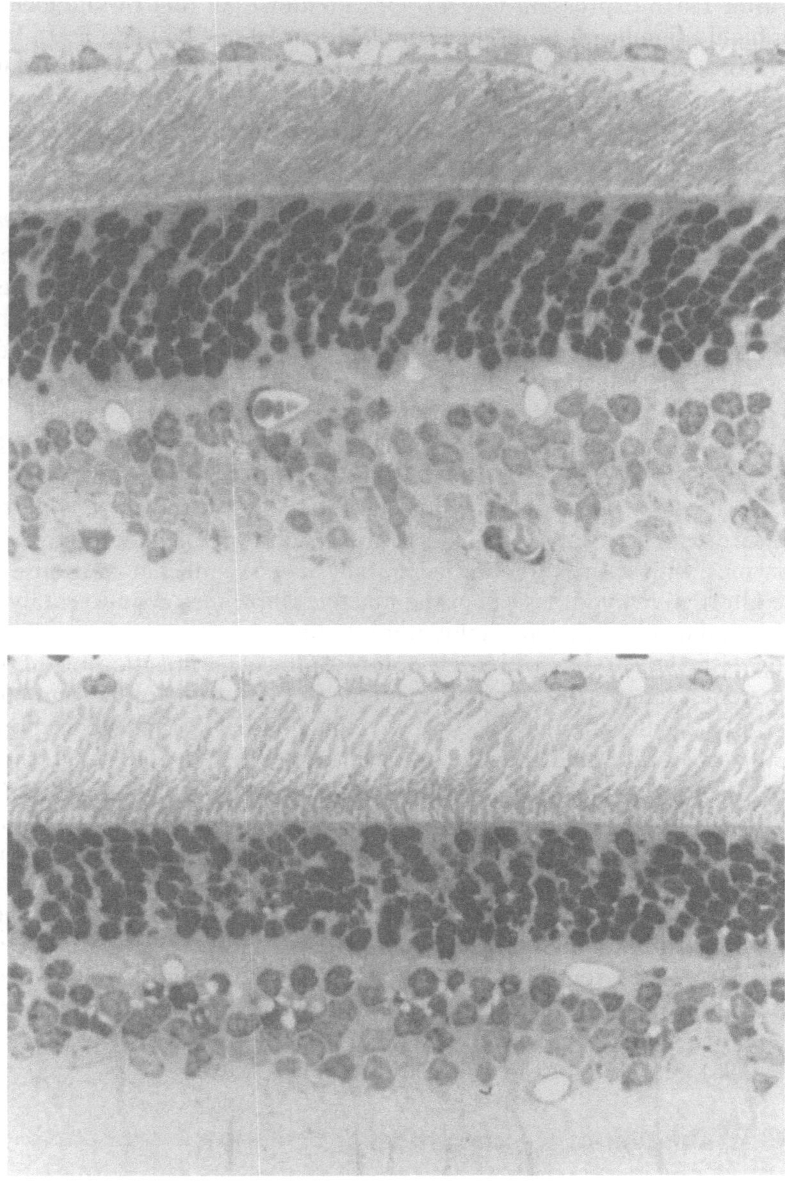

Figure 3. Light micrographs of photoreceptor rescue. Upper panel: Eye injected with Hh13 ribozyme at P14. Slightly greater than 50% more photoreceptor nuclei are present in the ONL than in the opposite, uninjected eye of the same animal (below), and the rod inner and outer segments are also longer. Lower panel: Uninjected eye of a P23H-3 mutant rhodopsin transgenic rat. In this example, the ONL has lost almost half of the normal complement of photoreceptor nuclei, and the rod inner and outer segments are significantly reduced in length from normal.
All are plastic-embedded rat retinas taken at P89 from the posterior inferior retina along the vertical meridian of the eye. Toluidine blue stain. Magnification, X600.

thickness, compared to about 60% in the uninjected controls (Fig. 3). Thus, the ONL thickness after Hh13 expression was 40–43% greater than that of uninjected P23H-3 controls, a highly significant difference ($p = 0.001$ or less at P60 and P90). Injection of the Hp11 ribozyme also resulted in significant rescue when compared to controls, with preservation of 77–83% of normal ONL thickness. Thus, the ONL thickness after Hp11 expression was 30–39% greater than that of uninjected P23H-3 controls, a highly significant difference ($p < 0.0005$ at all ages).

There was little or no rescue in PBS-injected control eyes ($p > 0.169$ in all cases). This is consistent with our finding that needle injury to the retina in young rats (P14–P15) does not rescue photoreceptors or up-regulate bFGF mRNA expression (M.M. LaVail and W. Cao, unpublished observations) as it does in older rats.[18,21] As a control for possible rescue by the expression of the bovine opsin promoter (BOPS), AAV-BOPS-*gfp* was injected at a titer of 1.75×10^8, similar to the titer used with the AAV-ribozymes. The injection of AAV-BOPS-*gfp* did not rescue photoreceptors. The inactive Hp11i did yield ONL thickness measures greater than uninjected control values, but they were consistently less than that resulting from the active Hp11 and Hh13 ribozymes. The partial rescue by delivery of Hp11i and Hh13i is likely explained by an antisense effect because this ribozyme binds and, without cleaving the target, could interfere with translation or lead to mRNA turnover. Others have noted that even ribozymes with short targeting arms may have significant antisense effects in cells.[22] The alternative hypothesis, that the inactive ribozymes acquire catalytic activity in the intracellular milieu, seems highly improbable.

Along with the survival of more photoreceptor cells, injection of the ribozymes resulted in greater lengths of rod inner segments (RIS) and rod outer segments (ROS). In the case of RIS, the uninjected control retinas had RIS that were about 90% of normal. Both the active and inactive ribozymes resulted in RIS lengths of 98% or greater of the normal length, and about 10–15% longer than uninjected controls. However, due to the short RIS length and degree of variance, only the Hh13 at P60 and P90, the Hp11 at P75 and the Hp11i at P60 were statistically different from uninjected controls ($p < 0.05$). The PBS and AAV-BOPS-*gfp* were indistinguishable from uninjected eyes. The ROS lengths were about 15–25% longer in the ribozyme-injected eyes compared to those in the uninjected control eyes. However, ROS in the ribozyme-injected eyes were, at greatest, only 65–75% of normal, compared to the virtually normal RIS lengths. The ROS of the active ribozymes differed significantly from the uninjected controls ($p < 0.005$ for all, except <0.02 for Hp11 at P90), as did the inactive Hp11i ($p < 0.05$), but none of the others did.

4.4. ERG Evaluation of Rescue Effect

We measured the electroretinographic (ERG) responses from a number of rats (Fig. 4) before some of the eyes were taken. The b-wave of the ERG, which arises from Müller cells and inner retinal neurons, is usually taken as a more reliable measure of retinal function, as the photoreceptor-derived a-wave is often more variable. When recording simultaneously from both eyes of normal rats, the responses measured within 5% of each other for both the b- and a-waves in almost every case (despite significant variation between different animals). Thus, it is highly significant that the active ribozymes resulted in b-waves 30–45% greater than in the opposite control eyes (Fig. 5). An example of the difference between two eyes in ERG amplitudes is shown in Fig. 4a. The inactive ribozymes, Hp11i and Hh13i, also resulted in greater b-waves,

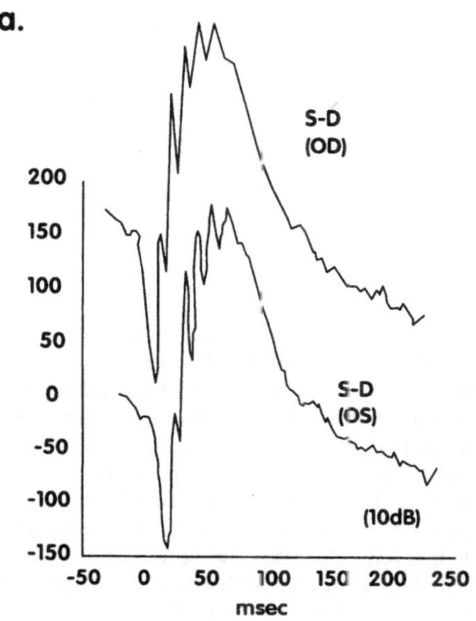

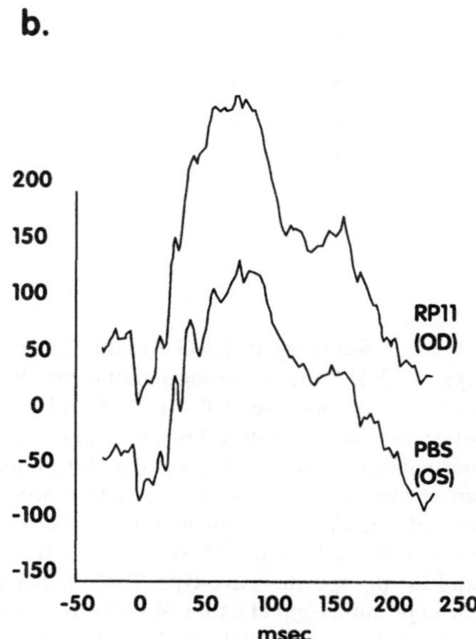

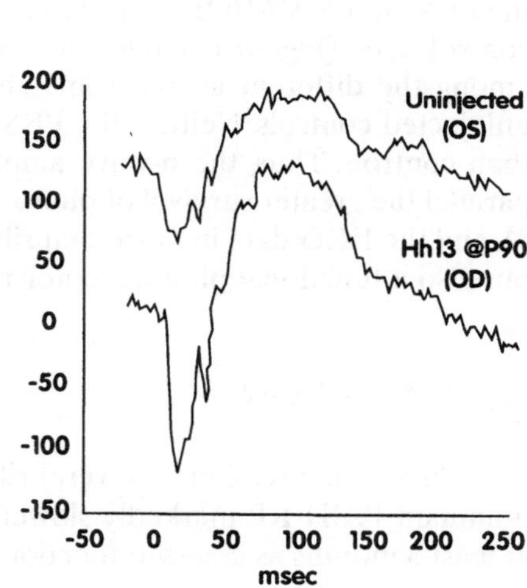

Figure 4. Electroretinographic analysis of P23H-3 animals. a. ERG responses recorded simultaneously from the two eyes of a normal Sprague-Dawley rat at P90. Waveforms from both eyes are very similar and reproducible, typically within 5%. Traces are the average of 3 responses to a stimulus intensity of 1.9 log cd m^{-2}. b. Simultaneous ERG responses from eyes of a P23H-3 transgenic rat at P90 that was injected in one eye with Hh11 hammerhead ribozyme. Erg resposes were not significantly different between the eyes at this stage. Traces are the average of 3 responses to a stimulus intensity of 1.9 log cd m^{-2}. c. ERG responses recorded simultaneously from the two eyes of a P23H-3 transgenic rat at P90 that was injected in one eye with Hh13 hammerhead ribozyme. Both the downward-deflecting a-wave and the upward-deflecting b-wave have much greater amplitudes in the Hh13-injected eye (right) than the uninjected eye (left). Traces are the average of 3 responses to a stimulus intensity of 1.9 log cd m^{-2}.

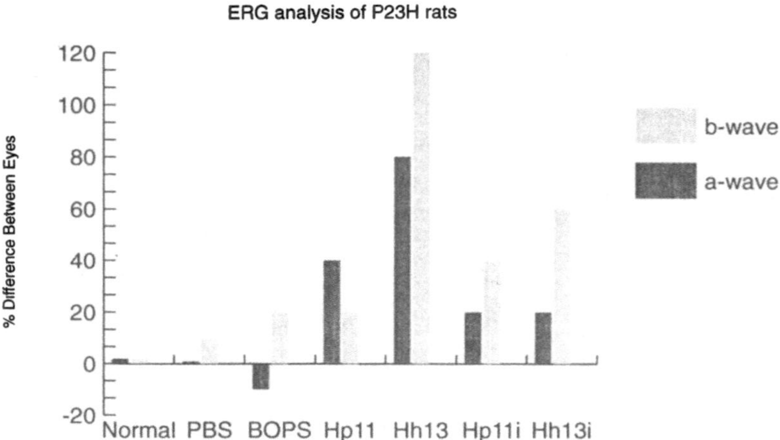

Figure 5. Summary of ERG Results. Percent differences between the uninjected control eye and injected eye of P23H-3 rats following simultaneous bilateral ERG measurements. Ribozymes or control agents were injected into one eye at P14 or P15 and recordings were made at P60, except for the hammerhead Hh13 ribozyme, where 2 rats at P60 and 2 at P90 were measured (data indistinguishable and thus combined). The normal (non-transgenic) rats had nothing injected in either eye and ranged in age from P30–P180. The values are the mean ± S.E.M. of percent differences between the injected and uninjected control eyes of individual recordings, which were made using stimulus intensities of –1.1, 0.9, and 1.9 log cd m^{-2} for the scotopic b-wave of each rat, and 0.9 and 1.9 log cd m^{-2} for the scotopic a-wave, all of which were combined for a given animal and for the specific wave-type. The number of rats is given below the b-wave values. BOPS, AAV-bovine opsin promoter-*gfp* at a titer of 1.75×10^8, similar to that used with the AAV-ribozymes (Hp11, Hh13, Hp11i, and Hh13i); P, postnatal day. (*$p < 0.05$, **$p < 0.01$, ***$p < 0.005$, ****$p < 0.001$ using a two-tailed, paired t-test.)

but the amplitudes were less than 25% greater than those of the control eyes, whereas the PBS- and AAV-BOPS-injected eyes were not significantly different from uninjected control eyes (Fig. 4b). While the a-wave amplitudes were somewhat more variable among the different agents, some showed amplitudes as much as 93% greater than uninjected controls. Neither the PBS- nor the AAV-BOPS-injected eyes were greater than controls. Thus, the b-wave amplitude changes for the active and inactive forms parallel the greater survival of photoreceptor cells as measured by ONL thickness (Fig. 3), and the ERG data indicate that ribozyme expression results in not only a structural, but also a functional photoreceptor rescue.

5. DISCUSSION

In vivo expression of several ribozymes in a transgenic rat model for autosomal dominant P23H RP markedly slowed the rate of rod photoreceptor degeneration for at least 3 months as assessed functionally by electroretinographic analysis, cellularly by preservation of retinal photoreceptors and molecularly by specific reduction in mutant mRNA levels in treated eyes.[23] Intracellular production of ribozymes in rod photoreceptors was achieved by transduction with an rAAV incorporating the same rod opsin promoter shown earlier to support rod specific expression.[9] This study establishes the initial proof-of-principle that a recombinant AAV vector containing the appropriate photoreceptor-specific promoter can mediate functional rescue of a genetically determined retinal photoreceptor degeneration and that ribozymes can be effective against an autosomal dominant genetic disease.

Although antisense oligonucleotides also have been tested as therapeutic agents for treatment of diseases resulting from inappropriate expression of gene products, antisense therapy requires rapid delivery of sufficient quantities of complementary mRNA in the cell to inhibit translation of the target mRNA. Ribozymes have an advantage in this context, because multiple copies of specific RNA molecules are cleaved at a catalytic rate. In addition, ribozymes hold promise for clinical applications because they are extremely allele-specific, exclusively cleaving mutated transcripts while leaving wild-type transcripts intact. Ribozyme catalyzed inhibition of gene expression has been tested in several *in vitro* systems, including plants,[1] mammalian cells,[2,3] and Xenopus oocytes.[4] Ribozymes also work *in vivo*. Adenovirus-delivered ribozymes led to 96% reduction in human growth hormone in transgenic mice expressing this protein,[5] and crosses between transgenic mice expressing bovine β-lactalbumin and mice expressing a ribozyme against that mRNA resulted in substantial reductions in mRNA levels in doubly transgenic progeny.[6] However, ribozymes to date have not been tested against a dominant negative allele *in vivo*.

Our design of rAAV vectors to deliver a functional P23H opsin-specific ribozyme to rod photoreceptors incorporated several features likely to be helpful in achieving photoreceptor rescue. AAV-packaged rod opsin promoter elements generate photoreceptor-specific expression in a significant portion of the rat retina after a single subretinal inoculation.[15] Such constructs result in persistent photoreceptor expression of the passenger gene (>21 months, WW Hauswirth and JG Flannery, unpublished observations) consistent with observations of rAAV-transduced skeletal muscle.[7,8]

The pan-retinal extent of photoreceptor rescue that resulted from a single two μl injection of the rAAV suspension was surprising. From photoreceptor counts, we estimate that there are approximately 10^7 photoreceptors in the rat retina. The number of rod photoreceptors that were rescued is within the limits predicted from our rAAV titers. Recombinant AAV titers were estimated using both an infectious center assay and a physical particle assay. Together they allow us to construct upper and lower bounds for the number of functional rAAV particles in a single two μl injection. The upper bound derives from the DNase resistant particle assay, indicating that two μl of the rAAV-ribozyme virus preparation contained 2×10^7 to 10^8 rAAV. This is an upper bound because not all particles counted are expected to be infectious. The lower bound for rAAV titer is generated by the infectious center assay, indicating 10^6 to $\times 10^7$ rAAV per μl. This titer is likely to be an underestimate of the true retinal infectivity, however, because the infectious particle titers were determined *in vitro* in another cell type (HEK 293). In this assay, the initial event, viral adherence to cells attached to culture plates, is diffusion limited.[24] Therefore, apparent infectious particle titers can be low by several orders of magnitude depending on viral concentration, transduction time and the volume of virus in the media above the cells.[25]

The lateral extent of rescue resulting from a single injection may also be explained by the unique nature of the retinal tissue. For *in vivo* delivery to the photoreceptors, rAAV is injected into extracellular space separating the photoreceptor and retinal pigment epithelium (RPE) layers. The initial volume of extracellular space, approximately 0.5 μl, increases greatly with the 2 μl injection. Following injection, the fluid transport function of the RPE dehydrates this space, reapposing the photoreceptors and RPE and concentrating the rAAV. The detachment of the photoreceptors from the RPE resolves within several hours. During the reattachment process, viral particles are spread laterally through the subretinal space. Such subretinal spreading of transducing rAAV particles has been experimentally confirmed using reporter gene constructs.[15] An additional factor favoring pan-retinal rescue is the increased transducibility of photoreceptors in neonatal rats compared to adults, as was recently shown for lentivirus vectors.[26] An alternative explanation is that rescued photoreceptors could support non-transduced photoreceptors by yet unidentified trophic factors.[27,28]

Our finding that ribozyme-targeted destruction of P23H mutant RNA markedly slows the rate of retinal degeneration in P23H transgenic rats, along with functional preservation of the retina, is the first demonstration of this therapeutic approach in an animal model of a dominantly inherited human disease, as far as we are aware. The successful slowing of inherited photoreceptor degeneration in an animal model of dominant retinitis pigmentosa encourages the further exploration and development of ribozymes as therapeutic agents for various forms of ADRP. Furthermore, because complete removal of mutant mRNA is not necessary to achieve phenotypic rescue (possibly due to the inferred ribozyme cleavage-independent antisense effect), this approach may be generally applicable to other dominantly inherited diseases as well.

ACKNOWLEDGMENTS

This work was supported by grants from the National Eye Institute, The Foundation Fighting Blindness, Research to Prevent Blindness, That Man Might See, Inc., and the March of Dimes Foundation. The authors are indebted to Kye Chestnut and the Vector Core Lab at the University of Florida for preparation of the recombinant viruses used in this study and to Dr. Michael Matthes for technical assistance. Correspondence and requests for reprints should be addressed to John G. Flannery (flannery@socrates.berkeley.edu).

REFERENCES

1. P. Steinecke, T. Herget, and P.H. Schreier, 1992, Expression of a chimeric ribozyme gene results in endonucleolytic cleavage of target mRNA and a concomitant reduction of gene expression in vivo., *EMBO. J.* **11**:1525–1530.
2. K.R. Birikh, P.A. Heaton, and F. Eckstein, 1997, The structure, function, and application of the hammerhead ribozyme., *Eur. J. Biochem.* **245**:1–16.
3. J.J. Rossi, 1997, Therapeutic applications of catalytic antisense RNAs (ribozymes)., *Ciba. Found. Symp.* **209**:195–204.
4. M. Cotten and M.L. Birnstiel, 1989, Ribozyme mediated destruction of RNA in vivo., *EMBO. J.* **8**:3861–3866.
5. A. Lieber, and M.A. Kay, 1996, Adenovirus-mediated expression of ribozymes in mice., *J. Virol.* **70**:3153–3158.
6. P. L'Huillier, S. Soulier, and M.G. Stinnakre, et al., 1996, Efficient and specific ribozyme-mediated reduction of bovine alpha-lactalbumin expression in double transgenic mice., *Proc. Natl. Acad. Sci.* **93**:6698–6703.
7. X. Xiao, J. Li, and R.J. Samulski, 1996, Efficient long-term transfer into muscle tissue of immunocompetent mice by adeno-associated virus vector., *J. Virol.* **70**:8098–8108.
8. K.J. Fisher, K. Jooss, and J. Alston, et al., 1997, Recombinant adeno-associated virus for muscle directed gene therapy., *Nat. Medicine.* **3**:306–312.
9. J.G. Flannery, S. Zolotukhin, M.I. Vaquero, M.M. LaVail, N. Muzyczka, and W.W. Hauswirth, 1997, Efficient photoreceptor-targeted gene expression in vivo by recombinant adeno-associated virus, *Proceedings of the National Academy of Sciences of the United States of America.* **94**(13):6916–6921.

10. S. Zolotukhin, M. Potter, W.W. Hauswirth, J. Guy, and N. Muzyczka, 1996, A "humanized" green fluorescent protein cDNA adapted for high-level expression in mammalian cells., *J. Virol.* **70**:4646–4654.
11. L.E. DesJardin and W.W. Hauswirth, 1996, Developmentally important DNA elements within the bovine opsin upstream region, *Investigative Ophthalmology and Visual Science.* **37**(1):154–165.
12. K.A. Drenser, A.M. Timmers, W.W. Hauswirth, and A.S. Lewin, 1998, Ribozyme-targeted destruction of RNA associated with autosomal-dominant retinitis pigmentosa, *Investigative Ophthalmology and Visual Science.* **39**(5):681–689.
13. M. Altschuler, R. Tritz, and A. Hampel, 1992, A method for generating transcripts with defined 5′ and 3′ termini by autolytic processing., *Gene.* **122**:85–90 (1992).
14. E. Bertrand, D. Castanotto, and C. Zhou, et al., 1997, The expression cassette determines the functional activity of ribozymes in mammalian cells by controlling their intracellular localization., *RNA.* **3**:75–88.
15. T.R. Flotte and B. Carter, 1995, Adeno-associated virus vectors for gene therapy., *Gene Therapy.* **2**:357–362.
16. R.H. Steinberg, J.G. Flannery, M.I. Naash, J. Chen, M. Matthes, D. Yasumura, and M.M. LaVail, 1996, Transgenic rat models of inherited retinal degeneration caused by mutant opsin genes., *Inv. Ophth. Vis. Sci.* **37**:S698.
17. M.M. LaVail and B.A. Battelle, 1975, Influence of eye pigmentation and light deprivation on inherited retinal dystrophy in the rat., *Exp. Eye Res.* **21**:167–192.
18. E.G. Faktorovich, R.H. Steinberg, D. Yasumura, M.T. Matthes, and M.M. LaVail, 1990, Photoreceptor degeneration in inherited dystrophy delayed by the basic fibroblast growth factor., *Nature.* **347**:83–86.
19. A.M. Timmers, B.R. Newton, and W.W. Hauswirth, 1993, Synthesis and stability of retinal photorecptor mRNAs are coordinately regulated during bovine fetal development., *Exp. Eye Res.* **56**:251–265.
20. J.J. Michon, Z.L. Li, S.N. Zhioura, R.J. Anderson, and M.O. Tso, 1991, A comparative study of methods of photoreceptor morphometry., *Invest. Ophthalmol. Vis. Sci.* **32**:280–284.
21. R. Wen, Y. Song, and T. Cheng, et al., 1995, Injury-induced upregulation of bFGF and CNTF mRNAS in the rat retina., *J. Neurosci.* **15**:7377–7385.
22. R. Hormes, M. Homann, and I. Oelze, et al., 1997, The subcellular localization and length of hammerhead ribozymes determine efficacy in human cells., *Nucleic. Acids. Res.* **25**:769–775.
23. A.S. Lewin, K.A. Drenser, W.W. Hauswirth, S. Nishikawa, D. Yasumura, J.G. Flannery, and M.M. LaVail, 1998, Ribozyme rescue of photoreceptor cells in a transgenic rat model of autosomal dominant retinitis pigmentosa, *Nature Medicine.* **4**(8):967–71.
24. N. Mittereder, K.L. March, and B.C. Trapnell, 1996, Evaluation of the concentration and bioactivity of adenovirus vectors for gene therapy., *J. Virol.* **70**:7498–7509.
25. C. Nyberg-Hoffman, P. Shabram, W. Li, D. Giroux, and E. Aguilar-Cordova, 1997, Sensitivity and reproducibility in adenoviral infectious titer determination., *Nature Medicine.* **3**:808–811.
26. H. Miyoshi, M. Takahashi, F. Gage and I. Verma, 1997, Stable and efficient gene transfer into the retina using an HIV-based lentiviral vector, *PNAS.* **94**:10319–10323.
27. S. Mohand-Said, D. Hicks, and M. Simonutti, et al., 1997, Photoreceptor transplants increase host cone survival in the retinal degeneration (rd) mouse., *Ophthalmic Res.* **29**:290–297.
28. P.C. Huang, A.E. Gaitan, Y. Hao, R.M. Petters, and F. Wong, 1993, Cellular interactions implicated in the mechanism of photoreceptor degeneration in transgenic mice expressing a mutant rhodopsin gene, *Proceedings of the National Academy of Sciences of the United States of America.* **90**(18):8484–8488.

28

TARGETING MÜLLER CELLS FOR GENE THERAPY

Gene Regulation Studies

Vijay Sarthy

Department of Ophthalmology
Northwestern University Medical School
300 E. Superior Street
Chicago, Illinois 60611

SUMMARY

Retinal support cells such as Müller cells and RPE are excellent candidates for targeting gene therapy vectors. However, the lack of knowledge of gene regulatory mechanisms in Müller cells has limited their use for designing cell type-specific expression vectors for gene therapy. As a first step in this direction, our recent research has focused on the identification of genetic regulatory elements for two Müller cell-specific genes—the glial intermediate filament protein (GFAP) gene and the cellular retinaldehyde-binding protein (CRALBP) gene. Experimental evidence from transfection assays suggest that promoter-proximal sequences are sufficient to drive expression of GFAP and CRALBP genes in Müller cells. However, studies with GFAP-transgenic mice suggest that GFAP gene transcription in Müller cells is complex, and that an inducible, Müller cell-specific enhancer is likely to control GFAP expression. Identification of the enhancer should provide a inducible, promoter system for expressing foreign proteins in the retina.

INTRODUCTION

Gene regulation studies are not only important for understanding basic biological processes such as development and differentiation, but are crucial for designing cell type-specific vectors for targeted delivery in gene therapy. In the retina, most recent

Retinal Degenerative Diseases and Experimental Therapy, edited by Hollyfield *et al.*
Kluwer Academic / Plenum Publishers, New York, 1999.

work on gene regulation has focused on genes expressed in photoreceptor cells.[1–5] Gene expression in other retinal cell types have not been well studied. The first studies of gene regulation in the retina were concerned with the role of cell-cell interactions in regulating gene expression in Müller cells.[6] These investigations showed that contact-interaction with neurons was essential for high level expression of the Müller cell-specific enzyme, glutamine synthetase.

Many Müller cell-specific genes are not normally expressed, or expressed at a low level during development. However, they are strongly expressed in response to trauma, retinal detachment, ischemia, and rod/cone degeneration.[7,8] Some of upregulated proteins are well-known cytokines and growth factors such as ciliary neurotrophic factor (CNTF) and basic fibroblast growth factor (bFGF) which might function to ameliorate neuronal damage, and thus serve a neuroprotective role.[9] Other proteins, particularly the cytoskeletal proteins, appear to participate in remodeling Müller cell membranes to seal off the sites of retinal damage.[10]

Understanding gene regulatory mechanisms in Müller cells is important for development of strategies for using Müller cell-specific promoters to test the biological effects of growth factors and cytokines in animals models of retinal degeneration. They are also essential for designing cell type-specific expression vectors for gene therapy. In recent years, our research has focused on the identification of genetic regulatory elements for two Müller cell-specific genes—the glial intermediate filament protein (GFAP) gene and the cellular retinaldehyde-binding protein (CRALBP) gene.

The GFAP gene was chosen for study because GFAP is strongly upregulated in glial cells in response to neuronal damage and is the best known marker for reactive gliosis.[7] Furthermore, there is substantial interest in the mechanisms regulating GFAP gene expression in glial cells in general, as this response is robust and widespread throughout the nervous system, and characteristic of a multitude of neurodegenerative diseases.[11] Indeed, as a molecular marker, GFAP turns out to be a more sensitive indicator of neuronal damage than either histology or electrophysiology. Moreover, identification of GFAP regulatory elements might offer an inducible promoter for gene therapy studies in retinal degenerative diseases.[12]

CRALBP gene was studied because it is expressed at a high level in Müller cells in the adult retina and its regulatory sequences might provide a means to target transgene-expression to Müller cells.[13,14] CRALBP gene regulation studies are also of interest because point mutations in this gene have been recently found to result in autosomal recessive retinitis pigmentosa.[15] In the present report, we describe some recent studies on GFAP and CRALBP gene regulation.

MATERIALS AND METHODS

Immunocytochemistry

The indirect imunofluorescence technique was employed to localize GFAP in frozen sections of the mouse retina.[16] After enucleation, eyes were fixed in fresh 4% paraformaldehyde, frozen in OCT compound, and sectioned on a cryostat. The sections were collected on gelatin-coated slides and processed for immunocytochemistry. After blocking with 1% BSA in phosphate buffered-saline, sections are treated with a polyclonal GFAP antibody (DAKO, Carpenteria, CA). GFAP was visualized by staining sections with FITC-goat antirabbit IgG. Preimmune rabbit serum was used as a control.

RNA Analysis

Total RNA was extracted from 10–12 eyes, and 10 μg aliquots were separated by electrophoresis on an agarose gel containing formaldehyde. After blotting RNA to nitrocellulose, blots were hybridized with ^{32}P-labeled GFAP probe. Following washes at a final stringency of 0.1 XSSC at 65°C, the blots were exposed for autoradiography.[17]

For transcription "run-on" assays,[18] nuclei were prepared from retinas, and nuclear transcription was carried out in the presence of ^{32}P-UTP. As a control, transcription was carried out in the presence of α-amanitin at 1 μg/ml concentration to inhibit Pol II transcription. ^{32}P-RNA was extracted and hybridized to slots bound with pBR 322, pRβA-1 (actin), and GFAP DNA as described elsewhere.[17] Blots were exposed for 1–3 days and the autoradiograms are scanned with a densitometer.

In Situ Hybridization

In situ hybridizations were carried out as described elsewhere.[17] ^{35}S-labeled DNA probes are prepared according protocols supplied by NEN Du Pont Co. (Boston, MA). Eyes were fixed in 4% paraformaldehyde, sectioned in a cryostat, and collected on gelatin-coated slides. After incubation with proteinase K, sections were hybridized with the GFAP probe overnight at 45°C. After several washes at high stringency, the slides were processed for autoradiography. Sections were counterstained with cresyl violet and examined.

Transfection Assays

Müller cells (rMC-1) and 3T3 cells were grown to 50–70% confluence, and co-transfected with pCRALBP-*luc* containing a 0.3 Kb, 5′ fragment of human CRALBP gene cloned into the promoter-less vector, pGL2B-*luc*, and pSV*lac* vector to normalize transfection assays (Promega Crop., Madison WI). Following electroporation, cells were harvested and assayed for luciferase and β-galactosidase.[19] Luciferase activities were expressed as photon counts per second X10^3 and normalized to β-gal activities in each case, to control for variations in transfection efficiency. All values were expressed as mean ± SEM for three separate determinations. Transfection with primary Müller cell cultures were carried out using Lipofectin according to instructions supplied Promega.

RESULTS

GFAP Is Upregulated in Retinal Degenerative Diseases

GFAP is a 54 kDa, type III intermediate filament protein that is the major constituent of glial filaments in astrocytes.[20] In order to examine whether GFAP is expressed in the mammalian retina, specifically in Müller cells, we treated sections of C57/B6 mouse retina with a polyclonal antibody against GFAP. As shown in Fig. 1, GFAP-containing structures were present mostly in the ganglion cell layer and are likely to represent the cell body and processes of retinal astrocytes. The radially-oriented Müller cells were not labeled. However, when GFAP-immunostaining was

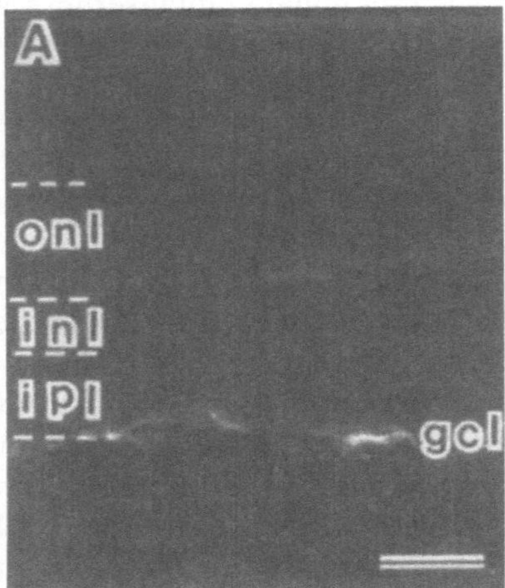

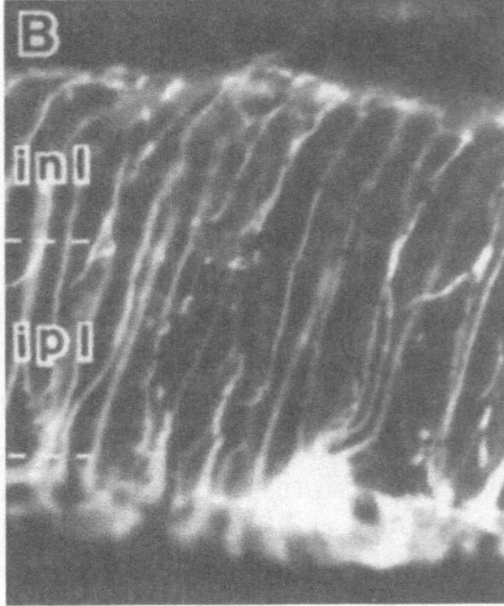

Figure 1. GFAP localization in normal and *rd* mouse retina. Immunostaining was carried out as described under "Methodology." A, normal retina and B, *rd/rd retina*. Note the radial GFAP-expression in Müller cells in the *rd/rd* retina.

examined in retinas from the retinal degeneration (*rd*) mouse or in retinas with constant light damage, there was strong labeling of Müller cells. These results indicate that both hereditary and acquired photoreceptor loss can lead to GFAP expression in Müller cells.

The increased GFAP-immunostaining is not unique to the *rd/rd* mouse retina. A similar increase in GFAP content has been reported in other retinal degenerative diseases (Table 1). Although detailed studies on the mechanism of GFAP induction have been carried out only in the mouse retina, it is likely that a common mechanism is used in all these cases.

Transcriptional Activation of GFAP Gene

The increase in GFAP-immunostainining could be due to a number of reasons—changes in antigenicity of existing GFAP filaments, epitope unmasking, transcriptional

Table 1. GFAP upregulation in retinal degenerative diseases

	Reference
Royal College of Surgeons (RCS) rat	16
Retina degeneration (*rd*) mouse	17
Retinal degeneration slow (*rds*) mouse	21
Vitilago (*vit*) mouse	22
Nervous (*nr*) mouse	unpublished
Purkinje cell degeneration (*pcd*) mouse	unpublished
Abyssinian cat	21
Retinitis Pigmentosa, Autosomal dominant	23
Retinitis Pigmentosa, Recessive	24
Age-related macular degeneration	25
Fundus flavimaculatus	26

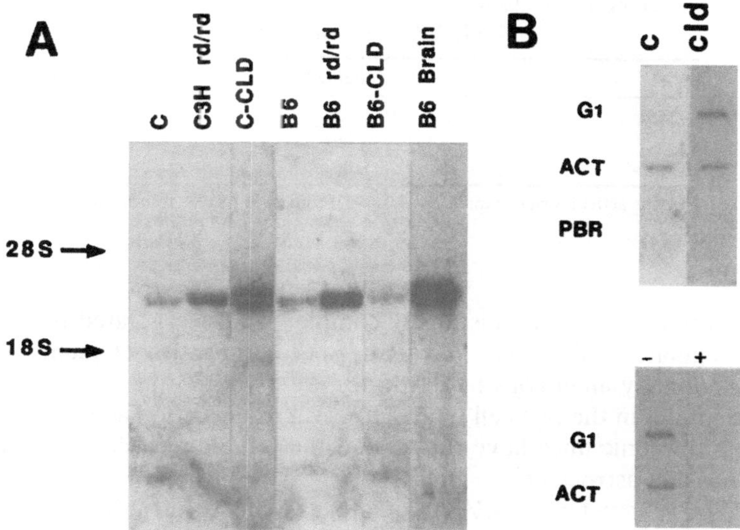

Figure 2. GFAP expression results from transcriptional activation of the GFAP gene. Steady state mRNA level measurements by northern blotting (A) and nuclear run-on assays (B) show that GFAP gene is activated in the retina. G1: GFAP probe; ACT: β-actin probe. In bottom of panel B: without (–) and with (+) α-amanitin.

activation of the GFAP gene or increased stability of GFAP mRNA. In a series of biochemical experiments, it was shown that GFAP induction in Müller cells is mainly regulated at the level of transcription.[17] Immunoblotting studies showed that there was a 3- to 5-fold increase in the GFAP content in retinas with photoreceptor loss; and therefore it is unlikely that GFAP immunostaining observed in Müller cells could arise from epitope-unmasking or changes in antigenicity.[17]

Northern blotting was carried out to examine whether there was a corresponding increase in GFAP mRNA level. We found a small amount of GFAP mRNA in the normal mouse retina, presumably derived from astrocytes. In *rd* retinas, there was an 8- to 10-fold increase, while in retinas with light damage, there was an even more dramatic increase by 15-fold. These results demonstrate that the GFAP mRNA content is elevated in retinas with photoreceptor degeneration.

In order to determine whether the increase in GFAP mRNA level was due to activation of GFAP gene transcription, or was the result of increased GFAP mRNA stability, we carried out nuclear transcription assays using nuclei isolated from normal and light-damaged retinas. As shown in Fig. 2B, we found a 10-fold increase in GFAP synthesis in nuclei from light-damaged retinas. Furthermore, this increase was blocked when the assay was performed in the presence of α-amanitin, a RNA polymerase II inhibitor.

Identification of GFAP Regulatory Elements

The mouse, rat and human GFAP genes has been isolated and characterized.[27] The regulatory sequences of the mouse and human GFAP genes have been extensively analyzed in astrocyte cultures and in transgenic mice.[27] The studies show that GFAP

Table 2. β-galactosidase activity in cells transfected with GFAP-lacZ constructs

	Müller cell	C6
C-259	5	33
C-339	5	40
pSVlac	100	100

Enzyme activities were expressed as % β-gal activity in pSVlac transfections.

gene transcription in astrocytes is highly complex, and is regulated by a combination of positive regulatory sequences and tissue-specific negative elements. Since Müller cells are functionally analogous to astrocytes, we expected the GFAP regulatory elements to be similar in the two cell types. However, transfection assays and studies with GFAP-lacZ transgenic mice have shown that the GFAP gene is regulated differently in Müller cells and astrocytes.

In order to characterize GFAP regulatory elements in Müller cells, cell transfection assays were carried out in primary Müller cell cultures that express GFAP.[28] Results of cell transfection assays are shown in Table 2. When a construct (C-259) containing ~2 kb of GFAP 5′ region was transfected, there was a small increase in the β-galactosidase (β-gal) reporter activity. The construct, however, showed a much higher β-gal activity when transfected into C6 cells. The Müller cell cultures appeared not to be a problem because a control vector, pSV lac, induced substantial β-gal activity. This result suggested that the 5′ region was not able to direct β-gal expression in Müller cells. The lack of reporter stimulation in Müller cells could be due to the absence of essential intronic or 3′ sequences in the constructs tested. To explore this possibility, we tested two other GFAP-lacZ fusion genes:[29] C-339 which carried the entire GFAP gene with 2 kb of 5′ and 1 kb of 3′ flanking sequence, and had lacZ inserted in the first exon; and C-445 which was similar to C-339 except that the first two ATGs were changed to TTGs. In C6 cells, C-339 showed a 40% stimulation whereas C-445 gave 39% stimulation of β-gal activity. Transfected Müller cell cultures, however, produced only a 5% change in β-gal activity. Therefore, the 5′-, 3′-flanking sequences and introns are not sufficient to direct GFAP gene expression in Müller cell cultures.

It could be argued that the lack of β-gal stimulation in Müller cells is due to unrecognized problems with transfection assays, and might not reflect the in vivo situation. To test this possibility, we carried out experiments with GFAP-lacZ transgenic mice carrying the C-339 fusion-gene. Following mechanical injury to the eye, we found that the endogenous GFAP gene was activated whereas the transgene was not expressed in Müller cells. In contrast, both genes were active in astrocytes.[29] These studies indicate that GFAP gene transcription in Müller cells is complex, and that an inducible, Müller cell-specific enhancer is likely control GFAP gene regulation. Identification of the enhancer should provide a inducible, promoter system for expressing foreign proteins in the retina.

CRALBP Gene Regulation in Müller Cells

CRALBP is a retinoid-binding protein highly expressed by retinal Müller cells and the pigment epithelium.[30] The human and mouse CRALBP genes have been

Table 3. Luciferase levels in rMC-1 and 3T3 cells transfected with a pCRALBP-*luc*

	rMC-1	3T3
pCRALBP-*luc*	180 ± 48	9.8 ± 0.27
pGL2B-*luc*	13.2 ± 4.1	8.3 ± 0.6

Experimental details are provided under "Methodology".

cloned, and CRALBP gene regulation has been studied in RPE.[13] The genetic elements that are responsible for CRALBP expression in Müller cells, however, have not been previously characterized. Recently, we examined the ability of 5′ flanking sequences of the human CRALBP gene to promote luciferase reporter expression in Müller cells.[19] DNA transfection was used to assay for gene activity in a Müller cell line (rMC-1) or in 3T3 cells. In Müller cells transfected with a construct carrying ~0.3 kb, 5′ sequences, there was a 20-fold increase in luciferase activity suggesting that the immediate 5′ flanking sequences of the human CRALBP gene contain the genetic elements sufficient for high level, cell-specific expression of the CRALBP gene in Müller cells. The in vivo specificity of this construct remains to be determined in CRALBP-lacZ transgenic mice.

DISCUSSION

Regulation of Müller Cell-Specific Genes

The pioneering studies of Moscona and his associates showed that cell-cell interactions are important for gene expression in the developing chick retina.[6] Both *in vivo* studies and work with reaggregated and dissociated cell cultures showed that interactions with neurons is crucial for induction of glutamine synthetase in Müller cells.[6] Although this phenomenon has been known for more than two decades, its molecular basis remains unknown.

Recent studies on gene expression in Müller cells have focused on genes expressed exclusively in Müller cells such as carbonic anhydrase II, glutamine synthetase, and CRALBP genes. In all these cases, cell transfection studies show that 5′ flanking sequences are sufficient for gene regulation.[13,31,32] Identification of the regulatory sequences will make it possible to express novel proteins in Müller cells and examine their biological effects in retina. The regulatory regions should also be useful in designing gene therapy vectors that are targeted for expression in Müller cells. However, one drawback is that both CA and CRALBP are expressed in the RPE, and inadvertent gene expression from RPE might be difficult to avoid.

In addition to the constitutively expressed proteins discussed above, several other proteins have been reported to be upregulated in Müller cells under pathological conditions (Table 4). In most of these cases, the supporting evidence comes from immunocytochemical data, and little is known about the gene regulatory mechanisms involved. A major reason for studying this class of genes is that they might provide cell type-specific, inducible, promoter systems suitable for gene therapy studies. There are certain problems, however. For example, GFAP is not only upregulated in Müller cells,

Table 4. Examples of proteins upregulated in Müller cells in the retina

Protein	Condition	Reference
Glial fibrillary acidic protein (GFAP)	Diverse	7
Ciliary neurotrophic factor (CNTF)	Trauma	9
Amyloid peptide (APP)	Degeneration	8
Bcl2	Degeneration	8
C-fos	Trauma	34
Glutamate transporter (GLAST)	Ischemia	35

but its expression is enhanced in astrocytes.[17] Also, GFAP gene activation occurs transiently, and expression of transgenes may also be affected similarly.[33] With a judicious choice of regulatory elements, it might be possible to design an inducible, Müller cell-specific GFAP vector for gene therapy. However, such an advance is critically dependent on a sound knowledge of GFAP genetic regulatory elements.

Targeting Müller Cells for Gene Delivery

Retinal support cells such as Müller cells and RPE are excellent candidates for targeting gene therapy vectors. First, Müller cells continue to function long after progressive rod loss,[7] and hence are well-suited for long-term delivery of neuroprotective agents to late-dying cells such as cones. Secondly, Müller cells span the entire retina; therefore, they are useful for delivering exogenous proteins to both the outer retina and the vitreous. Thirdly, there is evidence that some gene delivery vectors such as the adenovirus vectors are naturally targeted to Müller cells.[37] Lastly, Müller cells are endowed with mechanisms for synthesis and secretion of many neuroprotective cytokines and growth factors and hence are a natural site for expression of these substances.[9] In this context, it is worth mentioning that the availability of Müller cell cultures and Müller cell lines makes it possible to transplant transfected cells into the eye, an approach similar to that being tested with astrocyte cultures for neurodegenerative diseases.[38]

The work on gene regulation is also important for development of strategies for using Müller cell-specific promoters to test the biological effects of growth factors and cytokines in animal models of retinal degeneration.[39] In the future, it should be possible to create transgenic mice in which expression of cytokines and growth factors is under the control of GFAP-regulatory elements. When rod loss is induced in the transgenic animals by light-damage, GFAP gene activation is expected to result in synthesis and secretion of the cytokine or growth factor from the transgene. The neuroprotective or neurotoxic effects of the transgene products on rod and cone degeneration can be conveniently examined on a long-term basis in this system.

ACKNOWLEDGMENTS

I wish to thank many colleagues who participated in the studies described here. The research was supported by the NIH grant, R01EY03523, and an unrestricted grant from Research to Prevent Blindness, Inc.

REFERENCES

1. S. Chen and D.J. Zack, 1996, Ret 4, A positive-acting rhodopsin regulatory element identified using a bovine retina *in vitro* transcription system, *J. Biol. Chem.* **271**:28549–28577.
2. A.P. Dipolo, C.B. Rickman, and D.B. Farber, 1996, Isolation and initial characterization of the 5′ flanking region of the human and murine cyclic guanosine monophosphate-phosphodiesterase beta-subunit genes, *Invest. Ophthalmol. Vis. Sci.* **37**:551–560.
3. I. Ahmad, X. Yu, and C.J. Barnstable, 1994, A *cis*-acting element, T alpha-1, in the upstream region of rod alpha-transducin gene that binds a developmentally regulated retina-specific nuclear factor. *J. Neurochem.* **62**:396–399.
4. A.T. Morris, W.B. Fong, M.J. Ward, H. Hu, and S.L. Fong, 1997, Localization of upstream silencer elements involved in the expression of cone transducin α-subunit (GNAT2), *Invest. Ophthalmol. Vis. Sci.* **38**:196–206.
5. G.I. Liou, S. Matragoon, J. Yang, L. Geng, P.A. Overbeek, and D.P. Ma, 1991, Retina-specific expression from the IRBP promoter in transgenic mice conferred by 212 bp of the 5′-flanking region, *Biochem. Biophys. Res. Comm.* **181**:159–165.
6. P. Linser and A.A. Moscona, 1983, Induction of glutamine synthetase in the embryonic neural retina: localization in Müller fibers and dependence on cell interactions, *Proc. Natl. Acad. Sci.* **76**:6476–6480.
7. P.V. Sarthy, 1990, Reactive gliosis in retinal degenerations, in: *Retinal Degenerations* (R.E. Anderson, J.G. Hollyfield, and M.M. LaVail, eds.), pp. 109–115, CRC Press Inc., Boca Raton, FL.
8. W. Hartig, J. Gorsche, C. Distler, D. Grimm, E. El-Hifnawi, and A. Reichenbach, 1995, Alterations of Müller (glial) cells in dystrophic retinae of RCS rats, *J. Neurocytol.* **24**:507–517.
9. R. Wen, Y. Song, T. Cheng, M.T. Matthes, D. Yasumura, M.M. LaVail, and R.H. Steinberg, 1995, Injury-induced upregulation of bFGF and CNTF mRNAs in the rat retina, *J. Neurosci.* **15**:7377–7385.
10. G.E. Korte, G.S. Hageman, D.V. Pratt, S. Glusman, M. Marko, and A. Ophir, 1992, Changes in Müller cell plasma membrane specializations during subretinal scar formation in the rabbit, *Exp. Eye Res.* **55**:155–162.
11. M.D. Norenberg, 1996, Reactive astrocytosis, in: *The role of glia in neurotoxicity* (M. Ascher and H.K. Kimelberg, eds.), pp. 93–106, CRC Press Inc, Boca Raton, FL.
12. J.L. Ridet, S.K. Malhotra, A. Privat, and F.H. Gage, 1997, Reactive astrocytes: cellular and molecular cues to biological function, *Trends in Neurosci.* **20**:570–577.
13. B.N. Kennedy, S. Goldflam, M.A. Chang, P.A. Campochiaro, A.A. Davis, D.J. Zack, and J.W. Crabb, 1998, Transcriptinoal regulation of the human gene encoding cellular retinaldehyde-binding protein, *J. Biol. Chem.* **273**:5591–5598.
14. S.J. Brodjian, B.N. Kennedy, J.W. Crabb, and V. Sarthy, 1997, Regulation of cellular retinaldehyde-binding protein (CRALBP) gene in Müller cells, *Invest. Ophthalmol. Vis. Sci.* **37**:S336.
15. M.A. Maw, B. Kennedy, A. Knight, R. Bridges, K.E. Roth, E.J. Mani, J.K. Mukkadan, D. Nancarrow, J.W. Crabb, and M.J. Denton, 1997, R150Q mutation of cellular retinaldehyde-binding protein in autosomal recessive retinitis pigmentosa, *Nature Genetics*, **17**:198–200.
16. A.J. Eisenfeld, A.H. Bunt-Milam, and P.V. Sarthy, 1984, Müller cell expression of glial fibrillary acidic protein after genetic and experimental photoreceptor degeneration in the rat retina, *Invest. Ophthalmol. Vis. Sci.* **25**:1321–1328.
17. P.V. Sarthy and M. Fu, 1989, Transcriptional activation of an intermediate filament protein gene in mice with retinal dystrophy. *DNA*, **8**:437–446.
18. M.E. Greenberg and E.B. Ziff, 1985, Stimulation of 3T3 cells induces transcription of the c-fos proto-oncogene, *Nature* **311**:433–438.
19. V.P. Sarthy, S.J. Brodjian, K. Dutt, B.N. Kennedy, P.P. French, and J.W. Crabb, 1998, Establishment and characterization of a retinal Müller cell line, *Invest. Ophthalmol. Vis. Sci.* **39**:212–216.
20. A. Bignami and D. Dahl, 1995, Gliosis, in: *Neuroglia*, (H. Kettenmann and B.R. Ransom, eds.), pp. 843–858, Oxford University Press, New York, NY.
21. P. Ekstrom, S. Sanyal, K. Norfstrom, G.J. Chader, and T. van Veen, 1988, Accumulation of glial fibrillary acidic proterin in Müller radial glia during retinal degeneration, *Invest. Ophthalmol. Vis. Sci.* **29**:1363–1371.
22. S.B. Smith, S. Brodjina, S. Desai, and V. Sarthy, 1997, Glial fibrillary acidic protein (GFAP) is synthesized in the early stages of the photoreceptor cell degeneration of the mi^{vit}/mi^{vit} (vitilago) mouse, *Exp. Eye Res.* **64**:645–650.

23. Z-Y. Li, S.G. Jacobson, and A.H. Milam, 1984, Autosomal dominant retinitis pigmentosa caused by the threonine-17-methionine rhodopsin mutation: retinal histopathology and immunocytochemistry, *Exp. Eye Res.* **58**:397–408.
24. Z-Y. Li, I.J. Kljavin, and A.H. Milam, 1995, Rod photoreceptor neurite sprouting in retinitis pigmentosa, *J. Neurosci.* **15**:5429–5438.
25. M.C. Madigan, P.L. Penfold, J.M. Provis, T.K. Balind, and F.A. Billson, 1994, Intermediate filament expression in human retinal macroglia. Histopathologic changes associated with age-related macular degeneration, *Retina* **14**:65–74.
26. C.D. Birnbach, M, Jarvelainen, D.E. Possin, and Milam A.H., 1994, Histopathology and immunocytochemistry of the neurosensory retina in fundus flavimaculatus, *Ophthalmology* **101**:1211–1219.
27. M. Brenner, 1994, Structure and transcriptional regulation of the GFAP gene, *Brain Pathology* **4**:245–257.
28. P.V. Sarthy, 1985, Establishment of Müller cell cultures from adult rat retina, *Brain. Res.* **337**:138–147.
29. L.C. Verderber, L.C., W. Johnson, L. Mucke, and V. Sarthy, 1995, Differential regulation of a GFAP-lacZ transgene in retinal astrocytes and Müller cells, *Invest. Ophthalmol. Vis. Sci.* **36**:1137–1143.
30. J.C. Saari, 1994, Retinoids in photosensitive systems, in: *the Retinoids: Biology, Chemistry and Medicine* (M.B. Sporn, A.B. Roberts, and D.S. Goodman, eds.), pp. 351–385, Raven press, NY.
31. L.W. Lai, R.P. Erickson, P.J. Venta, R.E. Tashian, and Y.H. Lien, 1998, Promoter acitivty of carbonic anhydrase II regulatory regions in cultured renal proximal tubular cells, *Life Sci.* **63**:121–126.
32. Y-C. Li, D. Beard, S. Hayes, and A.P. Young, 1996, A transcriptional enhancer of the glutamine synthetase gene that is selective for retinal Müller glial cell, *J. Mol. Neurosci.* **6**:169–183.
33. V. Sarthy and H. Egal, 1994, Transient induction of the glial intermediate filament protein in Müller cells in the mouse retina, *DNA and Cell Biol.* **14**:313–320.
34. K. Yoshida, Y. Kuraki, K. Ohki, T. Harada, T. Ohashi, H. Matsuda, and J. Imaki, 1995, C-fos gene expression in rat retinal cells after focal retinal injury, *Invest. Ophthalmol. Vis Sci.* **36**:251–254.
35. Y. Otori, S. Shimada, K. Tanaka, I. Ishimoto, Y.T. Tano, and M. Tohyama, 1994, Marked increase in glutamate-aspartate transporter (GLAST/GluT-1) mRNA following transient retinal ischemia, *Mol. Brain Res.* **27**:310–314.
36. K. Clarke and E.E. Geisert, Jr, 1998, The target of the antiproliferative antibody (TAPA) in the normal and injured rat retina, *Mol. Vision* **4**:3.
37. M. Fukuhara, A. Suzuki, Y. Fukuda, and J. Kosaka, 1998, Adenovirus vector-mediated gene transfer into rat retinal neurons and Müller cells in vitro and in vivo, *Neurosci. Lett* **242**:93–96.
38. W.F. Blakemore, A.J. Crang, and R.J.M. Franklin, 1995, Transplantation of glial cells, in: *Neuroglia* (H. Kettenmann and B.R. Ransom, eds.), pp. 869–882, Oxford University Press, New York, NY.
39. M.M. LaVail, K. Unoki, D. Yasumura, M.T. Matthes, G.D. Yancopoulos, and R.H. Steinberg, 1992, Multiple growth factors, cytokines and neurotrophins rescue photoreceptors from the damaging effects of constant light, *Proc. Natl. Acad. Sci.* **89**:11249–11253.

Section III

THE INVOLVEMENT OF LIGHT IN RETINAL DEGENERATION

29

DARK ADAPTATION IS IMPAIRED IN DIABETICS BEFORE PHOTOPIC VISUAL LOSSES CAN BE SEEN

Can Hypoxia of Rods Contribute to Diabetic Retinopathy?

G. B. Arden,[1] J. E. Wolf,[1] J. Collier,[2] C. Wolff,[2] and M. Rosenberg[2]

[1] City University, London
[2] Queen Mary College, London

INTRODUCTION

The main theme of this volume is the inherited retinal degenerations, but the most important causes of blindness have a different aetiology, not directly related to genetic defects. Diabetes is the greatest cause of blindness in younger people, and even considering all age groups is as common a cause of blindness as Glaucoma and Age Related Maculopathy. Diabetes causes a retinopathy (DR) which is basically a vasculopathy[1,2] and the cellular biology of DR has been recently linked to cytokines.[3] Starting from this fact, it has been proposed that techniques of molecular biology which might be of use in the treatment of inherited degenerative diseases could also be applicable to DR. This paper is however concerned with a far simpler method of controlling DR, which is based on the physiology of the eye, the particular features of which have been known for half a century but not exploited in this connection. The basic idea is that anoxia early in DR will only develop during dark adaptation, occuring in long periods every night in sleep.

OXYGEN TENSION IN NORMAL RETINA AND EFFECTS OF LIGHT

DR is peculiar in that while all capillaries in the body are affected in diabetes, there are special and more severe effects on the vessels of the retina[4]—the vasculopathy develops because of a special retinal factor. Retinal neurones and glia are however

Retinal Degenerative Diseases and Experimental Therapy, edited by Hollyfield *et al.*
Kluwer Academic / Plenum Publishers, New York, 1999.

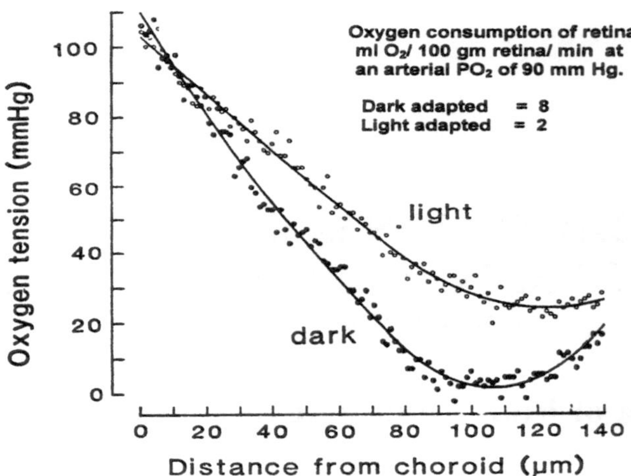

Figure 1. Adapted from.[9,10,12] Profiles of oxygen tension (PO_2) in the cat retina, recorded with an oxygen microelectrode inserted from the vitreous surface of the retina to the RPE, and withdrawn progressively by the distances indicated. Experiment performed in darkness and with background illumination of 8.6 log quanta/degree/s. Note the steep fall of PO_2 from the RPE to the region of the rod cell bodies. In darkness, the minimum oxygen tension 100 uM from the RPE becomes unmeasurable. The rise from this minimum vitreally indicates that photoreceptors in darkness are absorbing oxygen from the retinal circulation. The extra absorption in darkness can be calculated.

very similar to the relatively unaffected brain capillaries so the "special factor" is likely to involve those retinal elements not found in the brain. These are of course the photoreceptors, unusual since they are specialised for phototransduction. The outer limb grows and is shed, unlike any other neuronal cell. The contents turn over in about 10 days. This, and the rhodopsin cycle, require a very active metabolism in the rods (which greatly outnumber the cones). However the greatest energy requirement is due to the outer limb membrane in darkness having a high conductance, which causes the development of a dark current.[5] The photoreceptor signals light by reduction in this current. Very active "pumps" maintain the dark current and the result is that the oxygen consumption of the photoreceptors (especially rods) is greater than in any other cell.[6–13] Moreover, photoreceptors are in an avascular zone, and are supplied with oxygen by diffusion, mainly from the choroid. Although this is a very profuse blood supply, it is only just adequate. Figure 1 shows the profile of PO_2 in the cat retina: even with a background light the steep fall in oxygen tension from the retinal pigment epithelium, RPE, (100 mm. Hg.) to the outer nuclear layer, ONL, (18 mm. Hg.) is evidence of a high utilisation of oxygen by rods. In darkness, the tension at the OPL falls to zero, so even more oxygen is consumed, and the increase in PO_2 from the rod layer in a vitreous direction indicates that oxygen is flowing from the retinal circulation to supply the rods. If an oxygen microelectrode is placed in the OPL, a flash of light causes an "oxy-electrogram"[13] which is has a time course as fast as the well known changes in extracellular potassium, that produce the ERG.[14–16]

One consequence of the high consumption of oxygen which is not normally appreciated is that the normal eye (despite its huge choroidal blood supply), is in darkness, near the brink of anoxia. Small reductions of the oxygen in the inspired air—equiva-

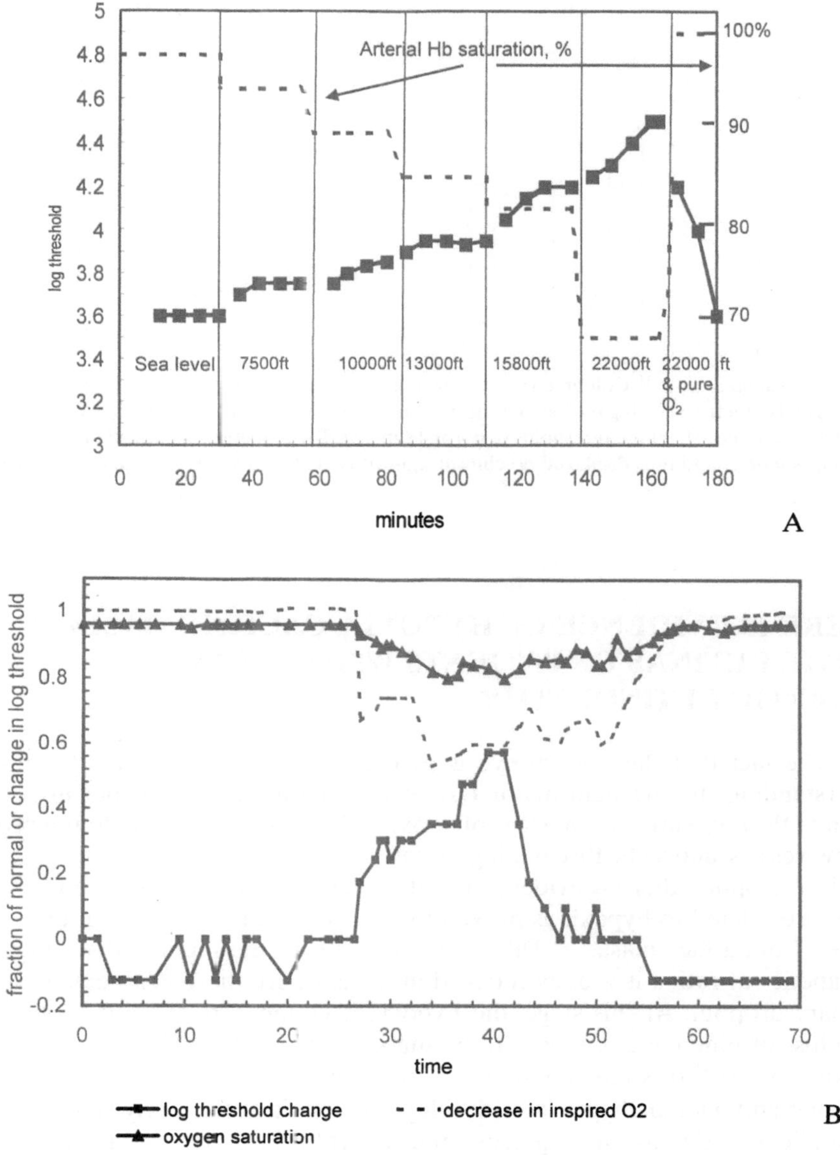

Figure 2. A: Adapted from.[17] The subject breathed various gas mixtures, in which oxygen was reduced to mimic the content at the altitudes indicated. Note the progressive decrease of rod sensitivity, and the recovery. B: Recent partial repeat of this experiment with continuous monitoring of arterial oxygen saturation and alveolar oxygen tension, showing time course of loss of rod sensitivity.

lent to an elevation of 1500m—cause the visual threshold to increase. This was first demonstrated over 50 years ago.[17,13] Figure 2a is taken from the original paper, and Fig. 2b shows a modern variant of this experiment, with continuous recordings of arterial and inspired oxygen. However in light-adaptation, much more hypoxia is required before vision (acuity, colour vision) becomes abnormal.

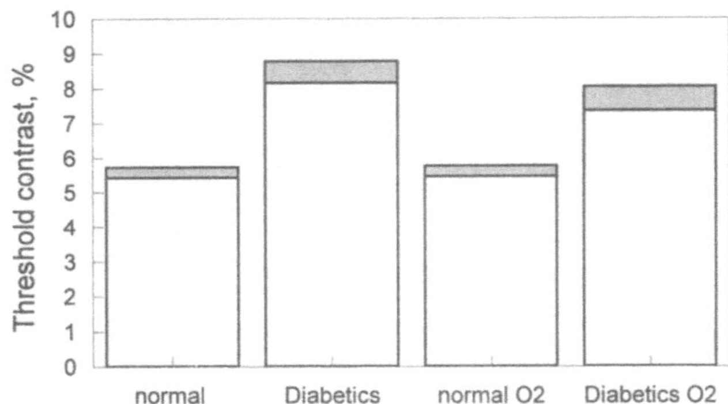

Figure 3. Adapted from.[34] Colour contrast sensitivity measured along a tritan colour confusion line in normals and diabetics, showing the loss of sensitivity in the patients and the partial recovery on oxygen inhalation, which does not affect the thresholds of normal subjects. The diabetics all had mild background retinopathy (stages 0 or 1) and thus displayed no clinical signs of retinal hypoxia. For technical details see.[66]

THERE IS EVIDENCE OF HYPOXIA CAUSING DAMAGE IN THE RETINAS OF PATIENTS WITH EARLY DIABETIC RETINOPATHY

The fact that the normal eye is so near its oxygen limit may be important in understanding the mechanisms of DR. It is generally accepted that in severe background DR, there are clinical signs of anoxia, which ultimately lead to the proliferation of new vessels, and sight-threatening disease.[1,2,19,20] However, before the stage of major capillary dropout, there is evidence of interference with visual function.[21–32] That such losses are related to hypoxia is proven by their partial reversal when diabetics breath oxygen from a face mask.[33,34] This is in keeping with the report[35] that oxygen tension in diabetic cat retina is grossly reduced in areas where there is no evidence of severe capillary dropout. At this stage, the hypoxia is confined to the outer retina, and the early loss of blue cone function in the diabetic human ERG has been located to the photoreceptors.[32] It is unlikely that the inner retinal layers, in which microaneurysms, exudates, and haemorrhages later develop are anoxic at this early stage, but the extra metabolic load due to raised glucose, the reduction in O_2 release associated with glycosylated haemoglobin, the rheological changes in red blood cells, the thickening of basement membrane, and swelling of endothelial cells can reduce the transport of oxygen[36–43] into the receptor layer which then becomes hypoxic.

It follows that the loss of visual function in DR should be worse in darkness, if it is related to mild anoxia. There are numerous prior reports of the difficulties diabetics experience in darkness, and many papers have demonstrated abnormal dark adaptation, which is incomplete, and may be slowed. We have recently analysed dark adaptation in early diabetics whose photopic visual function was normal.[44] All had abnormalities of dark adaptation. The patients simply stopped dark-adapting sooner than the normals i.e. the last 1–1.5 log unit of increase in rod sensitivity did not occur. This is consistent with the rods being unable to maintain a maximal dark current, and hence signaling an "equivalent background illumination".[45]

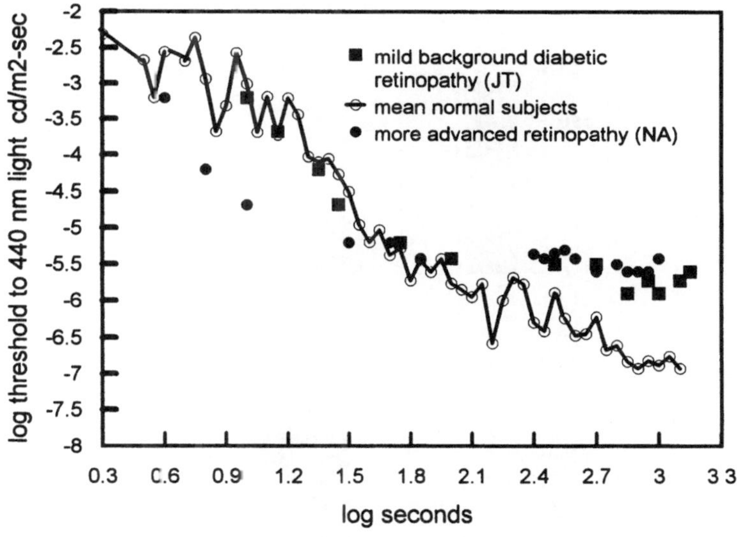

	Normals			Background DR		
number	25			4		
	Thresholds at 5 min.	10 min.	decrease 5-10min	Thresholds at 5 min	10 min	decrease 5-10 min
Mean	1.27	0.46	0.81*	1.56	1.47	0.097*
St. Error	0.10	0.15	0.11	0.45	0.57	0.06

* probability normals and DR are from same population (t) =0.0012

Figure 4. Adapted from[44] showing abnormal dark adaptation in 2 diabetics with early background retinopathy compared to the average result from 25 normals. The diabetics were chosen to have no measurable loss of photopic function. Note the logarithmic time scale, chosen to emphasise the abrupt termination of the diabetic's recovery. The panel shows that the effect, even in a small group, is highly significant.

A HYPOTHESIS LINKING MILD HYPOXIA TO DIABETIC RETINOPATHY

An important consequence of prolonged anoxia in the retina is the upregulation of cytokines.[3,46,47] One of these, VEGF, has aroused special interest because it is extremely potent, and it acts on capillaries to cause loss of pericytes, swelling and leakage of endothelial cells as well as provoking capillary proliferation—i.e., the histological changes of DR.[3] In man and in animal models of diabetes, VEGF levels are raised very early, before gross retinal damage to the capillaries.[47–51] VEGF controls the development of the retinal blood supply, and it has been suggested it is also a major causative factor in DR.[3] During prolonged periods of dark-adaptation, the increased retinal anoxia will provoke up-regulation of VEGF, and this could lead to a vicious circle in which the mild capillary swelling and leakage causes further anoxia, which further increases the rate of production of VEGF.

Of course, other cytokines could also be produced and be important:[52,53] other (unknown) factors apart from anoxia may also upregulate cytokines. However, the early and apparently oxygen dependent loss of function in the human diabetic retina

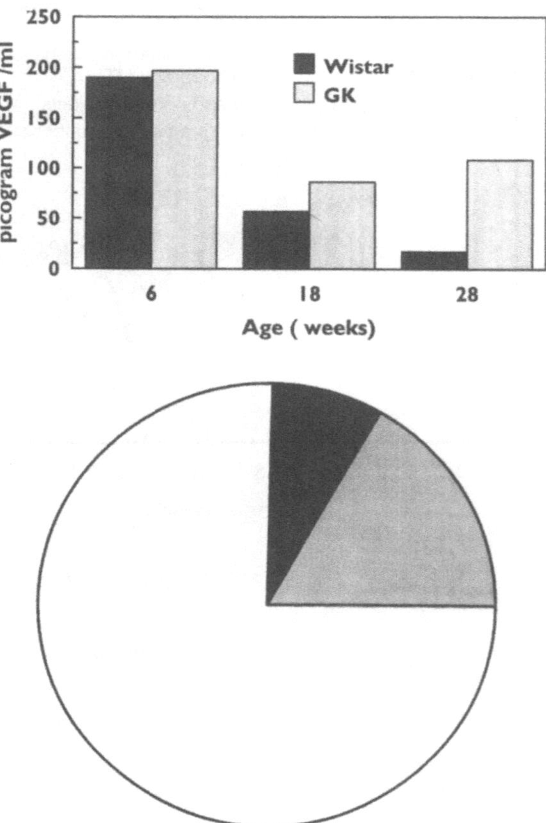

Figure 5. A: Adapted from.[50] Shows the concentration of VEGF in the spontaneously diabetic G-K rat as a function of age. Note the decrease of the cytokine in the normal Wistar controls, and the increase in the diabetics, which is pronounced by 6 months. B: Adapted from[49] showing the proportion of retinal capillaries of GK rats showing increased VEGF staining and histologic changes at the age of 6 months. Evidently, VEGF concentration increases before amy extensive "retinopathy".

which occurs before extensive capillary dropout—i.e. before the pre-proliferative stage of retinopathy, provides a linking hypothesis that explains the local severity of diabetic vasculopathy in the retina.

While this is an attractive theory, it requires to be tested in additional ways. Can the cytokine concentrations be modulated by dark adaptation in the way suggested? In an animal model of early DR, this appears to be the case. (in preparation)

Again, the production of VEGF has to be related to the histologic finding of early DR. In the embryo retina, immunocytochemistry shows VEGF is associated with glia, mostly astrocytes, but including Mueller cells, towards which blood vessels grow. As the vessels reach the cells, and oxygenation increases, the VEGF staining vanishes.[46,47,43] The vasculopathy of DR begins in the inner retinal capillaries, and also affects the RPE. But these regions are 100 uM from the most anoxic part of the retina. VEGF could diffuse from the Mueller cells to its receptors in the capillaries and RPE. However one well-studied function of the Mueller cell is "potassium funneling".[16] Retinal neuronal activity releases potassium into the extracellular spaces of the OPL and IPL where it is taken up by pumps, and released through conductances in specialised regions at the inner and outer limiting membranes. It is possible that VEGF is formed in the hypoxic nuclear region of the Mueller cell, and can only leave it at these specialised regions, where the anoxia is not found. Mueller cells become activated (e.g. for intracellular fibrillary substance—very early in DR.[1,2,47]

POSSIBLE REGIME FOR DIABETICS WHICH WOULD HELP PREVENT RETINOPATHY

Present medical methods of treating DR are: very close control of blood sugar levels and interference with the polyol pathway.[1,2,43] In established cases of sight threatening DR, the use of laser pan retinal photocoagulation (PRP) is protective. (The latter probably works by reducing retinal oxygen requirements[54,55] and would provide a reduction which is no larger than caused by light adaptation). It has been proposed that development of antisense mRNAs for VEGF injected into the vitreous could prevent DR by blocking the production of the cytokine.[3] There are obstacles to all these methods of treatment. Patients are often non-compliant, PRP is destructive, and in some types of DR, unsuccessful. In addition, surgical and future medical treatments are expensive, and not therefore not universally applicable—e.g. in developing countries. We therefore propose an entirely novel form of treatment that might avert DR, namely to prevent dark adaptation and if this is possible, there is a compelling body of evidence suggesting such a regime could be effective. It is prudent to consider if any there is are practical objections to it.

Normally people prefer to sleep in darkness, but in many cases of town dwellers, darkness is incomplete. However people do sleep under quite bright lights e.g. on beaches (or as every teacher knows, during lectures!). If patients without sight threatening DR are to comply with an instruction to sleep in the light, then it must be tolerable for them and their partners, or at least offer a disruption of the same order as using an insulin pump, or having repeated intravitreal injection of a molecular biological construct. (If patients are in imminent danger of becoming blind, they are more likely to be compliant with stringent therapies). Fortunately, the quantity of light required to light adapt patients to the level required does not seem to be large in early diabetes and much less than would affect melatonin and circadian rhythms.

THE QUANTITY OF LIGHT REQUIRED TO LIGHT ADAPT BEHIND CLOSED LIDS

From the work of Linsenmeier and his collaborators,[8–14] the quantity of light delivered to the retina required to completely suppress the extra oxygen requirements of the rod dark current is in the photopic range- 5×510^8 photons.degree^{-2}.sec^{-1}. However, a reduction of dark current requires a lesser light. In the case of the early diabetics we have studied, the elevation of threshold is only 1–1.5 log units (Fig. 3), and it is plausible that illumination which causes an increase in threshold ~2 log unit would prevent the anoxia which (it is hypothesised) is proving damaging. With the eyes open, this is a very low level of background illumination. However, we wish to achieve such a retinal illumination with the lids closed, with natural pupils and during sleep: in such a situation the eyes rotate upward (Bell's phenomenon). Thresholds and increment thresholds have rarely been investigated in such conditions. The attenuation of the lids is known,[56–60] but the effect of Bell's phenomenon on threshold has not been measured. In addition, light adaptation might affect pupillary diameter, and this would affect both light adaptation and apparent visual thresholds.

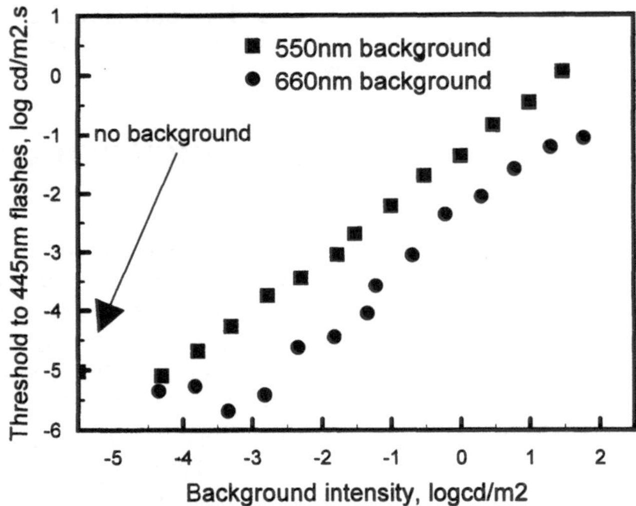

Figure 6. Two colour increment-thresholds in normal subject obtained through the closed lids, which act as a wavelength selective filter. The test flashes were from blue LEDs, peak output 445 nm, and dominant wavelength (through closed lids) of 495 nm. The backgrounds were either from green LEDs (peak output 550 nm, and dominant wavelength 560 nm) or red (peak output 660, dominant 645 nm).
The abscissa gives photopic intensities of the backgrounds at the cornea, after allowing for lid absorption/scatter. The red backgrounds are calculated to be 1.93 log units less effective than green in adapted rods. Thresholds begin to be determined by the cones at an elevation of rod threshold of 3 log units, and the light at this point is provided by LEDS with a junction current of 0.2 mA

Therefore we have measured the increment threshold of normal subjects through closed eyelids, both with the eyes in the primary position and also vertically rotated through the maximum sustainable angle, and with and without mydriasis. We have examined the effect of wavelength of the adapting light on rod increment threshold.

Figure 6 shows a representative increment threshold relationship for blue flashes on green or red backgrounds, obtained through closed lids. A "miniGanzfeld",—a translucent hemisphere 5 cm in diameter illuminated by LEDs from the convex rear side[61,62] was used as a stimulator. It was placed before the left eye, and fixation was maintained by providing a dim red LED in the visual axis, viewed by the right eye. The left eye was held closed with small pieces of transparent adhesive tape. The stimulator contained blue, green, amber, and red LEDs (peak wavelengths 445, 550, 610, and 660 nm respectively). Note that with eyes opened, the subjects saw a Ganzfeld: with the lids closed, the light falling on the lids was forward-scattered and thus the Ganzfeld stimulation was maintained, though the retinal illumination was reduced by the lid absorption and scatter. Thus, with this equipment direct measures of the effective reduction in retinal illumination by the lids can be obtained by comparing thresholds with eyes opened and closed. The absorption of the lids increases rapidly into the blue: an average is about 0.065 log unit/5 nm. though the individual value varies widely especially in the blue. Assuming the figure given, the dominant wavelength of light from LEDs reaching the retina is modified, so that of the blue LED is shifted significantly to 495 nm near the peak absorption of the rods, which are thus stimulated with a fair degree of selectivity. The green backgrounds are transmitted with less change in spectral composition (dominant wavelength 560 nm) and the light from the red LED attenuated about 6 fold has

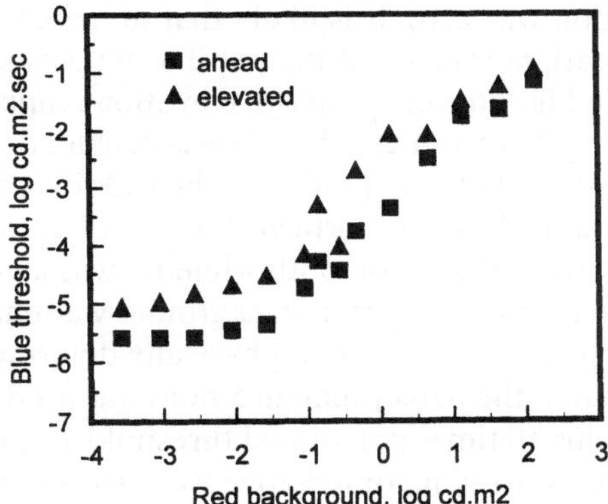

Figure 7. Effect of maximal voluntary upward elevation of the eyes on thresholds to blue flashes obtained through the closed lids (the fellow eye was provided by a fixation point). The thresholds were obtained alternately with eyes elevated and in the normal visual axis, at a variety of red light adapting backgrounds.

its dominant wavelength at 645 nm. In the figures, the intensities in the X-axis are in photopic units of background illumination, corrected for the individual's lid absorption. While very dim green backgrounds increase rod threshold, red backgrounds are ineffective until they are much more intense. This result shows it is possible to selectively light adapt rods through the lids with green LEDs. The horizontal separation corresponds to 1.8 log unit, which is close to the relative calculated rod-stimulating effectiveness of these sources when entering through the pupil (log 1.93). and thus it is probable that the effective light is entering through the pupil rather than through the sclera (and see below, Fig. 7). For backgrounds > log −1.5 cd.m^{-2} it is likely that the threshold with a green background begins to be determined by green cones. The rise of threshold here is about 3 log units, i.e. to the "Kohlrauch knee" on the standard dark adaptation curve, and well above the elevated level (Fig. 4) to which the scotopic thresholds of early diabetics can fall. The increment thresholds rise steadily with intensity, following the Weber law, and do not show the change from one to another of the π mechanisms that are indicated by abrupt decreases in slope characteristic of the Stiles 2-colour increment threshold technique with open eyes.[63] This feature of the present result is almost certainly due to the very large size of the stimulus field, and the variation of cone threshold as a function of retinal eccentricity. The secondary abscissal scale shows the LED current required to produce this light output on the lid surface: only 200 μA per junction (see below). It corresponds to about log 1 cd.m^{-2}, a light level in the mesopic region, which is certainly not perceptually disturbing. At such light levels, there is probably no significant pupillary constriction. When the backgrounds are still more intense the iris may constrict, and reduce the retinal flux of both the blue stimulus and the background to the same extent.

BELL'S PHENOMENON AND LIGHT ADAPTATION

In addition, in sleep the eyes rotate upwards, and this might reduce the light entering the eye through closed lids. We have therefore repeated the experiment of Fig. 6, but moving the fixation spot so that the eyes were elevated by 25° (which is the maximum possible for many people to maintain without acute strain) and is as large or larger than Bell's phenomenon.

Figure 7 shows the results obtained with red backgrounds in one male. The attenuation is fairly trivial and in 3 male observers measured 3.0 ± 0.3 2.6 ± 0.8 and 2.5 ± 0.3 for green backgrounds. In one female, with some degree of exophthalmos, the attenuation was zero. It is likely that in her case, the upward rotation did not cause the upper peripheral part of the pupil to be occluded by the orbital ridge. This explains also why in Fig. 7 the effect of eye elevation is apparently decreased with higher backgrounds: the pupil is constricted, so less is occluded. Hence it appears that Bell's phenomenon only affects light adaptation of the rods by a maximum of 0.5 log unit. We have also measured the rod-driven cortical visual evoked potential (VCEP)[64] in a subject using the same visual stimulator, both while he was awake (conditions as for Fig. 6), and also when he was asleep. A green background was employed (0.7 $cd.m^{-2}$ on the lids), which increased rod threshold, psychophysically determined, by 1 log unit. When a blue flash intensity 3 times the perceptual threshold was used, there was no measurable response. With a stimulus 10 times perceptual threshold, a very small VCEP of 2 μV was obtained, a complex wave with major negative wave (n1) at 120 msec. This is later than the response for photopic stimulation[65] but is characteristic of rods.[64] When the subject slept (signalled by changes in breathing), large slow waves appeared in the crude EEG record, but in the averaged response, a VCEP of similar amplitude and timing was obtained. If Bell's phenomenon had reduced the retinal illumination threefold, the VECP would have vanished. Moreover the background lights and the flashes used did not prevent sleep.

CONSTRUCTION OF A DEVICE FOR USE BY DIABETICS

The stimulator was powered by 10 superbright green LEDs each emitting 6 candela with a 20 mA current, which retro-illuminated an area of 10.4 cm^2, and in turn this illuminated the face and lids. If an eyeshield containing LEDs needs only to illuminate an area corresponding to the pupil, i.e. $<2\,cm^2$, 2 green LEDs each passing 0.2 mA would emit sufficient light to reduce rod dark current considerably, and more than is sufficient to prevent dark adaptation. Such a light source and power supply (a small rechargeable battery) can be incorporated into an eye shield, which the diabetic could wear at night, and which would only light adapt his/her own eyes.

REFERENCES

1. N. Frank, 1986, Diabetic retinopathy: current concepts of evaluation and treatment. Clinics in Endocrinology and Metabolism, 15:933–969.
2. N. Frank, 1995, Diabetic Retinopathy, In: Progress in Retinal and Eye Research, 14:361–392.
3. L.P. Aiello, 1997, Vascular Endothelial Growth Factors. 20th-century mechanisms, 21st-century therapies, Invest. Ophthalmol. Vis. Sci. 38:1847–1652.
4. T.S. Kern and R.L. Engerman, 1996, Capillary lesions develop in retina rather than cerebral cortex in diabetes and experimental galactosemia, Arch. Ophthalmol. 114:306–310.
5. W.A. Hagins, P.D. Ross, R.L. Tate, and S. Yoshikami, 1989, Transduction heats in retinal rods: Tests of the role of cGMP by pyroelectric calorimetry, Proc. Nat. Acad. Sci. USA. 86:1224–1228.
6. K.-W. Yau and D.A. Baylor, 1989, Visual Transduction, Annual Rev. Neurosci. 12:289–327.
7. V.A. Alder, E.S.J. Cringle, and I.J. Constable, 1983, The retinal oxygen profile in Cats, Invest. Ophthalmol. Vis. Sci. 24:30–36.
8. R.A. Linsenmeier, 1986, The effects of light and darkness on oxygen distribution and consumption in the cat retina, J. Gen. Physiol. 88:521–542.
9. R.A. Linsenmeier and R.D. Braun, 1992, Oxygen distribution and consumption in the cat retina during normotension and hypoxaemia. J. Gen. Physiol. 99:177–197.
10. L.M. Haugh, R.A. Linsenmeier, and T.K. Goldstick, 1990, Mathematical Models of the spatial distribution of retinal oxygen tension and consumption, including changes on illumination, Ann. Biomed. Eng. 18:10–36.
11. J. Ahmed, R.D. Braun, R. Dunn Jn., and R.A. Linsenmeier, 1993, Oxygen distribution in the macaque retina, Invest. Ophthalmol. Vis. Sci. 34:516–521.

12. R.D. Braun and R.A. Linsenmeier, 1995, Oxygen consumption in the inner and outer retina of the cat, Invest. Ophthalmol. Vis. Sci. 36:542–554.
13. L.M. Haugh, L.A. Scheidt, E.R. Griff, and R.A. Linsenmeier, 1995, Light evoked oxygen responses in isolated Toad retina, Exp. Eye Res., 61:73–81.
14. L.J. Frishman, F. Yamamoto, J. Bogucka, and R.H. Steinberg, 1992, Light-evoked changes in $[K(+)]_o$ in proximal portion of light-adapted cat retina, J. Neurophysiol. 67:1201–1212.
15. R.F. Miller and J.E. Dowling, 1970, Intracellular responses of the Muller (glial) cells of the mudpuppy retina: their relationship to the b-wave of the electroretinogram, J. Neurophysiol. 33:323–341.
16. E.A. Newman, 1985, Regulation of extracellular potassium by glial cells in the retina, Trends in Neuroscience 8:156–159.
17. R.A. McFarland, J.N. Evans, and M.H. Halperin, 1941, Ophthalmic aspects of acute oxygen deficiency, Arch Ophthalmol NY. 26:886–913.
18. J. Mandlebaum, 1941, Dark adaptation; physiologic and clinical considerations Arch. Ophthalmol. 26:203–239.
19. A.M.P. Hamilton, M.W. Ulbig, and P.J. Polkinghorne, 1996, Management of Diabetic Retinopathy, BMJ press London pp 136.
20. Diabetic Retinopathy Research Group. Second Report, 1978, Amer. J. Ophthalmol. 85:82–106.
21. S.E. Simonsen, 1965, Electroretinographic study of diabetics; Preliminary Report, Acta Ophthalmologica, 43:841–843.
22. T. Amemiya, 1977, Dark adaptation in diabetics. Ophthalmologica, 174:322–326.
23. D.B. Henson, and R.V. North, 1979. Dark adaptation in diabetes mellitus, Brit J. Ophthalmol. 63:539–541.
24. M.O. Scase, D.H. Foster, W.P. Honan, J.R. Heron, J.R., M.C. Guilliford, and J.H.B. Scarpello, 1990, Abnormalities in hue discrimination with very brief stimuli in diabetic patients, Clin. Vis. Sci. 6:49–57.
25. K. Frost-Larsen, H.W. Larsen, and S.E. Simonsen, 1981, The value of dark adaptation as a prognostic tool in diabetic retinopathy, Metabolic and Pediatric Ophthalmology, 5:39–44.
26. E.B. Roeker, E. Pulos, G.H. Bresnick, and M. Severns, 1992, Characterisation of the electroretinographic scotopic b-wave amplitude in diabetic and normal subjects, Invest. Ophthalmol. Vis. Sci. 33:1575–1583.
27. D.V. De Alwys, J.P. Reffin, S.J. Tregear, L.G. Ripley, and A.G. Caswell, 1993, Should the management of diabetic retinopathy be based on the measurement of visual function rather than observations of retinal morphology? Invest. Ophthalmol. Vis. Sci. 34: ARVO Absts. #80 pg. 719.
28. V.C. Greenstein, S.R. Thomas, H. Blaustein, K. Koenig, and R.E. Carr, 1993, Effects of early diabetic retinopathy on rod system sensitivity, Optom. Vis. Sci. 70:18–23.
29. J.V. Lovasik and H. Kergoat, 1993, Electroretinographic results and ocular vascular perfusion in type 1 diabetes, Invest. Ophthalmol. Vis. Sci. 34:1731–1743.
31. S.D. Tregear, P.J. Knowles, D.V. De Alwys, J.P. Reffin, L.G. Ripley, and A.G. Caswell, 1994, Colour vision deficits predict the development of sight-threatening disease in diabetic subjects with background retinopathy. Invest. Ophthalmol. Vis. Sci. 34: ARVO Absts. #81 pg. 719.
32. V.C. Greenstein, D.C. Hood, R. Ritch, D. Steinberger, and R.E. Carr, 1989, S(Blue) cone pathway vunerability in Retinitis pigmentosa, Diabetes, and Glaucoma. Invest. Ophthalmol. Vis. Sci. 30:1732–1737.
33. A. Harris, O. Arend, R.P. Danis, D. Evans, S. Wolf, and B.J. Martin, 1996, Hyperoxia improves contrast sensitivity in early diabetic retinopathy, Brit, J. Ophthalmol. 80:209–213.
34. F. Dean, A. Dornhorst, and G.B. Arden, 1997. Partial reversal of protan and tritan colour defects with inhaled oxygen in insulin dependent diabetic subjects, Brit. J. Ophthalmol. 81:27–30.
35. R.A. Linsenmeier, R.D. Braun, M.A. McRipley, L.B. Padnick, and D.L. Tatchell, 1997. Retinal hypoxia in long term diabetic cats. Invest. Ophthalmol. Vis. Sci. 38: ARVO Abs. S77 #3569.
36. J.S. Tiedman, S.E. Kirk, and J.M. Beach, Inner retinal oxygen consumption increases during hyperglycaemia in diabetic patients, Invest. Ophthalmol. Vis. Sci. 38: ARVO Abs. S714.
37. S. Konno, G.T. Feke, A. Yashida, N. Fujio, D.G. Goger, and S.M. Buzney, 1996, Retinal blood flow changes in Type I Diabetes, Invest. Ophthalmol. Vis. Sci. 37:1140–1148.
38. M.K. v.d. Enden, J.R. Nyengaard, E. Ostrow, J.H. Burgan, and J.R. Williamson, 1995, Elevated glucose levels increase retinal glycolysis and sorbitol pathway metabolism, Invest. Ophthalmol. Vis. Sci. 36:1675–1685.
39. G.T. Feke and S.M. Buzney, 1994, Retinal circulatory abnormalities in type 1 diabetes. Invest. Ophthalmol. Vis. Sci. 35:2968–2975.
40. J. Ditzel, 1979, Changes in red cell oxygen release capacity in diabetes mellitus, Fed. Proc. 38:2484–2488.

41. W.G. Robison, J.L. Jacot, J.P. Glover, M.D. Basso, and T.C. Hohman, 1997, Aldose reductase Inhibitor intervention after significant diabetic-like capillary basement membrane thickening. Invest. Ophthalmol. Vis. Sci. 38:ARVO Abs S715 #3305.
42. K.M. Reiser, 1990, Non-enzymatic glycation of collagen in ageing and diabetes, Proc. Soc. Exp. Biol. Med. 196:17–29.
43. W.D. Robison, N.M. Laver, and M.F. Lou, 1995, The role of aldose reductase in diabetic retinopathy: prevention and intervention studies, Progress in Retinal and Eye Research, 14:593–641.
44. G.B. Arden, J.E. Wolf, and Y. Tsang, 1998, Does dark adaptation exacerbate diabetic retinopathy? Evidence and a linking hypothesis, Vision Res. 38:1723–1729.
45. W.A.H. Rushton, 1963, Increment threshold and dark adaptation, J. Opt. Soc. Amer. 3:104–109.
46. J. Stone and J. Maslim, 1997, Mechanisms of retinal angiogenesis. Progress in Retinal and Eye Research, 16:157–181.
47. H. Tanihara, M. Inatani, and Y. Honda, 1997, Growth factors and their receptors in the retina and pigment epithelium. In Progress in Retinal and Eye Research. 16:271–301.
48. M.K. Mathews, C. Merges, D.S. McLeod, and G.A. Lutty, 1997, VEGFand vascular permeability changes in human DR, Invest. Ophthalmol. Vis. Sci. 38:2729–2741.
49. T. Murata, K. Nakagawa, A. Khalil, T. Ishibashi, H. Inomata, and K. Sueshi, 1996, The relation between the expression of VEGF and the breakdown of BRB in diabetic rat retinas, Lab. Invest. 74:819–825.
50. H. Sone, Y. Kawakami, Y. Okuday, Y. Sekine, S. Honmura, K. Matsuo, T. Seyawa, H. Suzuki, and K. Yamashita, 1997, Ocular VEGF levels in diabetic rats are elevated before observable retinal proliferative changes, Diabetologia, 40:726–730.
51. R.H. Amin, R.N. Frank, A. Kennedy, D. Eliott, J.E. Puklin, and G.W. Abrams, 1997, VEGF is present in glial cells of the retina and optic nerve of human subjects with non-proliferative diabetic retinopathy, Invest. Ophthalmol. Vis. Sci. 38:38.
52. J. Ambati, K.V. Chalan, D.K. Chawla, C.T. D'Angio, E.G. Guillet, S.S. Rose, R.E. Vanderlinde, and B.K. Ambati, 1997, Elevated gamma-aminobutyric acid, glutamate and VEGF levels in the vitreous of patients with PDR, Arch. Ophthalmol. 115:1161–624.
53. A. Khalik, D. Jarvis, D. McLeod, and D. Boulton, 1996, Oxygen modulates the response of the retinal pigment epithelium to Basic Fibroblast Growth Factor and Epidermal factor by receptor regulation. Invest. Ophthalmol. Vis. Sci. 37:436–443.
54. V.A. Alder, E.S.J. Cringle, and M. Brown, 1987, The effect of regional retinal photocoagulation on vitreal oxygen tension. Invest. Ophthalmol. Vis. Sci. 28:1078–1085.
55. C.J. Pournara, M. Tsacopoulos, K. Strommer, N. Gilodi, and P. Leuenberger, 1990, Scatter photocoagulation restores tissue hypoxia in experimental vasoproliferative microangiography in miniature pigs. Ophthalmology 97:1329–1333.
56. M.J. Mosely, S.C. Bayliss, and A.R. Fielder, 1988, Light transmission through the human eyelid: in vivo measurement, Ophthal. Physiol. Opt. 4:229–230.
57. K. Ando and D.F. Kripke, 1996, Light attenuation by the human eyelid. Biol. Psychiatry, 39:22–25.
58. J. Robinson, S.C. Bayliss, and A.R. Fielder, 1991, Transmission of light across the adult and neonatal eyelid in vivo, Vision Res. 31:1837–1840.
59. M.L. Crawford and R.E. Marc, 1976, Light transmission of cat and monkey eyelids, Vision Res. 16:323–324.
60. J.F. Cooper, 1996, A psychophysical method of measuring the attenuation of retinal illumination in humans caused by closing the eyelids, and the relationship of this to skin albedo. Project report submitted for B.Sc. City University, London, U.K.
61. W. Spileers, F. Falcao-Reis, C. Hogg, and G.B. Arden, 1993, A new Ganzfeld electroretinographic stimulator powered by red and green LEDs. Clin. Vis. Sci. 8:21–39.
62. G.B. Arden, J.E. Wolf, T. Berninger, C.R. Hogg, R. Tzekov, and G.E. Holder, 1998, S-cone ERGs elicited by a simple technique in normals and in tritanopes. Vision Res. in the press
63. W.S. Stiles, 1959, Color vision: The approach through increment threshold sensitivity. Proc. Nat. Acad. Sci. 45:100–128.
64. L. Adams, G.B. Arden, and Joan Behrman, 1969, Responses of human visual cortex following excitation of peripheral rods. Brit. J. Ophthalmol. 53:439–452.
65. D. Regan, 1972, Evoked Potentials in sensory physiology and clinical medicine, Chapman and Hall, Lond.
66. G.B. Arden and M. Hall, 1995, Does occupational exposure to Argon laser radiation decrease colour contrast sensitivity in UK Ophthalmologists? Eye, 9:686–696.

30

AGE AND MONOCULAR ENUCLEATION AS POTENTIAL DETERMINANTS OF LIGHT DAMAGE IN THE MOUSE RETINA

Matthew M. LaVail, Naveen N. Kumar, Gregg M. Gorrin, Douglas Yasumura, and Michael T. Matthes

Beckman Vision Center
University of California, San Francisco
San Francisco, California 94131-0730

1. ABSTRACT

Age and monocular enucleation have been examined as potential determinants of the susceptibility of the retina to light damage. BALB/c and C57BL/6J-c^{2J} mice were examined. BALB/c mice showed an initial increase in susceptibility to light damage between 8.5–10 weeks of age. This did not coincide with the time of sexual maturation, as reported previously for the rat. C57BL/6J-c^{2J} mice, which are significantly more resistant to light damage than most other albino strains, showed no age-related change in susceptibility. Monocular enucleation had no effect on the susceptibility of the intact eye to light damage in either strain.

2. INTRODUCTION

The degree of damage to the retina due to exposure to constant light is determined by properties of the exposing light, i.e., intensity, duration and wavelength,[1–3] as well as a number of conditions of the animal or retina. These include body temperature;[1,4] age;[5,6] species;[1,7] strain;[7–10] genetic constitution;[9,11] the dark-adapted and steady-state bleach rhodopsin levels of the animal prior to and during exposure;[1–3] and prior light history.[12,13] The determinants of light damage are reviewed more extensively elsewhere.[14–16]

Since Noell's landmark study demonstrating photoreceptors of the albino rat to be sensitive to the damaging effects of constant light,[1] most studies on light damage have been carried out in this species. The other most widely used rodent model in retinal research is the mouse, which has been shown to differ from the rat not only in

Retinal Degenerative Diseases and Experimental Therapy, edited by Hollyfield *et al.*
Kluwer Academic / Plenum Publishers, New York, 1999.

the sensitivity of the retina to light, but also in the cytopathlogic process that degenerating photoreceptors undergo, at least as viewed histologically.[7] Most of the determinants of light damage shown in the rat have not been analyzed in other species. For this reason, and because of differences seen in light damage between the rat and mouse, we have now explored the relationship of age to light-damage susceptibility in the mouse. Young rats have previously been shown to be more resistant to the damaging effects of light than adults,[5,6] with initial increased susceptibility occurring coincident with sexual maturation.[17,18]

We have also examined the effect of monocular enucleation on the susceptibility of the remaining retina to the damaging effects of light. The rationale for examining the effect of monocular enucleation is that the procedure is required for some studies (see Discussion), and it has been reported that the open eye of monocularly visually deprived rats (by lid suture) is more severely damaged than the eyes of undeprived rats.[19] It was assumed that an animal which has only one eye available keeps it open for longer periods of time than when both eyes are used. Moreover, there is president for unexpected changes in light-damage susceptibility following surgical manipulations such as optic nerve transection.[20]

At least one gene in mice has been implicated in determining the susceptibility of the retina to the damaging effects of light, and different presumptive alleles in different strains of mice impart remarkable differences in sensitivity to light.[9,11] For example, the retinas of albino C57BL/6-c^{2J} (B6-c^{2J}) mice are significantly more resistant to the damaging effects of light than are those of albino BALB/cByJ (BALB/c) mice.[9,11] Since these different strains offer opportunities to explore interactions among determinants of light damage, we have carried out the studies on the effects of age and monocular enucleation in both the light-sensitive BALB/c mice and the light-resistant B6-c^{2J} mice.

3. METHODS

BALB/c and B6-c^{2J} mice born in our laboratory were reared and maintained in cyclic fluorescent light at an illuminance level of less than 20 ft-c. For the study of age influences on light damage, mice at different ages (see Results) were placed into constant fluorescent light at an illuminance level of 115–200 ft-c for 2 or 3 weeks, as described elsewhere.[11] Following light exposure, the mice were then killed by overdose of carbon dioxide followed immediately by vascular perfusion with mixed aldehydes. The eyes were embedded in epoxy resin and sectioned at 1 µm thickness along the vertical meridian, so that superior-inferior hemispheric differences could be studied. These procedures are described elsewhere in detail.[21,22] The superior cornea was trimmed close to the ora serrata to distinguish the superior from the inferior hemisphere in the tissue sections, and the hemispheric identification was confirmed by visualizing the branches of the long posterior ciliary artery located in the choroid on the inferior side of the optic nerve head in the sections.[23]

For the study of monocular enucleation, mice at 4–5 months of age were used. The mice had either one eye removed with iris scissors under ether anesthesia immediately before being placed into light for 3 weeks (3 weeks of monocular exposure), after 2 weeks of light exposure (1 week of monocular exposure), after 1 week of light exposure (2 weeks of monocular exposure), or not at all (binocular control mice). After light exposure they were killed and their eyes processed the same way as for the studies on age influences.

The thickness of the outer nuclear layer (ONL) was taken as a measure of photoreceptor cell number.[24] For measurement of the ONL, 54 measurements were taken at defined points around the eye, 3 each in 9 sets of 250-μm lengths of retina in each of the superior and inferior hemispheres, with the aid of a Bioquant morphometry system (R and M Biometrics, Nashville, TN) as described elsewhere.[11] The measurements were made on one complete section from one eye of each mouse. From these 54 measurements a mean ONL thickness was obtained for each mouse. Statistical comparisons were made using a two-tailed, unpaired Student's *t*-test.

4. RESULTS

4.1. Age as a Potential Determinant of Light Damage

In the first experiment to examine the role of age as a potential determinant of light-damage susceptibility, mice were placed into constant light at various ages (Fig. 1) and examined histologically after 3 weeks. Based on ONL thickness, the BALB/c mice placed into the light at postnatal day (P) 30, P45, and P60 were indistinguishable from one another (Fig. 1). By contrast, the BALB/c mice placed into constant light at P105 or P150 were significantly more susceptible to the damaging effects of light, based on their greater loss of photoreceptor cells and smaller ONL thickness, than the younger mice ($p < 0.00001$; data from P105 and P150 statistically indistinguishable and thus pooled). By contrast, the ONL measurements of the B6-c^{2J} mice were virtually identical at all ages and showed no increased susceptibility to constant light at any age.

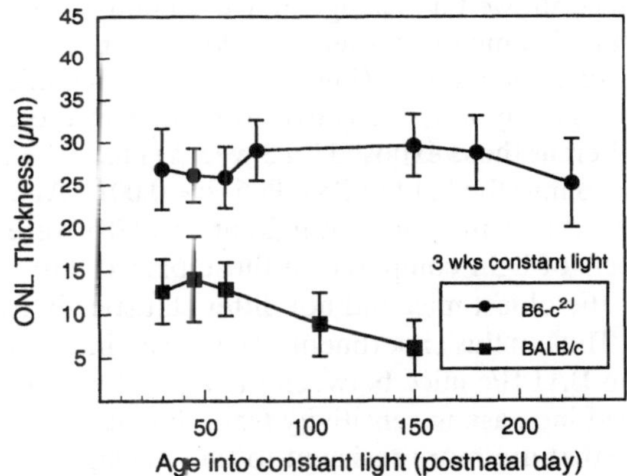

Figure 1. Measurements of the outer nuclear layer (ONL) thickness in mice placed into constant light at different ages and exposed for 3 weeks. The data points are the mean ± S.D. of 8–20 mice at ages P150 and younger and of 3–5 mice at each of the older mice (B6-c^{2J} only). The data points are plotted at the age the mice were introduced into constant light, but their chronological ages were 21 days greater than those represented by the data points. The mean ONL thickness of normal, cyclic light-reared control mice was 41–44 μm at ages up to P105 and about 40–41 μm at P150.

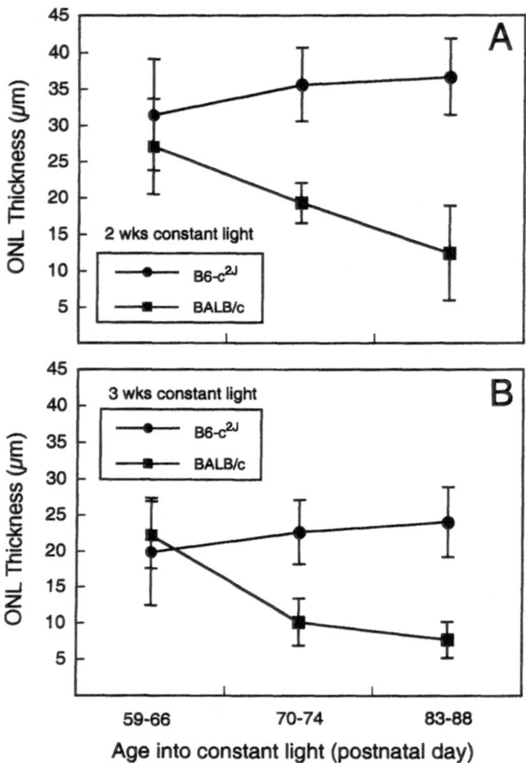

Figure 2. Measurements of the outer nuclear layer (ONL) thickness in mice placed into constant light at different ages and exposed for either 2 weeks (A) or 3 weeks (B). The data points are the mean ± S.D. of 7–15 mice at all ages, except for the BALB/c data points at P70–P74, which are based on 4 and 5 mice for the 2-week and 3-week exposure periods, respectively. The data points are plotted at the age the mice were introduced into constant light, but their chronological ages were 14 (A) or 21 (B) days greater than those represented by the data points.

A similar second experiment was carried out about one year later in an attempt to determine more precisely when the sensitivity in BALB/c mice changed and to confirm that B6-c^{2J} mice showed no change in age-related sensitivity to constant light. In this case, we examined some mice after 2 weeks of constant light and some after 3 weeks. After 2 weeks of constant light (Fig. 2A), the BALB/c mice placed into light at P70–P74 were just marginally more sensitive to light than those placed into light at P59–P66 ($p = 0.05$), whereas those exposed for 3 weeks (Fig. 2B) were significantly more sensitive to those placed into the light at P59–P66 ($p < 0.001$). At both exposure periods the BALB/c mice placed into the constant light at P83–P88 were even more sensitive to the damaging effects of light compared to those placed into light at P59–P66 (Fig. 2), but the values for the oldest mice did not differ statistically from those introduced into light at P70–P74. Thus, in this experiment a significant increase in sensitivity to constant light occurred in BALB/c mice between the ages of about P62 and P72, with an even more pronounced increase in sensitivity from that at P62 at about P85.

The B6-c^{2J} mice that are more resistant to the damaging effects of light showed no increase in susceptibility with either the 2- or 3-week exposure period (Fig. 2), consistent with the first experiment (Fig. 1). It should be noted that in the first experiment, the BALB/c mice were already more susceptible to light damage at P30–P60 than the B6-c^{2J} mice (Fig. 1), whereas in the second experiment the two strains were indistinguishable in their susceptibility at the younger ages of P59–P66 (Fig. 2). We cannot explain this difference, but the experiments were carried out about a year apart, and subtle lighting differences may have existed (e.g., with different spectra from fluores-

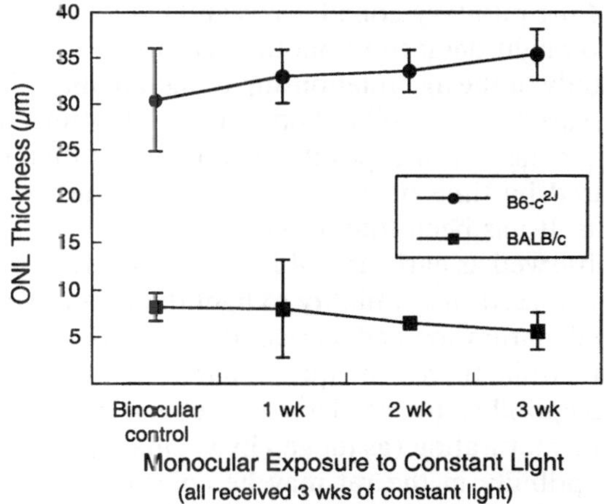

Figure 3. Measurements of the outer nuclear layer (ONL) thickness in adult mice placed into constant light for 3 weeks. The mice had either one eye removed before being placed into light (3 wk monocular exposure), after 2 weeks of light exposure (1 wk monocular exposure), after 1 week of light exposure (2 wk monocular exposure), or not at all (binocular control mice). The data points are the mean ± S.D. of 4–6, except for the BALB/c data point at 1 wk monocular exposure, which is based on 3 mice. No statistical differences were found among the 4 groups within either strain.

cent bulbs of different ages). Nevertheless, the results of each of the two experiments were consistent with one another with regard to the age-related changes in BALB/c mice and the absence of these changes in B6-c^{2J} mice.

4.2. Monocular Enucleation as a Potential Determinant

To examine the effect of monocular enucleation on the remaining eye, adult BALB/c and B6-c^{2J} mice were exposed to constant light for 3 weeks with one eye removed before the animals were placed into constant light. These were compared to age-matched binocular controls (many were littermates) that had both eyes intact with no surgical procedure. In neither strain were the ONL measurements of the monocularly enucleated mice statistically different from the binocular control mice (Fig. 3). In addition, some mice of both strains were exposed for a total of 3 weeks, but had both eyes intact for either 1 or 2 weeks of constant light, at which time one eye was removed. Thus, these mice had either 1 week or 2 weeks of monocular light exposure. In neither case were the ONL measurements of these mice statistically different from those of the binocular control mice or those that were monocularly enucleated for the full 3-week period of light exposure (Fig. 3).

5. DISCUSSION

In albino rats where age-related susceptibility to light damage was first seen,[5,6] the initial increased susceptibility occurred at 6–8 weeks of age. This was concurrent with the period of sexual maturation in the rat, as indicated by the vaginal opening age of P41–P42.[18] This suggested a possible relationship between the degree of light damage and the maturation of the pituitary-gonadal axis and led to a number of studies on the hormonal influences on light damage to photoreceptors.[17,18,25,26]

In the present study of the age relationship to light damage in mice, BALB/c mice showed an initial increased susceptibility between 8.5–10 weeks of age. Thus, BALB/c mice show the initial increase in susceptibility at a later age than rats. However, sexual maturation as measured by time of vaginal opening in mice generally occurs at an *earlier* age than in rats. It has been found to occur in albino mice at a median age of P35,[27] and has been observed as early as P24 in some C57BL/6J mice.[28] Thus, while our findings do show an increased susceptibility to light damage in BALB/c mice, as in the rat, the events of sexual maturation and increased susceptibility do not coincide as they do in the rat. Therefore, while there is no question that pituitary and gonadal hormones can influence the susceptibility to the damaging effects of constant light,[17,18,25,26] the precise timing of sexual maturation (as judged by vaginal opening), and initial increase in light damage susceptibility in the rat may be coincidental. This does not exclude some of the complex aspects of hormonal control in the regulation of age-related susceptibility to light damage.

The B6-c^{2J} mice showed no age-related change in susceptibility to light damage up to almost 8 months of age (Fig. 1). These animals presumably became sexually mature at or near the same time as pigmented (non-albino) coisogenic C57BL/6J mice, since they breed well in our laboratory. Thus, reproductive hormone differences presumably do not explain the lack of age-related change in susceptibility. A more likely explanation is that the influence of the light-resistance gene(s) in B6-c^{2J} mice[11] is so strong that it masks an age-related change in susceptibility. If so, we would predict that such a change might be seen if the mice were exposed to light of significantly higher intensity. This remains to be demonstrated.

Monocular enucleation had no effect on the susceptibility of the remaining eye to the damaging effects of constant light. As noted in the Introduction, Fifkovál reported that the open eyes of rats with monocular lid suture were more severely affected than the two open eyes of control rats.[19] There are several reasons why this observation may not be valid. First, only one measurement was made on either side of the optic nerve head by Fifková. It was later shown that a major superior/inferior asymmetry exists in the distribution of light damage,[2] and that it is important to obtain an ONL measurement in a way that integrates the thickness over the entire retinal section,[3] as we have done in the present study. Second, the eyes were embedded in paraffin, which does not always provide highly consistent thickness of retinal layers. Third, the differences reported were extremely small, less than 3 μm, i.e., less than two-thirds the thickness of a single photoreceptor nucleus. The combination of these factors makes the reported differences in the monocularly occluded rat eyes less than compelling. Moreover, in preliminary experiments with rats, Bush and Williams (R. Bush, personal communication) were unable to replicate the findings of Fifková. Thus, it seems highly unlikely that monocular enucleation has an effect on the susceptibility of the remaining, intact eye.

In the absence of molecular probes, monocular enucleation is often used when the genotype of a rodent must be determined for some breeding schemata. For instance, in the production of congenic strains of rats or mice with inherited retinal degenerations[29–31] or in the establishment of linkage between genetic loci,[31] it is often necessary to determine unequivocally the genotype of animals at genetic loci that cause photoreceptor degenerations while, at the same time, keeping the animals available for breeding purposes. Based on the present study, it seems clear that the monocular enucleation should have no effect on the susceptibility of the remaining eye to light damage.

ACKNOWLEDGMENTS

We thank Nancy Lawson and Gloria Riggs for technical and administrative assistance. This work was supported in part by NIH Research Grant EY01919 and Core Grant EY02162; and funds from the Foundation Fighting Blindness, Research to Prevent Blindness, and That Man May See, Inc.

REFERENCES

1. Noell, W.K., Walker, V.S., Kang, B.S., and Berman, S., 1966, Retinal damage by light in rats, *Invest. Ophthalmol.* **5**:450–473.
2. Rapp, L.M. and Williams, T.P., 1980, A parametric study of retinal light damage in albino and pigmented rats, in: *The Effects of Constant Light on Visual Processes*, (T.P. Williams and B.N. Baker, eds.), pp. 135–159, Plenum Press, New York.
3. Williams, T.P. and Howell, W.L., 1983, Action spectrum of retinal light-damage in albino rats, *Invest. Ophthalmol. Vis. Sci.* **24**:285–287.
4. Barbe, M.F., Tytell, M., Gower, D.J., and Welch, W.J., 1988, Hyperthermia protects against light damage in the rat retina, *Science* **241**:1817–1819.
5. Ballowitz, L. and Dammrich, K., 1972, Retinaschaden bei ratten nach einer fototherapie, *Z. Kinderheilk.* **113**:42–53.
6. O'Steen, W.K., Anderson, K.V., and Shear, C.R., 1974, Photoreceptor degeneration in albino rats: dependency on age, *Invest. Ophthalmol. Vis. Sci.* **13**:334–339.
7. LaVail, M.M., Gorrin, G.M., Repaci, M.A., and Yasumura, D., 1987, Light-induced retinal degeneration in albino mice and rats: strain and species differences, in: *Degenerative Retinal Disorders: Clinical and Laboratory Investigations*, (J.G. Hollyfield, R.E. Anderson, and M.M. LaVail, eds.), pp. 439–454, Alan R. Liss, Inc., New York.
8. O'Steen, W.K. and Donnelly, J.E., 1982, Chronologic analysis of variations in retinal damage in two strains of rats after short-term illumination, *Invest. Ophthalmol. Vis. Sci.* **22**:252–255.
9. LaVail, M.M., Gorrin, G.M., and Repaci, M.A., 1987, Strain differences in sensitivity to light-induced photoreceptor degeneration in albino mice, *Curr. Eye Res.* **6**:826–834.
10. Borges, J.M., Edward, D.P., and Tso, M.O., 1990, A comparative study of photic injury in four inbred strains of albino rats, *Curr. Eye Res.* **9**:799–803.
11. LaVail, M.M., Gorrin, G.M., Repaci, M.A., Thomas, L.A., and Ginsberg, H.M., 1987, Genetic regulation of light damage to photoreceptors, *Invest. Ophthalmol. Vis. Sci.* **28**:1043–1048.
12. Penn, J.S. and Anderson, R.E., 1987, Effect of light history on rod outer segment membrane composition in the rat, *Exp. Eye Res.* **44**:767–778.
13. Penn, J.S., Naash, M.I., and Anderson, R.E., 1987, Effect of light history on retinal antioxidants and light damage susceptibility in the rat, *Exp. Eye Res.* **44**:779–788.
14. Tso, M.O.M., 1989, Experiments on visual cells by nature and man: in search of treatment for photoreceptor degeneration, *Invest. Ophthalmol. Vis. Sci.* **30**:2430–2454.
15. Organisciak, D.T. and Winkler, B.S., 1994, Retinal light damage: practical and theoretical considerations, in: *Progress in Retinal and Eye Research*, (N. Osborne and G. Chader, eds.), pp. 1–29, Pergamon Press, Oxford.
16. Rapp, L.M., 1995, Retinal phototoxicity, in: *Handbook of Neurotoxicology*, (L.W. Chang, and R.S. Dyer, eds.), pp. 963–1003, Marcell Dekker, Inc., New York.
17. Olafson, R.P. and O'Steen, W.K., 1976, Hormonal influences on photoreceptor damage: the pituitary gland and ovaries, *Invest. Ophthalmol.* **15**:869–872.
18. O'Steen, W.K., 1980, Hormonal influences on retinal photodamage, in: *The Effects of Constant Light on Visual Processes*, (T.P. Williams, and B.N. Baker, eds.), pp. Plenum Press, New York.
19. Fifková, E., 1972, Effect of light and visual deprivation on the retina, *Exp. Neur.* **35**:450–457.

20. Bush, R.A. and Williams, T.P., 1987, The effect of unilateral optic nerve section on retinal light damage in rats, *Exp. Eye Res.* **52**:139–153.
21. LaVail, M.M. and Battelle, B.A., 1975, Influence of eye pigmentation and light deprivation on inherited retinal dystrophy in the rat, *Exp. Eye Res.* **21**:167–192.
22. Faktorovich, E.G., Steinberg, R.H., Yasumura, D., Matthes, M.T., and LaVail, M.M., 1992, Basic fibroblast growth factor and local injury protect photoreceptors from light damage in the rat, *J. Neurosci.* **12**:3554–3567.
23. LaVail, M.M., Matthes, M.T., Yasumura, D., and Steinberg, R.H., 1997, Variability in rate of cone degeneration in the retinal degeneration (*rd/rd*) mouse, *Exp. Eye Res.* **65**:45–50.
24. Michon, J.J., Li, Z.L., Shioura, N., Anderson, R.J., and Tso, M.O.M., 1991, A comparative study of methods of photoreceptor morphometry, *Invest. Ophthalmol. Vis. Sci.* **32**:280–284.
25. O'Steen, W.K. and Kraeer, S.L., 1977, Effects of hypophysectomy, pituitary gland homogenates and transplants, and prolactin on photoreceptor destruction, *Invest. Ophthalmol. Vis. Sci.* **16**:940–946.
26. O'Steen, W.K., 1979, Hormonal and dim light effects in retinal photodamage, *Photochem. Photobiol.* **29**:745–753.
27. Engle, E.T. and Rosasco, J., 1927, The age of the albino mouse at normal sexual maturity, *Anat. Rec.* **36**:383–388.
28. Bronson, F.H., Dagg, C.P., and Snell, G.D., 1966, Reproduction, in: *Biology of the Laboratory Mouse, 2nd Ed.*, (E.L. Green, eds.), pp. 187–204, McGraw-Hill Book Co., New York.
29. LaVail, M.M. and Sidman, R.L., 1974, C57BL/6J mice with inherited retinal degeneration, *Arch. Ophthalmol.* **91**:394–400.
30. LaVail, M.M., Sidman, R.L., and Gerhardt, C.O., 1975, Congenic strains of RCS rats with inherited retinal dystrophy, *J. Hered.* 242–244.
31. LaVail, M.M., 1981, Assignment of retinal dystrophy (rdy) to linkage group IV of the rat, *J. Hered.* **72**:294–296.

ULTRAVIOLET LIGHT DAMAGE AND REVERSAL BY RETINOIC ACID IN JUVENILE GOLDFISH RETINA

De-Mao Chen,[1] Guangjun Dong,[1,2] and William S. Stark[1]*

[1] Department of Biology
Saint Louis University
St. Louis, Missouri
[2] Institute of Microbiology
Chinese Academy of Sciences
Beijing, Peoples' Republic of China

1. SUMMARY

Because their spectral sensitivity shows that ultraviolet [UV] light reaches the retina,[1] we utilized juvenile goldfish [*Carassius auratus*] to study light damage induced by 300–400 nm light at about 30 $\mu W\,cm^{-2}$. Electron microscopy [EM] and the electroretinogram [ERG] were used to assay retinal health. Cone ellipsoids in the vicinity of mitochondria were tortuous and swollen at 3 days of such near-UV treatment. Double cones showed some outer segment damage. Curiously, at 6 days, some mitochondria were displaced toward the external limiting membrane in double cones. By 12 days, the double cone outer segments were nearly gone, and the mitochondrial displacement extended to single cones [including UV cones]. Inner segments were filled with a smooth reticulum. Rods were normal. Although there was ultimately [by 32 days] a preferential loss of UV visual sensitivity, at 12 days a UV sensitivity peak remained which could be isolated by chromatic adaptation, suggesting that UV cones continued to function. Widespread cone damage by UV light suggests a mechanism independent of visual pigment absorbance, though the delayed loss of UV sensitivity in UV cones may be mediated through their UV rhodopsin. Retinoic acid had been implicated in protection of photoreceptors from light damage, retinal development, and gene

* For correspondence at Department of Biology, Saint Louis University, 3507 Laclede Avenue, St. Louis, MO 63103-2010, U.S.A., E-mail: StarkWS@SLU.EDU, Home page: http://starklab.slu.edu

Retinal Degenerative Diseases and Experimental Therapy, edited by Hollyfield *et al.*
Kluwer Academic / Plenum Publishers, New York, 1999.

transcription. All-*trans* retinoic acid [1 μM] was added to the aquarium. Treatment of 3-day-UV fish with retinoic acid for 5 hr and of 12-day-UV fish for 7 hr reversed many of the signs of UV damage.

2. INTRODUCTION

The visual sensitivity of fish to UV light has been of substantial interest for nearly two decades.[2,3] We recently demonstrated that a UV sensitive cone mechanism could be detected and isolated by spectral analyses of the ERG in juvenile but not adult goldfish.[1] A similar UV sensitivity loss in developing retinas has been described in several other teleost species: brown trout,[4] rainbow trout,[5] sockeye salmon,[6] Atlantic salmon,[7] yellow perch,[8] and rudd.[9] Here we report that treatment with moderate intensities of UV light causes widespread damage to cones.

We also determined the effects of retinoic acid on UV-damaged goldfish retinas. Browman & Hawryshyn[10] reported that retinoic acid can induce precocial loss of UV photoreceptors in juvenile rainbow trout, but it also induces a reappearance of UV sensitivity in larger trout. The product of a retinoic acid responsive gene[11] [as well as several growth factors and neurotrophic agents][12,13] was shown to rescue rat photoreceptors from the light-damage. Retinoic acid and/or its binding protein is present in amacrine cells,[14] Muller cells,[15] pigment epithelium,[16] and interphotoreceptor matrix.[17] Retinoic acid and its receptors play an important role in morphogenesis in developing retinas.[16,18–21] We report here that UV-damaged cones of goldfish could be rescued by retinoic acid which adds to a growing literature on its relevance to visual system. This laboratory's interest in retinoic acid stems in large part from work on *Drosophila* where retinoic acid is neither essential nor toxic: total retinoid deprivation and replacement are thus possible *in vivo*, and retinoic acid was shown to regulate transcription of genes for rhodopsin[22,23] and a retinoid binding protein.[24,25]

3. METHODS

3.1. Animals and UV Light and Hormone Treatment

Juvenile common goldfish [*Carassius auratus*, General Pet Inc., Springfield, MO] [average body length = 34.9 ± 1 mm, average weight = 1.5 ± 0.14 g] were maintained in an aerated aquarium at 22–24°C on a 12 hr light/12 hr dark cycle of dim white room light. Animals were handled in accordance with the ARVO resolution in protocol #937 approved by Saint Louis University's Animal Care Committee.

For UV treatment, juvenile goldfish were illuminated by a "black light" [F 15 T 8/375 BLB, General Electric, USA]. Intensity was calibrated by a 27-5479 portable radiometer/photometer [Ealing Electro-Optics, Inc., Holliston, MA]. The average intensity at the water surface was 39 $\mu W\,cm^{-2}$. At the bottom of the aquarium, the average intensity was 25 $\mu W\,cm^{-2}$. The depth of the water was 10 cm. For retinoic acid treatments, 1 μM all- *trans* retinoic acid [Sigma, St. Louis, MO] was added to the aquarium for 5 hr at 3 days of UV treatment and 7 hr at 12 days of UV treatment before the UV treatment was terminated.

3.2. Electroretinograms

Goldfish were anesthetized with aerated water having 0.008% tricaine methanesulfonate [3-aminobenzoic acid ethyl, Sigma] and immobilized by intramuscular injection of gallamine triethiodide [Flaxedil, 1.5 mg/kg, Sigma]. Fish were placed in a chamber where water was continually circulated through the gills by a tube in the mouth. The recording eye faced into the air while the other side was in water. The recording side of the body was covered by wet sponges. All data were obtained from 12:00 pm to 6:00 pm with light onset at 8:00 am. ERGs were recorded by placing a NaCl glass microelectrode [pulled from inner filament glass with a Narashige puller, model PD-5, Scientific Instrument Lab, Tokyo, Japan] on the cornea and a reference electrode in the mouth. A Getting microelectrode amplifier [Model 5, Iowa City, Iowa] fed into a MacLab/2e [Milford, Mass]—Macintosh LC computer system.[1]

A dual beam optic system delivered stimuli and steady adapting background lights via beam splitter and lenses. Intensity was adjusted by glass neutral density filters. A 150 W Xenon arc lamp [XBO/CR, Osram, Germany], d.c. power supply [Model 1600 Opti-Quip, Highland Mills, NY], and a Bausch & Lomb 500 mm monochromator [Rochester, NY] provided monochromatic stimuli from 350 to 650 nm. Stimulus duration [100 ms] was controlled with an electromagnetic shutter [SD-10 shutter drive timer, Vincent, Rochester, NY]. Stimulus intensities were calibrated, as in previous papers,[1,26] by a photodiode [EG&G HUV 4000B, Salem, MA]. Steady adapting lights, calibrated by the radiometer/photometer [see above] were provided by an orange filter [K2, average full intensity = $138\,\mu W\,cm^{-2}$] and a blue filter [Kodak-Wratten-35, average full intensity = $117\,\mu W\,cm^{-2}$]. Stimuli and adapting background lights were superimposed for illumination onto the cornea. Spectral sensitivity was determined by the quanta required to elicit a $50\,\mu V$ ERG for different wavelengths of stimuli.

3.3. Microscopy

After goldfish were anesthetized, both eyes were enucleated under room light, slit along the lateral sclera, and fixed for about 30 min in a mixture of 4% formaldehyde, 2% glutaraldehyde, and 1% acrolein buffered with 0.1 M phosphate to pH = 7.4. The eyes were hemisected at the scleral margin with scissors. Anterior segments were discarded, and posterior eye cups were placed in fixative at 4°C overnight. The next day, eye cups were further divided into known portions and rinsed thoroughly with 0.1 M phosphate buffer. Postfixation was with 1% osmium tetroxide in 0.1 M phosphate buffer for 2 hr. A graded ethanol dehydration followed. The specimens were infiltrated and embedded with Spurr's low viscosity medium and 100% ethanol in ratios of 1:1 and 1:3 for 30 min each at room temperature. The specimens were moved into pure Spurr's medium for three days at room temperature with new mediim replaced every day. Polymerization took 15 hr at 70°C.

Blocks were trimmed, then sectioned with a glass knife [LKB glass and Knife breaker type 7801, Bromma, Sweden]. Thick sections of about 1 µm were cut with an LKB Ultratome III type 8802 and fixed with heat onto the back of a cover slip. If necessary, sections were stained with toluidine blue. Sections were viewed and photographed with an Olympus light microscope [BH-2], and determination was made as to whether a relevant area for thin sectioning had been reached. Blocks were then thin-sectioned for EM with a Diatome [Fort Washington, PA] diamond knife. Sections were picked up on uncoated grids and stained with 1% uranyl acetate for 10 min and with Reynold's lead citrate for 2 min. A JEOL 100S transmission EM was used for viewing and photography.

4. RESULTS

4.1. Morphology of Rods and Cones in Juvenile Goldfish

Our morphology of untreated goldfish photoreceptors are in agreement with previous light microscopy.[27,28,29] The retina of juvenile goldfish contains rods and four morphologically distinct classes of cones: double cones, a long single cone, a short single cone, and miniature UV cones. The principle cones and most of the long single cones contain a long wavelength sensitive [red] pigment with a maximum absorption at 579–625 nm; accessory cones and the remainder of the long single cones contain a middle wavelength sensitive [green] pigment with a peak at 509–537 nm; the short single cones are sensitive to short wavelength [blue] light [441–452 nm] or near UV light [350–380 nm]. Our microscopy is consistent with the available literature in that UV cones have the shortest myoids; they also are identified by their short outer segments and the location of their nuclei [close to or below the external limiting membrane].

In teleost fish retinas, cone outer segments are arranged in a regular geometric array called a mosaic.[8,29,30] Cone inner segments are known to consist of an ellipsoid and a myoid which contains abundant rough endoplasmic reticulum. Figure 1A is an oblique EM section of the juvenile goldfish retina before damage showing this mosaic of double cones, short and long single cones, UV cones, and rods. UV cones are absent in adult fish retinas.[8,10,30,31]

One hallmark of cone outer segments is the surround of thin calyces [Fig. 1B],[32] filled with closely packed actin filaments, devoid of microtubules and shown to be the major peanut agglutinin binding site.[33] The calycal processes which originate from the inner segment and are only in cones, not rods. An important factor in determining photoreceptor types is the fact that these calycal processes are not as obvious in UV cones as they are in other single cones and double cones.

4.2. The Effects of UV Light

When juvenile goldfish were treated with UV light, the ERG sensitivity to UV light, assayed at 0, 12,19, 32, 38, and 43 days, decreased by 0.35 log units relative to the 520 nm sensitivity. Thus, there may be treatment-induced loss of the UV cone's contribution to sensitivity. However, it should be noted that this loss is against the backdrop of a developmental loss of UV sensitivity.[1]

EM observations show that the cones, especially the double cones, were injured by UV exposure. At 3 days, outer segments were shorted and disorganized, and cone ellipsoids in the vicinity of mitochondria were tortuous and swollen. Interestingly, at 6 days of UV exposure, some mitochondria of double cones were displaced toward the external limiting membrane [Fig. 2A]. By 12 days, the double cone outer segments were nearly gone, and the mitochondrial displacement extended to all single cones includ-

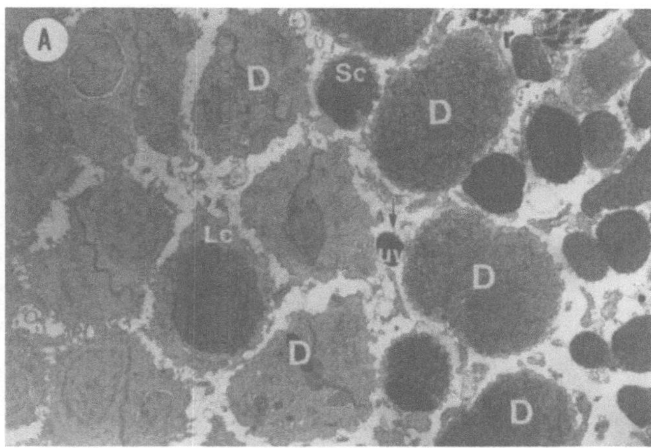

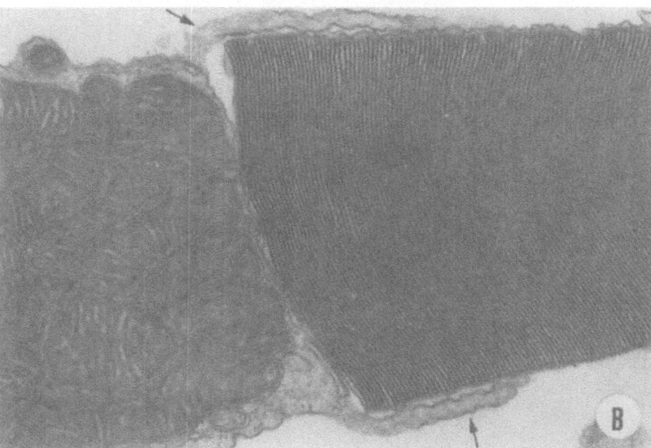

Figure 1. EMs with no UV treatment. A—Oblique section showing double cones [D], long single cones [Lc], short single cones [Sc], UV cones [uv, arrow], and rods [r]. ×3782. B—Longitudinally sectioned cone showing calycal processes [arrows] at the junction of the mitochondria-packed inner segment [left] and the disc filled outer segment [right]. ×22,600.

ing UV cones; inner segments were filled with a smooth reticulum [Fig. 2B]. A few islands of photoreceptor inner and outer segments survived in this damaged area. Seemingly normal UV cones can still be found [Fig. 2C] even though their ellipsoids and nuclei were damaged and vacuolized. Surprisingly, no rods showed obvious injury at 12 days of UV-exposure [Fig. 2D]. Although there was ultimately [by 32 days] a preferential loss of UV visual sensitivity, at 12 days a UV sensitivity peak remained which could be isolated by chromatic adaptation, suggesting that UV cones continued to function.

4.3. The Effects of Retinoic Acid

Two groups of juvenile goldfish were separately raised and exposed to UV light for 3 days and 12 days respectively. Before terminating the UV treatment, 1 μM retinoic acid was added to the aquaria for 5 hr [for the 3 day UV light fish group, Fig. 3A] and 7 hr [for the 12 day UV light fish group, Fig. 3B]. EM showed that most of UV light damage of photoreceptor cells was reversed. Mitochondria are located in the ellipsoid region as they should be. Outer segments are present, and the disorganization and vacuolization of inner segments is recovered.

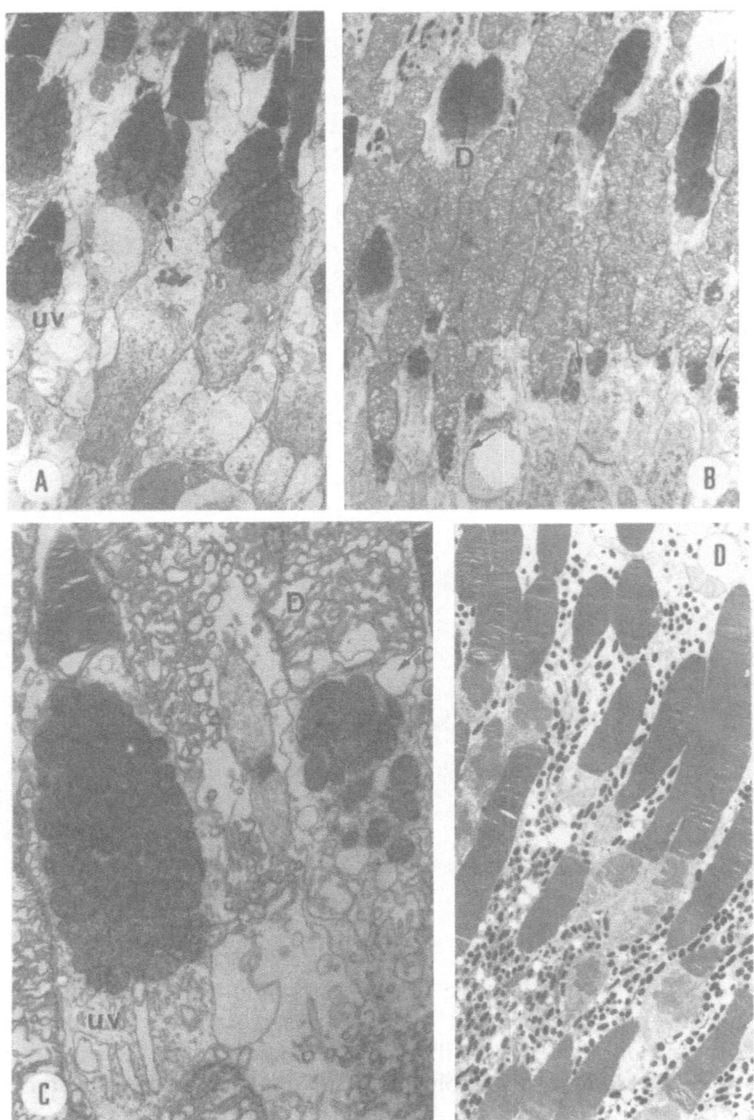

Figure 2. EMs after UV treatment. A—6 days of UV treatment. Note mitochondrial displacement in double cone [arrow] and normal UV cone [uv]. ×2860. B—12 days of UV treatment. Note widespread damage to cones, including double cones [D], and displaced mitochondria [arrow]. ×1380. C—12 days of UV treatment. A UV cone showing normal outer segment and mitochondria. Note displaced mitochondria [arrow] of damaged double cone [D]. ×9110. D—12 days of UV treatment. Lack of damage in rods. ×2590.

5. DISCUSSION

5.1. UV-Light Induced Damage

The conventional wisdom is that UV light is damaging to the eye in many ways.[34] This laboratory's interest in UV damage is based in part on work showing substantial

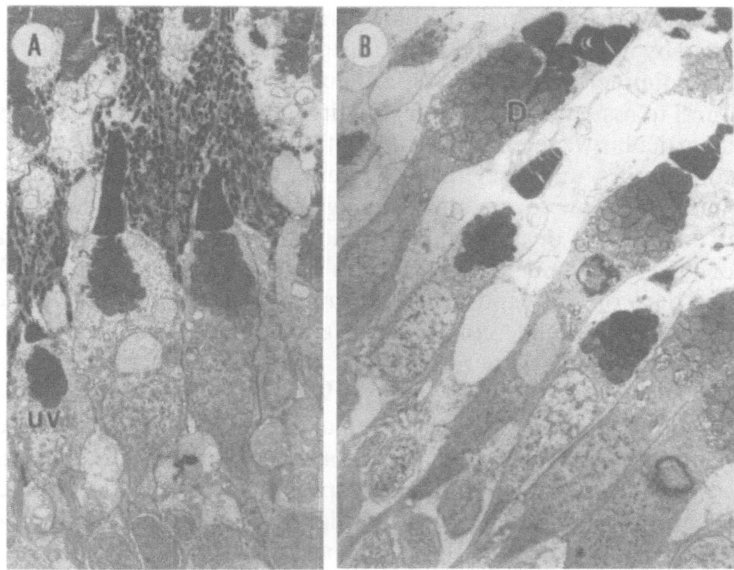

Figure 3. The effects of retinoic acid. A—3 day UV treatment ending with 5 hr of retinoic acid. Note UV cone [uv]. ×2380, B—12 day UV treatment ending with 7 hr of retinoic acid. Note double cone [D]. ×2000.

visual sensitivity of rods and cones to UV light in aphakic observers.[35,36,37] Importantly, the chronic retinal exposure to UV light in unprotected aphakic observers was shown to cause loss in short wavelength cone mediated sensitivity.[38] Work with animal models is thus very important, though these studies have varied substantially in intensity level, species and findings. Regarding intensity level, studies vary with respect to chronic vs. acute actinic exposure, and, especially in acute treatments, the quantity of light delivered [intensity multiplied by time] is probably the important independent variable.[39,40] Intensity levels in different studies are thus sometimes hard to compare.

A paper on rats by Noell et al.[41] is considered to be a pioneering study on retinal light-induced damage. With hindsight, their work relatively dim light in a nocturnal species, though eventually the albino rat model was used in studies of UV light-induced damage using intensities slightly brighter than those of this study.[42] A classic investigation of UV light-induced damage involves the aphakic monkey;[43,44] the intensities in this study were extremely high, perhaps four orders of magnitude brighter than those of the present study, and have only been matched in several studies of short wavelength-induced retinal damage in invertebrates.[39,40] In these studies, on rat[42] and monkey,[43,44] and *Drosophila*,[39,40] UV light was strikingly more efficient than visible light in inducing damage. However, only in the fly was it clearly shown that the UV damage was mediated through rhodopsin. Basinger and Matthes[45] reported that frogs exposed to constant fluorescent light for 14 months showed pigment epithelium swelling, while rod outer segments appeared normal. Summarizing retinal light damage studies in a variety of species, it is now well established that damage can be initiated by means either independent of or dependent on rod and cone visual pigment absorption of visible or UV light. The fish retina has been underutilized as a model of light damage.

Sperling et al.[46] reported that the blue cone sensitivity in rhesus monkeys is selectively eliminated by intense blue light exposure; although the blue stimuli were intense, they were much dimmer than the UV stimuli used in the study on aphakic monkeys.[39] Thus, it is natural to assume that the cell containing the chromophore that is maximally sensitive to a particular wavelength should be the cell that is initially damaged. In light injured photoreceptors, outer segments are usually the first to exhibit damage. For instance, albino trout show reversible "pruning" of rod outer segments upon exposure to solar light.[47] However, our experiments showed that not only UV cones but also other kinds of cones in juvenile goldfish retinas were damaged by UV light. We checked all fish at three days of UV exposure and found that UV light was very efficient for inducing damage to the double cone outer segments. At 12 days of UV exposure, all double cone outer segments were nearly gone, but a few UV cone outer segments can still be seen. ERG data also showed that a UV-sensitive cone mechanism can be isolated by chromatic adaptation. Obviously, retinal damage by UV light was not restricted to short-wavelength cones [UV cones and short single cones] of the juvenile goldfish. Widespread cone damage by UV light in the present work adds to the literature suggesting a mechanism independent of visual pigment absorbance.[42,43,44] However, later loss of UV sensitivity may involve the UV rhodopsin.

Sperling et al.'s[46] above-referenced work on rhesus monkeys suggests that rods are less susceptible to intense blue light than blue cones. We found that 3 days and 12 days of UV exposure did not affect the rods of juvenile goldfish retinas. At 40 days of UV exposure, some of the rods were disorganized, though most rods and even a few cones survived this exposure. However, when juvenile goldfish were treated with thyroxine for 23 days, we saw that most of the rod outer segments were separated and inner segments vacuolized [data not shown]. It is really interesting that rods in fish, frog, and monkey retinas show such high resistance to short wavelength exposure. It would be strategic to take advantage of the continuing development in fish retinas to determine whether the rod precursors, a specialized population of undifferentiated neuroepithelial cells, can develop into new rods and other cones.[10,29,48]

The mitochondrial changes in cone ellipsoids are particularly interesting. In normal cones, mitochondria are tightly packed. The mitochondria became disorganized by 3 days of UV exposure, and, by 6 days, some mitochondria in double cones were displaced toward the external limiting membrane. By 12 days, the mitochondrial displacement extended to all single cones including UV cones. To our knowledge, this is the first report of such mitochondrial displacement, and it is especially important that this can be dramatically reversed by retinoic acid treatment. Mitochondrial enzymes have been postulated to mediate short-wavelength retinal damage.[42] It has also been argued that the deterioration of mitochondrial structure and function appears to be primary in the development of pigment epithelium damage induced by blue light.[49] Since, in addition to mitochondria, the inner segment myoids contain abundant rough endoplasmic reticulum, with its mRNA in restricted locations,[29] the mitochondrial displacement may be an indirect consequence of changes in translation of photoreceptor cell transduction proteins induced by UV light.

5.2. Rescue of Light-Damaged Photoreceptors by Retinoic Acid

It has been shown that retinoic acid, a derivative of vitamin A, controls opsin gene transcription in the compound eye of *Drosophila* with northern[22] and promoter-reporter[23] analyses; retinoic acid also controls transcription of the gene encoding

Drosophila's retinoid and fatty acid binding glycoprotein.[24] These unique studies on *Drosophila* are possible because retinoic acid is neither essential, nor is it toxic, and hence it can be deprived and supplemented at will.

In the visual system, retinoic acid cannot participate in reactions of the visual cycle because it can not be biologically converted to retinaldehyde or retinol.[50] However, recent evidence shows that retinoic acid contributes to growth, morphogenesis, and differentiation in the vertebrate retina.[10,11,20,51,52] Retinoic acid is a necessary dietary supplement in vitamin A deprivation and replacement experiments in rats.[53,54]

Growth factors[12] and a retinoic acid-responsive gene product[11] can prevent light-induced retinal degeneration. Our result shows that the effect of retinoic acid in reversing UV damage is very fast and efficient. During UV exposure, fish were treated with retinoic acid for only 5 hr [3 day UV treatment] and 7 hr [12 days of UV light]. As a result of this treatment, cones appeared normal at the outer and inner segment levels. Presumably, only light-induced damage to cone inner and outer segments that did not result in cell death was repaired by retinoic acid. It would be interesting to verify the expectation that physiological function of the light damaged photoreceptor cells can also be recovered by retinoic acid. The effect of retinoic acid on the protection or rescue from light damage may be mediated by the retinal pigment epithelium where it has developmental effects.[16] Obviously, more work is needed to determine how retinoic acid rescues light damaged retinas in animal models.

ACKNOWLEDGMENTS

Supported by funding from the Provost's office at Saint Louis University and by NIH grant R01 EY07192 to WSS and a Chinese Academy of Sciences fellowship to GD. We thank Dr. Nancy Galvin and Dr. Vernon Fischer for technical help in EM.

REFERENCES

1. D.-M.Chen and W.S. Stark, 1994, Electroretinographic analysis of ultraviolet sensitivity and adult goldfish retinas, *Vision Res.* **34**:2941–2944.
2. F.I. Harosi and Y. Hashimot, 1983, Ultraviolet visual pigment in a vertebrate: a tetrachromatic cone system in the dace, *Science* **222**:1021–1023.
3. A.G. Palacios, T.H. Goldsmith, and G.D. Bernard, 1996, Sensitivity of cones from a cyprinid fish (*Danio aequipinnatus*) to ultraviolet and visible light, *Vis. Neurosci.* **13**:411–421.
4. J.K. Bowmaker and Y.W. Kunz, 1987, Ultraviolet receptors, tetrachromatic colour vision and retinal mosaics in the brown trout (*Salmo trutta*): age-dependent changes, *Vis. Res.* **27**:2101–2108.
5. D.J. Coughlin and C.W. Hawryshyn, 1994, Ultraviolet sensitivity in the torus semicircularis of juvenile rainbow trout (*Oncorhynchus mykiss*), *Vision Res.* **34**:1407–1413.
6. I.N. Flamarique and C.W. Hawryshyn, 1996, Retinal development and visual sensitivity of young Pacific sockeye salmon (*Oncorhynchus nerka*), *J. Exp. Biol.* **199**:869–882.
7. Y.W. Kunz, 1987, Tracts of putative ultraviolet receptors in the retina of the two-year-old brown trout (*Salmo trutta*) and the Atlantic salmon (*Salmo salar*), *Experienta* **43**:1202–1204.
8. E.R Loew and C.W. Wahl, 1991, A short-wavelength sensitive cone mechanism in juvenile yellow perch, *Perca flavescens*, *Vision Res.* **31**:353–360.
9. A.V. Whitmore and J.K. Bowmaker, 1989, Seasonal variation in cone sensitivity and short-wave absorbing visual pigments in the rudd *Scardinius erythrophthalmus*, *J. Comp. Physiol. A* **166**:103–115.
10. H.I. Browman and C.W. Hawryshyn, 1994, Retinoic acid modulates retinal development in the juveniles of a teleost fish, *J. Exp. Biol.* **193**:191–207.

11. K. Unoki, N. Ohba, H. Arimura, H. Muramatsu, and T. Muramatsu, 1994, Rescue of photoreceptors from the damaging effects of constant light by midkine, a retinoic acid-responsive gene product, *Invest. Ophthalmol. Vis. Sci.* **35**:4063–4068.
12. M.M. LaVail, K. Unoki, D. Yasumura, M.T. Matthes, G.D. Yancopoulos, and R.H. Steinberg, 1992, Multiple growth factors, cytokines, and neurotrophins rescue photoreceptors from the damaging effects of constant light, *Proc. Nat. Acad. Sci. USA* **89**:11249–11253.
13. E.G. Faktorovich, R.H. Steinberg, D. Yasumura, M.T. Mattes, and M.M. LaVail, 1992, Basic fibroblast growth factor and local injury protect photoreceptors from constant light damage in the rat, *J. Neurosci.* **12**:3554–3567.
14. V.P. Gaur, A.M. Leeuw, A.H. Milam, and J.C. Saari, 1990, Localization of cellular retinoic acid-binding protein to amacrine cells of rat retina, *Exp. Eye Res.* **50**:505–511.
15. R.B. Edwards, A.J. Adler, S. Dev, and R.C. Claycomb, 1992, Synthesis of retinoic acid from retinol by cultured Muller cells, *Exp. Eye Res.* **54**:481–490.
16. P.A. Compochiaro, S.F. Hackett, and B.P. Conway., 1991, Retinoic acid promotes density-dependent growth arrest in human retinal pigment epithelial cells, *Invest. Ophthalmol. Vis. Sci.* **32**:65–72.
17. G.S. Hageman, M.A. Kirchoff-Rempe, G.P Lewis, S.K. Fisher, and D.H. Anderson, 1991, Sequestration of basic fibroblast growth factor in the primate retina interphotoreceptor matrix, *Proc. Nat. Acad. Sci. USA* **88**:6706–6710.
18. G.A. Hyatt, E.A. Schmitt, J.M. Fadool, and J.E. Dowling, 1996, Retinoic acid alters photoreceptor development *in vivo*, *Proc. Nat. Acad. Sci. USA* **93**:13298–13303.
19. M.W. Kelley, J.K. Turner, and T.A. Reh, 1994, Retinoic acid promotes differentiation of photoreceptors *in vitro*, *Development* **120**:2091–2102.
20. M. Manns and B. Fritzsch, 1991, The eye in the brain: Retinoic acid affects morphogenesis of the eye and pathway selection of axons but not the differentiation of the retina in *Xenopus laevis*, *Neurosci. Let.* **127**:150–154.
21. E. Wagner, P. McCaffery, J. Mey, F. Farthangfar, M.L. Applebury, and U.C. Drager, 1996, Retinoic acid increases arrestin mRNA levels in the mouse retina, *FASEB J.* **11**:271–275.
22. W.L. Picking, D.-M. Chen, R.D. Lee, M.E. Vogt, J.L. Polizzi, R.G. Marietta, and W.S. Stark, 1996, Control of *Drosophila* opsin gene expression by carotenoids and retinoic acid: Northern and western analyses, *Exp. Eye Res.* **63**:493–500.
23. D. Sun, D.-M. Chen, A. Harrelson, and W.S. Stark, 1993, Increased expression of chloramphenicol acetyltransferase by carotenoid and retinoid replacement in *Drosophila* opsin promoter fusion stocks, *Exp. Eye Res.* **57**:177–187.
24. W. Samuel, R.K. Kutty, G. Kutty, T. Iwata, T. Duncan, W.S. Stark, and B. Wiggert, 1998, Transcriptional regulation of *Drosophila* retinoid- and fatty acid- binding glycoprotein gene by retinoic acid, *Invest. Ophthalmol. Vis. Sci. Suppl.* **39**:S423.
25. K. Shim, W.L. Picking, R.K. Kutty, C.F. Thomas, B.N. Wiggert, and W.S. Stark, 1997, Control of *Drosophila* retinoid binding protein expression by retinoids and retinoic acid: Northern, western, and immunocytochemical analyses, *Exp. Eye Res.* **65**:717–727.
26. D.-M. Chen and W.S. Stark, 1992, Electrophysiological sensitivity of carotenoid deficient and replaced *Drosophila*, *Vis. Neurosci.* **9**:461–469.
27. R.E. Marc and H.G. Sperling, 1976, Color receptor identities of goldfish cones, *Science* **191**:487–489.
28. W.K. Stell and F.I. Harosi, 1975, Cone structure and visual pigment content in the retina of the goldfish, *Vision Res.* **16**:647–657.
29. P.A. Raymond, L.K. Barthel, M.E. Rounsifer, S.A. Sullivan, and J.K. Knight, 1993, Expression of rod and cone visual pigment in goldfish and zebrafish: A rhosopsin-like gene is expressed in cones, *Neuron* **10**:1161–1174.
30. Y.W. Kunz, G. Wildenburg, L. Goodrich, and E. Callaghan, 1994, The fate of ultraviolet receptors in the retina of the Atlantic salmon (*Salmo salar*), *Vision Res.* **34**:1375–1383.
31. H.I. Browman and C.W. Hawryshyn, 1994, The developmental trajectory of ultraviolet photosensitivity in rainbow trout is altered by thyroxine, *Vision Res.* **34**:1397–1406.
32. M.M. Januschka, D.A. Burkhardt, S.L. Erlandsen, and R.L. Purple, 1987, The ultrastructure of cones in the walleye retina, *Vision Res.* **27**:327–341.
33. M. Ishikawa, Y. Hashimoto, A. Tonosaki, and S. Sakuragi, 1997, Preference of peanut agglutinin labeling for long-wavelength-sensitive cone photoreceptors in the dace retina, *Vision Res.* **37**:383–387.
34. W.S. Stark and K.E.W.P. Tan, 1982, Ultraviolet light: Photosensitivity and other effects on the visual system, *Photochem. Photobiol.* **36**:371–380.

35. W.S. Stark, R.H. Wagner, and C.M. Gillespie, 1994, Ultraviolet sensitivity of three cone types in the aphakic observer determined by chromatic adaptation, *Vision Res.* **34**:1457–1459.
36. M.S. Griswold and W.S. Stark, 1992, Scotopic spectral sensitivity of phakic and aphakic observers extending into the near ultraviolet, *Vision Res.* **32**:1739–1743.
37. W.S. Stark, 1987, Photopic sensitivity to ultraviolet and wavelengths and the effects of the macular pigments in human aphakic observers., *Curr. Eye Res.* **6**:631–638.
38. J.S. Werner, V.G. Steele, and D.S. Pfoff, 1989, Loss of human photoreceptor sensitivity associated with chronic exposure to ultraviolet radiation, *Ophthalmol.* **96**:1552–1558.
39. W.S. Stark and S.D. Carlson, 1984, Blue and ultraviolet light induced damage to the *Drosophila* retina: Ultrastructure., *Curr. Eye Res.* **3**:1441–1454.
40. W.S. Stark, K.D. Walker, and J.M. Eidel, 1985, Ultraviolet and blue light induced damage to the *Drosophila* retina: Microspectrophotometry and electrophysiology, *Curr. Eye Res.* **4**:1059–1075.
41. W.K. Noell, V.S. Walker, B.S. Kang, and S. Berman, 1966, Retinal damage by light in rats, *Invest. Ophthalmol.* **5**:450–473.
42. L.M. Rapp and S.C. Smith, 1992, Morphologic comparisons between rhodopsin-mediated and short-wavelength classes of retinal light damage, *Invest. Ophthalmol. Vis. Sci.* **33**:3367–3377.
43. W.T. Ham Jr., H.A. Mueller, J.J. Ruffolo Jr., and A.M. Clarke, 1979, Sensitivity of the retina to radiation damage as a function of wavelength, *Photochem. Photobiol.* **29**:735–743.
44. W.T. Ham Jr., H.A. Mueller, J.J. Ruffolo Jr., J.E. Millen, S.F. Cleary, R.K. Guerry, and D.I. Guerry, 1984, Basic mechanisms underlying the production of photochemical lesions in the mammalian retina, *Curr. Eye Res.* **3**:165–174.
45. S.F. Basinger and M.L. Matthes, 1980, The effect of long-term constant light on the frog pigment epithelium, *Vision Res.* **20**:1143–1150.
46. H.G. Sperling, R.S. Johnson, and R.S. Harwerth, 1980, Differential spectral photic damage to primate cones, *Vision Res.* **20**:1117–1125.
47. D.M. Allen and T.E. Hallows, 1997, Solar pruning of retinal rods in albino trout, *Visual Neurosci.* **14**:589–600.
48. P.A. Raymond, 1985, Cytodifferentiation of photoreceptors in larval goldfish: Delayed maturation of rods, *J. Comp. Neurol.* **236**:90–105.
49. E.L. Pautler, M. Morita, and D. Beezley, 1989, Reversible and irreversible blue light damage to the isolated, mammalian pigment epithelium, in: *Inherited and Environmentally Induced Retinal Degenerations*, (M.M. LaVail, R.E. Anderson, and J.G. Hollyfield, eds.), pp. 555–567, Alan R. Liss, New York.
50. J.E. Dowling and G. Wald, 1960, The biological activity of vitamin A acid, *Proc. Nat. Acad. Sci. USA* **46**:587–608.
51. R. Adler, 1993, Determination of cellular types in the retina, *Invest. Ophthalmol. Vis. Sci.* **34**:1677–1682.
52. M.W. Kelley, J.K. Turner, and T.A. Reh, 1994, Retinoic acid promotes differentiation of photoreceptor cells *in vitro*, *Development* **120**:2091–2102.
53. M.L. Katz, M. Kutryb, M. Norberg, C.-L. Gao, R.H. White, and W.S. Stark, 1991, Maintenance of opsin density in photoreceptor outer segments of retinoid-deprived rats, *Invest. Ophthalmol. Vis. Sci.* **32**:1968–1980.
54. M.L. Katz, D.-M. Chen, H.J. Stientjes, and W.S. Stark, 1993, Photoreceptor recovery in retinoid-deprived rats after vitamin A replenishment, *Exp. Eye Res.* **56**:671–682.

32

A COMPARISON OF LIGHT-INDUCED ROD DEGENERATION IN TWO TELEOST MODELS

Donald M. Allen,[1] Chris Pipes,[1] Kristi Deramus,[2] and Ted E. Hallows[3]

[1] Department of Biology
University of Texas of the Permian Basin
Odessa, Texas 79762
[2] Department of Biology
Odessa College Odessa Texas 79764
[3] Kamas State Fish Hatchery
2722 E. Mirror Lake Hwy
Kamas, Utah 84036

1. SUMMARY

Albino rainbow trout, *Oncorhynchus mykiss*, are resistant to light damage, losing only their rod outer segments (ROS) in full daylight (10,000 to 100,000 lux) at 11°C.[1] To compare light damage in albino trout with light damage in albino rodents, we analyzed central retinal structure in albino trout exposed to full daylight, indoor dim daylight (30 lux-30 days or longer) and strong constant light (3000 lux). In albinos remaining outdoors or placed in constant light, ROS volume was reduced but the number of photoreceptor nuclei did not decline. Albinos placed in dim daylight re-grew ROS to 60% or more of normal volume but when returned to outdoor raceways lost most of their ROS volume within 5 days. Outdoor albinos placed in dim daylight replaced ROS much more slowly. In neither case was there a change in number of photoreceptor nuclei. In affected albinos there is apparently little rod cell death during the initial phases of light insult to ROS or thereafter. This confirms that most rod cells with ROS damage survive and retain capacity to re-grow ROS, and any which undergo apoptosis are replaced by cells derived from rod progenitors.

To ascertain whether the 11°C temperatures at which trout were held was protective, we exposed albino and normal oscars, *Astronotus oscellatus*, to 3000 lux constant light at 28°C. Albinos lost ~50% of their rod nuclei over 14 days and normals appeared unaffected. Thus, retinal photo-degeneration in 3000 lux constant light in albino trout at 11°C was limited to ROS loss, whereas ROS destruction lead to loss of

Retinal Degenerative Diseases and Experimental Therapy, edited by Hollyfield *et al.*
Kluwer Academic / Plenum Publishers, New York, 1999.

rod cell bodies in albino oscars at 28°C. The ability to use ambient temperature to preclude or permit light damage to proceed to cell death in a cone rich, diurnal retina could be exploited to study mechanisms for susceptibility to and resistance to photo-degeneration.

2. INTRODUCTION

The nocturnal, rod-dominated retinas of rats and mice are damaged by forced exposures to moderate light and this has led to their choice as animal models for study of light damage to the human retina. Albino strains, which lack a melanized iris and retinal pigment epithelium (RPE), are even more susceptible to damage, where ~200 lux cyclic light or 100 lux constant light leads to death of photoreceptors in rats previously raised in 6 lux.[2–4] When albino rats are exposed to strong continuous light (1000–3000 lux), rods are damaged first and enter apoptosis.[5,6] Disruption of the rod outer segment (ROS) lamellae, loss in ROS volume, loss of visual pigment, loss of evoked scotopic visual responses (ERG), disappearance of photoreceptor nuclei and overall thinning of the ONL have been used, singly or in combination, to assess the progression and/or severity of light damage.[5,7]

For study of light damage to the diurnal human retina, the cone-rich retinas of squirrels or the miniature pig may be more relevant.[8,9] While the cone-rich retinas of normally pigmented diurnal mammals function well during daylight, light damage may occur in albinistic specimens. However, albino mammals are rare in nature, and may develop abnormal visual projections.[10,11] Diurnal fishes (*Teleostei*) have fully duplex retinas with spectrally distinct cones (including a uv-sensitive cone) sharing space with rods. Lacking mobile pupils, they employ photomechanical movements to protect rods from light damage.[12,13] At sunrise, rod myoids elongate and cones contract, while melanosomes within processes of the retinal pigment epithelium (RPE) move vitread to form a pigmented shield in front of the rods. At sundown, the movements are reversed, so that rods are exposed to innocuous light in front of cones.

Since albino teleosts lack functional melanosomes, they should susceptible to light damage. This was recently verified in hatchery-reared tyrosinase(−) albino and normally pigmented rainbow trout, *Oncorhynchus mykiss* by Allen and Hallows.[1] Albinos and normals were spawned in January and raised indoors until April, when they were put in full outdoor daylight (100,000 lux). Normal fish were behaviorally indifferent to sunlight, but albinos attempted to keep their heads shaded where the light level rarely exceeded 10,000 lux. Central portions of the retina were examined in June, July, August, October, November, and February. Normally pigmented trout showed no damage, whereas albinos had severely truncated or missing rod outer segments (ROS) and undamaged cone outer segments (COS). In spite of exposure to 10,000 lux daylight, the census of nuclei in the outer nuclear layer (ONL) of albinos did not decline as it surely would have in albino rats. Either rod somas survived and/or were lost via apoptosis and replaced from rod progenitors adjacent to the ONL. In albinos, there was generally a two-fold increase in numbers of macrophages in the outer retina distal to ROS, many of which contained phagosomes, suggesting that they assist the unpigmented RPE in removal of excess ROS fragments. Estimated ROS volume, derived from the numbers of ROS and ROS lengths observed in the plane of section over a 100 μm of retina, reached its lowest point in July and August, when ROS volume was only 6–8% of normal. This improved to 20% of normal by February and ROS attained near normal length and number when albinos were moved indoors in October and exposed to attenuated ambient daylight (10–30 lux) for ~30 days. Later, visual sensitivity tested during dark adaptation by an optomotor response verified that in August, outdoor albinos were scotopically blind, but those which had been exposed to 10–30 lux ambient daylight for 30 days previously had 2/3 of normal ROS volume and correspondingly improved scotopic visual sensitivity. Thus, rod cell somas persisted in spite of losing up to 94% of their ROS volume and were able to rebuild fully functional ROS.

We wanted to know why retinal damage did not proceed beyond destruction of ROS in albino trout exposed to 10,000 lux daylight. It was possible that the dark period was preventing apoptosis since Penn found cell losses due to constant light in golden shiners, *Notemigonus crysoleucas*.[14] Since apoptosis occurs in rats after brief exposures to constant light at doses about 1/3 as strong as those which are required in cyclic light, we judged that 3000 lux constant light should be comparable to 10,000 lux daylight. We compared the effects of this protocol to those of outdoor daylight and dim daylight on retinal structure of albino and normal trout. We also wanted to define the early trajectory of rod damage after placement in damaging light. Thus, we examined retinas of previously protected albino trout over the first 10 days after their transfer from innocuous light (10–30 lux daylight) to full outdoor light. Finally, since we had hypothesized that the low temperatures at which trout must be held may confer resistance to cell death following ROS damage, we followed the early trajectory of retinal damage in normally pigmented and albino oscars, *Astronotus oscellatus*, after transfer from 30 lux cyclic light into 3000 lux constant light at 28°C.

3. METHODS

3.1. Specimens and Light Measurement

Juvenile albino and normal rainbow trout were maintained in separate adjacent raceways at 11°C as described in Allen and Hallows.[1] Fish were spawned in January and brought into indoor light as fry after 1 month where they remained until being transferred to outdoor raceways on March 15. Albino and normal trout were mixed together after transfer indoors where they were exposed to dim ambient daylight (10–30 lux) or to 3,000 lux constant incandescent light (2–150 watt incandescent bulbs (Philips). Incandescent light is only slightly less effective than fluorescent light in inducing light damage to rat retinas.[15] Albino trout which had recovered their ROS indoors were transferred to outdoor raceways to check for cell losses during early exposure to damaging light. Albino and normal tiger oscars were obtained as 6–8 cm specimens (Crown Aquatics, Tampa, FL) and kept in separate 30 gallon aquaria at 28°C for 1 month in 30 lux cyclic light. They were then given 3000 lux constant light (2–25 watt florescent lamps and 4–150 watt incandescent sources) for 14 days. Trout were maintained on a standard hatchery diet, while oscars were maintained on cichlid pellets. Light was measured in lux with a digital photometer (Tektronix J-16).

3.2. Analysis of Cell Structure

Enucleated eyes were slit along the scleral margin and immersed in ice-cold fixative (2% glutaraldehyde-2% paraformaldehyde in 0.087M Sodium Phosphate, pH 7.2). After removal of the anterior segment, posterior eye cups were post-fixed in 1%

osmium tetroxide, dehydrated in an alcohol series to propylene oxide and embedded in an epoxy/araldyte mixture. Polymerized blocks were faced and cut at 1 µm on an ultramicrotome to expose a section of central retina with the long axes of ROS aligned tangentially to the plane of section. Each of several 100 µm section of central retina was scanned for profile counts, including total number of photoreceptor nuclei in the ONL, number of ROS, and mean length of ROS, as described previously.[1] Here we use the multiplicand [ROS length × No. of ROS/100 µm] as an estimate of ROS volume. ROS volume of albinos is compared to normals w/in the same treatment or referenced to pre-treatment albino controls. A priori statistical comparisons were based on the mean square of the appropriate ANOVA.

4. RESULTS

4.1. 3000 lux Constant Light Maintains the Low ROS Volume Established in Outdoor Daylight

Albino and normal trout were moved from outdoor raceways into 3000 lux constant incandescent light for 45 days (AL,NL) and compared to groups placed in indoor daylight (10–30 lux) for 65 days (AS,NS) or remaining in outdoor raceways (AO,NO). The 24 hours of 3000 lux was estimated to provide from 1/3 to 1/2 the lux available to outdoor albinos with their eyes shaded for 15 hours of daylight.

The effect of constant light was to maintain the reductions in ROS number and length that had been previously attained in outdoor light (AL < NL and AO < NO, Table 1. Fig.1), without decreasing the census of photoreceptor nuclei (ONL nuclei, Table 1). Thus, the effects of solar pruning observed in AO groups are similar to those of 3000 lux constant light. Albinos placed in indoor light showed significant increases in ROS number and length without an increase in the number of ONL nuclei (AS vs. AO, Table 1). Among normal trout, ROS were apparently lengthened in dim light (NS > NO or NL).

Damage to ROS in either AO or AL groups is clearly less severe than would be expected for albino rats exposed to this light, in which rhodopsin would disappear from the retina along with most rod cell nuclei.[5] ROS length appears to be reduced more than the number of existing ROS ellipsoids. Thus, the effect of damaging light is to

Table 1. Visual pigment and rod structure in dark-adapted albino (A) and normal (N) rainbow trout sampled after exposure to different light treatments

GROUP	ONL nuclei per 100 µm	ROS per 100 µm	ROS length (µm)	ROS Vol.	Body length (cm)	n
Sampled 8/23/95	Age: 220 days					
NO: 160d O	62.0 ± 1.12	33.6 ± 2.45	25.6 ± 2.62	860	10.1 ± 0.95	4
AO: 160d O	59.1 ± 3.41	12.1 ± 4.00	4.66 ± 1.15	56	11.0 ± 0.82	4
NS: 95d O, 65d S	61.2 ± 3.90	35.1 ± 2.41	33.7 ± 3.07	1183	11.0 ± 0.82	4
AS: 95d O, 65d S	58.1 ± 3.05	31.5 ± 3.41	22.5 ± 2.52	708	10.2 ± 1.04	4
NL: 114d O, 45d L	60.2 ± 2.95	35.0 ± 4.40	26.8 ± 4.85	938	9.87 ± 1.18	4
AL: 114d O, 45d L	58.0 ± 2.04	14.9 ± 5.19	6.55 ± 1.78	98	10.1 ± 1.03	4

O = outdoor daylight 10,000 to 100,000 lux; S = 10–30 lux daylight; L = 3000 lux constant light. Means ± S.D.
Significance: ROS/100 µm: AO < NO, AL < NL ($p < 0.001$) (AS = NS); ROS length: AO < NO, AS < NS, AL < NL ($p < 0.001$).

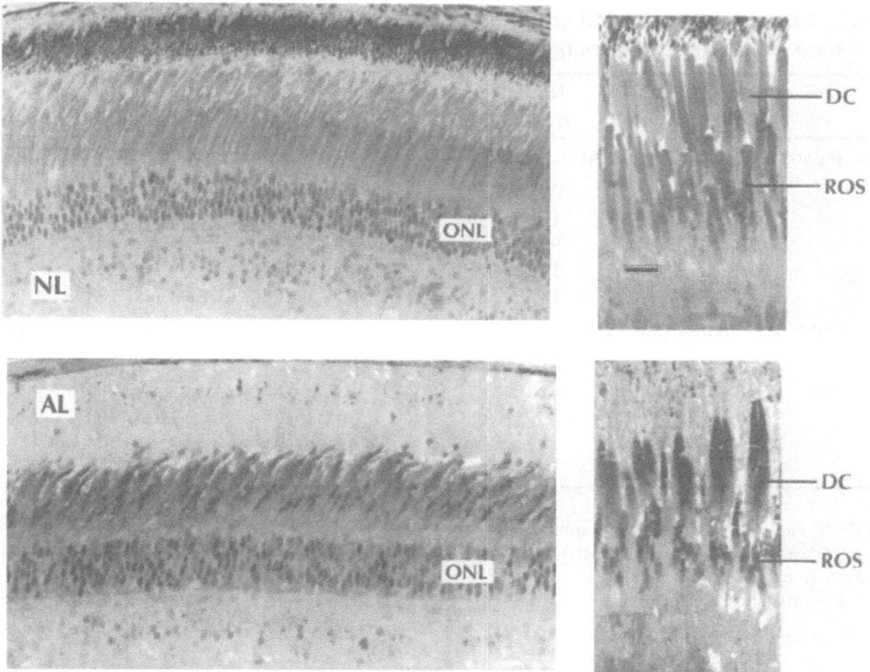

Figure 1. Retinas of normal (NL) and albino (AL) rainbow trout moved from an outdoor raceway and kept in 3000 lux constant light for 45 days. At right; DC = double cone; Scale bar = 10 μm.

"prune" ROS, leaving some ellipsoids in an apparently ROSless state. When albinos are protected (AO → AS), ROS elongate and most ellipsoids are no longer ROSless (Table 1: AS, ROS per 100 μm and ROS length).

4.2. Early Trajectory of ROS Damage and Repair in Albino Trout

Previous protocols featured sampling only after at least 30 days of exposure to experimental light conditions.[1] Early trajectories for ROS damage when indoor fish were moved outdoors (AS → AO) or ROS repair in protected albinos (AO → AS) were not defined. The former is important since there may be a transient loss in number of photoreceptor nuclei if cell death occurs early on which is not balanced by cell replacement. The latter is important also, because the rate at which shortened ROS are rebuilt in dim light may different than when ROS are of normal length. To measure the early phase of ROS damage we pre-treated albinos with 30 days of indoor ambient daylight. Then we returned these fish outdoors and sampled every two days for 10 days (AS → AO group, Table 2, Fig. 2). At the same time, we moved outdoor fish indoors and sampled them every two days for 10 days (AO → AS group, Table 2).

The loss of ROS volume present in protected albinos (AS) decreased rapidly following transfer into outdoor light. Significant declines in both ROS number and length occurred within 2 days after transfer (Fig. 2 and AS → AO, Table 2). Beyond day 4, there were no significant additional losses noted, and a mostly ROSless steady state in the central retina can be maintained thereafter as long as the fish remain outdoors.[1]

Table 2. Early changes in number of ONL nuclei, number ROS, and mean ROS length in albino trout after transfers between dim daylight and full daylight. Means ± S.D.

	ONL nuclei (per 100 μm)	No. of ROS (per 100 μm)	ROS length	ROS Volume	n
	Indoor → Outdoor (AS → AO)		Day 0: after 30 days in 10–30 lux		
Day 0	59.9 ± 3.71	29.5 ± 1.89	29.1 ± 7.14	858	4
Day 2	55.2 ± 2.46	17.8 ± 7.40	17.4 ± 7.00	310	4
Day 4	55.6 ± 8.65	9.86 ± 6.11	12.4 ± 8.17	122	3
Day 7	56.6 ± 2.78	12.3 ± 4.73	11.0 ± 2.97	135	4
Day 10	57.4 ± 3.33	12.5 ± 3.81	9.82 ± 1.08	123	4
	Outdoor → Indoor (AO → AS)		Day 0: after 180 days in 10,000 lux		
Day 0	57.7 ± 3.84	7.04 ± 3.10	8.90 ± 2.21	63	5
Day 2	56.7 ± 2.32	9.64 ± 8.45	9.41 ± 6.77	91	7
Day 4	58.4 ± 5.49	11.3 ± 6.86	10.6 ± 4.08	120	5
Day 7	58.3 ± 4.75	14.2 ± 6.26	12.0 ± 2.79	170	5
Day 10	58.3 ± 4.63	18.4 ± 6.48	14.4 ± 8.12	265	5

Statistics:
Day 0 ROS/100 and ROS length AS > AO initial controls ($p < 0.001$).
AS → AO: ROS/100: Day 2 < Day 0 ($p < 0.01$); Days 4,7,10 < Day 0 ($p < 0.001$); ROS length Day 2 < Day 0 ($p < 0.01$); Day 4,7,10 < Day 0 ($p < 0.001$).
AO → AS: ROS/100 and ROS length (ns.).

Importantly, during the trajectory of damage, the census of ONL nuclei did not change. Conversely, when damaged albinos were moved indoors, there was an evidence of recovery underway in both number of ROS and ROS length, but this was clearly much slower and had not reached significance by day 10. Again, there was no change in census of ONL nuclei. We believe that that significant increases in ROS length and number would have occurred within a few more days with these particular sample sizes, since

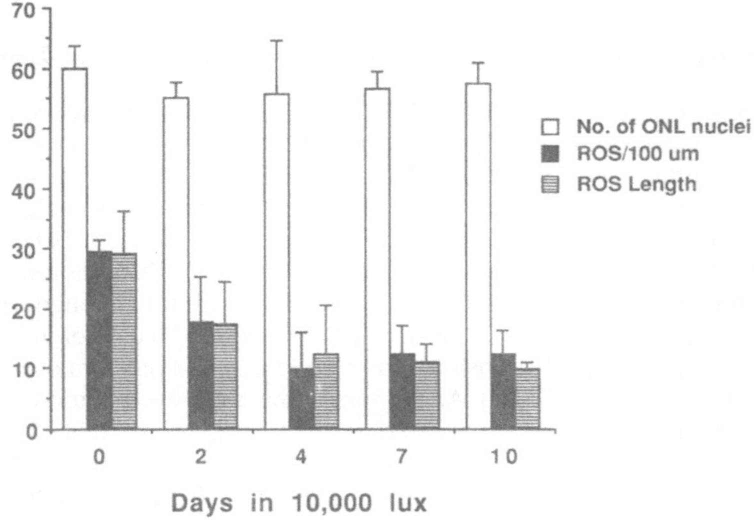

Figure 2. Trajectory of losses in ROS length, number of ROS per 100 μm in albino trout moved from indoor dim daylight into an outdoor raceway. Vertical bars = 1 S.D. as in Table 2. ONL = outer nuclear layer.

by day 30 both ROS number and ROS volume had increased significantly in the initial control protected group (Day 0 controls, AS > AO, Table 2). Clearly, if the rate of ROS loss is dependent on the dosage of noxious light, a change from 30 lux to 10,000 lux (AS → AO) is sufficient to inflict a high initial rate of destruction, and the process could clearly be encouraged to proceed more slowly at some intermediate dose. One the other hand, the rate at which ROS can be rebuilt after affected albinos are transferred into dim light (AO → AS) is probably limited by the rate at which new disc protein can be directed through the ellipsoids and incorporated into new ROS discs. The overall replacement rate must suffice to restore ROS to significantly toward normal volume within 30 days, as shown in the AS group (Table 1: 60% of normal) or as observed in the day 0 AS → AO group (Table 2).

4.3. Early Trajectory of Damage in Albino Oscars at 28°C

The relatively low temperature at which trout were held may have contributed to their ability to limit light damage to pruning of the ROS. Since acclimation of rainbow trout to temperatures higher than 17°C would have deleterious physiological effects, we used albino and normal forms of the oscar to examine the effects of 28°C on the light damage process. Fish were held in dim cyclic light (10–30 lux) for 30 days, then exposed to 3000 lux constant light for 14 days. One normal and one albino were sampled every two days, while 3–7 of each were sampled at day 0 and day 14 to provide statistical samples.

In the 1 μm sections of the oscar retina, it is easy to distinguish rod nuclei from cone nuclei so each class of nuclei was counted separately (note Fig. 3). At day 0, albinos and normals had the same number of rod and cone nuclei in their ONL, as well as having similar ROS volumes. However, though these did not change significantly in normally pigmented fish exposed to constant light, there was a steady decline in ROS

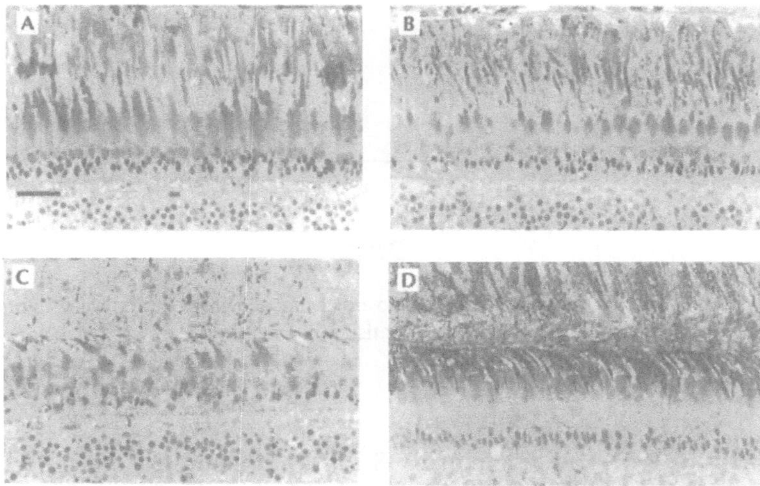

Figure 3. Damage to ROS and rod nuclei in albino oscars moved from 30 lux cyclic light to 3000 lux constant light (note loss of rod nuclei over time—rod nuclei are darkly stained and vitread of larger cone nuclei in ONL—see pointer). A: day 0, albino, B: day 8 albino, C: day 14 albino; D: day 14 normal. A few pyknotic nuclei can be seen (B,C). Scale bar = 25 μm.

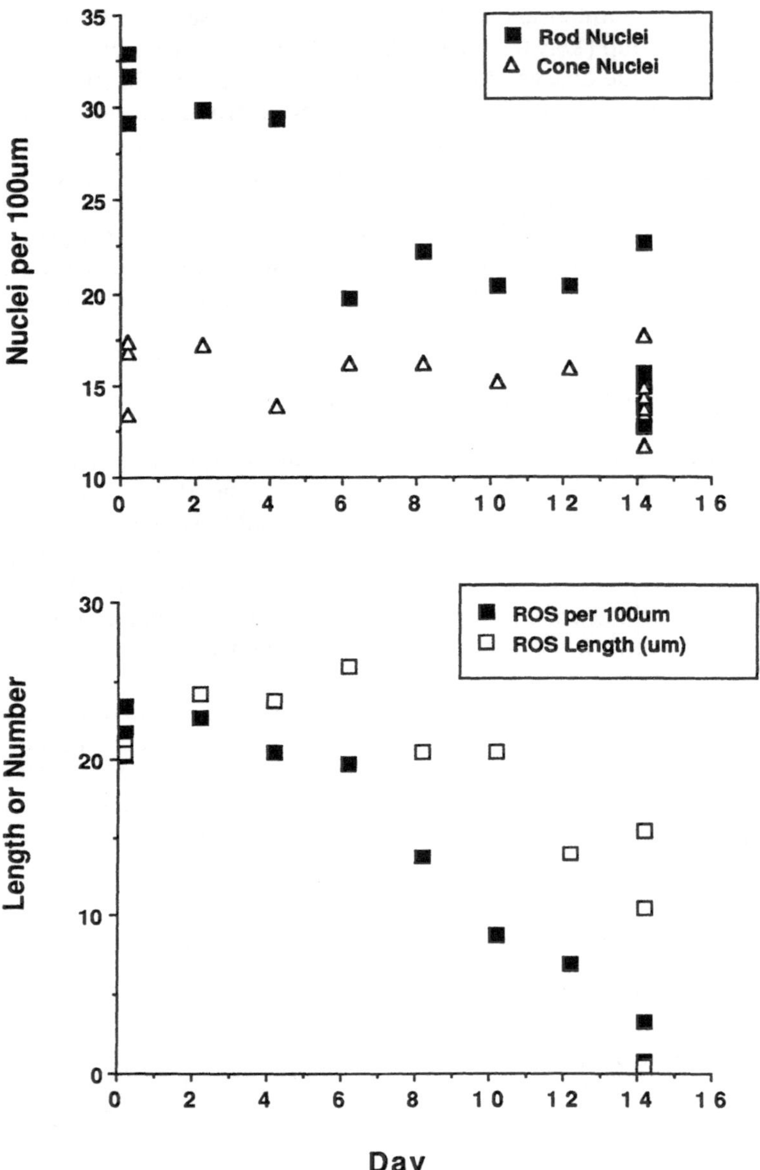

Figure 4. Trajectory of acute rod cell losses in albino oscars moved from 30 lux cyclic light to 3000 lux constant light. Groups at day 0 and day 14 provided statistics in Table 3.

volume (number of ROS and length of ROS, Fig. 4) in albinos. Most importantly, the number of rod nuclei also declined in albinos. By day 14, the ROS per 100 um, ROS length and number of rod nuclei per 100 μm were all significantly reduced ($p < 0.001$) in albinos, such that ROS had virtually disappeared. There were 50% fewer rod cells, and some cells were pyknotic, which is characteristic of apoptosis. The number of cones had not declined by day 14 (Figs. 3, 4 and Table 3).

Table 3. Cone and rod nuclei per 100 μm, ROS per 100 μm and mean ROS length in normal and albino oscars kept in 30 lux cyclic light (NS,AS) and moved to 3000 lux constant light for 14 days (NL,AL). (means ± 1 S.D.). Prior to day 0, both normals and oscars had received 30 days of <30 lux cyclic light

Cone nuclei/ 100 μm	Rod nuclei/ 100 μm	ROS per 100 μm	ROS length	(n)
NS: day 30:				
14.5 ± 0.89	29.7 ± 0.87	18.1 ± 0.75	21.7 ± 1.66	3
AS: day 30:				
15.4 ± 1.83	30.8 ± 1.88	21.2 ± 1.75	20.6 ± 0.17	3
NL: day 14:				
14.2 ± 1.14	30.4 ± 1.32	18.1 ± 2.85	21.6 ± 1.47	5
AL: day 14:				
15.2 ± 3.26	14.0 ± 1.81	0.53 ± 1 00	3.76 ± 5.16	7
ROS Volumes: NS: 537; AS: 652; NL: 545; AL: 7.42				

Statistics: Rod nuclei, ROS/100 um or ROS length: NS = AS = NL, AO < AS or NS or NO (p < 0.001).

5. DISCUSSION

5.1. Temperature, ROS Damage, and Rod Cell Death

Our results show that exposure of albino trout to 3000 lux constant light at 11°C maintains the solar pruning of ROS established previously without affecting the number of ONL nuclei (Table 1). Since this protocol causes apoptosis of rod cells in rats,[6] this confirms that albino trout are apoptosis-resistant. However, the 50% loss of rod cells in albino oscars at 28°C and 3000 lux constant light establishes the oscar as an apoptosis-permissive model. We have verified also that in both species, a significant loss of ROS volume occurs within a few days after previously protected fish are placed in damaging light. Cones remain undamaged in albinos of both species, and rods of normally pigmented fish appear undamaged.

Though effects of different temperatures on retinal light damage have not been tested on either species, our results argue that 28°C permits and 11°C prevents rod cells from entering apoptosis following destruction to ROS from of rod cells within the ONL. Even so, the 50% rod cell loss observed in albino oscars is mild in comparison to damage expected to the ONL in albino or pigmented rats. Interestingly, in albino rats, which maintain body temperatures near 37°C, 3° hyperthermia accelerates damage[2,16] while 7°C of hypothermia reduces the rate of damage.[17,18]

5.2.1. Low Temperature and Resistance to Light Damage. How might a low ambient temperature block cell death? First, if ROS membrane turnover can be managed so as to maintain ROS in a truncated state, this may stay the initiation of apoptosis. This appears to be the case in albino trout at 11°C, in which fish in 3000 lux continuous light for 45 days were still maintaining about 10% of normal volume (AL/NL, Table 1) and showed no loss of ONL nuclei after 30 days. Since dietary deficiencies in vitamin A or in docosahexaenoic acid (DHA) protect against light damage to ROS in rats,[19] it is plausible that as ROS shorten, substrates (DHA) for lipid peroxidation are reduced, eventually leading to a steady state at which truncated ROS can be maintained and/or retain the ability to regenerate new ROS membranes. One could also argue that the availability of opsins and membrane lipids may rise as they are released from disintegrating ROS membranes, and that this may up-regulate the rate at which new discs are added to the shortened ROS. This may be why COS are longer than normal in outdoor albinos.[20] However, this "auto-compensation" is speculative, since in pigmented rats, ROS disc synthesis is reduced proportionately by the degree of light damage.[21] The possible importance of maintaining a small functioning ROS volume at 11°C is emphasized by what happens when ROS volume collapses. Albino oscars in 3000 lux continuous light at 28°C for 14 days had almost no residual ROS volume (1.3% of normal) and 50% of normal rod nuclei in their ONL (Fig. 4, Table 3). In these fish, the balance of ROS synthesis and destruction is such that most rods are ROSless, and cell death is initiated.

5.2.2. Temperature and ROS Lipids. Another aspect of low temperature concerns possible changes in the lipid profile of the rods. Typically, during acclimitization to low temperatures in poikilotherms, the per cent of polyunsaturated fatty acids (PUFA) is increased to maintain membrane fluidity.[22–24] In the crayfish, *Procambarus clarkii*, rhabdomal unsaturated to saturated fatty acid ratios were 2.17 at 4°C and 1.46 at 25°C. When crayfish adapted to darkness and 4°C were placed suddenly into 5000 lux light and 25°C, the microvillar structural integrity was damaged, but this may have been due partly to peroxidation of PUFAs at the higher temperature.[25] An increase in PUFA would be expected to increase, rather than decrease opportunity for lipid peroxidation and generation of free radicals. At first glance this seems contradictory to the idea that low temperatures are protective of the rod cell bodies. However, recent studies have indicated that light-released docosahexaenoic acid (DHA) in rat retinas may inhibit formation of prostaglandins and other metabolites which may act to initiate apoptosis[26] and DHA has been shown to inhibit apoptosis in developing rat retinal cells *in vitro*.[27] Thus, the release of more DHA from ROS by light at 11°C in albino trout would remove substrate for lipid peroxidation from the ROS while its uptake into cell bodies could prevent cell death. Clearly, it would be of interest to examine the lipid profiles of ROS membranes in oscars and trout acclimatized to different temperatures.

5.2.3. ROS Fragments and RPE Lipofuscin. Increased phagosytosis during light damage of ROS could increase the peroxidizable material in the retinal pigment epithelium (RPE). In light damaged quail, *Coturnix coturnix*, rod damage was associated with increased RPE lipofuscin[28] and phagocytized photoreceptor outer segments are a source of lipofuscin in RPE cell cultures.[29] Lipofuscin itself has been shown to enhance lipid peroxidation in cultured human RPE cells.[30] Constant light produced a rapid increase in macrophages which may assist in processing of shed ROS fragments in albino mice.[31] In five different teleosts, Braekevelt found "wandering phagocytes" at the RPE-photoreceptor interface.[32] In albino trout, the volume of RPE phagosomes was 3–5 fold higher than normal. Macrophages, which appear distal to the tips of ROS, were increased significantly and the unpigmented RPE was still intact[1] (see also Figs. 1, 3). These findings indicate a protective role of macrophages in assisting in the removal of ROS fragments that might otherwise be the source of lipofuscin and/or lipid peroxidation within the RPE. Further study could employ inhibitors of macrophage function.

5.3. Natural Resistance to Light Damage: Diurnal Retinas

Though cell death is not blocked at higher at 28°C in albino oscars during 14 days of continuous light, normally pigmented fish appeared unaffected by 14 days of it

(Table 3). This is not surprising, since normal goldfish were unaffected by rearing up to 8 months in 340 lux constant light, but after 12 months exposure, rod nuclei were reduced by 37%.[33] Penn[14] reported that golden shiners sustained damage to ROS, thinning of the ONL and loss of ganglion cells when kept 96 hours in 850 or 1250 lux constant light. On the other hand, when trout or the cyprinid *Notropis cornutis*, were sampled from outdoor locations (100,000 lux or more), the total amount of extracted rod visual pigment showed variability, but not so much to suggest major damage to rods.[34] Such comparisons among normally pigmented diurnal fishes reveal the functional match of ocular pigmentation and natural daylight.

What makes a diurnal retina so resistant to light damage that daylight can be routinely tolerated? Of course, normally pigmented fish have melanin screening ROS from incoming light. Melanin may also act as an anti-oxidant adjacent to the ROS.[35,36] Another factor may be that diurnal retinas are cone rich. Cones appear to be unaffected by albinism, either by light exposure or during development.[1,37] In the rod-dominant nocturnal species, ganglion cell density of albinos is reduced 25% from normals, but albinos of cone rich grey squirrel, *Sciurus carolinensis*, have only 5% fewer ganglion cells than normal.[38] Albinism appears to have little effect on the retina of the budgerigar, *Melopsitticus undulatus*,[39] and experimental detachment of retinas in rhesis monkeys, *Macaca mulatta*, caused photoreceptor outer segments to be lost, but the inner segments remained intact to later re-grow outer segments.[40] This is analogous to the retention of rod cell bodies and recovery of ROS in dim light in albino trout (AS group, Table 1). Overall, these findings suggest that a cone-dominant retina retains a functional network of damage resistant cells sharing space with rods and interacting with them via inner retinal contacts (long wavelength cones have been shown to be active under scotopic conditions in the cyprinid, *Danio aequipinnatus*).[41] Such activity may preserve the function and inhibit apoptosis of rods. Whether COS positioned vitread of ROS during light adaptation offer any protection to rods has not been considered, but this may be of some minor value in albinos lacking the melanin barrier. This could be tested by inhibiting cone contraction in albino fishes prior to light exposure.

5.3.1. Rod Cell Replacement in Teleosts. A possible resistance factor peculiar to fishes is their ability to replace rod photoreceptors (for review, see Hoke and Fernald.[42] Rod cell replacement has been noted in goldfish, *Carassius aureatus*,[43] the cichlid *Haplochromis burton*,[44] and rainbow trout.[45] New rods are added during ocular growth at the retinal margins and centrally from rod progenitors residing adjacent to the ONL within the INL.[46] This mechanism could also provide new rods to replace those lost through mechanical or photic injury. Since a reversible loss in ROS volume is maintained for up to 220 days in albino trout without apparent losses in rod cells, rod cell replacement may occur only to maintain rod density during ocular growth.[1] However, albino oscars, which lose rod nuclei from the ONL, could be effectively used to study rod cell replacement during recovery from light damage. We have not exposed albino oscars to damaging light beyond 14 days, nor have we protected them once they have been damaged, but we would predict that recovery of the 50% loss in rods would require a much higher rate of cell division of rod progenitors adjacent to the ONL.

5.3.2. Growth Factors in Resistance to Light Damage. Growth factors could be protective by stimulating proliferation of new rods from rod progenitors.[47] In *H. burtoni*, insulin-like growth factors (IGF) are expressed in cones,[48] and in goldfish,

IGF-I has been localized to inner plexiform layer throughout the retina and in the circumferential zone, but it stimulates rod precursor proliferation only in the latter.[47,49] However, these goldfish were normally pigmented and their central retinas were not light-damaged. In a related finding, Allen and Hallows noted that rods in the far periphery of albino trout are of near normal length in spite of the tremendous damage sustained in the central retina.[50] This could result of a non-uniform distribution of growth factors which are supporting rod proliferation or ROS synthesis in the germinal zone, although a reflective iris pigment may also be a factor. The latter may not be important, since longer ROS are found in the peripheral retina of deep sea fish.[51] Basic fibroblast growth factor (bFGF) and ciliary neurotrophic factor (CNTF) are known to be up-regulated by mechanical or photic injury in mice,[52,53] and bFGF protects rat retinas from damage by constant light.[54,55] In normal humans, there is a central (less) to peripheral (more) gradient of bFGF.[56] If a similar gradient exists for a growth factor such as IGF in the trout retina, that may explain the occurrence of relatively normal rods in the peripheral margin of affected albinos. Because of it's greater susceptibility to light damage, the albino oscar would be perhaps more suitable for testing whether growth factors actually rescue damaged rod cells. A major point to be resolved is whether the refractory response to IGF in the central retina noted in normal goldfish would hold true in an affected albino, in which central rods are being pruned of ROS but not dying (trout) or are undergoing cell death (oscar).

The attributes of albino teleosts as experimental models for study of photodegeneration have been presented here. Much work remains to be done, but the ability to control ambient conditions (light, temperature) which preclude or permit rod cell death in a cone rich, diurnal retina can now be exploited to study mechanisms for susceptibility and resistance to retinal degeneration.

ACKNOWLEDGMENTS

We thank the State of Utah Fisheries Division of Natural Resources for permission to work at the Kamas hatchery. Travel for DMA was supported by an Ashbel Smith Professorship.

REFERENCES

1. Allen, D.M. and Hallows, T.E., 1997, Solar pruning of retinal rods in albino rainbow trout, *Visual Neurosci.* **14**:589–600.
2. Noell, W.K., Walker, V.S., Kang, B.S., and S. Berman, 1966, Retinal damage by light in rats, *Invest. Ophthalmol. Vis. Sci.* **5**:450–473.
3. Semple-Rowland, S.L., and W.W. Dawson, 1987, Retinal cyclic light damage threshold for albino rats, *Lab. Animal Sci.* **37**:289–298.
4. Williams, R.A., Howard, A.G., and T.P. Williams, 1985, Retinal damage in pigmented and albino rats exposed to low levels of cyclic light following a single mydriatic treatment, *Current Eye Res.* **4**:97–102.
5. Organisciak, D.T. and B.S. Winkler, 1994, Retinal light damage: practical and theoretical considerations, *Prog. in Retinal and Eye Res.* **13**:1–29.
6. Hafezi, F., Marti, A., Munz, K., and C. Reme, 1997, Light-induced apoptosis: differential timing in the retina and pigment epithelium, *Exp. Eye Res.* **64**:963–970.
7. Penn, J.S. and R.E. Anderson, 1991, Effects of light history on the rat retina, *Prog. in Retinal Res.* **11**:75–98.

8. Anderson, D.H., Fisher, S.K., and R. Steinberg, 1978, Mammalian cones: disc shedding, phagocytosis and renewal, *Invest. Ophthalmol. Vis. Sci.* **17**:117–133.
9. Dureau, P., Jeanny, J., Clerc, B., Dufier, J., and Y. Courtois, 1996, Long term light-induced degeneration in the miniature pig, *Molecular Vision* **2**:1–14.
10. Guillory, R.W., 1971, An abonormal retinogeniculate projection in the albino ferret (*Mustela furo*), *Brain Res.* **14**:482–485.
11. Zhang, H.Y. and K.-P. Hoffman, 1993, Retinal projections to the pretectum, accessory optic system and superior colliculus in pigmented and albino ferrets, *Eur. J. Neurosci.* **5**:486–500.
12. Douglas, R.H., 1982, The function of photomechanical movements in the retina of the rainbow trout (*Salmo gairdneri*), *Rev. Can. Biol. Exptl.* **42**:117–122.
13. Burnside, B. and B. Nagle, 1983, Retinomotor movements of photoreceptors and retinal pigment epithelium: mechanisms and regulation, *Prog. in Retinal. Res.* **2**:67–108.
14. Penn, J.S., 1985, Effects of continuous light on the retina of a fish, Notemigonas crysoleucas, *J. Comp. Neurol.* **238**:2121–2127.
15. Osteen, W.K. and K.V. Anderson, 1972, Photoreceptor degeneration after exposure of rats to incandescent illumination, *Z. Zellforsch.* **127**:306–313.
16. Organisciak, D.T., Darrow, R.M., Noell, W.K., and J.C. Blanks, 1995, Hyperthermia accelerates retinal light damage in rats, *Invest. Ophthalmol. Vis. Sci.* **36**:997–1008.
17. deLint, P.J., van Norren, D., and A.M.W. Toebosch, 1992, Effect of body temperature on threshold for retinal light damage, *Invest. Ophthalmol. Vis. Sci.* **33**:2382–2387.
18. Gorgels, T., van Beek, L., and D. van Norren, 1997, Effect of body temperature on retinal damage by 488 nm light in rat, *Microscopy Res. and Tech.* **36**:89–95.
19. Bush, R.A., Reme, C., and A. Malnoe, 1991, Light damage in the rat retina: the effect of dietary deprivation of n-3 fatty acids on acute structural alterations, *Exp. Eye Res.* **53**:741–752.
20. Allen, D.M. and Hallows, T.E., 1995, Reversible visual deficits in albino rainbow trout, abstract: published in the Proceedings of the Western Section of the American Fisheries Society, Park City UT, July 19, 1995.
21. Rapp, L.M., Fisher, P., and H. Dhindsa, 1994, Reduced rate of rod outer segment disk synthesis in photoreceptor cells recovering from UVA light damage. *Invest. Ophthalmol. Vis. Sci.* **35**:3540–3548.
22. Tiku, P.E., Gracey, A.Y., MacCartney, A.I., Beynon, R.J., and A. R. Cossins, 1996, Cold-induced expression of Δ9-desaturase in carp by transcriptional and posttranslational mechanisms, *Science* **271**:815–818.
23. Wallaert, C. and P.J. Babin, 1993, Circannual variation in the fatty acid composition of high-density lipoprotein phospholipids during acclimitization in trout, *Biochim. et Biophys. Acta.* **1210**:23–26.
24. Wallaert, C. and P.J. Babin, 1994, Thermal adaptation affects the fatty acid composition of plasma phospholipids in trout, *Lipids* **29**:373–376.
25. Kashiwagi, T. Meyer-Rochow, Nishimura, K., and E. Eguchi, 1997, Fatty acid composition and ultrastructure of photoreceptive membranes in the crayfish *Procambarus clarkii* under conditions of thermal and photic stress, *J. Comp. Physiol. B* **167**:1–8.
26. Reinboth, J., Clausen, M., and C.E. Reme, 1996, Light elicits the release of docosahexaenoic acid from membrane phospholipids in the rat retina in vitro, *Exp. Eye Res.* **63**:277–284.
27. Rotstein, N.P., Aveldano, M., Barrantes, F.J., Roccamo, A.M., and L.E. Politi, 1997, Apoptosis of retinal photoreceptors during development in vitro: protective effect of docosahexaenoic acid, *J. Neurochem.* **69**:504–513.
28. Fite, K.V., Bengston, L., and B. Donaghey, 1993, Experimental light damage increases lipofuscin in the retinal pigment epithelium of Japanese quail,(*Coturnix coturnix japonica*), *Exp. Eye Res.* **57**:449–460.
29. Wihlmark, U., Wrigstad, A., Roberg, K., Brunk, U.F., and S.E.G. Nilsson, 1996, Formation of lipofuscin in cultured retinal pigment epithelial cells exposed to pre-oxidized photorecetpor outer segments, *APMIS* **104**:272–279.
30. Wihlmark, U., Wrigstad, A., Roberg, K., Nilsson, S.E.V. and U.F. Brunk, 1997, Lipofuscin accumulation in cultured retinal pigment epithelial cells causes enhanced sensitivity to blue light irradiation, *Free Radical Biol. & Medicine* **22**:1229–1234.
31. Sanyal, S. De Ruiter, A., and Ch. Dees, 1984, Light dependent accumulation of macrophages at the photoreceptor-pigment epithelial interface in the retina of albino mice, *Experientia* **40**:851–854.
32. Braekevelt, C.R., 1980, Wandering phagocytes at the retinal epithelium-photoreceptor interface in the teleost retina, *Vision Res.* **20**:495–499.
33. Raymond, P.A., Bassi, C.J., and M.K. Powers, 1988, Lighting conditions and retinal development in goldfish: photoreceptor number and structure, *Invest. Ophthalmol. Vis. Sci.* **29**:27–36.

34. Allen, D.M., Loew, E.R., and W.N. McFarland, 1982, Seasonal change in the amount of visual pigment in the retinae of fish, *Can. J. Zool.* **60**:281–287.
35. Sanyal, S., and G.H. Zeilmaker, 1988, Retinal damage by constant light in chimaeric mice: implications for the protective role of melanin, *Exp. Eye Res.* **46**:731–743.
36. Corsaro, C., Scalia, M., Blanco A., Aiello, I., and G. Sichel, 1995, Melanins in physiological conditions protect against lipoperoxidation. A study on albino and pigmented *Xenopus, Pigment Cell Res.* **8**:270–282.
37. Jeffery, G., Darling, K., and A. Whitmore, 1994, Melanin and the regulation of mammalian photoreceptor topography, *Eur. J. Neurosci.* **6**:657–667.
38. Esteve, J. and G. Jeffery, 1998, Reduced retinal deficits in an albino mammal with a cone rich retina: a study of the ganglion cell layer at the area centralis of pigmented and albino grey squirrels, *Vision Res.* **38**:937–940.
39. Jeffery, G. and A. Williams, 1994, Is abnormal retinal development in albinism only a mammalian problem? Normality of a hypopigmented avian retina, *Exp. Brain Res.* **100**:47–57.
40. Guerin, C.J., Lewis, G.P., Fisher, S.K., and D.H. Anderson, 1993, Recovery of photoreceptor outer segment length and analysis of membrane assembly rates in regenerating primate photoreceptor outer segments, *Invest. Ophthalmol Vis. Sci.* **34**:175–183.
41. Van Roessel, P., Palacios, A.G., and T.H. Goldsmith, 1997, Activity of long wavelength cones under scoptopic conditions in the cyprinid fish, *Danio aequipinnatus, J. Comp. Physiol. A* **181**:493–500.
42. Hoke, K.L. and R.D. Fernald, 1996, Rod photoreceptor neurogenesis, *Prog. in Retinal and Eye Res.* **16**:31–49.
43. Johns, P.R. 1977, Growth of the aldult goldfish eye. III. Source of the new retinal cells, *J. Comp. Neurol.* **176**:343–358.
44. Johns, P.R., and R.D. Fernald, 1981, Genesis of rods in the retina of teleost fish, *Nature* **293**:141–142.
45. Julian, D., Ennis, K., and J. I. Korenbrot, 1998, Birth and fate of proliferative cells in the inner nuclear layer of the mature fish retina, *J. Comp. Neurol.* **394**:271–282.
46. Raymond, P.A. and P.F. Hitchcock, 1997, Retinal regeneration: common principles but a diversityof mechanisms, *Adv. Neurol* **72**:171–184.
47. Boucher, S-M. and P.F. Hitchcock, 1998, Insulin-related growth factors stimulate proliferation of retinal progenitors in the goldfish, *J. Comp. Neurol.* **394**:386–394.
48. Mack, A.F., Balt, S.L., and R.D. Fernald, 1995, Localization and expression of insulin-like growth factor in the telost retina, *Visual Neurosci.* **12**:457–461.
49. Boucher, S-M. and P.F. Hitchcock, 1998, Insulin-like growth factor-I binds in the inner plexiform layer and circumferential zone in the retina of the goldfish, *J. Comp. Neurol.* **394**:395–401.
50. Allen, D.M., 1997, Peripheral rods evade light damage in albino trout, *Invest. Ophthalmol. Vis. Sci.* **38**:Part II abstract # 4787 in ARVO suppl. page S1027.
51. Frohlich, E. and H-J. Wagner, 1996, Rod outer segment renewal in the retina of deep sea fish, *Vision Res.* **36**:3183–3194.
52. Gao, H. and J.G. Hollyfield, 1996, Basic fibroblast growth factor: increased gene expression in inherited and light-induced photoreceptor degeneration, *Exp. Eye Res.* **62**:181–189.
53. Cao, W., Wen, R, Feng, L, Lavail, M., and R. Steinberg, 1997, Mechanical injury increases bFGF and CNTF mRNA expression in the mouse retina, *Exp. Eye Res.* **65**:241–248.
54. LaVail, M.M., Yasumura, D., Matthes, M.T., Lau-Villacorta, C., Unoki, K., Sung, C-H., and R. Steinberg, 1997, Protection of mouse photoreceptors by survival factors in retinal degenerations, *Invest. Ophthalmol Vis. Sci.* **39**:592–602.
55. Matsuda, K., Watanabe, I., Unoki, K., Ohba, N., and T. Muramatsu, 1995, Functional rescue of photoreceptors from the damaging effects of constant light by survival-promoting factors in the rat, *Invest. Ophthalmol. Vis. Sci.* **36**:2142–2146.
56. Li, Z-Y., Chang, J.H., and A.H. Milam, 1997, A gradient of basic fibroblast growth factor in rod photoreceptors in the normal human retina, *Visual Neurosci.* **14**:671–679.

33

REGULATION OF LIGHT ABSORPTION IN THE PIGMENTED RAT RETINA

Theodore P. Williams, Barbara N. Baker, and Janice Dodge

Department of Biological Science
Program in Neuroscience
Florida State University
Tallahassee, Florida 32306-4340

SUMMARY

Earlier work from our laboratory showed that retinal rods of albino rats closely regulate the number of photons they absorb. This "photostasis number" is maintained at about 10^{16} per eye per day, regardless of the light intensity in the habitat or day length. We showed that the retina is plastic and that the regulation is achieved by modification of the optical density of rod cells.

In the present study, we showed that the eyes of pigmented rats also maintain photostasis. Total rhodopsin in the dark-adapted eye is higher in pigmented than in albino rats. Replicating the lower levels seen in albino rats requires either pupillary dilation or use of light intensity that approaches that of outdoor lighting. Under those conditions, rod cell count is also significantly lower than that at light intensities normally used in animal colonies.

Despite these variations in retinal light-absorbing properties, all animals absorbed the photostasis number of photons (1.4×10^{16} per eye per day), 13,000–15,000 photons per average rod per sec. This range of rates is well within that known to saturate mammalian rods and keep them unresponsive to further absorption. We therefore hypothesize that a function of photostasis is to render rods functionless during daylight hours, but this hypothesis requires in addition that most if not all rods absorb at these rates.

To test this possibility, we measured the rate of photon absorptions across a wide range of retinal areas with an ocular transmission photometer (OTP) and found that most, if not all, rods are bsorbing at these rates. An interesting aspect of the OTP studies

Retinal Degenerative Diseases and Experimental Therapy, edited by Hollyfield *et al.*
Kluwer Academic / Plenum Publishers, New York, 1999.

is that excised, pigmented rat eyes retain pupillary responses. These responses are circadian in nature: they anticipate light-period boundaries and shift only gradually after shifts in the photoperiod.

1. INTRODUCTION

The retinal rods of albino rats, regardless of habitat light intensity or day length, all absorb the same number of photons every day.[1,2] This is the case over a broad range of light intensities and day lengths in the habitats in which the rats were born and raised.[1–3] The "photostasis number" of photons absorbed per eye per day for albino rats is 1.2×10^{16} (± 20%).

This regulation is achieved by modification of the light-absorbing properties of the rods, adjusted largely by means of the renewal of the rods' outer segments (ROS). Rats living at low light intensities have long ROS and high rhodopsin concentrations in the ROS disk membranes. Those living at high intensities have short ROS and low rhodopsin concentrations in the disks. Renewal of the ROS provides the "coarse-tuning" of the regulation. It establishes R_o, the total rhodopsin content of the dark-adapted retina.

On the other hand, the rate of rhodopsin regeneration from the bleached state is known to depend on the extent of bleach: strong bleaches are associated with slow regeneration.[4] Regeneration from the bleached state determines the moment-to-moment availability of rhodopsin for absorbing light and, being variable, "fine-tunes" the regulation.

The retina is plastic. If rats born and raised at a given light intensity are moved from their "native" lighting regime to a new one, their ROS will lengthen or shorten, and their R_o levels will change, inversely with the intensity change. Renewal is the mechanism whereby these adaptations occur.[5–7]

An especially interesting interpretation of the "photostasis number" arises from consideration of the rate of absorption per rod that it implies: division of 10^{16} absorptions per eye per 12-h light period by the number of rods in the rat eye, 1.75×10^7 (± 15%),[8,9] yields 11,000–15,000 photons per rod per sec. These rates are in the middle range of those that reportedly saturate rods and keep them saturated.[10–12] We therefore hypothesize that a function of photostasis is to render the scotopic visual system saturated during the daylight hours.

The rate of absorptions per rod cited above is the average rate, which we have calculated from whole-eye rhodopsin measurements; in short, it says nothing about individual rods. To suggest, as we have, that photostasis results in the saturation of "the rod system" is to imply that all or at least most of the rods are absorbing at this rate. If our hypothesis is to be sustained, all rods should absorb at nearly the same rate, and this rate should lie in the range reported to saturate the rods, 8000–20,000 photons per rod per sec.[11]

We show here that pigmented rats maintain photostasis. The present study also supported our hypothesis that photostasis mechanisms are able to keep most if not all rods at saturation. The data come from the ocular transmission photometer (OTP), which measures local rhodopsin concentration and local intensity across the rat retina without interfering with the normal optics of the pigmented rat eye.[13]

The salient result is that most (perhaps all) rods absorb photons at nearly the same rate when the pigmented rat eye is exposed to *Ganzfeld* illumination. The data further suggest a new way to think about scotopic and photopic vision, namely that the duplex retina has evolved such that the number of photons absorbed by the rods over the daylight hours is regulated at just the "correct" (photostasis) level to saturate the scotopic system.

We also report here an unusual iridial response that causes problems in the OTP experiments: The iris of pigmented rats retains the ability to constrict in response to illumination even though the eye is excised.

2. METHODS AND MATERIALS

2.1. Animals

All experiments were conducted according to NIH guidelines for the care and use of animals. Long-Evans hooded (pigmented) rats were used in these studies. These rats were raised at one of three different illuminances: 3-, 200-, and 3000-lux light, cycled 12:12.

2.2. Dark-Adapted Rhodopsin Levels

Rats raised at the various light intensities were dark-adapted overnight and killed by CO_2 asphyxiation (and exsanguination) the next morning in dim red light. Their retinas were extruded through slits across the corneas. Some of the rats in the 200-lux habitat were given daily drops of mydriatic (2% atropine) in both eyes to reveal whether the pupillary response was reducing the light intensity in their eyes and thereby circumventing the need for lower rhodopsin concentration (which is readily observed in albino rats at this intensity).

We then immersed and macerated retinas in distilled water for 5–10 min to burst cells and release soluble proteins. After centrifugation at 10,000 × g for 10 min, the supernatant was removed, and 1.0 ml Emulphogene (1% in Tris buffer, pH 6.5) was added to the pellets, which were thoroughly disrupted for 2–3 min with a microspatula. These tubes were then set aside, with gentle agitation, at 4°C for 1 h. They were then spun down at 10,000 × g for 10 min. The supernatants were drawn off quantitatively, the total volume of each was measured, and they were scanned on a Hewlett-Packard diode array spectrophotometer. Spectra of all samples were obtained before and after complete bleaching in orange light, and difference spectra were generated. From these, the change in absorbance of the rhodopsin at 500 nm was determined. This value was then used to calculate from the Beer-Lambert law (molar absorbance coefficient = $42{,}000\,M^{-1}cm^{-1}$) the concentration of rhodopsin in the extract and, from that, the number of nanomoles per eye.

2.3. Regeneration Rate

At the end of their normal day, when the lights went off in their habitats, animals were killed at various intervals and the retinas assayed for rhodopsin as described above. We determined the rate constant for regeneration, k_r, by constructing rate plots and finding the slopes. More recently, we have shown that extrapolation of these plots to time zero gives R_{ss} values that agree well with those obtained by Rapp's method.[14] Thus, in the process of measuring k_r we also obtain R_{ss}, the other parameter needed for photostasis calculation.

2.4. Ocular Transmission Photometry

The OTP monitors the intensity of light transmitted through excised eyes.[13,15] Intense, actinic bleaching light, passing through the normal optics of an eye, traverses all of the ocular media including the photoreceptor/RPE/choroid complex. A fraction of the light that is not absorbed or scattered exits through the sclera, where it is sampled by a fine-tipped "optical pick-up," which touches the external surface of the sclera. The pick-up is, in turn, optically coupled to a photon-counting system where photon rates are measured, log-transformed, recorded, and displayed against time. The bleaching beam from a "green" He-Ne laser (543 nm) delivers up to 10^{15} photons per sec to the eye by way of a *Ganzfeld* whose intensity distribution has been shown to be very uniform.[13] Achieving this uniformity was important: we had to be sure that any nonuniformities in the intensity distributions found inside the eyes were not caused by nonuniform intensity distributions outside the eyes.

The mature rat retina is 80–90 mm^2 in area.[16] Scale drawings of the pick-up, sclera, choroid, and retina, done with appropriate refractive indices, showed that the optical pick-up collects light from an area of retina[13] of about 1.6 mm^2, corresponding to about 2% of the rods in the eye. The pick-up is attached to a goniometer calibrated in 10° steps out to 70° from the optic nerve.

For all OTP measurements in this study, an eye was suddenly exposed to the full intensity of the laser-lighted *Ganzfeld.* Simultaneously, the photon-counting system began to monitor the light transmitted through the sclera at the chosen location. Photons were collected into bin sizes of 800 ms and stored for later display as $\log_{10}$(cts/sec), change in which is directly proportional to rhodopsin loss due to bleaching. The intensity of this transmitted light increased with time as rhodopsin inside the eye bleached. The "total," asymptotic change in $\log_{10}$(cts/sec), Δ, is directly proportional to the rhodopsin concentration per unit area that had been present in the dark-adapted eye at the given location.[13,15] For the bleaching of rhodopsin to reach near-asymptotic levels, a run typically lasted 300–500 sec.

The data were stored as the log(cts/sec) vs. time, plotted, and then fit with a simple exponential of the form[13,15]

$$\Delta = \Delta_t(1 - \exp(-t/\tau))$$

From the curve fit, we obtained both the total absorbance change, Δ, and the time constant of bleaching, τ. The value of Δ, measured at 543 nm, when multiplied by 2 is equivalent to the absorbance change at 500 nm, the λ_{max} of rat rhodopsin. τ is inversely proportional to the intensity incident on the rods at the location. Thus, one run provides both parameters of interest: $1/\tau$, proportional to the local intensity, and Δ, proportional to the local, dark-adapted rhodopsin concentration per unit area. The product, $(1/\tau) \cdot \Delta$, which is proportional to the initial rate of photon absorptions, is then calculated for the retinal region being examined.

2.5. Pupillometry

The procedure was a modification of one described by Lau et al.[17] A rat was lightly anesthetized with Chloropent^R given IP and then oriented in a standard way beneath an operating microscope in dim red light. When the bright, white light was turned on, the minimal pupil diameter was determined by means of a calibrated graticule fitted

Table 1. Dark-adapted levels of rhodopsin in rats raised at different light intensities

	Light intensity in lux			
	3	200, without mydriatics	200, with mydriatics	3000
R_o (nmol/eye)	2.7	2.6	1.6	1.9

to the eyepiece of the microscope. Lamp intensity was kept constant for all measurements.

2.6. Rod Cell Counts

In each of two separate studies, two groups of Long-Evans rats were raised in cyclic light, one at 3 lux and the other at 3000 lux. When they were 12 weeks old, we killed two rats from each group and counted the nuclei in the outer nuclear layer columns at 48 locations in each retina; sections were made through the vertical meridian.

3. RESULTS

3.1. Dark-Adapted Rhodopsin Levels

Table 1 shows the dark-adapted levels of rhodopsin (R_o) as a function of light intensity. Note that, at 200 lux, when the pupillary response is defeated with mydriatic, the rhodopsin level is lower than when it is not. Rats raişed at 200 lux without mydriatics therefore apparently regulate the light intensity incident on the retina by means of their pupillary responses and thus need not reduce rhodopsin levels. On the other hand, in rats raised at 3000 lux, which approaches outdoor intensities, rhodopsin levels were low even without mydriatics.

3.2. Rod Cell Counts

Part of the difference in R_o between 3000-lux and 3-lux animals is due to a difference in rod cell count. In each of the two studies, a two-tailed, paired *t*-test revealed that the two treatment groups differed significantly ($p < 0.0001$) in mean number of nuclei. The difference was about 2 nuclei per column (Table 2).

3.3. Photostasis Calculation

To determine the photostasis number for pigmented rats, we used the 3000-lux group because only that group showed significant rhodopsin bleaching during the daylight hours, thus permitting the rate of regeneration to be measured when the lights are turned off at the end of the normal day. Figure 1 shows that, in pigmented rats, rhodopsin accumulated nearly linearly with time in the dark, in contrast to that in albino rats, which is regenerated in a first-order fashion.[1] The rate constant, simply determined from the slope of such a function, was 4.8×10^{-4} nmoles/sec. This value is

Table 2. Average outer nuclear layer thickness (nuclei/column) in 48 columns taken from each of two pigmented rats raised at each of two light intensities, in each of two studies. Significance of the difference between the values at the two light intensities was determined in each study by a two-tailed *t*-test

	Light intensity in lux		
	3	3000	p-value
Study 1	11.9 (0.11)	9.8 (0.14)	0.0001
Study 2	11.8 (0.13)	9.8 (0.14)	0.0001

similar to other rate constants we have measured for albino rats.[1] Using this rate constant, along with the steady-state bleach level (difference between the y-intercept [arrow in Fig. 1] and the R_o value, 1.9 nmol), in our equation for the photostasis calculation, we calculated the photostasis number for pigmented rats to be 1.4×10^{16} photons absorbed per eye per day, slightly higher than but within the range of that found for albino rats, 1.2×10^{16}.

3.4. Distribution of Absorption Rate

Figure 2 presents the distribution of absorption rates across part of the pigmented rat retina. This distribution is not as uniform as that in the albino eye, and the data are

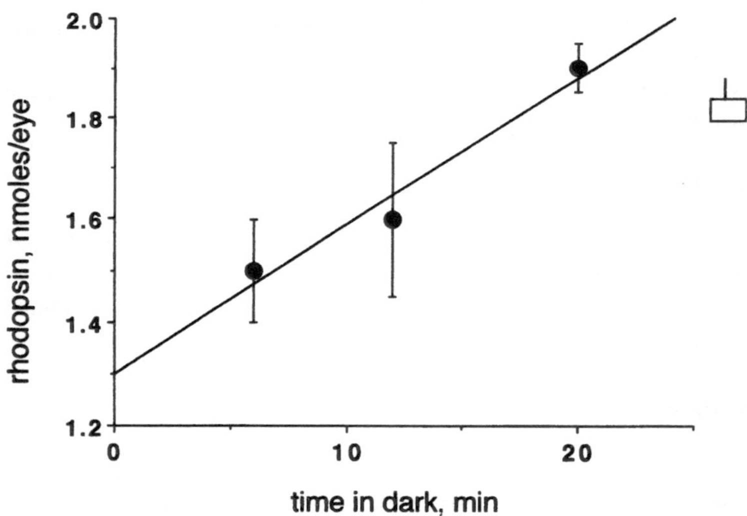

Figure 1. Rhodopsin accumulation with time in the dark. The rate at which rhodopsin is regenerated is linear with time in the dark under these conditions (least squares fit, $r^2 = 0.98$). The slope of this function, 4.8×10^{-4} nmoles/sec, is the rate constant for regeneration and is approximately the same as other rate constants found in albino rats with a similar extent of bleach. The single open symbol represents the totally dark-adapted rhodopsin level.

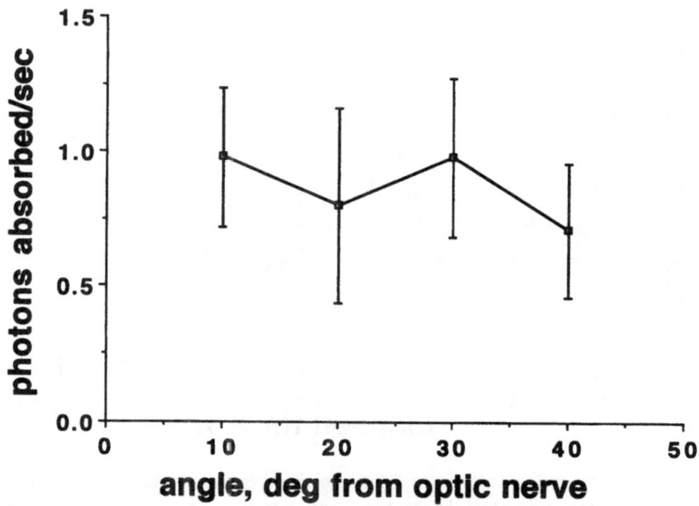

Figure 2. Distribution of photon-absorption rates across a restricted region of the superior half of the pigmented rat retina. These rates were determined by means of an ocular transmission photometer. The variability in the data was found to be due in part to an unusual pupillary response that persisted in excised pigmented rat eyes. Even with the variability, the rates of absorption differ by only 30% from 0° (the region of the optic nerve) to 50°.

more variable.[15] From 0° to 50°, the rate decreases approximately 30%; this, too, is unlike the more shallow decrease in the albino eye.

3.5. An Unusual Iridial Response

The higher variability of OTP results in the pigmented eye is due in part to an unusual pupillary response in the excised eyes: the iris continues to constrict in response to light despite lack of any central innervation. This phenomenon, discovered by Lau et al.[17] and rediscovered in our laboratory, causes problems with OTP measurements. In an attempt to circumvent the problems, we made some measurements on eyes excised during the night and found that the constriction responses were smaller and sometimes absent. Because this result hinted that the iridial response was under circadian control, we made measurements over a 24-h period. The results appear in Fig. 3 and are plotted twice to emphasize the rhythmic nature of the response. Note that the response anticipates lights-off and—on by about an hour.

Next we shifted the photoperiod in the rat habitat by 6 h and followed the shift of the pupillary-constriction response over the next several days (Fig. 4). No appreciable shift occurred immediately, but the response gradually shifted—the daylight responses became weak, taking on nightime character, while the nighttime responses became strong, taking on daytime character, as would a circadian process.

4. DISCUSSION

The present work shows that pigmented rats maintain photostasis. Their rhodopsin levels differ with habitat light intensity, and they lose rod cells at light inten-

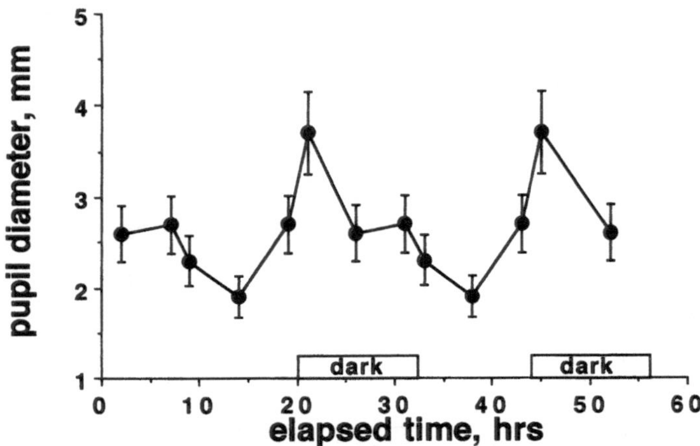

Figure 3. Pupillary constrictions in response to light in excised eyes as a function of time of day. Note that the strength of the constriction is greatest during the daytime hours and weakest during the night. Note also that the response anticipates both lights-on and lights-off, a characteristic of some circadian processes.

sities high enough to approach those outdoors and at somewhat lower intensities if pupillary reflexes are defeated with mydriatics. Until now, the photostasis response had been definitively demonstrated only in albino rats.

Although we have now demonstrated photostasis in them, two problems have slowed our progress toward studying all of the regulatory mechanisms in pigmented rats. First, only a few rats at a time can be raised at light intensities approaching those

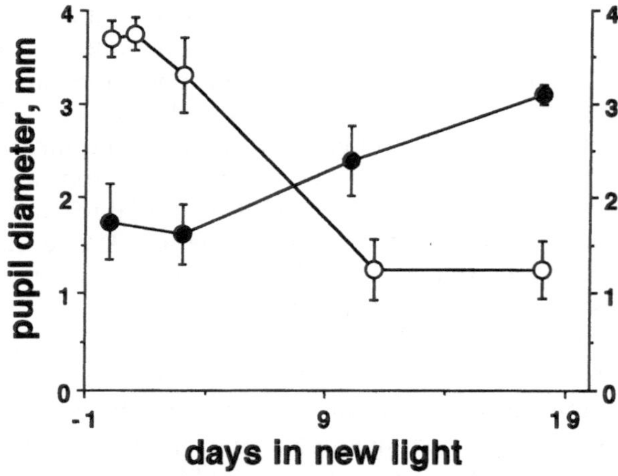

Figure 4. A further test of the circadian character of the unusual pupillary response. The photoperiod was shifted by 6 h, and the pupillary responses were measured as a function of time after the shift. Two points in time were monitored, one that had been during the light period before the shift and one that had been during the dark period before the shift. Both responses shifted gradually with days under the new photoperiod: the daytime response became weak, resembling those that occur at night, and the nighttime response became strong, resembling those that occur during the day. The gradual nature of the shift is a hallmark of circadian processes.

outdoors, because of the difficulty of achieving those intensities without raising the temperature in the cages. In fact, raising any number of rats from birth until 12 weeks of age at a fixed light intensity is extremely costly and labor-intensive. Second, at any but the problematic high light intensities, the pupillary response, the normal as well as the unusual one described here, necessitates use of multiple daily doses of mydriatics (side effects unknown).

One of the regulatory features not thoroughly studied here is the linear regeneration of rhodopsin. Rapp and Williams[14] have shown that rhodopsin in pigmented rats comes to steady state during the daylight hours but that details of this condition differ from that found in albinos. The difference seems not to be a simple intensity effect, although intensity is lower in the pigmented eye than in the albino.[14,15] The linear regeneration rates found here suggest that part of the difference could lie in the details of the regeneration mechanisms.

Finally, the pupillary response that persists without central neural connections is not understood. If this response is similar to the one found in amphibians, which is better understood, the implication is that the irises of pigmented rats contain rhodopsin and that absorption of light by this pigment causes iridial muscles to constrict in response to light.

ACKNOWLEDGMENTS

We thank Dr. Anne B. Thistle for editing the manuscript.

REFERENCES

1. J.S. Penn and T.P. Williams, 1986, Photostasis: Regulation of daily photon-catch by rat retinas in response to various cyclic illuminances, *Exp. Eye Res.* **43**:915–928.
2. C. Parker and T.P. Williams, 1995, The effects of daylength on photostasis, *Invest. Ophthalmol. & Visual Sci.* **36(Suppl.)**:s515.
3. T.P. Williams, B.N. Baker, and J. Dodge, 1994, Pigmented rats, pupillary reflex and photostasis, *Invest. Ophthalmol. & Visual Sci.* **35(Suppl.)**:1608.
4. I. Perlman, 1978, Kinetics of bleaching and regeneration in abnormal (RCS) and normal albino rats in vivo, *J. Physiol. (London)* **278**:141–159.
5. J.-L. Schremser and T.P. Williams, 1992, Photoreceptor plasticity in the albino rat retina following unilateral optic nerve section, *Exp. Eye Res.* **55**:393–399.
6. J.-L. Schremser and T.P. Williams, 1995a, Rod outer segment renewal as a mechanism for adaptation to a new intensity environment. I Rhodopsin levels and ROS length, *Exp. Eye Res.* **61**:17–24.
7. J.-L. Schremser and T.P. Williams, 1995b, Rod outer segment renewal as a mechanism for adaptation to a new intensity environment. II Rhodopsin synthesis and packing density, *Exp. Eye Res.* **61**:25–32.
8. K.S. Lashley, 1932, The mechanism of vision, V, *J. Comp. Psychol.* **13**:173–185.
9. R.A. Cone, 1963, Quantum relations of the rat electroretinogram, *J. Gen. Physiol.* **46**:1267–1286.
10. M. Aguilar and W.S. Stiles, 1954, Saturation of the rod mechanism of the retina at high levels of stimulation, *Optica Acta* **1**:59–65.
11. M.H. Pirenne, 1962, Liminal brightness increments, in: *The Eye,* (H. Davson, ed.), pp. 159–174, Academic Press, New York and London.
12. M. Alpern, W.A.H. Rushton, and S. Torii, 1969, Encoding of rod signals from rods, *Nature* **223**:1171–1172.
13. T.P. Williams and J.P. Webbers, 1995, Photometer for measuring intensity and rhodopsin distributions in intact eyes, *Appl. Optics* **34**:5720–5724.
14. L.M. Rapp and T.P. Williams, 1980, A parametric study of retinal light-damage in albino and pigmented rats, in: *Effects of Constant Light on Visual Processes* (T.P. Williams and B.N. Baker, eds.), pp. 135–159, Plenum Press, New York.

15. T.P. Williams, J.P.P. Webbers, L. Giordano, and R.P. Henderson, 1998, Distribution of photon absorption rates across the rat retina, *J. Physiol. (London)* **508**:515–522.
16. A. Hughes, 1979, A schematic eye for the rat, *Vision Res.* **19**:569–588.
17. K.C. Lau, K.-F. So, G. Campbell, and A.R. Lieberman, 1992, Pupillary constriction in response to light in rodents which does ont depend on central neural pathways, *J. Neurol. Sci.* **113**:70–79.

AN UNCOUPLING EFFECT OF REACTIVE OXYGEN SPECIES ON THE RETINAL HORIZONTAL CELLS

T. Matsukawa,[1] Z.-Y. Zhou,[1] K. Sugawara,[1] M. Devadas,[1] K. Sugitani,[2] and S. Kato[1]*

[1] Department of Molecular Neurobiology
Graduate School of Medicine
[2] Department of Laboratory Sciences
Faculty of Medicine University of Kanazawa
Kanazawa 920-8640, Japan

SUMMARY

Glutathione is a potent endogenous scavenger of oxygen free radicals in living cells. Decreased levels of glutathione are involved in aging and some neurodegenerative diseases. To investigate an effect of glutathione depletion in the horizontal cells of carp retina, the gap junctional intercellular communication between horizontal cells was studied using both techniques of intracellular recording of light-induced responses and dye coupling of the fluorescence, Lucifer Yellow (LY). The glutathione depletion was accomplished by single injection of 2.5 μmol L-buthionine sulfoximine (BSO), an inhibitor of glutathione synthesis. Intravitreal injection of BSO induced a dramatic reduction of retinal level of glutathione 2 days after treatment. The low level of glutathione continued for a further 4–5 days and thereafter was gradually recovered to about 70% of the control level by 50 days after injection. In normal retinas, the response amplitude of horizontal cells was monotonically enhanced as the diameter of the spot increased (0.5–4.0 mm) and correspondingly the dye diffusion area was wide, as the injected LY normally diffused to several neighboring cells. Depletion of glutathione by BSO significantly altered the spatial properties of horizontal cells by increasing the

* Correspondence should be addressed to S. Kato at Department of Molecular Neurobiology, Graduate School of Medicine, University of Kanazawa, 13-1 Takara-machi, Kanazawa 920-8640, Japan, TEL: +81 76 265 2450; E-mail: satoru@med.kanazawa-u.ac.jp

Retinal Degenerative Diseases and Experimental Therapy, edited by Hollyfield *et al.*
Kluwer Academic / Plenum Publishers, New York, 1999.

response amplitude to central spots and slightly decreasing that to peripheral annuli, which were observed by 4 days after treatment. It also restricted intracellular LY to one or two cells. Accompanying the recovery of the cellular level of glutathione, the spatial properties and dye coupling of horizontal cells were recovered to normal. These alterations of spatial and dye coupling properties of horizontal cells were well correlated with accumulation of reactive oxygen species (H_2O_2) during glutathione depletion. The present data have demonstrated for the first time that neuronal gap junctional channels can be modified by intracellular glutathione via oxidative stress.

INTRODUCTION

A tripeptide, glutathione (GSH), is a major thiol compound in living cells and is a substrate for scavenging reactive oxygen radicals or toxins through the enzymes glutathione peroxidase and glutathione reductase.[1] In the nervous system, GSH deficiency leads to neuronal and glial cell damage induced by oxidative stress in vitro and in vivo.[2–4] GSH has been found in high concentrations in the brain and it has been also found in primary astroglial and neuronal cultures.[5–7]

Horizontal cells in the vertebrate retina are electrically coupled through gap junctions.[8,9] Furthermore, a recent work has reported that radial glial Müller cells and horizontal cells in the rabbit retina have high levels of GSH.[10] We proposed an important role for GSH in the glial Müller cells with reference to the cystine/glutamate antiporter.[11,12] However, the role of GSH in the horizontal cells remained unclear. A large amount of evidence has been collected showing that GSH depletion leads to inhibition of gap junction-mediated intercellular communications in non-neuronal liver and ovarian granulosa cell lines.[13,14] Therefore, we examined the effect of GSH depletion on the intercellular communication of horizontal cells in the carp retina using electrophysiological methods; lateral spread of S-potential amplitude and dye diffusion of Lucifer Yellow (LY) injected into the cell.[15–17] GSH depletion was accomplished by intravitreal injection of an inhibitor of GSH synthesis, L-buthionine sulfoximine (BSO).[18]

METHODS

Chemicals

L-Buthionine(S.R) sulfoximine and Lucifer Yellow were purchased from Sigma (St. Louis). 2′,7′-Dichlorofluorescin diacetate (DCFH-DA) was purchased from Molecular Probes (Eugene, OR). Dopamine hydrochloride, 7-fluoro-4-sulfamoyl-2,1,3-benzoxadiazole (ABD-F), and 2′,7′-dichlorofluorescein (DCF) were obtained from Wako (Osaka). Carp (*Cyprinus carpio*, body weight 600-800 g) were purchased from a local dealer. GSH was depleted by single intravitreal injection of 2.5 μmol buthionine sulfoximine (BSO) dissolved in phosphate buffered saline.[11,19] The fish were reared for up to 50 days before eye enucleation. If required, 5 μl of dopamine (2 mM) plus pargyline was applied to the residual vitreous fluid beneath the isolated retina as a referred control.[16]

Optical System

A 500 W Xenon arc lamp was used as the source of light. Monochromatic light of various wave lengths (419, 460, 521, 578, 619, and 679 nm) was obtained by narrow-band interference filters in the light path. The size of the central spot on the retinal surface was varied from 0.5 to 4.0 mm in diameter. Two kinds of annulus of 1.0 mm (A1) or 2.0 mm (A2) in inner diameter with 4.0 mm in outer diameter were used in the present experiments. The intensity of light was attenuated by interposing neutral density filters which covered a range of 0–6.0 log units. All the monochromatic lights were equalized in quanta with a digital photometer (Tektoronix J-16), and the unattenuated (0 log) light was 8.3×10^{12} quanta/cm^2/sec. A red (619 nm) light spot of 3.0 mm in diameter at an intensity of −1.5 log unit was usually used for central illumination during microelectrode penetration. The light stimuli were 1.0 s in duration and were repeated at 4.0 s intervals.

Recording System

The fish were dark adapted for 1 h, and then the eyes were enucleated under MS222 (Sankyo Pharma Co.) anesthesia. The retina with or without treatment was isolated under dim light and placed receptor-side up in a transparent chamber. All the experiments were done under a similar mesotopic light condition.[19] Intracellular recordings from photopic L-type horizontal (H1) cells were made by glass micropipettes filled with 4% Lucifer Yellow CH (LY) dissolved in 0.1 M lithium chloride solution. The dye (LY) was iontophoretically injected into a recording cell with a sinusoidal current (4 c/s) of ±10 nA for 3 min.[17]

Biochemical Assay

Cellular levels of glutathione in the retina were assayed by HPLC method with a thiol group-specific fluorogenic reagent.[11] Reactive oxygen species (ROS) were assayed using a fluorescence probe, 2′,7′-dichlorofluorescin diacetate (DCFH-DA). ROS accumulation was measured by a modification of our previous method.[20] In brief, the retina was incubated with dispase (4 mg/ml) in L-15 medium at 30°C for 30 min. The retinal cells were dissociated by pipetting and then collected by centrifugation. The cells were washed in Hank's balanced salt solution (HBSS, pH7.4), and resuspended in a total volume of 2.0 ml of the same buffer. The cell suspension was divided into two tubes, one for ROS assay and the other for protein quantification. The cells were diluted in the same volume of HBSS and then incubated with 10 μM DCFH-DA at 25°C for 30 min. After loading, the cell suspension was centrifuged at 3000 xg for 5 min and the pellet was resuspended in 2.0 ml of the saline-cold buffer, and then the fraction was incubated for a further 60 min (25°C). At the beginning and end of incubation, fluorescence intensity was measured at 488 nm excitation and 525 nm emission (Hitachi, F-2000). ROS accumulation was quantitated from 2′,7′-dichlorofluorescein (DCF) standard curve (10–600 nM) and results were expressed as pmol DCF formed/h/μg protein.

Fluorescent Observation of Dye Coupling

Three to four cells were intracellularly recorded and then injected with LY for 3 min in each retinal preparation. The retina was fixed in 4% paraformaldehyde and

35% sucrose solution (FS solution) overnight and then flat-mounted with the vitreal side facing up on a glass plate. The LY marked cells were examined with half-dried flat mounts under a fluorescence microscope (Nikon EF).

Statistics

Results were expressed as the mean±SD and statistical differences were evaluated by Scheffe's method of one way ANOVA.

RESULTS

Spatial Property and Dye Coupling of L-Type Horizontal Cells during Glutathione Depletion

We measured retinal levels of GSH immediately after electrophysiological experiments in each preparation. The retinal level of GSH after a single injection of 2.5 µmol BSO was followed for 50 days. The BSO injection depleted the level of GSH in the carp retina 2 days after treatment. It reached 2.2 pmol/µg protein 3 days after treatment. The low level of GSH continued for a further 4–5 days and then the level of GSH was gradually recovered to 40% or 70% of the control value (12.4 pmol/µg protein) 20 days or 50 days after BSO treatment, respectively.

Figure 1A shows typical responses of photopic L-type horizontal (H1) cells in the control retina to red light (619 nm at −1.5 log units) spots and annuli with different configurations. When central spots with increasing diameter were flashed on the impaled element, they evoked hyperpolarizing responses which enlarged in amplitude as the diameter of the spot was increased (0.5–4.0 mm). The maximal response was observed with the largest spot (4.0 mm diameter). When annular light stimuli were used, the response decreased in amplitude with increasing inner diameter of the annulus. In retinas 4 days after BSO treatment, the response amplitude to the central spot suddenly increased 3-fold for small spots and 1.5-fold for large spots (Fig. 1B). In contrast, the response to the peripheral annulus decreased slightly compared to the control. The enhancement of H1 cell responses to the central spot continued for a further 2–3 days. In dopamine (DA) treated retina, the response amplitude to the small spots significantly increased (Fig. 1C), while the response to the peripheral annulus dramatically decreased compared to the control. By 20 days after BSO treatment, the response amplitude had almost returned to the control level (Fig. 1D). It is interesting to point out that a transient of responses at the onset of light stimuli was sharpened in GSH-depleted retina as compared to that in control and recovered retina. Furthermore, we made a graphical relationship between the response ratio of a central spot (1.0 mm in diameter) to an annulus (A2, 2 mm in inner diameter and 4 mm in outer diameter) and time periods after BSO treatment (Fig. 2). We can see a significant increase of the response ratio (c:a) 4–5 days after treatment and thereafter the ratio gradually returned to control levels by 20 days after treatment. In the DA treated retina, the response ratio of center to annulus is higher than that of GSH depleted retinas (Fig. 2).

On the 4th day after BSO treatment, the ratio of central to annular light stimuli in the H1 cells was dramatically enhanced. The receptive field properties of H1 cells in the BSO treated retina were very similar to those in the dopamine treated retina.

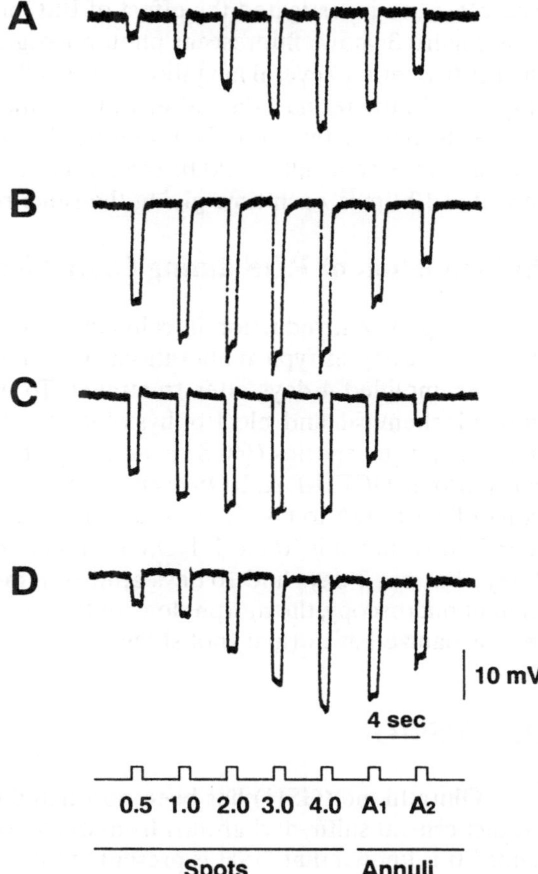

Figure 1. Representative recording from photopic L-type horizontal (H1) cells to spot and annular light stimuli with different sizes in differently treated retinas. A. Control. B. Four days after intravitreal injection of 2.5 μmol BSO. C. Thirty minutes after DA treatment. D. Twenty days after intravitreal injection of BSO. In this figure the upper trace shows electrophysiological records. The bottom trace indicates the light stimuli (619 nm at −1.5 log unit) and the numbers beside this trace indicate the diameter of light spots and the inner and outer diameter of annuli. Scale: 4 sec and 10 mV.

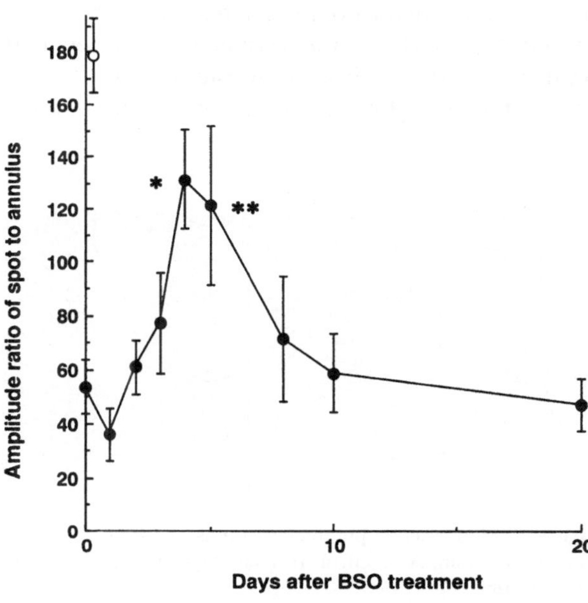

Figure 2. Graphic illustration of the amplitude ratio of spot to annulus in retinas at various times after BSO treatment (0–20 days), as indicated. The diameter of the spot was 1.0 mm. The inner diameter of the annulus was 2 mm and the outer diameter was 4 mm. The amplitude ratio was suddenly enhanced 4 days after BSO treatment and it continued until 6-7 days after BSO treatment (closed circles). Each point shows the mean±SD of 10–15 H1 cells. * and ** are $p < 0.01$, and $p < 0.05$, respectively as compared with the control (0 day). Note a similar high amplitude ratio of spot to annulus in the DA-treated retinas (open circle).

Therefore, we investigated the effect of BSO injection on diffusion of LY between H1 cells. Figure 3 shows fluorescent photomicrographs of intracellular markings of LY. In the control retina, several neighboring H1 cells including the injected cell could be seen (Fig. 3A). In the retina 4 days after BSO treatment, the number of fluorescent cells was limited to one or two (Fig. 3B). During the recovery phase 20 days after treatment, several fluorescent cells could be seen as in the control (Fig. 3C). The experiments were repeated 10 times, each time giving the same results.

Accumulation of ROS during Glutathione Depletion

A significant reduction in cellular levels of GSH was initiated 2 days after BSO treatment, whereas typical alterations of spatial or dye coupling properties of H1 cells were exemplified 4 days after treatment. To understand this time lag of initiation in both biochemical and electrophysiological phenomena, we measured formation of reactive oxygen species (ROS) generated after BSO treatment with a specific fluorescence probe, DCFH-DA. ROS accumulation was estimated utilizing a converting reaction of DCFH-DA to DCF oxidized by ROS. The rate of DCF formation in carp retinas ($n = 5$–6) is shown in Table 1. ROS accumulation was significantly enhanced (2.9-fold) 4 days, but not 2 days and 10 days after treatment with BSO, compared to the control. In light microscopy, the morphology of the carp retina 4 days after treatment with BSO was recognized as normal (not shown).

DISCUSSION

Glutathione (GSH) has been postulated to serve as a physiological reductant to protect critical sulfhydryl groups from oxidation by hydrogen peroxide and other oxidants.[1] It is known that GSH is present in the retina[21] and that this tissue also has enzymatic activity associated with GSH metabolism including glutathione peroxidase[22] and glutathione reductase.[23] Although it has been suggested that GSH is involved in defending the retina against oxidative damage,[24] there are few reports to prove the exact role of this tripeptide at the retinal cellular level. If the recent immunohistochemical finding of GSH localization in rabbit retina[10] is applied to our carp retina, horizontal cells in the fish retina give us a good material to investigate the functional role of GSH

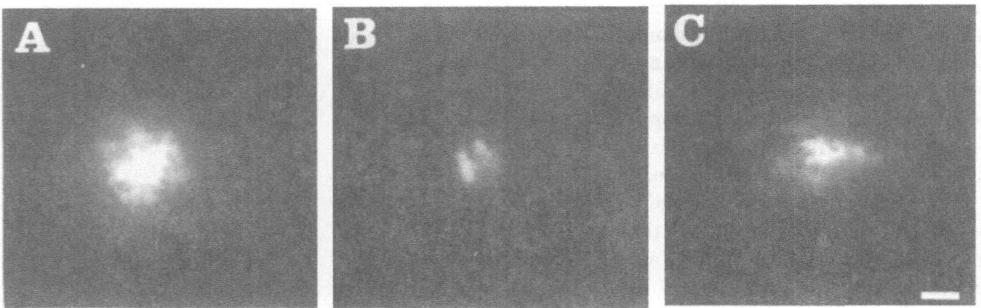

Figure 3. Fluorescence photomicrographs showing horizontal cells intracellularly marked with Lucifer Yellow for 3 min. A. Control. B. Four days after BSO treatment. C. Twenty days after BSO treatment. Scale: 50 μm.

Table 1. Accumulation of ROS during glutathione depletion

Days	DCF formed (pmol/hr/μg protein)
0	0.49 ± 0.21
2	0.70 ± 0.19
4	1.42 ± 0.40*
10	0.83 ± 0.16

A single dose (2.5 μmol) of BSO was injected into the eye. Retinal cells were dissociated by dispase and collected to measure DCF formation with DCFH-DA. The rate of DCF formation was significantly increased (2.9 fold) 4 days after BSO tratment, but not 2 days and 10 days after treatment. Each value is the mean ± SD (n = 5.6). *P < 0.05 compaed with control (0 days).

at the cellular level because of their big response (S-potential) to light stimuli. Therefore, we designed an electrophysiological approach to compare spatial and dye-coupling properties before and after GSH depletion in photopic L-type horizontal (H1) cells.

The spatial properties of the H1 cells response to central and annular light stimuli suddenly changed 4–5 days after BSO treatment; the response amplitude to the central spot was dramatically enhanced, but not that to the peripheral annulus. In general, an on transient response was found in GSH-depleted retina (Fig. 1B). Such transients are usually attributed to feedback from horizontal cells to cones.[25] ROS could facilitate such feedback within an illuminated area, because the responses to central spots were markedly enlarged in amplitude in the GSH depleted retina. The dye diffusion of LY was also changed 4 days after BSO treatment; the LY marked cells were restricted to one or two cells including the injected cell. The alterations of spatial and dye coupling properties continued for a further 3–4 days. Ten days after treatment with BSO, such typical changes of spatial responses and dye diffusion between H1 cells did not occur. Twenty days after treatment, they behaved like the control. In retinal GSH levels, there was no significant difference between 2 days (20% of control) and 4 days (17% of control) after BSO treatment. However, there was a big difference between accumulation of reactive oxygen species in retinas 2 days (140% of control) and 4 days (290% of control) after BSO treatment as shown in Table 1. This fact suggests that retinal GSH levels were almost depleted by 2 days after BSO treatment, while ROS accumulation was a little bit delayed until 4 days after treatment. Such an accumulation of ROS appears to uncouple the gap junctional intercellular communication between H1 cells through GSH depletion. The uncoupling action of GSH depletion in the present study is weaker than that of dopamine, which is a very strong uncoupler of intercellular communication of horizontal cells in the retina.[16,17,26] Generally, intracellular pH and Ca^{2+} are very important in the intercellular communication between cells.[27] In the retina, dopamine-cAMP[16,26,28] and NO-cGMP[29] systems are known as uncouplers of horizontal cells. At present, we do not know the exact reason why GSH depletion inhibited the intercellular communication between H1 cells. It is possible that the effect of ROS is the result of non-specific cellular damage, however it is very difficult to distinguish between specific actions and damaging effects at the neural cell membrane. Testing the acute effects of reagents which liberate ROS may discern the nature of this effect, but knowing the exact time to administer the reagent (ie after GSH depletion but before ROS accumulation) in each retinal sample would also be difficult. From the results of ROS accumulation, we speculate that oxidized stress induced by ROS accumulation during GSH depletion leads to uncoupling of the H1 cells. The fact that ROS had a reversible effect on the H1 cell response together with the observation that there was no significant change in retinal morphology suggests a regulatory role for ROS on retinal electrical synapses.[30]

ACKNOWLEDGMENTS

We thank Dr. T. Teranishi for his helpful suggestions throughout the experiments. We also thank Mr. K. Okamoto, Mr. M. Ohkawa, Ms T. Kano, and Ms Tami Urano for their technical and secretarial assistance. This study was supported in part by research grants from the ministry of Education, Science and Culture Japan (Nos 06808084 and 07558105), from the Sasakawa Health Scientific Foundation and from the Shimabara Scientific Foundation.

REFERENCES

1. A. Meister, 1988, Glutathione metabolism and its selective modification, *J. biol. Chem.* **263**:17205–17208.
2. J.T. Coyle and P.S. Puttfarcken, 1993, Oxidative stress, glutamate, and neurodegenerative disorders, *Science* **262**:689–694.
3. A. Jain, J. Martensson, E. Stole, P.A.M. Auld, and A. Meister, 1991, Glutathione deficiency leads to mitochondrial damage in brain, *Proc. natl. Acad. Sci. U.S.A.* **88**:1913–1917.
4. S. Kato, K. Negishi, K. Mawatari, and C.H. Kuo, 1992, A mechanism for glutamate toxicity in the C6 glioma cells involving inhibition of cystine uptake leading to glutathione depletion, *Neuroscience* **48**:903–914.
5. M. Orlowski and A. Karkowsky, 1976, Glutathione metabolism and some possible functions of glutathione in the nervous system, *Int. Rev. Neurobiol.* **19**:75–121.
6. E. Pileblad, P.S. Eriksson, and E. Hansson, 1991, The presence of glutathione in primary neuronal and astroglial cultures from rat cerebral cortex and brain stem, *J. Neural Transm.* **86**:43–49.
7. S.P. Raps, J.C.K. Lai, L. Hertz, and A.J.L. Cooper, 1989, Glutathione is present in high concentrations in cultured astrocytes but not in cultured neurons, *Brain Res.* **493**:398–401.
8. A. Kaneko, 1971, Electrical connexions between horizontal cells in the dogfish retina, *J. Physiol.* **213**:95–105.
9. K.I. Naka and W.A.H. Rushton, 1967, The generation and spread of S-potentials in fish (cyprinidae), *J. Physiol.* **192**:437–461.
10. D.V. Pow and D.K. Crook, 1995, Immunocytochemical evidence for the presence of high levels of reduced glutathione in radial glial cells and horizontal cells in the rabbit retina, *Neurosci. Lett.* **193**:25–28.
11. S. Kato, S. Ishita, K. Sugawara, and K. Mawatar, 1993, Cystine/glutamate antiporter expression in retinal Müller cells: Implications for DL-alpha-aminoadipate toxicity, *Neuroscience* **57**:473–482.
12. S. Kato, K. Mawatari, K. Sugitani, and Y. Yasui, 1996, DL-α-aminoadipate is a toxin to Müller cells, *Prog. Retinal and Eye Res.* **15**:435–456.
13. R. Barhoumi, H.R. Bailey, R.W. Hutchinson, J.A. Bowen, and R.C. Barghardt, 1995, Enhancement of melpalan toxicity by octanol in ovarian adenocarcinoma cell lines: effects of altered cell-cell communication, glutathione levels, and plasma membrane fluidity, *Fundam. Appl. Toxicol.* **25**:70–79.
14. R. Barhoumi, J.A. Bowen, L.S. Stein, J. Echols, and R.C. Burghardt, 1993, Concurrent analysis of intracellular glutathione content and gap junctional intercellular communication, *Cytometry* **14**:747–756.
15. W. Stewart, 1978, Functional connections between cells as revealed by dye-coupling with a highly fluorescent naphthalimide tracer, *Cell* **14**:741–759.
16. T. Teranishi, K. Negishi, and S. Kato, 1983, Dopamine modulates S-potential amplitude and dye-coupling between external horizontal cells in carp retina, *Nature* **301**:243–246.

17. T. Teranishi, K. Negishi, and S. Kato, 1984, Regulatory effect of dopamine on spatial properties of horizontal cells in carp retina, *J. Neurosci.* **4**:1271–1280.
18. O.W. Griffith and A. Meister, 1979, Potent and specific inhibition of glutathione synthesis by buthionine sulfoximine (s-n-butylhomocysteine sulfoximine), *J. biol. Chem.* **254**:7558–7560.
19. S. Kato, 1979, C-type horizontal cell responses to annular stimuli, *Exp. Eye Res.* **28**:627–639.
20. K. Mawatari, Y. Yasui, K. Sugitani, T. Takadera, and S. Kato, 1996, Reactive oxygen species involved in the glutamate toxicity of C6 glioma cells via xc antiporter system, *Neuroscience* **73**:201–208.
21. H. Hermann and S.G. Moses, 1945, Content and state of glutathione in the tissues of the eye, *J. biol. Chem.* **158**:33–45.
22. M. Reim, B. Heuvel, and H. Cattepoel, 1974, Glutathione peroxidase in some ocular tissues, *Ophthalmic Res.* **6**:228–234.
23. C. Castorina, A. Campisi, C. Di-Giacomo, V. Sorrenti, and A. Russo, 1992, Lipid peroxidation and antioxidant enzymatic systems in rat retina as a function of age, *Neurochem. Res.* **17**:599–604.
24. L. Feeney and E.R. Berman, 1976, Oxygen toxicity: membrane damage by free radicals, *Invest. Ophthalmol.* **15**:789–792.
25. W.K. Stell and D.O. Lightfoot, 1975, Color-specific interconnections of cones and horizontal cells in the retina of the goldfish, *J. Comp. Neurol.* **159**:473–502.
26. K. Negishi, T. Teranishi, and S. Kato, 1989, The dopamine system of the teleost fish retina, *Prog. Retinal Res.* **9**:1–48.
27. W.R. Loewenstein, 1981, Junctional intercellular communication: the cell-to-cell channel, *Physiol. Rev.* **61**:829–913.
28. M. Piccolin, J. Neyton, and H.M. Gerschenfeld, 1984, Decrease of gap junction permeability induced by dopamine and cyclic adenosine 3′,5′-monophosphate in horizontal cells of turtle retina, *J. Neurosci.* **4**:2477–2488.
29. E. Miyach, M. Murakami, and T. Nakaki, 1990, Arginine blocks gap junctions between retinal horizontal cells, *Neuroreport* **1**:107–110.
30. R.J. Ruch and J.E. Klaunig, 1988, Inhibition of mouse hepatocyte intercellular communication by paraquat-generated oxygen free radicals, *Toxic appl. Pharm.* **94**:427–436.

Section IV

ANIMAL MODELS WITH PHOTORECEPTOR DEGENERATION

ANIMAL MODEL FOR RETINITIS PIGMENTOSA

Retinal Degeneration Induced by N-Methyl-N-Nitrosourea in Rodents and Spontaneous Occurrence in *rd* Gene Carrying Mice

Hiroyuki Nambu,[1] Kenshi Yuge,[1] Motomaro Nakajima,[1] Kanji Takahashi,[1] Hirohiko Miki,[1] Masanobu Uyama,[1] Katsuhiko Yoshizawa,[2] Yoshiko Uemura,[2] and Airo Tsubura[2]

[1]Department of Ophthalmology
[2]Department of Pathology
Kansai Medical University
10–15 Fumizono-cho Moriguchi-shi
Osaka 570-8506, Japan

SUMMARY

N-methyl-N-nitrosourea (MNU) showed a toxic effect on the mouse and rat retinal photoreceptor cells leading to apoptosis. In agreement to human retinitis pigmentosa, the primary event was photoreceptor apoptosis. After photoreceptor cell loss, retinal pigment epithelial (RPE) cell migration to inner retina and accumulation around blood vessels were seen in humans; in rats, the migration was seen, but they were not accumulated around blood vessels, and in mice, RPE migration was not seen. In C_3H mice carrying the *rd* gene as well as C_3H mutants (Fas/Fas ligand-deficient mice), retinal degeneration occurred by an apoptotic mechanism. Therefore, Fas/Fas ligand do not play a primary role in the photoreceptor apoptosis in *rd* gene carrying mice.

1. INTRODUCTION

Retinitis pigmentosa is a human disease characterized by loss of photoreceptor cells leading to visual disturbance and eventually to blindness. Since it is difficult to

Retinal Degenerative Diseases and Experimental Therapy, edited by Hollyfield *et al.*
Kluwer Academic / Plenum Publishers, New York, 1999.

examine the early stage of the desease in humans, establishment of animal models are thus important. Various chemicals are known to cause retinal toxicity. Among the chemicals, N-methyl-N-nitrosourea (MNU), a nitroso compound presents widely in the environment, is known to cause retinal degeneration in several strain of animals.[1–3] We examined morphological changes after a single intraperitoneal injection of MNU in young adult mice and rats, and DNA fragmentation at single cell level was observed using TUNEL (TdT-mediated dUTP-biotin nick end labeling) methods to clarify the mechanism of the retinal degeneration caused by MNU.

In C_3H mice carrying *rd* gene, photoreceptor cell death occurs by an apoptotic mechanism. We examined the retinas of C_3H(– +/+) mice and C_3H mutants (*-lpr*, *-lpr^{cg}*, *-lpr/gld*) defective in apoptosis to clarify the involvement of Fas/Fas ligand system in retinal degeneration caused by *rd* gene.

2. MATERIALS AND METHODS

2.1. MNU-Induced Retinal Degeneration

2.1.1. Animals. C_{57}BL/6N mice and Brown-Norway(B-N) rats.

2.1.2. Chemical. MNU solution was freshly prepared by dissolving it in physiologic saline containing 0.05% acetic acid immediately before use.

2.1.3. Experimental Procedure. A single intraperitoneal(i.p.) injection of 60 mg/kg and 75 mg/kg body weight MNU solution was administered to C_{57}BL and B-N rats at 7 weeks of age, respectively. Control animals were injected with a comparable volume of physiologic saline containing 0.05% acetic acid.

At 1, 6, 12 hours, 1, 3, 5 days, and 1, 2, 3, 5, 8 weeks after the treatment, at least 4 mice and rats were anesthetized with ether and killed by cervical dislocation. After the animals were sacrificed, both eyes were quickly removed and examined.

2.2. rd Gene-Derived Photoreceptor Apoptosis in C3H Mice

2.2.1. Animals. C_3H/HeJ (– +/+) mice, C_3H/HeJ mutants (*-lpr/lpr*, *-lpr/gld*, *-lpr^{cg}/lpr^{cg}*), and C_{57}BL/6N mice.

2.2.2. Experimental Procedure. C_{57}BL and C_3H mice were killed by cervical dislocation at 8, 11, 14, 17, and 20 days of age. eight eyes from 4 individuals were examined at each point. C_3H mutants (*lpr* ; 3 mice, *lpr^{cg}* ; 4 mice, *lpr/gld* ; 4 mice) were randomly killed at >20 days of age. After the animals were sacrificed, both eyes were quickly removed and examined.

2.3. Tissue Fixation and Processing

One eye was fixed overnight in 10% neutral buffered formalin; another eye was fixed for 12 hours in methacarn or 0.5 × Karnovsky solution. The formalin- and methacarn-fixed tissues were routinely processed for paraffin embedding, sectioned at 4 μm, and stained with HE.

Formalin-fixed tissues were applied for TUNEL staining using an *in situ* apoptosis detection kit (Apop-Tag, Oncor, Gaithersburg, MD, USA).

2.4. Electron Microscopy

Tissue fixed in Karnovsky's solution was post-fixed 2%OsO4, dehydrated in graded alcohol, and processed for Luveak-812 embedding. Thin sections stained with uranyl acetate and lead citrate were examined with the electron microscope.

3. RESULTS

3.1. MNU-Induced Retinal Degeneration in Mice

Retinal degeneration was seen constantly in MNU-treated mice and rats. In all cases, the posterior pole of the retina was severely damaged, then proceeded to periphery of the retina. In contrast, control animals showed no morphologic changes in the retina at the end of the experiments (Fig. 1a).

3.1.1. Photoreceptor Apoptosis. No morphological change was evident in the retina at 1 hour after treatment.

At 3 hours, the vacuolization of photoreceptor inner segments and disorientation of photoreceptor outer segments were seen (Fig. 1b). Electron microscopically, these findings corresponded to dilatation of endoplasmic reticulum in photoreceptor inner segments, and the disorderly arranged photoreceptor outer segments. At this time, a slight degree of photoreceptor nuclei condensation, indicative of apoptosis, began and progressed in parallel to the passage of time (Fig. 2).

Thereafter, destruction of photoreceptor nuclei and widespread disruption of photoreceptor segments advanced (Fig. 1c).

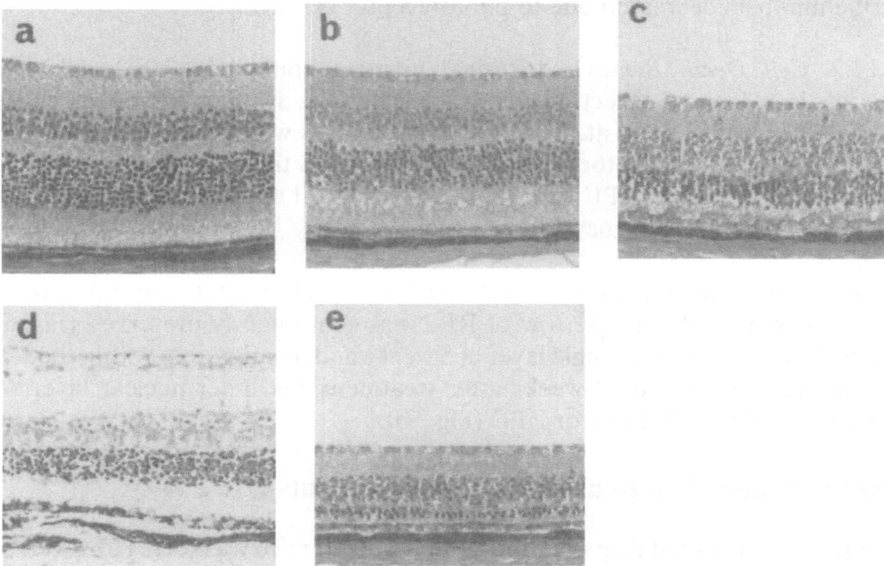

Figure 1. Morphologic change of MNU-induced photoreceptor apoptosis in mice. (a) control (b) 3 hours after treatment (c) day 3 (d) day 3 (TUNEL staining) (e) day 7.

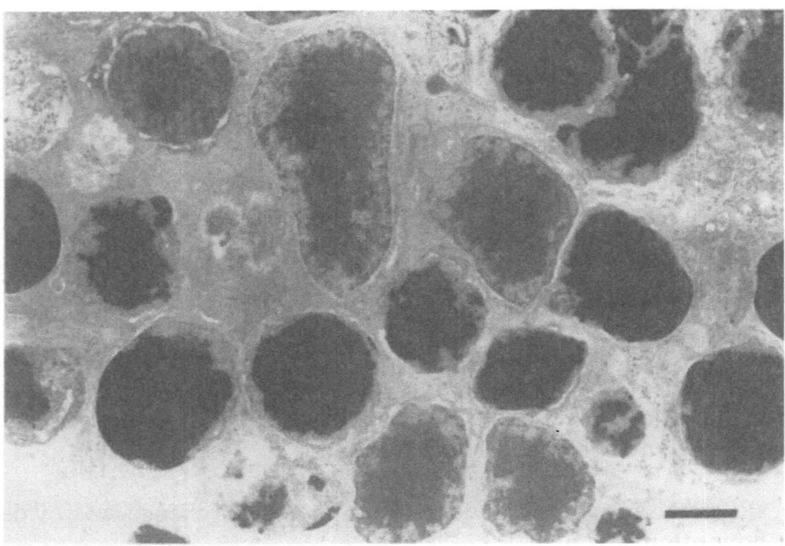

Figure 2. Outer nuclear layer in MNU-treated mouse 3 days after treatment (Bar = 2μm). Photoreceptor cells showed condensed nuclei while some adjacent photoreceptor nuclei appear unaffected.

At 5 days, the photoreceptor outer and inner segments disappeared, and at 1 week after treatment, the outer nuclear layer was lost or reduced to only a few layers of cells (Fig. 1e).

TUNEL-positive nuclear labeling began to appear in a few photoreceptor nuclei at 6 hours after dosing, which increased in number at days 1 and 3 (Fig. 1d). Then, at day 7, the numbers decreased due to photoreceptor cell loss.

3.1.2. Cell Death Removal. Retinal pigment epithelial (RPE) cells became swollen at days 1 and 3. Electron microscopically, the swollen RPE cells contained lamellar bodies (Fig. 3), and at this stage, macrophages with phagocytosed cell debris were seen in the photoreceptor layer (Fig. 4). After this time, when the photoreceptor cell death was completed, RPE cells returned to normal size, and neither macrophage infiltration or RPE cell phagocytosis was seen.

3.1.3. Late Retinal Change. At 2 and 3 weeks after treatment, bilayered RPE cells were seen, though a single row of RPE was continuously preserved (Fig. 5a, 6). However, they returned to a single layer at 5 weeks and no migration within the sensory retina was seen. Finally, at 12 weeks after treatment, the inner nuclear layer was in direct contact with single layer of RPE (Fig. 5b).

3.2. MNU-Induced Retinal Degeneration in Rats

In rat model, retinal degeneration comparable to mice was seen. However, at late stage, RPE migrated to inner retina without in contact with intraretinal vascular endothelial cells (Fig. 7).

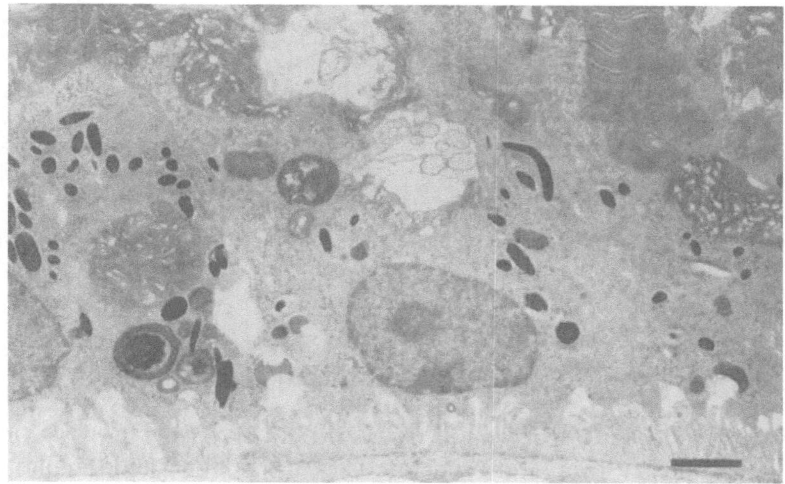

Figure 3. One day after MNU treatment (Bar = 2 μm).The swollen RPE cell contained phogocytosed lamellar bodies.

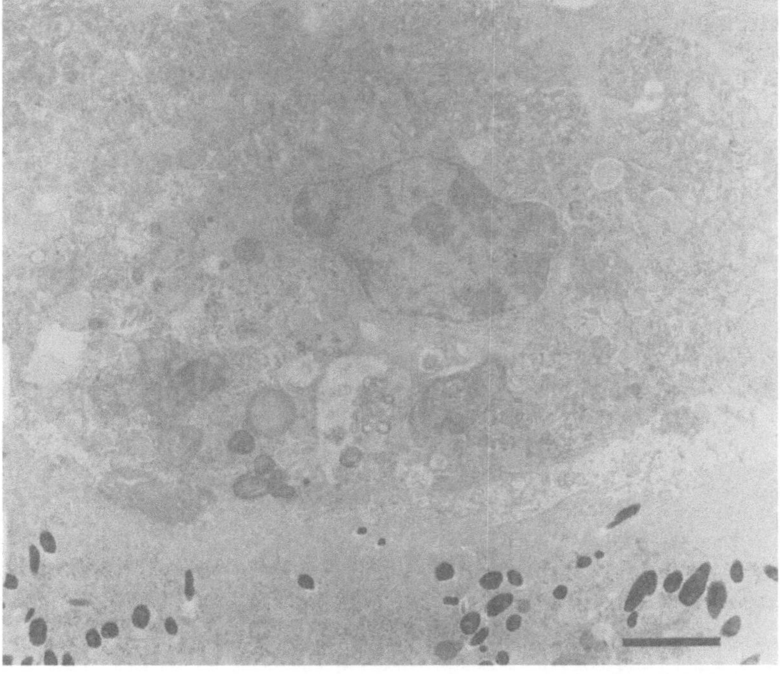

Figure 4. Three days after treatment (Bar = 2 μm). Macrophage with phagocytosed cell debris was seen.

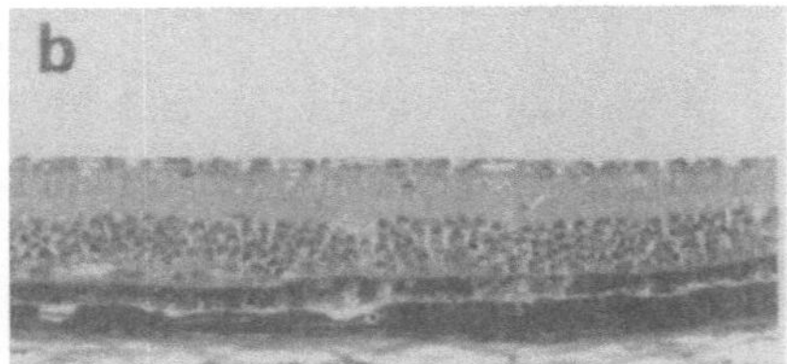

Figure 5. Late stage in MNU-induced retinal degeneration in mice. (a) 2 weeks (b) 12 weeks.

3.3. *rd* Gene-derived Photoreceptor Apoptosis in C_3H Mice

3.3.1. Morphogenesis of Retina in C_{57}BL and C_3H Mice. In C_{57}BL retina, the inner and outer nuclear layers were separated at 8 days of age, and the photoreceptor ayer began to grow between 8–11 days after birth with most rapid growth occurring between 14–17 days (Fig. 8). In C_3H retina, the development was comparable to that of C_{57}BL retina at 8 days. However, the reduction in thickness of the outer nuclear and photoreceptor layers was noted at 14 days. Finally, at 20 days of age the outer nuclear layer was lost or reduced to a single layers of cells (Fig. 9). The number of TUNEL-positive cells in the outer nuclear layer was high at 11–20 days, compared to C_{57}BL retina. Electron microscopically, pyknotic nuclei were seen in the outer nuclear layer at 11–20 days.

3.3.2. Retina in C_3H Mutant Mice. In all the *lpr/lpr*, *lpr/gld*, *lpr^{cg}/lpr^{cg}* mice, the retina was affected similarity to that of C_3H- +/+ mice.

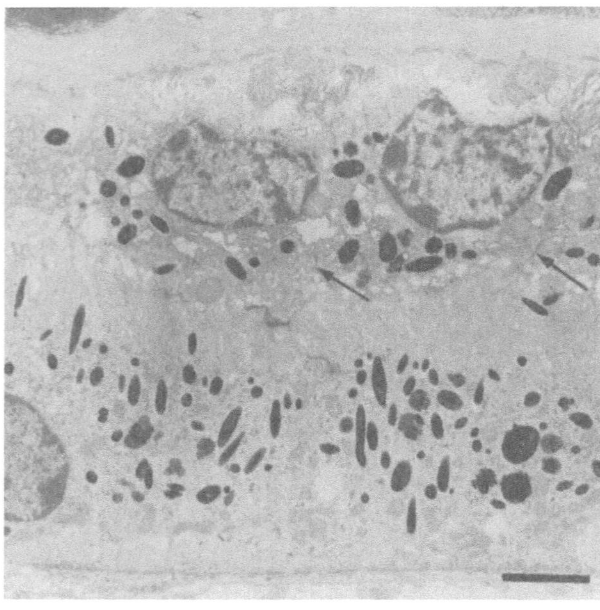

Figure 6. Three weeks after MNU treatment (Bar = 2μm). The dilated endoplasmic reticulum and swollen mitochondria in the RPE cell detached from Bruch's membrane were seen. However, RPE along Bruch's membrane appear normal. Note the tight junction (Arrow).

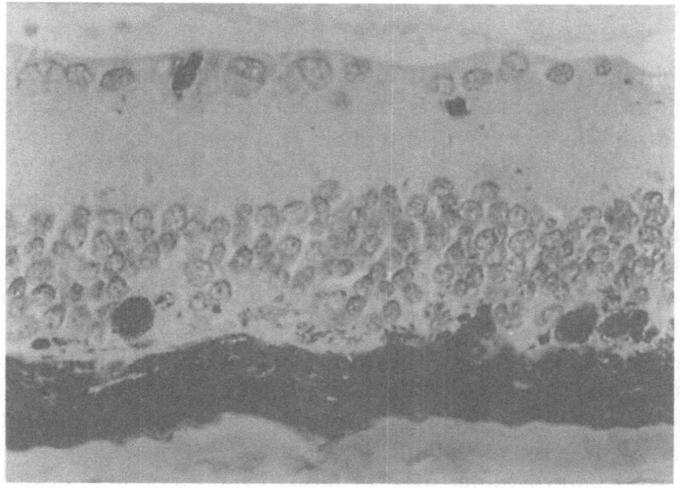

Figure 7. MNU-induced rat model (5 weeks after treatment). At late stage, migrated RPE cells to inner retina were seen.

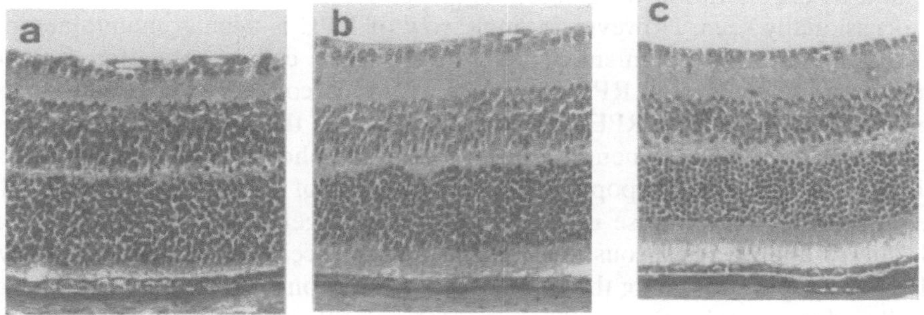

Figure 8. Developmental retinal changes in $C_{57}BL$ mice. (a) 8 days after birth (b) 14 days (c) 20 days.

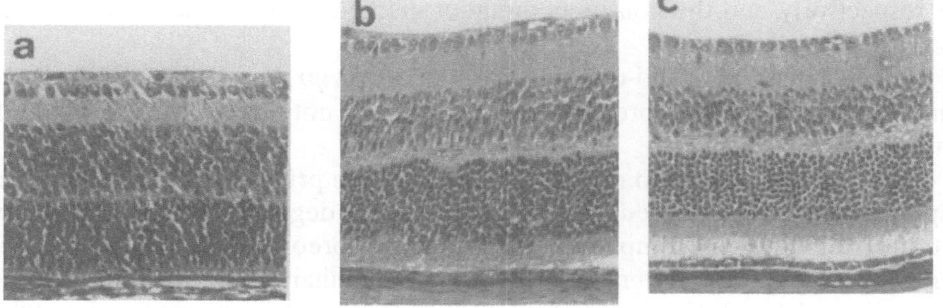

Figure 9. Developmental retinal changes in C_3H mice. (a) 8 days after birth (b) 14 days (c) 20 days.

4. DISCUSSION

A single intraperitoneal administration of MNU, a potent alkylating agent, induced progressive photoreceptor apoptosis in all treated mice[4] and rats.[5] Similar to human retinitis pigmentosa, photoreceptor apoptosis was confirmed by morphology, TUNEL method, and internucleosomal DNA fragmentation assay.[6] A previous experiment showed 40 mg/kg MNU did not produce retinal degeneration in all treated mice.[7] In addition, single administration of 60 mg/kg MNU at day 0 and 3 of neonatal mice resulted in retinal dysplasia characterized by the formation of rosettes, whereas MNU treatment at day 11, 14, 17 or 20 resulted in retinal degeneration; no retinal damage was seen when MNU was administered at days 5 or 8.[8] In rats, >50 mg/kg MNU administration at birth induced retinal dysplasia,[6] and >90 mg/kg, >50 mg/kg, >35 mg/kg MNU administration induced retinal degeneration at days 21, 50, 150 after birth, respectively.[6] Thus, the effect of MNU on the retina is time-specific and dose-dependent in mice and rats.

Compared with human retinitis pigmentosa, the MNU-induced adult model showed similarities and dissimilarities. In human retinitis pigmentosa, the damage is initiated in the equatorial zone, then extends to peripheral and central retina. In the inherited retinal degeneration *rd* mouse, the damage initiates from central to peripheral retina. In agreement with *rd* mouse, initial retinal lesion induced by MNU was posterior pole. In MNU-rat model, the RPE cells migrated within all layers of the retina at the late stage of the disease, but were not in contact with vascular endothelial cells.[9] In mice, RPE cells were swollen in regions where photoreceptor cell death was incomplete, and as degeneration progressed, detached RPE cells from Bruch's membrane were occasionally seen. However, a single row of RPE is usually maintained at late stage. In human retinitis pigmentosa, migration of RPE cells occurs after photoreceptor cell death. The migrated RPE cells were mainly in contact with vascular endothelial cells.[10] Thus, the fate of RPE was different between the species.

In human retinitis pigmentosa, primary event is photoreceptor cell apoptosis. In *rd* mice, photoreceptor cell apoptosis is primary events of retinal degeneration.[11] In the MNU-induced rat and mouse model, the primary target was the photoreceptor cell death due to apoptosis. Various mechanisms and intracellular signaling pathways are involved in apoptosis. Among them, the Fas antigen is one of the most well characterized cell surface proteins that mediate apoptotic signals into cell. The *lpr* (lymphoproliferation) locus is a defective allele of Fas characterized by no expression of the Fas antigen, and lpr^{cg} (*lpr* complementing *gld*) gene is a point-mutated nonfunctional Fas antigen. The *gld* (generalized lymphoproliferative disease) locus encodes a ligand for the Fas antigen and the *gld* is a point-mutation of locus leading to expression of non-functional Fas ligand. The absent and non-functioning Fas antigen in *-lpr/lpr*, $-lpr^{cg}/lpr^{cg}$ mice, respectively, and the non-functioning of the ligand in *-gld/gld* mice arrest apoptosis via the Fas pathway.

However, Fas/Fas ligand-deficient C_3H mice (*-lpr/lpr*, *lpr/gld*, lpr^{cg}/lpr^{cg}) all evoked retinal degeneration. Therefore, Fas and Fas ligand do not play a primary role in the *rd* disease.

In conclusion, similar to retinitis pigmentosa, the primary event was photoreceptor apoptosis in MNU-induced and *rd*-derived retinal degeneration in animals, but the onset site was different compared with humans. Moreover, the action of RPE was different among species. After understanding the similarity and dissimilarity, the experimental model of MNU-induced photoreceptor apoptosis may provide additional informations in the better understanding of human retinitis pigmentosa.

REFERENCES

1. K.M. Herrold, 1976, Pigmentary degeneration of the retina induced by N-methyl-N-nitrosourea: an experimental study in Syrian hamsters, *Arch. Ophthalmol.* **78**:650–653.
2. A.S.K. Murthy, G.F. Vawter, L. Kopito, and E. Rossen, 1972, Retinal atrophy and cataract following administration of N-methyl-N-nitrosourea, *Proc. Soc. Biol. Med.* **139**:84–87.
3. S.B. Smith and K.L. Yielding, 1986, Retinal degeneration in the mouse: a model induced transplacentally by methylnitrosourea, *Exp. Eye. Res.* **43**:791–801.
4. H. Nambu, K. Yuge, M. Nakajima, N. Shikata, K. Takahashi, H. Miki, M. Uyama, and A. Tsubura, 1997, Morphologic characteristics of N-methyl-N-nitrosourea-induced retinal degeneration in C_{57}BL mice, *Pathol. Int.* **47**:377–383.
5. M. Nakajima, K. Yuge, H. Senzaki, N. Shikata, H. Miki, M. Uyama, and A. Tsubura, 1996, Photoreceptor apoptosis induced by a single administration of N-methyl-N-nitrosourea in the rat retina, *Am. J. Pathol.* **148**:631–641.
6. K. Yuge, H. Nambu, H. Senzaki, I. Nakao, N. Shikata, K. Takahashi, H. Miki, M. Uyama, and A. Tsubura, 1996, N-methyl-N-nitrosourea-induced photoreceptor apoptosis in the mouse retina, *In Vivo* **10**:483–486.
7. H. Nambu, K. Yoshizawa, J. Yang, D. Yamamoto, and A. Tsubura, 1998, Age-specific and dose-dependent retinal dysplasia and degeneration induced by a single intraperitoneal administration of N-methyl-N-nitrosourea to rats, *Toxicol. Pathol.* **11**:127–131.
8. H. Nambu, M. Taomoto, E. Ogura, and A. Tsubura, 1998, Time-specific action of N-methyl-N-nitrosourea in the occurrence of retinal dysplasia and retinal degeneration in neonatal mice, *Pathol. Int.* **48**:199–205.
9. M. Nakajima, H. Nambu, N. Shikata, H. Senzaki, H. Miki, and A. Tsubura, 1996, Pigmentary degeneration induced by N-methyl-N-nitrosourea and the fate of pigment epithelial cells in the rat retina, *Pathol. Int.* **46**:874–882.
10. Z.Y. Li, D.E. Possin, and A.H. Milam, 1995, Histopathology of bone spicule pigmentation in retinitis pigmentosa, *Ophthalmology* **102**:805–816.
11. D.W. Caley, C. Johnson, and R.A. Liebelt, 1972, The postnatal development of the retina in the normal and rodless CBA mouse: A light and electron microscopic study, *Am. J. Anat* **133**:179–212.

36

DEVELOPMENT OF RAT MODELS FOR CHOROIDAL NEOVASCULARISATION (CNV)

A Comparison between Laser and Recombinant Adenovirus Induced CNV

P. E. Rakoczy,[1] W. Y. Shen,[2] K. Spilsbury,[2] M. Yu,[1] M. Lai,[1] C. J. Barry,[2] and I. J. Constable[1]

[1] Centre for Ophthalmology and Visual Science
University of Western Australia
[2] Lions Eye Institute
2 Verdun St. Nedlands, 6009, Australia

SUMMARY

This project aimed to investigate whether the overexpression of vascular endothelial growth factor (VEGF) in the retinal pigment epithelium (RPE), induced by recombinant adenovirus mediated delivery of $VEGF_{164}$, is sufficient to initiate the development of leaky choroidal blood vessels in a rat model. Eyes of pigmented rats were subjected to krypton laser photocoagulation or subretinal injection with recombinant adenovirus constructs Ad.RSV.VEGF or Ad.CMV.VEGF. Initially the animals were examined with fundus photography followed by fluorescein angiography. Retinal photographs demonstrated that the krypton laser produced moderate burns and that the subretinal injection resulted in the formation of a bleb in the subretinal space. Injection of control recombinant adenovirus Ad.RSV.βgal resulted in the expression of βgal reporter gene in the RPE layer. Of the 141 laser spots delivered into a total of 17 rat eyes, 52.4% developed fluorescein leakage that was detectable by angiography. Of seven eyes injected with Ad.RSV.VEGF, 5 developed leakage at 2 weeks post injection, and of 8 eyes which were injected with Ad.CMV.VEGF, 5 developed leakage at 1 week post injection. However, the development of leaky blood vessels was temporary and both Ad.RSV.VEGF and Ad.CMV.VEGF injected animals showed regression of leaky vessels by 4 and 5 weeks post injection, respectively. These results suggest that the upregulation of VEGF in the RPE layer may be a key factor in the series of events leading to the development of CNV.

INTRODUCTION

Choroidal neovascularisation (CNV) is the formation of new vessels between the retinal pigment epithelium (RPE) and the inner aspect of the Bruch's membrane[1]. Clinically CNV presents as serous or haemorrhagic detachment of the RPE and or sensory retina. CNV occurs in several ocular diseases including Best's disease, choroidal haemangioma, malignant melanoma, angiod streaks[1] and in the end stages of age related macular degeneration (AMD). The end stage of AMD, which is accompanied by rapid visual loss due to hemorrhage, is referred to as wet or exudative AMD. Although wet AMD occurs in ~38% of all AMD sufferers it accounts for the majority of legal blindness cases in the developed countries.[2,3] In spite of the significance of this condition, to date, our understanding of the pathomechanism of CNV is limited and there is no effective treatment for sub-macula CNV.

CNV in AMD is associated with the presence of basal linear deposit and soft, often confluent, drusen.[4–6] One of the intriguing questions of CNV research is to establish if the angiogenesis is a response to breaks in the Bruch's membrane or if it is an additional stimulus which facilitates the invasion of the subretinal space. Some experimental animal studies suggest that CNV can occur without prior breakage in the Bruch's membrane.[7–9] However, in the majority of animal models an initial breakage of the Bruch's membrane is required to generate CNV.[10–14]

Several studies have demonstrated the presence of chronic inflammatory cells including giant cells, macrophages and lymphocytes in CNV.[15–17] These observations were subsequently confirmed by animal models and it has been proposed that CNV formation can be linked to the transient overexpression of cellular adhesion molecules (CAMs) and vascular endothelial growth factor (VEGF) induced by the initial laser damage.[11,14] The overexpression of VEGF both in human CNV and in animal models is particularly significant as VEGF is a known ischemia induced glycoprotein with angiogenic and vasopermeability activity.[18] In addition, it is secreted by many cell types, is a highly specific mitogen for vascular endothelial cells[19,20] and is upregulated by hypoxia.[21] It has been suggested that hypoxia in the outer retina could be caused by the accumulation of debris in Bruch's membrane and may be involved in the stimulation of CNV.[22–25]

One of the most descriptive characteristics of CNV is that the new blood vessels penetrating Bruch's membrane remain immature and allow serum and blood to escape as an exudate. This leaky nature of CNV is characteristic not only of early CNV but also fully developed CNV membranes[26] and allows for a reliable clinical diagnostic tool, fluorescein angiography.[27]

This project aimed to investigate whether overexpression of VEGF, induced by adenovirus mediated delivery of a $VEGF_{164}$ gene into the RPE layer, is sufficient to induce the development of leaky blood vessels in a rat model. The presence and characteristics of leaky blood vessels was monitored by *in vivo* fluorescein angiography and compared to a krypton laser induced rat model.

MATERIALS AND METHODS

Krypton Laser Photocoagulation

Laser photocoagulation was performed as described by Shen et al. (1998). Briefly, dark Agouti pigmented rats were anaesthetised with an intramuscular injection of a mixture of xylazine and ketamine. The pupils were dilated with 2.5% phenylephrine and 1% tropicamide. Krypton laser irradiation (647.1 nm, Coherent Radiation System, California, USA) was delivered through a Zeiss slit lamp with a handheld coverslip serving as a contact lens. Krypton laser photocoagulation was applied using a setting of 100 μm diameter, 0.1 s duration, and 150 mW intensity, to the posterior pole around the optic nerve, 10 spots per eye.

Fluorescein Angiography

Fluorescein angiograms were performed as described earlier.[14] After krypton laser photocoagulation or subretinal injection, the eyes were followed by serial fluorescein angiography from intraperitoneal injection of 0.1–0.2 ml of 10% sodium fluorescein. High resolution 60 degree photography of the rat fundus was obtained by a simple modification of a Canon CF-60ZA retinal camera (Kawasaki, Kawagawa, Japan) as previously described. Fluorescein angiography was recorded on Kodak TMAX 100 film.

Histology

Eyes which had demonstrated fluorescein leakage during angiography were enucleated at either 2 weeks or 3 months post laser photocoagulation. The enucleated eyes were fixed in 2.5% glutaraldehyde in phosphate buffered saline (PBS) for 24 hours. The laser lesions associated with fluorescein leakage were selected under a dissecting microscope. The selected tissues were post fixed with 1% osmium tetroxide for 1 hour and embedded in epoxy resin after dehydration with a series of graded ethanols. Blocks were sectioned (4 μm) and stained with 1% toluidine blue.

Production of Recombinant Adenoviruses

Recombinant adenovirus Ad.RSV.βgal was produced as described previously[28] Recombinant adenovirus constructs Ad.RSV.VEGF and Ad.CMV.VEGF were generated by routine molecular biological techniques followed by homologous recombination[29] and high titre stocks were prepared as for Ad.RSV.βgal. Viral stocks were titrated and checked for the presence of wild type virus contamination.[30]

In vitro Adenovirus Transduction of Rat RPE Cells

Rat RPE cells (>18 passages) were maintained in Dulbecco's Modified Eagle's Medium (DMEM) supplemented with 10% fetal bovine serum (FBS; Gibco BRL, Grand Island, NY) at 37°C in 5% CO_2. For adenovirus transduction experiments, 2×10^5 rat RPE cells per well were seeded overnight in gelatin coated 24 well plates (Becton Dickinson, Lincoln Park, NJ). The following day, cell monolayers were washed in PBS then incubated in 200 μl serum free DMEM containing a range of Ad.RSV.βgal multiplicities of infection (MOIs). The virus was allowed to adsorb to the cells for 60 minutes at 37°C in 5% CO_2 before the addition of 1 ml DMEM/10% FBS. The cells were incubated for a further 24 hours before being washed in PBS, fixed for 10 minutes in 0.5% glutaraldehyde (BDH, Poole, England)/PBS, and stained in X-gal solution at 37°C overnight.

Subretinal Injections

The animals were anaesthetised and their pupils dilated as described earlier. The conjunctiva close to the limbus was cut and the sclera exposed. A shelving puncture of the sclera was then made with a 30 gauge needle. A 32 gauge needle was then passed through this hole in a tangential direction under an operating microscope. Two microlitres of Ad.RSV.VEGF containing 2.0×10^8 pfu or Ad.CMV.VEGF containing 6.0×10^8 pfu in 10% glycerol solution was delivered into the subretinal space. PBS or Ad.RSV.β-gal (2.0×10^8πfu) in 10% glycerol was injected in the same way as controls. The success of each subretinal injection was confirmed by a partial retinal detachment viewed under indirect ophthalmoscopy. The needle was kept in the subretinal space for 1 minute then withdrawn gently and antibiotic ointment was applied. At 3 and 7 days post injection, the injected eyes were observed using indirect ophthalmoscopy to assess any sign of injury or intraocular infection.

For X-gal staining, the eyes were enucleated at 7 days post injection. The enucleated eyes were fixed in 2% paraformaldehyde for 1 hour and washed in PBS for 15 minutes 3 times. The fixed eyes were bathed in X-gal staining solution at room temperature overnight. The eyes were washed with PBS and placed in 70% ethanol for 2 hours. After removing the anterior segment and the lens, the eye cups were embedded in paraffin and 6 μm sections were cut.

RESULTS

Krypton Laser Photocoagulation

There were 141 laser burns delivered into 17 rat eyes by krypton laser photocoagulation. In selected animals, the presence of laser burns was shown by fundus photography (Fig. 1A). Of the successfully delivered laser burns 74 (52.5%) developed easily recognizable fluorescein leakage by 3 weeks post photocoagulation (Fig. 1B). Some animals were followed with regular angiograms up to 3 months and they demonstrated that the laser spots which had developed leakage by 3 weeks post photocoagulation remained leaky during the length of the experimental period (Fig. 1C). Histological examination of one eye which presented leakage at 3 months post laser photocoagulation demonstrated the formation of a neovascular membrane (Fig. 1D).

Characterisation of Recombinant Adenoviruses

All recombinant adenovirus stocks were free of wild type virus contamination and produced titers of 1×10^{11} pfu/ml Ad.RSV.βgal, 1×10^{11} pfu/ml Ad.RSV.VEGF and 3×10^{11} pfu/ml Ad.CMV.VEGF. Rat RPE cells were shown to be transduced with recombinant adenovirus (Fig. 1A). Subretinal injection of Ad.RSV.βgal in a non pigmented rat eye demonstrated that approx. one quarter of the RPE cells were trans-

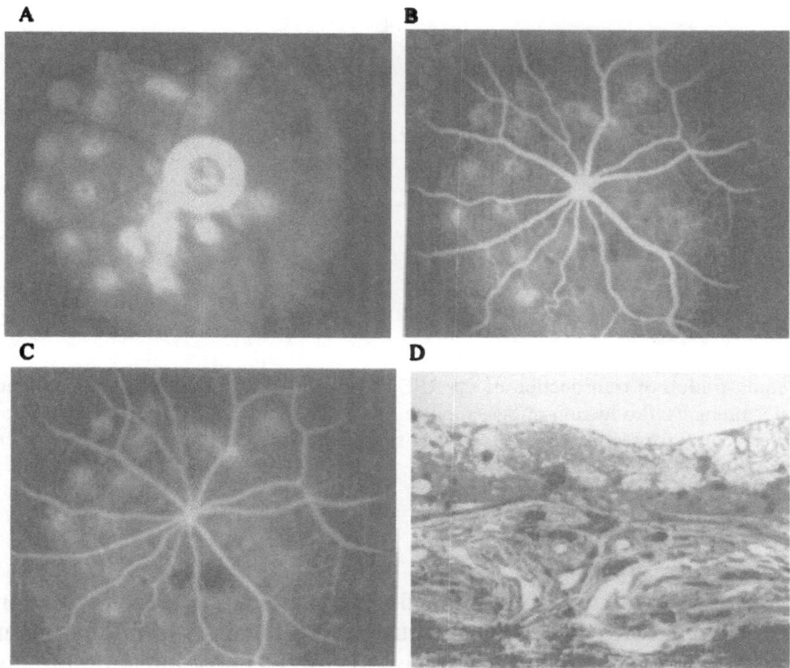

Figure 1. Choroidal neovascularisation induced by krypton laser photocoagulation. A: Fundus photograph at 2 weeks post photocoagulation. B and C: angiograms of the same animal at 2 weeks and 3 months post photocoagulation, respectively. D Histology of a retina.

duced at 10 days post injection (Fig. 1B). However the transduction was transient and by 3 weeks post injection only a scattered signal was visible (data not shown).

Fluorescein Angiograms of Recombinant Adenovirus Injected Rat Eyes

Indirect ophthalmoscopy of subretinally injected eyes demonstrated the formation of a bleb which covered approx. one quarter of the retina. A limited subretinal hemorrhage was observed in some of the eyes. Apart from these changes, no other retinal abnormalities were observed at the time of injections. Of the 6 animals injected with Ad.RSV.βgal, fundus photography did not reveal any retinal changes up to 2 weeks post injection. However by 3 or 4 weeks post injection, some changes in RPE pigmentation around the injection sites were observed. There were no fluorescein leakage's detected in any of the Ad.RSV.βgal injected eyes. Retinal photography of Ad.RSV.VEGF and Ad.CMV.VEGF injected animals also demonstrated changes in RPE pigmentation around the injection sites in some eyes (Fig. 3A arrow). Of the 7 animals injected with Ad.RSV.VEGF, 4 demonstrated signs of leakage (Fig. 3B) at 2 weeks post injection. The same eyes continued to show signs of leakage at 3 weeks post injection. (Fig. 3C). By 4 weeks post injection the intensity of leakage significantly weakened and by 5 weeks post injection all signs of fluorescein leakage had disappeared. Retinal photograph of Ad.CMV.VEGF injected eyes demonstrated a normal appearance (Fig. 3D) apart from occasional RPE pigmentary changes. Ad.CMV.VEGF

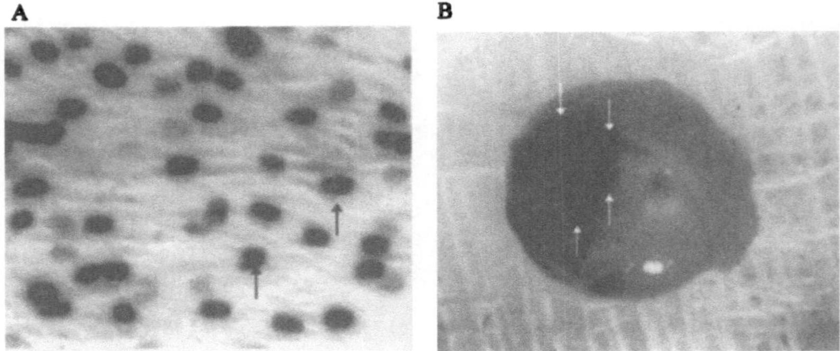

Figure 2. Demonstration of transduction of rat RPE cells with Ad.RSV.βgal recombinant adenovirus following X-Gal staining. A: Production of βgal reporter gene (black nucleus) in rat RPE cells. B: Production of βgal reporter gene (arrows, black area) following subretinal injection into a non-pigmented rat eye at 7 days post injection.

injected eyes demonstrated early signs of fluorescein leakage. Of the 8 eyes injected, 5 demonstrated leakage at 1 week post injection (Fig. 3E) and 4 of these remained leaky at 4 weeks post injection (Fig. 3F). However, by 5 weeks post injection the intensity of the leakages reduced. Three of these animals injected could not be analyzed by fluorescein angiography as they developed corneal neovascularisation obscuring the view of the retina.

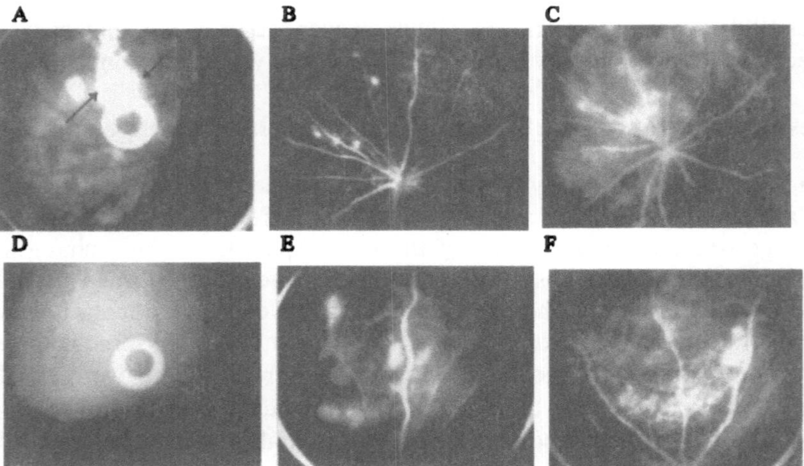

Figure 3. Fundus photographs and fluorescein angiograms of Ad.RSV.VEGF (A, B, and C) and Ad.CMV.VEGF (D, E, and F) injected rat eyes. A: Fundus photograph of a rat eye injected with Ad.RSV.VEGF at 3 weeks post injection *note some RPE pigment changes were present around the injection site (arrows). B: Fluorescein angiogram of the same rat eye at 2 weeks post injection. B: Fluorescein angiogram of the same eye at 3 weeks post injection. D: Fundus photograph of a rat eye injected with Ad.CMV.VEGF at 1 week post injection. E: Fluorescent angiogram of the same eye as in D at 1 week post injection. F: Fluorescein angiogram of the same eye at 4 weeks post injection.

DISCUSSION

In this study, we confirmed the relative reliability of the krypton laser induced CNV model. The development of leaky spots was consistent and histology demonstrated the presence of fully developed neovascular membranes similar to those seen in human CNV.[31] Although the appearance of leaky blood vessels was consistent, leakage was present only in 52.4% of all laser burns delivered. While these results are in agreement with previous results[9,12,14,32] further studies are necessary to investigate why this technique cannot achieve a higher success in generating fluorescein leaking laser burns.

Immunohistochemical analysis of human CNV samples have shown immunoreactivity to a variety of growth factors TGFβ, aFGF, bFGF, and VEGF[33–37] proliferating RPE cells, inflammatory cells, and vascular endothelial cells. Subsequent histological studies of developing CNV membranes in laser induced animal models demonstrated that the infiltrating cells and proliferating RPE cells also produce a wide range of angiogenic factors including VEGF.[12,14,38–40] The delivery of bFGF into the subretinal space has been shown to induce CNV in a rabbit model.[41] Although VEGF is considered to be a primary stimulus in inducing new vessel formation, VEGF induced CNV in an animal model has not been reported.

Recently, the development of retinal neovascularisation has been shown in a transgenic mouse model[42] which expresses VEGF under the opsin promoter. *In situ* hybridization demonstrated the overexpression of VEGF mRNA in the retina and histological studies showed wide spread retinal neovascularisation. However there was no sign of choroidal neovascularisation in any of these animals despite significant amounts of VEGF in the retina. These observations support the existence of two independent vascular beds in the eye, retinal, and choroidal. These two vascular beds are separated and the RPE layer prevents the traffic of proteins from the retina to the choroid. Supposing choroidal vascular endothelial cells are induced to proliferate by RPE produced factors, the VEGF must then be secreted basolaterally. A preliminary *in vitro* experiment have shown that VEGF expression induced by hypoxia in RPE cells is indeed basolaterally secreted.[43]

In this work we selected recombinant adenoviruses (Ad.RSV.VEGF, Ad. CMV.VEGF) to upregulate VEGF expression in the RPE cells. Adenovirus was selected as it has been shown to transduce RPE cells with high efficacy[44] and it is relatively easy to produce.[29] One of the disadvantages of the adenovirus delivery system is the transient transgene expression[45]. In our system the expression of βgal reporter gene was detected at a high level up to 2 weeks suggesting that a VEGF upregulation would occur for the same period of time. Previous studies on a krypton laser rat model[14] demonstrated that the upregulation of CAM's and VEGF expression in and around the laser spots for 10 days was associated with choroidal new vessel formation.[14]

In this work the temporary upregulation of VEGF expression in Ad.RSV. VEGF and Ad.CMV.VEGF injected eyes induced extended fluorescein leakage. These results suggest that in contrast to VEGF produced by cells of the retina, VEGF expressed in the RPE cells can reach the choroidal layer thus implying that recombinant adenovirus produced VEGF is secreted basolaterally. In addition, these results demonstrated that specific upregulation of VEGF is sufficient to induce fluoresceine leakage and suggest an important role for VEGF in the induction of CNV. However, in both Ad.RSV.VEGF and Ad.CMV.VEGF injected animals the leakages were temporary and regression of leaky vessels was observed at around 4 weeks post injection. As the development of leaky blood vessels in this model relied specifically on RPE secreted VEGF it can be proposed that the regression of leakages is due to the termination of VEGF overexpression in these cells. The regressive nature of these leaky choroidal vessels also suggest that there was no fully developed neovascular membranes formed. Further studies are required to establish the histological appearance of the leaky sites.

These results, suggest that the RPE specific upregulation of VEGF expression is an initiating factor in the development of leaky choroidal blood vessels. They therefore provide support for theories which propose that CNV in humans is initiated by hypoxia induced upregulation of VEGF expression in RPE cells.

ACKNOWLEDGMENT

The authors thank Hyal Pharmaceuticals Australia for providing funding for the project.

REFERENCES

1. W.R. Green and D.J. Wilson, 1986, Choroidal neovascularization, *Ophthalmology* **93**:1169–1176.
2. S.H. Sarks, 1973, New vessel formation beneath the retinal pigment epithelium in senile eyes, *Br J Ophthalmol* **57**:951–965.
3. W.R. Green, C. Enger, 1993, Age-related macular degeneration histopathologic studies. The 1993 Lorenz E. Zimmerman Lecture. *Ophthalmology* **100**:1519–1535.
4. N.M. Bressler, J.C. Silva, S.B. Bressler, S.L. Fine, and W.R. Green, 1994, Clinicopathologic correlation of drusen and retinal pigment epithelial abnormalities in age-related macular degeneration, *Retina* **14**(2):130–142.
5. S.H. Sarks, 1976, Ageing and degeneration in the macular region: a clinico-pathological study, *Br J Ophthalmol* **60**:324–341.
6. S.H. Sarks, D. van Driel, L. Maxwell, and M.C. Killingsworth, 1980, Softening of drusen and subretinal neovascularization, *Trans Ophthalmol Soc UK* **100**:414–422.
7. W.J. Heriot, P. Henkind, R.W. Bellhorn, and M.S. Burns, 1984, Choroidal neovascularization can digest Bruch's membrane—A prior break is not essential, *Ophthalmology* **91**(12):1603–1608.
8. A. Pollack, W.J. Heriot, and P. Henkind, 1986, Celluar processes causing defects in Bruch's membrance following krypton laser photocoagulation, *Ophthalmology* **93**:1113–1119.
9. R.N. Frank, A. Das, and M.L. Weber, 1989, A model of subretinal neovascularization in the pigmented rat, *Curr Eye Res* **8**:239–247.
10. X. Yi, N. Ogata, M. Komada, K. Takahashi, K. Omori, and M. Uyama, 1997, Vascular endothelial growth factor expression in choroidal neovascularization in rats. *Graefe's Arch Clin Exp Ophthalmol*:313–319.
11. T. Ishibashi, Y. Hata, H. Yoshikawa, K. Nakagawa, K. Sueishi, and H. Inomata, 1997, Expression of vascular endothelial growth factor in experimental choroidal neovascularization. *Graefe's Arch Clin Exp Ophthalmol* **235**:159–167.
12. E.T. Dobi, C.A. Puliafito, and M. Destro, 1989, A new model of experimental choroidal neovascularization in the rat, *Arch Ophthalmol* **107**:264–269.
13. L. Feeney-Burns, C. Gao, and E.R. Berman, 1988, The fate of immunoreactive opsin following phagocytosis by pigment epithelium in human and monkey retinas, *Invest Ophthalmol Vis Sci* **29**:708–719.
14. W. Shen, M. Yu, C. Barry, I.J. Constable, and P.E. Rakoczy, 1998, Expression of cell adhesion molecules and vascular endothelial growth factor in experimental choroidal neovascularisation in the rat, *Br J Ophthalmol* **82—in press**
15. M.C. Killingsworth, S.H. Sarks, 1982, Giant cells in disciform macular degeneration of the human eye, *Micron* **13**(3):359–360.

16. S.H. Sarks, P.L. Penfold, M.C. Killingsworth, and D. van Driel, 1985, Patterns in Macular Degeneration, in: Retinal Diseases, (S.J. Ryan, A.K. Dawson, eds.) pp. 87–93, Grune & Stratton. Inc. NY.
17. P.L. Penfold, M.C. Killingsworth, and S.H. Sarks, 1985, Senile macular degeneration: The involvement of immunocompetent cells, *Graefe's Arch Clin Exp Ophthalmol* **223**:69–76.
18. N. Ferrara, T. Davis-Smyth, 1997, The biology of vascular endothelial growth factor, *Endocr Rev* **18**(1):4–25.
19. D.W. Leung, G. Cachianes, W. Kuang, D.V. Goeddel, and N. Ferrara, 1989, Vascular endothelial growth factor is a secreted angiogenic mitogen, *Science* **246**:1306–1309.
20. H.F. Dvorak, L.F. Brown, M. Detmar, and A.M. Dvorak, 1995, Vascular permeanility factor vascular endothelial growth factor, microvascular hyperpermeability, and angiogenesis. *Am J Pathol* **146**:1029–1039.
21. A. Minchenko, T. Bauer, S. Salceda, and J. Caro, 1994, Hypoxic stimulation of vascular endothelial growth factor expression in vitro and in vivo, *Lab Invest* **71**: 374–379.
22. R.J. D'Amato, A.P. Adamis, 1995, Angiogenesis inhibition in age-related macular degeneration, *Ophthalmology* **102**:1261–1262.
23. P.A. D'Amore, 1994, Mechanisms of Retinal and Choroidal neovascularization, *Invest Ophthalmol Vis Sci* **35**:3974–3979.
24. H.P. Heidenkummer, 1991, Age-related macular degeneration: current aspects of pathogenesis and treatment, *Eye Science* **7**:6–20.
25. P.E. Rakoczy, C.M. Lai, M. Baines, S. Di Grandi, J.H. Fitton, and I.J. Constable, 1997, Modulation of cathepsin D activity in retinal pigment epithelial cells, *Biochem J* **324**:935–940.
26. W.R. Green. 1985, Pathology of the macula, in: Ophthalmic Pathology, Volume 2 (W.H. Spencer, ed.) pp. 924–1034, W.B. Saunders Company.
27. S.O. Sykes, N.M. Bressler, M.G. Maguire, and A.P. Schachat, 1994, Detecting recurrent chorodial neovascularization comparison of clinical examination with and without fluorescein angiography, *Arch Ophthalmol* **112**(12):1561–1566.
28. L. da Cruz, P. Rakoczy, M. Perricaudet, and I. Constable, 1996, Dynamics of gene transfer to retinal pigment epithelium, *Invest Ophthalmol Vis Sci* **37**(12):2447–2454.
29. T.C. Becker, R.J. Noel, W.S Coats, A.M. Gomez-Foix, T. Alam, R.D. Gerard, and C.B. Newgard, 1994, Use of recombinant adenovirus for metabolic engineering of mammalian cells, in: Methods in Cell Biology, Volume 43 pp. 161–189,.Academic Press.
30. W. Zhang, P.E. Koch, and J.A. Roth, 1995, Detection of wild-type contamination in a recombinant adenoviral preparation by PCR, *Biotechniques* **18**:444–446.
31. A.C. Bird, N.M. Bressler, S.B. Bressler, I.H. Chisholm, G. Coscas, and M.D. David, 1995, An international classification and grading system for age-related maculopathy and age-related macular degeneration. *Surv Ophthalmol* **39**(5):367–374.
32. S.J. Ryan. 1989, Subretinal neovascularization, in: Retina, (S.J. Ryan, E.J. Murray, and Patz, eds.) pp. 107–125, Mosby, St Louis.
33. R.N. Frank, R.H. Amin, D. Eliott, J.E. Puklin, and G.W. Abrams, 1996, Basic fibroblast growth factor and vascular endothelial growth factor are present in epiretinal and choroidal neovascular membranes, *Am J Ophthalmol* **122**:393–403.
34. R.N. Frank, 1997, Growth factors in age-related macular degeneration: pathogenic and therapeutic implications, *Ophthalmic Research* **29**:341–353.
35. P.F. Lopez, B.D. Sippy, H.M. Lambert, A.B. Thach, and D.R. Hinton, 1996, Transdifferentiated retinal pigment epithelial cells are immunoreactive for vascular endothelial growth factor in surgically excised age-related macular degeneration-related choroidal neovascular membranes, *Invest Ophthalmol Vis Sci* **37**:855–868.
36. A. Kvanta, P.V. Algvere, L. Berglin, and S. Seregard, 1996, Subfoveal fibrovascular membranes in Age-related macular degeneration express vascular endothelial growth factor, *Invest Ophthalmol Vis Sci* **37**(9):1929–1934.
37. G.A. Lutty, D.S. McLeod, C. Merges, A. Diggs, and J. Plouet, 1996, Localization of Vascular Endothelial Growth Factor in Human Retina and Choroid, *Arch Ophthalmol* **114**:971–977.
38. N.C. Tsang, P.L. Penfold, P.J. Snitch, and F. Billson, 1992, Serum levels of antioxidants and age-related macular degeneration. *Documenta Ophthalmologica* **81**(4):387–400.
39. B. Martini, S.J. Ryan, 1992, Argon laser lesion of the retina: occurrence and origin of macrophages. *Euro J Ophthalmol* **2**:51–57.
40. X. Yi, K. Takahashi, N. Ogata, and M. Uyama, 1996, Immunohischemical proof of origin of macrophages in laser photocoagulation lesion in the retina. *Jpn J Ophthalmol* **40**:192–201.

41. H. Kimura, T. Sakamoto, C. Spee, and et al. 1995, A new model for experimental subretinal neovascularisation in the rabbit, *Invest Ophthalmol Vis Sci* **35**:3429–3435.
42. N. Ogata, M. Matsushima, Y. Takada, T. Tobe, and et al. 1996, Expression of basic fibroblast growth factor mRNA in developing choroidal neovascularization, *Curr Eye Res*:1008–1018.
43. H.G.T. Blaauwgeers, G.M. Holtkamp, P. Koolwijk, van Hinsbergh, V, M. Kroon, A. Kijistra, and R. Schlingemann, 1998, Preferential basal secretion of VEGF by RPE *in vitro* and expression of its receptor KDR on the inner choriocapillaris *in vitro* suggest a trophic relation in the normal eye, *Invest Ophthalmol Vis Sci* **39**(4):S372
44. L. da Cruz, P. Rakoczy, M. Perricaudet, and I.J. Constable, 1996, The dynamics of gene transfer to retinal pigment epithelium. *Invest Opthalmol Vis Sci* **37**(12):2447–2454.
45. M.B. Reichel, R.R. Ali, A.J. Thrasher, D.M. Hunt, S.S. Bhattacharya, and D. Baker, 1998, Immune responses limit adenovirally medited gene expression in the adult mouse eye, *Gene Therapy* **5**:1038–1046.

37

IDENTIFICATION OF DEGENERATION-SPECIFIC GENES IN THE RCS RAT RETINA BY SUBTRACTIVE HYBRIDIZATION TECHNIQUES

Arman Tehrani, Thomas H. Wheeler-Schilling, Eberhart Zrenner, and Elke Guenther

Department of Experimental Ophthalmology
University Eye Hospital
Section for Pathophysiology of Vision and Neuro-Ophthalmology
Röntgenweg 11, D-72076 Tübingen
Germany

SUMMARY

Retinitis pigmentosa (RP) is one of the most frequent hereditary dystrophies. Multiple genetic defects have been found to be expressed in the photoreceptors and retinal pigment epithelium (RPE) cells of RP patients. The Royal College of Surgeons (RCS) rat is the animal model most commonly used to study RP. In the RCS rat, the photoreceptors begin to degenerate during the third postnatal week, and retinal degeneration continues toward the inner retina after their loss. We used a subtractive hybridization technique to identify differentially expressed genes that might be responsible for the induction of degenerative processes in the RCS retina. Until now, we have isolated about 30 clones with putative differentially expressed sequences. To date, two of these sequences were confirmed to be differentially expressed in the RCS rat but not in control rats. The method described here is therefore suitable for identifying differentially regulated genes, not only in RP retinas, but also in other retinal dystrophies and may help to further elucidate the pathophysiology of retinal diseases.

Retinal Degenerative Diseases and Experimental Therapy, edited by Hollyfield *et al.*
Kluwer Academic / Plenum Publishers, New York, 1999.

INTRODUCTION

Genetic causes have been attributed to more than 4000 human disorders, and a vast number of these genetic causes also affect ocular tissues and vision.[1,2] One of the most frequent retinal hereditary dystrophies, with a prevalence of 1:3000 to 1:5000, is retinitis pigmentosa (RP). RP is in reality a clinically heterogeneous group of disorders characterized by degeneration of photoreceptors and pigment epithelium with different modes of genetic transmission and different types of progression, usually leading to blindness in later life.[3–6] A hereditary pattern has been determined in approximately 50% of the patients.[7] Because multiple genetic defects have been found to be expressed in photoreceptors[8,9] and in RPE cells,[10] RP can be considered as a group of diseases caused by abnormal genes at various loci within the human genome.[11]

In recent years tremendous new insights have been gained by using molecular techniques for the investigation of retinal dystrophies. The molecular pathology of several ocular degenerative disorders is being studied in animal models, and such studies will hopefully provide new insights into the pathogenesis of these diseases. In this regard results from animal experiments over the past years leads to the assumption that the underlying pathophysiological mechanisms are not one but many.[12] The rat model most commonly used to study RP is the degenerative Royal College of Surgeons (RCS) strain. This strain, first described in 1938,[13] was initially maintained at the Royal College of Surgeons (London, UK).[14–16] RCS rats are characterized by an autosomal recessive retinal dystrophy resulting in loss of rod cells.[13,15] Due to the failure of retinal pigment epithelium cells to phagocytose the shed disks of the outer segments,[17,18] there is an accumulation of debris which changes the extracellular milieu and results in visual impairment. However, there are not only photoreceptors and RPE cells affected in RP. During the degeneration process, also the outer nuclear/plexiform layer (ONL/OPL) disappear and the inner nuclear layer (INL) becomes affected, too.[19]

Since most of these changes have to be accompanied by alterations in the gene expression of the cells affected, there should be detectable differences in the genes expressed in RP retinas compared to normal ones. Several methods can be used today to detect such changes in expression. All are based on the assumption that there are changes in the abundance of a small set of specific mRNAs. The difficulty lies in isolating these specifically altered transcripts from the thousands of constitutive transcripts within every eukaryotic cell. In higher eukaryotes, biological processes such as cellular growth, differentiation, development, and degeneration are mediated by programs of differential gene expression. Messengers of genes relevant for a specific developmental and/or degenerative process are thought to change their abundance due to induction or repression of transcription, although other levels of regulation may be involved as well.[20] To understand the molecular regulation of these processes, the relevant subsets of differentially expressed genes of interest must be identified, cloned, and studied in detail. Subtractive cDNA hybridization has proved to be a powerful tool for identifying and isolating cDNAs of differentially expressed genes.[21,22] Numerous cDNA subtraction methods have been reported. In general, they involve hybridization of cDNA from one population (tester) to an excess of mRNA (cDNA) from another population (driver) followed by separation of the unhybridized fraction (target) from hybridized common sequences.[23]

The intention of this chapter is mainly to provide a methodological description of the identification of up- and down-regulated genes in the retina/RPE of RCS rats by subtractive hybridization techniques. However, we wish to point out that any degenerative processes that depend on changes in transcript abundance can be examined with these methods. To date, we have been able to isolate about 30 clones; two of these sequences were found to be differentially regulated in the RCS rat. Further experiments will elucidate the cellular localization of their gene products.

To summarize, the approach described here is capable of identifying differentially expressed genes during retinal degeneration. It will thus hopefully help to deepen understanding of the pathophysiology of retinal disorders and promote new therapeutical concepts for the treatment of such diseases.

METHODS

Animals

Retinae were obtained from three RCS rats on postnatal day 18 (P18) and three control animals on postnatal day 20 (P20). After cervical dislocation and enucleation the retinae were removed from the eyecups and immediately frozen in liquid nitrogen. They were stored at –80°C until use. The treatment of animals in this study was in conformity with the ARVO Resolution On Animal Use in Research.

mRNA Isolation

mRNA was isolated using the oligo-dT based mRNA Isolation Kit (Boehringer Mannheim, Germany) according to the manufacturer's protocol (Fig. 1).

SMART cDNA PCR KIT

The cDNA was synthesized using the SMART cDNA PCR KIT (Clontech Laboratories, Palo Alto, USA). About 0.35 µg of mRNA was reverse transcribed in the presence of cDNA synthesis (CDS) primer and SMART II oligonucleotide (both provided by Clontech, Palo Alto, USA) using Superscript II reverse trancriptase (GIBCO/BRL, Karlsruhe, Germany). The reaction was then diluted with 90 µl TE buffer (10 mM Tris pH 7.6, 1 mM EDTA), heated to 72°C for 7 min and then stored at –20°C. To determine the optimal number of cycles for cDNA amplification two test tubes were set up according to the manufacturer's instructions, using the supplied PCR primer and the diluted first strand reaction as template. After 15, 18, 21, and 24 cycles, 15 µl of the PCR reaction was transferred into a extra tube and stored on ice. After the last cycle, these probes were run on a 1.2% agarose gel. By determining the size of the amplification product, we found that the optimal number of PCR cycles was 18 cycles. The cDNA amplifications were then carried out with a PCR consisting of 18 cycles. For purification, two tubes of each pool were mixed and phenol extracted according to standard procedure.[24] The aqueous phase of the phenol extraction was mixed with n-butanol to reduce the volume to about 60 µl. The remaining aqueous fraction was loaded onto CHROMA SPIN-1000 columns (Clontech, Palo Alto, USA) and eluted with TNE buffer (10 mM Tris pH 7.6, 10 mM NaCl, 0.1 mM EDTA). The presence of cDNA in the collected fractions was checked by agarosegel electrophoresis.

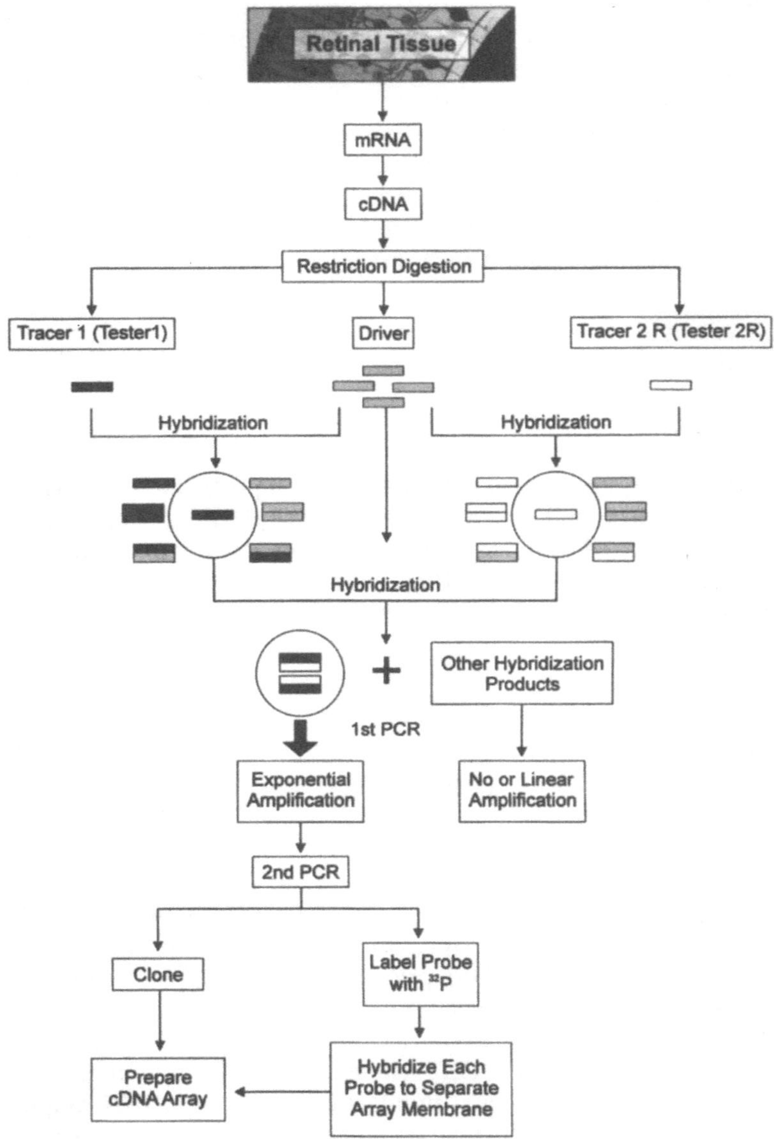

Figure 1. Schematic diagram of the PCR-Select cDNA subtraction. For simplicity, only the forward subtraction is shown. The restriction-digested cDNA of the RCS retina is ligated to Adaptor 1, forming Tracer 1, as shown in black. The same cDNA was also ligated to Adaptor 2R, resulting in tester 2R, as shown in white. The driver cDNA used (shown in gray) results from the *Rsa* I digestion of the control cDNA. The black/white molecules after the second hybridization are formed only if the sequence is up-regulated in tester cDNA. Only these molecules can be exponentially amplified in the following 1st PCR.

Rsa I Digestion

The purified cDNA was digested with *Rsa* I (Fig. 1). The digest was purified using the Qiagen PCR Purification Kit (Qiagen, Hilden, Germany). The digested cDNA was eluted with water. The concentration was checked spetrophotometrically and adjusted to 300 ng/μl.

Adaptor Ligation

The adaptors 1 and 2R (provided with the Clontech PCR Subtraction Kit) were ligated to both cDNA pools, using 400 units of T_4 DNA ligase per reaction in a total volume of 10 µl. The reactions were incubated at 16°C overnight. The reaction was stopped by adding 1 µl of 0.2 M EDTA; 1 mg/ml glycogen and heating the mixtures to 72°C for 5 min. Ligation products were stored at −20°C.

First Hybridization

In the following procedure, an excess of driver cDNA is added to each tester (e.g. adaptor ligated cDNA; Fig. 1) and the samples are heat-denatured and allowed to anneal. The concentrations of high- and low-abundance sequences that are left in single stranded (ss) form after annealing are equalized, because reannealing is faster for the more abundant molecules due to the second-order hybridization kinetics. Furthermore, the remaining ss cDNAs (available for the second hybridization) are dramatically enriched for differentially expressed sequences, since non-target cDNAs present in the tester and driver cDNA form hybrids. Control experiments provided with the kit were performed simultaneously.

In the forward subtraction, the cDNA of the RCS rat retina was used as tester and the cDNA of the control served as driver. 1.5 µl of both *Ras* I-digested driver cDNA and adaptor1-ligated tester were mixed with 4 µl hybridization buffer(4×) provided with the kit. The same was done for the adaptor2R-ligated tester in a separate tube. In the reverse subtraction, the cDNA of the RCS rat retina was used as driver and the cDNA of the control served as adaptor-ligated testers. The hybridizations were incubated in a thermal cycler at 98°C for 1.5 min. Incubation was continued at 68°C for 8 hr and then followed immediately by the second hybridization (Fig. 1).

Second Hybridization

The two samples of each subtraction from the first hybridization were mixed together and freshly denatured driver DNA was added to further enrich for differentially expressed sequences. New hybrid molecules corresponding to differentially expressed cDNAs with different adaptors on each end were formed. 1 µl of each driver was mixed with 1 µl hybridization buffer(4×) and 2 µl H_2O. This mixture was incubated in a thermal cycler at 98°C for 1.5 min. In the next step, this freshly denatured driver was mixed simultaneously with the two hybridization reactions from the first hybridization. The reactions were incubated at 68°C overnight. 200 µl of dilution buffer (20 mM HEPES-HCl pH 8.3, 50 mM NaCl, 0.2 mM EDTA pH 8.0) was added after hybridization was completed, and the mixture was heated to 68°C for 7 min.

PCR Amplification

Differentially expressed cDNAs were selectively amplified during the reactions described below. Before PCR amplification, the missing strands of the adaptors were filled in by a brief preincubation at 75°C. This created the binding site for PCR primer1. In the first amplification, only ds cDNAs with different adaptor sequences on each end were exponentially amplified. The second, nested PCR was used to further reduce background and enrich for differentially expressed sequences (all primer- and adaptor-sequences were provided by Clontech, Palo Alto, USA).

1 µl of each diluted cDNA was mixed with 2.5 µl 10× reaction buffer, 0.5 µl dNTP's (10 mM each), 19.5 µl H_2O, 1.0 µl PCR primer1 (10 µM), and 0.5 µl 50× Advantage cDNA Polymerase Mix (Clontech, Palo Alto, USA). The reactions were amplified with the following protocol: 75°C 5 min, 94°C 25 sec, 27 cycles of 94°C 10 sec, 66°C 30 sec, 72°C 1.5 min. After the PCR was completed, 3 µl of each reaction was diluted with 27 µl of H_2O. 1 µl of these diluted first PCR products were used as a template in the second, nested PCR. The composition of this PCR was as follows: 1 µl of diluted template, 18.5 µl H_2O, 2.5 µl 10× reaction buffer, 1.0 µl nested PCR primer1 (10 µM), 1.0 µl nested PCR primer2R (10 µM), 0.5 µl dNTP's (10 mM each), and 0.5 µl 50× Advantage cDNA Polymerase Mix. Reactions were cycled with the following protocol: 11 cycles of 94°C 10 sec, 68°C 30 sec, 72°C 1.5 min followed by 72°C for 5 min.

T/A Cloning

Cloning of the enriched cDNA fragments was done using the Clontech AdvanTAge PCR Cloning Kit (Clontech, Palo Alto, USA) according to the manufacturer's instructions.

cDNA Array

Clones that appeared positive in blue/white screening were selected and grown overnight at 37°C in a 96 well plate containing 100 µl of LB-Amp in each well. In parallel, each selected clone was grown on LB-Amp plates for later use. A PCR reaction mastermix was composed as follows with at least 96 reactions for each subtraction: 2.0 µl 10× reaction buffer, 0.6 µl nested primer 1, 0.6 µl nested primer 2R, 0.4 µl dNTP's (10 mM each) 15.2 µl H_2O, and 0.2 µl 50× Advantage cDNA Polymerase Mix. 19.0 µl of the master mix was aliquoted into a reaction tube, and 1 µl of each bacterial culture was added. Reactions were cycled with the following protocol: 94°C 30 sec, 23 cycles 95°C 30 sec, 68°C 3 min. Each reaction product was analyzed by agarose gelelectrophoresis. Preparation of cDNA dot blots for two sets of 96 PCR products resulting from forward and reverse subtraction: 5 µl of each PCR product was mixed with 5 µl of 0.6 N NaOH. 1.5 µl of each mixture was then spotted in double probes onto a positively charged nylon membrane (Boehringer Mannheim, Germany). Blots were neutralized for 2–4 min in 0.5 M Tris-HCl pH 7.5 and washed in H_2O. Cross-linkage was preformed in a UV linking device (UV Stratalinker, Stratagene, Heidelberg, Germany) under 120 mJ.

Restriction Digestion of Subtracted cDNA Probes

Before the forward- and reverse-subtracted cDNA can be used as probes for differential screening, the adaptor sequences had to be removed. Two tubes of secondary PCR product for each subtraction were purified using Qiagen PCR Purification Kit (Qiagen, Hilden, Germany). The cleaned product was then digested with *Rsa* I for 1 hr at 37°C, followed by a *Sma* I digestion for 1 hr at 37°C, according to standard procedures.[24] The adaptors were removed from the cDNA by using the Qiagen PCR Purification Kit (Qiagen, Hilden, Germany).

Random-Primer Labeling of cDNA Probes

3 µl of each adaptor-free forward- and reverse-subtracted cDNA were mixed with 6 µl of H_2O. Probes were denatured by heating for 8 min at 95°C and then chilled on ice. 3 µl reaction buffer (-dCTP), 2 µl random primer mix, 1 µl Klenow enzyme (exo-) (all provided by Clontech, Palo Alto, USA) and 5 µl [α-^{32}P]dCTP (50 µCi, 3000 Ci/mmol) were added. Reactions were incubated at 37°C for 30 min. The reaction was stopped by heating to 70°C for 5 min. Probes were purified from unincorporated dNTP's using CHROMA Spin-100 Columns (Clontech, Palo Alto, USA) according to the manufacturer's protocol.

Hybridization of the Dot Blots with Labeled Subtracted cDNAs

50 µl of 20× SSC were mixed with 50 µl of Blocking Solution (10 mg/ml sheared salmon sperm DNA, 0.3 mg/ml oligonucleotides corresponding to the nested primers and complementary sequences used to amplify cDNA), boiled for 5 min and chilled on ice. 5 ml of hybridization solution (ExpressHyb Hybridization Solution, Clontech, Palo Alto, USA) were added and membranes were prehybridized for 60 min at 72°C in rotating bottles. Hybridization probe was prepared by mixing 50 µl of 20× SSC, 50 µl blocking solution, and the purified radiolabeled probe. The probe was boiled for 5 min, chilled on ice and added to the hybridization container. Hybridization was done overnight at 72°C. The following washing steps were done with prewarmed buffers at 68°C: four times with 2× SSC/0.5% SDS, twice with 0.2× SSC/0.5% SDS, each step for 20 min. Blots were exposed to AGFA Curix films (AGFA, Leverkusen, Germany) overnight at –70°C using an intensifying screen.

Sequencing

Plasmids which seemed to be differentially expressed in the differentially screening procedure were sequenced. We used the DNA Sequencing Kit Big Dye Terminator (Perkin Elmer, Weiterstadt, Germany) according to the manufacturer's instructions for sequencing. Sequencing reactions were cleaned up and run on a ABI Prism 310 Sequencer (Perkin Elmer, Weiterstadt, Germany). The resulting sequences were aligned to Genbank and EMBL databases using programs of the GCG Package (Genetics Computer Group, Inc., Madison, USA).

Virtual Northern Blot

A virtual northern blot was prepared to confirm the differential expression of the sequences identified with the differential screening procedure. 750 ng of cDNA per lane, prepared by the SMART cDNA PCR procedure, was loaded onto a 1.0% agarosegel and run in 1× TAE. The gel was blotted onto positively charged nylon membrane (Boehringer Mannheim, Germany) using capillary blotting procedure.[24] Cross-linkage was performed in a UV linking device (UV Stratalinker, Stratagene, Heidelberg, Germany) under 120 mJ. The amplification product, which was also used for the cDNA array, was radiolabeled by random priming for each clone which had to be confirmed. 100 ng of amplified cDNA was heat denatured, chilled on ice, mixed with 4 µl of High Prime Solution (Boehringer Mannheim, Germany) containing all reagents. 2 µl of [α-^{32}P]dCTP (50 µCi, 3000 Ci/mmol) were added and the reaction was incubated at 37°C for 30 min. Unincorporated dNTP's were removed using the Qiagen Nucleotide Removal Kit. Blots were prehybridized in Hybridization Solution (Clontech, Palo Alto, USA) at 60°C for 30 min. The radiolabeled probe was heat-denatured, chilled on ice and added. Blots were incubated at 60°C for 60 min. The following washing steps were prepared: Four times with 2× SSC/0.05% SDS at 25°C, 10 min per washing, twice with 0.1× SSC/0.1% SDS at 50°C, 20 min per washing. Blots were exposed to AGFA Curix films (AGFA, Leverkusen, Germany) overnight at –70°C using two intensifying screens.

RESULTS

The subtractive hybridization strategy is intended to enrich differentially expressed sequences out of two gene pools. In the previously described procedure, a restriction digestion was made to reduce the average fragment size, so that the resulting fragments corresponded to digested cDNA fragments. The forward- and reverse subtracted pools are cloned into a vector. These cloned fragments are then amplified by colony PCR, using primers binding to the previous ligated linker sequences. These reactions provide the cloned inserts, which are later used in the differential screening procedure, in larger amounts. As Fig. 2 shows, the size of the PCR-fragments varied between approximately 200 and 1500 basepairs as was to be expected after a *Rsa* I restriction digestion. It can also be seen that some clones show no insert or have two inserts (Fig. 2; lane 3: clone without insert; lanes 10, 18, 22: clones with two inserts).

For the differential screening procedure, only clones were chosen which showed a single band of 200–1500 basepairs. The results of the differential screening procedure of the forward subtraction are shown in Fig. 3. The amplified cDNA fragments were spotted onto a nylon membrane and then hybridized to different probes. The membrane in Fig. 3A was probed with the radiolabeled forward subtracted fragments. The result of probing the same membrane with the reverse subtracted radiolabeled fragments is shown in Fig. 3B. A comparison of both shows cDNAs which hybridize to the forward subtracted probe but not to the reverse subtracted probe (Fig. 3A, B; indicated by the white arrowhead). These clones most likely correspond to differentially expressed mRNAs and call for further analysis. The intensity of the hybridization signal may vary significantly. The efficiency of enrichment for each particular

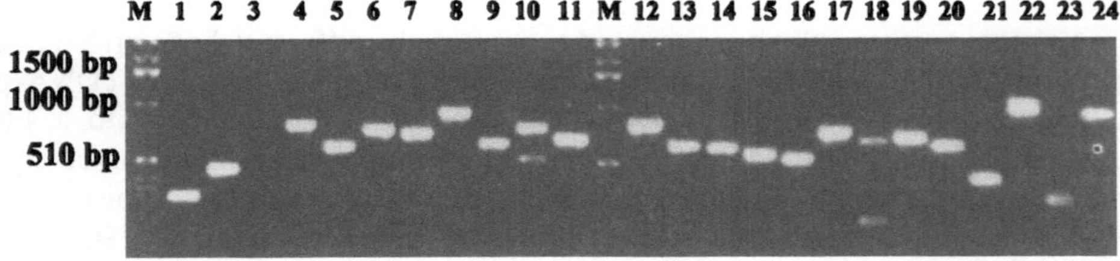

Figure 2. Ethidium bromide stained agarosegel. Lanes 1 to 24: amplification products resulting from cDNA Array PCR. Lane 3: no insert in vector. Lanes 10, 18, and 22: two inserts in vector. Each fragment represents a putative differentially regulated sequence. Only clones with a single band, corresponding to a single cloned fragment, were used for dot blotting and later on served as a probe in the virtual northern blot procedure (lanes 1, 2, 4–9, 11–17, 19–21, 23, 24). M: molecular weight marker, 1kbp-ladder, GIBCO/BRL.

fragment mainly depends on the abundance of the corresponding mRNA in tester versus driver.

In general, mRNAs that are much more abundant in tester than driver will be more efficiently enriched during subtraction. Nevertheless, some cDNA sequences cannot be rehybridized and/or amplified very efficiently. As a result, these cDNA fragments will generate fainter hybridization signals during differential screening. Clones that hybridize equally to both subtracted probes (Fig. 3A, B; indicated by the black arrowhead), do not represent differentially expressed mRNAs and were not further analyzed. Clones that hybridize to both subtracted probes, with an intensity of the hybridization signal that differs ≥ 5-fold, probably correspond to a differentially expressed gene and were also chosen for further analysis (Fig. 3A, B; white asterix). Clones that do not have a detectable hybridization signal for either of the subtracted probes usually represent nondifferentially expressed cDNAs present in the subtracted library. Some nondifferentially expressed cDNA fragments are randomly present in the subtracted library as single copies. Because these molecules are present at such low levels in the subtracted probe, they do not hybridize to the corresponding clone. In Fig. 3C and 3D the membranes were hybridized to the unsubtracted cDNA. In Fig. 3C the unsubtracted tester was used as radiolabeled probe. In Fig. 3D the unsubtracted driver was used as radiolabeled probe. Care was taken that the clones used for further analysis did not show a hybridization signal in Fig. 3C and no or only a very low hybridization signal in Fig. 3D. Clones that showed a signal in Fig. 3A, no signal in Fig. 3B, but a signal in Fig. 3C are false positives. The same is true for putative positive clones

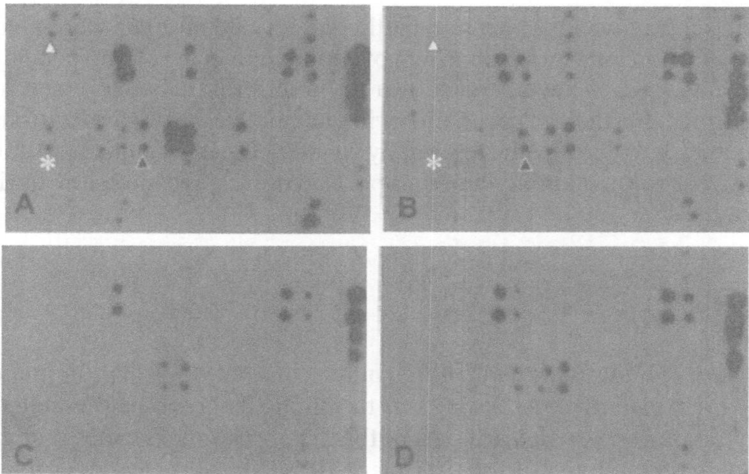

Figure 3. Autoradiographic results of the differential screening procedure. Four identical dot membranes were prepared for the 96 clones of the forward subtraction. (A) Probing with the forward subtracted cDNA. (B) Probing with the reverse subtracted cDNA. (C) Probing with unsubtracted driver cDNA. (D) Probing with unsubtracted tester cDNA. The white arrowhead indicates a (forward-subtracted) clone, which is enriched in the forward-subtraction, because it shows a signal in (A) but not in (B) and does not hybridize with the unsubtracted probes (C) and (D). The white asterix indicates a clone, which is also enriched in the forward-subtraction. Since its hybridization signal differs ≥ 5-fold, it also therefore corresponds to a differentially expressed gene. Only clones showing a hybridization signal as indicated by the white arrowhead or white asterix were chosen for further analysis. The black arrowhead indicates a false positive clone, because it is enriched in both blots (A) and (B), compared to the unsubtracted control.

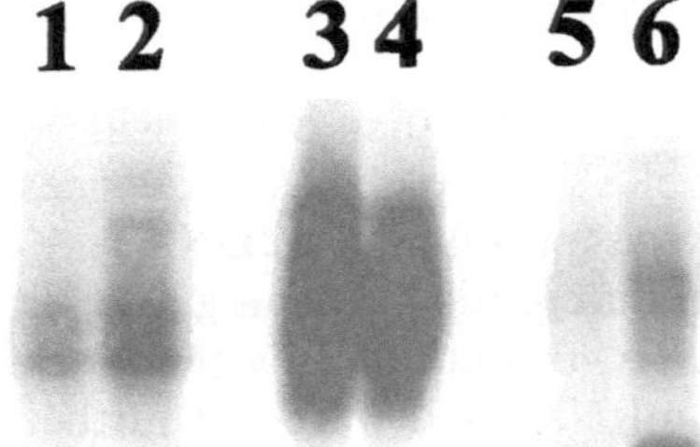

Figure 4. Autoradiogram of the virtual northern blot. Lanes 1, 3, 5: cDNA of the RCS retina. Lanes 2, 4, 6: cDNA of the control retina. Lane 3 and 4: Probed with β-actin as a positive control and to confirm that amounts of cDNA nearly equal were used per lane. The sequence used as a probe in lanes 1 and 2 is slightly repressed in the RCS retina. The same is true for the probe used in lanes 5 and 6.

showing a signal in Fig. 3D, because these sequences were certainly not enriched during the subtraction procedure.

Figure 4 shows the results of the virtual northern blotting, used to confirm the differentially expressed clones. In lanes 1, 3, and 5 the undigested cDNA resulting from the SMART cDNA PCR Kit was used as template, whereas in lanes 2, 4, and 6 the corresponding control cDNA served as template. Lanes 3 and 4 were probed with a radioactive labeled β-actin probe to check that equal amounts of cDNA were used per lane. In Fig. 4 (lanes 1 and 2), a differentially regulated cDNA fragment, identified by the differential screening procedure, was used as probe. As can be seen by the intensity of labeling, the analyzed sequence is expressed more intensely in the control retina (Fig. 4; lane 2) than in the RCS retina (Fig. 4; lane 1). The same is true for the cDNA fragment used as probe in Fig. 4 (lanes 5 and 6). This fragment is also strongly expressed in the control retina, but only a weak signal can be seen in the RCS cDNA.

With the subtractive hybridization method described here, we were able to isolate approximately thirty clones with an insert of putative degeneration-specific expressed genes; two of these genes have been shown to be differentially expressed in the RCS rat retina to date. Further experiments will include *in situ*-hybridizations and immunocytochemistry to localize differentially expressed gene products in the RCS rat retina and to learn more about their temporal pattern of expression during retinal degeneration.

DISCUSSION

Subtractive cloning is a powerful technique that allows isolation and cloning of differentially expressed mRNAs. Using this technique it should be possible to find still unknown candidate genes which are part of the degeneration cascades that are responsible for the functional defects in the retina of RCS rats. Here, we describe a possible methodological approach to identify such genes. The protocol is based on the assumption that retinal degeneration will be the result of changes in the abundance of a small set of specific mRNAs.

To date, different strategies have been used successfully to clone putative differentially expressed genes. For plus/minus screening, a cDNA library is generated from mRNA. To isolate the up- or down-regulated sequences within the library, cDNA probes are generated from mRNA isolated from the tissue of interest at each of two different stages.[25–27] The Differential Display protocol relies on arbitrary synthetic oligonucleotide primers that will anneal only to a subset of mRNA molecules and amplification by RT-PCR. The products are fractionated in DNA sequencing gels.[28,29] The PCR-based cloning strategy, i.e. subtractive hybridization Gene Expression Screen,[30] requires two mRNA populations; these are converted into short stretches of cDNA, to which adapters are ligated and amplified by PCR.[31] Despite successful identification of numerous important genes by these methods, they are usually inefficient for obtaining low abundance transcripts, often involve repeated subtraction steps, and are labor intensive. The difficulty lies in isolating specifically altered transcripts from the thousands of constitutive transcripts within the retina.

The PCR-based cDNA subtraction method, termed suppression subtractive hybridization[23] and based on the suppression PCR effect,[32,33] overcomes the problem of differences in the mRNA abundance by incorporating a hybridization step that normalizes sequence abundance during the course of subtraction by standard hybridization kinetics.[34] It eliminates any intermediate steps for physical separation of single-stranded and double-stranded cDNAs, requires only one subtractive hybridization round, and can achieve more than 1000-fold enrichment for differentially expressed cDNAs.[23]

In this study, we combined suppression subtractive hybridization methodology with three other molecular biological approaches to study differentially expressed genes in the RCS rat retina. The first approach, a cDNA amplification step, provided us with enough starting material to perform suppression subtracted hybridization, for which a certain amount of cDNA is needed to show reliable results. Suppression subtractive hybridization enriches the differentially expressed sequence, as mentioned above, but there are still nondifferential regulated fragments in the resulting reaction mixture. To overcome this problem, the differential screening procedure was applied, which allows differentiation between putative differentially expressed fragments and actually false positive sequences. This results in a dramatical reduction of the number of clones, excluding nearly all false positives. Furthermore, the differential expression of the remaining clones can be confirmed by a virtual northern blot, using the amplified cDNA from the first amplification (see Fig. 4).

The present study shows that this combination of existing techniques is efficient for generating cDNAs highly enriched for differentially expressed genes of both high and low abundance from the RCS rat retina. The high level of enrichment of rare transcripts has been achieved by the inclusion of a normalization step in the subtraction procedure. Future developments in these molecular techniques may play an increasing role for clinical classification and evaluation of retinal dystrophies. For example, the technique would be useful for evaluating retinal neurodegeneration induced by excitatory amino acids, the survival of cell populations after axotomy or retinal ischemia. It might reveal mechanisms which participate in the pathogenesis of glaucoma. Furthermore, modifications of this procedure will offer the possibility of identifying quantitatively regulated transcripts.

ACKNOWLEDGMENTS

Supported by a grant from the Federal Ministry of Education, Science, Research, and Technology (Fö. 01KS9602) and the Interdisciplinary Clinical Research Center (IKFZ) Tübingen.

REFERENCES

1. S. Merin, 1991, *Inherited Eye Disese: Diagnosis and Clinical Management*, Marcel Dekker, New York.
2. M.A. Musarella, 1992, Gene mapping of ocular diseases, *Surv. Ophthalmol.* **36**:285–312.
3. S. Merin, E. Auerbach, 1976, Retinitis pigmentosa, *Surv. Ophthalmol.* **20**:303–346.
4. J.R. Heckenlively, 1988, *Retinitis Pigmentosa*, J. B. Lippincott, Philadelphia.
5. R.A. Pagon, 1988, Retinitis pigmentosa, *Surv. Ophthalmol.* **33**:137–177.
6. E. Zrenner, E. Apfelstedt-Sylla, K. Rüther, 1995, Clinical aspects: retinitis pigmentosa, in: *Neurobiology and Clinical Aspects of the Outer Retina* (M.B.A. Djamgoz, S.N. Archer, and S. Vallerga, eds.), pp. 447–460, Chapman & Hall, London.
7. E.L. Berson, 1994, Retinitis pigmentosa and allied diseases, in: *Principles and Practice of Ophthalmology*, 2 (D.M. Albert and F.A. Jakobiec, eds.), pp. 1214–1237, WB Saunders, Philadelphia.
8. M.A. Maw, B. Kennedy, A. Knight, R. Bridges, K.E. Roth, E.J. Mani, J.K. Mukkadan, D. Nancarrow, J.W. Crabb, M.J. Denton, 1997, Mutation of the gene encoding cellular retinaldehyde-binding protein in autosomal recessive retinitis pigmentosa, *Nat. Genet.* **17**:198–200.
9. A.F. Wright, 1997, A searchlight through the fog [news], *Nat. Genet.* **17**:132–134.
10. S.M. Gu, D.A. Thompson, C.R. Srikumari, B. Lorenz, U. Finckh, A. Nicoletti, K.R. Murthy, M. Rathmann, G. Kumaramanickavel, M.J. Denton, A. Gal, 1997, Mutations in RPE65 cause autosomal recessive childhood-onset severe retinal dystrophy, *Nat. Genet.* **17**:194–197.
11. P. Stalmans, B. Himpens, 1998, A decreased Ca2+-wave propagation is found among cultured RPE cells from dystrophic RCS rats, *Invest. Ophthalmol. Vis. Sci.* **39**:1493–1502.
12. M.J. Voaden, 1988, Retinitis pigmentosa and its models, *Prog. Ret. Res.* **7**:293–331.
13. M.C. Bourne, D.A. Campell, K. Tansley, 1938, Hereditary degeneration of the rat retina, *Br. J. Ophthalmol.* **22**:613–623.
14. D.R. Lucas, M. Attfield, J.B. Davey, 1955, Retinal dystrophy in the rat., *J. Pathol. Bact.* **70**:469–474.
15. J.E. Dowling, R.L. Sidman, 1962, Inherited retinal dystrophy in the rat, *J. Cell Biol.* **14**:73–109.
16. M.M. LaVail, M. Sidman, R. Rausin, R.L. Sidman, 1974, Discrimination of light intensity by rats with inherited retinal degeneration: a behavioral and cytological study, *Vision Res.* **14**:693–702.
17. D. Bok, M.O. Hall, 1971, The role of the pigment epithelium in the etiology of inherited retinal dystrophy in the rat, *J. Cell Biol.* **49**:664–682.
18. R.J. Mullen, M.M. LaVail, 1976, Inherited retinal dystrophy: primary defect in pigment epithelium determined with experimental rat chimeras, *Science* **192**:799–801.
19. K. Kohler, J. Hartmann, S. Fischer, E. Zrenner, 1997, *Invest. Ophthalmol.* **38**:S38.
20. H.J. Balzer, H. Baumlein, 1994, An improved gene expression screen, *Nucleic. Acids. Res.* **22**:2853–2854.
21. J.R. Duguid, M.C. Dinauer, 1990, Library subtraction of in vitro cDNA libraries to identify differentially expressed genes in scrapie infection, *Nucleic. Acids. Res.* **18**:2789–2792.
22. E. Hara, T. Kato, S. Nakada, S. Sekiya, K. Oda, 1991, Subtractive cDNA cloning using oligo(dT)30-latex and PCR: isolation of cDNA clones specific to undifferentiated human embryonal carcinoma cells, *Nucleic. Acids. Res.* **19**:7097–7104.
23. L. Diatchenko, Y.F. Lau, A.P. Campbell, A. Chenchik, F. Moqadam, B. Huang, S. Lukyanov, K. Lukyanov, N. Gurskaya, E.D. Sverdlov, P.D. Siebert, 1996, Suppression subtractive hybridization: a method for generating differentially regulated or tissue-specific cDNA probes and libraries, *Proc. Natl. Acad. Sci. U.S.A.* **93**:6025–6030.
24. J. Sambrook, E.F. Fritsch, T. Maniatis, 1989, *Molecular Cloning*+Laboratory Manual*, 2nd, Cold Spring Harbor Laboratory Press, New York.
25. T.D. Sargent, I.B. Dawid, 1983, Differential gene expression in the gastrula of Xenopus laevis, *Science* **222**:135–139.
26. L.M. Schwartz, L. Kosz, B.K. Kay, 1990, Gene activation is required for developmentally programmed cell death, *Proc. Natl. Acad. Sci. U.S.A.* **87**:6594–6598.
27. Z.G. Liu, S.W. Smith, K.A. McLaughlin, L.M. Schwartz, B.A. Osborne, 1994, Apoptotic signals delivered through the T-cell receptor of a T-cell hybrid require the immediate-early gene nur77, *Nature* **367**:281–284.
28. P. Liang, A.B. Pardee, 1992, Differential display of eukaryotic messenger RNA by means of the polymerase chain reaction [see comments], *Science* **257**:967–971.
29. P. Liang, L. Averboukh, A.B. Pardee, 1993, Distribution and cloning of eukaryotic mRNAs by means of differential display: refinements and optimization, *Nucleic. Acids. Res.* **21**:3269–3275.

30. Z. Wang, D.D. Brown, 1991, A gene expression screen, *Proc. Natl. Acad. Sci. U.S.A.* **88**:11505–11509.
31. L. Buckbinder, D.D. Brown, 1992, Thyroid hormone-induced gene expression changes in the developing frog limb, *J. Biol. Chem.* **267**:25786–25791.
32. P.D. Siebert, A. Chenchik, D.E. Kellogg, K.A. Lukyanov, S.A. Lukyanov, 1995, An improved PCR method for walking in uncloned genomic DNA, *Nucleic. Acids. Res.* **23**:1087–1088.
33. N.G. Gurskaya, L. Diatchenko, A. Chenchik, P.D. Siebert, G.L. Khaspekov, K.A. Lukyanov, L.L. Vagner, O.D. Ermolaeva, S.A. Lukyanov, E.D. Sverdlov, 1996, Equalizing cDNA subtraction based on selective suppression of polymerase chain reaction: cloning of Jurkat cell transcripts induced by phytohemaglutinin and phorbol 12-myristate 13-acetate, *Anal. Biochem.* **240**:90–97.
34. S.O. von, W.G. Thies, M. Hofmann, 1997, A high throughput screening for rarely transcribed differentially expressed genes., *Nucleic. Acids. Res.* **25**:2598–2602.

38

ANALYSIS OF FIELD POTENTIALS AND SPIKE PATTERNS EVOKED BY LOCAL ELECTRICAL STIMULATION OF THE CHICKEN RETINA

S. Weiss,[1,2] T. Herrmann,[1] A. Stett,[1,2] E. Zrenner,[2] and H. Haemmerle[1]

[1] Natural and Medical Sciences Institute
Reutlingen, Germany
[2] University Eye Hospital, Dept. II, Tübingen
Germany

1. INTRODUCTION

It is well known that the retina can be stimulated *in vitro* and *in vivo* with monopolar transretinal or local bipolar applied electric current.[1–4] These results indicate that it might be possible to develop a retina implant that electrically stimulates the remaining cells of the retinal network electrically, even if the photoreceptors are completely degenerated. Current research focuses on two different approaches for the development of an intraocular prosthesis for patients suffering from photoreceptor degeneration. Several groups favour direct electrical stimulation of retinal ganglion cells[5–7] from the epiretinal side while Tassiker and Chow proposed in patents to stimulate neurons of the inner nuclear layer[8–10] from the subretinal side (see also Zrenner et al. in this volume).

The aim of our study was to analyze slow retinal potentials and ganglion cell activity evoked by *local* sub- or intraretinal applied current pulses *in vitro*. There are, in our opinion, three main reasons for this type of experiments: First, much of the previously published data were either obtained under global transretinal stimulation[3,11,12] or only the summed activity of large retinal areas like electrically evoked cortical potentials was measured.[2,9] Therefore, it is almost impossible to gain information about the spatial resolution which can be achieved with focal subretinal electrical stimulation of

Retinal Degenerative Diseases and Experimental Therapy, edited by Hollyfield *et al.*
Kluwer Academic / Plenum Publishers, New York, 1999.

the retina. Second, some groups have used transretinal current pulses to investigate the lateral and horizontal signal transmission in the inner retina by recording the intracellular responses of single neurons.[1,12,13] Their results clearly indicate that it is possible to electrically evoke a transient change in the membrane potential of for example bipolar or amacrine cells. Unfortunately, it is not clear whether these graded potentials lead to stimulus correlated ganglion cell activity, which is a prerequisite for a subretinal implantable retinal prosthesis. Third, Humayun et al.[4] published results from bipolar stimulation of retinas *in vitro*, but the authors have restricted their investigations to *epiretinal* application of current pulses.

To obtain more insight in the encoding of sub- and intraretinal applied electrical stimuli, we used current pulses of varying strength and duration to elicit slow potentials and spikes. This allowed us to define a useful stimulation range for a subretinal implantable visual prosthesis. Furthermore we used multichannel recordings of slow retinal potentials and ganglion cell activity to define the retinal area that is stimulated by a locally applied current pulse.

2. LIGHT EVOKED ELECTRORETINOGRAM OF THE CHICKEN RETINA

We established an *in vitro* model to record local electroretinograms (ERG) and ganglion cell spikes from a piece of chicken retina adhered to a microelectrode array[14] (MEA). The MEA has 60 substrate integrated microelectrodes (electrode diameter: 10 μm, electrode spacing: 100 μm or 200 μm). Light evoked ERGs were recorded from up to 8 MEA electrodes to evaluate whether ERGs measured with the MEA are comparable to standard ERGs. We used a custom made diode stimulator consisting of a voltage-current converter and a light-emitting diode (RS 564-015, dominant wavelength: 660 nm) to project spot stimuli (diameter: 100 μm) through the optics of an inverted microscope and the MEA onto the photoreceptors. The voltage-current converter was driven by a freely programmable waveform generator (analogue output of a VSG2/3 graphics card, Cambridge Research Systems Ltd). A representative ERG recorded from a MEA electrode is shown in Fig. 1. It shows the typical a- and b-waves after the onset and d-waves after the offset of light in the chicken photopic ERG.

The b-wave is relatively small because a relatively high concentration of Mg^{++} was used to suppress the development of spreading depressions.[15] Regarding the waveforms it is obvious that the ERG recorded with the MEA is comparable to standard ERGs.[15–17] In addition to the a, b, and d-wave one can typically record prominent population spikes from the MEA-electrodes (Fig. 1). The population spikes reflect correlated spiking of ganglion cells. These results demonstrate that one can record focal ERGs and multi-unit ganglion cell activity with a MEA.

3. ELECTRICALLY EVOKED SLOW RETINAL POTENTIALS OF THE CHICKEN RETINA

Knighton[3,18] has shown that it is possible to elicit slow retinal potentials electrically by global application of transretinal current to an isolated frog eycup. We used metal microelectrodes to apply monopolar charge balanced current pulses locally to the distal retina. The microelectrodes were made of platinum iridium with parylene coating and had an exposed tip length of about 30 μm with the tip sharpened to 3 μm.

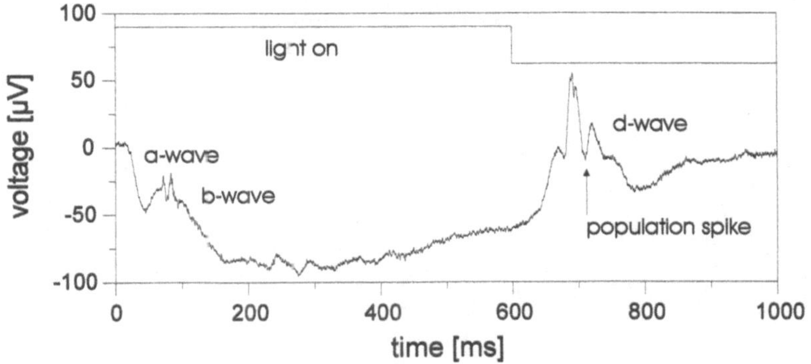

Figure 1. Local light evoked electroretinogram (ERG) of a chicken retina recorded with a microelectrode array (MEA) *in vitro*. The ERG was evoked by a light spot projected through the MEA onto the photoreceptors (spot diameter: 100 µm, retinal irradiance: 44 W/m^2). It is composed of the typical a-, b-, and d-wave and population spikes, which reflect correlated ganglion cell activity.

Monopolar injection of charge into the distal retinal network evoked a complex stimulus correlated signal. It consisted of the stimulus artifact, slow retinal potentials and ganglion cell spikes. Blockade of synaptic transmission by switching to a bath solution with higher Mg^{++} content (8–10 mM) resulted in the suppression of the retinal responses, while the passive electrical current spike remained unchanged. This enabled us to separate the biological signal from the stimulus artifact by subtracting the signal obtained under high Mg^{++} from the signal recorded with normal ringer solution (Fig. 2, upper panel). The result of the subtraction is shown in the lower panel of Fig. 2.

This difference signal is composed of three waves, a fast positive wave with a peak latency of about 10 ms, a long lasting negative wave and a sharp negative peak with a latency of about 80 ms. Quantitative analysis of the early positive wave revealed that the amplitude of the signal depends on the strength of the applied current pulse. The relationship between current strength and the mean signal amplitude (n = 5 experiments) is shown in Fig. 3.

The data shown in Fig. 3 clearly indicate that the working range for transmitting luminance information by electrically evoked slow potentials is relatively small. However, we are sure that for a subretinal implantable visual prosthesis it is appropriate to code luminance information by changing the amplitude of the electrical stimulation pulse. To test how the electrically evoked signals depend on the intraretinal position of the stimulation electrode we advanced the electrode into the retina up to 160 µm while we recorded retinal responses from the MEA electrode below the stimulation site. Original traces from such an experiment are shown in Fig. 4.

Retinal responses could be evoked with positive monopolar current pulses at stimulation depths varying from 20 µm to 120 µm (Fig. 4, left panel) while pulses with negative polarity were effective only when they were applied at a retinal depth of 80 µm and deeper (Fig. 4, right panel). It is obvious that the network response of an intact retina to a monopolar current pulse strongly depends on the polarity of the stimulus. Other authors[1,3,12,13,19] have adressed this question as well and found also that a retinal response was elicited by current flowing from photoreceptors to ganglion cells, while current flowing in the other direction did not evoke any response. Their results are comparable to the ones we obtained with positive monopolar charge injection. However,

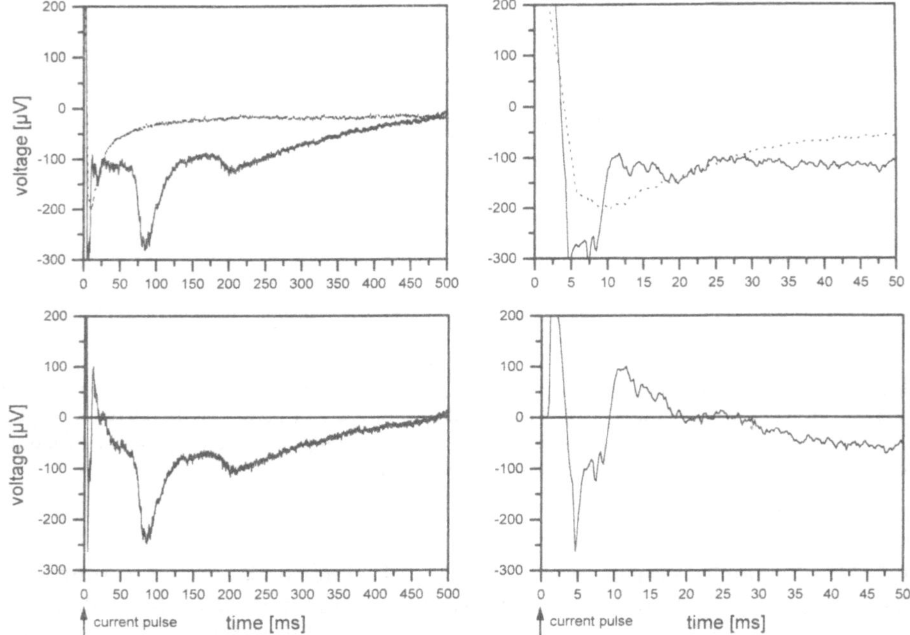

Figure 2. Upper panel: Electrically evoked responses of the chicken retina (solid lines). A negative monopolar current pulse (time: 0 ms, 30 µA, 1 ms duration) was applied at a retinal depth of 120 µm. The retinal response diminished (dotted lines) after we changed to a ringer solution with higher Mg^{++}-content (8 mM nstead of 1 mM). Lower panel: Difference signal between the traces recorded with a ringer solution with ow and high Mg^{++}-content. The difference signal is composed of positive and negative waves with different ımplitudes and timings.

when we lowered the electrode deeply into the retina, an injection of negative charge was also effective in our e. ·eriments (Fig. 4, right panel). This result can be explained by the fact that the negative pole of the electrical field was situated into or below the inner nuclear layer (INL) which resulted in a positive current flowing from the pho-

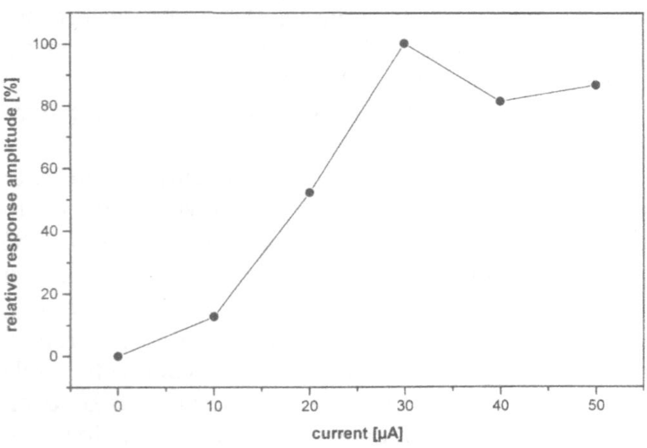

Figure 3. Mean relative amplitude (n = 5) of the electrically evoked retinal field potential with positive polarity (see Fig. 2) vs. strength of the current pulse. Between 0 µA and 30 µA an almost linear relationship between pulse strength and response amplitude was found. Saturation of the response occured above 30 µA.

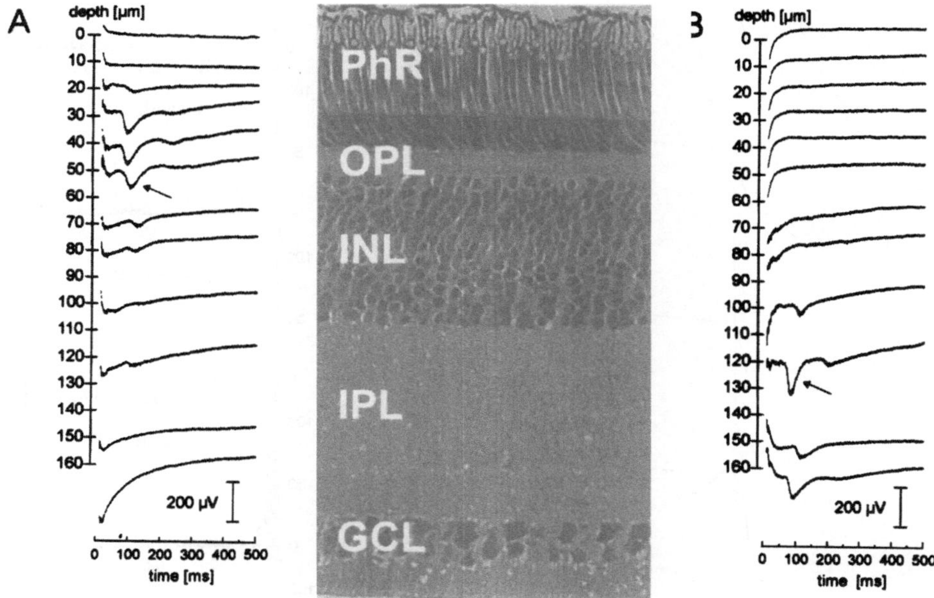

Figure 4. Retinal responses to a monopolar current pulse (30 µA, 1 ms duration) applied at time 0 ms with a PtIr needle electrode at different retinal depths. The vertical positions of the response traces indicate the position of the electrode tip in relation to the distal surface of the retina. For comparison a histological section through a chicken retina is shown. The traces in the left panel (A) were recorded after charges with positive polarity were injected, while the right traces (B) were obtained after negative charge injection. The prominent negative potentials (arrows) could either be elicited with a positive pulse at stimulation depths of 30 µm to 70 µm or at stimulation depths of 100 µm to 160 µm with a negative pulse. Abbreviations: PhR: photoreceptors; OPL: outer plexiform layer; INL: inner nuclear layer; IPL: inner plexiform layer; GCL: ganglion cell layer.

toreceptors along the bipolar cells to the border between the inner nuclear and the inner plexiform layer. We conclude, that a subretinal visual prosthesis with planar microelectrodes should be most effective if positive charges were injected at the distal side of the retina. Parallel recordings of retinal field potentials from up to eight MEA electrodes enabled us to evaluate the spatial extent of the electrically evoked retinal excitation. The results of such an experiment are shown in Fig. 5.

After a negative monopolar current pulse was injected in the INL at time zero, an electrically evoked slow retinal potential with a positive and negative component was recorded from the MEA electrode just below the stimulation electrode (Fig. 5, left panel, electrode 44, solid line). For comparison a baseline measurement recorded from the same electrode under high Mg^{++}-ringer is shown for this electrode too (Fig. 5, left panel, electrode 44, dotted line). On the neighbouring electrodes 43, 42, and 41 no slow potential could be evoked which indicates that the retinal excitation in terms of field potentials is locally defined. However, on electrodes 45 and 47 small but significant negative field potentials with different peak-timing characteristics were recorded. Because they had a different waveform and timing in comparison to the slow potentials that were recorded on electrode 44 we conclude, that they might reflect a spatial spread of

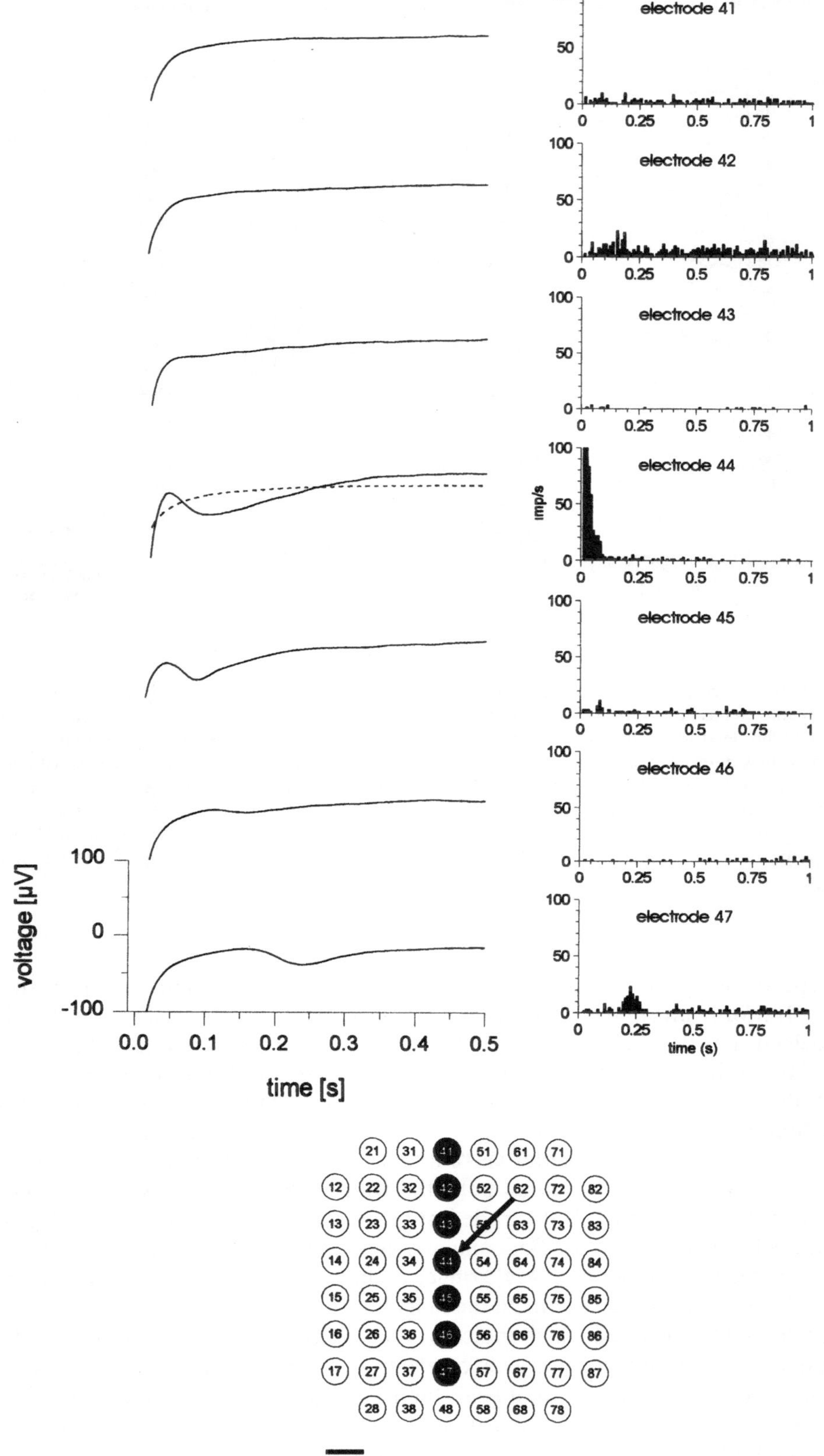

electrode 41
electrode 42
electrode 43
electrode 44
electrode 45
electrode 46
electrode 47
imp/s
time (s)
voltage [µV]
time [s]
200µm

the electrically evoked excitation over the retinal network. If this is really the case one can speculate that the electrically evoked depolarization of for example photoreceptor terminals leads, via the retinal pathways, to a de- or hyperpolarisation of some wide field amacrine cell which spreads over a distance of up to 600 μm. Overall, we conclude that distally applied current pulses evoke long lasting locally defined retinal field potentials which reflect the excitation of the retinal network.

4. ELECTRICALLY EVOKED GANGLION CELL ACTIVITY OF THE CHICKEN RETINA

For a subretinal visual prosthesis it is necessary that the electrically evoked slow retinal potentials lead to distinguishable spatio-temporal patterns of ganglion cell activity which are then transmitted to higher visual centers via the optic nerve. Therefore, we examined whether the electrically evoked slow potentials are related to a modulation of ganglion cell activity. The data presented in Fig. 5 are suitable for such a comparison. In the right panel of Fig. 5 peristimulus time histograms (PSTH) of multi-unit ganglion cell activity are shown. The spikes were recorded from the same electrodes as the slow retinal potentials displayed in the left panel of the figure. A transient increase in the spike rate of the ganglion cells was observed on the recording electrode below the stimulation site (Fig. 5, PSTH) which was highly correlated with the timing of the positive field potential on the same electrode (Fig. 5 left panel). The negative slow potential which lasted up to 250 ms was not correlated with any modulation of ganglion cell activity. However, in other cases there was also a significant correlation between the occurrence of negative field potentials and an increase of the ganglion cell spike rate. A typical example of a temporal correlation between a negative field potential and the modulation of ganglion cell activity was found in the same experiment on electrode 47 (Fig. 5). If one compares, in Fig. 5, the spatial distribution of the electrically evoked slow retinal potentials with the ganglion cell activity recorded from the same electrodes, it is obvious that there is a clear relationship between the occurrence of slow retinal potentials and a stimulus correlated modulation of ganglion cell activity. This result indicates that the electrically evoked and locally defined network activity was translated into a spatio-temporal firing pattern of ganglion cells which might transmit useful information about the electrical stimulus to the brain. The data presented so far were obtained by averaging the retinal responses from up to 60 stimulus representations. However, for a person who has a retinal implant and wants to visually navigate through an unknown environment it is a prerequisite that a continous stream of information about light modulations is transmitted by the ganglion cells to the brain. There will be no time for averaging the retinal output to support, for example, object detec-

Figure 5. Multichannel recordings of electrically evoked slow retinal potentials (left panel) and multi unit ganglion cell activity (right panel). The seven recording sites (black dots, electrode 41 to electrode 47) and the position of the stimulation electrode (arrow) are shown in the schematic drawing of the MEA. On electrode 44 electrically evoked retinal potentials with positive and negative polarity were recorded (solid line), which diminished after a ringer solution with high MG^{++}-content was superfused (dotted line). Stimulus correlated ganglion cell activity was recorded from the same electrode as it is shown in the peristimulus time histogram in the right panel. The ganglion cell firing rate was increased during the occurence of the positive retinal field potential. On electrodes 41 to 43 no electrically evoked retinal responses were recorded while on electrode 45 and 47 negative field potentials were elicited which, in the case of electrode 47, was correlated with a small but significant increase in ganglion cell spiking rate.

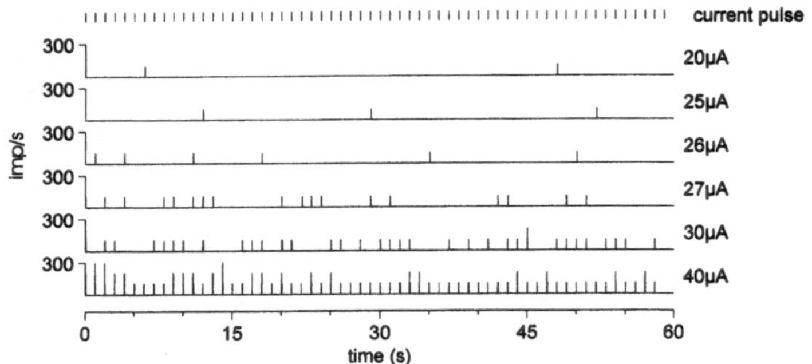

Figure 6. Rate histograms of multi unit ganglion cell activity recorded from a MEA electrode situated below a stimulation electrode. The latter was positioned in the INL over the recording electrode. The application of a negative monopolar current pulse (1 ms duration, strength as indicated) is depicted on top of the histograms. The ganglion cells respond with a short burst of spikes to the electrical stimulus which can be seen in the histograms. Only at the highest stimulation strength (40 μA) the increase of the ganglion cells firing rate is correlated with *every* current pulse over the whole recording time. With decreasing current strength the ganglion cell more often does not signal the occurence of an electrical stimulus with an increase in its firing rate.

tion in the environment. Therefore, we tested whether *every* current pulse injected into the retinal network elicits a short but significant answer of the ganglion cells (Fig. 6).

We applied negative monopolar current pulses (1 ms duration, strength as indicated) at an interstimulus intervall of 1 s to the distal side of the retina and recorded multi unit spikes. Inspection of the rate histograms revealed that much larger amount of charge (40nC per pulse) was needed for a reliable coding of every current stimulus

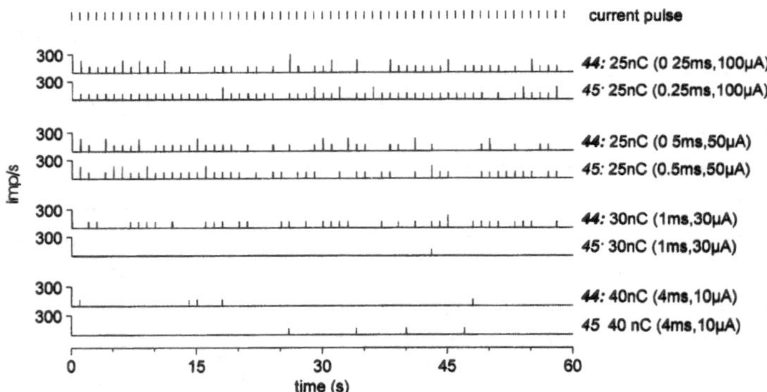

Figure 7. Rate histograms of multi unit ganglion cell activity recorded from a MEA electrode located below the stimulation electrode (44) and a MEA electrode located 200 μm away from the stimulation site (45). The time of injection of a negative monopolar current pulse (1 ms duration, strength as indicated) in the INL is shown on top of the histograms. A stimulus correlated increase in ganglion cell spike rate was evoked by a charge injection of 25nC when the duration of the current pulse was shorter than 1 ms. With a stimulus duration of 1 ms a ganglion cell response could be elicited by a charge transfer of 30nC while with longer pulse durations even an increase of the charge transfer to 40nC failed to elicit stimulus correlated activity. A shorter but stronger pulse (upper 4 histograms) elicited an increase in firing rate on electrode 44 and 45 which could be attributed to the wider extent of the electrical field generated by the injected charge, while a longer pulse with moderate current strength (row 5 and 6) evoked a stimulus correlated response only at the electrode located below the stimulation electrode.

than would be expected from excitation thresholds obtained by averaging retinal responses. However, the spatial extension of the electrical field increases too with stimulation strength which limits spatial resolution. Therefore we checked the possibility to facilitate the local electrical stimulation of retinal neurons without decreasing spatial resolution by applying pulses with longer durations and less strength (Fig. 7).

Unfortunately the threshold for evoking retinal responses increased with increasing pulse duration, although the total amount of injected charge was the same (Fig. 7, lower traces). Because the response threshold decreased with shorter pulse durations (Fig. 7, upper traces), we found that the spatial extension of the electrical field was wide enough to elicit stimulus correlated ganglion cell responses 200 µm away from the stimulation electrode (Fig. 7, electrode 45). This demonstrates that a careful adjustment of pulse parameters is needed for optimal electrical stimulation of the retinal network.

5. SUMMARY

We conclude that it is possible to evoke local stimulus correlated retinal activity in terms of slow potentials and ganglion cell activity by applying current pulses with different strengths and durations to the distal side of an intact chicken retina. The spatial extension of the retinal excitation is well defined and depends on pulse parameters. However, a lateral spread of the electrically evoked responses over the retinal network cannot be excluded and might lead to certain spatial noise. Whether this noise is relevant for the adequate functioning of a subretinal visual prosthesis needs further evaluation. The working range for transmitting luminance information with electrical pulses to the retinal network is relatively small. It is limited at the lower end by the fact that a threshold must be overcome to elicit a retinal response. The upper limit is set by the safe charge injection limit which depends on the electrode materials and geometry. Overall, we believe that for a subretinal visual prosthesis it is neccessary that the pulse parameters can be adjusted to find a compromise between a large working range for luminance coding and an optimized spatial resolution, respectively.

ACKNOWLEDGMENT

The authors wish to thank K. Kohler for contributing the histological section of the chicken retina. The authors were supported by BMBF Grants 01IN502 A,C and *f*ortüne Grant Nr. 233.

REFERENCES

1. S. Barnes and F. Werblin, 1987, Direct excitatory and lateral inhibitory synaptic inputs to amacrine cells in the tiger salamander retina, *Brain Res.*, **406**:233–237.
2. W.W. Dawson and N.D. Radtke, 1977, The electrical stimulation of the retina by indwelling electrodes, *Invest. Ophthalmol. Vis. Sci.*, **16**:249–252.
3. R.W. Knighton, 1975, An electrically evoked slow potential of the frog's retina. I. Properties of response, *J. Neurophysiol.*, **38**:185–197.
4. M. Humayun, R. Probst, E. de Juan, K. McCormick, and D. Hickingbotham, 1994, Bipolar surface electrical stimulation of the vertebrate retina, *Arch. Ophthalmol.*, **112**:110–116.

5. R. Eckmiller, 1997, Learning retina implants with epiretinal contacts, *Ophthalmic Res.*, **29**:269–280.
6. M.S. Humayun, E. de Juan Jr., G. Dagnelie, R.J. Greenberg, R.H. Propst, D.H. Phillips, 1996, Visual perception elicited by electrical stimulation of retina in blind humans, *Arch. Ophthalmol.* **114**:40–46.
7. J. Wyatt and J. Rizzo, 1996, Ocular implants for the blind, *IEEE Spectrum*, **33**:47–53.
8. A.Y. Chow, 1993, Electrical stimulation of the rabbit retina with subretinal electrodes and high density microphotodiode array implants, *Invest. Ophthalmol. Vis. Sci.*, **34**:835 (Abstract).
9. A.Y. Chow and V.Y. Chow, 1997, Subretinal electrical stimulation of the rabbit retina, *Neurosci. Lett.*, **225**:13–16.
10. E. Zrenner, K.D. Miliczek, V.P. Gabel, H.G. Graf, E. Guenther, H. Haemmerle, B. Hoefflinger, K. Kohler, W. Nisch, M. Schubert, A. Stett, and S. Weiss, 1997, The development of subretinal microphotodiodes for replacement of degenerated photoreceptors, *Ophthalmic Res.*, **29**:269–280.
11. D.R. Crapper and W.K. Noell, 1963, Retinal excitation and inhibition from direct electrical stimulation, *J. Neurophysiol.*, **26**:924–947.
12. A. Kaneko and T. Saito, 1983, Ionic mechanisms underlying the responses of off-center bipolar cells in the carp retina, *J. Gen. Physiol.*, **81**:603–612.
13. J.-I. Toyoda and M. Fujimoto, 1984, Application of transretinal current stimulation for the study of bipolar-amacrine transmission, *J. Gen. Physiol.*, **84**:915–925.
14. W. Nisch, J. Böck, H. Haemmerle, and A. Mohr, 1994, A thin film microelectrode array for monitoring extracellular neuronal activity in vitro, *Biosensors & Bioelectronics*, **9**:737–741.
15. R.P. Gallemore, E.R. Griff, and R.H. Steinberg, 1988, Evidence in support of a photoreceptoral origin for the "light-peak substance", *Invest. Ophthalmol. Vis. Sci.*, **29**:566–571.
16. V. Porciatti, P. Bagnoli, and R. Alesci, 1987, On and off activity in the retinal and tectal responses to focal stimulation with uniform or patterned stimulation, *Clin. Vision Sci.*, **2**:93–102.
17. P.A. Sieving, K. Murayama, and F. Naarendorp, 1994, Push-pull model of the primate photopic electroretinogram: A role for hyperpolarizing neurons in shaping the *b*-wave, *Vis. Neurosci.*, **11**:519–532.
18. R.W. Knighton, 1975, An electrically evoked slow potential of the frog's retina. II. Identification with PII component of electroretinogram, *J. Neurophysiol.*, **38**:198–209.
19. P.D. Lukasiewicz and F.S. Werblin, 1990, The spatial distribution of excitatory and inhibitory inputs to ganglion cell dendrites int the tiger salamander retina, *J. Neurosci.*, **10**:210–221.

Section V

MOLECULAR AND CELL BIOLOGICAL STUDIES

39

ANTISENSE INACTIVATION OF *rds*/PERIPHERIN IN *XENOPUS LAEVIS* EMBRYONIC RETINAL CULTURES

Vikas Kancherla,[1] Wojciech Kedzierski,[2] Gabriel H. Travis,[2] and Monica M. Jablonski[3]

[1] Department of Ophthalmology and Visual Science
University of Texas Health Science Center at Houston
Houston, Texas 77030
[2] Department of Psychiatry
University of Texas Southwestern Medical Center
5323 Harry Hines Boulevard NB4.202
Dallas, Texas 75235
[3] Department of Ophthalmology
University of Tennessee Memphis
956 Court Avenue D228
Memphis, Tennessee 38163

1. SUMMARY

To investigate the role of *rds*/peripherin in photoreceptor outer segment membrane assembly, we have transfected embryonic *Xenopus laevis* retinas with phosphorothioated antisense or sense RNA complementary to specific regions of the three *Xenopus rds*/peripherin homologs. We have demonstrated that using antisense oligonucleotides complementary to the three homologs of *rds*/peripherin, we can significantly alter the structure of photoreceptor outer and inner segments. In addition, the immunolocalization patterns using xrds35 and –38 anti-peptide sera and an anti-opsin monoclonal antibody were significantly reduced, suggestive of aberrant outer segment membrane assembly. We have successfully generated a model upon which to study the functional role of *rds*/peripherin, the product of the *rds* gene, by down-regulating the expression of this protein in developing photoreceptors. A better understanding of *rds*/peripherin gene expression will shed light on the role of this molecule in rod and cone photoreceptor outer segment development, structure, function and survival. This will also provide additional insights into the mechanisms of, and possibly open new therapeutic avenues for, the retinal degenerative processes taking place in the heterogeneous clinical patterns so far described in association with peripherin/RDS human gene mutations in association with autosomal dominant retinitis pigmentosa.

2. INTRODUCTION

The *rds* gene encodes an integral membrane glycoprotein of 346 residues, known as *rds*/peripherin. This protein is postulated to belong to a group of adhesion molecules[1–3] and has been localized to the rim region of the outer segments of both rods and cones.[4–7] The *rds*/peripherin product is similar in size and topology, and is partially homologous to rom-1, another rim-specific protein with which *rds*/peripherin is tightly associated to form a putative disc membrane protein complex in mammals.[8]

On experimental grounds, an insertion in the *rds*/peripherin gene produces a truncated protein that, in the *rds*/*rds* homozygous mouse, precludes the normal development of outer segments by impairing disc folding, leading to a slowly progressive degeneration of both rod and cone photoreceptors.[9] This is believed to be due to a null (-/-) effect of the *rds* mutant. The *rds*/+ heterozygote mouse has a much milder form of the disease, most likely due to haploinsufficiency, rather than a dominant negative effect.[10] In the *rds*/+ mouse, outer segments are formed, but they are disorganized, with swollen and vacuolated discs.[11] Further evidence of the essential role of *rds*/peripherin in disc folding comes also from the complete rescue of photoreceptors achieved by transgenic insertion of wild-type *rds* gene in homozygous *rds*/*rds* mice. An incomplete protection was instead obtained in a transgenic line with a low expression of the wild-type gene, showing a histopathologic pattern closely resembling that seen in the *rds*/+ mouse.[12]

Mutations in the orthologous human gene, peripherin/RDS, cause a wide spectrum of retinal manifestations that are inherited in an autosomal dominant fashion. Disease expression is extremely variable, both in terms of extension (i.e., from diffuse RP-like degeneration or retinitis punctata albescens, to macular pattern dystrophy or AMD-like manifestations), and in terms of the pattern of photoreceptor dysfunction (either rod- or cone-predominant). Not only has this phenotypic variability been accounted for by allelic heterogeneity, but it has also been observed within the same pedigree (clinical heterogeneity).[13–16] Explanations for this puzzling clinical behavior remain elusive. In addition, heterozygous peripherin/RDS mutations have been also shown to determine RP in a digenic fashion, i.e., only when coupled to further heterozygosity for rom-1 gene mutations.[17] An early histopathologic study on a retina from a maculopathy due to a peripherin/RDS gene mutation showed RPE damage, drüsen-like deposits and minimal loss of photoreceptors, although more detailed information on outer segment disc morphology was not available.[18] Outer segment shortening and disorganization have been documented in retinas from adRP patients,[19] in close analogy to what has been observed in *rds*/+ mice.[11]

We have recently isolated and characterized three homologs of mammalian *rds*/peripherin from *Xenopus* retinas. This is in contrast to mammals that have only one *rds* gene. The protein products of the xrds genes vary in their distribution among rods and cones: xrds35 and xrds36 are localized to the disc-rims of rod photoreceptors while xrds38, the *Xenopus* ortholog, is distributed in the disc periphery of both rods and cones.[7] In rod photoreceptors, the xrds molecules are hypothesized to form heterotrameric or higher-order complexes containing up to all three xrds proteins, whereas in cones complexes composed of only xrds38 subunits are located. In both photoreceptor types, the *rds*/peripherin complexes are believed to stabilize disc rims by bridging the intradiscal space in the terminal loop region of outer segment discs.[7] In the present study we sought to examine the function of the *rds* gene by inhibiting its expression using antisense oligonucleotides.

3. METHODS

3.1. Oligonucleotide Synthesis

Phosphorthioated oligonucleotides derived form the 5′ end of the coding sequences of the three isoforms of *Xenopus rds*/peripherin were synthesized by Operon Technologies, Inc. (Alameda, CA). Oligos were fully deprotected and desalted to remove organic contaminants. Two 20 base oligos for xrds35/36 were used in these experiments: [5′AGA GGA CCA TTT TTA CAG GG-3′ and 5′-TGG CCT TAA AGA GGA CCA TT-3′ (both antisense)], corresponding to nucleotide positions 28–47 and 37–56, respectively. Two 20 base oligos for xrds38 were also used: [5′-AGT TTT CAT CAG GGC CAT TG-3′ and 5′-TTC ACT CTC CTC TTC AGG TT-3′ (both antisense)], corresponding to nucleotide positions 62–81 and 89–108, respectively. One 20 base oligo was utilized as a negative control: 5′-AAT GGT CCT CTT TAA GGC CA-3′ (sense).

3.2. Culture of Developing Retinas

The culture preparation used in these experiments has been previously described.[20–26] *Xenopus laevis* embryos were obtained through induced breeding of adult *Xenopus* by injection of human chorionic gonadotropin (Sigma Chemical Co., St. Louis, MO). Embryos were staged by external morphological criteria[27] after which time eye rudiments from stage 35/36 embryos were isolated. The RPE was atraumatically removed by nicking the cell layer with forceps and allowing the RPE to retract away from the surface of the differentiating neuroepithelium. Rudiments were cultured in Niu-Twitty medium containing 5 mM lactose to permit organization of the nascent photoreceptor outer segments[25,26] and were maintained at 23°C for 3 days under cyclic lighting conditions as previously described.[24–26,28]

3.3. Exposure to Oligonucleotides

To increase the cellular uptake and decrease the toxicity of the oligos, Lipofectin (Invitrogen Corporation, San Diego, CA), was used as a transfecting agent. Optimal exposure conditions to the oligos and Lipofectin were determined empirically with a titration protocol. Possibly toxicity was monitored by examining retinal morphology at the light microscopy level, while cellular uptake was monitored with confocal microscopic (Biorad MC1024, Hercules, CA) analysis of an FITC-conjugated sense oligo. The final concentrations of oligo and Lipofectin added to the Niu-Twitty medium containing lactose were 1.5 μM and 12 μg/ml, respectively. The exposure paradigm that maintained the cellular organization and ultrastructure of the retina while allowing for maximum cellular uptake was a pulse of four hours followed by a 20 hour resting period for a duration of three and one-half days. This exposure paradigm was followed in all subsequent experiments. Four antisense and one sense oligo were studied. Additional controls included retinas without an adherent RPE, both with and without lactose added to the medium.

3.4. Morphological Assessment and Immunocytochemistry

Prior to harvesting for morphological or biochemical analysis, each rudiment was grossly examined under a dissecting microscope for rudiment integrity and smoothness of the neuroepithelial surface to insure that all rudiments were undamaged and intact. Any rudiments that exhibited an uneven surface or had many loose cells associated with it were discarded.

Standard light microscopy was used to monitor the morphology of retinas developing in vitro with emphasis on photoreceptor structure. Tissues were fixed in Tucker fix on ice and embedded as previously described.[29] One micron sections were stained with Toluidine blue-O and viewed on a light microscope. Eyes to be immunocytochemically evaluated were fixed in 4% paraformaldehyde in 0.05 M phosphate buffer, dehydrated, embedded in Unicryl (Electron Microscopy Sciences, Fort Washington, PA), cut at one micron thickness and collected on microscope slides. The sections were incubated in 5% goat serum (Vector Laboratories, Burlingame, CA) in phosphate buffered saline (PBS), rinsed in PBS and incubated overnight in primary antibody. The following antibodies were used: anti-opsin, 1:2000 dilution (generously provided by Dr. Paul Hargrave, University of Florida, Gainesville); xrds35 antiserum, 1:3000 dilution;[7] and xrds38 antiserum, 1:4000 dilution.[7] Gold-conjugated secondary antibodies followed by silver enhancement were applied to the tissue, as described by the manufacturer (Electron Microscopy Services, Fort Washington, PA). Retinal sections were viewed on a Nikon Axiophot microscope.

4. RESULTS

4.1. Structural Analysis of Control Retinas

When placed in culture without a closely apposed RPE, the intact undifferentiated retinas produced extensive amounts of outer segment membrane. However, the membranes lacked the normal stacked disc conformation and were arranged as large whorls with little evidence of normal disc stacking. Most of the membrane profiles did not appear to be associated with any particular photoreceptor, and formed a dense membranous mat at the outer retinal surface, yet the amount of membrane produced was substantial (Fig. 1A). However, the addition of 5 mM lactose to the culture medium permitted the organization of nascent outer segment membranes. As shown in Figure 1B, outer segment membranes that were elaborated in the absence of the RPE, yet in the presence of lactose ultrastructurally resembled eyes that matured in the presence of the RPE.[25,26] The addition of sense oligonucleotide had no effect on the ultrastructure of photoreceptors (Fig. 1C). Inner and outer segments appeared very similar to those that differentiated in the presence of lactose only. Outer segments were organized with the expected architecture of stacked flattened membranous saccules.

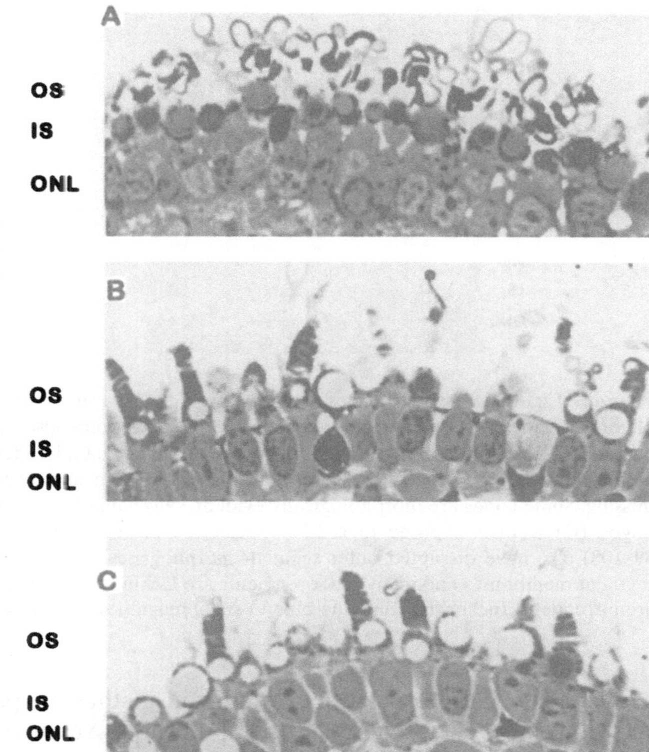

Figure 1. Light microscopic images of control *Xenopus laevis* retinas removed from stage 35/36 embryos and cultured for 3.5 days. (A) Retinas maintained in culture without an adherent RPE in Niu-Twitty medium elaborate outer segment membranous material that is disorganized and whorl-like in appearance. (B) Retinas maintained without an adherent RPE in medium containing 5 mM lactose elaborate outer segments that are arranged in an orderly array of stacked membranous saccules. (C) Sense oligonucleotide exposure of retinas that differentiate in the absence of the RPE, yet in the presence of lactose does not disrupt outer segment organization. Tucker fix; Toluidine blue-O stain; magnification bar = 1 μm.

4.2. Structural Analysis of Retinas Exposed to Antisense Oligonucleotides

The addition of antisense oligonucleotides complementary to identical regions of xrds35 and xrds36 prevented the formation of outer segments in the majority of the rods (Figs. 2A and 2B). In the few photoreceptors that possess a full photoreceptor complement, the outer segments have a less organized structure than the control tissue shown in Fig. 1. There was some evidence of membranous blebs and whorls of membrane. The inner segments of many photoreceptors were also disrupted and missing, suggesting severe photoreceptor damage. Some cone outer segments were evident and organized, therefore cones did not appear to be affected by these particular oligos.

The antisense oligos complementary to xrds38 had similar dramatic alterations in photoreceptor structure. The majority of photoreceptors were lacking inner and outer segments. In Figs. 2C and 2D, areas with some photoreceptor inner and outer segments are shown, however, in the remainder of the retina, these photoreceptor components are lacking. Both rod and cone photoreceptors were affected by these antisense oligos.

4.3. Immunolocalization Analysis of Control Retinas

Immunolocalization analysis of opsin, xrds35 and xrds38 in control retinas are shown in Fig. 3. At the light microscopy level, opsin immunolabeling was seen over the outer segment regions of the photoreceptors. The entire outer segment was heavily

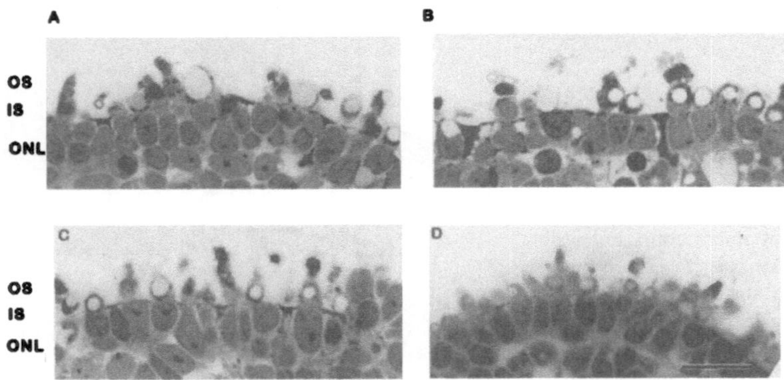

Figure 2. Light microscopic images of embryonic *Xenopus* retinas exposed to xrds35/36 and xrds38 antisense oligonucleotides. (A) Retinas exposed to xrds35/36 (nucleotide positions 28–47) or (B) xrds35/36 (nucleotide positions 37–56), did not elaborate outer segment material. Only a few scattered photoreceptors produced membranes that, however, were somewhat disorganized in appearance. Also, many inner segments were missing. Some cones are morphologically evident, suggesting that cones are not affected by these particular oligos. (C) Retinas exposed to xrds38 (nucleotide positions 62–81) or (D) xrds38 (nucleotide positions 89–108) also have disrupted outer segment morphogenesis. Many photoreceptors have not formed outer segment membranes and many inner segments are lacking. Some outer segments are present, although they are infrequent. Tucker fix; Toluidine blue-O stain; magnification bar = 1 μm.

labeled. The anti-opsin antibody utilized in these experiments has been previously shown to label exclusively the principal rod of *Xenopus* retina.[29] Although not shown in Figure 3A, other retinal areas had absence of labeling indicating the presence of cones. Immunolabeling of *rds*/peripherin was performed using xrds35 and –38 anti-peptide sera. Xrds35 gave a labeling pattern within the outer segments corresponding to the disc periphery and incisures. Not all outer segments were immunolabeled (Fig. 3B), however, and are most likely cone photoreceptors which do not contain xrds35.[7] Xrds38 gave a very defined labeling pattern corresponding to the disc periphery and incisures (Fig. 3C). All photoreceptor outer segments were labeled with this antibody which is consistent with our previous findings that this homolog is present in both rods and cones.[7]

4.4. Immunolocalization Analysis of Retinas Exposed to Antisense Oligonucleotides

In contrast to control eyes represented in Fig. 3, retinas exposed to the antisense oligos did not label heavily with any of the antibodies utilized. In retinas exposed to the an antisense oligo complementary to xrds35 and –36, the labeling of opsin is undetectable (Fig. 4A). No opsin-immunopositive outer segment membranous material was detected. Additionally, the inner segments of the photoreceptors were very disrupted and along much of the outer retinal surface, the inner segments were missing. In similarly treated eyes, the immunolabeling patterns using xrds35 antisera were indistinguishable from background (Fig. 4B). Retinas exposed to the xrds38 antisense oligo also showed a disrupted morphology of the inner segment and lack of outer segment membranous material. These retinas showed no immunolabeling with the xrds38 antiserum (Fig. 4C) or with our anti-opsin antibody (data not shown).

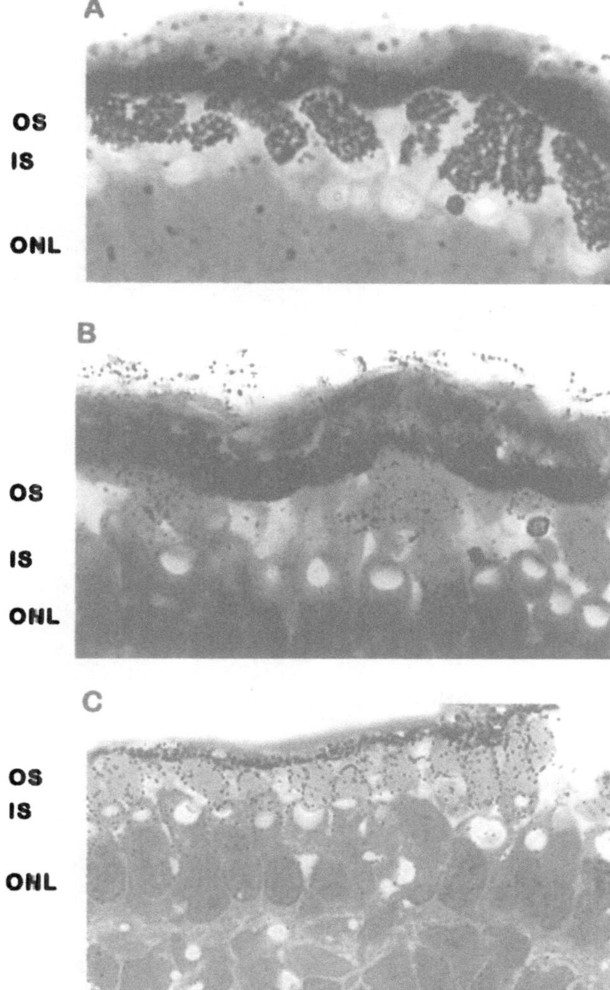

Figure 3. Immunocytochemical analysis of control retinas removed from stage 35/36 and cultured for 3.5 days. (A) Outer segments immunolabel heavily with the anti-opsin antibody. The xrds35 peptide antiserum (B) labels the disc periphery and incisures of rod outer segments, and the xrds38 antiserum (C) labels heavily the disc periphery and incisures of all photoreceptors. Paraformaldehyde fix; Azure II/methylene blue stain; magnification bar = 1 μm.

5. DISCUSSION

The results presented herein demonstrate that we are able to significantly alter the structure of photoreceptor outer segments and also reduce the immuolabeling of photoreceptor outer segments using antisense oligonucleotides. Cone photoreceptors were preserved during exposure to the xrds35/36 antisense oligonucleotides, which is consistent with our previous finding that cones do not express the xrds35 and –36 homologs. Both rods and cones were affected by the xrds38 antisense oligos, both of which express the xrds38 isoform. The sense oligonucleotide did not affect photoreceptor morphology or immunolabeling patterns. This cell-specific targeting confirms the specificity of the antisense oligos.

With the experimental conditions described in the present report, we have mimicked end-stage photoreceptor degeneration due to downregulation of *rds*/peripherin. In large expanses of the retina, the photoreceptors lack both an outer and the inner

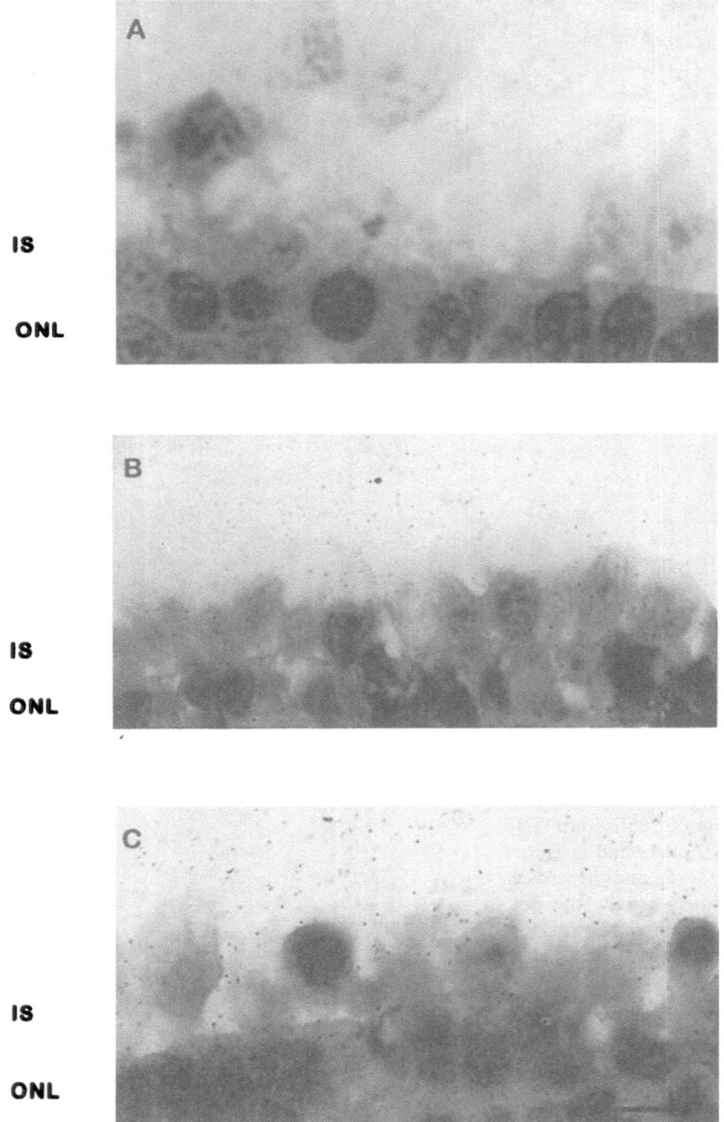

Figure 4. Immunocytochemical analysis of retinas exposed to antisense oligos. (A) Utilizing the anti-opsin antibody, there was no labeling of retinas exposed to the xrds35/36 antisense oligos, suggesting that no membranous material containing opsin was present. (B) In these same eyes, there also no labeling detected with the xrds35 antiserum. (C) In eyes exposed to the xrds38 antisense oligonucleotides, the level of immunolabeling was not above background using anti-xrds38 antiserum. Paraformaldehyde fix; Azure II/methylene blue stain; magnification bar = 1 µm.

segment, although the outer nuclear layer appears to be intact. The experimental system we have described in the present study will allow us to carefully evaluate the mechanisms involved in retinal degenerations due to mutations in the *rds*/peripherin gene. By titrating the antisense oligos, we will be able to determine the minimal amount of *rds*/peripherin necessary to allow outer segments to form and perhaps develop a better understanding of the role of *rds*/peripherin in outer segment membrane folding.

This system will allow us to determine the function of the rds/peripherin gene by loss-of-function analysis.

It has been proposed that the *rds*/peripherin protein is essential for normal disc morphogenesis.[1,2,3,6] Based upon the model of outer segment disc morphogenesis originally postulated by Steinberg et al.,[30] Arkikawa proposed a model for disc elaboration accounting for the localization of *rds*/peripherin.[6] They suggested that outer segment discs are formed via a two-step process in which there is an initial evagination of membrane from the connecting cilium followed by a secondary formation of the disc rim in a zipper-like fashion. It is this second step in which *rds*/peripherin is proposed to play a key role. This model provides a viable explanation of the cause of the lack of outer segment formation in both the *rds/rds* homozygous mouse and also in the present study. Although opsin may be the most abundant protein within the outer segment, it alone is insufficient to produce outer segments of the proper size, conformation and folding patterns. Indeed, it has been indicated that in the *rds* mouse, opsin mRNA levels and synthesis rates are relatively high, 70% and 92% of normal, respectively. However the amount of opsin protein levels are very low (~3% of normal), due to a high turnover rate of opsin.[31] A complete absence of the *rds*/peripherin protein, as in the *rds/rds* mouse[32] results in lack of outer segment formation and ultimately in cell death.[11] The results from the present study parallel these phenomenon closely. In our experimental paradigm, we precluded normal outer segment membrane formation and promoted the loss of inner segments using antisense oligonucleotide technology.

The experimental system utilizing *Xenopus laevis* embryonic eyes in culture possesses many advantages in the analysis of the functional role of *rds*/peripherin in outer segment disc structure and assembly. Firstly, we are able to efficaciously deliver the antisense oligonucleotide to the cells of study with great ease. The oligonucleotides and transfecting agent are simply added to the culture medium in which the intact eyes are differentiating. The Niu-Twitty medium is very simple, consisting of salts and water with no confounding serum or other additives. However, in this media the intact eye rudiment is able to differentiate from a neuroblastic layer to a stratified retina complete with outer segment membranes of stacked membranous saccules in perfect register. Because the retinas remain intact, photoreceptors, the targets of the xrds antisense oligonucleotides in the present study, maintain their contacts with second order neurons, thereby eliminating any potential secondary effects of cellular isolation. In addition, the use of embryonic retinas is advantageous because it circumvents the need for depleting an existing pool of mRNA, as in the case of fully developed retinas. Using a model in which the retina is readily accessible to the antisense oligonucleotide, while maintaining the normal organized structure of the photoreceptor outer segment will allow us to readily evaluate the effect of the antisense inactivation of *rds*/peripherin on photoreceptor outer segment membrane assembly.

ACKNOWLEDGMENTS

The authors would like to thank Dr. Paul Hargrave (University of Florida, Gainesville) for supplying us with the anti-opsin antibody. We would also like to acknowledge the skillful technical assistance of Cynthia S. Ervin, Amira Wohabrebbi, and Michael Lee. This work was supported by grants from the National Eye Institute, Fight for Sight and Research to Prevent Blindness (RPB). M.M.J. is the RPB William and Mary Greve Special Scholar.

REFERENCES

1. G.H. Travis, J.G. Sutcliffe, and D. Bok, 1991, The *retinal degeneration slow (rds)* gene product is a photoreceptor disc membrane-associated glycoprotein, *Neuron* **6**:61–70.
2. R.A. Bascom, S. Manara, L. Collins, R.S. Molday, V.I. Kalnins, and R.R. McInnes, 1992, Cloning of the cDNA for a novel photoreceptor membrane protein (rom-1) identifies a disk rim protein family implicated in human retinopathies, *Neuron* **8**:1171–1184.
3. P.K. Bhatia and G.H. Travis, 1994, Rds/peripherin and rom-1: A new class of adhesion molecules, *Invest. Ophthalmol. Vis. Sci* **35 (Suppl.)**:2675.
4. R.S. Molday, D. Hicks, and L. Molday, 1987, Peripherin. A rim-specific membrane protein of rod outer segment discs., *Invest. Ophthalmol. Vis. Sci.* **28**:50–61.
5. G. Connell, R. Boscom, L. Molday, D. Reid, R. McInnes, and R.S. Molday, 1991, Photoreceptor peripherin is the normal product of the gene responsible for retinal degeneration in the rds mouse., *Proc. Natl. Acad. Sci. USA* **88**:723–726.
6. K. Arikawa, L.L. Molday, R.S. Molday, and D.S. Williams, 1992, Localization of peripherin/rds in the disk membranes of cone and rod photoreceptors: Relationship to disk membrane morphogenesis and retinal degeneration., *J. Cell Biol.* **116**:659–667.
7. W. Kedzierski, W.N. Moghrabi, A.C. Allen, M.M. Jablonski-Stiemke, S. Azarian, D. Bok, and G.H. Travis, 1996, Three homologs of rds/peripherin in Xenopus laevis photoreceptors that exhibit covalent and non-covalent interactions., *J. Cell Sci.* **109**:2551–2560.
8. O.L. Moritz and R.S. Molday, 1996, Molecular cloning, membrane topology, and localization of bovine rom-1 in rod and cone photoreceptor cells., *Invest. Ophthalmol. Vis. Sci.* **37**:352–365.
9. S. Sanyal, G. Chader, and G. Aguirre, 1985, Expression of retinal degeneration slow (rds) gene in the retina of the mouse, in: p. 239–256. Alan R. Liss Inc., New York, 1985.
10. J. Ma, J.C. Norton, A.C. Allen, J.B. Burns, K.W. Hasel, J.L. Burns, J.G. Sutcliffe, and G.H. Travis, 1995, Retinal degeneration slow (rds) in mouse results from simple insertion of a haplotype-specific element into a protein-coding exon II., *Genomics* **28**:212–219.
11. R.K. Hawkins, H.G. Jansen, and S. Sanyal, 1985, Development and degeneration of retina in rds mutant mice: photoreceptorabnormalities in the heterozygotes, *Exp. Eye Res.* **41**:701–720.
12. G.H. Travis, K.R. Groshan, M. Lloyd, and D. Bok, 1992, Complete rescue of photoreceptor dysplasia and degeneration in transgeneic retinal degeneration slow (rds) mice., *Neuron* **9**:113–119.
13. R.G. Weleber, R.E. Carr, W.H. Murphey, V.C. Sheffield, and E.M. Stone, 1993, Phenotypic variation including retinitis pigmentosa, Pattern dystrophy, and fundus flavimaculatus in a single family with a deletion of codon 153 or 154 of the peripherin/RDS gene, *Arch. Ophthalmol.* **111**:1531–1542.
14. C.M. Kemp, S.G. Jacobson, A.V. Cideciyan, A.E. Kimura, V.C. Sheffield, and E.M. Stone, 1994, RDS gene mutations causing retinitis pigmentosa or macular degeneration lead to the same abnormality in photoreceptor function, *Invest. Ophthalmol. Vis. Sci.* **35**:3154–3162.
15. M.B. Gorin, K.E. Jackson, R.E. Ferrell, V.C. Sheffield, S.G. Jacobson, J.D. Gass, E. Mitchell, and E.M. Stone, 1995, A peripherin/retinal degeneration slow mutation (Pro-210-Arg) associated with macular and peripheral retinal degeneration, *Ophthalmology* **102**:246–255.
16. M. Nakazawa, E. Kikawa, Y. Chida, Y. Wada, T. Shiono, and M. Tamai, 1996, Autosomal dominant cone-rod dystrophy associated with mutations in codon 244 (Asn244His) and codon 184 (Tyr184Ser) of the peripherin/RDS gene., *Arch. Ophthalmol.* **114**:72–78.
17. K. Kajiwara, L.B. Hahn, S. Mukai, G.H. Travis, E.L. Berson, and T.P. Dryja, 1991, Mutations in the human retinal degeneration slow gene in sutosomal dominant retinitis pigmentosa., *Nature* **354**:480–483.
18. J.D.M. Gass, 1974, A clinicalpathologic study of a peculiar foveo-macular dystrophy., *Trans. Am. Ophthalmol. Soc.* **72**:139–156.
19. J.G. Flannery, D.B. Farber, A.C. Bird, and D. Bok, 1989, Degenerative changes in a retina affected with autosomal dominant retinitis pigmentosa, *Invest. Ophthalmol. Vis. Sci.* **30**:191–211.
20. R.N. Lolley, D.B. Farber, M.E. Rayborn, and J.G. Hollyfield, 1977, Cyclic GMP accumulation causes degeneration of photoreceptor cells: simulation of an inherited disease, *Science* **196**:664–666.
21. J.G. Hollyfield and P. Witkovsky, 1974, Pigmented retinal epithelium involvement in photoreceptor development and function, *J. Exp. Zool.* **189**:357–378.
22. J.G. Hollyfield, 1976, In vitro elimination of oocyte melanosomes from the retinal rudiment of Rana pipiens, *Exp. Eye Res.* **22**:126–140.
23. J. Hollyfield, M. Rayborn, D. Farber, and R. Lolley, 1982, in: *The Structure of the Eye* (J. Hollyfield, ed.), pp. 97–114. Elsevier Biomedical, Inc., New York.

24. M.M. Stiemke, R.A. Landers, M.R. Al-Ubaidi, and J.G. Hollyfield, 1994, Photoreceptor outer segment development in *Xenopus laevis*: Influence of the pigment epithelium, *Dev. Biol.* **162**:169–180.
25. M.M. Stiemke and J.G. Hollyfield, 1994, Outer segment disc membrane assembly in the absence of the pigment epithelium: The effect of exogenous sugars, *Dev. Brain Res.* **80**:285–289.
26. M.M. Stiemke and J.G. Hollyfield, 1995, in: Degenerative Diseases of the Retina. (R.E. Anderson, J.G. Hollyfield, and M.M. LaVail, eds.), pp. 129–137, Plenum Publishing Corp., New York.
27. P.D. Nieuwkoop and J. Faber, 1956, Normal Table of Xenopus laevis (Daudin). North Holland Publishing Co., Amsterdam.
28. J.G. Hollyfield and M.E. Rayborn, 1979, Photoreceptor outer segment development: Light and dark regulate the rate of membrane addition and loss, *Invest. Ophthalmol. Vis. Sci.* **18**:117–133.
29. M.M. Stiemke, R.A. Landers, M.R. Al-Ubaidi, and J.G. Hollyfield, 1992, Rod photoreceptor development in the Xenopus laevis embryo, *Invest. Ophthalmol. Vis. Sci. (Suppl.)* **33**.
30. R.H. Steinberg, S.K. Fisher, and D.H. Anderson, 1980, Disc morphogenesis in vertebrate photoreceptors, *J. Comp. Neurol.* **190**:501–518.
31. N. Agarwal, I. Nir, and D.S. Papermaster, 1990, Opsin synthesis and mRNA levels in dystrophic retinas devoid of outer segments in retinal degeneration slow (*rds*) mice., *J. Neurosci.* **10**:3275–3285.
32. J. Usukura and D. Bok, 1987, Changes in the localization and content of opsin during retinal development in the rds mutant mouse: Immunocytochemistry and immunoassay, *Exp. Eye Res.* **45**:501–515.

PHOSDUCIN-LIKE PROTEIN 1 (PhLP1) AND PHOSDUCIN-LIKE ORPHAN PROTEIN 1 (PhLOP1)

Chromosomal Localization to 1q25.3 and Genomic Structure Reveal Alternative Splicing of the Human Phosducin Gene

Cheryl M. Craft,* Xuemei Zhu, Jaji Murage, and Xiankui Li

Mary D. Allen Laboratory for Vision Research
Doheny Eye Institute
Department of Cell and Neurobiology
University of Southern California School of Medicine
Los Angeles, California 90033

SUMMARY

Phosducin (Phd) is a soluble phosphoprotein, selectively binding to the beta-gamma (βγ) complex of G-proteins, playing a role in intracellular signaling processes. We have molecularly identified 3 Phd isoforms expressed in the human retina. Based on their ability or inability to interact in our assay system with Gβγ proteins, we named them Phosducin-Like Protein 1 (PhLP1) or Phosducin-Like Orphan Proteins (PhLOP1 & PhLOP2), respectively. To further study these Phd isoforms and their potential involvement in normal and abnormal visual processes through genetic alterations, we identified and characterized an 85 kb P1-PhLP human genomic clone that contains PhLP1 and PhLOP1. With fluorescence *in situ* hybridization (FISH) to human metaphase chromosomes, this P1-PhLP genomic clone was mapped to human chromosome 1q25.3. Previous work identified a gene locus for phosducin (PDC) to 1q25–q31.1 by somatic cell hybridization and *in situ* hybridization. Within the P1-PhLP genomic clone,

* To whom correspondence and reprint requests should be addressed: Cheryl M. Craft, Ph.D., Mary D. Allen Professor for Vision Research, Doheny Eye Institute, Professor and Chair, Department of Cell and Neurobiology, BMT 401, 1333 San Pablo Street, Los Angeles, CA 90033, Telephone: (323) 442-1794; FAX: (323) 442-2709; Email: ccraft@hsc.usc.edu

Retinal Degenerative Diseases and Experimental Therapy, edited by Hollyfield *et al.*
Kluwer Academic / Plenum Publishers, New York, 1999.

the previously characterized complete PDC gene was identified; furthermore, the 5′-flanking region and a potential promoter region of PhLOP1 was identified between Exon 2 and Exon 3 of the PDC gene. Initial transfection experiments with luciferase activity reporter PhLOP1 constructs, ranging in size from 386 to 852 basepairs (bp), suggest that no tissue specific retinal promoter activity is contained in this 5′-flanking region of the PhLOP1 when compared to the high retinal specific promoter activity observed for a 123 bp construct for Interphotoreceptor Retinoid-Binding Protein (IRBP). Alternatively, a repressor element is still present and preventing promoter activity in the PhLOP1 constructs. These results correspond to the low levels of mRNA of the PhLOP1 observed in the retina compared to Phd. These data suggest that both PhLP1 and PhLOP1 are created through alternative splicing of the Phd gene and that a single PDC gene at 1q25.3 is responsible for these three retinal isoforms.

1. INTRODUCTION

Heterotrimeric G-proteins with their alpha (α), beta (β), and gamma (γ)-subunits play a major role in transmembrane G-protein receptor coupled (GPRC) signaling processes.[1,2] Phosducin (Phd) binds to the G-protein βγ complex and prevents the reassociation of Gα.[3–5] Originally identified in retina and pineal gland,[6–9] Phd, phosducin-like protein ($PhLP_{Long(L)/Short(S)}$, PhLP1), and phosducin-like orphan protein (PhLOP1 and PhLOP2) isoforms are expressed at low levels in other tissues examined.[10–17] PhLPs, like Phd, regulate Gβγ signaling; however, without the amino terminal domain, the binding and affinity for Gβγ decreases, suggesting alternative physiological functions for the truncated, orphan Phd isoforms.[18]

To investigate the Phd isoforms in the human retina, we identified three variants of Phd from a human retina cDNA library, named PhLP1, PhLOP1, and PhLOP2, respectively (Accession #AF076463, AF076464, and AF076465).[18] One Phd isoform, PhLP1, has the complete Phd sequence plus an additional amino terminal domain of 37 predicted amino acids. Like Phd, *in vitro* studies with recombinant PhLP1 and Gβγ proteins reveal similar Gβγ binding capacity. PhLOP1 has the carboxy terminal part of Phd; however, it lacks the first 52 amino acid terminal residues of Phd. PhLOP2 is also truncated, sharing limited amino acid sequence identity to the other isoforms, but its nucleotide sequence identity is significantly homologous with the other three isoforms. In our reconstitution assays with glutathione S-transferase/phosducin isoform fusion proteins, both PhLOP1 and PhLOP2 fail to bind to Gβγ because each lacks the major 11 amino acid domain within the amino terminus, which by crystallography analysis reveals structural interaction with Gβ.[19] In search of other functional partners for the PhLOP1 isoform, we discovered through a yeast 2-hybrid screen of a retinal cDNA library a protein-protein interaction with a proteasomal 26S ATPase protein, SUG1.[20]

2. METHODS

2.1. Fluorescent *In Situ* Hybridization

The human genomic P1-PhLP clone was labeled with digoxigenin dUTP by nick translation and combined with sheared human DNA, hybridized to normal metaphase chromosomes derived from PHA stimulated peripheral blood lymphocytes from a male donor in a solution containing 50% formamide, 10% dextran sulfate, and 2X SSC. Specific hybridization signals were detected by incubating the hybridized slides in fluoresceinated antidigoxigenin antibodies followed by counterstaining with DAPI. For two color experiments the detection was the same except for the inclusion of avidin Texas red.

2.2. Restriction Endonuclease Enzyme Digestions and Southern Blot Analysis

The P1-PhLP genomic DNA was digested with each of the following enzymes: *Hin*DIII, *Eco*RI, *Bam*HI, *Xba*I, *Cla*I / *Sma*I, and *Cla*I / *Sac*I (Boehringer Mannheim and New England BioLabs) and electrophoresed on an agarose gel. Typically, 5 μg of DNA was digested with 10 units of the enzyme in appropriate buffer in a final volume of 20 μl. After 4 hr incubation, the digestions were applied onto a 0.65% agarose gel (12 × 20 × 0.6 cm) in 1x TBE (89 mM Tris base, 89 mM Boric Acid, 2 mM EDTA) and electrophoresed at 80 V for 12 hrs. The DNA bands were transferred onto Hybond-N+ nucleic acid transfer membrane (Amersham). The transferred DNA was subsequently cross-linked to the membrane by UV irradiation. The membrane was incubated at least 4 hr at 42°C in 50% deionized formamide, 1x Denhardt's (100x is 10 g of Ficoll 400, 10 g of polyvinylpyrolidone, and 10 g of Bovine Serum Albumin), 2X SSC (20X SSC is 0.3 M Sodium Citrate; 3 M NaCl, pH 7.0), 1% SDS, 25 M NaH_2PO_4, 0.01 mg/ml of Salmon Sperm DNA. The total cDNA encoding PhLOP1 and Phd were isolated by gel purification after double enzyme *Xba*I/*Xho*I digestion of the original λMAX clones.[18] The probes were labeled with 100 μCi of α-[P^{32}]-dCTP by random priming, using the DNA Labeling System (Amersham). Hybridization was performed in 50% formamide buffer at 42°C overnight after prehybridization for 4 hr. The membrane was then washed 2 times for 20 min at room temperature (2X SSC, 0.1% SDS). The final wash was done for 20 min at 62°C (0.2X SSC, 0.5% SDS). The membrane was briefly dried and exposed to X-ray film or phosphoimager for 12 hrs.

2.3. DNA Sequence Analysis

DNA sequence analysis was performed by automated DNA sequencing with an ABI-310 Genetic Analyzer (Applied Biosystem/Perkin-Elmer), according to the manufacturer's recommendation. Both vector and Phd isoform primers were used to obtain DNA sequence information.

2.4. Cell Culture and Transfections

WERI retinoblastoma (Rb-1) cells (ATCC) were grown in RPMI 1640 containing Glutamax (Gibco, BRL), penicillin, streptomycin, and 10% fetal bovine serum (inactivated) in a 5% CO_2 humidified atmosphere. Transfection was performed using Superfect (Qiagen) in accordance to the manufacturer's recommendation. Cells seeded on a 3 cm plate (10^6 in 2.0 ml), were co-transfected with 1.7 μg test plasmids (PhLOP1/IRBP/CMV constructs) and 0.8 μg pGL3 control plasmid as an internal control for transfection efficiency, using 8 ml of Superfect in serum free medium to a total 60 μl volume.

Cells were collected 40 h later, washed twice with phosphate buffered saline, resuspended in 200 μl of PLB buffer (Promega) and lysed by two cycles of freezing and thawing. *Renilla* and firefly activities were assayed on the same 20 μl cell extract, using the Dual Kit Luciferase Assay (Promega). *Renilla* luciferase activity was normalized to the corresponding firefly luciferase activity for each sample. The data obtained represent the average ± S.E.M. of at least three independent experiments, each performed in duplicate.

3. RESULTS

3.1. Chromosomal Localization of a P1-PhLP Genomic Clone

To characterize the chromosomal localization and the genomic structure of these human phosducin isoforms, we isolated a genomic DNA clone (P1-PhLP) from a human P1 genomic library with a radiolabeled cDNA probe for the human retinal PhLP1, synthesized by polymerase chain reaction (PCR) using vector primers (λMAX vector, sense and antisense primers, +Gal4 and -T7, respectively), as previously described.[18] The purified PCR product was used for the human P1 genomic library hybridization screen using previously published protocols[5] (Genome Systems Inc. St. Louis, MO). The result of the screen yielded a positive genomic DNA clone, P1-PhLP (F322), about 85 kilobases long. F322 was subsequently used for fluorescence *in situ* hybridization (FISH). The initial experiment resulted in specific labeling of the long arm of a group A chromosome. A second experiment was conducted in which a biotin labeled probe which is specific for the heterochromatic region of chromosome 1 was cohybridized with clone P1-PhLP. This experiment resulted in the specific labeling of the heterochromatin in red and the long arm in green of chromosome 1 (data not shown). Measurement of 10 specifically hybridized chromosomes 1 demonstrated that P1-PhLP is located at a position, which is 42% of the distance from the heterochromatic-euchromatic boundary to the telomere of chromosome arm 1q, an area that corresponds to band 1q25.3 (Fig. 1). A total of 80% metaphase cells were analyzed with 68 exhibiting specific labeling. Confirming the previously report that the human phosducin gene was mapped to chromosome 1q25–1q32.1,[23–25] we have further narrowed the location to 1q25.3 and mapped the PhLP1 isoform to this identical chromosomal region.

3.2. Genomic Structural Relationship of Phd, PhLP1, and PhLOP1

To study the genomic structural relationship among the retinal Phd isoforms in P1-PhLP genomic clone, we analyzed the restriction endonuclease digestions of the P1 DNA by Southern blot hybridization and compared the fragment pattern obtained with cDNA probes for either Phd, PhLP1, or PhLOP1 isoform. The P1-PhLP genomic DNA was digested with each of the following enzymes: *Hin*DIII, *Eco*RI, *BamH*I, *Xba*I, *Cla*I / *Sma*I, and *Cla*I / *Sac*I (Boehringer Mannheim and New England BioLabs), electrophoresed on an agarose gel, transferred to membrane, hybridized with each appropriately labeled cDNA encoding each Phd isoform, washed and exposed to a phosphoimager. The results on this study are shown in Fig. 2. As expected because of a high degree of nucleotide sequence identity among the three isoforms, shared, common fragments are hybridized across all three isoforms; however, unique bands hybridize with either the labeled Phd or PhLP1 probes. The fragments present in Phd but not PhLOP1 are indicated with a diamond to the left side of the fragments (Fig. 2,

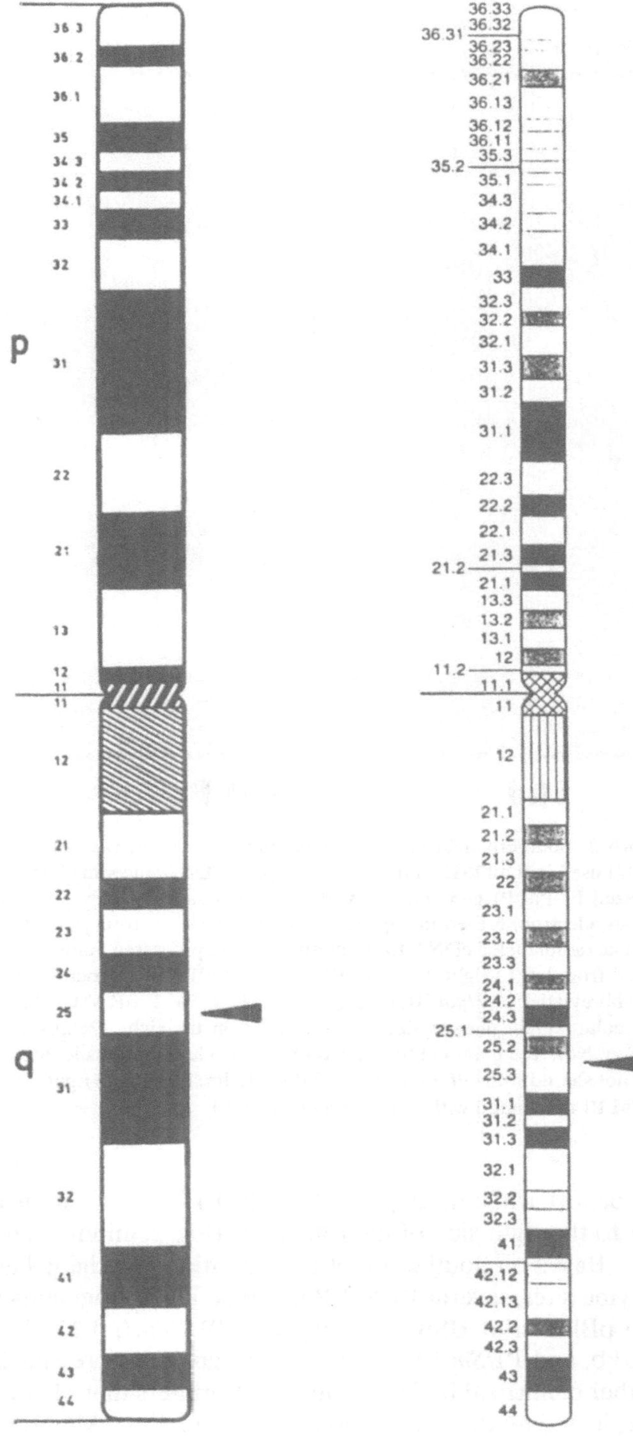

Figure 1. Ideogram illustrating the chromosomal position of P1-PhLP1 at 1q25.3 from the International System for Human Cytogenetic Nomenclature 1995. The pointers identify the location of the fluorescent *in situ* hybridization (FISH) results with the P1-PhLP, 85 kb genomic clone (F322), which maps to human 1q25.3 band, which refines the location of the PDC gene (1q25–1q31).[21,22]

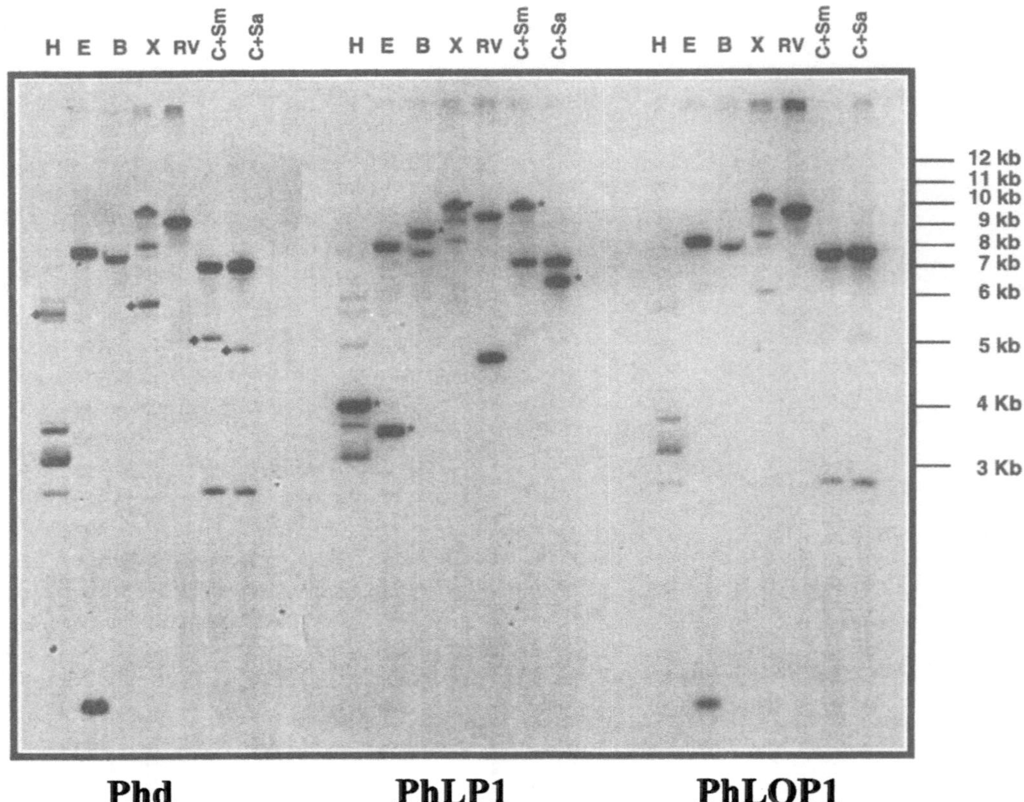

Figure 2. Southern hybridization of restriction endonuclease digestions of the P1-PhLP genomic clone (F322) used in the FISH with either a complete cDNA encoding human retina Phd, PhLP1 or PhLOP1 with digested P1-PhLP1 genomic DNA. P1-PhLP genomic DNA was digested with specific restriction endonucleases, electrophoresed in triplicate through 0.65% agarose gels, transferred and hybridized with random primed, radiolabeled cDNA for each isoform, and processed before exposure to a phosphoimager overnight. Listed from left to right, Phd, PhLP1, and PhLOP1. The restriction endonucleases indicated across the top are abbreviated: H, *Hind* III; E, *Eco*R I; X, *Xba* I; RV, *Eco*R V; C + Sm, *Cla* I/*Sma* I; C + Sa, *Cla* I/*Sac* I. The molecular weight marker sizes are indicated on the right. Details of the procedures are in the Methods section. Note the diamond markers on the left side fragments identify the bands unique to Phd hybridization not shared with either PhLP1 or PhLOP1; the stars on the right side fragments identify the bands unique to PhLP1 not shared with either Phd or PhLOP1.

left panel), while those present in PhLP1 but not Phd or PhLOP1 are indicated with a star to the right side of the fragments (Fig. 2, middle panel).

Based on Southern blot hybridization and the gel analysis of selected restriction enzyme sites, several P1-PhLP genomic DNA fragments were selected for subcloning into pBluescript IISK+ (Stratagene). P1/*BamH* I-11.5 kb, P1/*Cla* I-12.6 kb, P1/*Sma* I-12.3 kb, and P1/*Sma* I-8.3 kb DNA fragments were subcloned. These fragments were further confirmed by PCR using PCR primers that identify the entire PDC gene structure, including the coding, noncoding, and 5′-flanking region previously published.[23–25] The schematic diagram of the predicted P1-PhLP gene structure, its restriction map and the location of the subcloned genomic fragments are shown in Fig. 3.

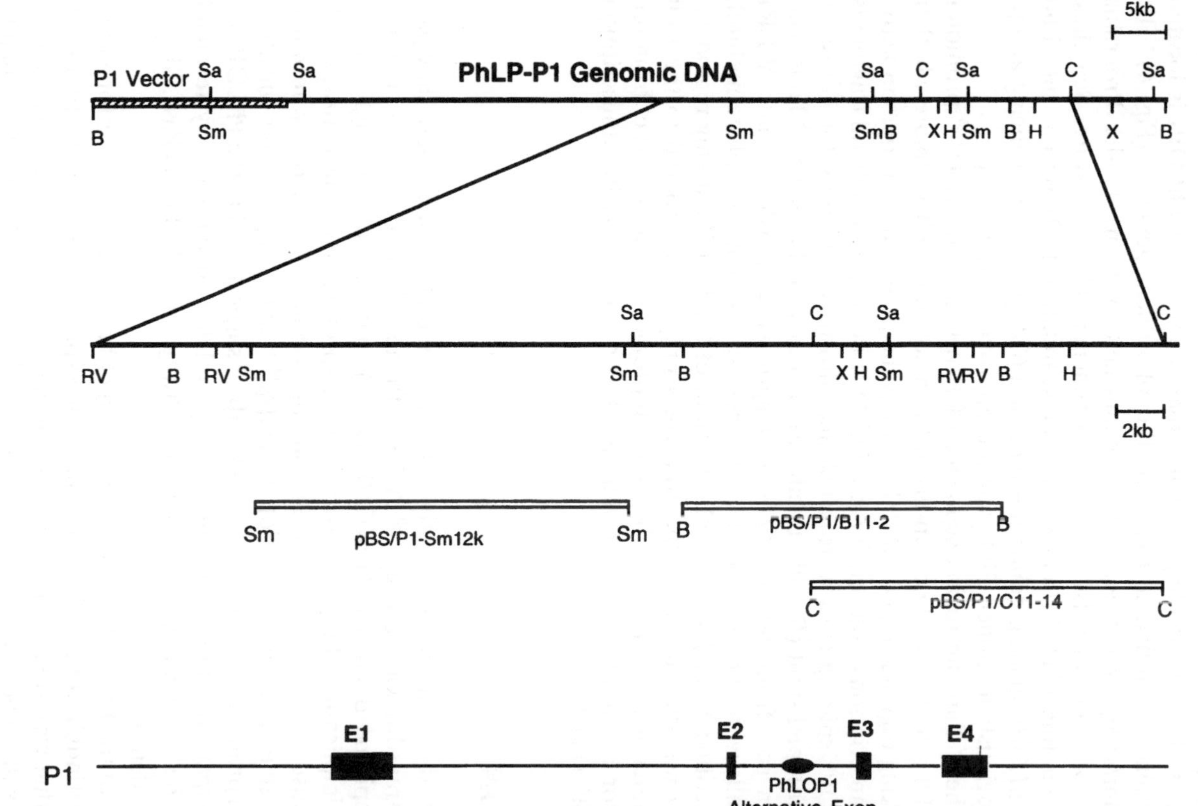

Figure 3. Diagram of Phd gene structure, including the alternative exon for PhLOP1 and the restriction map of the P1-PhLP genomic clone, including the position of the subclones. The top line represents the entire P1-PhLP (F322) genomic DNA fragment of approximately 85 kb in length, 5 kb indicator length above the line. Based on the Southern blot hybridization results of the cDNA of the Phd isoforms (Fig. 2), fragments that cover the Phd cDNA were selected, subcloned into pBluescript (pBS), and analyzed further: *Sma* I/12.3 kb = pBS/P1-Sm12k, *Bam*H I/12 kb = pBS/P1/B11-2, and *Cla* I/12.5 kb = pBS/P1/C11-14. The enzyme abbreviations are identical to Fig. 2. The lower line, 2 kb indicator length above the subclones, represents the exon (E)/intron locations of the PhLP gene verified by sequence analysis in the subcloned fragments, E1–E4 of the previously identified Phd gene,[23–25] including the alternative exon for PhLOP1 located within the intron between E2/E3.

3.3. Identification of an Alternative Exon and 5′-Flanking Region for PhLOP1

DNA sequence analysis using internal primers in the Phd coding region and the P1-PhLP genomic subclones as templates, the sequence was analyzed across each predicted exon/intron junction.[23–25] Sequence analysis of the P1/*BamH*I-11.5 kb subclone revealed that the 103 bp 5′-noncoding sequence that is unique for PhLOP1 is located within the previously identified intron 2 of the phosducin gene[23–25] (Fig. 4). The splicing site donor/acceptor sequences were confirmed by Gene Finder Programs (Baylor College of Medicine; BCM Search Launcher: Gene Feature Searches, http://dot.imgen.bcm.tmc.edu:9331/seq-search/gene-search.html). Between the Phd exon 2 and the PhLOP1 splicing donor site there are 678 bp, from the splicing acceptor site to the previously identified Phd exon 3 is approximately 1200 bp.

Although the 5′ flanking region sequence of Phd is documented,[23–25] promoter activity for the Phd gene has not been observed (data not shown). To explore whether or not the 5′ flanking region isolated for PhLOP1 has the potential for promoter activity to drive the transcription of this mRNA for PhLOP1 in the retina, several fragments around the PhLOP1 splicing sites were subcloned into pGL-3 (Firefly Luciferase as reporter, Promega), pRL-null (Renilla Luciferase as reporter, Promega).[26] Four constructs ranging in size from 386 to 852 bp in length were transfected into WERI retinoblastoma cells with SuperFect method (Qiagen) or Calcium Phosphate method, as previously described.[26] Similar to our observations of the Phd 5′ flanking region, no promoter activity was detected from the constructs prepared compared to the well established promoter activity of the 124 bp minimal promoter gene construct from the Interphotoreceptor Retinoid-binding protein, used as a retinal specific transcription activation control[27] (Fig. 5).

4. DISCUSSION

Phosducin and its isoforms are widely distributed in body tissues where they participate in signal transduction pathways.[3–9] Phd, PhLP1, and $PhLP_L$ possess an 11 amino acid conserved epitope domain (TGPKGVINDWR) that controls the high affinity binding of Gβγ, implicating these isoforms in the G-protein signaling pathway.[9–11,18] Several reports also suggest that even without this domain, the carboxy terminal domain is still capable of limited interaction with Gβγ.[14–16] In our experiments, the phosducin-like orphan proteins (PhLOPs) fail to bind Gβγ,[18] suggesting that the PhLOP isoforms may participate in other signaling pathways related to the carboxy terminal domain, such as the proteasomal ubiquination pathway through the ATPase, SUG1, that was recently identified.[20]

The mRNA of the Phd isoforms is expressed in either human retina or forskolin-stimulated Y79 retinoblastoma cells, verifying their expression independently of the cDNAs for each that was isolated through a human retina library screen.[18] The present data demonstrates that Phd and at least two isoforms, PhLP1 and PhLOP1, all map to 1q25.3 and that alternative spicing is responsible for the truncated isoform, PhLOP1. Although no retinal specific promoter activity is currently measured from constructs of the 5′-flanking region of PhLOP1 with WERI retinoblastoma cells (Fig. 5), we have observed significant reporter activity of the 852 bp fragment in a non-retinal cell line (data not shown). Because low mRNA levels of PhLOP1 are only detectable by

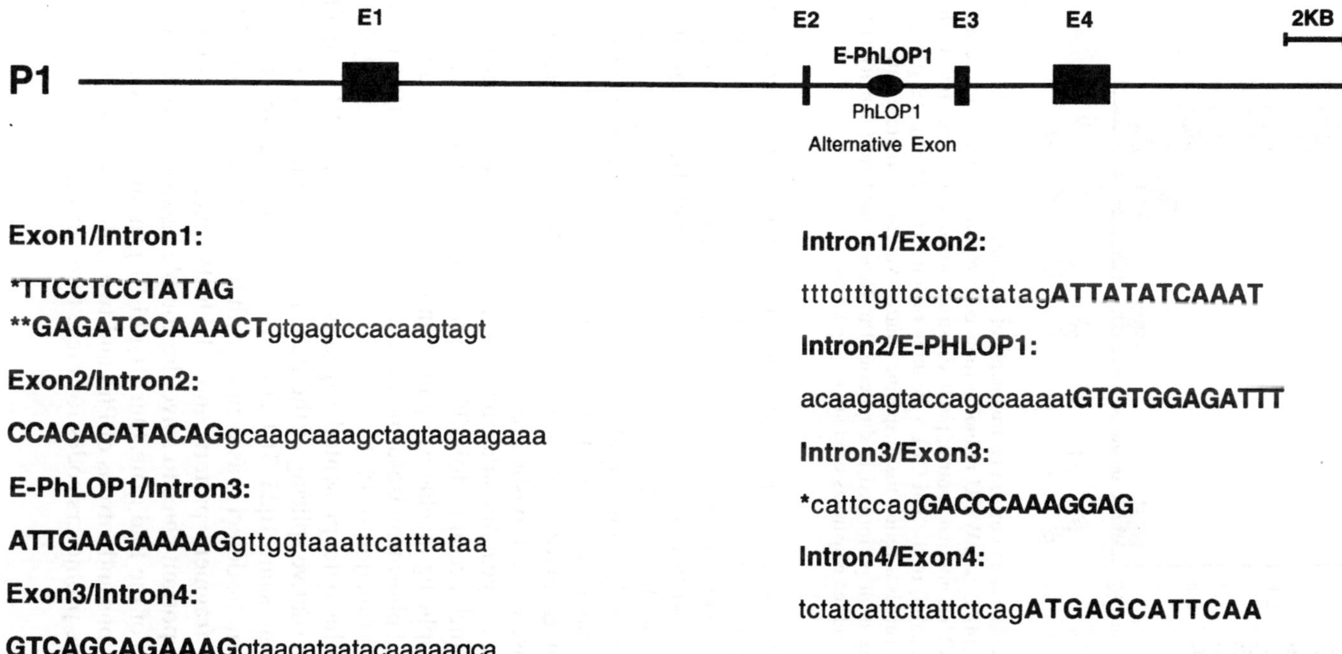

Figure 4. Splicing junctions and the exon/ intron nucleotide boundary sequences of Phd and PhLOP1. The lower line from Fig. 3 is identical. The nucleotide sequence at each exon/intron boundary is listed. The asterisks *, ** indicate verification of previously published boundaries from reference 23 and 24, respectively.

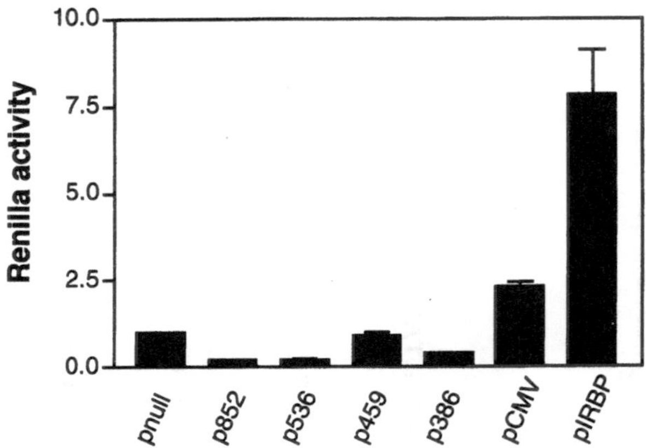

Figure 5. Reporter gene expression driven by fragments of decreasing size of the human PhLOP1 and the 123 bp IRPB promoter (pIRBP) in WERI retinoblastoma cells. *Renilla* luciferase expression driven by the CMV promoter (pCMV), an internal control; pnull, construct without promoter inserts, decreasing base-pair (bp) lengths (852–386 bp) of the PhLOP1 5′-flanking region (p852, p536, p459, p386) and the pIRBP (123 bp) promoter construct is a retina tissue specific control. Values are representative of at least three transfection experiments, each performed in duplicate. The *Renilla* luciferase activity arbitrary units are on the y axis (0.0–10.0) and the constructs used are on the x axis.

RT-PCT in the retina sample, perhaps an inhibitory element is blocking higher levels of tissue specific expression.[18] Additional studies are necessary to resolve this issue.

One plausible explanation for the alternative PhLOP1 splice variant would be the existing splice site at the PDC exon 2/intron 2 junction. The donor site at this junction is GC, not the predicted GT (Fig. 4), while all other donor/acceptor sites follow the predicted GT/AG rule, including the new donor site for the PhLOP1 exon. Altered splicing sites can lead to aberrant RNA splicing and is frequently associated with human disease, such as amyotrophic lateral sclerosis.[28] Although not common, an aberrant splicing database was established recently to track these altered variants (Web site: http://www/imcb.oaka-u.ac.jp/nakai/asdb.html), including "cryptic" splice sites that lead to exon skipping, which accounts for 50% of the genetic mutations listed.[29] It is possible that an altered splicing mechanism and exon skipping could result in an improper balance of the retinal phosducin isoforms, which could ultimately effect the Gβγ binding capacity and normal function of Phd.

In summary, the data presented suggests that both PhLP1 and PhLOP1 are created through alternative splicing of the PDC gene, and the P1-PhLP genomic fragment maps within the same 1q25–31.1 chromosomal region as the PDC gene.[21,22] At present, no visual genetic defect is associated with the 1q25.3 chromosomal region. The unique hybridizing fragment pattern and P1-PhLP subclones for Phd, PhLP1, and PhLOP1 provide important clues to solve the complex genomic structure and to identify the promoter(s) of the Phd gene. Our P1-PhLP1 genomic DNA covers all the previously identified exons and introns of Phd and two isoforms, suggesting that only one gene with alternative promoters codes for these three isoforms.

ACKNOWLEDGMENTS

This work is dedicated to Mary D. Allen for her generous and continued support of vision research. Additional support provided, in part, by NIH/EY00395 (R.N. Lolley

& C.M. Craft), EY03042 Core Vision Research Center grant (Doheny Eye Institute), Michael P. Connell Foundation, L.K. Whittier Foundation and the Neurogenetic Analysis Core (Hans-Jürgen Fülle) supported by the Howard Hughes Medical Institute Resources Grant. We wish to thank Dr. R.N. Lolley for critical discussions and suggestions pertaining to this research and Dr. Nicoletta Bobola for the IRBP promoter construct. C.M. Craft is the Mary D. Allen Professor for Vision Research.

REFERENCES

1. G. Gilman, 1987, G proteins: transducers of receptor-generated signals, *Annu. Rev. Biochem.* **56**:615–649.
2. M.I. Simon, M.P. Strathmann, and N. Gautam, 1991, Diversity of G proteins in signal transduction. *Science* **252**:802–808.
3. R.H. Lee, B.S. Lieberman, and R.N. Lolley, 1987, A novel complex from bovine visual cells of a 33,000-dalton phosphoprotein with β- and γ-transducin: purification and subunit structure, *Biochemistry* **26**(13):3983–3990.
4. T. Abe, H. Nakabayashi, H. Tamada, T. Takagi, S. Sakuragi, K. Yamaki, and T. Shinohara, 1990, Analysis of the human, bovine and rat 33-kDa proteins and cDNA in retina and pineal gland, *Gene* **91**:209–215.
5. C.M. Craft, R.N. Lolley, M.F. Seldin, and R.H. Lee, 1991, Rat pineal gland phosducin: cDNA isolation, nucleotide sequence, and chromosomal assignment in the mouse, *Genomics* **10**:400–409.
6. R.N. Lolley, C.M. Craft, and R.H. Lee, 1992, Photoreceptors of the retina and pinealocytes of the pineal gland share common components of signal transduction, *Neurochem. Res.* **17**:81–89.
7. S. Danner, and M.J. Lohse, 1996, Phosducin is a ubiquitous G-protein regulator, *Proc. Natl. Acad. Sci. USA* **93**(19):10145–10150.
8. K. Sunayashiki-Kusuzaki, T. Kikuchi, E.F. Wawrousek, and T. Shinohara, 1997, Arrestin and phosducin are expressed in a small number of brain cells, *Mol. Brain. Res.* **52**(1):112–120.
9. H. Tanaka, C. Iwami, C.H. Kuo, Y. Ding, E. Do, Y. Irie, and N. Miki, 1997, Analysis of the Tβγ-binding domain of MEKA/phosducin, *Neurochem. Int.* **31**(4):625–634.
10. J. Xu, D. Wu, V.Z. Slepak, and M.I. Simon, 1995, The N terminus of phosducin is involved in binding of βγ subunits of G protein, *Proc. Natl. Acad. Sci. USA* **92**(6):2086–2090.
11. B.E. Hawes, K. Touhara, H. Kurose, R.J. Lefkowitz, and J. Inglese, 1994, Determination of the Gβγ-binding domain of phosducin. A regulatable modulator of Gβγ signaling, *J. Biol. Chem.* **269**(47):29825–29830.
12. I. Boekhoff, K. Touhara, S. Danner, J. Inglese, M.J. Lohse, H. Breer, and R.J. Lefkowitz, 1997 Phosducin, potential role in modulation of olfactory signaling, *J. Biol. Chem.* **272**(7):4606–4612.
13. M.F. Miles, S. Barhite, M. Sganga, and M. Elliott, 1993, Phosducin-like protein: an ethanol-responsive potential modulator of guanine nucleotide-binding protein function. *Proc. Natl. Acad. Sci. USA* **90**(22):10831–10835.
14. S. Schroder, and M.J. Lohse, 1996, Inhibition of G-protein βγ-subunit functions by phosducin-like protein, *Proc. Natl. Acad. Sci. USA* **93**(5):2100–2104.
15. S. Schroder, K. Bluml, C. Dees, and M.J. Lohse, 1997, Identification of a C-terminal binding site for G-protein βγ-subunits in phosducin-like protein. *FEBS Lett.* **401**(2–3):243–246.
16. C. Thibault, M.W. Sganga, and M.F. Miles, 1997, Interaction of phosducin-like protein with G protein βγ subunits, *J. Biol. Chem.* **272**(19):12253–12256.
17. T. Abe, H. Tomita, and M. Tamai, 1998, Gene expression of the phosducin-like protein in the retina, *Ophtha. Res.* **30**:74–83.
18. C.M. Craft, J. Xu, V.Z. Slepak, X. Zhan-Poe, X. Zhu, B. Brown, and R.N. Lolley, 1998, PhLPs and PhLOPs in the phosducin family of Gβγ binding protein, *Biochemistry*, in press.
19. R. Gaudet, A. Bohm, and P.B. Sigler, 1996, Crystal structure at 2.4 A resolution of the complex of transducin-βγ and its regulator, phosducin, Cell 87:577–588.
20. X. Zhu and C.M. Craft, 1998, Interaction of phosducin and phosducin isoforms with a 26S proteasomal subunit, SUG1, *Molecular Vision* **4**:13, <http://www.molvis.org/molvis/v4/p13>
21. R.S. Sparkes, R.H. Lee, T. Shinohara, C.M. Craft, T. Kojis, I. Klisak, C. Heinzmann, and J.B. Bateman, 1993, Assignment of the phosducin (PDC) gene to human chromosome 1q25–1q32.1 by somatic cell hybridization and *in situ* hybridization, *Genomics* **18**(2):426–428.

22. C. Ding, X. Li, C.A. Griffin, E.W. Jabs, A.L. Hawkins, and M.A. Levine, 1993, The gene for human phosducin (PDC) a soluble protein that binds G-protein βγ dimers, maps to 1q25–q31.1, *Genomics* **18**(2):457–459.
23. Y. Watanabe, K. Kawasaki, N. Miki, and C.H. Kuo, 1990, Isolation and analysis of the human MEKA gene encoding a retina-specific protein, *Biochem. Biophys. Res. Commun.* **170**:951–956.
24. T. Abe, T. Kikuchi, and T. Shinohara, 1994, The sequence of the human phosducin gene (PDC) and its 5′-flanking region, *Genomics* **19**(2):369–372.
25. T. Abe, T. Kikuchi, and T. Shinohara, 1993, The sequence of the mouse phosducin gene and its 5′ flanking region, *Gene* **133**:179–186.
26. N. Bobola, X. Zhu, R. Ravazzolo, and C.M. Craft, Characterization of two AT-rich sequences in a minimal IRBP promoter by transient transfections in the WERI-Rb1 retinoblastoma cell line, submitted.
27. N. Bobola, E. Hirsch, A. Albini, F. Altruda, D. Noonan, and R. Ravazzolo, 1995, A single cis-acting element in a short promoter segment of the gene encoding the Interphotoreceptor Retinoid-binding protein confers tissue-specific expression, *J. Biol. Chem.* **170**:1289–1294.
28. G. Bai, and S.A. Lipton, 1998, Aberrant RNA splicing in sporadic amyotrophic sclerosis, *Neuron* **20**:363–366.
29. K. Nakai, and H. Sakamoto, 1994, Construction of a novel database containing aberrant splicing mutations of mammalian genes, *Gene* **141**:171–177.

41

ORGANIZATION OF THE INTERPHOTORECEPTOR MATRIX

Binding of SPACR to Hyaluronan

Joe G. Hollyfield, Mary E. Rayborn, and Shreeta Acharya

Cole Eye Institute and The Lerner Research Institute
The Cleveland Clinic Foundation
Cleveland, Ohio 44195

1. SUMMARY

The distribution of hyaluronan (HA) in the interphotoreceptor matrix (IPM) was evaluated in human tissues using a specific probe for HA (bHABC, biotinylated hyaluronan binding complex). In tissue sections, bHABC heavily decorates HA in the matrix surrounding rods and shows lighter binding to the cone associated matrix. Both HA and the glycoprotein SPACR are present in insoluble IPM isolates removed from unfixed retinas with distilled water treatment. To determine whether SPACR associates with HA, precipitation experiments were performed using the detergent cetylpyridinium chloride (CPC), in conjunction with hyaluronidase digestions. SPACR is present in the CPC pellet when an undigested IPM sample is precipitated, but is not present in the CPC pellet when the IPM is first digested with *Streptomyces* hyaluronidase. These results indicate that SPACR binds to HA, and implicates HA as the primary scaffold onto which IPM molecules form associative interactions in the IPM.

2. INTRODUCTION

Hyaluronan (HA) is a nonsulfated member of the glycosaminoglycan (GAG) family of polysaccharides, composed of repeating disaccharide units of β-(1,4)-D-glucuronic acid-β-(1,3)-N-acetyl-D-glucosamine.[1,2] It is a near ubiquitous constituent of connective tissues, as well as epithelia, where it is found in high concentration in intercellular spaces.[3,4,5,6] For over three decades, HA has been recognized as a minor constituent of the IPM. Earlier descriptions of macromolecules that can be rinsed from the

Retinal Degenerative Diseases and Experimental Therapy, edited by Hollyfield *et al.*
Kluwer Academic / Plenum Publishers, New York, 1999.

outer retina report that approximately 2% of the aqueous extractable IPM macromolecules are GAGs and the remainder glycoprotein. Of the GAGs present, chondroitin sulfate comprised approximately 60%, and HA represented approximately 15%.[7,8,9]

Because only minor amounts of HA can be rinsed from the IPM, coupled with the more recent observations that an extensive aqueous insoluble matrix remains associated with the outer retina following rinses,[10,11] we wished to re-examine the issue of HA distribution in this compartment. In this analysis, we employed a specific probe for HA which can be applied to tissue sections prepared with methods which retain HA, thus avoiding the use of extraction procedures which may remove only a portion of the HA or could contribute to cross contamination of HA from other tissue compartments (from the inner retinal surface or vitreous). We then used this HA probe to determine whether HA is contained in the insoluble IPM isolated from the human retina and whether the newly described molecule, SPACR,[12,13] also present in the insoluble IPM, is retained in this matrix through attachment to HA.

3. METHODS

Human eyes were obtained from a local eye bank between 1 to 6 hrs postmortem. For histology, after removal of the anterior segment, the posterior globe was immersion fixed in a solution containing 2.5% glutaraldehyde, 2% formaldehyde (prepared from paraformaldehyde), in 0.1 M phosphate buffer, (pH 7.2) with 0.1% to 0.5% cetylpyridinium chloride (CPC) to precipitate GAGs and prevent their extraction during subsequent processing. After overnight fixation, tissues were dehydrated, embedded in paraffin, sectioned at 10 µm, and mounted on glass microscope slides for further processing.

The bHABC probe, which contains the HA binding region of aggrecan and link protein ternary complex, was isolated from bovine articular cartilage proteoglycan aggregates following digestion with trypsin type XI (Sigma)[5,14] and biotinylated[4,15] as described previously. Deparaffinized tissue sections were incubated with 6% bovine serum albumin (BSA) in phosphate buffer (0.1 M, pH 7.2) for 30 min to block nonspecific binding. Sections were incubated with biotinylated bHABC (protein concentration 5 µg/ml diluted in 6% BSA/0.1 M phosphate buffer overnight at 4°C). Specificity of bHABC for HA binding was determined by evaluating the degree of bHABC binding to tissues predigested with *Streptomyces* hyaluronidase (100 TRU/ml sodium acetate buffer, pH 5.0 at 37°C) before applying the bHABC probe or the degree of bHABC binding to tissues using the bHABC preincubated overnight at 4°C with excess HA oligosaccharides (average length 11 disaccharides, 10 µg/2 µg bHABC).[16]

IPM isolates were prepared from human retinas using the distilled water method described previously.[10,11,17] The insoluble IPM was collected[12] after centrifugation, the pellet was solubilized in 0.1 M TBS (Tris buffered saline) pH 8.0 containing 5 mM dithiothreitol. Aliquots of the TBS extract were immobilized on nitrocellulose membranes by the application of 25 µl to 100 µl into the wells of a BioRAd 96 well dot blot apparatus.[18] Proteins were adsorbed to the membrane for 1 hr at room temperature before removing buffer and unbound molecules. The membrane was then washed with Tris-acetate buffer, pH 7.3, for 5 min and then incubated for 1 hr in 5% nonfat milk powder in TBS before applying the bHABC probe. HA (Healon™) and a TBS extract of a rat swarm cell chondrosarcoma tumor were used as positive controls.

To evaluate HA-SPACR interactions we used the protocol described previously.[19] The IPM extract (100 µg total protein) was incubated in 50 mM sodium acetate buffer pH 6.0 for 2 h at 37°C with or without *Streptomyces* hyaluronidase (2–5 TRU). IPM samples were then incubated at 37°C with 1.25% cetylpyridinium chloride (CPC) for 1 h. The preparation was centrifuged (12,000 rpm) for 15 min at room temperature. The pellets were washed with 1% CPC in Tris-acetate buffer, solubilized in Laemmli buffer[20] and both pellets and supernatants analyzed by SDS-PAGE, followed by staining with Gel Code. BSA was included as a negative control.

4. RESULTS AND DISCUSSION

The IPM showed intense binding of the bHABC probe, from the apical surface of the RPE to the outer limiting membrane of the retina (Fig. 1A). The most intense bHABC binding was present surrounding rods. In the IPM surrounding cones, bHABC was reduced or absent. These radial gaps in bHABC binding were clearly associated with the IPM surorunding cones. In the hyaluronidase digests, some resistance to HA digestion was noted (not shown), but all binding of bHABC to the IPM was eliminated when the probe was preincubated with HA oligosaccharides (Fig. 1B). These results indicate that HA is present in the IPM, and that it is more abundant in the rod-associated matrix than in the matrix surrounding cones.

Our previous analysis of the components present in the IPM indicates that the glycoprotein SPACR is also present in this complex and is retained in the complex when the IPM is isolated with distilled water.[12,21] To determine whether HA is also present in the IPM isolate, we used dot blot analysis of the isolated IPM samples probed with bHABC. As is evident in Fig. 2A, bHABC binds to the blotted IPM samples in a dose dependent manner as well as to HA positive controls, Healon,™ and to an extract of rat chondrosarcoma, which is known to contain HA.[22] When IPM samples are probed with bHABC and an anti-SPACR antibody, both HA and SPACR are identified in the

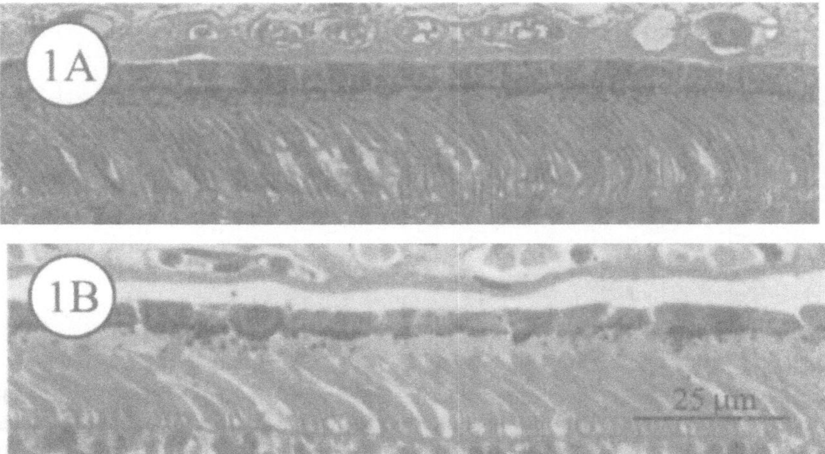

Figure 1. Photoreceptor/IPM/RPE complex from humnan retina treated with (A) native bHABC or (B) bHABC which was preincubated with HA oligosaccharides. Note the heavy staining of HA in (A) and the absence of HA staining in (B).

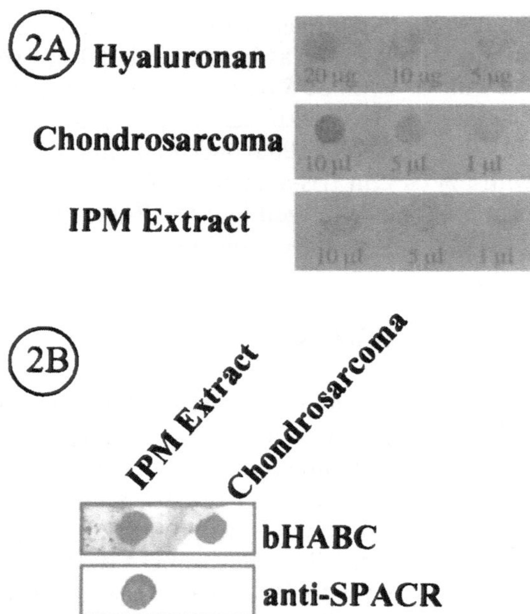

Figure 2. Dot blots of IPM and control samples demonstrating the presence of HA and SPACR in the insoluble IPM. In (2A), bHABC was used to probe the samples for the presence of HA. The hyaluronan sample (Healon™) and chondrosarcoma extract show strong HA signals in a dose dependent manner. The IPM also shows bHABC binding with an intensity suggesing a concentration of HA in the extract of approximately 0.5 µg/µl. In (2B), parallel dot blots of IPM and chondrosarcoma extracts were probed with bHABC (upper panel) and a polyclonal anti-SPACR antibody. Note that both HA and SPACR are present in the IPM extract, but only HA is present in the chondrosarcoma sample.

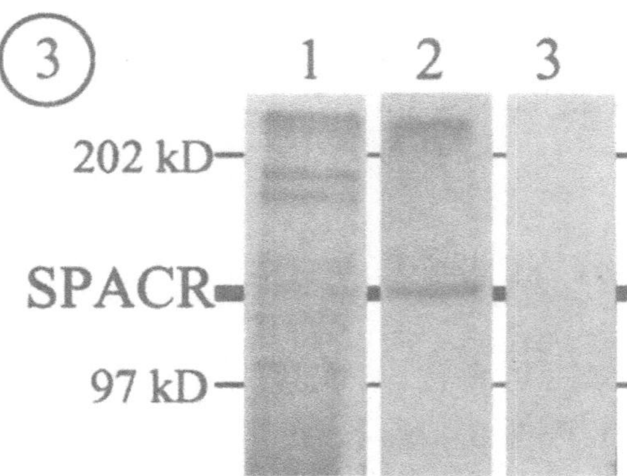

Figure 3. Precipitation of SPACR and hyaluronan from IPM extracts using CPC. IPM samples in TBS were incubated with 1.25% CPC in 0.1 M Tris-acetate buffer pH 7.3 for 1 h at room temperature as described in the methods. Lane 1 contains 25 µg of IPM protein from the original extract without CPC. The sample in lane 2 is from the CPC pellet that was incubated in the absence of enzyme, whereas the sample in lane 3 was predigested with *Streptomyces* hyaluronidase to degrade HA before CPC precipitation. Note the precipitation of SPACR (at the level of the thick marker) as well as the high molecular weight components at the top of the lane (lane 2) and the failure of these proteins to precipitate when HA is degraded (lane 3).

IPM samples (Fig. 2B). These results clearly demonstrate that HA and SPACR are present in the insoluble IPM extract as relatively stable constituents, since both are retained in this compartment during the hypotonic isolation procedure.

Considering the importance attributed to HA in the organization of other matrices,[23,24] HA present in the IPM may provide the basic framework to which other IPM molecules are associated. Since SPACR, the recently characterized IPM glycoprotein, is also retained as a part of the insoluble IPM, we evaluated the possibility that SPACR might be stabilized in the IPM by binding to HA. To test this hypothesis, we used the detergent cetylpyridinium chloride (CPC) in precipitation studies. Under non-dissociative conditions, CPC can efficiently precipitate HA and proteoglycans, along with any associated protein.[25] When an extract of the IPM is treated with CPC, SPACR, along with some higher molecular weight bands, is observed in the pellet (Fig. 3). When the IPM sample is pretreated with *Streptomyces* hyaluronidase (which specifically degrades HA) no SPACR is evident in the CPC pellet. The ability of CPC to precipitate SPACR from the IPM sample, and the absence of SPACR in the CPC pellet after hyaluronidase treatment, suggests that SPACR forms associative interactions with HA. These HA associations may be of fundamental importance to the stability of macromolecules comprising the insoluble IPM.

HA is well suited to serve as the fundamental framework for the IPM. It is extremely large, with molecular weights ranging between 1 and 10 million (between 2500 and 25,000 repeat disaccharides), supporting an extended molecular length of 2.5 to 25 µm. Specific associative interactions allow HA molecules to align and associate into more complex fiber bundles, which branch to form expansive three dimensional networks, contributing to the basic framework of matrices in which HA is found.[5] Further support for the involvement of HA as a candidate in organization of matrices comes from the findings that a number of molecules present in extracellular matrices can bind to HA, including fibronectin,[23] type VI collagen,[26] proteoglycans,[24] and plasma membrane receptors, such as CD44.[5,24] It is of interest that the deduced amino acid sequence of SPACR contains a linear, RHAAM-type HA binding domain.[13] The results of the coprecipitation studies of the IPM isolates suggests that this HA binding domain in SPACR is functional.

ACKNOWLEDGMENTS

We thank the Cleveland Eye Bank for the human tissues used in this study. The study was supported by grants from the National Institutes of Health (EY 02362), The Foundation Fighting Blindness, Hunt Valley, MD, The Retina Research Foundation, Houston, TX and funds from The Cleveland Clinic Foundation.

REFERENCES

1. Hascall, V. and Hascall, G, 1981, Proteoglycans. In *Cell Biology of the Extracellular Matrix.* (Ed E. Hay). Pp. 39–63. Plenum Press: New York.
2. Laurent, T. and Fraser, J, 1992, Hyaluronan. *FASEB J.* **6**:2397–2404.
3. Meyer, K, 1947, The biological significance of hyaluronic acid and hyaluronidase. *Physiol. Rev.* **27**:333–359.
4. Tammi, R., Ripellino, J., Margolis, R., and Tammi, M, 1988, Localization of epidermal hyaluronic acid using the hyaluronate binding region of cartilage proteoglycan as a specific probe. *J. Invest. Dermatol.* **92**:326–332.

5. Wang, C., Tammi, M., and Tammi, R, 1992, Distribution of hyaluronan and its CD44 receptor in the epithelia of human skin appendages. *Histochem.* **98**:105–112.
6. Wang, C., Tammi, M., Guo, H., and Tammi, R, 1996, Hyaluronan distribution in the normal epithelium of esophagus, stomach, and colon and their cancers. *Amer. J. Path.* **148**:1861–1869.
7. Berman, E.R. and Bach, G, 1968, The acid mucopolysaccharides of cattle retina. *Biochem. J.* **108**:75–88.
8. Adler, A.J. and Klucznik, K.M, 1982, Proteins and glycoproteins of the bovine interphotoreceptor matrix: composition and fractionation. *Exp. Eye Res.* **34**:423–434.
9. Kaneko, M, 1987, Interphotoreceptor matrix glycosaminoglycans in bovine eyes. *Ophthalmic Res.* **19**:330–337.
10. Johnson, L.V. and Hageman, G.S, 1989, Characterization of isolated cone matrix sheath substructure. *Invest. Ophthalmol. Vis. Sci. (Suppl)* **30**:490.
11. Hollyfield, J.G., Rayborn, M.E., Landers, R.A., and Myers, K.A, 1990, Insoluble interphotoreceptor domains surround rod photoreceptors in the human retina. *Exp. Eye Res.* **50**:107–110.
12. Acharya, S., Rayborn, M.E., and Hollyfield, J.G, 1998, Characterization of SPACR, a sialoprotein associated with cones and rods present in the interphotoreceptor matrix of the human retina: immunological and lectin binding analysis. *Glycobiology* **8**:997–1006.
13. Acharya, S., Rodriguez, I.R., Moreira, E.F., Midura, R.J., Misono, K., Todres, E., and Hollyfield, J.G, 1998, SPACR: a novel interphotoreceptor matrix glycoprotein in human retina that interacts with hyaluronan. *J. Biol. Chem.* **273**:31599–31606.
14. Tammi, R., Ronkko, S., Agren, U.M., and Tammi, M, 1994, Distribution of hyalronan in bull reproductive organs. *J. Histochem. Cytochem.* **42**:1479–1486.
15. Ripellino, J., Klinger, M., and Margolis, R, 1985, The hyaluronic acid binding region as a specific probe for the localization of hyaluronic acid in tissue sections. *J. Histochem. Cytochem.* **92**:1060–1066.
16. Hollyfield, J., Rayborn, M., Tammi, M., and Tammi, R, 1998, Hyaluronan in the interphotoreceptor matrix of the eye: Species differences in content, distribution, ligand binding and degradation. *Exp. Eye Res.* **66**:241–248.
17. Hollyfield, J.G., Rayborn, M.E., and Landers, R.A, 1990, A technique for isolation of the photoreceptor layer from other neurons in the human retina. *Exp, Eye Res.* **50**:335–338.
18. Plaas, A., Wong-Palms, S., Roughley, P., Midura, R., and Hascall, V, 1997, Chemical and immunological assay of the nonreducing terminal residues of chondroitin sulfate from human aggrecan. *J. Biol. Chem.* **272**:20603–20610.
19. Lee, T., Wisniewski, H., and Vilcek, J, 1992, A novel secretory tumor necrosis factor-inducible protein (TSG-6) is a member of the family of hyaluronate binding proteins closely related to the adhesion receptor CD44. *J. Cell Biol.* **116**:545–557.
20. Laemmli, U, 1970, Cleavage of structural proteins during assembly of the head of bacteriophage. *Nature* **227**:680–685.
21. Tien, L., Rayborn, M.E., and Hollyfield, J.G, 1992, Characterization of the interphotoreceptor matrix surrounding rod photoreceptors in the human retina. *Exp. Eye Res.* **55**:297–306.
22. Oegema, T., Hascall, V., and Dziewiatkowski, D, 1975, Isolation and characterization of proteoglycans from the Swarm rat chondrosarcoma. *J. Biol. Chem.* **250**:6151–6159.
23. Toole, B.P, 1990, Hyaluronan and its binding proteins, the hyaladherins. *Curr. Opin. Cell Biol.* **2**:839–844.
24. Knudson, C. and Knudson, W, 1993, Hyaluronan-binding proteins in development, tissue homeostasis, and disease. *FASEB J.* **7**:••.
25. Scott, J.E, 1973, Affinity competition and specific interactions in the biochemistry and histochemistry of polyelectrolytes. *Biochem. Soc. Trans.* **1**:787–806.
26. McDevitt, C., Marcelino, J., and Tucker, L, 1991, Interaction of intact type VI collagen with hyaluronan. *FEBS Lett.* **294**:167–170.

DIVERSE LOCALIZATION OF CYCLIC NUCLEOTIDE GATED CHANNELS IN THE OUTER SEGMENTS OF RODS AND CONES

Marion S. Eckmiller*

C. and O. Vogt Brain Research Institute
Heinrich Heine University of Düsseldorf School of Medicine
Postfach 101007, D-40001 Düsseldorf
Germany

ABBREVIATIONS USED

COS = cone outer segment; cGMP = cyclic guanosine monophosphate; CNG = cyclic nucleotide gated; IS = inner segment; OS = outer segment; PBS = phosphate buffered saline; ROS = rod outer segment

SUMMARY

The spatial distribution of cyclic nucleotide gated (CNG) channel molecules in photoreceptor outer segments (OS) dissociated from amphibian retinas was investigated by performing immunofluorescent localization of spectrin and using the known spectrin immunoreactivity of the beta subunit of the channel in rods to infer the location of CNG channels. In the OS of rods and cones, anti-spectrin immunoreactivity occurred as a bright streak of fluorescence at the ciliary axoneme. Rod OS displayed an additional pattern of staining not present in cone OS, namely a series of thin, discrete, longitudinal lines of fluorescence that extended the entire length of the OS and coincided with incisures. Thus, the location of immunoreactivity to spectrin in the OS of both photoreceptor cell types coincided with locations known to contain arrays of longitudinally-oriented microtubules. These findings provide strong evidence that

* Phone: 49 211 811 2767, Fax: 49 211 811 2336, E-mail:eckmill@hirn.uni-duesseldorf.de

Retinal Degenerative Diseases and Experimental Therapy, edited by Hollyfield *et al.*
Kluwer Academic / Plenum Publishers, New York, 1999.

CNG channel molecules are confined within OS membranes to specific restricted locations in the immediate vicinity of microtubules, eg., CNG channel molecules may be tethered to microtubules via the spectrin-like portion of their beta subunits. Because the localization of CNG channels within photoreceptor OS is expected to influence the spatiotemporal dynamics of phototransduction and adaptation, the diverse localization of channels within the OS could contribute to the different functional properties of rods and cones. Because evidence suggests that the OS of human and amphibian photoreceptors have similar microtubule-containing cytoskeletal systems at similar locations, the spatial distribution of CNG channel molecules described here for amphibian photoreceptor OS is also expected to occur in human photoreceptor OS. A disturbance in the localization of CNG channels, or in their associations with other molecules or microtubules, within photoreceptor OS is expected to disturb OS structure and function, which may be relevant for some human retinal degenerations.

INTRODUCTION

The light sensitive OS of vertebrate rod (R) and cone (C) photoreceptor cells[1–3] comprise hundreds of well-ordered parallel layers of membrane, supported along one side by the axoneme of a cilium, which extends from a basal body in the inner segment (IS). Between the photoreceptor IS and OS, this connecting cilium consists of a primary ciliary axoneme with nine microtubule doublets.[4] Within the OS these microtubule doublets become reduced to singlets, which in amphibian photoreceptors apparently disappear at more distal heights of the OS in rods[5] but extend all the way to the tip of the OS in cones.[6] In the retinas of amphibians, primates and many other vertebrates, ROS have multiple longitudinal indentations or incisures,[1,2,5,7] which have been shown in amphibian rods to be the site of an additional separate cytoskeletal system with longitudinal microtubules.[3,8–10] COS do not have multiple incisures or an associated microtubule-containing cytoskeleton.[1–3] It has been hypothesized that a disturbance of the cytoskeletal system at ROS incisures may be involved in some forms of human retinal degeneration.[3,8,9]

The biochemical events of phototransduction are known in considerable detail for rods but are less well known for cones.[11–14] In rods, the absorption of light by rhodopsin molecules within ROS disc membranes activates a cascade of enzymatic reactions, which decreases the concentration of cyclic guanosine monophosphate (cGMP) in the cytoplasm, which lowers the ionic permeability of cyclic nucleotide gated (CNG) channel molecules in the ROS surface plasma membrane, which reduces the circulating current entering the OS, which hyperpolarizes the OS and then the rest of the cell. In the OS of rods, CNG channels[15,16] are integral membrane proteins that are known from immunolocalization and biochemical studies to be located near the edge of the ROS and to be primarily localized in the plasma membrane, rather than the disc membranes.[17–19] In native ROS, the CNG channel molecules exist as tetramers composed of two alpha and two beta subunits, and the opening and closing of CNG channels is regulated both directly by the binding of cGMP and the calcium binding protein calmodulin and indirectly by the intracellular concentration of free Ca++.[16,19–21] Electrophysiological studies have found the cGMP-activated currents through membrane patches excised from ROS to be highly variable, suggesting that the distribution of CNG channels in the ROS plasma membrane may be inhomogeneous.[22,23] Immunocytochemical studies have shown stripes of immunoreactivity to actin, calmodulin, myosin, and fodrin (a spectrin-like protein found in brain tissue) at the cilium of ROS in various vertebrate retinas.[24–27] Suspecting that the CNG channels within ROS might all occur along the cilium, one group compared the currents at different positions around the circumference of amphibian ROS, failed to observe markedly different current densities at different positions, and concluded that the circumferential distribution of channels was fairly uniform.[23]

Spectrin denotes a family of actin-binding proteins commonly found in association with the plasma membrane of mature cells that are involved in membrane-cytoskeleton interactions, bestow stability on the membrane, and link transmembrane proteins to the cytoskeleton.[28,29] In erythrocytes, spectrin forms a fibrous network lining the cytoplasmic surface of the plasma membrane; in nonerythroid cells, proteins that are immunologically closely related to spectrin also occur in association with other cytoplasmic cytoskeletal structures. In neurons, the cytoskeletal proteins spectrin, actin, and myosin have been found to occur at restricted intracellular locations, where they may be associated with each other and with plasma membrane ion channels, whose activity can thereby be modified. For example, neuronal spectrin and actin can be closely associated with a high membrane density of voltage gated sodium channels at axon hillocks and at nodes of Ranvier, locations with a low threshold for action potential excitation.[30,31] Except for the studies showing fodrin immunoreactivity at the ROS cilium and another study described below, spectrin-immunoreactive proteins have not been reported within vertebrate photoreceptor OS.

Previous research has established that CNG channels are integral membrane proteins that are present in the surface membranes of ROS and COS, and that are immunoreactive to spectrin. Biochemical and immunochemical studies have indicated that bovine rods contain a 240 kDa protein showing crossreactivity to polyclonal sera that recognize erythrocyte spectrin.[32] Subsequent work has identified this 240 kDa spectrin-immunoreactive protein as the rod CNG channel beta subunit, whose N-terminal is on the cytoplasmic side of the OS plasma membrane and contains the calmodulin binding site.[16,17,19–21,33]

The recent findings that Usher syndrome and retinitis pigmentosa in humans can be caused by gene defects in myosin VIIA, a cytoskeletal protein associated with the photoreceptor connecting cilium,[34,35] has rekindled keen interest in the cytoskeleton of vertebrate photoreceptor OS. The present study has investigated the location of CNG channels, as inferred from the location of spectrin immunoreactivity, within the OS of rods and cones from amphibian retinas. Discrete lines of specific anti-spectrin immunoreactivity (and therefore presumably CNG channel molecules) were found at locations within OS known to contain microtubules, i.e., at the ciliary axoneme of ROS and COS, and also at the multiple incisures of ROS. Thus, the different microtubule-containing cytoskeletal systems that provide structural support for the membranous OS of rods and cones apparently also function to confine the CNG channels at diverse locations within their OS (portions of this work have been published in abstract form).[36]

METHODS

Neural retinas from African clawed toads (*Xenopus laevis*) were mechanically dissociated and the resulting cell suspension was fixed, applied to glass slides, and the slides processed for indirect immunofluorescent staining of spectrin by methods similar to those used previously for immunolocalization of other antibodies.[6,10] In the present study the primary antibody incubation was with antisera against chicken and human spectrins, i.e., antisera developed in rabbit using (alpha and beta chains of) spectrins purified from chicken and human erythrocytes as immunogens (Sigma, St. Louis, MO, U.S.A.); the secondary antibodies used were the appropriate fluorochrome-conjugated goat anti-rabbit IgG and/or IgM (Dianova, Hamburg, Germany). The phosphate buffered saline (PBS) solution used for rinsing and fixing the cells, and for diluting the antibodies and sera contained 2.7 mM KCl, 1.5 mM KH_2PO_4, 137 mM NaCl, 8 mM Na_2HPO_4, 2.0 mM $MgCl_2$, and 10 mg/ml sucrose and had pH 6.9. The photoreceptor cells and OS on experimental and control slides were examined and photographed in a Zeiss light microscope using bright field, differential interference contrast optics, and epifluorescent illumination.[6,10]

RESULTS

The cell preparations dissociated from *Xenopus* retinas comprised photoreceptor cell fragments, isolated OS attached to IS, and solitary OS, which could be readily identified by their size, shape, and appearance in the light microscope. The preparations contained many wide cylindrical ROS from red rods, some narrower cylindrical ROS from green rods, and also some smaller tapered COS.

On experimental slides treated for indirect immunofluorescent staining of spectrin, COS had a bright streak of fluorescence along one side of the OS that extended from the base to the tip of the COS (Fig. 1a,b). This streak of anti-spectrin immunofluorescence in COS closely resembled the streak of anti-tubulin immunofluorescence previously described at the COS ciliary axoneme.[6,8] COS showed no other staining. This immunofluorescent streak was taken to represent labelling closely associated with the COS axoneme. These observations suggested that the fluorescence in COS produced by anti-spectrin antibodies reflected the presence of spectrin, or an immunologically similar protein, in close association with the microtubules of the COS ciliary axoneme.

Dissociated ROS on experimental slides had a bright longitudinal streak of fluorescence (Fig. 1c) at one position of their circumference that was generally visible at a single focal depth. This streak arose at the base of the ROS and continued part way to the distal end of the ROS, resembling the ciliary axoneme in previous studies.[3,5,6,8] In addition, ROS (Fig. 1c–e) displayed a series of long, discrete fluorescent lines in a plane of focus close to the plasma membrane that were aligned parallel to the longitudinal axis of the OS and generally extended from the base to the distal end of the OS. The intensity of these fluorescent lines was quite similar at all lines in an individual ROS and at all heights of a given line, but it was less than that at the axoneme. The position, number, and spacing of these lines of anti-spectrin fluorescence coincided with those of the longitudinal indentations in the ROS plasma membrane that are termed incisures, which were visible by differential interference contrast optics. A similar pattern of immunolabelling was found in the OS of green (Fig. 1c) and red rods (Figure 1d,e). These observations suggested that the fluorescence in ROS produced by anti-spectrin antibodies demonstrated that spectrin, or an immunologically closely related protein, was present in close association with the microtubules in ROS, i.e. with the axoneme microtubules at the cilium and with the cytoplasmic microtubules at the multiple incisures.[3,8]

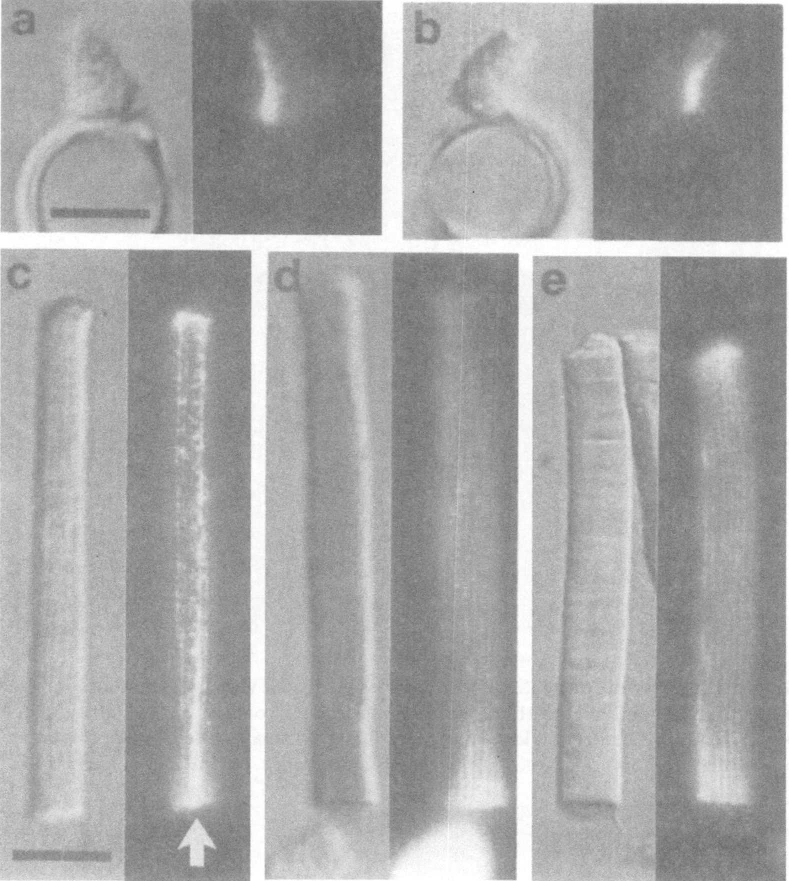

Figure 1. Photoreceptor cells and OS dissociated from *Xenopus* retinas on experimental slides processed for immunofluorescent staining of spectrin, each photographed with differential interference contrast optics (left) and epifluorescence (right). a, b The small tapered OS of cones had one bright streak of fluorescence that extended the entire COS length and resembled the ciliary axoneme.[6] c The large cylindrical OS of rods likewise had a single bright fluorescent streak at one position (white arrow) around their circumference that arose at the base of the ROS and extended part way to the distal end of the ROS, resembling the ROS ciliary axoneme.[6] d,e ROS on experimental slides also displayed a series of long, thin, discrete lines of anti-spectrin fluorescence that extended the entire length of the OS and coincided with their multiple incisures; these lines resembled lines of anti-tubulin fluorescence that are associated with microtubules at ROS incisures.[3,8] Magnification bar = 5 µm in a,b, 10 µm in c–e.

There was no fluorescence in photoreceptor OS on control slides incubated with PBS instead of anti-spectrin antibodies. ROS and COS on control slides incubated with diluted normal rabbit serum (Fig. 2), instead of anti-spectrin antibodies, sometimes showed a slight nonspecific fluorescence (particularly near the base of OS), but they had no bright lines of fluorescence. These findings indicated that the fluorescent streak associated with the ciliary axoneme of photoreceptor OS and the fluorescent lines at the multiple incisures of ROS on the experimental slides were specific for antibodies to spectrin, or an immunologically closely related protein.

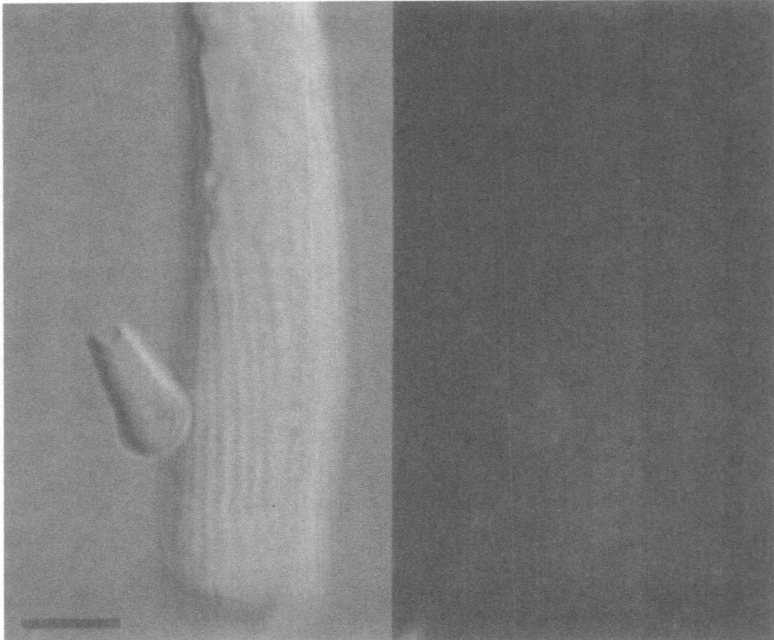

Figure 2. Dissociated photoreceptor OS on control slides incubated with normal mouse serum instead of an antibody to spectrin, photographed with differential interference contrast optics (left) and epifluorescence (right). COS and ROS on control slides sometimes showed a small amount of non-specific background fluorescence, but they had no bright lines or streaks of fluorescence. Magnification bar = 5 µm.

CONCLUSIONS

This study demonstrates that spectrin, or a protein that is immunologically closely related to spectrin, is associated with the microtubules in the OS of amphibian photoreceptors and, together with several other lines of evidence, provides strong evidence that this spectrin-like protein is the beta subunit of the CNG channel molecule. These findings are the first experimental evidence that CNG channels are not distributed homogeneously in the plasma membrane of vertebrate photoreceptor OS, but occur at specific positions in association with the cytoskeleton. Because the locations of microtubules are different in the OS of rods and cones,[3,6,8] the locations of their CNG channels likewise differ: CNG channel molecules are present along a longitudinal stripe at the ciliary axoneme in both ROS and COS, and they also occur along a separate series of thin longitudinal lines that coincide with microtubules in the cytoplasm at the multiple incisures in ROS. These observations suggest that CNG channels occur at these restricted locations within OS membranes because they are tethered to microtubules in the adjacent cytoplasm by filamentous connections, perhaps involving the spectrin-like beta subunit of the CNG channel. These findings reveal a novel association between photoreceptor OS structure and function, i.e., a direct interaction between microtubules, which provide structural support for the fluid OS membranes, and CNG channel molecules, which play a central functional role in phototransduction. Besides having implications for normal phototransduction in rods and cones, these findings may have clinical relevance for human retinal degenerations because CNG channels are expected to occur at the same restricted locations in the OS of human photoreceptors.

Previous immunocytochemical studies[17,18,32] of CNG channnels in ROS used antibodies raised against the the 240 kDa protein and the channel and found immunogold labelling around the ROS edge and near the ciliary and plasma membranes of bovine ROS in the electron microscope, but did not notice a possible higher labelling density in the membrane adjacent to incisures, which would be difficult to distinguish in single ultrathin sections. Earlier studies have also reported a stripe of immunoreactivity to fodrin and calmodulin at the ciliary axoneme of amphibian ROS[25] and a stripe of immunoreactivity to actin at the axoneme of ROS from several vertebrate species.[24,26,27] CNG channels in ROS bind calmodulin[16,19–21] and ion channels in various neurons are often closely associated with spectrin and related cytoskeletal proteins.[30,31] Accordingly, these earlier findings are entirely consistent with the present observation of spectrin immunofluorescence, and inference of CNG channels, at the axoneme of ROS.

Although the observation of lines of spectrin immunoreactivity and the inferred presence of CNG channels at the multiple incisures of *Xenopus* ROS as described here have not been previously reported, several lines of evidence suppport the validity of these findings. By performing light microscopic immunofluorescence on whole photoreceptor OS dissociated from amphibian retinas (in which the ROS are wide and have multiple incisures) and using methods that preserve cytoskeletal stability and immunoreactivity,[6,10] we have been able to visualize the immunoreactivity throughout whole ROS and to determine its subcellular distribution with respect to detailed structures, such as the position of the incisures. Electrophysiological studies have found that the CNG current through different patches excised from ROS membranes varies greatly, suggesting that CNG channels are not homogenously distributed in the membrane but might occur with higher density along the cilium. One group[23] that investigated this possibility by measuring the local electrical currents at 2.5 μm intervals around the circumference of ROS from the salamander retina found some variation in current densities but concluded that the circumferential distribution of channels was fairly uniform on a 1 μm scale. Although these results were taken as evidence for a homogeneous channel distribution, the intervals examined may not have been sufficiently small to distinguish currents through adjacent incisures. Additional evidence that the spectrin immunofluorescence reflects immunoreactivity to CNG channel molecules comes from the report[37] that a fluorescent cGMP conjugate, which is expected to bind to the CNG channel, preferentially binds at the base of amphibian ROS along longitudinal stripes that coincide with calycal processes, but also with incisures. The strongest evidence that the spectrin immunofluorescence observed in the present study reflects immunoreactivity to CNG channel molecules is that the CNG channel of ROS is the only protein within photoreceptor OS that has been reported to cross react with spectrin.[17,32] Thus, although in the present study immunolocalization was performed using sera containing antibodies that recognize erythrocyte spectrin, there is convincing evidence that the fluorescence that we obtained in ROS actually represents immunoreactivity to the beta subunit of CNG channel molecules.

Previous studies have not described the subcellular localization of CNG in the surface membranes of COS. By using appropriate methods,[6,10] we have been able to observe and localize the spectrin immunoreactivity throughout whole COS, and found it confined to a streak at the ciliary axoneme. Except for the possibility that the axoneme may not be renewed in mature ROS,[3,6] no major differences are known between the cilium in rod and cone photoreceptors. Accordingly, it is not surprising that we have found a streak of immunoreactivity to spectrin, demonstrating localization of CNG channels, in close association with the axoneme in COS, as in ROS.

In the present study the pattern of immunofluorescence to spectrin was different in the OS of rods and cones from the *Xenopus* retina, and resembled the pattern of immunofluorescence to tubulin in these OS in previous studies.[3,6,8–10] In ROS from amphibian retinas, longitudinal microtubules have been shown to occur both at the ciliary axoneme and within the narrow longitudinal strips of cytoplasm that occur between the edge of discs and the surface membrane at each of the multiple incisures.[3,6,8,9] In vertebrate COS, the only microtubules are those at the ciliary axoneme, which is consistent with the general observation that COS lack incisures.[1–3,6] Thus, the location of CNG channels, as inferred from the location of anti-spectrin immunofluorescence, in ROS and COS from amphibian retinas in the present study precisely coincided with OS locations known to contain microtubules.

Ultrastructural studies of ROS and COS have revealed filamentous connections linking the axoneme microtubules to the adjacent ciliary membrane and linking the microtubules at ROS incisures to the overlying ROS plasma membrane.[3–5,8,9,38,39] Although the biochemical identity of these filaments is not known, they have been assumed to play an important cytoskeletal role in maintaining the normal structural integrity of photoreceptor OS. Because the location of presumptive CNG channels in photoreceptor OS in the present study coincided with locations known to contain cytoskeletal microtubules, our findings suggest that the CNG channel molecules are held at these restricted locations within OS plasma membranes via filamentous attachments to the microtubules. It follows that one component of the filaments mediating these connections may be the beta subunit of the CNG channel.

In addition to their close association with one another, CNG channels and microtubules are expected to have specialized associations and/or interactions with other photoreceptor OS components. Based on research in photoreceptors and other cells, these microtubules can be expected to interact with motor and cytoskeletal proteins, eg., actin, kinesins, myosins, and various microtubule associated proteins.[4,10,27,28,35] Ultrastructural studies have demonstrated close associations (in some cases mediated by filamentous connections) between the microtubules (at the photoreceptor axonemes and at ROS incisures) and the highly curved membrane regions found at the open margins of lamellae in COS and at the incisures and rims of discs in ROS.[3,5,8,9,39,40] Furthermore, a recent study[16] has noted that under certain conditions, CNG channels isolated from ROS become directly associated with Na+/Ca++—K+ exchanger molecules; in combination with the present study this suggests that the exchanger molecules may likewise be confined in the plasma membrane to locations in the vicinity of microtubules. Because the microtubules at photoreceptor axonemes and at ROS incisures form highly organized associations with many different OS components and molecules, these OS cytoskeletons comprise specialized multicomponent cytoskeletal systems.[3,4,8–10]

A non-homogenous distribution of CNG channel molecules within the plasma membrane of vertebrate OS has not been previously reported but is consistent with certain findings in rods[22,23] and in other cells.[30,31] The subcellular localization of the CNG channels is expected to have a direct influence on photoreceptor function, eg., on the spatiotemporal dynamics of phototransduction and adaptation,[11–14] because the only route by which the circulating photocurrent enters OS is via the CNG channels. This heterogeneous distribution of channels in photoreceptor OS is expected to give rise to microdomains within the OS cytoplasm having very different concentrations, eg., of

Na+, Ca++, cGMP.[41] Accordingly, the close association between CNG channel molecules and microtubules in each of these several cytoskeletal systems within photoreceptor OS is expected to be important not only for the normal structure of vertebrate photoreceptors, but also for their normal function.

A diverse subcellular distribution of CNG channels in ROS and COS could contribute to the known functional differences between rods and cones, such as the faster response to light and recovery from light in cones.[11,12,14] Recent research has shown that the same events underlie phototransduction in both cell types and that the proteins involved are similar although not identical, but that the quantitative features of Ca++ homeostasis in ROS and COS, which mediate adaptation to changes in light intensity,[13] differ profoundly: the changes in cytoplasmic Ca++ concentration are much larger and faster in COS than ROS.[12] The two routes for information flow from CNG channels in the OS to the IS, i.e., along the membrane for electrotonic propagation of the photovoltage and through the cytoplasm for diffusion and/or transport of cytosolic molecules, are different for the cytoskeletal systems at photoreceptor axonemes and ROS incisures. The CNG channels in both cones and rods that are associated with the axoneme (and presumably present in the ciliary membrane) are located along a short, direct route to the connecting cilium, and hence to the IS and on to the synaptic terminal. By contrast, the channels in the ROS plasma membrane that are associated with the incisural microtubules are located along a longer, more tortuous route to the cilium, IS, and synapse. Accordingly, the channels at the ROS axoneme are expected to have an effect on the voltage along the connecting cilium membrane and on the OS free Ca++ concentration that is larger and faster than that of the channels at ROS incisures. Lacking incisures, COS apparently have all of their CNG channels positioned along the direct ciliary route, which may contribute to their fast response to changes in light intensity and their fast changes in Ca++ concentration. Having some CNG channels at the cilium and others at their incisures, ROS having multiple incisures would be expected to exhibit slower overall responses to changes in light intensity and slower overall changes in Ca++ concentration. A diverse localization of CNG channels, as described here, is thus expected to contribute to the different kinetics of the photocurrents and changes in free Ca++ concentration in the OS of rods and cones.

Photoreceptor OS in human and amphibian retinas differ in their dimensions and in the extent to which the lamellae in their COS are surrounded by an external membrane but they are otherwise very similar morphologically—in both cases the OS of rods have multiple incisures, the OS of cones lack incisures, and each OS is supported by a ciliary axoneme.[1–3,6,7,42] A cytoskeletal system containing longitudinal microtubules is present in the cytoplasm at the multiple incisures of ROS in amphibian rods[3,6,8,9] and an analogous system is expected to occur at the morphologically similar multiple incisures of ROS in human rods. Accordingly, the present findings that the CNG channels in photoreceptors from amphibian retinas are closely associated with microtubules at the axoneme of ROS and COS and the incisures of ROS strongly suggest that CNG channels are likewise closely associated with microtubules at the same locations in ROS and COS in the human retina.

These findings are likely to be clinically relevant because a disturbance involving CNG channel localization within the OS of human photoreceptors is expected to result in a pathological disturbance of photoreceptor structure and function. Defects in cilia or microtubules have long been suspected to play a role in human retinal disturbances because patients with Usher syndrome have retinitis pigmentosa plus hearing and vestibular defects, and mutations in the ciliary cytoskeletal protein myosin VIIA have now been shown to cause this syndrome.[34,35] Researchers have also discovered that mutations in three proteins (peripherin/rds, ROM 1, and rim protein) that are confined to the highly curved membrane found at the rims and incisures of ROS discs can cause human retinitis pigmentosa and/or macular degeneration.[40,43–47] These findings show that human retinal degenerations can result from a disturbance of cytoskeletal systems within photoreceptor OS. In addition, mutations involving an absence or paucity of functional CNG channels can cause retinitis pigmentosa.[48,49] Because in human photoreceptor OS the CNG channel molecules are expected to be localized in close association with microtubule-containing cytoskeletal systems as described here in amphibian photoreceptor OS, a disturbance in the localization of CNG channels and/or the associations between these channels and other molecules or microtubules within photoreceptor OS may contribute to some forms of human photoreceptor degeneration.

ACKNOWLEDGMENTS

The author thanks B. Rohbeck for technical assistance. This research was supported by the Deutsche Forschungsgemeinschaft (Ec 85/4–2), Ministry for Science and Research of Northrhine-Westfalia (400 017 92), and Deutsche Retinitis Pigmentosa Vereinigung.

REFERENCES

1. B. Borwein, 1981, The retinal receptor: A description, in: *Vertebrate Photoreceptor Optics* (J.M. Enoch and F.L. Tobey, eds.), pp. 11–81, Springer Verlag, Berlin.
2. J.M. Corless and D.R. McCaslin, 1984, Vertebrate visual photoreceptors, in: *The Receptors* (P.M. Conn, ed.), Vol.1, pp. 315–376, Academic Press, New York.
3. M.S. Eckmiller, 1997, Morphogenesis and renewal of cone outer segments, *Prog. Ret. Eye Res.* **16**:401–441.
4. J.C. Besharse and C.J. Horst, 1990, The photoreceptor connecting cilium, in: *Ciliary and Flagellar Membranes* (R. Bloodgood, ed.), pp. 389–417, Plenum Press, New York.
5. D. Roof, M. Adamian, D. Jacobs, and A. Hayes, 1991, Cytoskeletal specializations at the rod photoreceptor distal tip, *J. Comp. Neurol.* **305**:289–303.
6. M.S. Eckmiller, 1996, Renewal of the ciliary axoneme in cone outer segments of the retina of *Xenopus laevis*, *Cell Tissue Res.* **285**:165–169.
7. R.W. Young, 1971, Shedding of discs from rod outer segments in the rhesus monkey, *J. Ultrastruct. Res.* **34**:190–203.
8. M.S. Eckmiller, 1993, A cytoskeletal system at incisures of rod outer segments and its implications for renewal and retinal degeneration, *Invest. Ophthalmol. Vis. Sci. (Suppl.)* **34**:1359.
9. M.S. Eckmiller, 1995, Renewal of incisural microtubule cytoskeleton in rod outer segments of a mphibian and primate retinas, *Invest. Ophthalmol. Vis. Sci. (Suppl.)* **36**:514.
10. M.S. Eckmiller and A. Toman, 1998, Association of kinesin with microtubules in diverse cytoskeletal systems in the outer segments of rods and cones, *Acta Anatomica* (in press).
11. T.D. Lamb and E.N. Pugh, 1990, Physiology of transduction and adaptation in rod and cone photoreceptors, in: *The Neurosciences*, pp. 3–13.
12. J.L. Miller and J.I. Korenbrot, 1994, Differences in calcium homeostasis between retinal rod and cone photoreceptors revealed by the effects of voltage on the cGMP-gated conductance in intact cells, *J. Gen. Physiol.* **104**:909–940.
13. Y. Koutalos and K.W. Yau, 1996, Regulation of sensitivity in vertebrate rod photoreceptors by calcium, *Trends Neurosci.* **19**:73–81.
14. R.W. Rodieck, 1998, *The First Steps in Seeing*, Sinauer Associates, Sunderland, Massachusetts.

15. W.N. Zagotta and S.A. Siegelbaum, 1996, Structure and function of cyclic nucleotide-gated channels, *Annu. Rev. Neurosci.* **19**:235–263.
16. R.S. Molday and L.L. Molday, 1998, Molecular properties of the cGMP-gated channel of rod photoreceptors, *Vision Research* **38**:1315–1323.
17. L.L. Molday, N.J. Cook, U.B. Kaupp, and R.S. Molday, 1990, The cGMP-gated cation channel of bovine rod photoreceptor cells is associated with a 240-kDa protein exhibiting immunochemical cross-reactivity with spectrin, *J. Biol. Chem.* **265**:18690–18695.
18. R.S. Molday, L.L. Molday, A. Dosé, I. Clark-Lewis, M. Illing, N.J. Cook, E. Eismann, and U.B. Kaupp, 1991, The cGMP-gated channel of the rod photoreceptor cell characterization and orientation of the amino terminus, *J. Biol. Chem.* **266**:21917–21922.
19. T.-Y. Chen, Y.-W. Peng, R.S. Dhallan, B. Ahamed, R.R. Reed, and K.-W. Yau, 1993, A new subunit of the cyclic nucleotide-gated cation channel in retinal rods, *Nature* **362**:764–767.
20. M.E. Grunwald, W.-P. Yu, H.-H. Yu, and K.-W. Yau, 1998, Identification of a domain on the beta-subunit of the rod cGMP-gated cation channel that mediates inhibition by calcium-calmodulin. *J. Biol. Chem.* **273**:9148–9157.
21. D. Weitz, M. Zoche, F. Müller, M. Beyermann, H.G. Körschen, U.B. Kaupp, and K.-W. Koch, 1998, Calmodulin controls the rod photoreceptor CNG channel through an unconventional binding site in the N-terminus of the beta-subunit, *EMBO J.* **17**:2273–2284.
22. K.-W. Yau and D.A. Baylor, 1989, Cyclic GMP-activated conductance of retinal photoreceptor cells, *Ann. Rev. Neurosci.* **12**:289–327.
23. J.W. Karpen, D.A. Loney, and D.A. Baylor, 1992, Cyclic GMP-activated channels of salamander retinal rods: spatial distribution and variation of responsiveness, *J. Physiol.* **448**:257–274.
24. B.W. Nagle and B. Burnside, 1984, Calmodulin-binding proteins in teleost retina, rod inner and outer segments, and rod cytoskeletons, *Europ. J. Cell Biol.* **33**:248–257.
25. D. Roof, M. Applebury, and J. Kirsch, 1984, Localization of calmodulin and characterization of calmodulin binding proteins in the vertebrate rod outer segment, *Biophys. J.* **45**:1a.
26. M.H. Chaitin and D. Bok, 1986, Immunoferritin localization of actin in retinal photoreceptors, *Invest. Ophthalmol. Vis. Sci.*, **27**:1764–1767.
27. D.S. Williams, M.A. Hallett, and K. Arikawa, 1992, Association of myosin with the connecting cilium of rod photoreceptors, *J. Cell Science* **103**:183–190.
28. L.A. Amos and W.B. Amos, 1991, *Molecules of the Cytoskeleton*, Macmillan Education Ltd, London.
29. T. Kreis and R. Vale, 1993, *Guidebook to the Cytoskeletal and Motor Proteins*, Oxford University Press, New York.
30. Y. Srinivasan, L. Elmer, J. Davis, V. Bennett, and K. Angelides, 1988, Ankyrin and spectrin associate with voltage-dependent sodium channels in brain, *Nature* **333**:177–180.
31. S.J. Wood and C.R. Slater, 1998, Beta-spectrin is colocalized with both voltage-gated sodium channels and ankyrinG at the adult rat neuromuscular junction, *J. Cell Biol.* **140**:675–684.
32. S. Wong and R.S. Molday, 1986, A spectrin-like protein in retinal rod outer segments, *Biochemistry* **25**:6294–6300.
33. H.G. Körschen, M. Illing, R. Seifert, F. Sesti, A. Williams, S. Gotzes, C. Colville, F. Müller, A. Dosé, M. Godde, L. Molday, U.B. Kaupp, and R.S. Molday, 1995, A 240kDa protein represents the complete beta subunit of the cyclic nucleotide-gated channel from rod photoreceptor, *Neuron* **15**:627–636.
34. D. Weil, S. Blanchard, J. Kaplan, P. Guilford, F. Gibson, J. Walsh, P. Mburu, A. Varela, J. Levilliers, M.D. Weston, P.M. Kelley, W.J. Kimberling, M. Wagenaar, F. Levi-Acobas, D. Larget-Piet, A. Munnich, K.P. Steel, S.D.M. Brown, and C. Petit, 1995, Defective myosin VIIA gene responsible for Usher syndrome type 1B, *Nature* **374**:60–61.
35. X. Liu, G. Vasant, I.P. Udovichenko, U. Wolfrum, and D.S. Williams, 1997, Myosin VIIa, the product of the Usher 1B syndrome gene, is concentrated in the connecting cilia of photoreceptor cells, *Cell Motil. Cytoskeleton* **37**:240–252.
36. M.S. Eckmiller, 1998, Diverse localization of cyclic nucleotide gated channels in rods and cones inferred from spectrin immunolocalization, *Invest. Ophthalmol. Vis. Sci. (Suppl.)* **39**:678.
37. A. Caretta and H. Saibil, 1989, Visualization of cyclic nucleotide binding sites in the vertebrate retina by fluorescence microscopy, *J. Cell Biol.* **108**:1517–1522.
38. R.D. Fetter and J.M. Corless, 1987, Morphological components associated with frog cone outer segment disc margins, *Invest. Ophthalmol. Vis. Sci.* **28**:646–657.
39. R.S. Molday, 1988, Monoclonal antibodies to rhodopsin and other proteins of rod outer segments, *Prog. Ret. Res.* **8**:173–209.

40. R.S. Molday, 1994, Peripherin/rds and rom-1: Molecular properties and role in photoreceptor cell degeneration, *Prog. Ret. Eye Res.* **13**:271–299.
41. R. Rizzuto, M. Brini, M. Murgia, and T. Pozzan, 1993, Microdomains with high Ca^{2+} close to IP3-sensitive channels that are sensed by neighboring mitochondria, *Science* **262**:744–747.
42. D.H. Anderson, S.K. Fisher, and R.H. Steinberg, 1978, Mammalian cones: Disc shedding, phagocytosis, and renewal, *Invest. Ophthalmol. Vis. Sci.* **17**:117–133.
43. R. Allikmets, N.F. Shroyer, N. Singh, J.M. Seddon, R.A. Lewis, P.S. Bernstein, A. Peiffer, N.A. Zabriskie, Y. Li, A. Hutchinson, M. Dean, J.R. Lupski, and M. Leppert, 1997, Mutation of the Stargardt disease gene (ABCR) in age-related macular degeneration, *Science* **277**:1805–1807.
44. R. Allikmets, N. Singh, H. Sun, N.F. Shroyer, A. Hutchinson, A. Chidambaram, B. Gerrard, L. Baird, D. Stauffer, A. Peiffer, A. Rattner, P. Smallwood, Y. Li, K.L. Anderson, R.A. Lewis, J. Nathans, M. Leppert, M. Dean, and J.R. Lupski, 1997, A photoreceptor cell-specific ATP-binding transporter gene (ABCR) is mutated in recessive Stargardt macular dystrophy, *Nature Genetics* **15**:236–246.
45. S.M. Azarian and G.H. Travis, 1997, The photoreceptor rim protein is an ABC transporter encoded by the gene for recessive Stargardt's disease (ABCR), *Fed. Europ. Biochem. Soc. Letters* **409**:247–252.
46. H. Sun and J. Nathans, 1997, Stargardt's ABCR is localized to the disc membrane of retinal rod outer segments, *Nature Genetics* **17**:15–16.
47. A. Martinez-Mir, E. Paloma, R. Allikmets, C. Ayuso, T. del Rio, M. Dean, L. Vilageliu, R. Gonzàlez-Duarte, and S. Balcells, 1998, Retinitis pigmentosa caused by a homozygous mutation in the Stargardt disease gene ABCR, *Nature Genetics* **18**:11–12.
48. T.L. McGee, D. Lin, E.L. Berson, and T.P. Dryja, 1994, Defects in the rod cGMP-gated channel gene in patients with retinitis pigmentosa, *Invest. Ophthalmol. Vis. Sci. (Suppl.)* **35**:1716.
49. T.P. Dryja, J.T. Finn, Y.W. Peng, T.L. McGee, E.L. Berson, and K.W. Yau, 1995, Mutations in the gene encoding the alpha subunit of the rod cGMP-gated channel in autosomal recessive retinitis pigmentosa, *Proc. Natl. Acad. Sci. USA.* **92**:10177–10181.

Section VI

RETINAL PROSTHESIS AS A REPLACEMENT FOR PHOTORECEPTORS

43

DEVELOPMENT OF AN EPIRETINAL ELECTRONIC VISUAL PROSTHESIS

The Harvard Medical School—Massachusetts Institute of Technology Research Program

Joseph F. Rizzo, John Loewenstein, and John Wyatt

Massachusetts Eye and Ear Infirmary
243 Charles St.
Boston, Massachusetts 02114

SUMMARY

Diseases of the retina and optic nerve are common causes of irreversible blindness. Given the lack of effective treatments, several laboratories are utilizing microelectronic technology to develop either a cortical or retinal prosthesis. Each strategy offers certain advantages, but both face numerous and formidable challenges. Consequently, a clinically useful device of either type is still conceptual. The technological means to build prostheses are available but the ultimate obstacle is the integration of the technology with the brain. This review primarily focuses on our efforts to develop a retinal prosthesis. In particular we address the two problems that we believe to be most challenging: 1) need to demonstrate that retinal stimulation can produce "useful" vision in a blind patient, and 2) need to demonstrate long-term biocomopatibility of an implanted device in an animal.

Brindley produced crude visual sensations by stimulating the visual cortex in a blind patient 32 years ago.[1] This early success coupled with the broad therapeutic potential of a cortical prosthesis to treat blindness caused by damage of either the retina or optic nerve has made this pursuit attractive to many scientists. Developing a cortical prosthesis that remains functional and biocompatible for prolonged periods of time is a difficult task and progress has not surprisingly been slow.[2] The convoluted surface of the mobile brain and the need for penetration of the subsurface layer of the visual cortex increase the difficulty of maintaining a stable interface. The potential risks of neurosurgery are also a formidable if not simply a psychological barrier for patients.

Retinal Degenerative Diseases and Experimental Therapy, edited by Hollyfield *et al.*
Kluwer Academic / Plenum Publishers, New York, 1999.

Although few patients have been tested, the research teams have concentrated on improving the electrode arrays and other tools that will hopefully improve results in the future. Alternative sites for a visual prosthesis include the lateral geniculate body, which has the disadvantage of being located deep within the brain which creates even greater surgical risk, and the optic nerve. The latter is more accessible but the density of axons would make it difficult to activate them in small clusters that might potentially generate focal perceptions of light.

Partly because of these considerations nearly ten years ago we initiated an effort to develop a retinal prosthesis. Our interest in working more peripherally was stimulated in part by the occurrence of relatively common causes of blindness that spare the retinal ganglion cells,[2] by the increasing sophistication and miniaturization of microelectronics, and by the success of cochlear implants.[3] Working within the eye, however, is a formidable challenge and there are numerous limitations. Compared to the inner ear, the eye is prone to chronic inflammatory reactions, its nerve tissue (i.e. the retina) can relatively easily detach and become non-functional and the potential for a devastating infection make it untenable to have wires that penetrate the eye. Mechanical stabilization of a prosthesis on the retinal surface is difficult and fraught with potential danger to the retina. There is also the added possibility that epiretinal implantation may stimulate formation of surface membranes that could dislodge the device and create a barrier of high electrical resistance between the electrodes and the cells we intend to stimulate. Lastly, many blinding diseases destroy the output (retinal ganglion) cells or their axons, which would preclude use of a retinal prosthesis and therefore reduce the number of potential patients.

Nevertheless we believe the quest for a retinal prosthesis is justifiable. Two major blinding diseases, retinitis pigmentosa and macular degeneration, currently have little effective treatment. In both diseases there is evidence of relative preservation of retinal ganglion cells,[4,5] at least until the end-stage in the case of retinitis pigmentosa. There are large numbers of patients with these diseases who could potentially benefit from a retinal prosthesis, which unlike other forms of developing therapies might be able to restore function rather than halting or slowing progression.

1. RATIONALE FOR AN EPIRETINAL PROSTHESIS

Our choice to pursue an epiretinal rather than subretinal prosthesis is based partly on several compelling facts. The diseases we hope to treat relatively spare output cells that are easily accessible with standard surgical approaches. The ganglion cells lie close to the inner surface of the retina where they could be easily stimulated, and the topographic relationship of the ganglion cells to the underlying photoreceptors (and therefore to visual field coordinates) is reasonably straightforward. The latter feature provides hope that an array of stimulating electrodes might create visual perceptions that correspond to the pattern of delivered stimulation.

On the other hand there are theoretical benefits to a subretinal prosthesis.[6,7] Stimulating electrodes placed in the subretinal space would lie closer to surviving horizontal and bipolar cells, thus making it possible to more easily benefit from intraretinal neural processing. A subretinal device would also be easier to stabilize than an epiretinal device. These possible advantages are counter-balanced by concern that a foreign body in the subretinal space might stimulate the pigment epithelium to migrate into the retina and cause destructive fibrosis. A subretinal device might also impair diffusion of nourishment from the choriocapillaris and trophic influences of the pigment epithelium, and thus compromise retina that had survived the blinding insult. Research on subretinal and epiretinal devices is progressing in parallel and presently it is not possible to decide which approach might be most useful.

2. ENGINEERING CONSIDERATIONS

In designing the prosthesis we attempt to minimize the size and complexity of the implanted electronics. This serves two purposes. First, the smaller and lighter the device, the easier it will be to fixate and the better it will be tolerated in the eye. Second, modifications to the prosthesis will be easier to make if most of the active electronics are outside the eye. Our current concept is to use a miniature electronic camera to capture images (Fig. 1). The output of the camera will be used to modulate the intensity of an external laser or a radiofrequency beam delivered from a pair of glasses. Either form of transmission could be used to supply both power and imaging signal to the implanted microchip device. The laser has the advantages of wide bandwidth, low heat dissipation, good transmission through the ocular media, small size and low cost. Disadvantages include need for accurate targeting onto the implant and interruption of transmission by eyelid blinking or opaque media. Radiofrequency transmission does not require accurate targeting or clear media but would be susceptible to interference from other electronic sources.

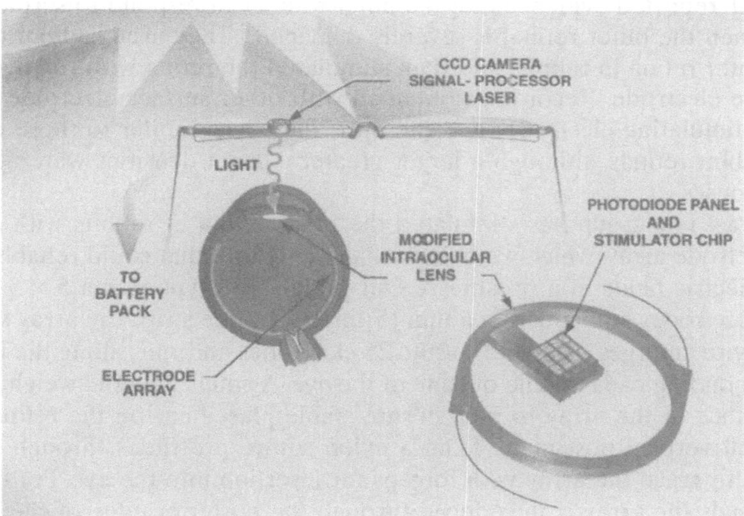

Figure 1. Patients with retinitis pigmentosa and macular degeneration become blind when the rods and cones degenerate and no longer convert incoming light into electrical impulses. While other parts of the retina remain healthy, the brain stops receiving impulses required to provide vision. The retinal prosthesis is designed to bypass the lost rods and cones and directly stimulate the surviving ganglion cells that are connected to the brain through the optic nerve. Specially designed glasses capture the visual scene and transmit this information into the eye through an invisible laser beam. The laser strikes a solar panel (photodiode array) located within the pupil to generate internal power. An ultra-thin electrode array carries the power to the retinal surface where it stimulates the ganglion cells.

If a laser is used, the internal device will have a photodiode array to convert laser light into electrical current. Microelectronics within the implant will control the distribution of current to stimulating electrodes on the undersurface of the device. The internal components will function only in response to external signals that will dictate strength, duration, polarity, timing, and location of stimulation to the retina. Adopting this design strategy reduces the number of internal components. This not only reduces the size of the internal device but reduces the likelihood that electronic failure will render the prosthesis non-functional. Our system design would also permit updated stimulation paradigms to be easily installed externally, affording the opportunity for external optimization based on experience and individual patient response.

The engineering limit for information transfer with this system is the restiction of speed based upon the power demand of the intraocular stimulator chip. The processing speed of microelectronic circuits grows linearly with its power consumption at any given operating voltage. The possibility of tissue damage places a limit on the power of the laser aimed at the internal chip. The design of a highly efficient photodiode array and low-power stimulator chip are significant engineering challenges.

3. ANIMAL MODELS

Cortical responses to local inner surface electrical stimulation of the retina have previously been reported in experiments on cats.[8] The cortical responses were similar to visually evoked potentials in the same animals, except for shorter latencies, which may be evidence that inner retinal cells are responding to the electrical stimulus. Humayun et al. reported evidence that the inner retina can respond to surface stimulation even when the outer retinal is severely damaged.[9] They used sodium iodate to destroy the outer retina in rabbits and then stimulated the retina with relatively large bipolar surface electrode. Recordings obtained with other surface electrodes located between the stimulating electrodes and the optic disc were similar to those obtained from intact rabbit retinas, although a larger greater current densities were needed to drive the responses.

By contrast our group has stimulated the inner retina of rabbits with a microfabricated electrode array, which we believe is the only tool that could reliably deliver highly focal electric fields in a prosthesis. Our initial prototype was a 5×5 array of bipolar gold electrodes embedded in a thin (5 μm) polyimide strip. The array spans 1×1 mm. A thin wire emerges from each of the 25 electrodes and runs along the length of the ribbon to bonding pads that lie outside of the eye. A small, flat gold weight is glued to the top surface of the array to help ensure stable placement on the retina during testing. A small vertical post made from a nylon suture protrudes through the gold weight is used to grasp the array with forceps for insertion into the eye. For electrical testing in animals the array is introduced through the posterior edge of the pars plicatta and placed on the retinal surface. The insertion is observed through the dilated pupil with an operating microscope. Coaxial illumination of the microscope is sufficient to visualize the retina in the rabbit eye, so endoillumination is unnecessary. If the array does not remain in contact with the retina, viscoelastic can be injected over it with a cannula to force it down into position.

Using this arrangement, we have recorded electrically evoked cortical potentials (EECP) from extra-dural electrodes over the visual cortex of 39 Dutch-belted rabbits. Surface electrical stimulation is delivered through the array, and computer-averaged extradural recordings of the activity over the visual cortex are made. Typically responses to 100 consecutive stimulations are computer-averaged. Electroretinograms are also recorded in all animals. Electrical stimulation of the retina was performed using either monopolar (with distant return) or bipolar electrode configuration. A switchbox outside the eye with wire traces to each stimulation electrode permitted selection of active electrodes without altering the position of the array. Responses to 100–500 consecutive stimulations are computer averaged.

Our EECP waveforms are similar to those elicited by light although as expected the latency is shorter (since presumably surface stimulation bypasses a large proportion of the relatively slow intraretinal synapses). Verification that "cortical" responses reflected brain activity rather than passively conducted potentials from the eye was made by recording over the brain before and after intravitreal injection of tetrodotoxin, which disrupts neural connection between the eye and brain. The lowest current threshold with monopolar stimulation was found with 75 μA and 200 μsec pulses, which yielded a 45 μV cortical response. For bipolar stimulation using four positive and four negative electrodes, the lowest current threshold was obtained with 100 μA and 100 μsec pulses, which yielded a 87 μV response. Increasing stimulus duration for a given current gave larger cortical responses, but this effect seemed to show saturation with very high charge (>90 mC). Over the 0.5 to 16 Hz stimulus frequencies tested, ERG responses varied relatively little but both evoked potential responses showed a significant monotonic decline in amplitude with higher frequencies. EECP amplitudes varied significantly with the orientation of stimulating electrodes on the retina, although in the three animals in which this effect was studied there was not a constant relationship between the orientation of the stimulating electrodes and the orientation of the nerve fiber layer. Vertical displacement of the stimulating electrode array above the retina produced a linear decline in evoked cortical amplitudes.

Maximizing evoked cortical responses will facilitate testing of the effectiveness and biocompatibility of a prosthesis since larger amplitudes make it easier to detect small declines in the induced cortical activity. The electrical charge needed to reliably produce EECP recordings in the rabbit is within the output capability of our photodiode array. Hence we are now prepared to begin testing of the efficacy of a prosthesis implanted in a rabbit eye.

4. HUMAN TESTING

Although animal work has demonstrated cortical responses to electrical stimulation from microelectrodes, we cannot gauge the perceptual quality an array can produce. Vision consists of many qualitative aspects, including color, contrast, form, and motion. The most basic of these elements is spatial detail. Visual prostheses are based on the premise of generating an image with multiple stimulation points. The anatomy of the retina poses several problems that might limit our ability to produce perceptions with useful spatial detail.

An epiretinal prosthesis will lie over both ganglion cell axons and cell bodies. The goal of our prosthesis is to stimulate ganglion cell bodies, because of their topographic relationship to the visual field. We would hope that a given microelectrode would produce very localized excitation and initiate action potentials within a small cluster of neighboring cell bodies. Ganglion cell axons, however, lie closer to the stimulating electrodes and are more numerous in a given area than ganglion cell bodies. If the electrical stimuli excite axons, which originate from a wide area of ganglion cell bodies, diffuse rather than focal images may be produced by individual electrodes.

It may be possible to design electrodes and stimulation strategies to favor activation of cell bodies rather than axons. Andrew Grumet of our group has studied models of the interaction of electrical fields with geometrically simplified axons and ganglion cell bodies. The likelihood of initiating an action potential is directly related to the depolarizing voltage induced across the cell membrane. An electric field (i.e. voltage gradient) perpendicular to an axon (modeled as a thin cylinder) leaves all points within that axon at nearly equal electrical potential. The electric field would therefore have to be very large to create sufficient depolarization to generate an action potential. An electrical field parallel to the axon, however, produces potential gradients over a much longer length, creating a greater depolarizing voltage. In contrast, the electrical field will have a similar effect on a small, spherical ganglion cell body regardless of its orientation. We may be able to exploit this strategy for selective stimulation.

Humayun et al. performed acute stimulation experiments that suggests focal electrical stimulation in the human eye may produce localized perceptions.[5] Their subjects had little or no light perception due to a variety of conditions expected to damage mainly the outer retina. They used hand-held bipolar electrodes positioned just above the inner retinal surface in the vitreous. All subjects reported perception of small spots of light in response to stimulation. This is very encouraging with regard to the viability of the inner retina in such cases, and in regard to the focal nature of the perception produced.

Encouraged by these human tests and the results of our surface contact stimulation experiments in rabbits, our group has begun human trials. A logical next step is to determine if subjects can reproducibly perceive patterns in response to stimulation by a larger number of smaller electrodes, which would presumably be required to generate spatially detailed perceptions. We using a microfabricated array that contains 100 electrodes distributed over a 2×2 mm area, employing techniques we developed in rabbits to work safely on the retinal surface.

5. CONCLUSIONS

This review describes selected aspects of our research effort to develop an epiretinal prosthesis. Our primary focus is to perform additional trials in humans and test the biocompatibility and efficacy of a chronically implanted device in an animals. Both of these are necessary "proof-of concept" endeavors. Success must be achieved in both realms prior to considering long-term implants in humans.

Research programs to develop retinal and cortical visual prostheses are progressing in parallel, and it is too early to determine if either will provide therapeutic benefit. The cortical projects are more mature and their experiences and successes provide a path for the younger retinal projects to follow. Collectively our groups must still demonstrate that blind humans can experience perceptions with enough geometric detail to be deemed "useful", perhaps if only to provide more independent mobility. Success lasting only a few hours, however, must not be taken as evidence that vision could be sustained for months or ultimately years, which would be required if a prosthesis were to be used clinically. We believe that long-term biocompatibility is the major impediment to this emerging field of research.

REFERENCES

1. Brindley G and Lewin W. The sensations produced by electrical stimulation of the visual cortex. *J. Physiol (Lond)*. 1968;196:479–493.
2. Rizzo JF Wyatt J. Prospects for a visual prosthesis. *Neuroscientist*. 1997;3:251–262.
3. Balkany T, Hodges A, and Luntz M. Update on cochlear implantation. *Otolaryngol Clin North Am*. 1996;29:277–289.
4. Stone J Barlow W. Morphometric analysis of macular photoreceptors and ganglion cells in retinas with retinitis pigmentosa. *Arch Ophthalmol*. 1992;110:1634–1639.
5. Humayun M, de Juan E, Dagnelie G, Greenberg RJ, Propst RH, and Phillips DH. Visual perception elicited by electrical stimulation of retina in blind humans. *Arch Ophthalmol*. 1996;114:40–46.
6. Chow A Chow V. Subretinal electrical stimulation of the rabbit retina. *Neuroscience Letters*. 1997;225:13–16.
7. Zrenner E, Miliczek K, Gabal V, et al. The development of subretinal microphotodiodes for replacement of degenerated photoreceptors. *Ophthalmic Res*. 1997;29:269–280.
8. Doty RW Grimm FR. Cortical responses to local electrical stimulation of retina. *Experimental Neurology*. 1962;5:319–334.
9. Humayun M, Propst R, de Juan E, McCormick K, and Hickingbotham D. Bipolar surface electrical stimulation of the vertebrate retina. *Arch Ophthalmol*. 1994;112:110–116.

44

RESPONSE CHARACTERISTICS OF SUBRETINAL MICROPHOTODIODE-BASED IMPLANT-MEDIATED CORTICAL POTENTIALS

Neal S. Peachey,[1,2] Alan Y. Chow,[3,4] Machelle T. Pardue,[1,2] Jay I. Perlman,[2,5] and Vincent Y. Chow[3]

[1] Research Service (151)
Hines VA Hospital, Hines, Illinois 60141
[2] Stritch School of Medicine
Loyola University Chicago
Maywood, Illinois 60153
[3] Optobionics Corporation
191 Palamino Place
Wheaton, Illinois 60187
[4] Louisiana State University Eye Center
2020 Gravier Street
New Orleans, Louisiana 70112
[5] Ophthalmology Section
Surgical Service
Hines VA Hospital
Hines, Illinois 60141

SUMMARY

Presently, there is no treatment by which to restore vision to patients blinded by outer retinal diseases such as retinitis pigmentosa and age-related macular degeneration. We have developed a retinal prosthetic that is designed to be placed in the subretinal space and thereby stimulate second-order retinal neurons spared by disorders that induce selective photoreceptor degeneration. The device is semiconductor-based and does not require an external power supply for operation. Previous studies have characterized the device with respect to biocompatibility and function. The present study concerns cortical potentials elicited when the implant is stimulated with infrared light, to which the native retina is insensitive. The results indicate that these potentials share several response characteristics with conventional visual evoked potentials and support the conclusion that the implant is capable of electrically activating the visual system.

1. INTRODUCTION

Degeneration of photoreceptors is known to underlie the blindness induced by disorders such as retinitis pigmentosa (RP) and age-related macular degeneration (AMD). Although there is currently no means to restore vision to patients blinded by these disorders, a number of possible therapies are presently being evaluated.

This chapter concerns an approach directed towards this goal that is grounded in semiconductor technology. The essential idea is to implant into the subretinal space a microphotodiode-based device which can respond to light with an electrical current.[1–3] In the subretinal location, the device would be in position to contact and thereby stimulate the inner retinal layers, which may be spared in RP[4] and in AMD.[5] As a consequence, the device could be used to prosthetically replace the function of the degenerated photoreceptor layer and thereby restore some degree of vision to affected patients.

There are substantial hurdles involved in the development of such a prosthetic device, including prototype fabrication and function, the development of subretinal surgical techniques, evaluation of implant biocompatibility and function *in vivo*, as well as establishing that the implant will make a functional connection with the neural retina. Many of these issues have been addressed in our previous work in which implants were placed in the subretinal space of normal adult rabbits[6] or cats.[7] These studies demonstrated (*i*) that implants can be placed in a stable location within the subretinal space; (*ii*) that these devices will respond electrically to incident light with a response that is graded with intensity; (*iii*) that implants will continue to function within the subretinal space for many months without any external power supply; (*iv*) that retinal locations away from the implant continue to respond normally to white light and retain a normal histologic appearance; (*v*) that the photoreceptor layer overlying the implant was lost, possibly due to the solid disk implant blocking choroidal nourishment to the outer retina; and (*vi*) that the inner retinal layers overlying the implant were relatively spared, and to a greater degree in the cat model which possesses a dual retinal circulation while the rabbit retina receives circulation only from the choroid. In addition, in the cat model, infrared stimulation of the implant evoked a slow cortical potential in each cat where the surgery was successful.[7] The work described in this chapter is concerned with a series of studies that were carried out to further define the response characteristics of this cortical potential.

2. MATERIALS AND METHODS

2.1. Implant Design

Figure 1 presents photographs of the type of device used here. At low power (Fig. 1A), the implant appears as a thin disk which is 2.5 mm in diameter, ~50 μm thick, and incorporates no external connections. At high power (Fig. 1B), it is possible to discern

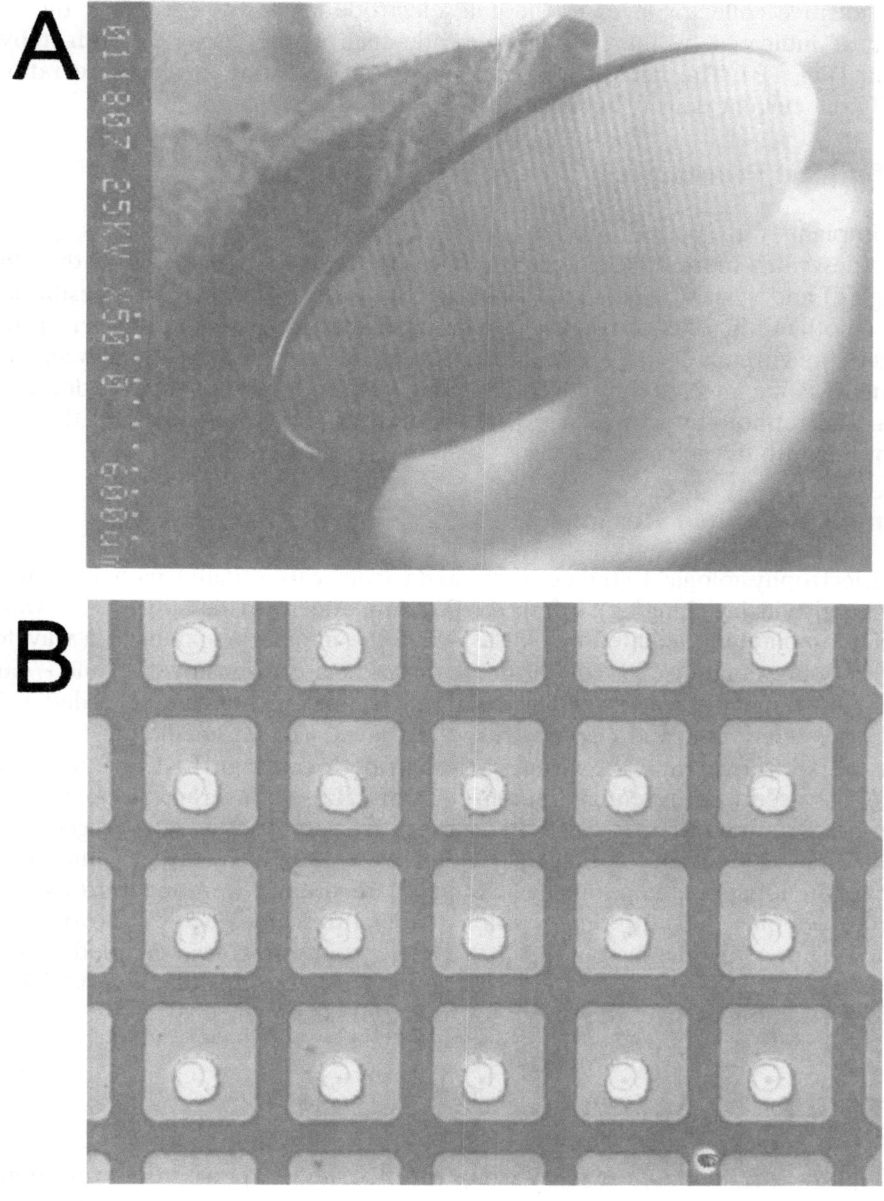

Figure 1. Implants photographed at low power (A) and at high power (B). The individual microphotodiode subunits that comprise the 2.5 mm diameter implant are clearly seen in the high power image.

the array of individual photodiode subunits that comprise the device. Each of these are 20 μm by 20 μm, and are separated on all sides by a 10 μm gap. In cross-section, the device is comprised primarily of silicon that has been doped using standard semiconductor techniques. The orientation of these doped layers with light stimuli determines whether the device operates in the negative-intrinsic-positive (NiP) or positive-intrinsic-negative (PiN) configuration. The electrode is composed of a thin layer of gold bound to the silicon layer by titanium. In previous work[6,7] the ratio of the photosensitive collector layer to the gold electrode layer was 1:1. Here, the ratio is higher, as indicated by the smaller electrode seen at the center of each individual subunit (Fig. 1B). The net result of this increased collector to electrode ratio is to amplify the current density achieved at the electrode surface.

2.2. Surgical Procedures

Implants were placed into the subretinal space of normal adult cats, using techniques described more fully elsewhere.[7] In brief, animals were sedated with ketamine (11 mg/kg) and xylazine (2 mg/kg), and brought to a surgical plane of anesthesia with 0.9% halothane in 100% oxygen. An incision made in the temporal sclera provided access to the vitreous cavity. A retinotomy was made near the *area centralis* and a small amount of saline injected through this opening was used to form a bleb underneath the retina. The retinotomy was then expanded, and the implant was placed through the retinotomy into the subretinal space.

2.3. Implant-Evoked Responses

Electrophysiological studies were made from cats sedated with ketamine (11 mg/kg) and xylazine (2 mg/kg). Infrared (IR) stimulation provided by an LED was used to isolate implanted-mediated responses; while the implant responds to a wavelength range of approximately 500 to 1100 nm, the cat retina is insensitive beyond ~750 nm.[8] Implant-mediated responses produced at the retinal level were recorded using a contact lens electrode, and also at the cortical level, using a needle electrode placed under the scalp overlying the visual cortex. In both cases, an LKC signal averaging system was used to differentially amplify (1–1500 Hz) and average responses evoked by the presentation of an IR pulse; 400 consecutive responses were averaged together in each trial. Retinal and cortical potentials were usually recorded simultaneously. However, in a mapping experiment (see Fig. 5), recordings were made from an array of electrodes placed at different positions under the scalp. Since only two recording channels were available, the entire set of electrode positions was sampled by recording from two leads in a given trial and alternating the electrode input across trials. Each electrode position was recorded from for at least 3 trials.

3. RESULTS

Figure 2 presents representative implant-mediated responses that were obtained in response to the presentation of a 200 msec IR stimulus (middle tracing). At the retinal level (lower tracings) there is a clear implant spike at stimulus onset which has a negative polarity, indicating that the implant is an NiP device. At stimulus offset, the implant generates a smaller spike of opposite polarity. At the cortical level (upper tracings), it is possible to see the implant spike as a far-field potential. In addition, there is a distinct slow wave that peaks approximately 100 msec after stimulus onset or offset (arrows).

Figure 3 presents retinal (lower three tracings) and cortical (upper three tracings) recordings made at three different stimulus intensities. The amplitude of the implant spike recorded at the retinal level decreased slightly as the flash was decreased from the intensity usually used (top retinal tracing) to half of that level (middle retinal

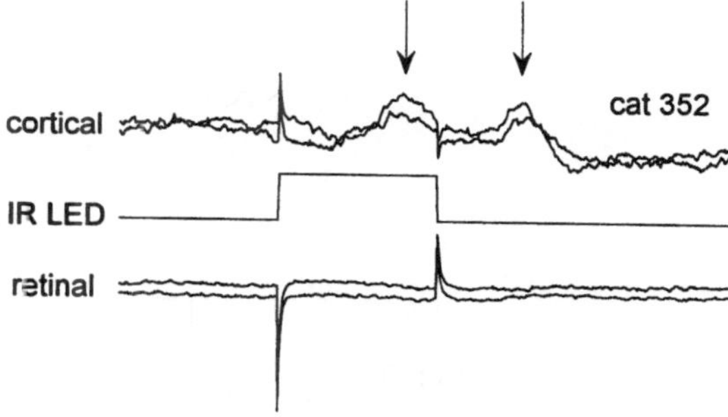

Figure 2. Implant-evoked electrophysiological responses, recorded at the retinal (lower tracing) and cortical (upper tracing) levels. Two repeats are shown for each condition. The tracing labeled 'IR LED' indicates the time of presentation of a 200 msec IR stimulus; the amplitude of this trace represents 10 μV for the retinal recordings and 5 μV for the cortical responses. Cat 352; 6 weeks post-implantation.

tracing). This reduction was accompanied by a modest reduction in cortical response amplitude (compare upper two cortical tracings). An additional two-fold decrease in stimulus intensity reduced the amplitude of both the implant spike (lower retinal tracing) and the cortical response (lower cortical tracing), and also delayed the peak time of the cortical response. These changes were verified in multiple recordings.

Figure 4 presents a series of cortical responses that were obtained to IR LED stimuli that ranged in duration from 10 to 100 msec. A clear cortical response was evoked by the 100 msec stimulus. As the duration was shortened to 50 and 33 msec, cortical responses were also obtained. In Figure 4, the tracings are aligned so that the LED offsets coincide. Therefore, the observation that the peak of the cortical responses stay in general alignment indicates that these responses are time-locked to stimulus offset. In addition, it is interesting to note that the amplitude of the response declined with shorter stimulus duration. In fact, the cortical response evoked by a 10 msec duration flash approached the baseline noise.

To better define the generators of the cortical response, an array of eight electrodes was positioned over the scalp. Figure 5 presents the responses recorded from this array. Position 4 overlies the cat visual cortex and is the electrode location used to routinely follow implanted cats.[7] The other electrodes are located at varying distances from this location. It is clear that the amplitude of the cortical response declines for electrode positions away from the visual cortex, and that this decline appears greater

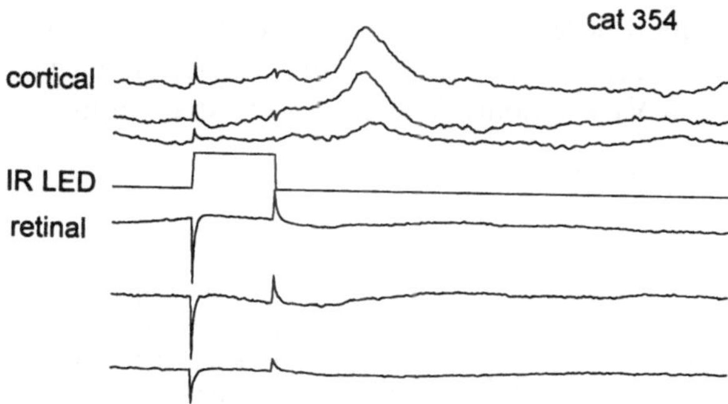

Figure 3. Relationship between stimulus intensity and implant-mediated response amplitude. Shown are cortical (upper three tracings) and retinal (lower three tracings) recordings made to the presentation of a 100 msec IR LED stimulus (middle tracing). The stimulus intensity was maximal for the upper of each set, and was decreased by a factor of two for each of the lower two records. The amplitude of the stimulus trace indicates 10 μV for the retinal recordings and 5 μV for the cortical responses. Cat 354; 3.5 months post-implantation.

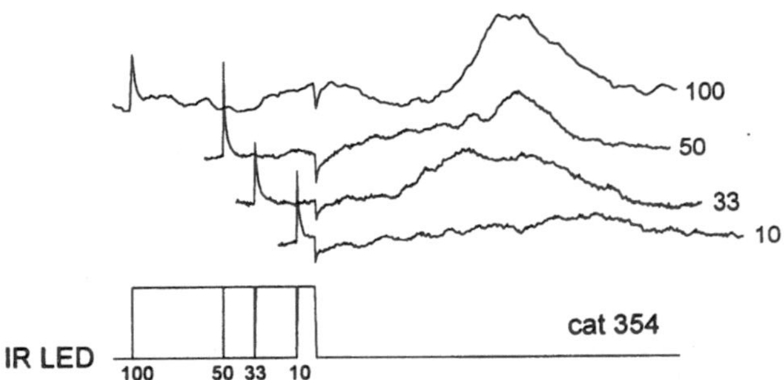

Figure 4. Duration dependence of implant-mediated cortical response. Shown are cortical recordings made to IR LED stimuli of 10, 33, 50 or 100 msec duration. The small number to the right of each tracing indicates the stimulus duration used to evoked that response. The small numbers on the lower tracing indicate the time of onset for each duration LED stimulus; the height of this record represents 5 μV. Cat 354; 3 months post-implantation.

for more distant electrode positions. In addition, these data indicate that the implant spike has a scalp distribution which differs substantially from that for the slow cortical response. Specifically, the implant spike is greatest at electrode position 6, which is located nearest the implanted eye. This spike diminishes at electrode positions of progressively greater distance from this location.

4. DISCUSSION

The electrophysiological results presented here indicate several important features of the cortical responses evoked by IR stimulation of the subretinal implant. Taken together, the results provide an initial basis with which to compare this response with the conventional visual evoked potential (VEP) that has been studied in a variety of species. First, the response is largely evoked by flash offset. This feature is in keeping with the observation that a visible light LED pulse (i.e., green or red) presented to cats under these conditions evokes a VEP which is dominated by the response to LED offset. Second, the amplitude of the response changes with stimulus intensity, at least over the range examined here. This is consistent with an extensive literature demonstrating that, within a several log unit operating range, increasing stimulus intensity leads to an increase in VEP amplitude.[9] Finally, the maximum response maps to the

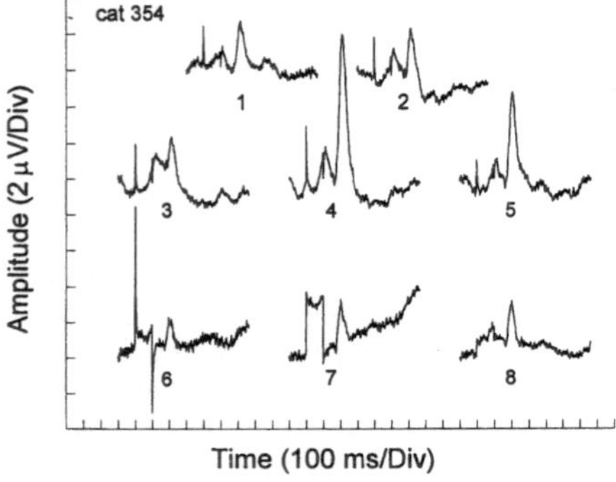

Figure 5. Cortical mapping of implant-mediated cortical response. The number below each waveform indicates the electrode position used to obtained that tracing. Electrode position 4 corresponds to the scalp overlying the cat visual cortex; other electrode positions were posterior (1, 2), lateral (3, 5) or anterior (6, 7, 8) to that location. The distances from electrode position 4 was 3 cm (1, 2), 2 cm (3, 5), and 4 cm (6, 7, 8), in approximately the orientation shown by the waveform placement. Stimulation was provided by a 100 msec presentation of an IR-emitting LED. Cat 354; 3.5 months post-implantation.

scalp overlying the visual cortex. This distribution indicates that this location most probably overlies the generators of this response and provides the strongest indication to date that the implant-mediated cortical potentials actually reflect activity of the visual system, and support the conclusion that the implant is capable of establishing a functional connection with the neural retina.

Histological data are not available from the animals described here because we are still in the process of following them post-operatively. However, previous work has found that the implants are well-tolerated by the retina, and there is no reason to expect that to be different here. In fact, examination of the implanted eye did not reveal funduscopic abnormalities beyond that present at the surgical site, and the electroretinogram of the implanted eye resembled that of the unoperated control eye in all respects. The overall healthy appearance of the retina resembled closely that reported previously.[7]

Although cortical potentials have been obtained from every cat with a successful surgery, in each case this potential is lost at about 3–5 months post-operatively. At present, the reason for this loss is unclear. One possibility that is addressed by the present data concerns the materials used for the bonding layer. Our earlier studies[7] used a chromium bonding layer, which raises the possibility that the loss of the cortical potentials might be caused by leaching of neurotoxic chromium ions from the implant into the retina. However, since the bonding layer of the implants used here was comprised of titanium, this factor alone cannot account for the disappearance of the implant-evoked cortical responses. Addressing alternative explanations for this phenomenon is an important goal of our present work, and we have made several modifications in implant design which will allow us to address these possibilities in the future.

ACKNOWLEDGMENTS

Supported by the Department of Veterans Affairs, NIH grant R43 EY1182, and by the MidWest Eye Banks and Transplantation Center.

REFERENCES

1. Chow, A.Y, 1991, Artificial retina device. United States Patent #5,016,633, issued May 21, 1991.
2. Chow, A.Y, 1991, Artificial retina device. United States Patent #5,024,223, issued June 18, 1991.
3. Chow, A.Y. & Chow, V, 1995, Independent photoelectric artificial retina device and method using same. United States Patent #5,397,350, issued March 14, 1995.
4. Santos, A., Humayun, M.S., de Juan, E. Jr., Greenberg, R.J., Marsh, M.J., Klock, I.B., & Milam, A.H, 1997, Preservation of the inner retina in retinitis pigmentosa. A morphometric analysis. *Archives of Ophthalmology* 115, 511–515.
5. Curcio, C.A., Medeiros, N.E., & Millican, C.L, 1996, Photoreceptor loss in age-related macular degeneration. *Investigative Ophthalmology & Visual Science* 37, 1236–1249.
6. Peyman, G.A., Chow, A.Y., Liang, C., Chow, V.Y., Perlman, J.I., & Peachey, N.S, 1998, Subretinal semiconductor microphotodiode array. *Ophthalmic Surgery & Lasers* 29, 234–241.
7. Chow, A.C., Pardue, M.T., Chow, V.Y., Perlman, J.I., & Peachey, N.S, 1998, Implantation of semiconductor-based photodiodes into the cat subretinal space. *Vision Research*, submitted.
8. Schneider, T., Olsen, B.T., & Zrenner E, 1986, Characteristics of the rod-cone transition in electroretinogram and optic nerve response. *Clinical Vision Sciences* 1, 81–91.
9. Brigell, M.G. & Marchese, A.L, 1993, The luminance-amplitude function of simultaneously recorded flash evoked potential and electroretinogram. *Clinical Vision Sciences* 8, 41–46.

45

ELECTRICAL STIMULATION OF THE HUMAN RETINA

Mark S. Humayun, James D. Weiland, and Eugene de Juan Jr.

Wilmer Ophthalmological Institute
Johns Hopkins University
Baltimore, Maryland

1. INTRODUCTION

The visual prosthesis being developed will provide useful vision to people blind because of photoreceptor loss due to retinal degenerative diseases such as age related macular degeneration (AMD) and retinitis pigmentosa (RP). A retinal prosthesis will replace photoreceptor function with an electronic device. In a healthy retina the photoreceptors initiate a neural signal in response to light. Photoreceptors are almost completely absent in the retina of end-stage RP patients and in the macular region of AMD patients. However, cells to which photoreceptors normally synapse, (i.e. bipolar cells) survive at high rates. In blind volunteers, we have shown that controlled electrical signals applied with a microelectrode positioned near the retina elicit the perception of a spot of light that correlates both spatially and temporally to the applied stimulus.[1,2] When multiple electrodes are activated in a two dimensional electrode array, a number of small spots of light are perceived by the patient(s) which when viewed together form an image representative of the pattern of active electrodes. Thus similar to how an image is formed by a dot-matrix printer, when controlled pattern electrical stimulation of the remaining retinal neurons is coupled with an extraocular image acquisition and transmission system, it could allow blind patients to regain form vision.

2. BACKGROUND

An illustration of a prototype retinal prosthesis system is shown below (Fig. 1). A video camera worn in a glass frame will capture an image and convert it to an electric signal. The electric signal will be appropriately coded and then sent by a radio frequency transmitting coil to a receiver coil implanted in the eye. Receiver electron-

Retinal Degenerative Diseases and Experimental Therapy, edited by Hollyfield *et al.*
Kluwer Academic / Plenum Publishers, New York, 1999.

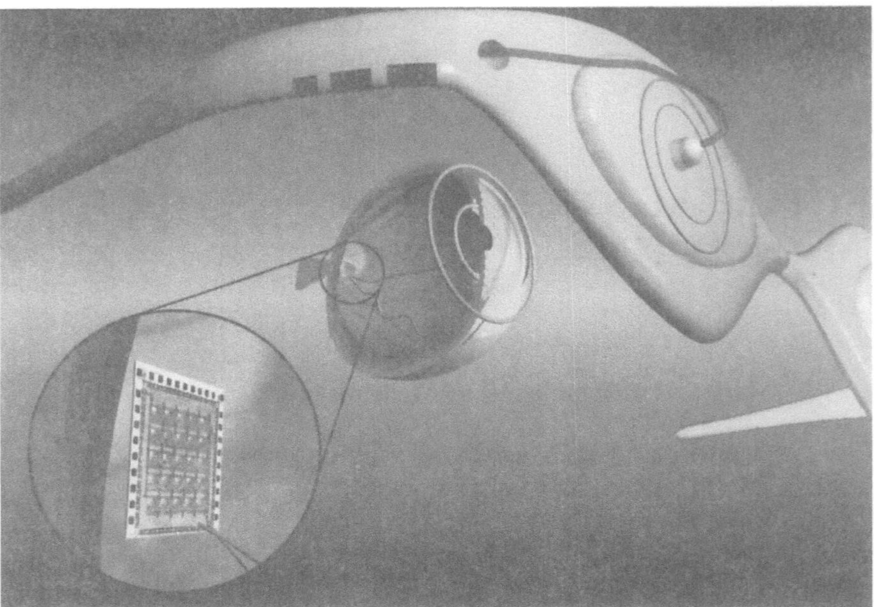

Figure 1. Retinal Prosthesis Concept. The external components (camera, video processor, power, and data transmitter) will be mounted on a glasses frame. The implanted components would include a receiver coil and the intraocular electronics, positioned over the retina, to decode the received signal and produce the appropriate pattern of electrical stimulus at the electrode array.

ics will decode the signal and stimulation electronics will generate the desired current pulse pattern for the stimulating array. The engineering of the system is considered feasible with available technology, much of which is used in today's cochlear implants. The clinical studies discussed below have begun to address the question of whether or not such a device will produce a usable image in a blind individual. While a final answer will not be available until someone is implanted with a retinal prosthesis, animal and human experiments to date have produced encouraging results.

The retina is a complex neural system consisting of several cell layers. The image signal that is produced by the photoreceptors is processed significantly by the remaining layers of the retina, before the final signal is sent to higher neural centers. However, in RP and AMD patients the photoreceptor layer is severely damaged. The success of the retinal prosthesis is predicated on the survival of other retinal cells in spite of almost complete photoreceptor degeneration. Post-mortem studies of human eyes of RP patients show significant survival of both inner nuclear layer cells and ganglion cells.[3,4] This identifies both the retinal ganglion cell (RGC) and the bipolar cell (BC) as possible targets for an electrical stimulation pulse. Additionally, both of these cell types have several locations where an action potential can be initiated (e.g. dendritic trees, soma, axon). The location of the signal initiation will be influenced by at least two factors: 1) the relative stimulation thresholds of the various cells/cell structures and 2) the location of the stimulating electrode in relation to the neural structures. The source of the electrical pulse, the stimulating electrode, in our prosthesis will lie directly on the inner limiting membrane. Intuitively, one might think that this positioning makes the RGC axons the most easily stimulated of all the neural elements, since the RGC axons would be closest to the stimulating electrode. The experiments conducted in our lab and by others run contrary to that conclusion. In fact, our experiments in isolated retina have shown that by modifying the pulse parameters, it is possible to selectively stimulate bipolar cells over ganglion cell soma or axons, even though the ganglion cell structures are closer to the stimulating electrode.[5,6] These results agree with the "size principal" (i.e. larger cells have lower thresholds to electrical stimulation), which has been well-established in electrophysiological studies of other systems.[7]

Despite the results that suggest we can selectively stimulate bipolar cells, using an electronic device to restore the sense of vision, produced by signals from 100 million photoreceptors, remains a seemingly impossible task. However, a similar task has already been accomplished in the auditory system. The cochlear implant electrically stimulates the auditory nerve, bypassing damaged cochlear hair cells, to produce the sensation of sound. Using only six electrical inputs to the auditory nerve, which contains approximately 30,000 fibers, the cochlear implant can produce sound perception with sufficient fidelity to enable a deaf individual to speak on an ordinary telephone. Additionally, cochlear implant patients improve their ability to hear with practice. This suggests that the brain can adapt to the reduced input and, with training, restore significant sensory function. Research in the visual system has shown that reducing visual input to a 25×25 array of pixels still allows a good degree of mobility.[8] These are encouraging examples when one considers that a retinal implant will provide input to a system designed to process information from 1 million ganglion cells.

Electrical stimulation of the human visual system can produce the perception of vision (as opposed to tactile or pain sensation). Stimulation of the visual cortex has been shown to produce phosphenes.[9,10] In our lab, human volunteers are evaluated prior to surgery with electrical stimulation of the globe with a contact lens electrode. The subjects have been able to "see flashes" when the electrical current is applied. These results make an argument that electrical stimulation can result in excitation of the visual system, but the extraocular positioning of the stimulating electrode significantly limits the available acuity of electrically stimulated vision. Using an intraocular probe, our group has run experiments (to be discussed in more detail below) that provides direct evidence that electrical stimulation of a non-seeing retina can provide usable vision.

3. METHODS

Human volunteers (13 end stage RP, 2 AMD) were tested to assess the perception experienced when a small metal electrode, in close proximity to the retina, was used to stimulate the cells of the retina. Two types of electrodes were used. A device consisting of platinum wires attached to a hand piece (Fig. 2) was used to determine the stimulation threshold, flicker fusion, and three-point discrimination. A multielectrode array of platinum disks arranged in a silicone substrate (Fig. 3) was used to stimulate in patterns on the retina. Protocols established for safe stimulation of neural tissue were followed.[11]

The stimulating probe and an endoilluminator were inserted into the subject's vitreous cavity through small sclerotomies. In order not to compromise the optic nerve function, 4% Lidocaine anesthesia was injected intrasclerally only at the site of the intended sclerotomy. A vitrectomy was performed only in the tests with the larger,

Figure 2. Discrete wire electrode. A photomicrograph of the tip of a compound electrode array is shown. Three concentric electrode sites of different sizes are visible. The electrode material is platinum with the center mantle diameters of the 2 large electrodes sites measuring 100 μm and the smaller mantle diameter being 50 μm. Also housed in with the electrodes is a fiberoptic sleeve, which is not visible due to the reflections off the metal electrode sites. The outer metal canula is connected to a hand piece. A cable extends from the end opposite to the electrode sites to allow connection to the electronics.

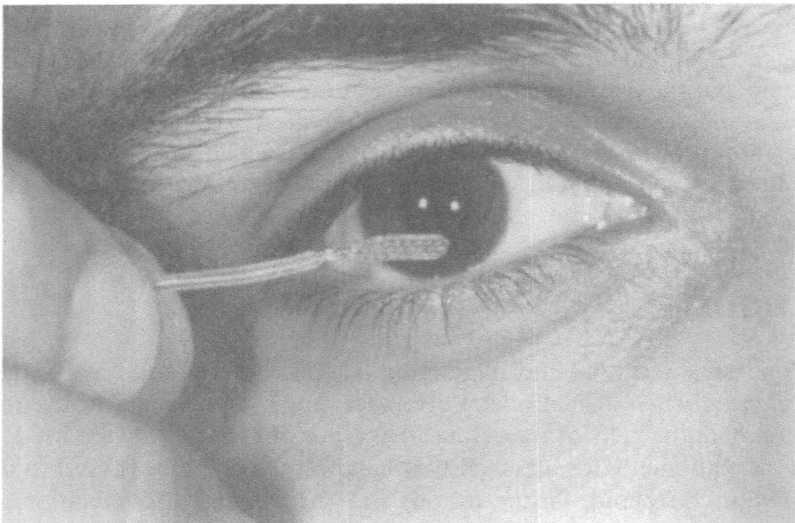

Figure 3. A multi-electrode array of platinum disks in a silicone matrix held near an eye. The array is approximately 3 × 7 mm. A cable extends from the end opposite to the electrode sites to allow connection to the electronics.

multi-electrode array. The surgical procedure, including the conversation between the subject and the experimenters, was videotaped via a camera and microphone built into the surgical microscope. The video running time was noted to correlate the subjects' description of the visual percepts with the electrical stimulation parameters. The stimulating probe was placed closely over the retinal surface, without touching (we estimate the distance at 0.5 mm) as shown in Fig. 4. Charge balanced, constant current electrical pulses were applied at a rate of 1 Hz or higher. Subjects were asked to describe the size, shape, color, and position of any visual perception. Additionally, as the rate of stimulation was increased, the subject was asked to define the point of flicker fusion (i.e. when the electrically elicited visual percept was no longer flashing). Flicker-fusion is important, as a strobe-like visual input is far less appealing than a stable non-flickering image.

4. RESULTS

All of the 15 human subjects were able to perceive phosphenes with electrical stimulation. All but one of these patients were able to localize the position of the stimulating electrode and could track the stimulus as it was moved along the retina. Stimulation threshold varied slightly, dependent on the proximity of the electrode to the retina and the condition of the retina. Flicker fusion typically occurred between 30–40 Hz. Most of the subjects reported seeing a white or yellow flashing light, though some of the subjects saw a black dot surrounded by a white or yellow ring. Patient perceptions tabulated (Table 1) and some of the more interesting results are described below.

A 61-year-old male with RP (no light perception vision) described the phosphene to be as small as the head of a match, indicating that electrical stimulation can be very

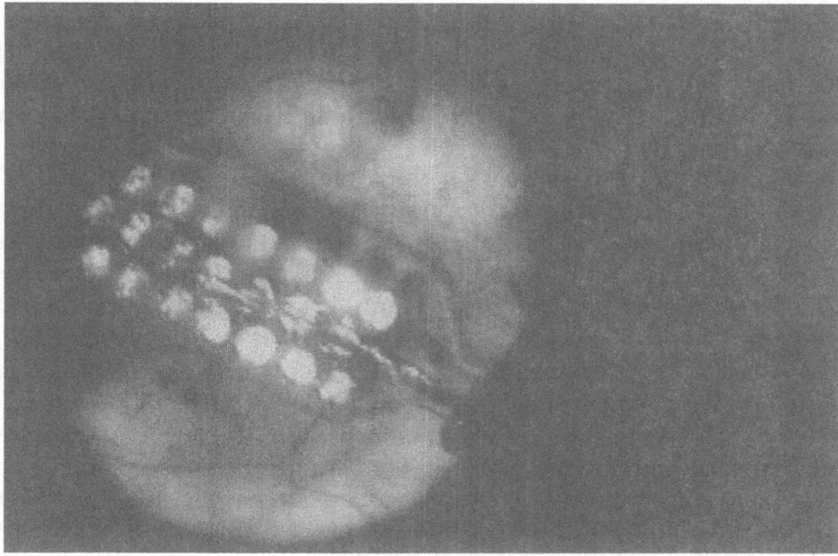

Figure 4. A multi-electrode array of platinum disks in a silicone matrix about to be placed on the retina of an RP patient. Multi-electrode probes allowed subjects to see simple shapes and letters. A cable extends from the electrode sites to allow connection to the electronics.

Table 1. Summary of clinical test results

Subject	Diagnosis	Pre-op Visual Acuity	Percept. shape	Percept. color
HC	RP	NLP	Letter H	Yellow-green
CC	RP	NLP	Match-head	Yellow
RS	RP	LP	Pin	Yellow
WG	AMD	NLP	Pencil	White
PS	RP	LP	Pea	Yellow
BC	RP	LP	Pin	Yellow
RJ	RP	LP	Pin	Yellow
BH	RP	NLP	Pin	White
AB	AMD	20/400	Pin	White
CS	RP	LP	Pin	Blue
VO	RP	LP	Pin	Yellow
HW	RP	LP	Pin	White
JT	RP	LP	Box	White

focal. He also discriminated between two points 435 µm apart, corresponding to 4/200 (crude ambulatory) vision.[1] Tests with a 5 × 5 array yielded the following results: a) when electrodes were stimulated in rows or columns the patient saw horizontal and vertical lines respectively, b) when electrodes in a 'U' like pattern were activated, the patient saw the letter H.[2] Although the patient's responses were not always correct, this result is very encouraging for a number of reasons. First, the stimulating electrode is hand-held and the eye is fully mobile resulting in movement of the electrodes relative to the retina and creating a non-uniform electrical field. Second, the tests were conducted for periods of less than 1 hour, which is a very short period to get accustomed to a new mode of visual input. From the cochlear implant experience, we know that usually it takes weeks to begin to use the artificial auditory input. Third, the choice of the letter 'U' was not optimal as it is frequently confused with the letter 'H'. This may have been especially true in our test, since our electrodes created a straight horizontal bottom for the letter 'U' rather than the slightly curved shape we are accustomed to seeing.

For this next patient, we selected a more distinct pattern than the letter 'U'. A 73-year-old female with RP (bare light perception vision) saw lights that outlined the shape of a 'matchbox' when pattern stimulation was applied through electrodes positioned on the perimeter of a square shaped electrode array. Every electrode on the perimeter of a 3 × 3 electrode array was activated to form this perception. The combination of electrodes that formed the box covered a square area of retina 1.2 mm on a side.

A 52-year-old female with RP (bare light perception vision) described the perception as a firefly; again indicating that the perception produced by electrical stimulation is focused on a small part of the visual field. Stimulation thresholds were lower for macular (central) regions and higher in the peripheral retina. In a crude approximation of a visual function test, we were able to map out her electrically elicited threshold by moving the stimulating electrode over the surface of the retina. The phosphene seen by the subject would disappear as the probe was moved peripherally, and then reappear as the probe was moved back into the macula. We believe that this difference in threshold to elicit visual perceptions indicates more damage in the peripheral retina, a result consistent with the progression of retinal degeneration in RP.

5. CONCLUSION

We have shown that electrical signals can create the perception of vision in individuals blind from photoreceptor degeneration. The fact that only single spots and shapes have been created is due to the simplicity of the device used to create the perceptions. The next task facing the retinal prosthesis project is building an electronic device with hundreds of electrodes that can be implanted for a long period of time. This work is progressing, and as more sophisticated devices become available, more human tests will be required to evaluate the ability of a degenerated retina to process a complex electrical stimuli.

REFERENCES

1. M.S. Humayun, E. de Juan, Jr., G. Dagnelie, R.J. Greenberg, R.H. Propst, and D.H. Phillips, 1996, Visual perception elicited by electrical stimulation of retina in blind humans. *Arch. Ophthalmol.* 114:40–46.
2. M.S. Humayun, E. de Juan, Jr., R.J. Greenberg, G. Dagnelie, R.S. Rader, and S.J. Katona, 1997, Electrical Stimulation of the Retina in Patients with Photoreceptor Loss, *Inves. Opthalmol. & Vis. Sci.*, 38(4):S39.
3. J.L. Stone, W.E. Barlow, M.S. Humayun, E. de Juan, Jr., and A.H. Milam, 1992, Morphometric analysis of macular photoreceptors and ganglion cells in retinas with retinitis pigmentosa, *Arch. Ophthalmol.* 110:1634–1639.
4. A. Santos, M.S. Humayun, E. de Juan, Jr., R.J. Greenberg, M.J. Marsh, and A.H. Milam, 1998, Inner retinal preservation in RP: a morphometric analysis, *Arch. Ophthalmol.*, in press.
5. R.J. Greenberg, 1997, Electrical stimulation of the vertebrate retina, Ph.D dissertation, Johns Hopkins University.
6. S.J. Katona, 1997, Excitation thresholds for the electrical stimulation of the normal and degenerate mouse retina, master's thesis, Johns Hopkins University.
7. R. Baratta, M. Ichie, S.K. Hwang, and M. Solomonow, 1989, Orderly stimulation of skeletal muscle motor units with tripolar nerve cuff electrodes, *IEEE Trans. on Biomed. Eng.*, 36(8):836–843.
8. K. Cha, K.W. Horch, and R.A. Normann, 1992, Simulation of a phosphene-based visual field: visual acuity in a pixelized vision system, *Ann Biomed. Eng.*, 20:439–449.
9. M. Bak, J.P. Girvin, F.T. Hambrecht, C.V. Kufta, G.E. Loeb, and E. M. Schmidt, 1990, Visual sensations produced by intracortical microstimulation of the human occipital cortex, *Med. & Biol. Eng. & Comput.*, 28:257–259.
10. G.S. Brindley and W.S. Lewin, 1968, The sensations produced by electrical stimulation of the visual cortex, *J Physiol (London)*, 196:479–493.
11. S.B. Brummer, L.S. Robblee, and F.T. Hambrecht, 1983, Criteria for selecting electrodes for electrical stimulation: theoretical and practical considerations, *Ann. N.Y. Acad. Sci.*, 405:159–171.

GOALS, CONCEPTS, AND CURRENT STATE OF THE RETINA IMPLANT PROJECT

EPI-RET

Rolf Eckmiller*

University of Bonn, Department of Computer Science VI
D-53117 Bonn
Germany

1. ABSTRACT

For blind subjects with retina degenerative diseases (especially: retinitis pigmentosa and macular degeneration) to regain visual perception, a team of 14 expert groups develops a partially implantable, learning visual prosthesis (retina implant). This team with experts from several biological, medical, and technological areas is supported by the German research ministry (BMBF) and is coordinated by the author. Retina implants consist of a learning retina encoder (RE)—to be mounted on a frame of glasses or embedded in a contact lens—for the approximate simulation of parts of the retina by transforming light patterns into impulse trains similar to the receptive field properties of ganglion cells, a microcontact foil as retina stimulator (RS) to be implanted adjacent to the ganglion cell layer, and a wireless signal- and energy transmission between RE and RS. The function of the various spatiotemporal filters of the RE, which is being implemented by learning neural nets, will be tuned individually in a dialog with the implant-carrying subject for optimal visual perception.

The development and successful test of retina implant prototypes in animals is expected at the end of the first 4-years research phase in 1999. In a subsequent research phase with participation from industry, the next step for adaptation of the retina implant system for application in humans and first trial tests with a small number of volunteers will follow. It is expected that implant-carrying subjects will be able to recognize position and 'gestalt' of larger objects (e.g. window, door, chair, table) based on RE and RS with about 500 microcontacts in connexion with retinal ganglion cells, and

* Email: eckmiller@nero.uni-bonn.de; URL: http:/www.nero.uni-bonn.de

Retinal Degenerative Diseases and Experimental Therapy, edited by Hollyfield *et al.*
Kluwer Academic / Plenum Publishers, New York, 1999.

that they will be able to walk and orient themselves without help in most unknown environments. This hope is partly based on recent findings that simple gestalt perceptions could already be elicited in several blind subjects by temporary microstimulation of retinal ganglion cells. Furthermore, anatomical studies have recently shown that a significant portion of the ganglion cells and the optic nerve in this group of blind subjects remain intact, even though the layer of photoreceptors is degenerated.

2. INTRODUCTION

A considerable number of visually impaired human subjects suffers from retinal degenerative defects (especially: retinitis pigmentosa, RP, and macular degeneration, MD), which in many cases of RP begins with night blindness (loss of rod photoreceptors), deteriorates into tunnel vision, and finally leads to total blindness (additional loss of cone photoreceptors, even within the fovea).[1,2,3] However, a significant number of retinal ganglion cells forming the optic nerve at the 'retinal output' as well as subsequent parts of the central visual system often remain intact.[4] In a recent study[5,6] it could be demonstrated that local electrical stimulation of the retinal ganglion cell layer in blind RP-patients yielded useful, localized visual sensations such as small dots or lines at defined locations in the visual space. Thus in principle it seems possible to bypass the defective retina under the following conditions:

"Eq. 1." development of a soft microcontact foil as retina stimulator (RS) for long-term implantation adjacent to the ganglion cell layer with a sufficient number of contacts for stimulation of individual ganglion cells/fibers;

"Eq. 2." development of a retina encoder (RE) for functional simulation of the defective 5-layered retina by mapping visual patterns onto spike trains by means of spatiotemporal filters with receptive field properties (RF-filters), and

"Eq. 3."development of a wireless signal- and energy transmission system (SE) for signal transmission from RE outside the eye to the individually addressable, microcontacts inside the eye.

Several groups in the USA[5,6,7] and Germany[8–16] are currently developing components for such retina implants. It is hoped to complete the development of entire retina implant prototypes and experimental tests in animals within the next 3 to 5 years and to apply this technology to humans within the next 10 years.

3. STRUCTURE OF THE RESEARCH CONSORTIUM

In October 1995, the development and animal test of an 'intelligent' retina implant was started in a consortium of 14 expert groups funded by the German federal research ministry (BMBF) with 10 Mill. DM over four years. Our retina implant consortium (the three coordinators are underlined) with about 40 scientists comprises expertise in: biomaterials science (C. Mittermayer / Aachen), microelectronics (B. Hosticka / Duisburg), microsystems technology & biomedical engineering (J.-U. Meyer / St. Ingbert, W. Mokwa / Duisburg, B. Wagner / Itzehoe), neural computation (R. Eckmiller / Bonn), optoelectronics (D. Jäger / Duisburg), retinal cell biology (M. Eckmiller / Düsseldorf), vitreoretinal surgery (N. Bornfeld / Essen, H. Gerding / Münster, K. Heimann / Köln, L. Hesse / Marburg, P. Wiedemann / Leipzig), and visual neurophysiology (R. Eckhorn / Marburg).

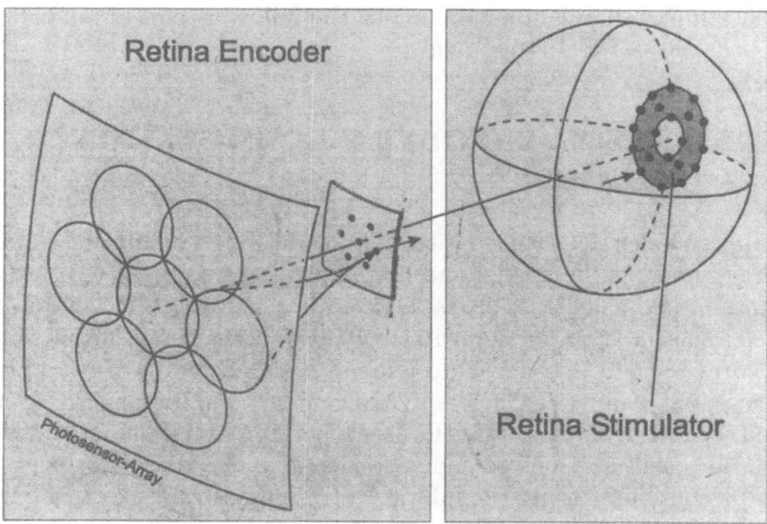

Figure 1. Schema of the retina implant with a retina encoder outside the eye and an implanted retina stimulator adjacent to the ganglion cell layer. Circle and arrow structures in the retina encoder depict the individual adaptive spatiotemporal filters with receptive field properties (RF-filters) of the primate retina.

The considerable challenge of forging an interdisciplinary partnership for this mission-oriented project is met by a number of novel organizational measures, including:

1. specification of a net plan with research topics: retina encoder-RE, signal and energy transmission-SE, retina stimulator-RS, implantation and wound healing-IH, material compatibility and tissue reaction-MT, and systems integration and test-ST;
2. assignment of working packages and mile stones for the different partners and sub-teams with a tight time schedule in order to successfully develop a prototype of the retina implant system and test its function over an extended period of time in different animal species within four years;
3. communication with visually impaired humans throughout the implant development in order to properly consider various related ethical and acceptance questions of the possible future recipient group for these visual prostheses.

Figure 1 depicts the retina implant schematically. The retina encoder (RE), which will be located outside the eye initially in a frame of glasses and later imbedded in a contact lens, has a photosensor array with about 100,000 smart pixels at the input and about 100 to 1000 technical ganglion cell (G-cell) outputs generating impulse sequences for elicitation of spike trains. Information processing within RE simulates the receptive field type filter operations for each G-cell individually. The G-cell output is subsequently encoded and transmitted via an electromagnetic and/or optoelectronic wireless transmission channel (SE) to the implanted retina stimulator (RS). RS will be implanted adjacent to the retinal ganglion cell layer and consists of an array of 100 to 500 microcontacts for localized, bi-phasic electrical stimulation of ganglion cells/fibers as well as a receiver and signal distributor. Considerable progress has been made recently with regard to alternative SE and RS[13,15] developments, RS implantation[12,15] as well as specification of stimulation parameters. The following part of the paper emphasizes recent progress in RE development.

4. LEARNING RETINA ENCODER FOR SIMULATION OF RETINAL FUNCTIONS

The retina encoder (RE) will consist of about 100 to 500 tunable RF-filters.[8,9,10,11,15] Each RF-filter simulates the spatial and temporal receptive field (RF) properties of typical primate retinal ganglion cells as depicted in Fig. 2. Each RF-filter receives its light input via a selectable set of about 100 to 1000 adjacent photosensor elements and generates a corresponding asynchronous impulse sequence at the G-cell output.

Various spatial parameters such as size of center C and periphery P of the receptive field input or resting impulse rate IR_0 as well as temporal parameters such as time t_p to impulse rate peak or time constant τ for impulse rate decay during a light step function can be continuously adjusted during a learning process. For this purpose, the individual RF-spatial filters are implemented as adaptive neural networks in order to simulate various receptive field characteristics of the primate retina.[17–20]

The implementation of the retinal information processing by a number of tunable spatiotemporal RF-filters emphasizes the input-output features of the retina and considers only a small portion of the large body of anatomical[18,21–24] and physiological data on intra-retinal structure and function. Rather, the adaptive RF-filters have to be flexible enough to be tunable in a multi-dimensional parameter space to the typical M-cell or P-cell receptive field properties of G-cells within the primate retina.[17–20]

As indicated schematically in Fig. 3, a typical RF-filter of the retina encoder (RE)[8,10,11,26] has an array of neighboring photosensors at the input and one G-cell output to map spatiotemporal events onto an asynchronous train of stimulation pulses for one-to-one elicitation of neural action potentials (spikes) at a given contacted retinal ganglion cell/fiber. The information processing of the RF-filter is flexible enough to assign a selectable set of photosensors to the excitatory (+) or inhibitory (–) receptive field center and another selectable set of adjacent photosensors to the corresponding inhibitory (–) or excitatory (+) receptive field periphery of a typical antagonistic On-

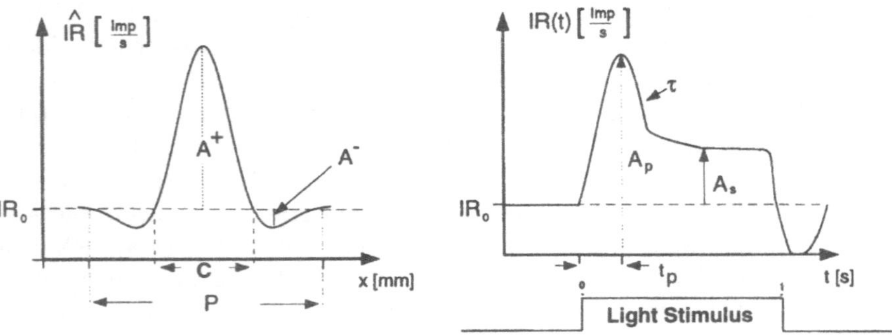

Figure 2. Spatial and temporal properties of the learning spatiotemporal RF-filters with receptive field (RF) properties.

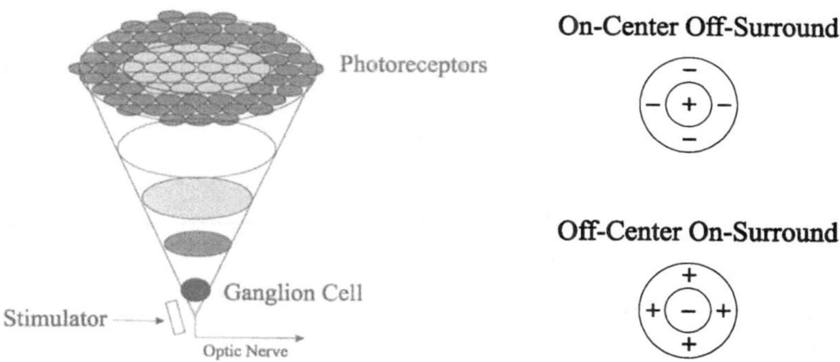

Figure 3. Scheme for a tunable RF-filter with antagonistic receptive field properties similar to those of retinal ganglion cells in the primate retina. Filter properties can be tuned to various spatial and temporal characteristics for example of M-cells or P-cells with On-centers or Off-centers.

Center or Off-Center receptive field. The receptive fields of different spatiotemporal RF-filters can widely overlap as already indicated in Fig. 1. Although their shape will typically be concentric circular, the receptive fields can also be changed to other shapes such as ellipses for example in a pre-defined set or a learning process. In principle, such tunable RF-filters can also be tuned to spatiotemporal receptive field properties of neurons in the central visual system (e.g.: in the LGN of the thalamus or in area V1 of the visual cortex.[27]

5. TECHNICAL IMPLEMENTATION OF LEARNING RETINA ENCODERS

For the first phase of our retina implant project, we have developed a Mark I RE as hardware simulator for real-time implementation of four RF-filters attached to an array of discrete photosensors The mixed analog and digital electronic circuits on several circuit boards are integrated in a portable system, which can be controlled and monitored via a PC.

Typical stimuli are moving or temporally modulated light patterns. The corresponding G-cell outputs are membrane potential time courses, which lead to impulse trains for simulation of a P-cell, as indicated in Fig. 4. System identification to specify the dynamic properties of individual RF-filters was performed with neural networks. For this purpose, a homogeneous light spot was moved with constant velocity over the photosensor array. RF-filter output signals were fed into a multilayer perceptron with several hidden layers. The current filter parameter vector was used as teacher for the back propagation learning process. The Mark I RE as portable system may also be valuable for collaborations with other labs.[8–11]

Simultaneously, we are developing the Mark II RE (Fig. 5) , which currently simulates about 200 tunable RF-filters in real time on a digital signal processor, DSP (C80,TI), each with an array of several thousand photosensors at the input. The algorithm implemented in assembler, which used only a fraction of the DSP processing power, was able to adjust various spatial and temporal properties of each RF-filter by means of 10 parameters. Various spatiotemporal patterns were processed by the RF-filters, yielding new values of the corresponding ganglion cell membrane potentials

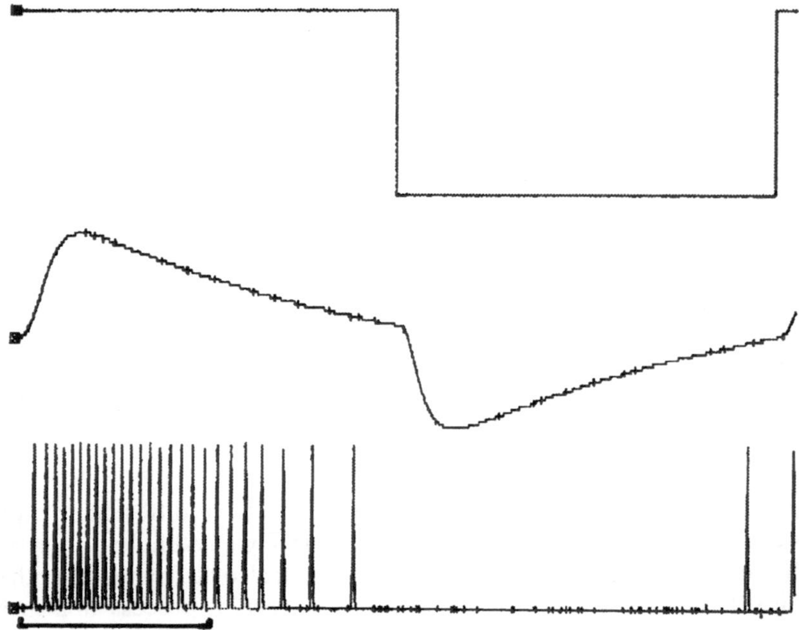

Figure 4. Typical recording from a Mark I RE RF-filter. Top trace: light spot time course centered on the On-center of the receptive field. Middle trace: membrane potential time course of the technical retina ganglion cell with a given setting of spatial and temporal filter parameters. Bottom trace: corresponding spike train at the ganglion cell output for stimulation of a contacted ganglion cell/fiber. Time scale: 100 ms.

every 13 ms. These quasi-continuous membrane potential time courses were subsequently transformed into asynchronous pulse trains, which could be used to stimulate ganglion cells in retina implant applications. The functional range of operation of the RF-filters as measured by the spatiotemporal amplitude spectra, included the spatial and temporal properties of primate retinal M and P cells with good approximation in the photopic range.[10,11]

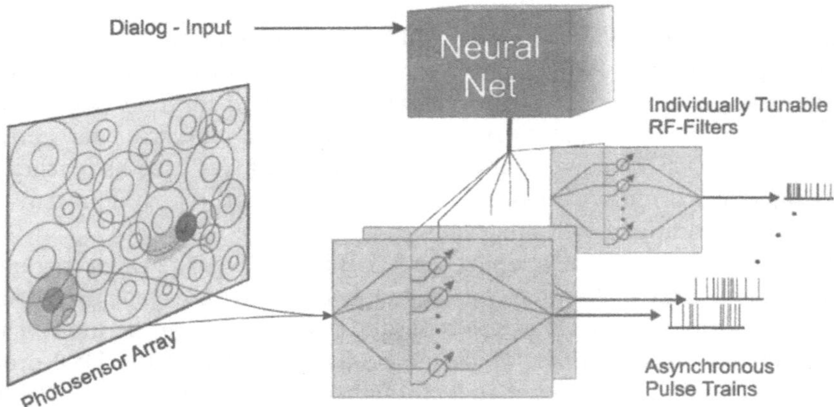

Figure 5. Schema of information processing in the Mark II RE.

6. DIALOG BETWEEN HUMAN AND RETINA ENCODER VIA VISUAL PERCEPTION

Bi-directional communication in a dialog between the learning RE and the implant-carrying human subject was recently proposed.[9,10,28] In principle, the various tunable RF-filters need to be optimized with regard to the spatiotemporal function, which is 'expected' by the central visual system from the randomly contacted ganglion cell. This tuning task in a dialog has to be based on the actual, regained visual perception of the initially blind subject. We propose a procedure in preparation of this dialog, while implant-carrying subjects are not yet available. Fig. 6 gives the schema of the proposed dialog concept. The monitor picture on the left depicts pattern P as rectangular edge moving upwards to the right. The right monitor shows an ellipse moving downwards to the right as simulation of the actually perceived pattern P'. RE is depicted as a module of nine tunable RF-filters with receptive field input surfaces as open circles on the left and the corresponding outputs on the right (dots). Two of the RF-filters are emphasized with connections to the assumed ganglion cell (1) as P-cell and ganglion cell (9) as M-cell. The central visual system with nine contacted ganglion cells on the left represents two structures for two different cases:

1. Simulation of the central visual system during a simulated dialog of a subject with normal vision, and
2. Real central visual system of an implant-carrying subject perceiving pattern P' during presentation of pattern P.

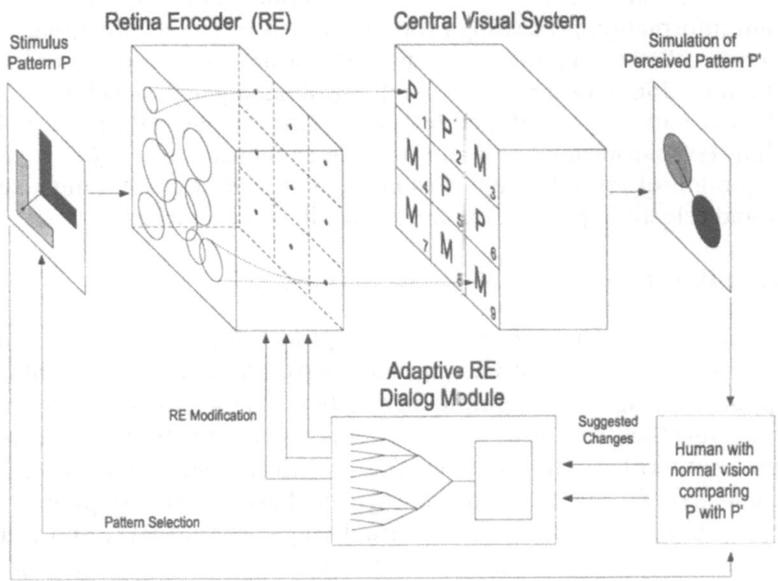

Figure 6. Dialog schema for a learning retina encoder with a human visual system. The resulting RE and dialog module following an extensive dialog phase in case 1) have to be sufficiently pre-trained to allow a successful tuning during a dialog phase in case 2). The implant-carrying, initially blind subject will provide the input to the dialog module (Fig. 6) based on a comparison between the actually perceived pattern P' and the desired pattern P, which is not visible but made known to the subject via another sensory system. Accordingly, the envisioned dialog in case 2) may be thought of as a combination of 'tuning' a multi-dial radio to a station and producing desired visual patterns with a functionally cryptic visual pattern generator.

The learning RE dialog module in Fig. 6 depicts a combined structure with a neural network and a knowledge based component. It is assumed here that in both cases 1) and 2), a human subject indicates suggestions to change the RE function (e.g. by changing a number of hand dials). This set of changed values is processed by the dialog module, which in turn generates either RE modification signals to modify some of the parameters of the various tunable RF-filters, or command signals to select another stimulus pattern P.

It should be noted that at least in case 1), the function of RE can be interpreted as a mapping operation from the pattern space P onto a ganglion cell activity space G (represented by the neural activity of stimulated ganglion cells). The central visual system module in Fig. 4 is positioned to map the ganglion cell activity space G onto the perceived pattern space P'. The goal of the dialog is to approximate P' as closely as possible to P. In the ideal case, P' becomes identical to P, which implies that the simulated central visual system module has to perform an inverse mapping with regard to the RE mapping.

7. FUTURE DEVELOPMENTS AND STEPS TOWARD COMMERCIALIZATION

7.1. Synchronization of the Spike Activity of Different Ganglion Cells

A growing body of recent neurophysiological data suggests the temporary coupling of the neural impulse activity of different retinal ganglion cells.[29,30] This may be an important information processing property caused by intra-retinal (e.g. spike activity of amacrine cells or coupling between neighboring ganglion cells) or extra-retinal (e.g. signal interaction between extra-thin unmyelinated ganglion cell fibers within the eye) events. The currently developed learning retina encoder (RE) offers the temporary coupling (synchronization) of the impulse sequences of a given set of G-cells. During adaptation of the RE properties in a dialog with a given human subject, this temporary and selective spike synchronization will be considered.

7.2. Information Processing of Color Signals

The primate retina uses cone photoreceptors with three different absorption spectra in the photopic and mesopic range and maps color patterns onto different types of color-opponent ganglion cell output signals (e.g. yellow-blue or red-green P-cells). One has to assume that the central visual system of humans with RP still 'expects' color information from most P-cells and interprets electrically induced signals accordingly.[6] The retina encoder (RE) in an advanced stage will have an array of photosensors with three different absorption spectra in the visual range as input layer and will allow for the simulation of color-opponent cells with the characteristic spatial and temporal properties of P-cells.

7.3. Transfer of Research Results into Medically Useful Prosthetic Products

There exist large gaps between the different scientific communities of research fields in biology, medicine, and technology on the one hand and between research and industry on the other hand. The first gap is successfully being minimized in our consortium by its cooperative structure and its mission-oriented applied research agenda. With regard to the gap between research and industry, we have identified some of the challenges as:

- Combining science interests with industry interests,
- Combining long term planning (5 to 20 years) in research with short term planning (1 to 5 years) in industry,
- Combining public investment risks for research with private investment risks for production and marketing.

Especially scientists have to learn that commercial exploitation of their research results is not only desirable but necessary. Accordingly, scientists have to focus on patented or at least patentable know how. Under this condition, the research results of scientists become sufficiently interesting for industry including venture capital and corporate venture to consider a required commercial risk investment based on patents (exploitation monopolies) and to actually build and market future products such as intelligent visual implants.

In order to facilitate the transfer of scientific and technological know-how into visual prosthesis products, patent applications[28] as well as the foundation of a high-tech company, which was generated as part of a major neurotechnology initiative,[31] have been implemented.

ACKNOWLEDGMENT

Supported in part by the German Federal Ministry for Education, Science, Research and Technology (BMBF).

REFERENCES

1. D.J. D'Amico, 1994, Diseases of the retina, *New England J. Med.* **331**:95–106.
2. T.P. Dryja and E.L. Berson, 1995, Retinitis pigmentosa and allied diseases, *Invest. Ophthal. & Vis. Sci.* **36**:1197–1200.
3. R.W. Massof and D. Finkelstein, 1987, A two-stage hypothesis for the natural course of retinitis pigmentosa", in: *Adv. in the Biosciences* Volume 62, pp. 29–58, Pergamon Press.
4. J.L. Stone, W.E. Barlow, M.S. Humayun, E. de Juan, A.H. Milam, and 1992, Morphometric analysis of macular photoreceptors and ganglion cells in retinas with retinitis pigmentosa, *Arch. Ophthalmol.* **110**:1634–1639.
5. M.S. Humayun, R.H. Propst, E. de Juan K. McCormick, and D. Hickingbotham, 1994, Bipolar surface electrical stimulation of the vertebrate retina, *Arch. Ophthal.* **112**:110–116.
6. M.S. Humayun, E. de Juan, G. Dagnelie, R. Greenberg, and R. Propst, 1996, Artificial vision, *Invest. Ophthal. & Vis. Sci.* **37**:S451.
7. J.F. Rizzo, S. Miller, T. Denison, T. Herndon, J.L. Wyatt, 1996, Electrically evoked cortical potentials from stimulation of rabbit retina with a microfabricated electrode array, *Invest. Ophthal. & Vis. Sci.* **37**:S707.
8. M. Becker, M. Braun, and R. Eckmiller, 1998, Retina Implant adjustment with reinforcement learning, in: *IEEE Int. Conf. Acustics, Speech, Signal Processing, ICASSP '98*, Seattle, Volume 2, pp. 1181–1184.
9. R. Eckmiller, 1996, Concerning the development of retina implants with neural nets, in: *Proc. Int. Conf. Neural Inf. Proc., ICONIP'96*, Hong Kong, Vol. 1, pp. 21–28.
10. R. Eckmiller, 1997, Learning Retina Implants with epiretinal contacts, *Ophthalmic Res.* **29**:281–289.

11. R. Hünermann, M. Becker, and R. Eckmiller, 1997, Towards real time implementation of a learning retina encoder, *Invest. Ophthal. & Vis. Sci.* **38 (Suppl.)**:191.
12. H. Gerding, C. Uhlig, and U. Thelen, 1998, The retina implant project: development of techniques for implantation and epiretinal fixation of stimulators, *Invest Ophthal. & Vis. Sci.* **39 (Suppl.)**:991.
13. W. Mokwa, H.K. Trieu, and L. Ewe, 1998, Implantable retina stimulator for a retina implant, in: *EUFIT '98*, Aachen, pp. 1788–1792.
14. N. Peixoto, S. Straburger, R. Hornig, P. Walter, P. Szurmann, and R. Eckmiller, 1998, Evaluation of implanted epiretinal microcontacts in the mammalian retina, *Invest. Ophthal. & Vis. Sci.* **39 (Suppl.)**:902.
15. M. Schwarz, B.J. Hosticka, R. Hauschild, W. Mokwa, M. Scholles, and H.K. Trieu, 1996, Hardware architecture of a neural net based retina implant for patients suffering from retinitis pigmentosa, in: *Proc. IEEE ICNN'96*, Washington, pp. 653–658.
16. P. Walter, P. Szurmann, N. Peixoto, S. Stra burger, H.K. Trieu, L. Ewe, T. Stiglitz, J.U. Meyer, and K. Heimann, 1998, Evoked cortical potentials after electrical surface stimulation of the rabbits retina, *Invest. Ophthal. & Vis. Sci.* **39 (Suppl.)**:990.
17. B.B. Lee, J. Pokorny, V.C. Smith, and J. Kremers, 1994, Responses to pulses and sinusoids in macaque ganglion cells, *Vision Res.* **34**:3081–3096.
18. R.W. Rodieck, and M. Watanabe, 1993, Survey of the morphology of macaque retinal ganglion cells that project to the pretectum, superior colliculus, and parvicellular laminae of the lateral geniculate nucleus, *J. Comp. Neurol.* **338**:289–303.
19. M. Watanabe and R.W. Rodieck, 1989, Parasol and midget ganglion cells of the primate retina, *J. Comp. Neurol.* **289**:434–454.
20. T. Yeh, B.B. Lee and J. Kremers, 1996, The time course of adaptation in macaque retinal ganglion cells, *Vision. Res.* **36**:913–931.
21. C.A. Curcio, K.R. Sloan, R.E. Kalina, A.E. Hendrickson, 1990, Human photoreceptor topography, *J. Comp. Neurol.* **292**:497–523.
22. D.M. Dacey and M.R. Peterson, 1992, Dendritic field size and morphology of midget and parasol ganglion cells of the human retina, *Proc. Natl. Acad. Sci.* **89**:9666–9670.
23. J.B. Jonas, U. Schneider, and O.H. Naumann, 1992, Count and density of human retinal photoreceptors, *Graefes Arch. Clin. Exp. Ophthal.* **230**:505–510.
24. H. Kolb, 1994, The architecture of functional neural circuits in the vertebrate retina, *Invest. Ophthal. & Vis. Sci.* **35**:2385–2404.
25. R. Eckmiller, 1975, Electronic simulation of the vertebrate retina, *IEEE Trans. Biomed. Eng.* **BME-22**:305–311.
26. P. Gaudiano, 1992, Toward a unified theory of spatiotemporal processing in the retina, in: *Neural Networks for Vision and Image Processing* (G. Carpenter, S. Grossberg, eds.), pp. 195–220, MIT Press, Cambridge MA.
27. J. Bullier and L.G. Nowak, 1995, Parallel versus serial processing: New vistas on the distributed organization of the visual system, *Curr. Opin. Neurobiol.* **5**:497–503.
28. R. Eckmiller, 1998, Lernfähiger sensomotorischer Encoder für Seh- und Hörprothesen, *International Patent Application with 28 claims*, PCT/EP98/00968, 1998.
29. M. Meister, L. Lagnado, and D.A. Baylor, 1995, Concerted signaling by retinal ganglion cells, *Science* **270**:1207–1210.
30. S. Neuenschwander, W. Singer, 1996, Long-range synchronization of oscillatory light responses in the cat retina and lateral geniculate nucleus, *Nature* **379**:728–733.
31. R. Eckmiller and S. Suchert, in press, Strategy for the foundation of a neurotechnology complany, in: *Int. Conf. Neural Inf. Proc., ICONIP '98*, Kitakyushu, November 1998, 8 pages.

ARE SUBRETINAL MICROPHOTODIODES SUITABLE AS A REPLACEMENT FOR DEGENERATED PHOTORECEPTORS?

E. Zrenner,[1] S. Weiss,[1,2] A. Stett,[1,2] B. Brunner,[2] V.P. Gabel,[4] M. Graf,[3] H.G. Graf,[3] H. Haemmerle,[2] B. Hoefflinger,[3] K. Kobuch,[4] K.-D. Miliczek,[1] W. Nisch,[2] H. Sachs,[4] and M. Stelzle[2]

[1] University Eye Hospital, Dept. II, Tübingen, Germany
[2] Natural and Medical Sciences Institute, Reutlingen, Germany
[3] Institute for Microelectronics, Stuttgart, Germany
[4] University Eye Hospital, Regensburg, Germany

1. INTRODUCTION

The feasibility of a retinal prosthesis for the blind has been discussed since the 1950s. Early attempts to realize a subretinal implantable prosthesis suffered from technical and microsurgical limitations at that time. Nowadays these problems can be solved because of the amazing progress that has been made in the fields of intraocular surgery, silicon microfabrication and our knowledge of the silicon/neuron junction. Therefore, a new attempt to develop a retinal prosthesis seems justified. While other research groups[1–3] have concentrated their efforts on the development of an *epiretinal* visual prosthesis, which receives energy and preprocessed visual information via a relatively complex telemetric system, Chow and coworkers[4–7] published the concept of a retinal implant consisting of thousands of microphotodiodes. The principle of this concept is based on an implant that is placed in subretinal space and stimulates the underlying retinal cells with the photocurrents generated solely from the incident light (Fig. 1).

Besides the technological development of an optimized subretinal implantable microphotodiode array (MPDA) which was, at the beginning in 1995, based on previous work of Chow and his associates we have aimed some of our experiments in the last three years at answering the following crucial questions:

Is it possible to stimulate a degenerated retinal network electrically from the distal side under conditions of chronic and safe charge injection?

What are the appropriate spatiotemporal stimulation patterns?

Retinal Degenerative Diseases and Experimental Therapy, edited by Hollyfield *et al.*
Kluwer Academic / Plenum Publishers, New York, 1999.

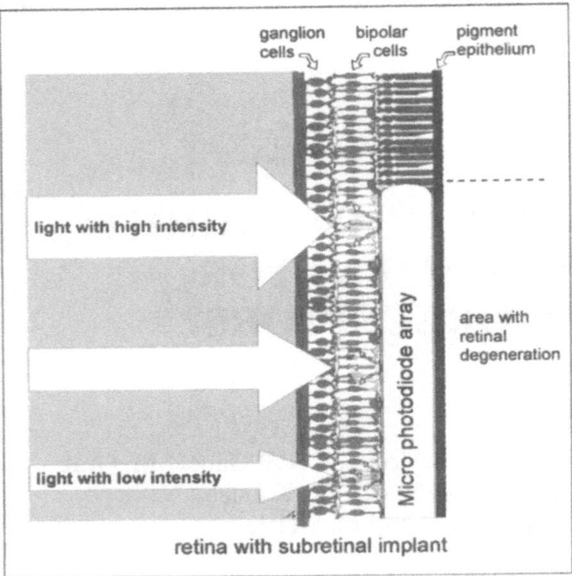

Figure 1. Illustration of the subretinal approach of a retinal prosthesis. A silicon chip containing a large number of individual microphotodiodes is implanted in the subretinal space in an area with degenerated photoreceptors. Incident light (arrows from the left) is absorbed from the photodiodes which generate currents in direct proportion to the local intensity of the light. Via the metallic stimulation electrodes the current is injected into the retinal tissue (small bundles of arrows from the right).

What minimal invasive surgical techniques can be utilized for the subretinal implantation of MPDAs?

Is the implant tolerated by the overlying retina over an extended period of time?

In the following sections we will present results that provide some answers to the above questions, which are in some aspects not only important for a subretinal but also for an epiretinal visual prosthesis.

2. MICROPHOTODIODE ARRAYS

MPDA prototypes with several different designs of microphotodiodes were produced using standard CMOS technology[8] which will be described in more detail elsewhere. Because we learned from several *in vitro* and *in vivo* experiments that a solid MPDA in the subretinal space disturbs the diffusion of nutrients between choriod and neuroretina, we modified the production process in such a way that a regular array of perforation holes in the silicon allows for fluid exchange between pigment epithelium and retina. The structure of the light sensitive area of the photodiodes is not disturbed by these holes. A SEM picture of such a perforation hole is shown in Fig. 2.

One of the main objectives in fabrication of MPDAs was to develop stimulation contacts with a maximized charge transfer capacity. Basically, the capacity increases with electrode area. However, the geometric area is limited since there has to be a certain lateral resolution in the stimulation of the retina. In contrast, the effective electrode area may be increased by several orders of magnitude by deposition of nanoporous rather than planar thin film electrodes. Therefore titanium nitride (TiN) was chosen to fabricate stimulation contacts. When deposited at high pressure, this type of electrode material exhibits similar nano-porosity as platinum black while showing superior mechanical stability (Fig. 3).

Further details about fabrication, morphological and electrical properties of this type of electrodes have previously been published.[9]

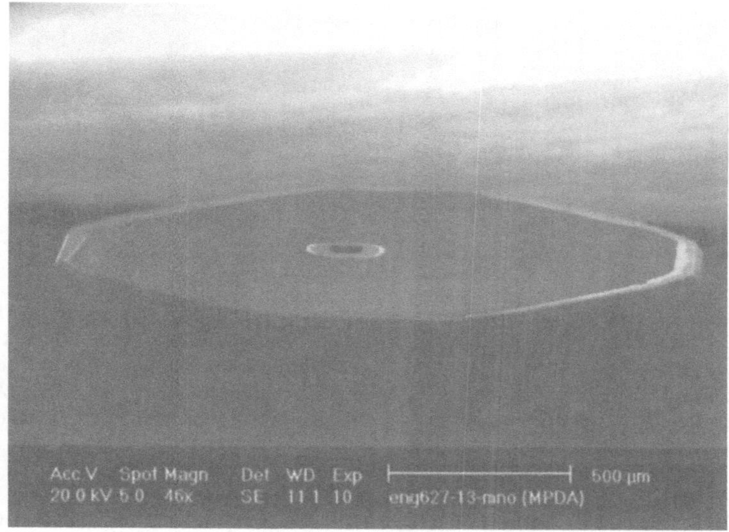

Figure 2. SEM-image of a MPDA with its individual photodiodes. In the middle of the MPDA a perforation hole was obtained by anisotropic etching of the silicon bulk.

3. FUNCTIONAL ELECTRICAL MULTISITE STIMULATION OF DEGENERATED RETINAS

To evaluate the optimal stimulation parameters for subretinal electrical stimulation of degenerated retinas, we established an *in vitro* model for investigating the basic principles of electrical retina stimulation. We used retinas of royal college of surgeons (RCS) rats, a well established animal model for human retinal degenerations,[10,11] which were adhered to a microelectrode array (MEA) for stimulation or recording.[12,13] The

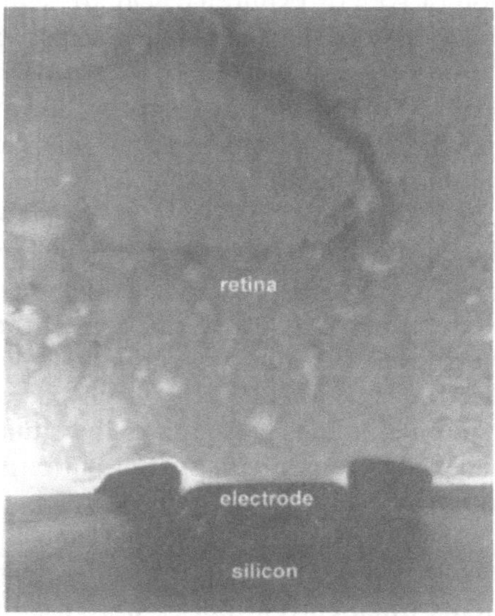

Figure 3. TEM-image of a cross-section of the chip retina interface. The retinal tissue has close contact to the TiN-electrode.

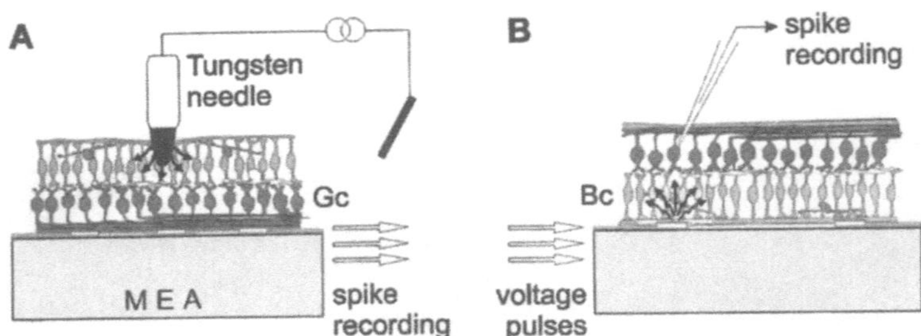

Figure 4. Retina-MEA assemblies for in vitro recording or stimulation of degenerated retinas. (A) The retina is put on the MEA with the ganglion cells (Gc) on the MEA electrodes. The distal retinal network is stimulated with current pulses delivered via a tungsten needle. During stimulation ganglion cell activity can be recorded from various MEA electrodes. (B) The retina is adhered with the distal side on the MEA. Bipolar cells (Bc) are stimulated via the MEA electrodes with voltage pulses while spikes can be recorded from a single ganglion cell with a glass pipette.

microelectrode array consists of 60 substrate integrated planar microelectrodes (diameter: 10 μm) which can be used either for multichannel recordings of ganglion cell activity when the proximal side of the retina is on the electrodes, or for multichannel stimulation, when the distal side of the retina is put on the MEA (Fig. 4).

We first tested whether it is possible to elicit stimulus correlated ganglion cell activity by applying monopolar charge balanced current pulses to the distal side of a degenerated retina via a tungsten needle. A typical result from these experiments is shown in Fig. 5.

The ganglion cells respond to a current pulse delivered to the retinal network with either an increase or a decrease in firing rate. To exclude the possibility that the ganglion cells were stimulated directly by the currents injected, we used pharmacological agents to specifically block retinal pathways. For example, the glutamate agonist 2-amino-4-phosphono-butyric acid[14] (AP-4) or glutamate antagonist kynurenic acid[15] was used to selectively block the "light-ON" or "light-OFF" pathway in the retina. The changes in response characteristics of the ganglion cells shown in Fig. 5 after application of AP4 or kynurenic acid are summarized in Fig. 6.

The transient initial increase of the ganglion cells firing rate elicited by the application of a positive monopolar current pulse (upper right) is inhibited by kynurenic acid (100 μM), as shown on the lower right, while AP-4 (100 μM) seems to have no significant effect (middle row, right). Negative current pulses elicited an inhibition (left column) of spontaneous activity that is only little influenced by pharmacological actions. However, it is obvious that it was the retinal network that was stimulated by the current pulse, because the ganglion cells response was modulated by the pharmacological agents. These results demonstrate that it is possible to stimulate a degenerated retinal network electrically from the subretinal side. Because the MPDA is fabricated with planar stimulation electrodes we have, in addition, used the MEA-electrodes to stimulate the distal side of a degenerated retina with voltage pulses. This experimental design mimicks the situation when an implanted MPDA stimulates the underlying retina with spatial patterns of photocurrents. A typical result from those experiments is shown in Fig. 7.

The recorded ganglion cell shows the same characteristic response features when a voltage pulse was applied to the retinal network, as it was found in previous experiments when current pulses were used (Fig. 5). Furthermore, the firing rate of the cell

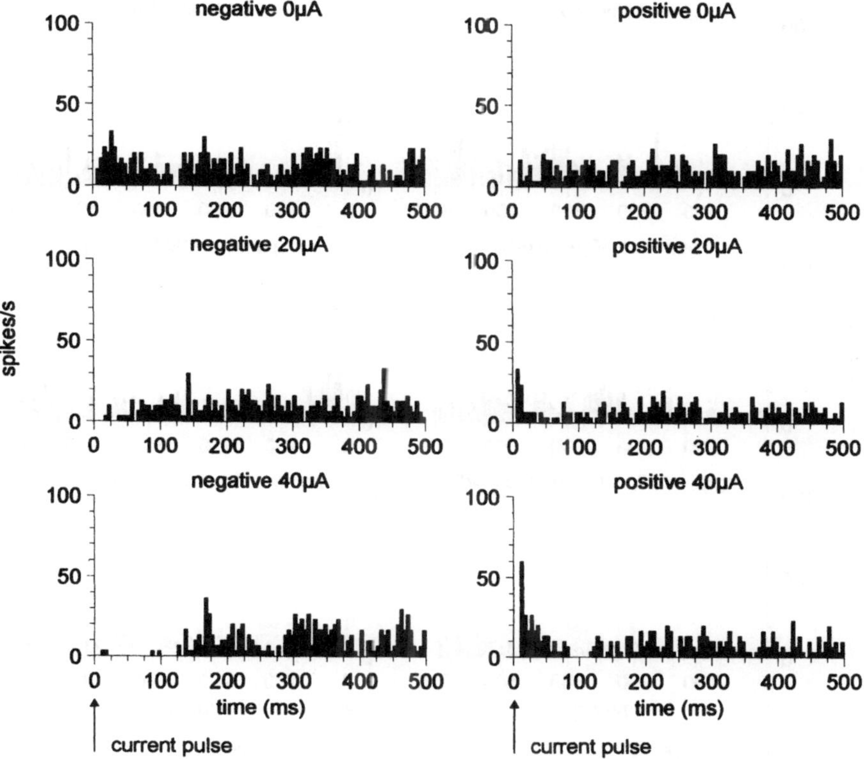

Figure 5. Peristimulus time histograms (PSTHs) of multi-unit ganglion cell activity recorded with a MEA electrode (60 stimulus presentations, binwidth 5 ms). A monopolar charge balanced current pulse with 1 ms duration and the indicated strength and polarity was injected in the distal retina at time 0 ms. Spike activity was either increased or decreased after the current pulse, depending on the strength and polarity of the stimulus. Retina of a 80 days old RCS rat.

depends not only on the stimulation strength (Fig. 7, right panel), but also on the spatial pattern of the electrical field (Fig. 7, left panel). Overall, we conclude that it is possible to electrically stimulate a degenerated retinal network with either voltage or current pulses, which evoke spatio-temporal distinguishable ganglion cell response patterns.

4. IMPLANTATION TECHNIQUES: A NOVEL APPROACH TO THE SUBRETINAL SPACE

As an alternative to the more conventional *ab interno* implantation technique previously described[8] we developed an *ab externo* transchoroidal access to the subretinal space in order to avoid the risks and late complications of an intraocular procedure. In contrast to the *ab interno* implantation technique the MPDA is put into the subretinal space via a transscleral and transchoroidal incision (Fig. 8).

A scleral flap is prepared 7 mm away from the corneo-scleral limbus in the upper quadrant of the eye. Intraocular pressure is reduced by paracentesis. After drug-induced vasoconstriction the choroid is incised following the course of the big vessels.

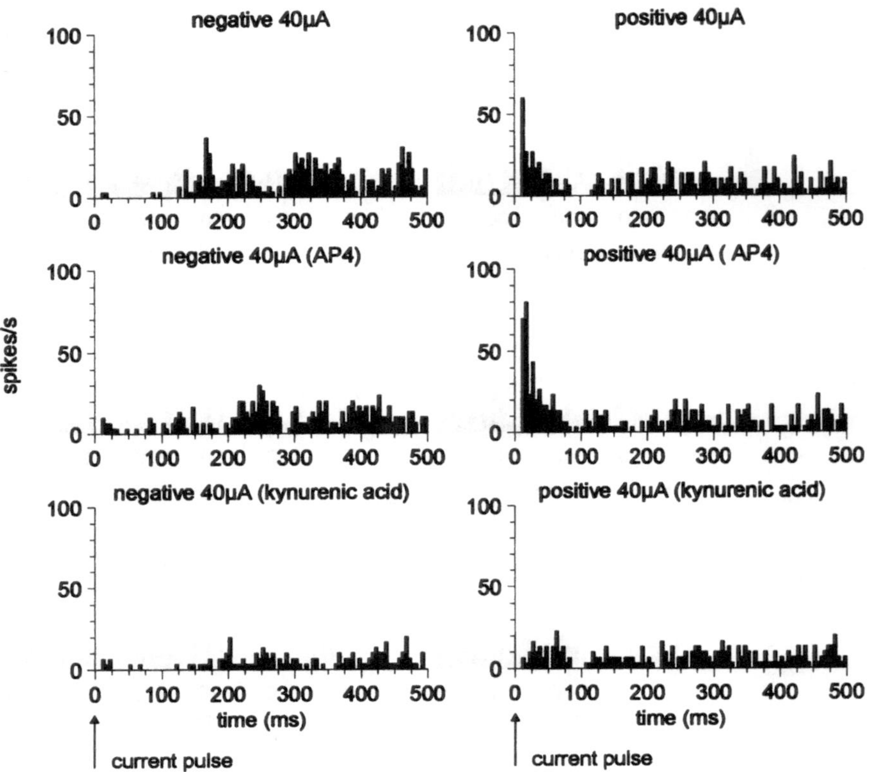

Figure 6. PSTHs of multi-unit ganglion cell activity recorded with a MEA electrode (60 stimulus presentations, binwidth 5 ms). A suprathreshold monopolar charge balanced current pulse with 1 ms duration and the indicated strength and polarity was injected in the distal retina at time 0 ms. The retina was either superfused with normal ringer solution (top row), ringer containing 100 µM AP-4 (middle row) or ringer containing 100 µM kynurenic acid (bottom row). Kynurenic acid abolished the ganglion cell excitation after the positive stimulus pulse (arrow). Retina of a 80 days old RCS rat.

Through this incision a soft foil strip is placed in the subretinal space to protect the retina and as a slide rail for the MPDA. The chip is pushed along this foil towards the regio macularis with a special designed curved ring forceps. The position of the MPDA can be estimated from the direction and length of the implanted foil without subsequent direct intraocular visual control. The *ab externo* technique was successfully tested in enucleated porcine eyes and in rabbits. The main risk of this technique is a perforation of the retina which, in the rabbits eye, is only about 100 µm thick. However, perforation may be less of a problem in another animal model with a thicker retina (e.g. minipig) or in the human eye. The novel transchoroidal access to the subretinal space using a thin foil strip for stabilisation of the neurosensory retina and as a slide rail into the subretinal space represents a surgical technique, which may help to minimize the intraoperative risks and long term complications of implantation of MPDAs. Furthermore, such a surgical procedure may also be valuable for other clinical applications. So far the implants in rabbits were very well tolerated for more than 14 months, as tested by biomicroscopy and electroretinography. In some rabbits a

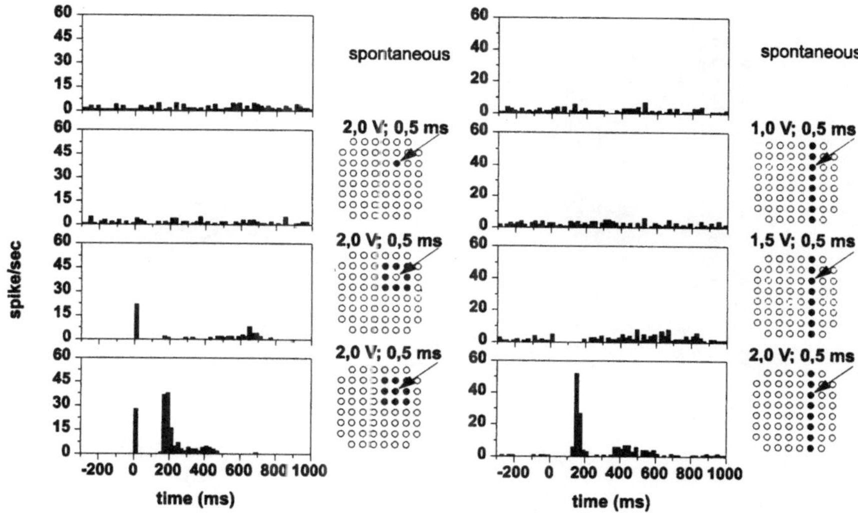

Figure 7. PSTHs of single-unit ganglion cell activity recorded with a patch pipette (50 stimulus presentations, binwidth 20 ms). Voltage pulses (parameters as indicated) were applied to the distal side of the retina via the MEA-electrodes (TiN-electrodes with 10 μm diameter and 100 μm interelectrode spacing) at time 0 ms. The recording site is indicated by the arrow and the electrodes that were used for stimulation are shown in black. Retina of a 82 days old RCS rat. The multisite electrical stimulation of the distal retina results in typical ganglion cell activity response patterns which consists of a fast burst, transient inhibition, and delayed discharge. The ganglion cell response depends on the strength and pattern of the voltage pulses.

evoked cortical potential could be elicited as well by infrared stimulation of implanted MPDAs.

5. BIOCOMPATIBILITY OF IMPLANTED MPDAS

Since the rabbit is one of the "classical" animal models in experimental ophthalmology, most of the implantations of chip prototypes for *in vivo* biocompatibility and functional tests were done in rabbit eyes. Although this animal model is suitable for developing intraocular surgical procedures, it has certain drawbacks if one wants to test the long-term biocompatibility of a subretinal implanted MPDA.[16] The implanted MPDA may act as an unwanted barrier in metabolic processes between the inner and outer retinal layers. This is especially a problem in the rabbit's avascular retina, where the whole nutritional supply of the inner retina derives from the choroid. Therefore, at present, intraretinal oxygen measurements are performed over different types of implanted chips, unperforated and perforated ones, in order to judge the influence of the MPDA on retinal oxygen supply and on retinal degenerative processes in different animal models: in the rabbit with the avascular retina and in the minipig with a human-like vascularized retina. For these measurements a new oxygen-specific optical microsensor, based on dynamic fluorescence quenching, is used. First results in rabbit eyes indicate that the oxygen tension of the inner retina is reduced over an unperforated MPDA as compared to a perforated chip, both implanted in the same eye. The relevance of these findings for a vascularized retina still has to be evaluated by measurements in minipigs.

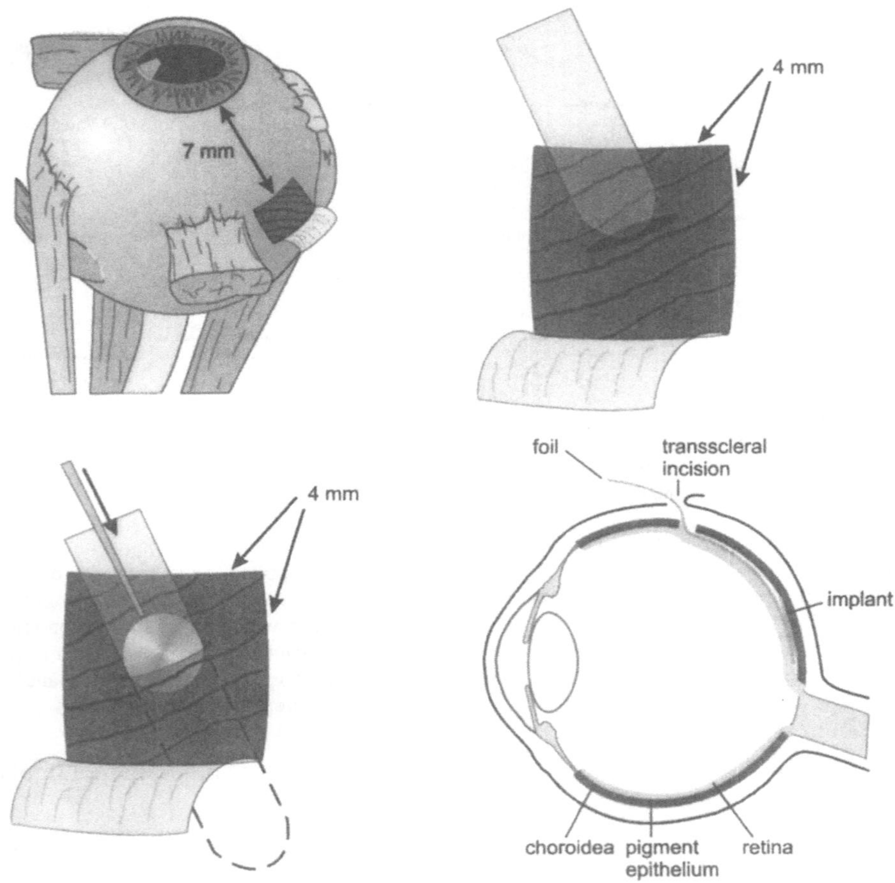

Figure 8. Schematic drawing of the *ab externo* implantation technique. A sclera incision is made about 7 mm away from the corneo-scleral limbus. Then the choriod is incised and a soft foil strip inserted into the subretinal space. A chip is pushed along this foil strip towards the macula. The final position of the chip in the subretinal space can be estimated from the direction and length of the foil strip.

6. SUMMARY

Is it really possible to replace degenerated photoreceptors with subretinal implanted microphotodiodes?

The experiments on the electrical excitability of degenerated retinas, the successful surgical techniques and the long-term results are pointing to "yes". Overall, we think that many of the basic questions regarding the development of a subretinal implantable visual prosthesis are now more or less solved. We know the range of stimulation parameters for safe and chronic subretinal electrical stimulation of degenerated retinas and which implant materials are most suitable in terms of biocompatibility and long-term stability. We have developed new implantation techniques for minimal invasive subretinal implantation and added major improvements to the original design of the MPDAs. However, there are still several technological problems to be solved before one can think of implanting MPDAs in patients suffering from photoreceptor loss. For example, both *in vitro* and *in vivo* we found that photocurrents generated from the naturally occurring retinal illuminances are not strong enough for successful electrical stimulation of the retinal network. To solve this problem various concepts are currently being developed to provide additional energy to a subretinal implant.

ACKNOWLEDGMENTS

The authors are very much indebted to the following collaborators: W. Barth, V. Bucher, P. Gnauck, T. Herrmann, S. Scholz, M. Tepfenhart. For further information see http://www.uni-tuebingen.de/subret.

This study is supported by grants from the German Federal Ministry of Education, Science, Research and Technology (BMBF) No. 01 IN 502 A–D; *fortüne* grant Nr.233, University Tuebingen; the Pro Retina Germany Foundation; by technical device support by Hewlett Packard GmbH, Boeblingen/Germany.

REFERENCES

1. J.F. Rizzo and J. Wyatt, 1997, Prospects for a visual prosthesis, *The Neuroscientist*, **3**:251–262.
2. M.S. Humayun, E. de Juan Jr., G. Dagnelie, R.J. Greenberg, R.H. Propst, D.H. Phillips, 1996, Visual perception elicited by electrical stimulation of retina in blind humans, *Arch. Ophthalmol.* **114**:40–46.
3. R. Eckmiller, 1997, Learning retina implants with epiretinal contacts, *Ophtalmic Res.*, **29**:281–289.
4. A.Y. Chow, 1993, Electrical stimulation of the rabbit retina with subretinal electrodes and high density microphotodiode array implants, *Invest. Ophthalmol. Vis. Sci.*, **34**:835 (Abstract).
5. A.Y. Chow and V.Y. Chow, 1997, Subretinal electrical stimulation of the rabbit retina, *Neurosci. Lett.*, **225**:13–16.
6. A.Y. Chow, V.Y. Chow, M.T. Pardue, J.I. Perlman, and N.S. Peachey, 1998, Retinal and cortical potentials induced by subretinally implanted microphotodiode arrays, *Invest. Ophthalmol. Vis. Sci.*, **39**:565 (Abstract).
7. G. Peyman, A.Y. Chow, C. Liang, V.Y. Chow, J.I. Perlman, and N.S. Peachey, 1998, Subretinal semiconductor microphotodiode array, *Ophthalmic Surg. Lasers*, **29**:234–241.
8. E. Zrenner, K.-D. Miliczek, V.P. Gabel, H.G. Graf, E. Guenther, H. Haemmerle, B. Hoefflinger, K. Kohler, W. Nisch, M. Schubert, A. Stett, and S. Weiss, 1997, The development of subretinal microphotodiodes for replacement of degenerated photoreceptors, *Ophthalmic Res.*, **29**:269–280.
9. M. Janders, U. Egert, M. Stelzle, and W. Nisch, 1996, Novel thin film micro-electrodes with excellent charge transfer capability for cell stimulation and sensing applications, *Proc. 18th Ann. Int. Conf. IEEE Eng. Med. Biol. Soc.*, #364.
10. K. Kohler, E. Guenther, and E. Zrenner, 1997, Tiermodelle in der Retinitis-pigmentosa-Forschung, *Klin. Monatsbl. Augenheilkd.*, **211**:84–93.
11. H.J. Sheedlo, V. Gauer, L.X. Li, A.D. Seaton, and J.E. Turner, 1991, Transplantation to the diseased and damaged retina, *TINS*, **14**:347–350.
12. A. Stett, K. Kohler, S. Weiss, H. Haemmerle, and E. Zrenner, 1998, Electrical stimulation of degenerated retina of RCS rats by distally applied spatial voltage patterns, *Invest. Ophthalmol. Vis. Sci.*, **39**:162 (Abstract).
13. S. Weiss, T. Herrmann, K. Kohler, A. Stett, and H. Haemmerle, 1998, Spontaneous and electrically evoked spatio-temporal firing patterns in the RCS rat retina, *Invest. Ophthalmol. Vis. Sci.*, **39**:991 (Abstract).
14. M.M. Slaughter and R.F. Miller, 1981, 2-amino-4-phosphonobutyric acid: A new pharmacological tool for retina research, *Science*, **211**:182–184.
15. P.A. Coleman, S.C. Massey, and R.F. Miller, 1986, Kynurenic acid distinguishes kainate and quisqualate receptors in the vertebrate retina, *Brain Res.*, **381**:172–175.
16. K.-D. Miliczek, S. Scholz, R.B. Aramant, M.J. Seiler, R.P. Tornow, K. Kohler, and E. Zrenner, 1998, Subretinal microphotodiodes in rat eyes: Biocompatibility, electroretinography and implantation technique, *Invest. Ophthalmol. Vis. Sci.*, **39**:565 (Abstract).

Section VII

NEUROPROTECTIVE AGENTS PROMOTE PHOTORECEPTOR SURVIVAL

48

RODS PRODUCE A DIFFUSIBLE FACTOR PROMOTING CONE PHOTORECEPTOR SURVIVAL *IN VIVO* AND *IN VITRO*

David Hicks,* Saddek Mohand-Said, Thierry Leveillard, Henri Dreyfus, and Jose Sahel

Laboratoire de Physiopathologie Retinienne
Clinique Ophtalmologique
Centre Hospitalier Universitaire Régional, BP. 426
1 Place de l'Hopital
67091 Strasbourg Cedex
France

1. SUMMARY

1.1. Purpose

To investigate possible trophic effects of transplanted rod photoreceptors upon recipient cone photoreceptors in mutant retinal degeneration (*rd*) mice retinas.

1.2. Methods

Rod-rich photoreceptor layers were isolated from postnatal 8 day normal sighted C57 mouse retinas through planar vibratome sectioning, and grafted into the subretinal space of 5 week old *rd* mice. Two weeks after surgery, the numbers of residual cone photoreceptors within the central posterior pole region were evaluated by counting of peanut lectin labelling of whole-mounted host retinas. In another set of experiments, 5 week *rd* retinas were cultured alone or co-cultured for one week in the presence of primary cell cultures prepared from dissociated postnatal 8 day C57 or *rd* mice, or equivalent masses of explanted retinas from 5 week old C57 (all preparations

* Author for correspondence and reprint requests: Tel.: (33) 388 24 33 15; FAX: (33) 388 24 33 14; e-mail: hicks@neurochem.u-strasbg.fr

Retinal Degenerative Diseases and Experimental Therapy, edited by Hollyfield *et al.*
Kluwer Academic / Plenum Publishers, New York, 1999.

containing large numbers of rods) or 5 week *rd* mice (containing very few rods). These feeder cultures were separated from the target *rd* retina by a semi-permeable membrane. The number of residual cones within the target retina at the end of the culture period was then determined through unbiased stereological counting of lectin or anti-arrestin antibody labelled retinas.

1.3. Results

Examination of flat-mounted, lectin-labelled retinas revealed the presence of numerous fluorescent cones scattered in a monolayer throughout the retinal surface. The visualization of rod-rich photoreceptor grafts was accomplished by opsin immunohistochemistry. Enumeration of cones within the central field revealed highly significantly greater numbers (38% more) in grafted compared to sham-operated paired control retinas ($p < 0.01$), irrespective of the location of the rod-rich graft. Analysis of organ-cultured *rd* retinas revealed highly significantly greater numbers of total surviving cones when cultured in the presence of rod-containing feeder cultures (8 day *rd* and C57, 5 week C57) compared to controls or to feeder cultures lacking rods (5 week *rd*). Indeed, when figures were corrected for non-specific cell death prevention of cone loss compared to expected numbers *in vivo* was approximately 80%.

1.4. Conclusions

These are the first quantitative data indicating the existence of diffusible trophic factors released by rods and affecting cone survival. As these latter are responsible for high acuity vision and degenerate secondarily to rods in many retinal pathologies, such data have important implications for both eventual therapeutic approaches to human retinal degeneration and for the understanding of cellular interactions within the retina.

2. INTRODUCTION

Despite considerable recent advances, our current level of understanding of the events leading to visual malfunction and cell losses in retinitis pigmentosa (RP), a group of untreatable inherited blinding diseases, has not yet provided clues to improve retinal cell survival.[1–3] Identified human mutations affect genes encoding for both structural (eg. peripherin/rds[4]) and functional (eg. rhodopsin[5]) proteins of rod photoreceptors. RP is characterized largely by early features of rod photoreceptor impairment, yet at various stages of the disease a loss of the photopic function reflecting degeneration of cone photoreceptors is consistently found.[6] Hence although of extreme importance for the patient, cone cell death and loss of function appear to be secondary to rod cell degeneration. No current data exist to account for cone cell death, but observations made on animal models of RP provide arguments for cone survival being dependent on the presence of rods. In several models with selective elimination of rods, secondary loss of cones is observed.[7–10] The cause of rod cell death in these animals, either transgenic or spontaneous mutants, cannot account for direct cone cell loss. Putative mechanisms of cone cell death include the liberation of endotoxins by degenerating rods, environmental alteration, or deprivation of rod-derived trophic factor(s).

Rod photoreceptor degeneration caused by a mutation in the beta subunit of rod cGMP phosphodiesterase has been described in both the retinal degeneration (*rd/rd*) mouse[11] and in human RP.[12] In the *rd/rd* mouse, a differential effect of the mutation on rods and cones was demonstrated by Carter-Dawson and co-workers,[8] i.e. cones die after rods. This model hence appeared appropriate to test the hypothesis of the dependence of cones on the viability of rods. To search for rod-derived trophic factors influencing cone survival, we used two experimental approaches: transplantation of rod-rich grafts into rodless *rd* hosts,[13] and *in vitro* models permitting a more controlled and reproducible assessment of photoreceptor cell interactions in the *rd* mouse retina.[14] We demonstrate the existence of such a factor(s). In view of the importance of cones for high acuity vision and colour perception,[15] such a beneficial effect has important implications for both the understanding of retinal degeneration and the development of therapies.

3. MATERIALS AND METHODS

Experimental protocols are given in detail elsewhere,[13,14] and will be summarized here.

3.1. Transplantation Studies

Recipient mice (C3H/He, *rd/rd*) aged 35 days, an age at which the rods have degenerated leaving a monolayer of cones,[8] were anesthetized. Dissection and isolation of the photoreceptor layer from donor normal C57 mice (aged 8 days postnatal) were performed by vibratome sectioning as previously described.[16] The photoreceptor preparation was loaded by aspiration into a polyethylene catheter, and positioned in the sub-retinal space of receivers. All transplantations were performed in the right eye. All left eyes were sham-operated, i.e. injection of culture media, to induce comparable retinal detachment.

The animals were killed with an overdose of anesthetic two weeks after surgery, the eyes were enucleated and fixed, and the retinas were processed for immunofluorescent labelling. Grafts were visualized using rho-4D2 rod opsin-specific antibody,[17] and host cones were detected by incubation in the cone-specific marker peanut agglutinin lectin (PNA).[18] The retinas were then flat mounted with the photoreceptor layer facing up and viewed with a Nikon Optiphot 2 photomicroscope.

Cell counts were performed on four quadrants centered on the optic nerve, selected in view of previous studies[8] showing that photoreceptor cell loss was maximal in that location, i.e. absence of residual rods and less than 50% remaining cones. Areas where cell counts could not be performed (e.g. radial incisions, presence of transplant, residual RPE) were subtracted and the overall countable surface determined for each observer after digitization.

3.2. Tissue Culture Studies

Retinas destined for organ culture experiments were isolated from 5 week old *rd* mice and flattened gently at the bottom of culture wells, the photoreceptor layer facing up, in 0.5 ml Dulbecco's Modified Eagle's Medium (DMEM). Control retinas were cultured in DMEM only for 7 days, while treated retinas were co-cultured for the same period under Falcon minicell inserts containing dissociated cell cultures prepared from retinas of 8 postnatal day normal or *rd* mice, or intact explanted retinas from 5 week old normal or *rd* mice respectively. All controls included the retina from the fellow eye.

After 7 days in culture, the retinas were fixed and treated as above. Labelling of cones in cultured retinas was performed either by lectin cytochemistry as above, or with specific anti-arrestin polyclonal antibody (generous gift of Dr. I. Gery, NIH, USA); or rho-4D2 anti-opsin monoclonal antibody.[17] Following extensive rinsing the retinas were flat mounted and examined as above.

Total numbers of labelled cones throughout the entire retinal surface were estimated using a stereological approach in order to obtain unbiased samples.[19] The cells were counted using a "systematic random" sampling procedure consisting of an initial random choice made within the first measurement interval, with all subsequent sampling performed at predetermined intervals throughout the entire retinal surface.

4. RESULTS

4.1. Transplantation Studies

As murine retinas are heavily rod-dominant (> 95% rods), photoreceptor grafts were primarily composed of rods. Within flattened wholemounts of *rd* retinas removed at two weeks post-operation, photoreceptor grafts were visualized either through their prior incorporation of the fluorescent marker PKH26 (data not shown) or by immunolabelling with anti-opsin antibody. Transplants were of variable size and shape, routinely appearing as a large sheet or scattered islands of rods covering an estimated 5–20% of the total retinal surface (Fig. 1). Host cones were visualized through the binding of PNA, which at low magnification appeared as discrete spots distributed throughout retinal wholemounts (Fig. 2). The quantitative estimates of lectin-labelled cone numbers within the central field of grafted and sham-operated retinas are summarized in Fig. 3. Means of 1700 versus 1235 cones/mm^2 or an increase in density of 38% were scored for grafted compared to control retinas respectively. Statistical differences between grafted and sham-operated animals were highly statistically significant ($p < 0.01$). The additional cones present in grafted retina did not arise by migration from the grafts as they did not exhibit red fluorescence. Sham-operated retinas did not possess significantly different cone numbers from age-matched non-operated *rd* mice.

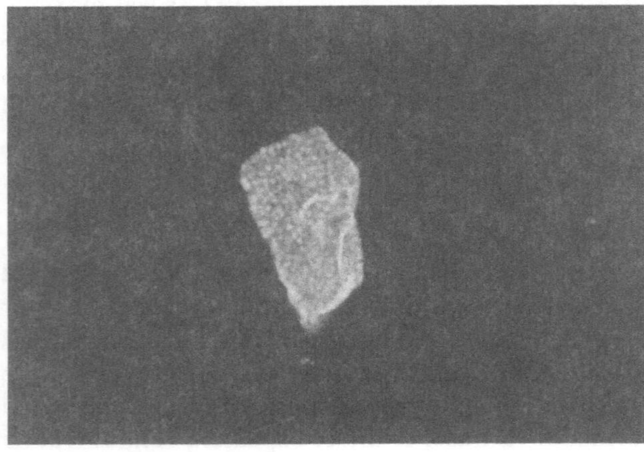

Figure 1. Fluorescently labelled photoreceptor transplant positioned on fixed, flat-mounted host *rd* mouse retina (black background), 2 weeks post-surgery. Magnification = ×250. Taken from ref. 13.

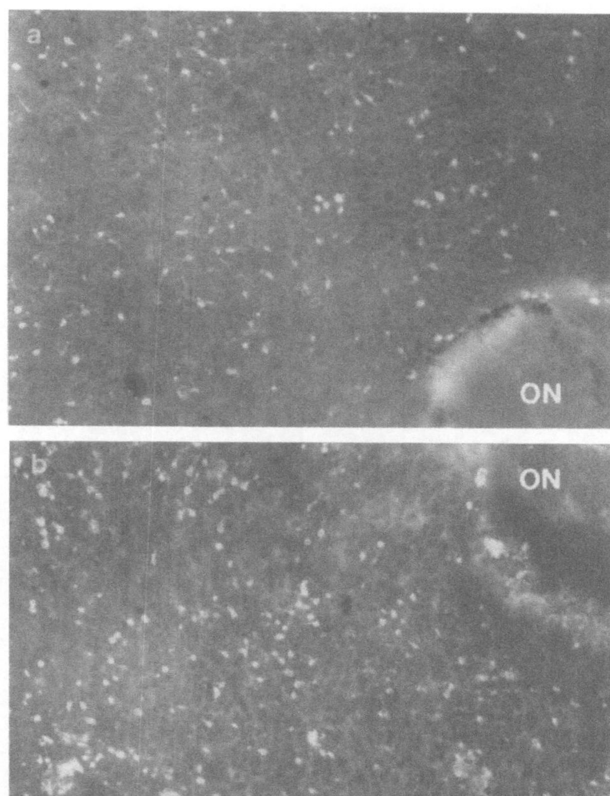

Figure 2. Representative sectors photographed from central quadrants of flat-mounted grafted (a) and sham-operated (b) *rd* mice retinas. At this low magnification, the cones are visible as brightly fluorescent points scattered across the surface. Greater numbers of cones are visible in the grafted compared to the control retina. ON = optic nerve head. Magnification = × 250. Taken from ref. 13.

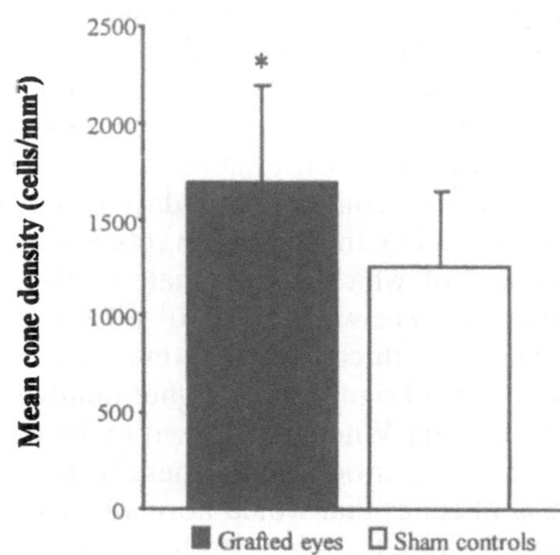

Figure 3. Quantitative estimates of cone density in the central field of grafted and sham-operated *rd* mice retinas (n = 41), two weeks post-operation survival time. Highly significantly different numbers of cones (*$p < 0.01$) are observed in grafted host retinas. Taken from reference 13.

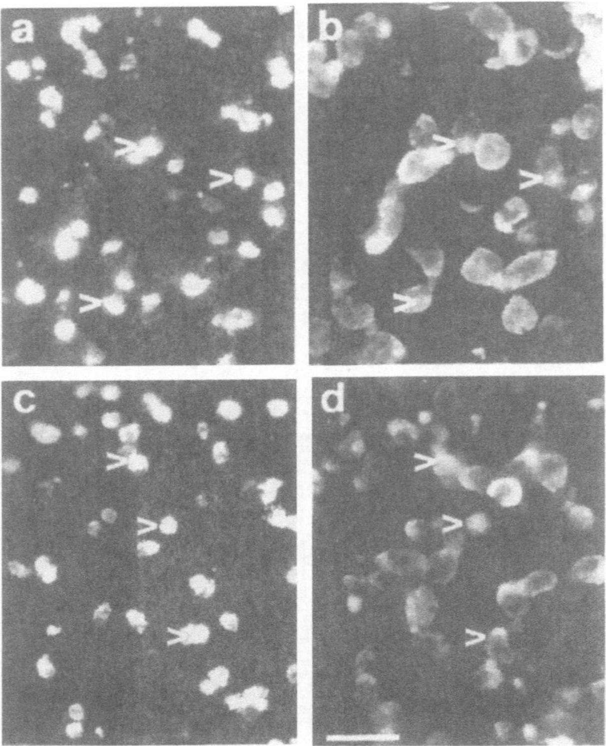

Figure 4. Explanted *rd* mice retinas cultured for one week in DMEM alone (a, b) or in the presence of dissociated young retinal cells (c, d), then flat-mounted and double immunolabelled with PNA lectin (a, c) and arrestin antiserum (b, d). The same fields are shown in a and b, and in c and d, and these higher power micrographs reveal that PNA lectin labels rounded matrix-associated structures protruding from the retinal surface, while arrestin is present throughout the photoreceptor cell bodies and outer segments. The differing focal plane of the two labels makes superimposition difficult, but the overlapping nature of the staining pattern can be observed (corresponding arrowheads in paired fields a and b, and paired fields c and d). Scale bar = 10 μm. Taken from ref. 14.

4.2. Tissue Culture

After 7 days *in vitro*, explants of *rd* retinas that had been cultured alone or in the presence of rod-containing "feeder cultures" were fixed and processed for immunolabelling. PNA labelling of explanted *rd* retinas revealed small rounded structures, corresponding to cone outer segment matrix sheaths, present exclusively at the scleral surface (Figs. 4a, c). Double immunolabelling of retinas with arrestin antiserum revealed in addition the presence of underlying cell bodies and small irregular outer segments overlapping with the PNA stain (Figs. 4b, d).

Quantitative estimates of PNA labelled cones of 5 week old *rd* mouse retinas cultured alone (control) for 8 days revealed absolute numbers of $10.2 \times 10^4 \pm 6.4 \times 10^3$ (mean ± S.D.). In contrast, in retinas co-cultured with cells from 8 day old *rd* or normal retinas (of which approximately 30–50% are rods), the average number of PNA-labelled cones was $11.8 \times 10^4 \pm 7.6 \times 10^3$ and $12.3 \times 10^4 \pm 7.4 \times 10^3$ respectively ($p < 0.0001$ in both cases). *Rd* retinas co-cultured with 5 week old normal retinal explants also showed significantly higher numbers of PNA-labelled cones: $12.0 \times 10^4 \pm 7.5 \times 10^3$ ($p < 0.0001$). When normalized to the expected cone loss in this animal strain over the same time period *in vivo*, these increases represented maintenance of approximately 40% of cones that would normally degenerate; compensation for cone loss due to the tissue culture technique itself revealed that these increases represented actually around 80–90%. No such difference was found when the mutant retina was co-cultured with 5 week old *rd* mice retinal explants ($10.1 \times 10^4 \pm 5.4 \times 10^3$ in controls versus $10.2 \times 10^4 \pm 6.4 \times 10^3$) ($p > 0.5$) (Fig. 5). Residual rods present mainly at the periphery of 5 week *rd*

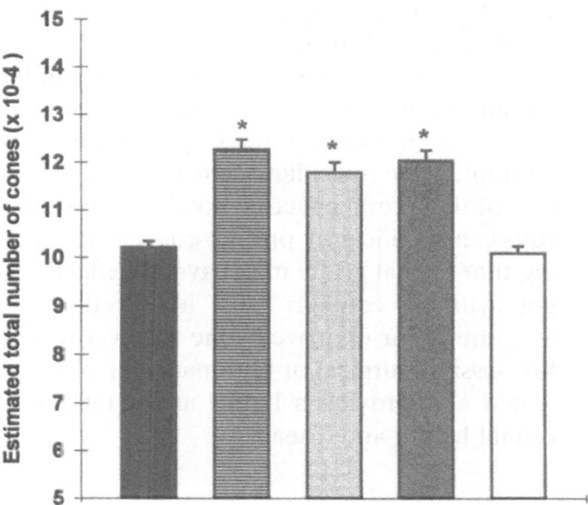

Figure 5. Estimated total number of cones (Mean +/− SD) in 5 week old *rd* (C3H/He/J) mice retinas cultured in DMEM alone (black bar), and from 5 week old *rd* mice retinas co-cultured with cells from 8 day old normal mice retinas (horizontal striped bar), with cells from 8 day old *rd* mice retinas (finely stippled bar), with retinal explants from 5 week old normal mice (diagonally striped bar) and with retinal explants from 5 week old *rd* mice (white bar). *$p < 0.0001$. Taken from ref. 14.

retinas (visualized through binding of rho-4D2) represented ~10% cone photoreceptor numbers, and in contrast to cones their numbers were not altered significantly by co-culture (controls: $12.0 \times 10^3 \pm 3.4 \times 10^2$; co-cultures: $12.7 \times 10^3 \pm 4.3 \times 10^2$).

5. DISCUSSION

This review summarizes data from two lines of investigation implicating rod photoreceptors in cone survival. The improved survival of *rd* mouse cone cells following rod photoreceptor transplantation, irrespective of the distance from the graft site, indicated the release of soluble factors from the transplant. Co-culture experiments involving different rod-containing populations and target *rd* explants confirmed the existence of such a factor(s). These studies represent, to our knowledge, the first reports of any objective rescue effects on cones in genetic retinal degenerations affecting primarily rods.

Research aimed at alleviating human retinal degenerations is currently exploring three main avenues: gene therapy, pharmacology, and transplantation. The first includes attempts at replacing defective genes,[20,21] the second at slowing photoreceptor cell loss using growth factors[22–24] or vitamin supplements[25] and the third by grafting different retinal cell populations.[16,26–31] The present studies suggest overlap between these categories, since diffusible pharmacologically active molecules appear to be released either directly by transplanted rods or by supportive cells such as Müller glia stimulated by them. We have shown in an *in vitro* model of aged human retina that glial cells are a prerequisite for long term photoreceptor survival.[32] The existence of diffusible trophic factors influencing photoreceptor survival was initially suggested from studies of chimeric normal and dystrophic rat retina in which regions of rod survival overgrew normal RPE.[33] In chimeric retinas composed of patches of normal and transgenic photoreceptors expressing a mutant rhodopsin gene, uniform retinal degeneration independent of the genotype was observed.[34] The authors proposed that each photoreceptor contributes to and draws from a common pool of trophic factor. Taken together these studies support a role for cellular interactions in cone survival within dystrophic and possibly normal retina.

It is interesting to note that rods were not protected under the experimental conditions used in our study, presumably because by this stage their apoptotic cycle has already been activated. Such data indicate that timing of treatments will be of paramount importance in protecting one or the other photoreceptor population.

In conclusion, the data presented here suggest that rescue of cones could be attempted by transplantation of photoreceptor cells, and eventually by administration of the "trophic factor" involved. Transplantation performed after loss of almost all rods can significantly protect cones even at a distance from the surgical area, suggesting that retinal grafts may have beneficial effects other than through directly reforming synaptic contacts with host retinas. Identification of a diffusible substance accounting for improved cone survival would not only provide a significant incentive for possible surgical or pharmacological therapeutic approaches to treatment of RP. It would also provide a better understanding of cone survival and death pathways in retinal health and disease.

REFERENCES

1. E.L. Berson, 1993, Retinitis pigmentosa: The Friedenwald Lecture, *Invest. Ophthalmol. Vis. Sci.* **34**:1659–1676.
2. R. Adler, 1996, Mechanisms of photoreceptor death in retinal degeneration, *Arch. Ophthalmol.* **114**:79–83.
3. G.Q. Chang, Y. Hao, and F. Wong, 1993, Apoptosis: final common pathway of photoreceptor death in rd, rds, and rhodopsin mutant mice, *Neuron* **11**:595–605.
4. J. Farrar, P. Kenna, S.A. Jardan, R. Kumar-Singh, M.M. Humphries, E.M. Sharp, D.M. Sheils, and P. Humphries, 1991, A three base pair deletion in the peripherin/rds gene in one form of retinitis pigmentosa, *Nature* **354**:478–480.
5. T. Dryja, T.L. McGee, E. Reichel, L.B. Hahn, G.S. Cowley, D.W. Yandell, M.A. Sandberg, and E.L. Berson, 1990, A point mutation of the rhodopsin gene in one form of retinitis pigmentosa, *Nature* **343**:364–366.
6. A.V. Cideciyan, D.C. Hood, Y. Huang, E. Banin, Z.Y. Li, E.M. Stone, A.H. Milam, and S.G. Jacobson, 1998, Disease sequence from mutant rhodopsin allele to rod and cone photoreceptor degeneration in man, *Proc. Natl. Acad. Sci. USA*. **95**:7103–7108.
7. D.B. Farber, 1995, From mice to men: The cyclic GMP phosphodiesterase gene in vision and disease, *Invest. Ophthalmol. Vis. Sci.* **36**:261–275.
8. L. Carter-Dawson, M.M. Lavail, and R.L. Sidman, 1978, Differential effect of the rd mutation on rods and cones in the mouse retina, *Invest. Ophthalmol. Vis. Sci.* **17**:489–498.
9. J. Usukura, W. Khoo, T. Abe, M.L. Breitman, and T. Shinohara, 1994, Cone cells fail to develop normally in transgenic mice showing ablation of rod photoreceptor cells, *Cell Tiss. Res.* **275**:79–90.
10. M.A. McCall, R.-G. Gregg, K. Merriman, Y. Goto, N.S. Peachey, and L.R. Stanford, 1996, Morphological and physiological consequences of the selective elimination of rod photoreceptors in transgenic mice, *Exp. Eye Res.* **63**:35–50.
11. C. Bowes, T. Li, M. Danciger, L.C. Baxter, and D.B. Farber, 1990, Retinal degeneratioon in rd mouse is caused by a defect in the β-subunit of rod cGMP-phosphodiesterase, *Nature* **347**:667–680.
12. M.E. McLaughlin, M.A. Sandberg, E.L. Berson, and T.P. Dryja, 1993, Recessive mutations in the gene encoding the β-subunit of rod phosphodiesterase in patients with retinitis pigmentosa. *Nature Genetics* **4**:130–133.
13. S. Mohand-Said, D. Hicks, M. Simonutti, D. Tran-Minh, A. Deudon-Combe, H. Dreyfus, M.S. Silverman, J.M. Ogilvie, T. Tenkova, and J.A. Sahel, 1997, Photoreceptor transplants increase host cone survival in the retinal degeneration (rd) mouse, *Ophthalmic Res.* **29**:290–297.
14. S. Mohand-Said, A. Deudon-Combe, D. Hicks, M. Simonutti, V. Forster, A.C. Fintz, T. Léveillard, H. Dreyfus, and J.A. Sahel, 1998, Normal retina releases a diffusible factor stimulating cone survival in the retinal degeneration (rd) mouse, *Proc. Natl. Acad. Sci. USA*, **95**:8357–8362.
15. J.E. Dowling, 1987, *The Retina-An Approachable Part of the Brain*, Harvard University Press, Cambridge, MA, USA.

16. M.S. Silverman and S.E. Hughes, 1989, Transplantation of photoreceptors to light-damaged retinas, *Invest. Ophthalmol. Vis. Sci.* **30**:1684–1690.
17. D. Hicks and R.S. Molday, 1986, Differential immunogold-dextran labeling of bovine and frog rod and cone cells using monoclonal antibodies against bovine rhodopsin, *Exp. Eye Res.* **42**:55–71.
18. J.C. Blanks and L.V. Johnson, 1984, Specific binding of peanut lectin to a class of retinal photoreceptor cell, *Invest. Ophthalmol. Vis. Sci.* **25**:546–557.
19. R.E. Coggeshall and H.A. Lekan, 1996, Methods for determining numbers of cells and synapses: a case for more uniform standards of review, *J. Comp. Neurol.* **364**:6–15.
20. J. Lem, J.G. Flannery, T. Li, M.L. Applebury, D.B. Farber, and M.I. Simon, 1992, Retinal degeneration is rescued in transgenic rd mice by expression of the cGMP phosphodiesterase beta subunit, *Proc. Natl. Acad. Sci. USA* **89**:4422–4426.
21. J. Bennett, T. Tanabe, D. Sun, Y. Zeng, H. Kjeldbye, P. Gouras, and A.M. Maguire, 1996, Photoreceptor cell rescue in retinal degeneration (rd) mice by in vivo gene therapy, *Nature Medicine* **2**:649–654.
22. R.H. Steinberg, 1994, Survival factors regulating retinal degenerations, *Curr. Opin. Neurobiol.* **4**:515–524.
23. E.G. Faktorovich, R.H. Steinberg, D. Yasumura, M.T. Matthes, and M.M. Lavail, 1990, Photoreceptor degeneration in inherited retinal dystrophy delayed by basic fibroblast growth factor, *Nature* **347**:83–86.
24. M.M. LaVail, D. Yasumura D, M.T. Matthes, C. Lau-Villacorta, K. Unoki, C.H. Sung, and R.H. Steinberg,1998, Protection of mouse photoreceptors by survival factors in retinal degenerations. *Invest. Ophthalmol. Vis. Sci.* **39**:592–602.
25. E.L. Berson, B. Rosner, M.A. Sandberg, K.C. Hayes, B.W. Nicholson, C. Weigel-DiFranco, and W. Willett, 1993, A randomized trial of vitamin A and vitamin E supplementation for retinitis pigmentosa, *Arch. Ophthalmol.* **111**:761–772.
26. M. Del Cerro, 1989, Retinal Transplants, in: *Progress in Retinal Research*, Volume 9 (N.N. Osborne and G.J. Chader, eds.), pp 229–272, Pergamon Press, Oxford, UK.
27. L. Qi Jian, M. Jorquera, and J.W. Streilein, 1993, Subretinal space and vitreous cavity as immunologically privileged sites for retinal allografts, *Invest. Ophthalmol. Vis. Sci.* **34**:3347–3354.
28. R.D. Lund, J.D. Radel, and P.J. Coffey, 1991, The impact of intracerebral retina transplants on types of behaviour exhibited by host rats, *Trends Neurosci.* **14**:358–362.
29. R. Aramant, M. Seiler, B. Ehinger, A. Bergstrom, B. Gustavii, P. Brundin, and A.R. Adolph, 1990, Transplantation of human embryonic retina to adult rat retina. *Restorative Neurol. Neurosci.* **2**:9–22.
30. L. Li and J.E. Turner, 1988, Inherited retinal dystrophy in the RCS rat: prevention of photoreceptor degeneration by pigment epithelial transplantation, *Exp. Eye Res.* **47**:911–917.
31. R. Lopez, P. Gouras, H. Kjeldbye, B. Sullivan, V. Reppucci, M. Brittis, F. Wapner, and E. Goluboff, 1989, Transplanted retinal pigment epithelium modifies the retinal degeneration in the RCS rat, *Invest. Ophthalmol. Vis. Sci.* **30**:586–588.
32. D. Hicks, V. Forster, H. Dreyfus, and J.A. Sahel, 1994, Survival and regeneration of adult human photoreceptors in vitro, *Brain Res.* **643**:302–305.
33. R.J. Mullen and M.M. LaVail, 1976, Inherited retinal dystrophy: primary defect in pigment epithelium determined with experimental rat chimeras, *Science* **192**:799–801.
34. P.C. Huang, A.E. Gaitan, Y. Hao, R.M. Petters, and F. Wong, 1993, Cellular interactions implicated in the mechanism of photoreceptor degeneration in transgenic mice expressing a mutant rhodopsin gene, *Proc. Natl. Acad. Sci. USA* **90**:8484–8488.

PIGMENT EPITHELIUM-DERIVED FACTOR (PEDF) IN THE RETINA

Protective Effect against Photoreceptor Cell Degeneration

S. Patricia Becerra,[1]* Elena Alberdi,[1] Alfredo Martinez,[2] Luis M. Montuenga,[2,3] Michel Cayouette,[4] and Claude Gravel[4]

[1] Laboratory of Retinal Cell and Molecular Biology
National Eye Institute
National Institutes of Health, Bethesda, Maryland
[2] CCBD, DCS, National Cancer Institute
Bethesda, Maryland
[3] Department of Histology and Pathology
School of Medicine
University of Navarra
Pamplona, Spain
[4] Laboratoire de Transfert de Gènes
Centre de Recherche
Université Laval Robert-Giffard
Beauport, Quebec, Canada

1. INTRODUCTION

Pigment epithelium-derived factor (PEDF) was first described as a secreted product of human fetal retinal pigment epithelial (RPE) cells in culture with a potent neuronal differentiating activity on human retinoblastoma Y-79 cells.[1] It was reported that PEDF at nanomolar concentrations induces a morphological differentiation of

* Correspondence should be addressed to S. Patricia Becerra, National Eye Institute, National Institutes of Health, Building 6, Room 308, 6 Center Drive, MSC 2740, Bethesda MD 20892-2740. Telephone: 301-496-6514; fax: 301-402-9211; e-mail: pbecerra@helix.nih.gov

Retinal Degenerative Diseases and Experimental Therapy, edited by Hollyfield *et al.*
Kluwer Academic / Plenum Publishers, New York, 1999.

Y-79 cells, characterized by the elongation of neurite-like processes and the expression of neuronal markers, e.g., neuron-specific enolase and the 200-kDa neurofilament subunit.[2] It was shown later that PEDF can also induce the differentiation of cells from another human retinoblastoma, the Weri cells.[3] While its role in vivo on retina cells is not known yet, its origin and these activities suggest that PEDF may play a role as a neurotrophic factor for the retina. In this regard, evidence of its effects on neurons from the CNS and PNS is accumulating rapidly and supports the notion that PEDF is a multipotent neurotrophic factor that acts upon various types of neurons and from different species. For example, it promotes the survival of rat cerebellar granule cell (CGC) neurons in culture,[4] including protection against glutamate neurotoxicity[5] and against both natural and induced apoptosis in vitro;[6] it promotes survival and differentiation of developing avian and murine spinal motor neurons.[7] Given the important biological activities of PEDF, it is of interest to investigate its mechanism of action on the retina. Therefore, the overall objective of this review will be to discuss the biochemistry of PEDF in the context of a potential use as a relevant neurotrophic factor for the retina.

2. STRUCTURE-FUNCTION RELATIONSHIPS OF PEDF

Isolation, cloning, and sequencing of the full length cDNA for human PEDF was the first step for the study of the structure-function relationships of this protein.[2,8] The derived amino acid sequence predicts a polypeptide of 418 residues (about 46,700 Da) bearing an amino-terminal signal peptide and a consensus glycosylation site, both of which are characteristics common to secreted proteins. Sequence alignment and comparison with GenBank databases revealed that it is a unique sequence but with strong homology to members of the serine protease inhibitor (serpin) superfamily. The serpins are structurally related proteins of 400–500 amino acids, and most of them behave as extracellular, irreversible serine-protease-inhibitors. They have one well-defined structural-functional characteristic: the serpin reactive region located toward the carboxy-end of their polypeptide sequence, which in the folded protein forms an exposed peptide loop and acts as a bait for an appropriate serine protease. Examples of serpins include those proteins with inhibitory activities against serine proteases, such as antitrypsin, antithrombin III, and protease nexin-1; and proteins that based on strong sequence similarities are said to belong to this family, but lack a demonstrable inhibitory activity, e.g., ovalbumin, angiotensinogen, and maspin.[9] In addition to PEDF, the other serpin with demonstrated neurotrophic activity is protease nexin-1. It is believed that this serpin induces its activity by inhibiting thrombin, its physiological target.[10,11]

Extensive biochemical and biophysical analyses on the PEDF protein have revealed that, although it has the structural characteristics of serpins, PEDF behaves as a noninhibitory serpin, like ovalbumin, maspin and angiotensinogen.[12,13] As opposed to protease nexin-1, PEDF does not inhibit thrombin in the absence or presence of heparin, nor several other known serine proteases. Furthermore, PEDF cleaved at its serpin exposed peptide loop or even without large segments from its carboxy-end, retains its neurotrophic activity.[6,7,12] These observations have demonstrated that a region spanning amino acid positions 44–121 confers a neurotrophic function to the PEDF polypeptide.[12] Thus, the neurotrophic activity of the PEDF serpin is not mediated by protease inhibition.

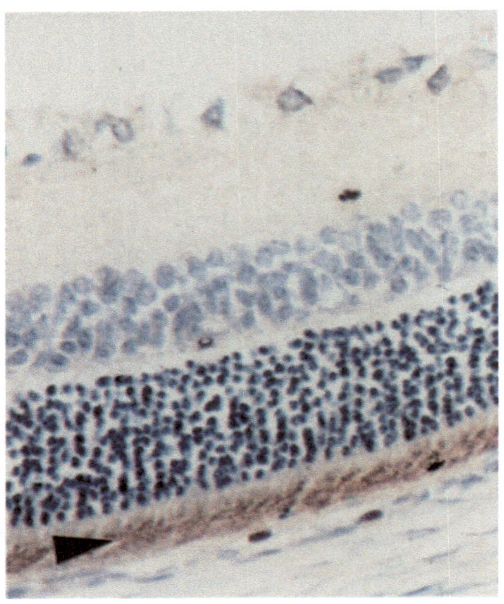

Figure 1. Cross-section of the retina showing immunoreactivity to the polyclonal antiserum to PEDF, Ab-rPEDF[18], in the IPM region from an albino rat (22 days old). The arrow head points to the IPM region and photoreceptor outer segments.

3. PEDF IS A NATURAL PROTEIN OF THE RETINA

To elucidate the biochemical activities by which the PEDF neurotrophic property occurs, we are studying the interactions of PEDF with components of its natural microenvironment. As a first step, we have investigated the expression pattern of the PEDF gene in the eye. In all ocular tissues examined so far (bovine, human, monkey, and murine), we found high levels of PEDF mRNA in the retinal pigment epithelium (RPE) relative to the lower or undetectable levels found in the neural retina.[14–16] In addition to the fact that cultured RPE cells secrete soluble PEDF protein to the media,[17] these observations have proposed that the RPE is the major source of PEDF as a diffusible factor of the retina. Indeed, in the native eye, the PEDF protein has been identified as an extracellular protein of the interphotoreceptor matrix (IPM), and also in vitreous and aqueous.[14,17–19] It is readily purified by biochemical fractionation of saline lavages of the bovine IPM and characterized as a soluble glycoprotein of 50,000 MW[18] with an amino-terminus corresponding to position 21 of the bovine PEDF precursor.[15] To support these findings, immunocytochemistry of the rat retina with a polyclonal antiserum to human PEDF shows that the most significant immunostaining is found to the IPM (Fig. 1). These observations imply that PEDF may bind to components of the insoluble IPM and/or cell-surface receptors on retina cells.

4. INTERACTIONS WITH IPM COMPONENTS

The IPM contains glycosaminoglycans (GAGs) and proteoglycans that confer a polyanionic nature or a negative charge to this matrix.[20–25] GAGs are highly sulfated and negatively charged polysaccharide chains, e.g., heparin, heparan and chondroitin sulfates, and proteoglycans are covalently linked GAGs to proteins. Since PEDF coexists with GAGs in the IPM, and is associated with this matrix by ionic interactions, we have examined their interaction at a molecular level.[26] We found that PEDF has binding affinity for heparin, heparan sulfate, and chondroitin sulfates A, B, and C. As an example, direct binding to heparin using affinity column chromatography showed that bovine PEDF protein, either from IPM or vitreous, has heparin-binding affinity mediated by ionic interactions (see Fig. 2). The negative nature of GAGs implies that positively charged residues of PEDF are involved in the binding. We have shown that chemical modification of primary amines of PEDF by biotinylation prevents the binding to heparin, revealing the involvement of positively charged lysine residues of PEDF in heparin binding. In addition, cleavage of the serpin-exposed loop does not affect the heparin binding property, but denaturation of the PEDF protein with urea eliminates this binding affinity. Therefore, the folded protein structure is essential for these interactions, and the determinants for this interaction can be identified in its tertiary conformation.

To map the GAG binding domain we developed homology molecular models for PEDF based on the X-ray crystal structures of ovalbumin (a noninhibitory serpin), and antithrombin III (a heparin dependant inhibitory serpin), as described previously.[26] Using computer graphics we replaced the amino acid sequence of these serpins with that of PEDF, and calculated the surface electrostatic properties to reveal clusters of basic or positively charged areas on the surface of PEDF. The most striking feature is an extended positively charged region at the center of β-sheet A /helix F that has basic electrostatic surface potential. This region is densely populated with lysines exposed to

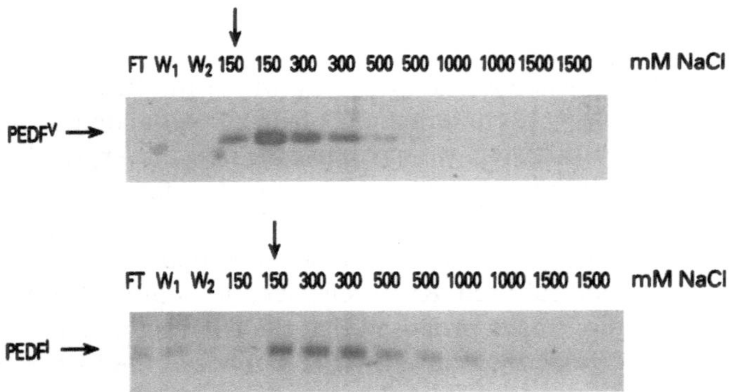

Figure 2. Heparin-binding property of PEDF. PEDF purified from bovine vitreous,[19] PEDF^V, and IPM,[18] PEDF^I, was subjected to heparin affinity column chromatography, as described previously.[26] A total of 15 μg of PEDF^V was mixed with 0.5 ml heparin acrylic beads, and 20 μg PEDF^I with 0.25 ml heparin acrylic beads in binding buffer. After incubation at 4°C for 30 min, the flow-through was collected, the beads were washed extensively with binding buffer and the bound protein eluted with a step gradient of NaCl. Proteins in fractions were resolved by SDS-12.5% polyacrylamide gel electrophoresis. Photographs of silver stained gels are shown. Flow-through (FT), washes (W_1 and W_2), and eluted fractions with increasing NaCl concentrations are indicated at the top of each photograph. Top arrow indicates the salt concentration at which the bound PEDF starts eluting from the column.

the surface (K134, K137, K189, K191, K214) that are available to interact with various GAGs and other polyanions. We believe that the negatively charged GAGs bind by anchoring onto this positively charged cleft of PEDF. The spatial relationship of the PEDF GAG-binding area and the neurotrophic active region results in interesting observations. The region spanning 44–121 of the human PEDF sequence, which we have previously shown to contain the neurotrophic active region,[4] is oriented toward the side opposite to the GAG-binding domain, and both sites are distant from the serpin-exposed loop. These two regions, the GAG-binding region and the neurotrophic active site, are thus distinct not only at the sequence, but most likely also at the conformational level, in the native PEDF protein.

5. PEDF BINDING TO RETINA

First, we investigated whether PEDF binds to retinoblastoma cells, since they constitute a functional model for the study of potential PEDF receptors.[27] Bovine PEDF purified from IPM[18] was radiolabeled with ^{125}I, and obtained in a biologically active form. Binding activity was determined in retinoblastoma membrane extracts or cells under identical conditions as for the differentiation assay. As shown in Fig. 3, increasing concentrations of unlabeled PEDF used as competitor, increasingly competed for the binding of the radiolabeled ligand to plasma membrane extracts, demonstrating a specific and saturable binding of ^{125}I-PEDF protein to the cell-surface of Y-79 cells. While the polyclonal antiserum to PEDF, which we have shown previously to block the PEDF-mediated activity on Y-79 cells,[18] inhibited binding, an excess of ovalbumin did not, revealing further specificity for the ^{125}I-PEDF-binding to these cells. In a similar fashion as above, we then examined the binding to the native retina. Binding assays were performed with plasma membrane extracts of the neural retina from bovine eyes, and

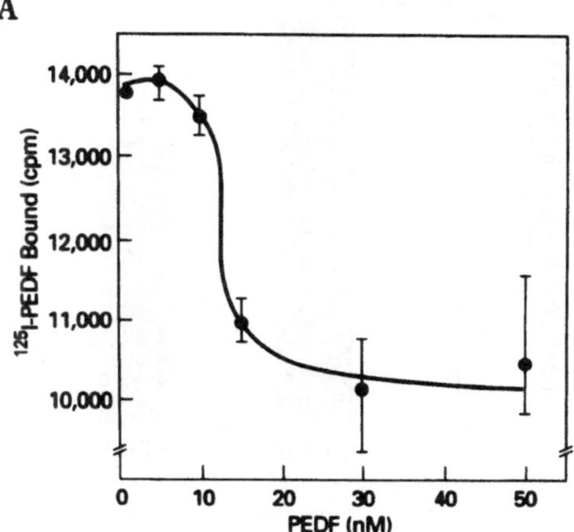

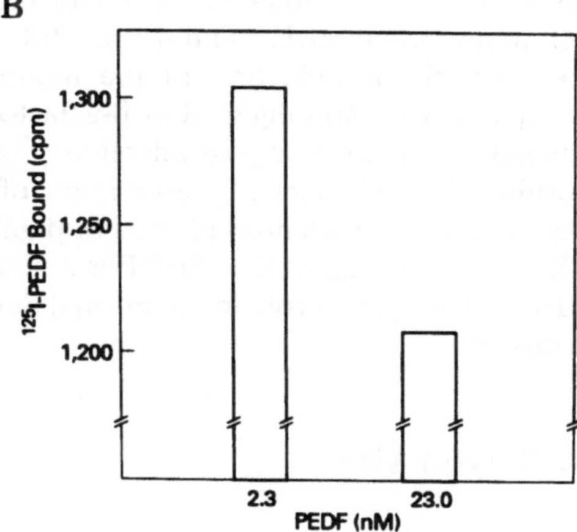

Figure 3. PEDF-binding to human retinoblastoma Y-79 cell membranes. Cells were cultured in fresh serum-free media for 16 hours before lysis of cells. Plasma membrane extracts were prepared by differential centrifugation, and a detergent soluble membrane fraction was obtained using 0.5% CHAPS in phosphate buffered saline. Binding reactions were performed with radiolabeled ^{125}I-PEDF and with increasing concentrations of unlabeled PEDF as competitor. Mixtures were incubated at 4°C for 90 min, and free ligand was separated from bound PEDF by filtration through glass fiber filters under vacuum. Panel A, membrane extracts were resuspended in conditioned media, and mixed with ^{125}I-PEDF at 0.6 nM. Each binding reaction mixture contained extracts derived from about 1.7×10^6 cells in a final volume of 0.2 ml. Panel B, each reaction contained solubilized membranes derived from 3.2×10^6 cells in 0.5 ml, and a total of 2.3 nM radiolabeled PEDF. Plots of bound radioactivity versus total PEDF concentration are shown. All experimental points are given as the average of triplicates.

demonstrated that the radiolabeled PEDF binds specifically to a protein(s) on the surface of retina cells. Using PEDF-affinity column chromatography, we found that cell membranes of human retinoblastoma and bovine retina have a PEDF-binding protein of about 80,000-MW. These results provide evidence for a cell-surface receptor for PEDF, which may be used to trigger the PEDF-mediated neuronal activities in the retina, and add further support to the notion that PEDF is a neurotrophic factor for retinal neurons.

6. PEDF EFFECT ON PHOTORECEPTORS OF RETINAL DEGENERATION MOUSE MODELS

Retinitis Pigmentosa (RP) is a class of genetic diseases causing photoreceptor cell degeneration. The availability of mice bearing mutations in RP genes allows the testing of potential factors for therapeutics strategies. Since PEDF can induce neuronal

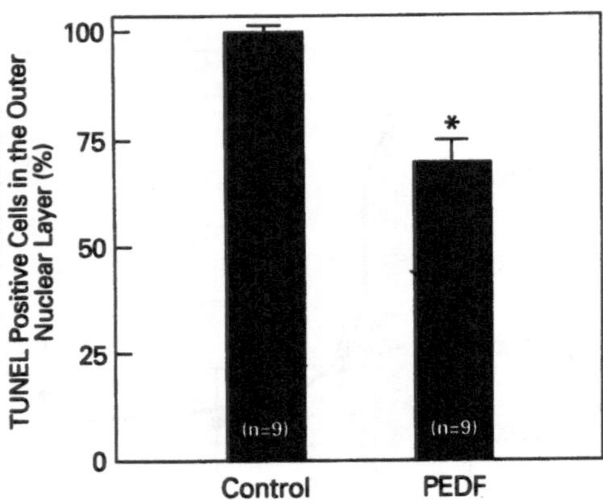

Figure 4. Effect of intravitreal PEDF injection in rd mouse on the number of TUNEL positive cells of the outer nuclear layer (ONL). Injections of 1 µg of purified recombinant human PEDF[13] were on the right eye of PD17 rds mice, while controls correspond to the left eye of each animal either injected with recombinant β-galactosidase or non-injected. TUNEL assays were performed three days post-injection. The percentage of TUNEL positive cells averaged from the right eye (PEDF) compared to left eye (control) is illustrated. Number in each bar corresponds to number of treated animals treated.

differentiation on retinoblastoma cells and rescue cerebellar neurons from natural cell death and glutamate-induced neurotoxicity, we have tested whether exposure to recombinant human PEDF could reduce photoreceptor cell degeneration in the RP mouse models *rd/rd* and *rds/rds*. Intravitreal injection into PD17 *rds* mice resulted in a significant reduction of the number of TUNEL positive photoreceptor cells compared with non-injected or β-galactosidase-injected contralateral eye of the same animals by three days post-injection (Fig. 4). Intravitreal injection into PD14 *rd* mice resulted in a significant preservation of the mean ONL column height after three days compared with non-injected or β-galactosidase injected contralateral eye (Fig. 5). These results suggest that PEDF is a neurotrophic factor for photoreceptors, and that PEDF could play a role in the normal development, survival, and homeostasis of these neurons.

7. SUMMARY

PEDF is an extracellular glycoprotein (50 kDa) identified as a soluble component of the interphotoreceptor matrix. While its role in the retina in vivo is not known yet, its biological activities in vitro suggest it may play a role as a diffusible neurotrophic

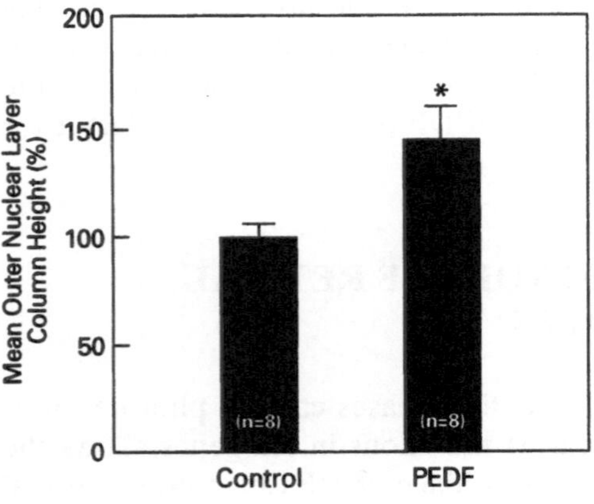

Figure 5. Effect of intravitreal PEDF injection in rds mouse on the mean ONL column height. Injections of 1 µg of recombinant human PEDF[13] were on the right eye of PD14 rd mice, while controls correspond to the left eye of each animal either injected with recombinant β-galactosidase or non-injected. The ONL column height was measured three days post-injection. The percentage of the mean ONL column height from the right eye (PEDF) compared to left eye (control) is illustrated. Number in each bar corresponds to number of treated animals treated.

factor for the retina. It is a member of the serpin superfamily, but does not display a serpin-related- inhibitory activity to explain its mechanism of action. To explore its role on the retina we have examined its interaction with components of its natural microenvironment. PEDF-immunoreactivity is significantly and specifically detected in the IPM area, suggesting that PEDF may be bound to extracellular matrix components and/or receptors on photoreceptor cells where it may exert its action. PEDF is associated with the IPM by ionic interactions, and has binding affinity to several GAGs, which represents the likely molecular basis for its association with the IPM. Human retinoblastoma cells and bovine neural retinae contain a specific cell-surface binding protein for PEDF, which may serve as a receptor for its neurotrophic activity. Finally, PEDF can protect photoreceptors in the rd and rds mice models. In addition to its localization and binding affinity to the retina, this novel protective effect reveals that PEDF can play a role as a neurotrophic factor for the retina.

REFERENCES

1. J. Tombran-Tink, G.J. Chader, and L.V. Johnson, 1991, PEDF: a pigment epithelium-derived factor with potent neuronal differentiative activity, *Exp. Eye Res.* **53**:411–414.
2. F.R. Steele, G.J. Chader, L.V. Johnson, and J. Tombran-Tink, 1993, Pigment epithelium-derived factor: neurotrophic activity and identification as a member of the serine protease inhibitor gene family, *Proc. Natl. Acad. Sci. USA* **90**:1526–1530.
3. S.P. Becerra, 1997, Structure-function relationships of PEDF: A noninhibitory serpin with neurotrophic activity, *Adv. Exp. Med. Biol.* **425**:223–237.
4. T. Taniwaki, S.P. Becerra, G.J. Chader, and J.P. Schwartz, 1995, Pigment epithelium-derived factor is a survival factor for cerebellar granule cells in culture, *J. Neurochem.* **64**:2509–2517.
5. T. Taniwaki, N. Hirashima, S.P. Becerra, G.J. Chader, R. Etcheberrigaray, and J.P. Schwartz, 1997, Pigment epithelium-derived factor protects cultured cerebellar granule cells against glutamate-induce neurotoxicity, *J. Neurochem.* **68**:26–32.
6. T. Araki, T. Taniwaki, S.P. Becerra, G.J. Chader, and J.P. Schwartz, 1998, Pigment epithelium-derived factor (PEDF) differentially protects immature but not mature cerebellar granule cells against apoptotic cell death, *J. Neurosci. Res.* **53**:7–15.
7. L.J. Houenou, A.P. D'Costa, L. Li, V.L. Turgeon, C. Enyadike, E. Alberdi, and S.P. Becerra, Pigment epithelium-derived factor (PEDF) promotes the survival and differentiation of developing spinal motor neurons (in preparation).
8. R.J. Pignolo, V.J. Cristofalo, and M.O. Rotenberg, 1993, Senescent WI-38 cells fail to express EPC-1, a gene induced in young cells upon entry into G_0 state, *J. Biol. Chem.* **268**:8949–8957.
9. R. Huber and R.W. Carrell, 1989, Implications of the three-dimensional structure of α_1-antitrypsin for structure and function of serpin, *Biochemistry* **28**:8951–8966.
10. J. Guenther, H. Nick, and D. Monard, 1985, A glia-derived neurite-promoting factor with protease inhibitory activity, *EMBO Journal* **4**:1963–1966.
11. S. Gloor, K. Odink, J. Guenther, H. Nick, and D. Monard, 1986, A glia-derived neurite promoting factor with protease inhibitory activity belongs to the protease nexins, *Cell* **47**:687–693.
12. S.P. Becerra, A. Sagasti, P. Spinella, and V. Notario, 1995, Pigment epithelium-derived factor (PEDF) behaves like a non-inhibitory serpin: neurotrophic activity does not require the serpin reactive loop, *J. Biol. Chem.* **270**:25,992–25,999.
13. E. Stratikos, E. Alberdi, P.G.W. Gettins, and S.P. Becerra, 1996, Characterization of human PEDF: overexpression in eukaryotic cells, *Protein Science* **5**:2575–2582.
14. J. Ortego, J. Escribano, S.P. Becerra, and M. Coca-Prados, 1996, Gene expression of the neurotrophic factor PEDF in the human ciliary epithelium: Synthesis and secretion into the aqueous humor, *Invest. Ophthalmol. Vis. Sci.* **37**:2759–2767.
15. L.A. Perez-Mediavilla, C. Chew, P.A. Campochiaro, R.W. Nickells, V. Notario, D.J. Zack, and S.P. Becerra, 1998, Sequence and expression analysis of bovine pigment epithelium-derived factor (PEDF), *Biochim. Biophys. Acta* **1398**:203–214

16. V.K. Singh, G.J. Chader, and I.R. Rodriguez, 1998, Structural and comparative analysis of the mouse gene for pigment epithelium-derived factor (PEDF), *Mol. Vis.* **4**:7–••.
17. J. Tombran-Tink, S.M. Shivaram, G.J. Chader, L.V. Johnson, and D. Bok, 1995, Expression, secretion, and age-related downregulation of pigment epithelium-derived factor, a serpin with neurotrophic activity, *J. Neurosci.* **15**:4992–5003.
18. Y.-Q. Wu, V. Notario, G.J. Chader, and S.P. Becerra, 1995, Identification of pigment epithelium-derived factor in the interphotoreceptor matrix of bovine eyes, *Protein Express. Purif.* **6**:447–456.
19. Y.-Q. Wu and S.P. Becerra, 1996, Proteolytic activity directed towards pigment epithelium-derived factor in vitreous of bovine eyes: Implications of proteolytic processing, *Invest. Ophthalmol. Vis. Sci.* **37**:1984–1993.
20. J.E. Morris, M. Yanagishita, and V.C. Hascall, 1987, Proteoglycans synthesized by embryonic chicken retina in culture: composition and compartmentalization, *Arch. Biochem. Biophys.* **258**:206–218.
21. M. Kaneko, 1987, Interphotoreceptor matrix glycosaminoglycans in bovine eye, *Ophthalmic Res.* **19**:330–337.
22. G.S. Hageman and L.V. Johnson, 1987, Chondroitin 6-sulfate glycosaminoglycan is a major constituent of primate cone photoreceptor matrix sheaths, *Curr. Eye Res.* **6**:639–646.
23. F. Murillo-Lopez, L. Politi, R. Adler, and A.T. Hewitt, 1991, Proteoglycan synthesis in cultures of murine retinal neurons and photoreceptors, *Cell Mol. Neurobiol.* **11**:579–591.
24. Y. Chu, L.N. Walker, S.L. Vijayasekaran, R.L. Cooper, K.V. Porrello, and I.J. Constable, 1992, Developmental study of chondroitin-6-sulphate in normal and dystrophic rat retina, *Graefes Arch. Clin. Exp. Ophthalmol.* **230**:476–482.
25. R.A. Landers and J.G. Hollyfield, 1992, Proteoglycans in the mouse interphotoreceptor matrix. VI. Evidence for photoreceptor synthesis of chondroitin sulfate proteoglycan using genetically fractionated retinas, *Exp. Eye Res.* **55**:345–356.
26. E. Alberdi, C.C. Hyde, and S.P. Becerra, 1998, Pigment epithelium-derived factor (PEDF) binds to glycosaminoglycans: Analysis of binding site, *Biochemistry* **37**:10,643–10,652.
27. E. Alberdi and S.P. Becerra, 1997, PEDF-binding activity to retinolastoma cells, *Adv. Exp. Med. Biol.* **425**:307.

50

RETINAL NEURONS IN PRIMARY CELL CULTURE

Inhibition of Apoptosis by Pigment Epithelial Derived Factor (PEDF)

J. F. McGinnis,[1]* W. Chen,[1] J. Tombran-Tink,[2] D. Mrazek,[1] V. Lerious,[1] and W. Cao[1]

[1]Department of Ophthalmology
Dean McGee Eye Institute
608 Stanton L. Young Blvd.
University of Oklahoma Health Science Center
Oklahoma City, Oklahoma, 73104
[2]Center for Neuroscience Research
Children's National Medical Center
111 Michigan Avenue N.W.
Washington, DC 20010

ABSTRACT

We have been interested in cellular regulatory mechanisms operating in mammalian retinal neurons, especially those determining sensitivity to factors which either promote or inhibit cell death. The production of reactive oxygen species through oxidative stress is believed to be an important mediator of neuronal cell death although the precise mechanism by which this occurs is unknown. Pigment epithelia derived factor (PEDF) has been shown to promotes neurotrophic differentiation and the survival of neurons of the central nervous system. Using a cell culture system, we demonstrate that multiple populations of retinal neurons can differentiate and be maintained in a chemically defined environment and that cell death induced by reactive oxygen species can be quantitated. The data also show that this induced death is dose dependent and occurs by an apoptotic mechanism which can be inhibited by PEDF. The mecha-

* email: james-mcginnis@ouhsc.edu

Retinal Degenerative Diseases and Experimental Therapy, edited by Hollyfield *et al.*
Kluwer Academic / Plenum Publishers, New York, 1999.

nism by which PEDF provides this protection may be very important for inhibiting the apoptotic death of retinal neurons which occurs in retinitis pigmentosa, macular degeneration, glaucoma or other neurodegenerative diseases. These cultured neurons will also be useful for analyzing apoptotic responses of each of the retinal neuron cell types to a variety of "cytotoxic" agents and to the rescue effect of a number of "factors" either alone or in combination.

INTRODUCTION

The mammalian retina is composed of an orderly layered arrangement of neurons whose positions reflect general functions with respect to the process of vision. The most exquisite cellular morphology belongs to the rods and cones, the first cells in the signal transduction pathway. Perhaps, because of their morphology or their function, these cells appear to be the most sensitive to genetic and/or environmental insults and are usually among the first cells to die in response to these perturbations. Apoptosis and necrosis are two alternative mechanisms by which cells can die, yet apoptosis, the active form of cell death, is the mechanism by which blindness occurs in all types of retinal degeneration so far investigated including degeneration of cells within the inner retina.[1–4]

We established an *in vitro system* which would enable us to identify and characterize agents which promote the death of retinal neurons and to use this system to screen for growth factors, cytokines, hormones, etc. which can either prevent the cells from choosing death or possibly rescue them after having made that choice. One of the major assets of our cell culture system is that it contains representatives of each of the retinal neuron cell types present *in vivo*. The advantages of using this *in vitro* system of mixed retinal neurons include: 1) differential susceptibility of multiple populations of retinal neurons to the toxins and/or factors can be simultaneously studied and compared; 2) the survival environment for neurons can be chemically defined and controlled; 3) various agents and/or factors can be screened alone or in combinatorial arrays for their ability to promote neuron survival or neuron death; 4) specific signal transduction pathways can be easily studied and cellular regulatory mechanisms identified and/or manipulated.

Oxidative stress is believed to be an important mediator of neuronal cell death but the precise mechanism by which this occurs is unknown. Oxidative stress refers to the cytotoxic consequences of oxygen radicals: superoxide, hydroxy radical, and hydrogen peroxide, which are generated as byproducts of normal and aberrant metabolic processes that utilize molecular oxygen.[5] Oxygen radicals can attack proteins, DNA and lipid membranes, thereby disrupting cellular functions and integrity. The attractive feature of the oxidative stress hypothesis is that it can account for cumulative damage associated with the delayed onset and progressive nature of neurodegenerative disorders, including retinal degeneration.[6]

Pigment epithelia derived factor (PEDF),[7] a protein synthesized by the pigment epithelia cells of the eye, promotes neurotrophic differentiation and the survival of neurons of the central nervous system[8] It has been shown to protect cultured cerebellar neurons from serum withdrawal induced cell death[9] as well as from glutamate cytotoxicity.[10] For these reasons, because of its ability to promote differentiation of Y79 retinoblastoma cells,[11] and because it is present in high concentration in the interphotoreceptor matrix of the retina,[11] we decided to evaluate its potential to protect photoreceptor cells from oxidative stress.

In this study, we demonstrate that our model retina cell culture system is capable of sustaining all of the multiple populations of retinal neurons in a chemically defined environment and that cell death can be quantitated. The data also show that free radical damage, induced by H_2O_2, results in a dose-dependent apoptotic cell death which can be inhibited by PEDF. The mechanism by which PEDF provides this protection may be very important for inhibiting the apoptotic death of retinal neurons which occurs in retinitis pigmentosa, macular degeneration, glaucoma or other neurodegenerative diseases.

METHODS

Primary Retinal Neuron Cell Cultures

This procedure represents a modification of that which we have used for mammalian brain cells[12] and has been presented previously.[13] Retinas of 10–15 rat pups, 0–2 days old, are dissected under sterile conditions in a tissue culture hood. The retinas are suspended in 25 ml of DMEM-F12 plus 10% fetal calf serum in a plastic bag, mechanically dissociated and the suspension is filtered through two sieves. The cells are centrifuged, resuspended and the concentration of cells is determined with a cell counter or hemocytometer, and the suspension. The cells, diluted with medium to 1×10^5 cells per ml are plated (1 ml) in 24 well tissue culture plates on 12 mm coverslips that had been pre-treated overnight with poly-D-lysine (10 µg/ml). The cells are maintained in the synthetic serum-free media developed by Bottenstein and Sato[14] as modified by Lillien and Cepko.[15]

Antibodies

We used monospecific antibodies to recognize individual retinal neuron populations. Rods were identified using a monoclonal antibody against rhodopsin (mAb Rho1D4)generously provided by Robert Molday[16,17] Rods, cones, and flat midget cone bipolar cells were visualized with rabbit anti-recoverin IgG[13] while anti-arrestin monoclonal antibody A9C6[18] was generously provide by Larry Donoso. Rod bipolar cells are labeled with mAb anti-PKC (Sigma). Amacrine cells are visualized with anti syntaxin from Sigma. Retinal ganglion cells are identified with mAb anti-Thy-1 (Chemicon, Temecula, CA). The monospecificity of the rabbit anti-mouse recoverin antibodies was recently demonstrated[13] and the others have been extensively characterized. PEDF was purified from human retina.[19]

Immunocytochemistry

For immunocytochemical staining, the cells grown in culture are fixed for 30 minutes in 4% paraformaldeyhde fixative in Tris-HCL (100 mM, pH 7.5) buffered saline (TBS), rinsed 3 times with 0.1 M Tris-HCL, pH 7.5 and maintained at 4°C in that solution until processed for immunocytochemistry. Routinely, the coverslips in the 24 well plates are incubated overnight with diluted antibody (1/100–1/4000) in TBS with 1% bovine serum albumin and 1% goat serum , rinsed 3× with 1.0 ml of same buffer, incubated with biotinylated secondary antibody for 1 hour; rinsed as above, followed by a one hour incubation with streptavidin conjugated- FITC or -Texas Red. For double labeling, recoverin was detected using goat anti rabbit IgG conjugated to Texas Red while rhodopsin was visualized using biotinylated goat anti-mouse IgG followed by incubation with streptavidin conjugated to FITC. For triple labeling, rhodopsin was detected using goat anti mouse IgG conjugated to FITC, while recoverin was visualized using biotinylated goat anti rabbit IgG followed by incubation with streptavidin conjugated to AMCA (blue); and arrestin was detected using biotinylated goat anti-mouse IgG followed by incubation with streptavidin conjugated to Texas Red. The cells on cover slips are mounted with antifade mounting medium and viewed and photographed with an Eclipse 800 microscope equipped with fluorescence and Nomarski optics and a digital camera.

TUNEL Method

Staining for the TUNEL technique[20] is done using a commercially available kit (Oncor, Apoptag). Coverslips are fixed in 4% paraformaldehyde for 10 min at room temperature and washed in two changes of phosphate-buffered saline (PBS) for 5 min each at room temperature. The positive cells are visualized using diaminobenzidine as substrate for the kit's horseradish peroxidase. The results are viewed with a Nikon Eclipse 800 microscope.

MTT Method

MTT (3,(4,5-dimethylthiazol-2-yl)2,5-diphenyl-tetrazolium bromide) is purchased from Sigma. The general principle involved is that cell death reduces the conversion of the yellow tetrazolium salt to the purple colored product, formazan, the concentration of which can be measured with a spectrophotometer.[21] A decrease in number of living cells results in a decrease in the total metabolic activity in the sample and the decrease directly correlates with the amount of purple formazan formed, as monitored by its spectral absorbance. For the assay 25 µl of the 5 mg/ml stock solution of MTT is added to each well, and after 2 hours of incubation at 37°C, 100 µl of the lysing buffer is added. After an overnight incubation at 37°C, the absorbance at 550 is measured using a microtiter plate ELISA reader.

RESULTS

The retina consists of multiple populations of neurons with the phototransduction cascade being initiated by the absorption of a photon of light within the photoreceptor cells and the transmission of this information to the neurons in the inner layers. The regulation of the expression of genes in photoreceptor cells has been of major interest to us and we have focused on the protein products of three genes which are retina specific. These proteins, recoverin, rhodopsin, and arrestin are differentially expressed *in vivo*. Immunocytochemical staining of normal adult rat retina show (Fig. 1A) that recoverin is expressed in rods, cones and a small population of bipolar cells and that some of these cells have higher concentrations of recoverin than others (arrows Fig. 1A). Recoverin is also expressed in a rare population of cells in the ganglion cell layer although not seen in this section. Rhodopsin (Fig. 1B) is confined to the rods and is most highly concentrated in the outer segments. Arrestin (Fig. 1C) is present in both rods and cones although not in any of the cells of the inner retina. Using

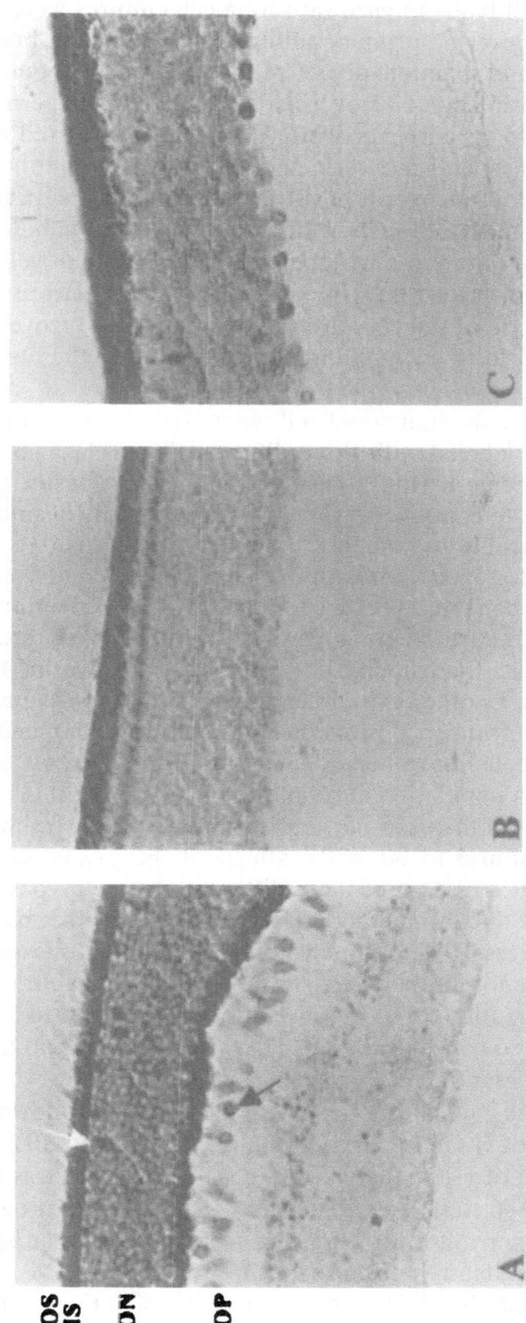

Figure 1. Immunocytochemical localization of A) recoverin, B) rhodopsin, and C) arrestin in normal adult rat retina using monospecific antibodies, horseradish peroxidase, and diaminobenzidine. (See Methods) Arrows in A indicate cells which have higher concentrations of recoverin.

antibodies which are specific for these three antigens, it is possible to identify at least three classes of cells. Rods will express all three antigens; cones will express recoverin and arrestin; and the subpopulation of bipolar cells will only express recoverin.

To visualize these cell types, *in vitro* rat retina cell cultures were established. We modified our method for growing primary cultures of mammalian brain neurons and applied it to the growth and maintenance of retinal neurons. To identify cells in our culture containing rhodopsin and/or recoverin, double labeling experiments were performed using fluorescein to visualize recoverin⊕ cells and Texas red for rhodopsin⊕ cells. The application of these anti-recoverin and anti-rhodopsin antibodies to our cultures demonstrated numerous recoverin-positive and rhodopsin positive cells (Fig. 2A) among a background of unlabeled cells which are visualized with Nomarski optics. Rods which are positive for both proteins appear yellowish green or yellowish red while cones and bipolar cells which only have recoverin are green. Surprisingly, under our culture conditions, about 3% of the cells are red [rhodopsin⊕, recoverin⊖], a phenotype not previously observed *in vivo*. At higher magnification, these cell types can be clearly seen (Fig. 2B).

To specifically identify the individual cell types, fixed cultures were reacted with three different antibodies labeled with three different fluorochromes: anti-rhodopsin (green) (Fig. 3B); anti-recoverin (blue) (Fig. 3C), and anti-arrestin (red) (Fig. 3D). Although it is not possible to combine all three colors and easily distinguish the resulting mix of colors, it is possible to compare the same field visualized with Nomarski optics and a filter specific for each fluorochrome. Under these conditions, rods are positive foı all three antigens; cones are positive for arrestin and recoverin; and the bipolar cells are only positive for recoverin. The arrows point out cells representative of each of these cell types as well as the [rhodopsin⊕, arrestin⊕, recoverin⊖] cells among a number of unlabeled cells. In other experiments (data not shown), anti-Thy1 positive cells (ganglion cells), anti-syntaxin positive cells (amacrine cells), and anti-calbindin positive cells (horizontal cells) have been detected in our cultures. Collectively these data demonstrate that each population of retinal neurons is present in our cultures and that they can be identified on the basis of the presence of specific retinal antigens. This supports the use of the cultures to assess the affects of changes in the chemical environment on the maintenance of each of these differentiated cell types.

H_2O_2 is one of the products of metabolic reactions which utilize oxygen and which also produces oxidative stress. To determine whether retinal neurons under our culture conditions were vulnerable to this oxidative stress, the MTT assay for cell viability was established and used to quantitate the response of retinal neurons to H_2O_2. Cells were incubated with the indicated concentration of H_2O_2 for 24 hours and then assayed for viability. Data from a representative experiment are shown in Fig. 4 and document a progressive decrease in the optical density in the MTT assay following exposure to increasing concentrations of H_2O_2. In parallel cultures treated with PEDF at 100 ng/ml, one hour prior to addition of H_2O_2, the change in optical density was less for each of the concentrations of H_2O_2 tested. These data show that retinal neurons in culture die in a dose dependent manner following exposure to reactive oxygen species and that they can be protected from death by prior exposure to PEDF.

To determine whether the H_2O_2-dependent death of retinal neurons is due to apoptosis or necrosis, the TUNEL assay was applied to our cultures following treatment with H_2O_2 for 24 hours (Fig. 5). Control cultures exhibited very little cell death (Fig. 5A) whereas cultures treated with H_2O_2 contained large numbers of cells undergoing apoptotic cell death (Fig. 5B). To test the ability of PEDF to prevent the apop-

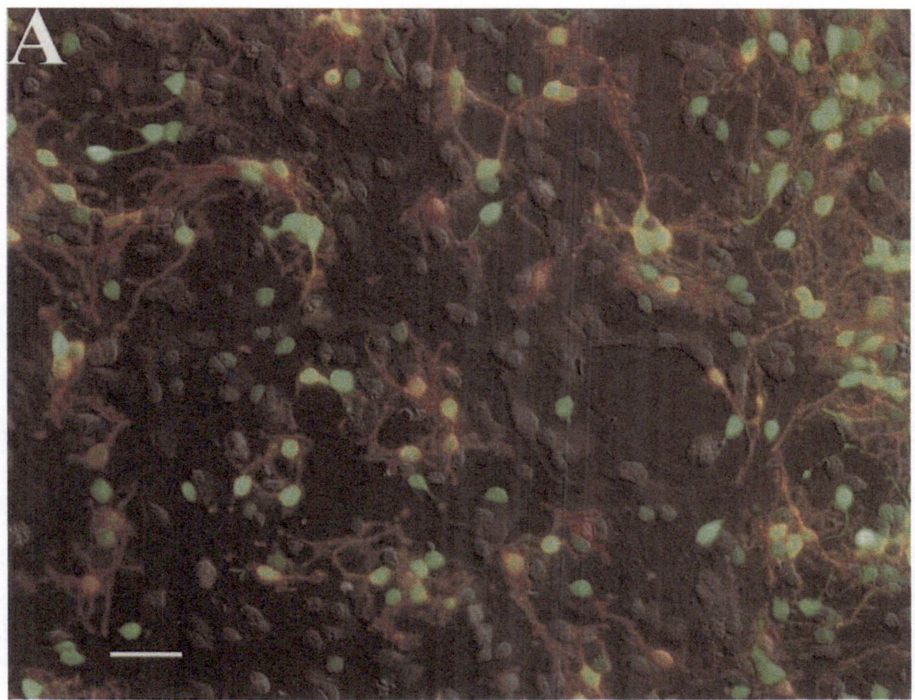

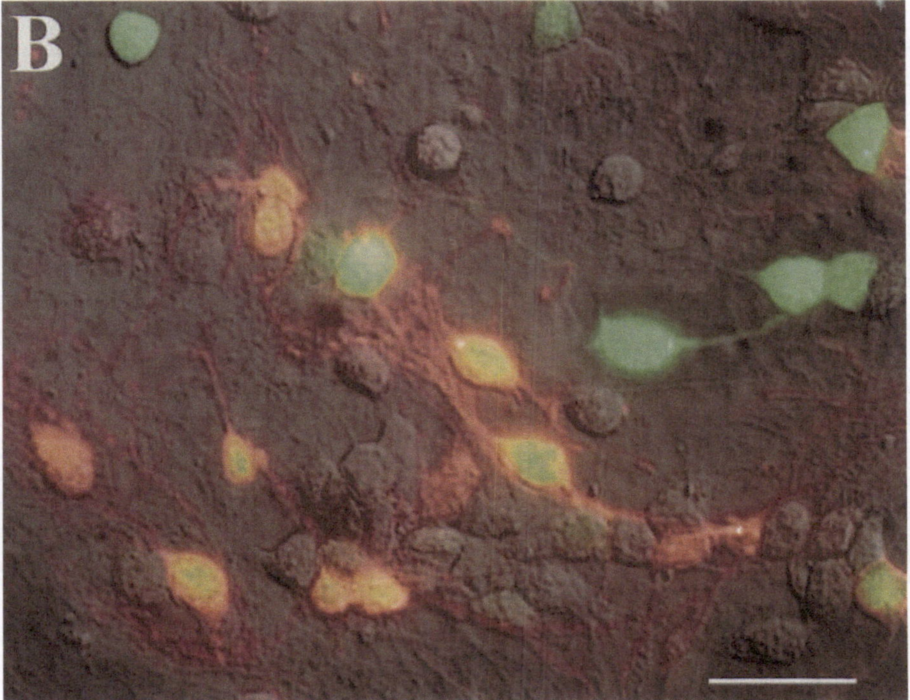

Figure 2. Cultured postnatal retinal neurons labeled with anti-recoverin and rhodopsin antibodies. Fields were viewed with filters which allow the Nomarski image, fluorescein (recoverin, green) and Texas-red (rhodopsin, red) stained cells to be seen simultaneously. (A) Low magnification: bar = 30 μm. (B) high magnification: bar = 30 μm.

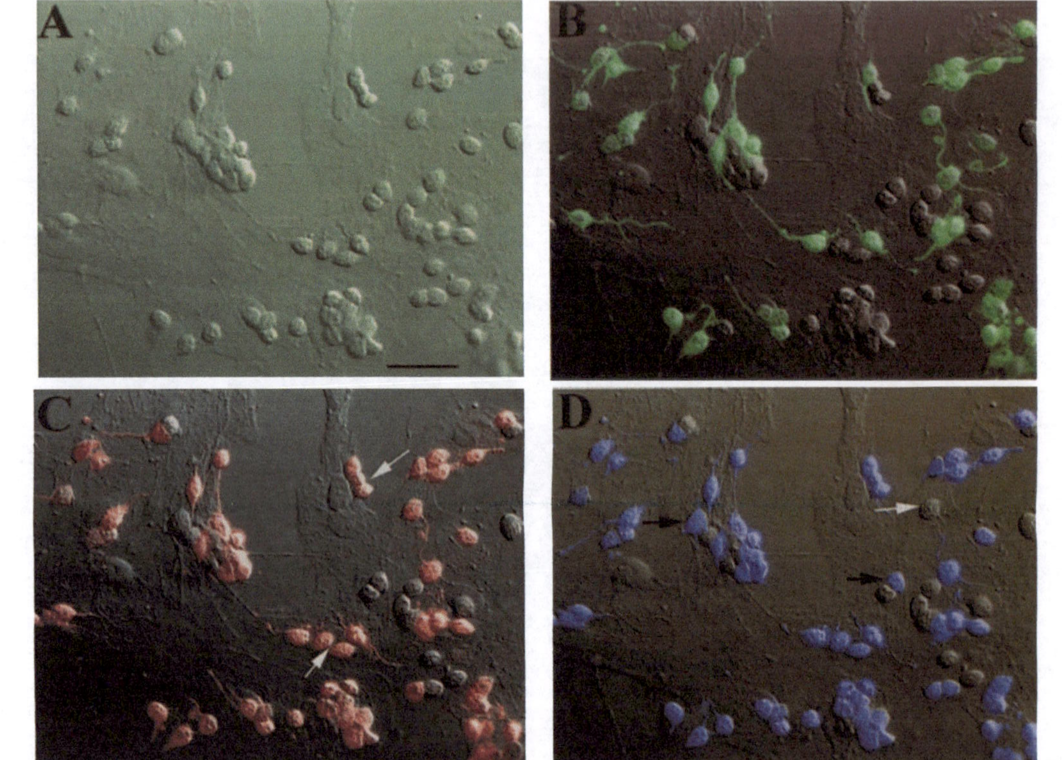

Figure 3. Identification of retinal cells in culture with antibodies against recoverin, arrestin, and rhodopsin. (A) Nomarski image: bar = 30 μm. (B) Superimposition of Nomarski image on the same field of fluorescein stained cells (rhodopsin positive, rod photoreceptor cells). (C) Superimposition of Nomarski image on the same field of Texas-red stained cells (arrestin positive, both rods and cones). The arrows indicate cells that are cone photoreceptor cells (rhodopsin negative and arrestin positive). (D) Superimposition of Nomarski image on the same field of AMCA (Blue) stained cells (recoverin positive, showing rods, cones, and some bipolar cells). The black arrows indicate cells that are rhodopsin and arrestin negative but recoverin positive (bipolar cells). The white arrow indicates a cell that is rhodopsin and arrestin positive but recoverin negative.

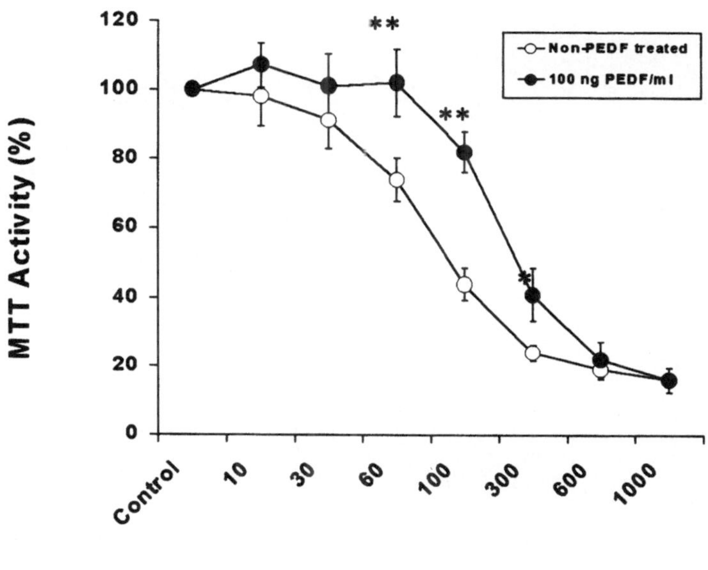

Figure 4. Protective effect of pigment epithelium-derived factor (PEDF) on cultured retinal neurons against H_2O_2-induced cell death, as determined by MTT assay. Cultures were pretreated with PEDF (100 ng/ml) for 1 hr before exposed to different doses of H_2O_2 (from 10–1000 µM) for 24 hrs. $*p < 0.05$; $**p < 0.01$ versus the same dose of H_2O_2 exposure without PEDF pretreatment (mean ± S.E., n = 3).

totic death of these neurons, parallel cultures were treated with PEDF. When control cultures were treated only with PEDF (Fig. 5C), there was essentially no effect on the number of apoptotic cells but cultures which were treated with PEDF one hour prior to exposure to H_2O_2 (Fig. 5D) exhibited a dramatic decrease in the number of apoptotic cells. These data support the conclusions that H_2O_2 causes the apoptotic death of retinal neurons and that PEDF is very effective in providing protection from the effects of this agent.

DISCUSSION

The process of vision is initiated with the passage of light through the layers of neurons in the retina and the absorption of photons by photopigments within the rod or cone photoreceptor cells. Within the photoreceptor cell there are a number of signal transduction enzymes whose amino acid sequences are unique to either the rod enzyme or the cone enzyme and whose activities result in the biochemical amplification of this initial event and the eventual transmission of an electrical signal to the brain. Hereditary mutations in the genes for these unique proteins can result in a family of diseases which may be classified broadly under the name retinal degeneration. The common feature of diseases within this group is the death of the rods, cones and the eventual loss of vision. A number of animal models for retinal degeneration have been developed and in all cases examined, the photoreceptor cells die by an apoptotic mechanism.[2,22–25] These include models for hereditary

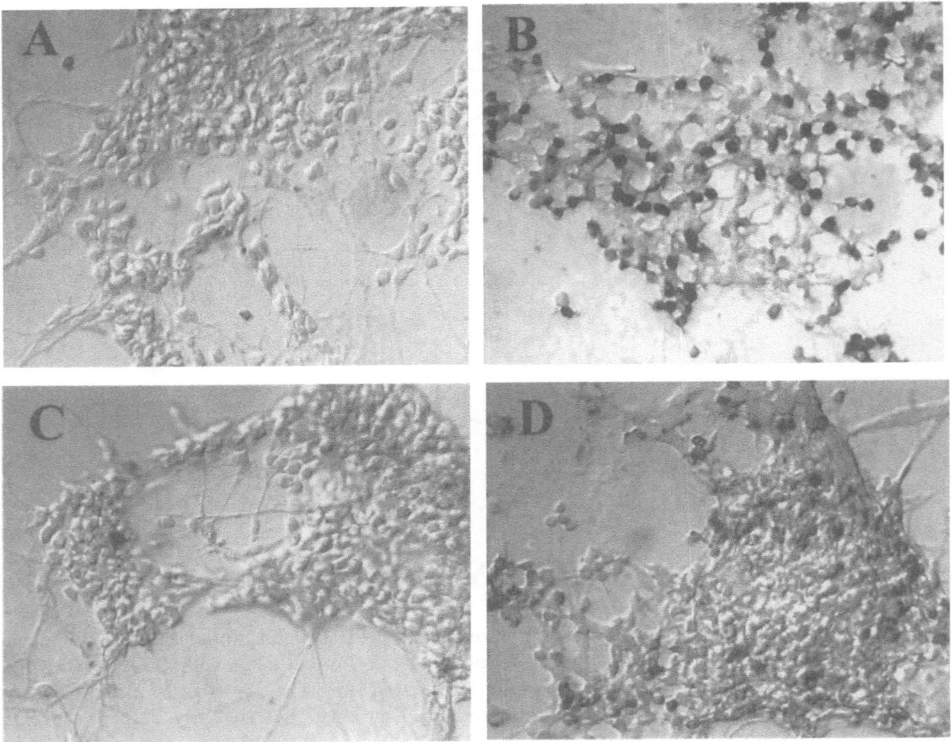

Figure 5. Pigment epithelium-derived factor (PEDF) inhibits H_2O_2-induced apoptosis in cultured retinal neurons. A. Untreated control showing 2% TUNEL-positive cells; B. H_2O_2 (100 μM, 24 hr) treated culture showing 50–60% TUNEL-positive cells; C. PEDF (100 ng/ml, 25 hr) treated culture showing 3–4% TUNEL positive cells; D. Culture pre-treated with PEDF (100 ng/ml) for 1 hr before H_2O_2 (100 μM, 24 hr) treatment, showing 26% TUNEL positive cells.

blindness, retinal detachment,[26] light induced degeneration,[27] ischemia,[28] and glaucoma.[29,30] Although the molecular pathways by which any of these degenerations occur are at present unknown, it is thought that light injures photoreceptors by increasing the formation of reactive oxygen species, and this is supported by evidence that antioxidants rescue light-damaged photoreceptors.[31]

In vitro, the rods and cones do not display the exquisite morphology that so characterizes them *in vivo* and without the positional location, identification is dependent on the detection of cell-specific gene products. Initial studies demonstrated that rat retinal neurons could be grown in a chemically defined environment and that each type of retinal neuron was represented in our cultures including rods, cones, bipolar, amacrine, horizontal, and ganglion cells. Because of our interest in the regulation of photoreceptor cell genes, rhodopsin, recoverin, and arrestin were chosen as specific antigens to be used to demonstrate the presence of rods, cones, and recoverin positive bipolar cells in our cultures. Recently,[13] using double labeling with either rhodopsin and recoverin antibodies or arrestin and recoverin antibodies, we had found two novel photoreceptor cell phenotypes i.e., [rhodopsin⊕, recoverin⊖] and [arrestin⊕, recoverin⊖] which were not found *in vivo* and which we hypothesized represented the same population of photoreceptor cells. Using triple labeling, our present experiments demonstrate cells which are [rhodopsin⊕, arrestin⊕, recoverin⊖] and confirm our hypotheses. These data also support the conclusion from recoverin knockout mice studies,[32] that recoverin is not necessary for photoreceptor cell survival. Although the mechanistic origin of these photoreceptor cells in our cultures is unknown, their existence suggests that signals which turn the recoverin gene off and on can be investigated and these features may be important towards understanding how the recoverin gene is illicitly turned on in small cell lung carcinomas[33,34] in Cancer Associated Retinopathy (CAR).

The data presented show that our culture conditions support the development and maintenance of representatives of each of the differentiated retinal neurons present *in vivo* and that H_2O_2 is capable of inducing apoptosis in these neurons in a dose dependent manner. In addition, PEDF can prevent the apoptotic death of these neurons. Because this manner of cell death requires the active participation of the cell and because of the multitude of signal transduction pathways which may be involved, multiple therapies can be tested for their efficacy in preventing the death of these neurons. These data provide the basis for further studies to characterize the mechanisms by which specific populations of retinal neurons are committed to die, not only by H_2O_2, but also by a variety of agents known to induce neuronal apoptosis in degenerative diseases of the central nervous system. Similarly, the exact mechanism by which PEDF protects these cells can be investigated. This *in vitro* model system will also enable combinatorial arrays of rescue factors to be qualitatively and quantitatively evaluated for their ability to rescue retinal neurons from cellular suicide induced by specific agents and/or conditions.

ACKNOWLEDGMENTS

The authors thank Na Wei for her expertise in generating and maintaining our primary cell cultures. This work was supported by a Jules and Doris Stein Professorship from the Research to Prevent Blindness Foundation to JFM, a National Institutes of Health grant EY06085 to JFM, by a grant from the University of Oklahoma Medical Alumni Association to Wei Cao, and by an unrestricted grant from RPB to the Department of Ophthalmology.

REFERENCES

1. Barber, A.J.; Lieth, E.; Khin, S.A.; Antonetti, D.A.; Buchanan, A.G.; Gardner, T.W., and Penn State Retina Res Grp, 1998 Neural apoptosis in the retina during experimental and human diabetes—Early onset and effect of insulin. *J. Clin. Invest.* **102**:783–791.
2. Lolley, R.N.; Rong, H., and Craft, C.M., 1994 Linkage of photoreceptor degeneration by apoptosis with inherited defect in phototransduction. *Invest. Ophthalmol. Vis. Sci.* **35**:358–362.
3. Papermaster, D.S. and Windle, J., 1995 Death at an early age: Apoptosis in inherited retinal degenerations. *Invest. Ophthalmol. Vis. Sci.* **36**:977–983.
4. Wong, P. 1994 Apoptosis, retinitis pigmentosa, and degeneration. *Biochem.Cell Biol.* **72**:489–498.
5. Davis, A.A.; Bernstein, P.S.; Bok, D.; Turner, J.; Nachtigal, M. and Hunt, R.C., 1995 A human retinal pigment epithelial cell line that retains epithelial characteristics after prolonged culture. *Invest. Ophthalmol. Vis. Sci.* **36**:955–964.
6. Anderson, G.P., 1997 Bcl-2 related proteins, apoptosis, and disease. *Trends Pharmacol.Sci.* **18**:51.
7. Tombran-Tink, J. and Johnson, L.V. 1989 Neuronal differentiation of retinoblastoma cells induced by medium conditioned by human RPE cells. *Invest. Ophthalmol. Vis. Sci.* **30**:1700–1707.

8. Steele, F.R.; Chader, G.J.; Johnson, L.V., and Tombran-Tink, J., 1993 Pigment epithelium-derived factor: neurotrophic activity and identification as a member of the serine protease inhibitor gene family. *Proc. Natl. Acad. Sci. U.S.A.* **90**:1526–1530.
9. Taniwaki, T.; Becerra, S.P.; Chader, G.J., and Schwartz, J.P., 1995 Pigment epithelium-derived factor is a survival factor for cerebellar granule cells in culture. *J. Neurochem.* **6**:2509–2517.
10. Taniwaki, T.; Hirashima, N.; Becerra, S.P.; Chader, G.J.; Etcheberrigaray, R., and Schwartz, J.P., 1997 Pigment epithelium-derived factor protects cultured cerebellar granule cells against glutamate-induced neurotoxicity. *J. Neurochem.* **68**:26–32.
11. Seigel, G.M.; Tombran-Tink, J.; Becerra, S.P.; Chader, G.J.; Diloreto, D.A.J.; del, C.C.; Lazar, E.S., and del, C.M., 1994 Differentiation of Y79 retinoblastoma cells with pigment epithelial-derived factor and interphotoreceptor matrix wash: effects on tumorigenicity. *Growth Factors.* **10**:289–297.
12. Breen, G.A.; McGinnis, J.F., and de Vellis, J., 1978 Modulation of the hydrocortisone induction of glycerol phosphate dehydrogenase by N6,O2′-dibutyryl cyclic AMP, norepinephrine, and isobutylmethylxanthine in rat brain cell cultures. *J. Biol. Chem.* **253**:2554–2562.
13. McGinnis, J.F.; Stepanik, P.; Chen, W.; Elias, R.; Cao, W., and Lerious, V., 1999 Unique retina cell phenotypes revealed by immunological analysis of recoverin expression in rat retina cells. *J. Neurosci. Res.* In Press.
14. Bottenstein, J.E. and Sato, G.H., 1979 Growth of a rat neuroblastoma cell line in serum-free supplemented medium. *Proc. Natl. Acad. Sci. U.S.A.* **76**:514–517.
15. Lillien, L. and Cepko, C., 1992 Control of proliferation in the retina: temporal changes in r esponsiveness to FGF and TGF alpha. *Development* **115**:253–266.
16. Reid, D.M.; Laird, D.W., and Molday, R.S., 1992 Characterization and application of an in vitro detection system for studying the binding and phagocytosis of rod outer segments by retinal pigment epithelial cells. *Exp. Eye Res.* **54**:775–783.
17. Hsu, Y.-T.; Wong, S.Y.C.; Connell, G.J., and Molday, R.S., 1993 Structural and functional properties of rhodopsin from rod outer segment disk and plasma membrane. *Biochim. Biophys. Acta Bio-Membr.* **1145**:85–92.
18. Donoso, L.A.; Merryman, C.F.; Edelberg, E.; Naids, R., and Lalsow, C., 1985 S-antigen in the developing retina and pineal gland: a monoclonal antibody study. *Invest. Ophthalmol. Vis. Sci.* **26**:561–567.
19. Tombran-Tink, J.; Chader, G.G., and Johnson, L.V., 1991 PEDF: a pigment epithelium-derived factor with potent neuronal differentiative activity [letter]. *Exp. Eye Res.* **53**:411–414.
20. Gavrieli, Y.; Sherman, Y., and Ben-Sasson, S.A., 1992 Identification of programmed cell death in situ via specific labeling of nuclear DNA fragmentation. *J. Cell Biol.* **119**:493–501.
21. Mosmann, T., 1983 Rapid colorimetric assay for cellular growth and survival: application to proliferation and cytotoxicity assays. *J. Immunol. Methods* **65**:55–63.
22. Chang, G.Q.; Hao, Y., and Wong, F., 1993 Apoptosis: final common pathway of photoreceptor death in rd, rds, and rhodopsin mutant mice. *Neuron* **11**:595–605.
23. Portera-Cailliau, C.; Sung, C.H.; Nathans, J., and Adler, R., 1994 Apoptotic photoreceptor cell death in mouse models of retinitis pigmentosa. *Proc. Natl. Acad. Sci. U.S.A.* **91**:974–978.
24. Tso, M.O.M.; Zhang, C.; Abler, A.S.; Chang, C.-J.; Wong, F.; Chang, G.-Q., and Lam, T.T., 1994 Apoptosis leads to photoreceptor degeneration in inherited retinal dystrophy of RCS rats. *Invest. Ophthalmol. Vis. Sci.* **35**:2693–2699.
25. Wong, P., 1994 Apoptosis, retinitis pigmentosa, and degeneration. *Biochem.Cell Biol.* **72**:489–498.
26. Cook, B.; Lewis, G.P.; Fisher, S.K., and Adler, R., 1995 Apoptotic photoreceptor degeneration in experimental retinal detachment. *Invest. Ophthalmol. Vis. Sci.* **36**:990–996.
27. Wong, P.; Kutty, R.K.; Darrow, R.M.; Shivaram, S.; Kutty, G.; Fletcher, R.T.; Wiggert, B.; Chader, G., and Organisciak, D.T., 1994 Changes in clusterin expression associated with light-induced retinal damage in rats. *Biochem. Cell Biol.* **72**:499–503.
28. Rosenbaum, D.M.; Rosenbaum, P.S.; Gupta, A.; Michaelson, M.D.; Hall, D.H., and Kessler, J.A., 1997 Retinal ischemia leads to apoptosis which is ameliorated by aurintricarboxylic acid. *Vision Res.* **37**, 3445–3451.
29. Garcia-Valenzuela, E.; Shareef, S.; Walsh, J., and Sharma, S.C., 1995 Programmed cell death of retinal ganglion cells during experimental glaucoma. *Exp. Eye Res.* **61**:33–44.
30. Kerrigan, L.A.; Zack, D.J.; Quigley, H.A.; Smith, S.D., and Pease, M.E., 1997 TUNEL-positive ganglion cells in human primary open-angle glaucoma. *Arch. Ophthalmol.* **115**:1031–1035.
31. Steinberg, R.H., 1994 Survival factors in retinal degenerations. *Curr.Opin.Neurobiol.* **4**:515–524.
32. Lem, J. and Makino, C.L., 1996 Phototransduction in transgenic mice. *Curr. Opin. Neurobiol.* **6**:453–458.

33. Polans, A.S.; Witkowska, D.; Haley, T.L.; Amundson, D.; Baizer, L., and Adamus, G., 1995 Recoverin, a photoreceptor-specific calcium-binding protein, is expressed by the tumor of a patient with cancer-associated retinopathy. *Proc. Natl. Acad. Sci. U.S.A.* **92**:9176–9180.
34. Thirkill, C.E.; Tait, R.C.; Tyler, N.K.; Roth, A.M., and Keltner, J.L., 1992 The cancer-associated retinopathy antigen is a recoverin-like protein. *Invest. Ophthalmol. Vis. Sci.* **33**:2768–2772.

51

MELATONIN RECEPTOR BLOCKERS ENHANCE PHOTORECEPTOR SURVIVAL AND FUNCTION IN LIGHT DAMAGED RAT RETINA

Ronald A. Bush,[1] Takeshi Sugawara,[1] P. Michael Iuvone,[2] and Paul A. Sieving[1]

[1] Department of Ophthalmology
Center for Retinal and Macular Degeneration
University of Michigan
Ann Arbor, Michigan 48105
[2] Department of Pharmacology
Emory University School of Medicine
Atlanta, Georgia 30322

1. SUMMARY

Melatonin is an important retinal neurohormone involved in rhythmic photoreceptor metabolism and in regulating the release of retinal dopamine. Systemic administration of melatonin and conditions supporting its synthesis in the retina seem to be related to increased light damage susceptibility. We found that intraocular injection of competitive melatonin receptor antagonists prior to the dark period before bright light exposure resulted in substantial protection of photoreceptors from light damage. Photoreceptor cell structure was much better preserved in antagonist injected eyes and the ERG threshold was approximately 2 log units of intensity lower. Intraocular injection of a number of other compounds which are known to affect endogenous dopamine levels or stimulate or block dopamine receptors did not have a significant effect on the ERG. However, some showed a tendency to reduce or enhance the effect of light damage on the ERG depending on their expected effect on dopamine levels. These results indicate that blocking the action of endogenous melatonin at retinal receptors can protect photoreceptors from damage by light. Further work is needed to investigate a possible role for dopamine in this protection.

Retinal Degenerative Diseases and Experimental Therapy, edited by Hollyfield *et al.*
Kluwer Academic / Plenum Publishers, New York, 1999.

2. INTRODUCTION

Retinal neuromodulators or neurohormones regulate metabolic functions in photoreceptor cells and can modulate light damage susceptibility (Reviews see:[1,2]). Melatonin (N-acetyl-5-methoxytryptamine) is a retinal neurohormone that is synthesized in both the pineal gland and the retina at low rates during daylight hours and at higher rates at night. In the retina melatonin plays a key role in rhythmic photoreceptor metabolism, e.g. disk shedding, and in regulating the synthesis and release of dopamine from post photoreceptor neurons.[2–4] It may also influence the phagocytic activity of the retinal pigment epithelium by modulating cellular levels of c-AMP.[1]

Light-damage susceptibility is enhanced under conditions that would be expected to elevate retinal melatonin levels, such as night time light exposure[5] or previous exposure to darkness.[6,7] When retinal melatonin levels are low, such as during mid-daylight hours or immediately following a period of light adaptation, damage from bright light exposure is reduced.[5,6] Conversely, systemically injected melatonin enhances susceptibility to light-induced photoreceptor damage in rats.[8,9] However, it is unknown whether systemic melatonin acts directly on receptors in the retina or indirectly by affecting other circulating factors.

We are currently investigating the role of endogenous retinal neurohormones and neuromodulators in retinal degenerations. The previous evidence indicates systemic levels of melatonin can modulate light damage susceptibility. Indirect evidence suggests that retinal levels of melatonin are involved in both light damage (see above) and inherited retinal degeneration in the RCS rat.[10,11] In order to more directly test the hypothesis that endogenous melatonin can affect photoreceptor survival in light damage, we injected specific melatonin receptor antagonists into the vitreous of rats prior to light exposure. Since blocking melatonin receptors can also enhance the release retinal dopamine[4,12] we also tested substances which have been shown to modulate dopamine levels or activity for their effect on light damage susceptibility.

3. METHODS

3.1. Animals

These studies were conducted in accordance with the ARVO Statement for the Use of Animals in Ophthalmic and Vision Research. Male Sprague-Dawley albino rats were obtained from Charles River Labs (Wilmington, MA) at 7 weeks of age. They were housed in our colony room for at least 7 days with 50 lux illumination (VitaLite) on a 12-hour light/dark cycle with lights on at 7:00 AM. Food and water were provided *ad lib* at all times.

3.2. Light Exposure

At the time of light exposure the rats were between 8 and 9 weeks of age. Our earlier experiments indicated that rats of this age and rearing conditions received a substantial amount of retinal light damage when exposed to1000–3000 lux for 48 hours.[13] In this study two different degrees of light induced damage were achieved by exposing rats to either 2500 lux (severe) or 1500 lux (mild) of diffuse white fluorescent light (GE, Cool White) for 48 hr. During exposure, rats were housed individually in clear Plexiglas cages in which the light intensity did not vary by more than 30% in any direction. The temperature was kept at 22.5 ± 1°C. After light exposure rats were returned to the colony room under normal cyclic lighting for 7 days.

3.3. Drugs and Injections

Melatonin receptor antagonists Luzindole (2-benzyl-N-acetyltryptamine) and 4-P-PDOT (4-phenyl-2-propionamidotetralin) were obtained from Tocris Cookson, Inc. (Ballwin, MO). Both are competitive ML_1 receptor antagonists with higher affinity for the ML_{1b} receptor shown to mediate melatonin suppression of dopamine release in the rabbit retina.[14] These compounds were dissolved in 50% DMSO for injection (luzindole, 40 mM; 4-P-PDOT, 80 mM). Because light damage results are often much more variable between animals than between eyes in the same animal,[13,15,16] we injected drug into one eye of each animal and used the other eye for comparison. One μl of drug solution was injected into the vitreous of the right eye using a Hamilton syringe and 30 gauge needle, and 1 μl of DMSO was injected into the left eye of each animal as a control. From estimates of the vitreal volume of the rat[17] the concentration at the retina, assuming complete mixing, was estimated to be about 1 mM for luzindole and 2 mM for 4-P-PDOT. Injections were carried out as previously described under dim red light[18] 16–18 hours before bright light exposure (between 4–6 PM) as shown in Figure 1. The rats were kept in darkness until the onset of bright light exposure at 10:00 AM the following day. Rats with either luzindole or 4-P-PDOT injected in one eye and DMSO injected in the other eye as well as a group with no injection (n = 4–5 each) were light damaged under the severe conditions. Five uninjected animals and 5 animals injected in both eyes with DMSO were also light damaged.

In a second series of experiments using mild light damage (1500 lux, 48 hr.) was conducted on groups of rats injected with drugs which have been shown to modulate the action or release of dopamine *in vivo*, or *in vitro* (n = 5 per drug). In addition luzindole and 4-P-PDOT were injected in 5 rats each using the same schedule as in Fig. 1. Most of the other substances were injected 2–3 hours before light exposure as they act directly on the dopaminergic system which is most active just prior to and after light onset.[19,20] Animals were kept in darkness from 7 PM until bright light exposure at 10 am the next day except for a brief period of dim red light exposure during injections. Drug and control injections were made as described above. Table 1 is a list of substances which were injected in this series of experiments in addition to luzindole and 4-P-DOT.

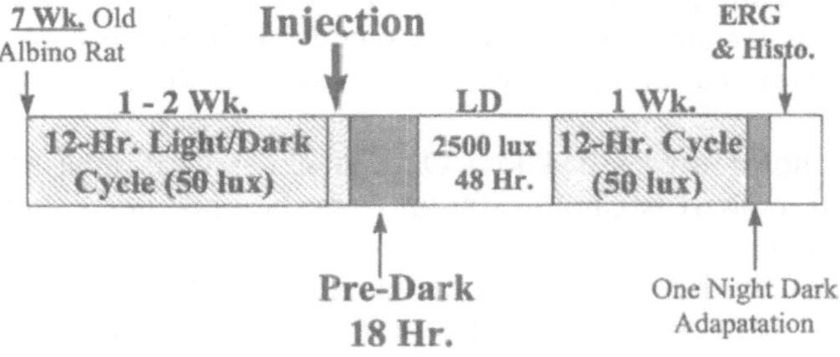

Figure 1. Time sequence of experiments.

Table 1. Additional drugs injected in second series of experiments

Substance	Vitreal concentration	Vehicle	Time before light damage	Action
Melatonin	2.5 mM	DMSO	3 hr.	Reduce dopamine release 1983[3] Enhance light damage[9]
Sulpiride	2 mM	DMSO	3 hr.	Standard D_2 antagonist
GDNF (glia-derived neurotrophic factor)	1 μg/μl	PBS	18 hr.	Enhances dopamine release[21]
Naloxone	1 mM	DMSO	3 hr.	Enhances dopamine release[22]
Bicuculline	1 mM	PBS	3 hr.	Enhances dopamine release[22]
PDA (2,3 piperidine dicarboxylic acid)	5 mM	PBS	3 hr.	Enhances dopamine release[23]
Apomorphine	1 mM	DMSO	3 hr.	Standard D_2 agonist

3.4. ERG Recordings

Before ERG recordings animals were dark adapted for 12 hours. They were prepared under dim red light, and both pupils were dilated with 0.1% atropine and 0.1% phenylephrine HCL. Animals were anesthetized intramuscularly (IM) with a loading dose of xylazine (10 mg/kg) and ketamine (14 mg/kg) and then maintained by IM infusion via pump (Razel Instruments). After this preparation, they were dark adapted for an additional 30 minutes. Animals were held steady by a bite bar and nose clamp during recordings, and a heating pad maintained body temperature at 36.8°C.

Dark adapted ERGs were recorded bilaterally as previously described.[18] White (2800°K) flashes of 50 ms were presented in Ganzfeld bowl. The maximal stimulus intensity of 2.3 log cd/m^2 was attenuated with neutral density filters. Recording began at intensities below ERG threshold and progressed upward in 0.3 to 0.4 log unit steps. The criterion amplitude was 10 μV for the b-wave thresholds, determined by plotting b-wave amplitude against light intensity on log-log plots. The difference in b-wave threshold between the drug treated and the vehicle treated eyes of each rat was used as an indicator of rescue of photoreceptor function by the drug.

3.5. Histology

One day after ERG recordings the rats were killed with an overdose of sodium pentobarbital and the eyes removed for retinal morphology, using overnight fixation with 4% paraformaldehyde and 2.5% glutaraldehyde in 0.1 M cacodylate buffer at 4°C. Eyes were trimmed and postfixed in 1% osmium for 1 hr. Epon embedded tissue was sectioned at 1 μm and stained with toluidine blue for light microscopy. All sections for light microscopy were cut along the vertical meridian of the eye passing through the optic nerve.

To assess photoreceptor cell loss ONL thickness was measured at ten locations, spaced 400 μm apart beginning 200 μm from the optic nerve, in the inferior and superior halves of retinal sections. Thickness was measured in μm using an ocular micrometer or by counting the number of cell nuclei in the column at each location. Photomicrographs were taken at corresponding points along the retinas of the control and drug injected eyes of several animals using an image capture system consisting of an Optronics video camera on an Olympus Vanox microscope. Images were processed in Adobe Photoshop 4.0.

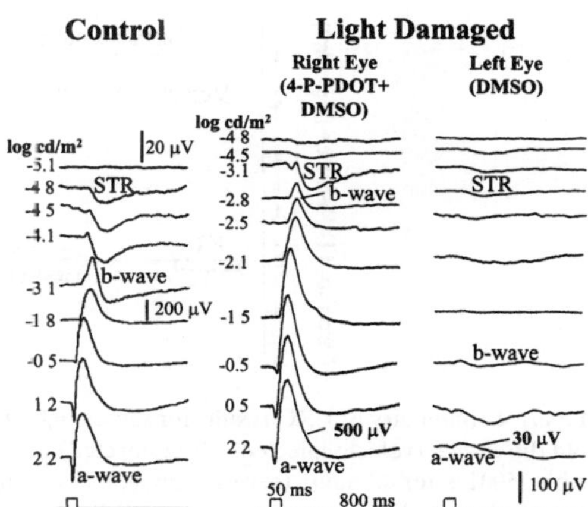

Figure 2. Dark adapted ERG waveform series. The control rat was maintained in 50 lux cyclic light and received no injections. The light damaged rat received a 1 µl injection of 1 mM (vitreal concentration) 4-P-PDOT in DMSO in one eye about 18 hours before being exposed to 2500 lux for 48 hours. The left eye was injected with DMSO alone. Recordings were done 1 week after light damage. Flash intensity is indicated at the left of each series. The 100 µV scale bar refers to the light damaged rat series.

4. RESULTS

Injection of melatonin receptor antagonists 16–18 hr. before exposure to severe light damage conditions, resulted in substantially better retinal function, as reflected in the ERG, than in vehicle injected eyes. Figure 2 shows an ERG response series from a light damaged rat treated in one eye with 4-P-PDOT and the other eye with vehicle. The eye injected with DMSO alone shows a near complete loss of the ERG response. The maximum b-wave response is decreased to about 30 µV from over 1000 µV in control. In contrast the b-wave of the 4-P-PDOT treated retina is only about 50% reduced at the maximum flash intensity. The threshold for the b-wave's appearance, an indication of the sensitivity of the response, was increased several log units by the light damage in the control eye, but was much less affected in the drug treated eye. The scotopic threshold response (STR), though reduced in amplitude, is relatively little affected.[13]

The log-log plots of response amplitude at various intensities for these 2 animals are shown in Fig. 3. The degree of protection is illustrated by the fact that the b-wave threshold (the flash intensity needed to give a 10 µV response) and the maximum

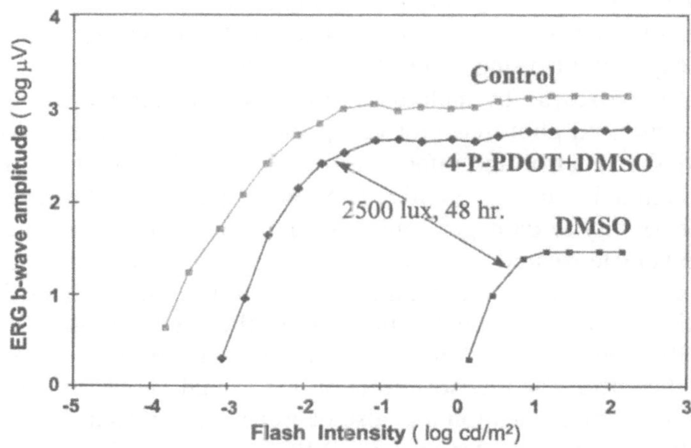

Figure 3. ERG b-wave log V-log I plots. The b-wave amplitude is plotted against flash intensity for the light damaged and control animal in Fig. 2.

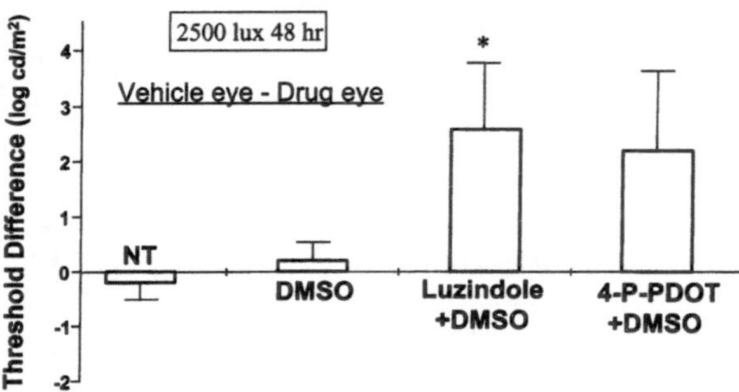

Figure 4. Summary of ERG results for severe light damage. The bars show the average ERG b-wave threshold difference (vehicle injected—drug injected) between the eyes of the same animal in rats light damaged with 2500 lux for 48 hours (severe damage). For controls the difference was left—right. N = 5 for the luzindole and control groups and 4 for the 4-P-PDOT group. * = statistically significant, $p < 0.02$, using 2-tailed Student t-test.

b-wave amplitude of the 4-P-PDOT treated eye was much closer to the undamaged control than to the control eye of the same animal. The sensitivity difference between the drug injected and control eye of the experimental animal was approximately 3 log units of intensity and the maximum amplitude difference was about 1.5 log μV. There was only about 1 log unit of intensity difference in threshold and less than 0.5 log μV difference in maximum amplitude between the 4-P-PDOT eye and the undamaged control animal.

A summary of the ERG b-wave threshold results from 2 groups of animals injected with the melatonin antagnost luzindole or 4-P-PDOT and 2 groups of controls is shown in Fig. 4. This graph plots the average difference between the left and right eyes of injected and uninjected rats one week after exposure to 2500 lux for 48 hr. In both groups injected with antagonists the antagonist injected eye had a much lower threshold than the vehicle injected eyes indicating a greater sensitivity in these eyes. For luzindole the difference was 2.5 ± 0.6 (S.E.M.) log cd/m^2 ($p < 0.02$) and for 4-P-PDOT the difference was 2.3 ± 0.8 (S.E.M.) log cd/m^2. The variability and fact that there were only 4 animals in the 4-P-PDOT group kept the average difference from being statistically significant ($p < 0.09$), but the drug injected eye in these 4 animals was always at least 10 times more sensitive. The protective effect of luzindole has been confirmed in 3 separate experiments using severe light damage.

The functional rescue by melatonin antagonists, illustrated in Figs. 2–4, can be explained by enhanced photoreceptor cell preservation seen in histological sections. Figure 5 shows the superior and inferior retinas of the right and left eyes of the same light-damaged animal as in Figs. 2 and 3. The central 1–2 mm of the superior retina has the greatest sensitivity to damage[24] and this region of the retina is shown in Figs. 5A and 5B. In the vehicle treated eye only a single layer of photoreceptor nuclei with no discernable outer segments remain. Since cones survive rods in light damage[25] these remaining cells may be cone nuclei. In contrast the ONL of the 4-P-PDOT retina has 5 to 6 cells remaining and outer segments which are approximately $^1/_2$ that of the undamaged control. The inferior retina of this animal, shown in Figs. 5D and 5E, is displayed at a lower magnification and also demonstrates substantially greater preservation of photoreceptor structure in the 4-P-PDOT treated retina. In contrast to the vehicle treated eye in this region, substantial outer segment length and well defined

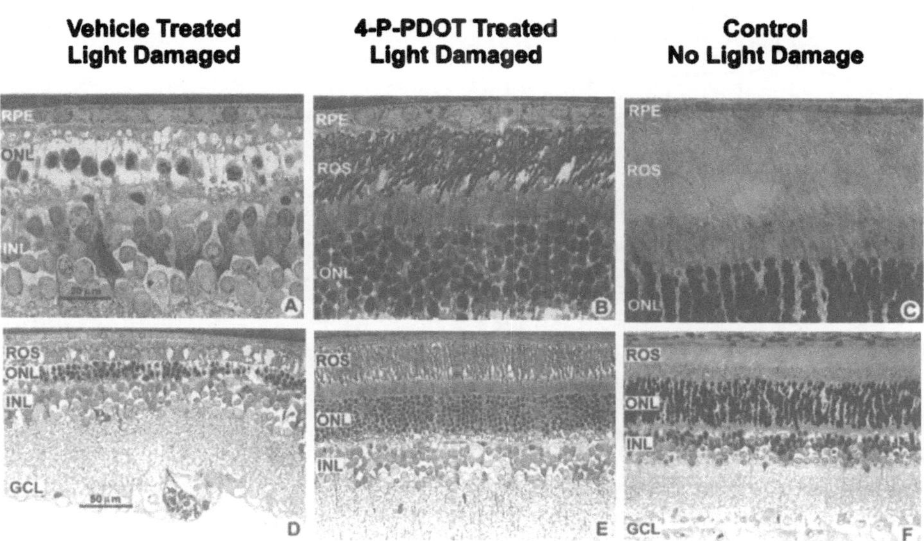

Figure 5. Retinal histology of the light damaged and undamaged control rat from Figs. 2 and 3. Panels A, B, and C are from the superior central retina. Panels D, E, and F are from the inferior central retina and are shown at a lower magnification (see scale bars). Some stretching of outer segments has occurred in processing in panel E.

inner segments remain. These results indicate that the preservation of retinal function in the 4-P-PDOT injected eye is correlated with the wide spread preservation of photoreceptor structure. This is further demonstrated by Figs. 6A and 6B which plot ONL thickness along the vertical meridian of the retina in luzindole and 4-P-PDOT treated animals. Because of the inter-animal variability in light damage, averaging across animals, as in Fig. 6A, may obscure the effect to some degree. Thus, the overall difference between eyes of the single animal in Fig. 6B indicates a greater degree of rescue.

These drugs and several others were tested for their protective effect under conditions of mild light damage (see Methods section). Many of the drugs had been previously shown to affect retinal or brain dopamine release or activity (see Methods section, Table 1) and were used as a preliminary test of the hypothesis that melatonin antagonists achieve photoreceptor rescue by their effects on retinal dopamine.[1,9] The ERG b-wave threshold difference between the left and right eyes is plotted in Fig. 7. In each case the right eye was injected with drug and the left eye with the vehicle (see Methods section). Only Luzindole and 4-P-PDOT, melatonin receptor antagonists, showed a statistically significant effect on the ERG threshold. Injection of these compounds, performed as in the previous severe light damage experiment, resulted in a lower threshold (greater sensitivity) in the drug injected eye than in the vehicle injected eyes. Because the overall damage is less under the light damage conditions of Fig. 7 than in those of Fig. 4, the difference in threshold between drug injected and control eyes is also less.

Most other compounds were injected 2 to 3 hours before bright light exposure to increase their chances of interacting with endogenous dopamine in the light. Injection of Sulpiride, a dopamine D2 receptor antagonist, resulted in tendency towards increased damage in that eye, though statistically significant only if a one-tailed t-test

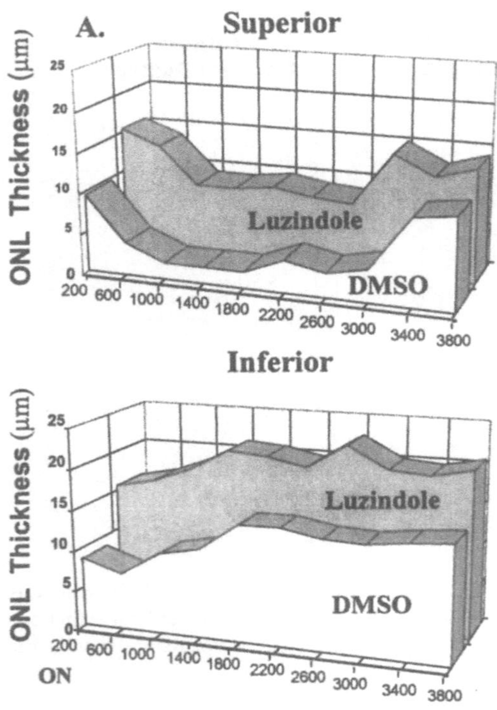

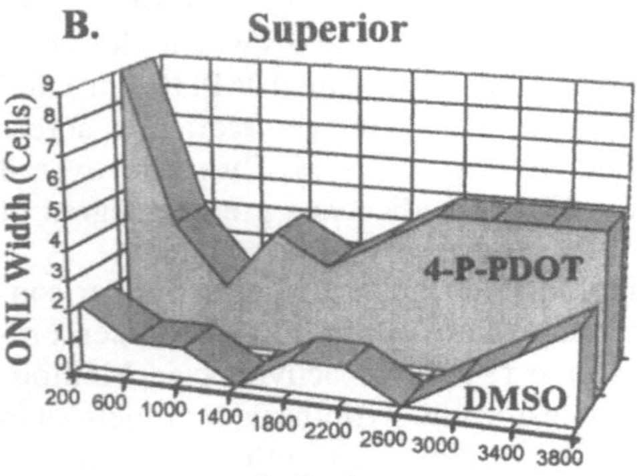

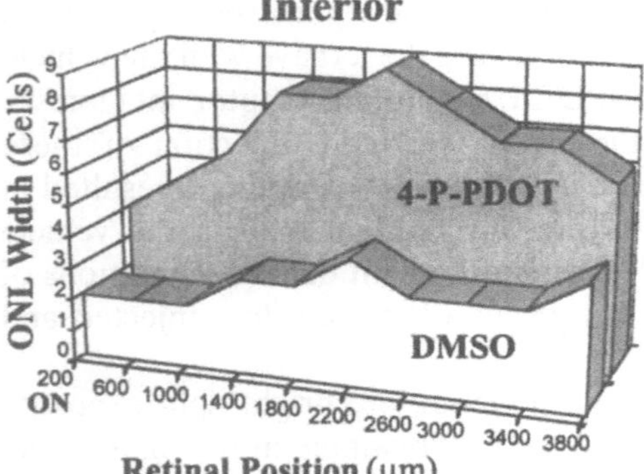

Figure 6. Plots of ONL width along the vertical meridian of light damaged retinas from rats pretreated with a melatonin antagonist in one eye and vehicle in the other. The value at each location in (A) is the average of 5 rats. ONL widths were measured in μm (Normal = approx. 40 μm). The right eye was injected with luzindole and the left eye with DMSO. The values at each location in (B) are from a single animal with 4-P-PDOT injected into the right eye and DMSO injected into the left eye. ONL widths are expressed as number of cells (Normal = approx. 11). Injections were made during the pre-exposure dark period prior to light damage as described in the Methods section. Light damage was produced with 2500 lux for 48 hours.

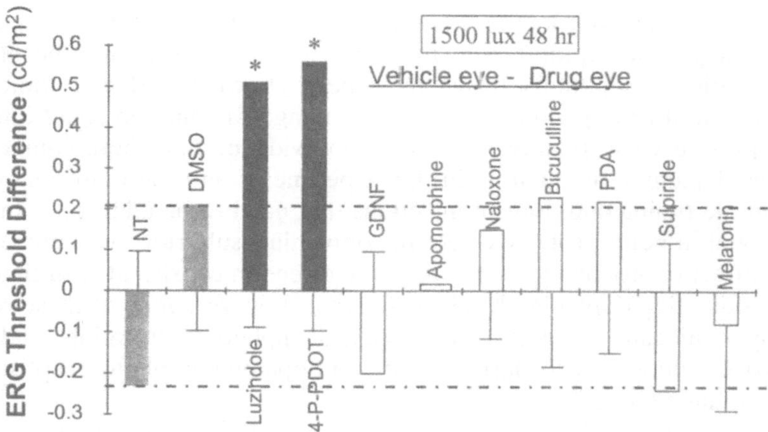

Figure 7. ERG b-wave threshold differences between the right and left eyes of light damaged rats. The right eye was injected with the drug shown and the left eye with vehicle (see Table 1). N = 5 for each group. * = statistically significant, $p < 0.05$, using two-tailed matched pair t-test. NT = no treatment. DMSO = DMSO in both eyes.

was applied. In eyes injected with bicuculline and PDA, shown to enhance retinal dopamine release, there was a tendency toward increased cell survival though, again, not statistically significant using the 2-tailed test. These results suggest that retinal dopamine may modulate photoreceptor survival in light damage. Injecting melatonin itself, which might be expected to suppress retinal dopamine release, had no significant effect on cell survival. None of the other compounds resulted in a b-wave threshold differences greater than that seen in no treatment rats or those injected with DMSO in both eyes.

5. DISCUSSION

The preservation of photoreceptor structure and function in light damage by local application of melatonin antagonists is consistent with previous findings that systemic melatonin enhances damage.[8,9] Recent evidence indicates that in rat, as in other species,[1] melatonin is synthesized in photoreceptors at night under regulation by a circadian clock.[26] Since melatonin is produced and released in the rat retina within minutes of exposure to darkness[27] the application of melatonin receptor antagonists at the beginning of the dark period prior to light exposure could block the action of endogenous melatonin which may normally enhance light damage susceptibility. Since endogenous melatonin production is suppressed in the light, this supports the idea that the protective effect of antagonists results from inhibiting the night time action of melatonin. Experiments we conducted in which melatonin antagonists were given shortly before light exposure showed no protection (unpublished results).

One of the well known actions of melatonin is the suppression of retinal dopamine release from inner retinal neurons.[1] Since the D2 agonist, bromocriptine, injected systemically, was protective in light damage it has been hypothesized that the enhancement of light damage susceptibility by melatonin is actually the result of the reduction in retinal dopamine.[1,9] Since luzindole has been shown to enhance the release of dopamine in darkness from the Xenopus retina *in vitro*,[4] our data are consistent with the idea that photoreceptor rescue by melatonin antagonists may be mediated by dopamine. Other compounds, which have been shown to affect retinal release of dopamine, as well as dopamine agonists and antagonists had no significant effect on photoreceptor survival. However, we have no evidence that these compounds were effective as dopamine modulators in our experiments and there are many questions concerning the timing and dose of injections that need to be addressed. Interestingly, consistent with a neuroprotective role for dopamine, sulpiride, bicuculline, and PDA tended to affect photoreceptor survival in the direction consistent with their expected effect in modulating dopamine levels or activity. However, none of these results were statistically significant using the same criteria applied to luzindole and 4-P-DOT. Further work is underway to clarify the role of dopamine in photoreceptor protection by melatonin antagonists.

The intracellular signaling pathways that might link melatonin or dopamine receptors to photoreceptor survival in light damage or other retinal degenerations has not been studied. Dopamine receptors of the D2/D4 subtype are negatively coupled to c-AMP levels in photoreceptor cells in mice.[28] Melatonin inhibits c-AMP formation in cultured RPE cells from human and rat by binding to pertussis sensitive receptors[29] and produces membrane conductance changes in chick RPE.[30] This evidence suggests that melatonin antagonists could affect normal photoreceptor-RPE interactions. We noted that one such interaction, the process of dark adaptation, which involves the regeneration of visual pigment in the RPE and return to the photoreceptors (data not shown), was unaffected by luzindole. In addition, a direct affect on photoreceptor transduction seems unlikely since we observed no change in the ERG after injection of luzindole solution or DMSO alone. The linkage of melatonin and dopamine receptors to c-AMP indicate that it is a candidate mediator of the neuroprotective effects. Therefore, other means of modulating intracellular levels or activity of c-AMP might be effective.

Another important question raised by these results is whether melatonin antagonists protect photoreceptors in inherited retinal degenerations. The early study by Bubenik and Purtill[8] found that systemic administration of bromocriptine, a D2 receptor agonist, which inhibits retinal melatonin synthesis, increased photoreceptor survival in the RCS rat. The RCS rat retina has elevated levels of melatonin[11] and reduced dopamine release[10] prior to the onset of cell death. The authors of the former study suggested that melatonin antagonists might ameliorate the degeneration in RCS rat. In a preliminary study, we have found reduced photoreceptor death and reduced sensitivity loss at 8 weeks in RCS rats after a single intravitreal injection of luzindole at 3 weeks of age. These data need to be confirmed and extended to other animal models of inherited retinal degeneration before a general role for melatonin antagonists in photoreceptor protection can be established.

REFERENCES

1. P.M. Iuvone, 1996, Circadian rhythms of melatonin biosynthesis in retinal photoreceptor cells: Signal transduction, interactions with dopamine, and speculations on a role in cell survival, in: *Retinal Degeneration and Regeneration*, (S. Kato, N.N. Osborne, and M. Tamai, eds.), pp. 3–13, Kugler Publications, Amsterdam/New York.

2. J.C. Besharse, P.M. Iuvone, and M.E. Pierce, 1988, Regulation of rhythmic photoreceptor metabolism: a role for post-receptoral neurons, in: *Progress in Retinal Research*, Volume 7 (N.N. Osborne and G.J. Chader, eds.), pp. 21–61, Pergamon Press, Oxford.
3. M.L. Dubocovich, 1983, Melatonin is a potent modulator of dopamine release in the retina, *Nature* **306**:782–784.
4. J.H. Boatright, N.M. Rubim, and P.M. Iuvone, 1994, Regulation of endogenous dopamine release in amphibian retina by melatonin: The role of GABA, *Vis. Neurosci.* **11**: 1013–1018.
5. T.E. Duncan and W.K. O'Steen, 1985, The diurnal susceptibility of rat retinal photoreceptors to light-induced damage, *Exp. Eye Res.* **41**:497–507.
6. M.P. White and L.J. Fisher, 1987, Degree of light damage to the retina varies with time of day of bright light exposure, *Physiol. Behav.* **39**:607–613.
7. D.T. Organisciak, Y.L. Jiang, H.M. Wang, M. Pickford, and J.C. Blanks, 1989, Retinal light damage in rats exposed to intermittent light. Comparison with continuous light exposure, *Invest. Ophthalmol. Vis. Sci.* **30**:795–805.
8. G.A. Bubenik and R.A. Purtill, 1980, The role of melatonin and dopamine in retinal physiology, *Can. J. Physiol. Pharmacol.* **58**:1457–1462.
9. A.F. Wiechmann and W.K. O'Steen, 1992, Melatonin increases photoreceptor susceptibility to light-induced damage, *Invest. Ophthalmol. Vis. Sci.* **33**:1894–1902.
10. M. Hankins and H. Ikeda, 1994, Early abnormalities of retinal dopamine pathways in rats with hereditary retinal dystrophy, *Doc. Ophthalmol.* **86**:325–334.
11. M. Hawlina, H.G. Jenkins, and H. Ikeda, 1992, Diurnal variations in the electroretinographic c-wave and retinal melatonin content in rats with inherited retinal dystrophy, *Doc. Ophthalmol.* **79**:141–150.
12. M.L. Dubocovich, 1988, Luzindole (N-0774): a novel melatonin receptor antagonist, *J. Pharmacol. Exp. Ther.* **246**:902–910.
13. T. Sugawara, P.A. Sieving, and R.A. Bush, 1998, ERG threshold elevation is quantitatively related to cell loss in light damaged rat:b-wave by a log-linear and STR by a log-log function, ARVO Abstract, *Invest. Ophthalmol. Vis. Sci.* **39**:S978.
14. M.L. Dubocovich, M.I. Masana, S. Iacob, and D.M. Sauri, 1997, Melatonin receptor antagonists that differentiate between the human Mel1a and Mel1b recombinant subtypes are used to assess the pharmacological profile of the rabbit retina ML1 presynaptic heteroreceptor, *Naunyn-Schmiedebergs Archives of Pharmacology* **355**:365–375.
15. D. Birch and G.H. Jacobs, 1977, Effects of constant illumination on vision in the albino rat, *Physiol. Behav.* **19**:255–259.
16. M.M. LaVail, M.T. Matthes, D. Yasumura, G. Faktorovich, and R. Steinberg, 1997, Histological method to assess photoreceptor light damage and protection by survival factors, in: *Degenerative Retinal Diseases*, (M.M. LaVail, J.G. Hollyfield, and R.E. Anderson, eds.), pp. 369–384, Plenum Press, New York.
17. R.A. Bush, K.W. Hawks, and P.A. Sieving, 1995, Preservation of inner retinal responses in aged royal college of surgeons rat: Evidence against glutamate excitotoxicity in photoreceptor degeneration, *Invest. Ophthalmol. Vis. Sci.* **36**:2054–2062.
18. T. Sugawara, P.A. Sieving, P.M. Iuvone, and R.A. Bush, 1998, The melatonin antagonist luzindole protects retinal photoreceptors from light damage in the rat, *Invest. Ophthalmol. Vis. Sci.* **39**: In press.
19. A. Wirz-Justice, M. Da Prada, and C. Remé, 1984, Circadian rhythm in rat retinal dopamine, *Neuroscience Letters* **45**:21–25.
20. R.A. Bush, C.E. Remé, M. Terman, and A. Malnoë, 1990, Disk shedding and dopamine rhythms under simulated dawn and dusk, ARVO Abs, *Invest. Ophthalmol. Vis. Sci.* **31**:6.
21. P.A. Lapchak, P.J. Miller, S. Jiao, D.M. Araujo, D. Hilt, and F. Collins, 1996, Biology of glial cell line-derived neurotrophic factor (GDNF): implications for the use of GDNF to treat Parkinson's disease. [Review] [39 refs], *Neurodegeneration* **5**:197–205.
22. W. Kolbinger and R. Weiler, 1993, Modulation of endogenous dopamine release in the turtle retina: effects of light, calcium, and neurotransmitters, *Vis. Neurosci.* **10**:1035–1041.
23. S.D. Critz and R. Marc, 1992, glutamate antagonists that block hyperpolarizing bipolar cells increase the release of dopamine from turtle retina, *Vis. Neurosci.* **9**:271–278.
24. L.M. Rapp and T.P. Williams, 1980, A parametric study of retinal light damage in albino and pigmented rats, in: *The Effects of Constant Light on Visual Processes* (T.P. Williams and B.N. Baker, eds.), pp. 135–159, Plenum Press, New York.
25. M.M. LaVail, 1976, Survival of some photoreceptor cells in albino rats following long-term exposure to continuous light, *Invest. Ophthalmol. Vis. Sci.* **15**:64–70.

26. G. Tosini, S. Doyle, and M. Menaker, 1998, Melatonin synthesis in the rat retina: cellular localiziation and circadian regulation, ARVO ABSTRACTS, *Invest. Ophthalmol. Vis. Sci.* **39**:s236.
27. D.A. Redburn and C.K. Mitchell, 1989, Darkness stimulates rapid synthesis and release of melatonin in rat retina, *Visial Neurosience* **3**:391–403.
28. A.I. Cohen, R.D. Todd, S. Harmon, and K.L. O'Malley, 1992, Photoreceptors of mouse retinas possess D4 receptors coupled to adenylate cyclase, *Proc. Natl. Acad. Sci., U.S.A.* **89**:12093–12097.
29. M.S. Nash and N.N. Osborne, 1995, Pertussis toxin-sensitive melatonin receptors negatively coupled to adenylate cyclase associated with cultured human and rat retinal pigment epithelial cells, *Invest. Ophthalmol. Vis. Sci.* **36**:95–102.
30. N. Nao-i, S.E. Nilsson, R.P. Gallemore, and R.H. Steinberg, 1989, Effects of melatonin on the chick retinal pigment epithelium: membrane potentials and light-evoked responses, *Exp. Eye Res.* **49**:573–589.

Section VIII

TRANSPLANTATION STUDIES

52

IRIS PIGMENT EPITHELIAL CELL TRANSPLANTATION IN MONKEY EYES

Toshiaki Abe,* Hiroshi Tomita, Toshifumi Ohashi, Katsura Yamada, Yoshiyuki Takeda, Keiko Akaishi, Madoka Yoshida, and Makoto Tamai

Department of Ophthalmology
Tohoku University
School of Medicine
Sendai 980-8574, Japan

ABSTRACT

To establish auto iris pigment epithelial (IPE) cell transplantation, we examined monkey IPE and performed auto IPE transplantation. IPE cells of the monkeys were obtained by peripheral iridectomy and cultured as we previously reported. Immunocytochemical study was also performed to confirm that they were epithelial in origin. Cultured condition were also studied with bovine, rabbit, mouse, auto, or human serum. Transplantation study was performed by transvitreal approach in the monkey eyes. Mouse and rabbit serums were extremely toxic to the monkey IPE cell culture. Conversely, the cells grew well in the medium with bovine, monkey, and human serum. The difference was not statistically significant among bovine, monkey, and human serum. We demonstrated the presence of the transplanted cultured IPE cells after 60 days of transplantation by ophthalmoscopical and histochemical examination. Several reports also have demonstrated that IPE cells may have about 80% phagocytic function and may also be able to form tight junctions in the subretinal space. The establishment of auto IPE cell transplantation may improve the problem of the rejection.

* Corresponding author; Department of Ophthalmology, Tohoku, University, School of Medicine, 1-1 Seiryomachi Aobaku, Sendai, Miyagi, 980-8574, Japan. Fax: 81-22-717-7298; Tel: 81-22-717-7294; e-mail: toshi@oph.med.tohoku.ac.jp

Retinal Degenerative Diseases and Experimental Therapy, edited by Hollyfield *et al.*
Kluwer Academic / Plenum Publishers, New York, 1999.

INTRODUCTION

Transplantation of neural retina[1] retinal pigment epithelium (RPE),[2–4] and iris pigment epithelium (IPE)[5] have been performed using animal model, such as Royal College of Surgeons (RCS) rats, whose original condition contributed to dysfunction of the phagocytosis of the RPE.[6] The results of the transplantation have shown good preservation of photoreceptor cells and have reported no evidence of rejection. However, xenograft transplantation of RPE with collagen sheets in the subretinal space or anterior chamber demonstrated well-preserved histochemistry of the retina, even though the electroretinographic (ERG) findings reported decreased wavelengths.[5,7] Recently, the transplantation of fetal RPE cells has also been evaluated in human age-related macular degeneration (ARMD), and some of the findings reported with cystoid macular edema may be due to graft rejection.[8] We also observed that several cytokine genes were transcribed at the transplanted area, when we performed transplantation of human RPE or IPE to the subretinal space of the RCS rats.[9] These results indicated that even though, the subretinal space have reported to be an immunologically privileged site, some local reactions may be generated.

IPE cells have been reported to have function of specific phagocytic capacity against rod outer segment and may have potential possibility to be used as alternative to RPE cells.[10] IPE cells also have the possibility that IPE cells would form tight junctions in the subretinal space, thereby substituting for the blood-retinal barrier normally formed by RPE cells.[11] As described above, the IPE cells also were transplanted at the subretinal space by using cultured IPE cells[12] or using basic fibroblast growth factor gene transfected IPE cells.[5] These study delayed the photoreceptor degeneration of RCS rat. In order to prevent graft rejection, we performed auto cell transplantation using monkey IPE.

MATERIALS AND METHODS

Preparation of Cultured Monkey IPE Cells

Monkey (*Macaca fuscata*) IPE cell culture was established as we previously reported.[13] In brief, after we obtain the iris by peripheral iridectomy, the iris was incubated with 0.25% trypsin in Dalbeco's solution at 37 °C for 40 minutes. After incubation, the IPE was mechanically detached from the stroma and pipetting was performed using a glass pipette. By performing the pipetting under visualization of a dissecting microscope, we could collect IPE in 20% fetal bovine serum (FBS), F-12 medium (Gibco BRL, Bethesda, MD, U.S.A.). Next, the IPE cells were seeded at 37 °C in 5% CO_2 in modified polystyrene dishes (Becton-Dickinson Labware, Franklin Lakes, NJ, U.S.A.) in 20% FBS in F-12 medium. Instead of FBS, we also used mouse, rabbit, monkey (auto), and human serums in order to examine the effect of each serum. Monkey serums were collected by venous puncture from the monkeys, who were performed peripheral iridectomy, followed centrifugation at 3500 ×g for 15 min at RT. We also changed the concentration of the serum between 5 to 20%. These cells also were examined with anti-cytokeratin pan (a monoclonal antibody mixture, Sigma, St. Louis, MO, U.S.A.) to confirm that the cells were epithelial in origin. Cell number was determined with use of a Burke-Turk hemocytometer. Phase-contrast photographs of each cells was taken with OLYMPUS IMT-2 (Olympus Tokyo, Japan).

Immunocytochemistry

Cytokeratin immunocytochemistry was performed as described previously.[13] Briefly, the cells at the indicated passage was pre incubated with 0.3% hydrogen peroxide to remove endogenous peroxidase activity. Incubation with the anti-cytokeratin pan (a monoclonal antibody mixture, Sigma) was performed overnight at 4 °C. Following incubation with the second antibody, the color development was achieved by 3-amino-9-ethylcarbazole (Dako). Control slides were made by adding mouse IgG instead of the first antibody in the incubation mixture containing Tw-PBS-2% FBS.

The Methods for Transplantation

IPE cells of the monkey were transplanted in the subretinal space as we reported previously.[5] In brief, monkeys were anesthetized with an injection of pentobarbital natrium (60 mg/kg) followed by an injection of atropine (0.4 mg/kg). Tree-port vitrectomy was performed. Then, micro incision was made at the upper region of the macula and mid periphery of the retina and micro bleb detachment was made by injecting Balanced Salt Solution (BSS) at the subretinal space with 30-gauge needle through the micro incision. Through that incision, 6 μl suspension of IPE cells (approximately 6×10^5 cells) was injected into the subretinal space using a 30-gauge needle. The surgery was finished with the substitution of the vitreous cavity with 20% Sulfur hexafluoride (SF6) and antibiotics instillation into the eye.

Microscopic Examination

For light microscopic examination, the eyeballs used for transplantation were enucleated after the monkeys were anesthetized with the standard protocol, and fixed with periodate-lysine-paraformaldehyde at 4 °C overnight. Sections 3 μm thick were treated with xylene and alcohol and washed with PBS at pH7.4 and PBS containing 0.05% Tw after embedding in paraffin. The prepared sections were stained with hematoxylin and eosin.

Statistical Analysis

Statistical significance was determined by using *t*-test. P value less than 0.05 was considered to be statistically significant.

RESULTS

Figure 1 showed the phase-contrast photographs of primary cultured IPE (Fig. 1 (left)) and at passage 6 (Fig. 1 (right)). Immunocytochemistry revealed that the cells used for the study were in epithelial origin. The proliferation rate of the IPE cells were also less proliferative if compared to that of RPE cells (data not shown).

25,000 cells of monkey IPE at passage 7 were seeded at 37 °C in 5% CO_2 in modified polystyrene dishes in F-12 medium with 20% bovine, mouse, rabbit, monkey (auto), or human serum and without serum. Three days after incubation, the cells were counted. The serums of mouse and rabbit were extremely toxic to the monkey IPE cell culture and a statistically significant decrease in cell numbers was observed with mouse

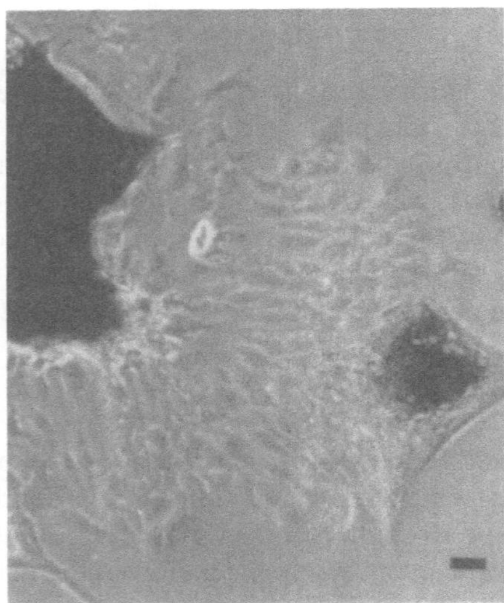
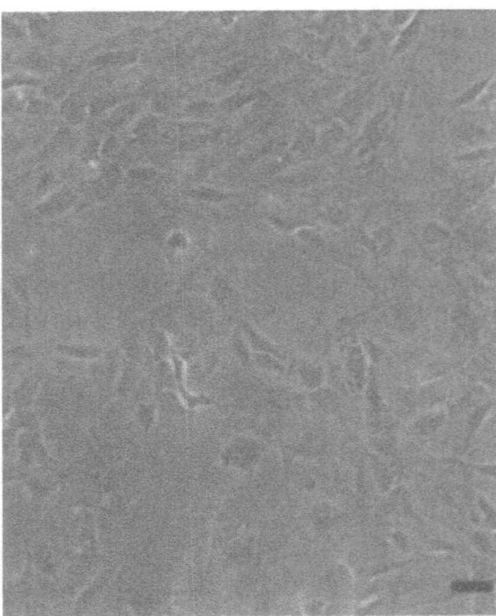

Figure 1. Phase-contrast photographs of iris pigment epithelial (IPE) cells. Primary cultured IPE cells (left) and cultured IPE cells (right) of passage 6 were shown. These sells were confirmed that they were in epithelial origin. Bar is 50μ.

and rabbit serums, when compared to bovine, monkey, and human serums ($P < 0.01$). A statistically significant decrease in cell numbers was also observed with no serum, when compared to bovine, monkey, and human serum ($P < 0.05$). Conversely, statistical analysis was not significant among monkey, bovine, and human serums. Maximum effect was also observed between 10 to 15% of the auto serum of the monkey in the F-12 medium.

The IPE cells at passage 2 were transplantated in the monkey eye after vitrectomy as described in the materials and methods. After 60 days of transplantation, the transplanted eyes were enucleated and fixed in the 4% paraldehyde solution. Rejection was not observed in the operated eye by ophthalmoscopic examination and fluorescein angiography. Histological examination revealed that the transplanted cells were observed without rejection at the subretinal space (Fig. 2).

DISCUSSION

Although transplantation have been evaluated in human disease of ARMD using RPE from fetus[8,14] or RPE from eye bank,[15] the authors also suggested that focal rejection may play an important role for the clinical results. We also observed that local cytokine gene transcriptions, such as IL-1α, -β, and -6 were detected at the transplanted region, if we used xeno- or allograft for transplantation. These results suggest that the transplantation of xeno- or allograft may be affected by the immunological reaction, even though the subretinal space have reported to be an immunologically privileged site.[16,17] In order to prevent rejection, we performed auto IPE transplantation using monkey eyes.

The methods of IPE culture was established using auto serum. Interestingly, the serums of mouse and rabbit were toxic to the monkey IPE cells, conversel bovine,

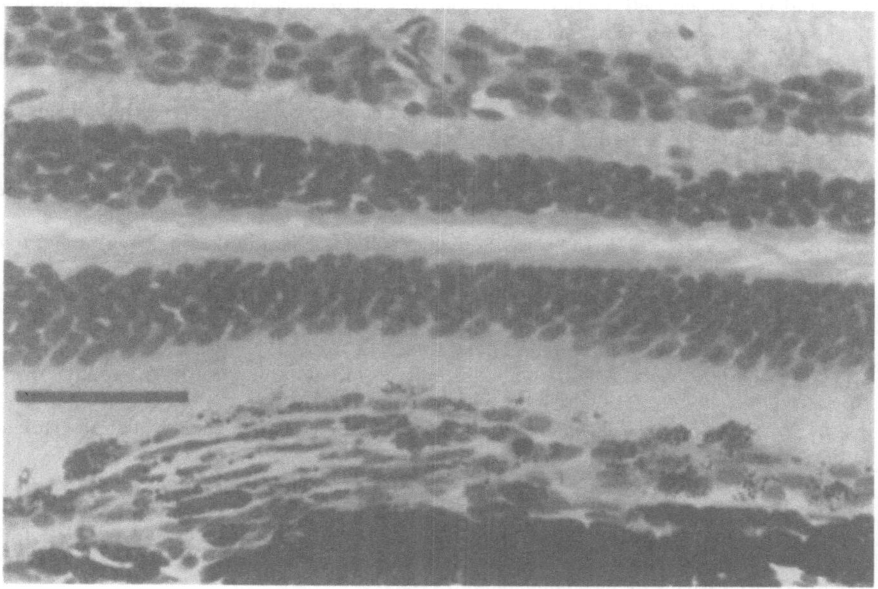

Figure 2. Auto IPE transplantation in monkey eye. Cultured monkey IPE cells at passage 2 were transplanted to the subretinal space of the monkey. Transplanted IPE cells with pigment were observed at the subretinal space. Bar is 50μ.

human, and auto serums were not toxic. As we reported elsewhere,[9] the gene transcription of human RPE cells at the subretinal space of rats have also extensively decreased after 3 days of transplantation. In performing subretinal transplantation, especially throug a transscleral approach, the diseased background conditions or surgical trauma may damage the blood ocular barriers and the transplanted cells may be influenced by the blood of the recipient.

Although we do not know whether the cultured IPE cells may do function as that of RPE cells at the subretinal space, several reports have demonstrated that the IPE cells may have about 80% phagocytic function. The transplantation of IPE cells as RPE cells also delayed the retinal degeneration of RCS rats.[12] The auto IPE cell transplantation in the subretinal space of the monkey demonstrated that there was no sign of rejection and the transplanted cells were observed after 60 days of transplantation. The establishment of the methods of auto IPE cell transplantation may improve the problem of the rejection of the transplanted cells.

ACKNOWLEDGMENT

The authors thank Miss Y. Sasaki for her technical assistance.

REFERENCES

1. L.Q. Jiang, M. Jorquera, and J.W. Streilein, 1993. Subretinal space and vitreous cavity as immunologically privileged sites for retinal allografts. *Invest Ophthalmol Vis Sci* **34**:3347–3354.

2. L. Li and J.E. Turner, 1991, Optimal conditions for long-term photoreceptor cell rescue in RCS rats: the necessity for healthy RPE transplants. *Exp Eye Res* **52**:669–679.
3. M.M. Lavail, L. Li, J.E. Turner, and D. Yasumura, 1992, Retinal pigment epithelial cell transplantation in RCS rats: normal metabolism in rescued photoreceptors. *Exp Eye Res* **55**:555–562.
4. H.J. Sheedlo, L. Li, and J. E. Turner, 1991, Photoreceptor cell rescue at early and late RPE-cell transplantation periods during retinal disease in RCS dystrophic rats. *J NeuralTransplant Plast* **2**:55–63.
5. M. Tamai, 1996, Retinal pigment epithelial cell transplantation: perspective. *Acta Societatis Ophthalmologicae Japonicae* **100**:982–1006.
6. D. Bok and M.O. Hall, 1971, The role of the pigment epithelium in the etiology of inherited retinal dystrophy in the rat. *J Cell Biol* **49**:664–682.
7. N.S. Bhatt, D.A. Newsome, T. Fenech et al., 1994, Experimental transplantation of human retinal pigment epithelial cells on collagen substrates. *Am J Ophthalmol* **117**:214–221.
8. P.V. Algvere, L. Berglin, P. Gouras, Y. Sheng, and E.D. Kopp, 1997, Transplantation of RPE in age-related macular degeneration: observations in disciform lesions and dry RPE atrophy. *Graefes Arch Clin Exp Ophthalmol* **235**:149–158.
9. T. Abe, Y. Takeda, K. Akaishi, H. Tomita, K. Yamada, and M. Tamai, Transcription of cytokine genes at the subretinal space after transplantation. Submitted.
10. K.A. Rezai, A. Lappas, L. Farrokh-siar, L. Kohen, P. Wiedemann, and K. Heimann, 1997, Iris pigment epithelial cells of Long Evans rats demonstrate phagocytic activity. *Exp. Eye Res.* **65**:23–29.
11. K.A. Rezai, A. Lappas, L. Farrokh-siar, L. Kohen, P. Wiedemann, and K. Heimann, 1997, Comparison of tight junction permeability for albumin in iris pigment epithelium and retinal pigment epithelium in vitro. *Graefes Arch Clin Exp Ophthalmol* **235**:48–55.
12. K.A. Rezai, L. Kohen, P. Wiedemann, and K. Heimann, 1997, Iris pigment epithelium transplantation. *Graefes Arch Clin Exp Ophthalmol* **235**:558–562.
13. Y.K. Durlu and M. Tamai, 1997, Transplantation of retinal pigment epithelium using viable cryopreserved cells. *Cell Transplant.* **6**:149–162.
14. P.V. Algvere, L. Berglin, P. Gouras, and Y. Sheng, 1994, Transplantation of fetal retinal pigment epithelium in age-related macular degeneration with subfoveal neovascularization. *Graefes Arch Clin Exp Ophthalmol* **232**:707–716.
15. G.A. Peyman, K.J. Blinder, C.L. Paris, W. Alturki, N.C. Jr. Nelson, and U. Desai, 1991, A technique for retinal pigment epithelium transplantation for age-related macular degeneration secondary to extensive subfoveal scarring. *Ophthalmic Surg* **22**:102–108.
16. J.Y. Niederkorn and J.W. Streilein, 1986, Lymphoma allografts abrogate immune privilege within the anterior chamber of the eye. *Invest Ophthalmol Vis Sci* **27**:1235–1243.
17. J.W. Streilein, 1990, Anterior chamber associated immune deviation: the privilege of immunity in the eye. *Surv Ophthalmol* **35**:67–73.

53

THE RETINAL PIGMENT EPITHELIAL CELL DIFFERENTIATION AND CELL MARKER EXPRESSION FOLLOWING CRYOPRESERVATION AT –80°C AND UNDER LIQUID NITROGEN AT –196°C

Y. K. Durlu,[1]* S-I. Ishiguro,[1] K. Akaishi,[1] T. Abe,[1] Y. Chida,[1]
S. Shibahara,[2] and M. Tamai[1]

[1] Department of Ophthalmology
[2] Department of Molecular Biology
1-1 Seiryo-machi, Aoba-ku
Tohoku University School of Medicine
Sendai, Miyagi 980, Japan

SUMMARY

Cryopreserved retinal pigment epithelial (RPE) cells at –80°C can be used for transplantation studies. Long-term viability at –196°C has been used for many other cell lines. The origin and purity of human RPE cell lines were assessed by immunocytochemistry and reverse-transcriptase polymerase chain reaction (RT-PCR) using RPE cell markers [cytokeratin, cellular retinaldehyde binding protein (CRALBP), tyrosinase, tyrosinase-related proteins I and II (TRP-I and II), Na,K-ATPase α_1 and β_1]. Cultured human RPE cell lines were cryopreserved at –80°C and –196°C. After cryopreservation, human RPE cells were re-cultured on laminin-coated polystyrene dishes and laminin-coated collagen sheets. Differentiation was evaluated by the expression level of RPE marker genes using RT-PCR. It was found that cryopreserved human RPE cells on laminin-coated collagen sheets disclosed relatively strong re-expression of their marker genes if compared to human RPE cells cultured on laminin-coated polystyrene dishes. Cryopreservation of RPE cells at both temperatures, thawing and subsequent

* Current address: Adakale Sokak 25/34, Adakale Ishani Kat 2, Yenisehir 06420, Ankara, Turkey

Retinal Degenerative Diseases and Experimental Therapy, edited by Hollyfield *et al.*
Kluwer Academic / Plenum Publishers, New York, 1999.

culturing on laminin-coated collagen sheets can enhance the feasibility of using cryopreserved RPE cells for transplantation.

INTRODUCTION

Retinal pigment epithelial (RPE) cell transplantation may have potential clinical applications for the surgical treatment of RPE-specific retinal degeneration.[1,2] Suitable donor tissue is an important issue for the prognosis of successful transplantation. The ideally preserved tissue contributes to the success of the surgery and final aim of the transplantation. Inadequate techniques of preservation may lead to the failure transplantation surgery.

In our previous study, we reported that cryopreserved RPE cells at –80°C can be used for transplantation studies.[3] Cryopreservation temperature at –196°C (under liquid nitrogen) has been frequently used for other cell lines. The vulnerability or resistance of human RPE cells under liquid nitrogen has not been known. We would like to assess the differentiated features of human RPE cells after cryopreservation at –80°C and –196°C using RPE cell markers like cellular retinaldehyde binding protein (CRALBP), tyrosinase, tyrosinase-related proteins I and II (TRP-I and TRP-II), Na,K-ATPase α_1 and β_1.

MATERIALS AND METHODS

Two different adult human RPE cell cultures (TEB-0915 and TEB-1116/1) were established according to our previously published procedure.[4] 10 ng/ml basic-fibroblast growth factor (b-FGF, a gift of Takeda Chemical Industries, Ltd., Japan) was included to the culture medium 10% fetal bovine serum (FBS, supplied from GIBCO) in minimum essential medium (MEM, supplied from GIBCO). The purity and origin of the RPE cultures were checked by cytokeratin immunocytochemistry and Western blotting, and by RT-PCR of CRALBP, tyrosinase, TRP-I and TRP-II, Na,K-ATPase α_1 and β_1.[5] The protocols of immunocytochemistry,[3] Western blotting,[3] RT-PCR,[6] and cryopreservation[3] techniques were applied as reported previously. Laminin (human, supplied from GIBCO)-coated collagen (rat-tail tendon collagen, type I, supplied from Collaborative Research) gels were prepared as follows; in each 8-microwell plate (supplied from Nunc), collagen (4 mg/ml) at a volume of 100 µl was inserted, and after treating the wells by ammonia-water under laminar air-flow, the gels were washed and neutralized with Hank's balanced salt solution (HBSS-, supplied from GIBCO). Then, the collagen gels were soaked in laminin at a concentration of 2 µg/ml in HBSS- overnight at cold room. The excess laminin was tapped off and human RPE cells were cultured on these laminin-coated collagen gels.

RESULTS

Adult human RPE cell cultures (TEB-0915 and TEB-1116/1) were free of other cell contamination (Figs. 1 and 2).

The cytokeratin immunoreactivity of human RPE cell cultures (TEB-0915 and

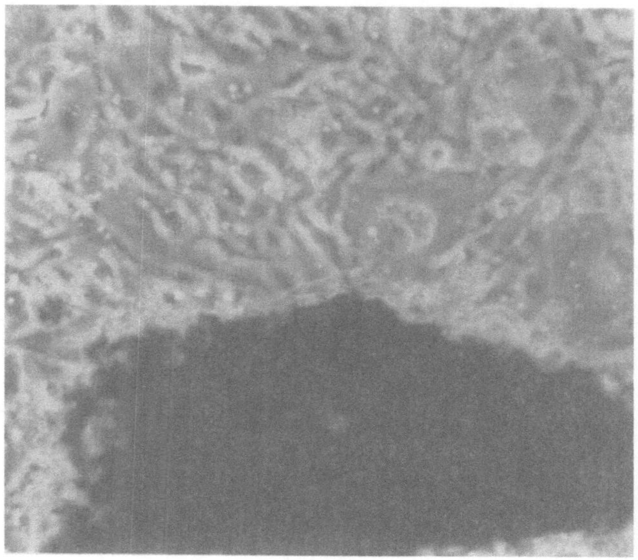

A

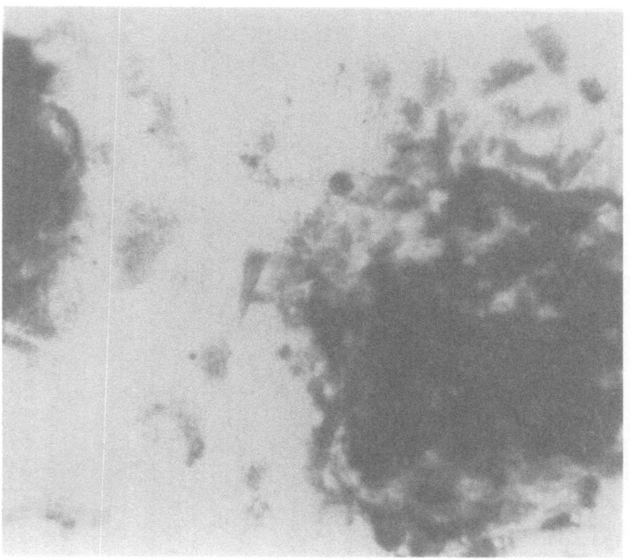

B

Figure 1. TEB-0915 adult human RPE cell culture. A. Primary culture (phase microscopy, ×240). B. Primary culture (light microscopy, ×240).

TEB-1116/1) was found positive (Figs. 3A and 4A) where the controls lacked reaction product (Figs. 3B and 4B).

Phase and light microscopy before and nine months after cryopreservation at –80°C showed the presence of melanin granules in TEB-1116/1 human RPE cells (Fig. 5).

Following cryopreservation of human RPE cells for four months under liquid nitrogen, strong re-expression of RPE cell markers was observed in human RPE cell cultures established on laminin-coated collagen gels but not human RPE cell cultures established on plain collagen gels as checked by RT-PCR (Fig. 6).

The results of RT-PCR are summarized at Table 1.

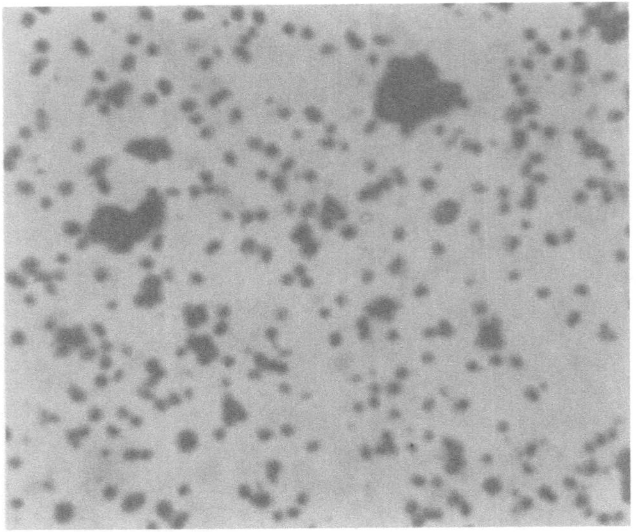

A

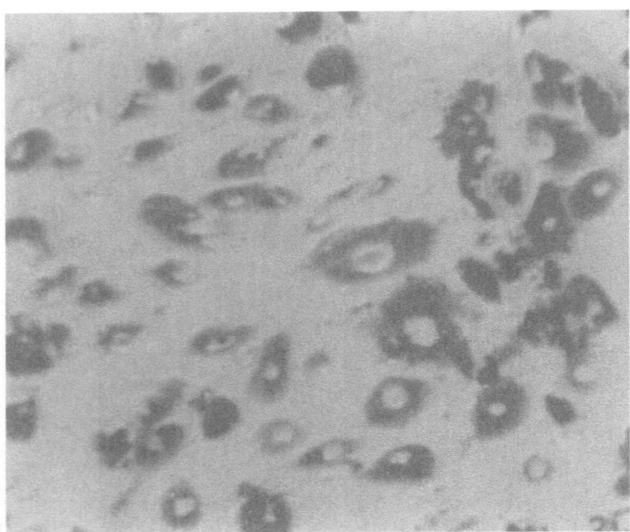

B

Figure 2. TEB-1116/1 adult human RPE cell culture. A. After isolation (light microscopy, ×120). B. Passage 2 (light microscopy, ×240).

DISCUSSION

Age-related macular degeneration, a disease of the RPE cell layer, is the major cause of blindness in developed countries.[7] It affects a third of the population older than age 75.[7] RPE transplantation is now considered as an alternative method for the treatment of this disease.[1] The initial reports by Algvere et al.,[1,2] indicated that rejection phenomenon after RPE transplantation in human subjects limited the rate of success. Systemic immunity might have significant influence on RPE allografts in the subretinal space of RCS rats.[8]

Our previous[3] and this study indicates that, the analysis for MHC expression, HLA subtyping, and rejection markers in cryopreserved human RPE cells may be useful to reduce the rate of rejection after RPE transplantation.

A

B

Figure 3. A. The cytokeratin immunocytochemistry of TEB-0915 (×120), and its control (B) by substituting human IgG for the first antibody (×120).

Donor RPE cell culture conditions may effect the photoreceptor cell rescue in transplanted RCS rats.[9] RPE cells are known to loose their differentiated features after several passages in culture.[4,6] Transplantation of tissue cultured neonatal Long Evans rat RPE cells was less effective in promoting photoreceptor cell rescue in RCS rat retina if compared to the freshly-isolated neonatal Long Evans rat RPE cells.[9] Transplanted RPE cells should express their highly-differentiated feature that's sufficient to function as healthy RPE to restore vision. We thought that the cryopreservation period of cultured human RPE cells under liquid nitrogen might give advantages to search the differentiated features and function of RPE cells suitable for transplantation.

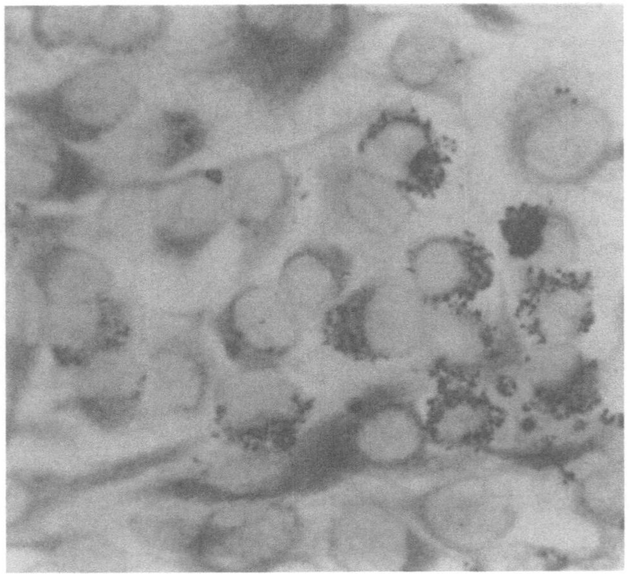
A

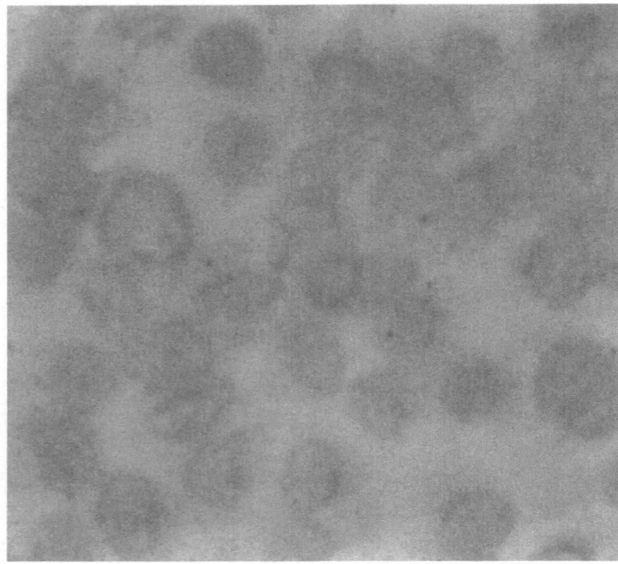
B

Figure 4. A. The cytokeratin immunocytochemistry of TEB-1116/1 (×480), and its control (B) by substituting human IgG for the first antibody (×480).

In this study we have shown that human RPE cells can be cryopreserved under liquid nitrogen at –196°C as well as at –80°C. The re-differentiation of human RPE cells could be achieved after cryopreservation as checked by morphology and the RT-PCR of RPE cell-specific markers. The extracellular matrix proteins (both laminin[10] and collagen, an integral part of Bruch's membrane) might have critical roles for re-differentiation of RPE cells after cryopreservation.

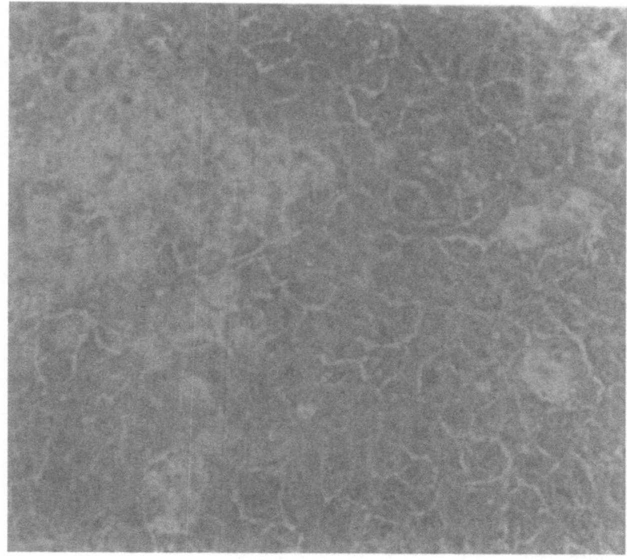

A

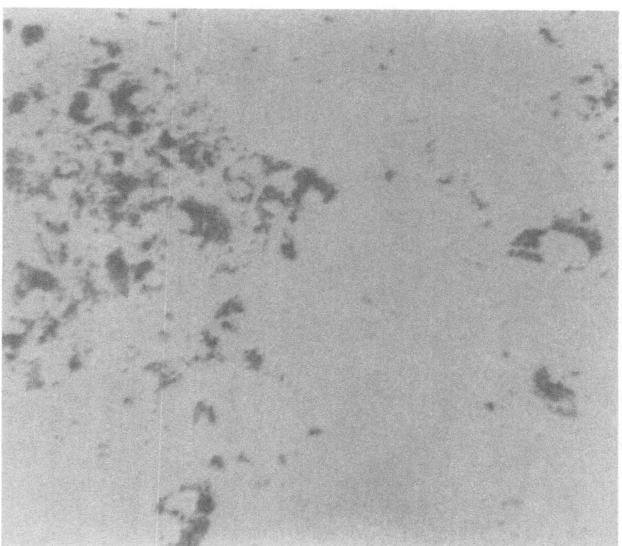

B

Figure 5. Phase (A) and light (B) microscopy of TEB-1116/1 at passage 4 (×240). C and D showed phase and light microscopy of TEB-1116/1 at passage 4 cryopreserved for nine months at –80°C (×240).

CONCLUSION

Human RPE cells can be cryopreserved at –80°C and at –196°C, thawed and subsequently used for transplantation studies. There might be several advantages of cryopreservation of human RPE cells for transplantation; the viability, origin, function, differentiation, and contamination (viral, fungal or bacterial) of the RPE cells could be checked during storage. Re-differentiation of cryopreserved human RPE cells on laminin-coated collagen gels may enhance the feasibility of this method.

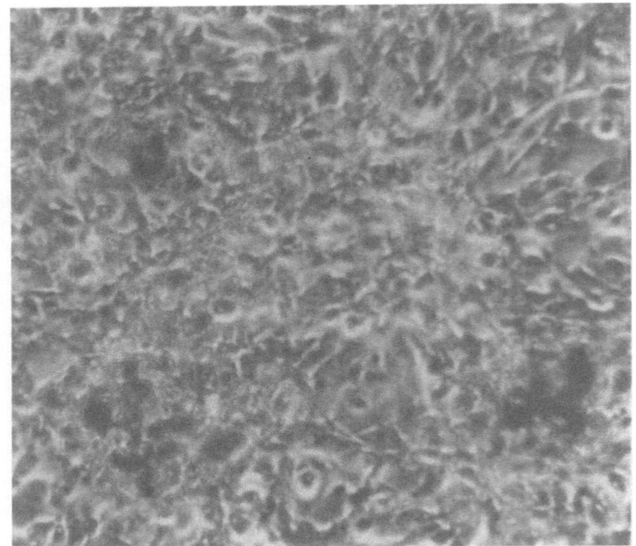
C

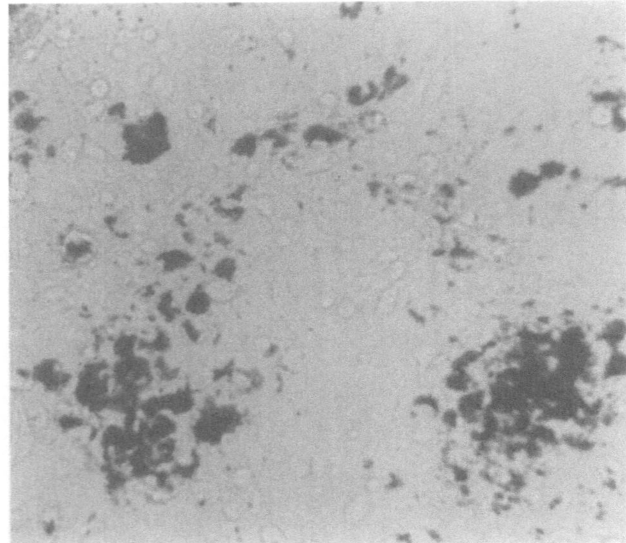
D

Figure 5. *Continued.*

Table 1. RPE cell markers

	Tyrosinase	TRP-I	TRP-II	CRALBP	Na,K-ATPase	
					α_1	β_1
in situ	+	+	+	+	+	+
in vitro:						
Passage 3	+	+	+	+	+	+
Passage 9	+	+	−	+	+	+
Cryopreserved:						
−80°C (9 months)	+	+	−	+	+	+
−196°C (4 months)	+	+	−	+	+	+

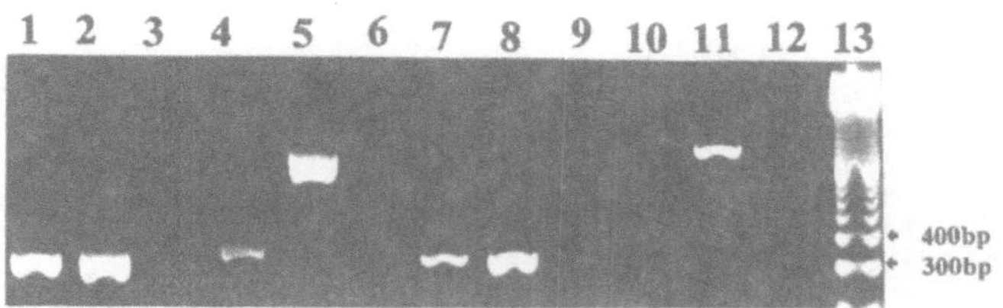

Figure 6. RT-PCR of TEB-1116/1 (four months cryopreserved at –196°C and re-cultured). Lane 1-Cells on aminin-coated collagen gel, Tyrosinase (359 bp). Lane 2-Cells on laminin-coated collagen gel, TRP-I (364 bp). Lane 3-Cells on laminin-coated collagen gel, TRP-II (345 bp). Lane 4-Cells on laminin-coated collagen gel, CRALBP (434 bp). Lane 5-Cells on laminin-coated collagen gel, Na,K-ATPase α_1 (918 bp). Lane 6-Cells on laminin-coated collagen gel, Na,K-ATPase β_1 (764 bp). Lane 7-Cells on collagen gel, Tyrosinase. Lane 8-Cells on collagen gel, TRP-I. Lane 9-Cells on collagen gel, TRP-II. Lane 10-Cells on collagen gel, CRALBP. Lane 11-Cells on collagen gel, Na,K-ATPase α_1. Lane 12-Cells on collagen gel, Na,K-ATPase β_1. Lane 13-Standard, 100 bp ladder.

REFERENCES

1. P.V. Algvere, L. Berglin, P. Gouras, and Y. Sheng, 1994, Transplantation of fetal retinal pigment epithelium in age-related macular degeneration with subfoveal neovascularization, *Graefe's Arch. Clin. Exp. Ophthalmol.* **232**:707–716.
2. P.V. Algvere, L. Berglin, P. Gouras, Y. Sheng, and E.D. Kopp, 1997, Transplantation of RPE in age-related macular degeneration: observations in disciform lesions and dry RPE atrophy, *Graefe's Arch. Clin. Exp. Ophthalmol.* **235**:149–158.
3. Y.K. Durlu and M. Tamai, 1997, Transplantation of retinal pigment epithelium using viable cryopreserved cells, *Cell Transplantation*, **6**:149–162.
4. Y.K. Durlu and M. Tamai, 1995, *In vitro* expression of epidermal growth factor receptor by human retinal pigment epithelium cells, in: *Degenerative Diseases of the Retina* (R.E. Anderson, M.M. LaVail, J.G. Hollyfield, eds.), pp. 69–76, Plenum Press, New York.
5. A. Ruiz, S.B. Bhat, and D. Bok, 1995, Characterization and quantification of full length and truncated Na,K-ATPase α_1 and β_1 RNA transcripts expressed in human retinal pigment epithelium, *Gene*, **155**:179–184.
6. T. Abe, Y.K. Durlu, and M. Tamai, 1996, The properties of retinal pigment epithelial cells in proliferative vitreoretinopathy compared with cultured retinal pigment epithelial cells, *Exp. Eye Res.*, **63**:201–210.
7. S.H. Sarks and J.P. Sarks, 1989, Age-related macular degeneration: atrophic form, in: *Retina* (S.J. Ryan, ed.), Volume 2, pp. 149–173, The C. V. Mosby Co., St. Louis.
8. X. Zhang and D. Bok, 1998, Transplantation of retinal pigment epithelial cells and immune response in the subretinal space, *Invest. Ophthalmol. Vis. Sci.*, **39**:1021–1027.
9. L. Li and J.E. Turner, 1991, Optimal conditions for long-term photoreceptor cell rescue in RCS rats: the necessity for healthy RPE transplants. *Exp. Eye Res.*, **52**:669–679.
10. P.A. Campochiaro and S.F. Hackett, 1993, Corneal endothelial cell matrix promotes expression of differentiated features of retinal pigment epithelial cells: implication of laminin and basic fibroblast growth factor as active components, *Exp. Eye Res.*, **57**:539–547.

ABOUT THE EDITORS

Joe G. Hollyfield, Ph.D., is Director of Research in The Eye Institute at The Cleveland Clinic Foundation, Cleveland, Ohio, and Professor of Cell Biology, Neurobiology and Anatomy, Ohio State University, Columbus, Ohio. He received his Ph.D. degree from the University of Texas at Austin and was a postdoctoral fellow at the Hubrecht Laboratory in Utrecht, The Netherlands. He has held faculty positions at Columbia University College of Physicians and Surgeons in New York City and at the Cullen Eye Institute, Baylor College of Medicine in Houston, TX. He was Director of the Retinitis Pigmentosa Research Center in The Cullen Eye Institute from 1978 until his move to The Cleveland Clinic Foundation in 1995, and is currently Director of the Foundation Fighting Blindness Research Center at The Cleveland Clinic Foundation. Dr. Hollyfield has published over 150 papers in the area of cell and developmental biology of the retina and retinal pigment epithelium in both normal and retinal degenerative tissue. He has edited eight books, seven on retinal degenerations and one on the structure of the eye. Dr. Hollyfield has received the Marjorie W. Margolin Prize (1981, 1994), the Sam and Bertha Brochstein Award (1985) and the Award of Merit in Retina Research (1998) from the Retina Research Foundation; the Olga Keith Wiess Distinguished Scholars' Award (1981), two Senior Scientific Investigator Awards (1988, 1994) from Research to Prevent Blindness, Inc.; an award for Outstanding Contributions to Vision Research from the Alcon Research Institute (1987); the Distinguished Alumnus Award (1991) from Hendrix College, Conway, Arkansas; and the Endre A. Balazs Prize (1994) from the International Society for Eye Research (ISER). He is currently Editor-in-Chief of the journal, *Experimental Eye Research* published by Academic Press. Dr. Hollyfield has been active in the Association for Research in Vision and Ophthalmology (ARVO) serving on the Program Committee, as a Trustee and as President. He is also a past President and former Secretary of the International Society of Eye Research. He currently serves on the Scientific Advisory Boards of The Foundation Fighting Blindness, Research to Prevent Blindness, The Helen Keller Eye Research Foundation, The South Africa Retinitis Pigmentosa Foundation, and is Co-Chairman of the the International Retinitis Pigmentosa Foundation Medical and Scientific Advisory Board.

Robert E. Anderson, M.D., Ph.D., is Dean A. McGee Professor of Ophthalmology; Professor of Cell Biology, Biochemistry & Molecular Biology, and Geriatric Medicine; Chairman of the Cell Biology Department; and Director of the Oklahoma Center for

Neuroscience at The University of Oklahoma Health Sciences Center in Oklahoma City, Oklahoma. He received a Ph.D. in Biochemistry (1968) from Texas A&M University and M.D. from Baylor College of Medicine in 1975. In 1968, he was a postdoctoral fellow at Oak Ridge Associated Universities. At Baylor, he was appointed Assistant Professor in 1969, Associate Professor in 1976, and Professor in 1981. He joined the faculty of the University of Oklahoma in January, 1995. Dr Anderson's research has concentrated on the retina where he has published extensively in the areas of glycerolipid and fatty acid metabolism, lipid second messengers, and the biochemistry of retinal degenerations. He has edited eight books, seven on retinal degenerations and one on the biochemistry of the eye. Dr. Anderson has received the Marjorie Margolin Prize, and the Sam and Bertha Brochstein Award for Outstanding Achievement in Retina Research from the Retina Research Foundation (1980); the Dolly Green Award (1982), two Senior Scientific Investigator Awards (1990 and 1997) from Research to Prevent Blindness, Inc.; an Award for Outstanding Contributions to Vision Research from the Alcon Research Institute (1985). He has served on the editorial boards of Investigative Ophthalmology and Visual Science, Journal of Neuroscience Research, Neurochemistry International, and Current Eye Research, and is currently on the editorial board of Experimental Eye Research. He has been an active participant in the program committees of the Association for Research in Vision and Ophthalmology (ARVO) and was a trustee representing the Biochemistry and Molecular Biology section. He has served on the Vision Research Program Committee and Board of Scientific Counselors of the National Eye Institute and the Board of the Basic and Clinical Science Series of The American Academy of Ophthalmology. Dr. Anderson is a past Councilor and Treasurer for the International Society for Eye Research.

Matthew M. LaVail, Ph.D., is Professor of Anatomy and Ophthalmology at the University of California, San Francisco School of Medicine. He received a Ph.D. degree in Anatomy (1969) from the University of Texas Medical Branch in Galveston and was subsequently a postdoctoral fellow at Harvard Medical School. Dr. LaVail was appointed Assistant Professor of Neurology-Neuropathology at Harvard Medical School in 1973. In 1976, he moved to UCSF, where he was appointed Associate Professor of Anatomy. He was appointed to his current position in 1982, and in 1988, he also became director of the Retinitis Pigmentosa Research Center at UCSF, later named the Kearn Family Center for the Study of Retinal Degeneration. Dr. LaVail has published extensively in the research areas of photoreceptor-retinal pigment epithelial cell interactions, retinal development, circadian events in the retina, genetics of pigmentation and ocular abnormalities, inherited retinal degenerations and light-induced retinal degeneration. He is the author of more than 110 research publications and has edited 8 books on inherited and environmentally induced retinal degenerations. Dr. LaVail has received the Fight for Sight Citation (1976); the Sundial Award from the Retina Foundation (1976); the Friedenwald Award from the Association for Research in Vision and Ophthalmology (1981); a MERIT Award from the National Eye Institute (1989); an Award for Outstanding Contributions to Vision Research from the Alcon Research Institute (1990); the Award of Merit from the Retina Research Foundation (1990); the first John A. Moran Prize for Vision Research, University of Utah; (1997); and the first Trustee Award, The Foundation Fighting Blindness (1998). He has served on the editorial board of Investigative Ophthalmology and Visual Science and is currently on the editorial board of Experimental Eye Research. Dr. LaVail has been an active participant in the program committees of the Association for Research in Vision and Ophthalmology and the International Society for Eye research, and he is currently a Trustee of the Association for Research in Vision and Ophthalmology representing the Retinal Cell Biology Section.

INDEX